Research Trends
in Combinatorial Optimization

William J. Cook · László Lovász · Jens Vygen

Editors

Research Trends
in Combinatorial Optimization

Bonn 2008

 Springer

William Cook

Industrial and Systems Engineering
Georgia Tech
765 Ferst Drive
Atlanta, Georgia 30332-0205
USA
bico@isye.gatech.edu

László Lovász

Eötvös Loránd University
Institute of Mathematics
Pázmány Péter sétány 1/C
H-1117 Budapest
Hungary
lovasz@cs.elte.hu

Jens Vygen

Research Institute for Discrete Mathematics
University of Bonn
Lennéstr. 2
53113 Bonn
Germany
vygen@or.uni-bonn.de

ISBN 978-3-540-76795-4   e-ISBN 978-3-540-76796-1

DOI 10.1007/978-3-540-76796-1

Library of Congress Control Number: 2008932935

Mathematics Subject Classification (2000): 90CXX

© 2009 Springer-Verlag Berlin Heidelberg

Cover design: WMXDesign GmbH

Printed on acid-free paper

9 8 7 6 5 4 3 2 1

springer.com

Dedicated to Bernhard Korte

Preface

The editors and authors dedicate this book to Bernhard Korte on the occasion of his seventieth birthday. We, the editors, are happy about the overwhelming feedback to our initiative to honor him with this book and with a workshop in Bonn on November 3–7, 2008. Although this would be a reason to look back, we would rather like to look forward and see what are the interesting research directions today.

This book is written by leading experts in combinatorial optimization. All papers were carefully reviewed, and eventually twenty-three of the invited papers were accepted for this book.

The breadth of topics is typical for the field: combinatorial optimization builds bridges between areas like combinatorics and graph theory, submodular functions and matroids, network flows and connectivity, approximation algorithms and mathematical programming, computational geometry and polyhedral combinatorics.

All these topics are related, and they are all addressed in this book. Combinatorial optimization is also known for its numerous applications. To limit the scope, however, this book is not primarily about applications, although some are mentioned at various places.

Most papers in this volume are surveys that provide an excellent overview of an active research area, but this book also contains many new results. Highlighting many of the currently most interesting research directions in combinatorial optimization, we hope that this book constitutes a good basis for future research in these areas.

We owe sincere thanks to all authors for their valuable contributions. We also thank all referees for carefully reviewing the papers and making many suggestions for improvements. Special thanks go to Ina Prinz for her portrait and cover design, and to Klaus Radke for technical help. Moreover, we thank Springer-Verlag for the efficient cooperation. Last, but not least, our most important thanks go to Bernhard Korte, without whom the field of combinatorial optimization would not be the same.

Atlanta, Budapest, and Bonn,   *William Cook*
June 2008   *László Lovász*
                                    Jens Vygen

Contents

List of Contributors

Mourad Baïou
CNRS, LIMOS Complexe scientifique
des Cezeaux
CUST – Campus des Cezeaux
BP 206 – 63174 Aubière Cedex
France
mourad.baiou@isima.fr

Francisco Barahona
IBM T.J. Watson Research Center
Yorktown Heights, NY 10589
USA
barahon@us.ibm.com

Sylvia Boyd
School of Information Technology
and Engineering
University of Ottawa
800 King Edward Ave.
Ottawa, Ontario, K1N 6N5
Canada
sylvia@site.uottawa.ca

Vašek Chvátal
Canada Research Chair in
Combinatorial Optimization,
Department of Computer Science
and Software Engineering
Concordia University
1455 De Maisonneuve Blvd. West
Montréal, Québec, H3G 1M8
Canada
chvatal@cse.concordia.ca

Jack Edmonds
Equipe Combinatoire et
Optimisation
Université Pierre et Marie Curie (Paris 6)
4, place Jussieu
75252 Paris Cedex 05
France
jackedmonds@rogers.com

Jean Fonlupt
Equipe Combinatoire et Optimisation
CNRS et Université Pierre et Marie
Curie (Paris 6)
4, place Jussieu
75252 Paris Cedex 05
France
jean.fonlupt@math.jussieu.fr

András Frank
Department of Operations Research
Eötvös Loránd University
Pázmány Péter sétány 1/C
1117 Budapest
Hungary
frank@cs.elte.hu

Satoru Fujishige
Research Institute for Mathematical
Sciences
Kyoto University
Kyoto 606-8502
Japan
fujishig@kurims.kyoto-u.ac.jp

Frank Göring
Fakultät für Mathematik
TU Chemnitz
Reichenhainer Str. 39
09107 Chemnitz
Germany
frank.goering@
mathematik.tu-chemnitz.de

Jochen Harant
Institut für Mathematik
TU Ilmenau
Postfach 100565
98684 Ilmenau
Germany
jochen.harant@tu-ilmenau.de

Stefan Hougardy
Research Institute for Discrete
Mathematics
University of Bonn
Lennéstr. 2
53113 Bonn
Germany
hougardy@or.uni-bonn.de

T.C. Hu
Department of Computer Science
and Engineering
University of California, San Diego
9500 Gilman Drive
La Jolla, CA 92093-0404
USA
techianghu@gmail.com

Tamás Király
Department of Operations Research
Eötvös Loránd University
Pázmány Péter sétány 1/C
1117 Budapest
Hungary
tkiraly@cs.elte.hu

Leo Landa
Department of Computer Science
and Engineering
University of California, San Diego

La Jolla, CA 92093-0404
USA
leo@leolan.com

Thomas M. Liebling
EPFL – Ecole Polytechnique
Fédérale de Lausanne
Mathematics Institute
1015 Lausanne
Switzerland
Thomas.Liebling@epfl.ch

Kazuo Murota
Department of Mathematical Informatics
Graduate School of Information
Science and Technology
University of Tokyo
Tokyo 113-8656
Japan
murota@mist.i.u-tokyo.ac.jp

Guyslain Naves
Laboratoire G-SCOP
CNRS, INPG, UJF
46 avenue Félix Viallet
38031 Grenoble Cedex 1
France
Guyslain.Naves@g-scop.inpg.fr

Jaroslav Nešetřil
Department of Applied Mathematics
(KAM) and
Institute of Theoretical Computer
Science (ITI)
Charles University
Malostranské nám 25
11800 Praha 1
Czech Republic
nesetril@kam.mff.cuni.cz

Michael D. Plummer
Department of Mathematics
Vanderbilt University
Nashville, TN 37240
USA
michael.d.plummer@
vanderbilt.edu

Yves Pochet
Center of Operations Research and
Econometrics (CORE) and
Institut d'Administration et de
Gestion (IAG)
Université catholique de Louvain
34, Voie du Roman Pays
1348 Louvain-la-Neuve
Belgium
Yves.Pochet@lhoist.com

Lionel Pournin
EPFL – Ecole Polytechnique
Fédérale de Lausanne
Mathematics Institute
1015 Lausanne
Switzerland
Lionel.Pournin@epfl.ch

Myriam Preissmann
Laboratoire G-SCOP
CNRS, INPG, UJF
46 avenue Félix Viallet
38031 Grenoble Cedex 1
France
Myriam.Preissmann@
g-scop.inpg.fr

William R. Pulleyblank
Center for Business Optimization
IBM Global Business Services
Route 100
Somers, NY 10589
USA
wp@us.ibm.com

Dieter Rautenbach
Institut für Mathematik
TU Ilmenau, Postfach 100565
98684 Ilmenau
Germany
dieter.rautenbach@tu-ilmenau.de

András Recski
Budapest University of Technology
and Economics,
Department of Computer Science

and Information Theory
1521 Budapest, P.O.B. 91
Hungary
recski@cs.bme.hu

Ingo Schiermeyer
Institut für Diskrete Mathematik
und Algebra
Technische Universität Bergakademie
Freiberg
Prüferstr. 1
09596 Freiberg
Germany
schierme@math.tu-freiberg.de

Andreas S. Schulz
Massachusetts Institute of Technology
77 Massachusetts Avenue
Cambridge, MA 02139
USA
schulz@mit.edu

András Sebő
Laboratoire G-SCOP
CNRS, INPG, UJF
46 avenue Félix Viallet
38031 Grenoble Cedex 1
France
Andras.Sebo@g-scop.inpg.fr

F. Bruce Shepherd
James McGill Professor
Department of Mathematics and
Statistics
McGill University
805 Sherbrooke West
Montreal, Québec, H3A 2K6
Canada
bruce.shepherd@mcgill.ca

Man-Tak Shing
Department of Computer Science
Naval Postgraduate School
1411 Cunningham Road
Monterey, CA 93943
USA
shing@nps.edu

Alexandre Skoda
Equipe Combinatoire et
Optimisation
CNRS et Université Pierre et Marie
Curie (Paris 6)
4, place Jussieu
75252 Paris Cedex 05
France
askoda@math.jussieu.fr

Martin Skutella
Institut für Mathematik
Technische Universität Berlin
Straße des 17. Juni 136
10623 Berlin
Germany
skutella@math.tu-berlin.de

Zoltán Szigeti
Laboratoire G-SCOP
CNRS, INPG, UJF
46 avenue Félix Viallet
38031 Grenoble Cedex 1
France
zoltan.szigeti@g-scop.inpg.fr

Mathieu Van Vyve
n-Side
Chemin du Cyclotron 6
1348 Louvain-la-Neuve
Belgium
Mathieu.VanVyve@n-side.be

Dominic Welsh
Mathematical Institute
University of Oxford
24-29 St Giles'
Oxford, OX1 3LB
Great Britain
dwelsh@maths.ox.ac.uk

Laurence A. Wolsey
Center of Operations Research and
Econometrics (CORE) and
Département d'Ingénierie
Mathémathique (INMA)
Université catholique de Louvain
34, Voie du Roman Pays
1348 Louvain-la-Neuve
Belgium
laurence.wolsey@uclouvain.be

1

On the Location and p-Median Polytopes

Mourad Baïou and Francisco Barahona

Summary. We revisit classical systems of linear inequalities associated with location problems and with the p-median problem. We present an overview of the cases for which these linear systems define integral polytopes. We also give polynomial time algorithms to recognize these cases.

1.1 Introduction

Facility location and p-median are among the most well-studied problems in combinatorial optimization. They are both NP-hard, so there is not much hope of having a complete polyhedral characterization of them. The linear programming relaxations that we use have been known since the 60's and have been the basis for many heuristics, branch and bound algorithms, and approximation algorithms. Despite all this work, very little is known about special cases where these formulations give integral polytopes, and also there are not many special cases where the associated polytope has been completely characterized. We have found a characterization of the graphs for which these linear relaxations define polytopes with all extreme points being integral. Here we present an overview of all these cases. We also give polynomial time algorithms to recognize these classes of graphs. Our characterization shows the basic structures that a graph contains when the polytope has fractional extreme points.

We first deal with location problems, we show that the linear relaxation gives an integral polytope if and only the graph does not contain a certain type of "odd" cycles. Then we deal with the p-median problem. We show that there are five configurations that should be forbidden in order to have an integral polytope. Here the proof consists of three parts as follows. First we show the result for the so-called Y-free graphs. We denote by Y some basic configuration in the graph. The result on Y-free graphs is used to start an induction proof for oriented graphs. These are directed graphs where between any two nodes u and v, at most one of the arcs (u, v) and (v, u) exists. Here the induction is done on the number of Y configurations. The third part consists of extending our result to general directed graphs. Here the induction is done on the number of pairs of nodes u and v such that both (u, v) and (v, u) exist. The initial step of the induction is given by the result on oriented graphs.

This paper is organized as follows. Section 1.2 contains some definitions. Section 1.3 deals with location problems. Section 1.4 covers the p-median problem. In Sect. 1.5 we give an algorithm to recognize the graphs defined in Sect. 1.4. Section 1.6 is devoted to some extensions.

1.2 Preliminary Definitions

A directed graph $G = (V, A)$ is called *oriented* if $(u, v) \in A$ implies $(v, u) \notin A$. For a directed graph $G = (V, A)$ and a set $W \subset V$, we denote by $\delta^+(W)$ the set of arcs $(u, v) \in A$, with $u \in W$ and $v \in V \setminus W$. Also we denote by $\delta^-(W)$ the set of arcs (u, v), with $v \in W$ and $u \in V \setminus W$. We write $\delta^+(v)$ and $\delta^-(v)$ instead of $\delta^+(\{v\})$ and $\delta^-(\{v\})$, respectively. If there is a risk of confusion we use δ_G^+ and δ_G^-. A node u with $\delta^+(u) = \emptyset$ is called a *pendent* node.

A simple cycle C is an ordered sequence

$$v_0, a_0, v_1, a_1, \ldots, a_{p-1}, v_p,$$

where

- $v_i, 0 \le i \le p - 1$, are distinct nodes,
- $a_i, 0 \le i \le p - 1$, are distinct arcs,
- either v_i is the tail of a_i and v_{i+1} is the head of a_i, or v_i is the head of a_i and v_{i+1} is the tail of a_i, for $0 \le i \le p - 1$, and
- $v_0 = v_p$.

By setting $a_p = a_0$, we associate with C three more sets as below.

- We denote by $\hat{C}$ the set of nodes v_i, such that v_i is the head of a_{i-1} and also the head of a_i, $1 \le i \le p$.
- We denote by $\dot{C}$ the set of nodes v_i, such that v_i is the tail of a_{i-1} and also the tail of a_i, $1 \le i \le p$.
- We denote by $\tilde{C}$ the set of nodes v_i, such that either v_i is the head of a_{i-1} and also the tail of a_i, or v_i is the tail of a_{i-1} and also the head of a_i, $1 \le i \le p$.

Notice that $|\hat{C}| = |\dot{C}|$. A cycle will be called *odd* if $p + |\dot{C}|$ (or $|\tilde{C}| + |\dot{C}|$) is odd, otherwise it will be called *even*. A cycle C with $\dot{C} = \emptyset$ is a *directed* cycle. The set of arcs in C is denoted by $A(C)$.

If we do not require $v_0 = v_p$ we have a *path* P. In a similar way we define $\dot{P}$, $\hat{P}$ and $\tilde{P}$, excluding v_0 and v_p. We say that P is *odd* if $p + |\dot{P}|$ is odd, otherwise it is *even*. For the path P, the nodes $v_1, \ldots, v_{p-1}$ are called *internal*.

If G is a connected graph and there is a node u such that its removal disconnects G, we say that u is an *articulation point*. A graph is said to be *two-connected* if at least two nodes should be removed to disconnect it. For simplicity, sometimes we use z to denote the vector (x, y), i.e., $z(u) = y(u)$ and $z(u, v) = x(u, v)$. Also for $S \subseteq V \cup A$ we use $z(S)$ to denote $z(S) = \sum_{a \in S} z(a)$.

A *polyhedron P* is a set defined by a system of linear inequalities, i.e., $P = \{x \mid Ax \leq b\}$. A *face* of P is obtained by setting into equation some of these inequalities. An *extreme point* of P is given by a face that contains a unique element. In other words, some inequalities are set to equation so that this system has a unique solution. A *polytope* is a bounded polyhedron. A polyhedron is called *integral* if all its extreme points are integral.

1.3 Location Problems

Let $G = (V, A)$ be a directed graph, not necessarily connected, where each arc and each node has weight associated with it. We study a "prize collecting" version of a *location problem* (LP) as follows. A set of nodes is selected, usually called *centers*, and then each non-selected node can be assigned to a center. The weight of a node is the revenue obtained by opening a facility at that location, minus the cost of building the facility. The weight of an arc (i, j) is the revenue obtained by assigning the location i to the location j, minus the cost originated by this assignment. The goal is to maximize the sum of the weights of the selected nodes plus the sum of the weights yielded by the assignment. The linear system below defines a linear programming relaxation.

$$\max \sum w(u, v) x(u, v) + \sum w(v) y(v)$$

$$\sum_{(u,v) \in A} x(u, v) + y(u) \leq 1 \quad \forall u \in V, \tag{1}$$

$$x(u, v) \leq y(v) \quad \forall (u, v) \in A, \tag{2}$$

$$0 \leq y(v) \leq 1 \quad \forall v \in V, \tag{3}$$

$$x(u, v) \geq 0 \quad \forall (u, v) \in A. \tag{4}$$

For each node u, the variable $y(u)$ takes the value 1 if the node u is selected and 0 otherwise. For each arc (u, v) the variable $x(u, v)$ takes the value 1 if u is assigned to v and 0 otherwise. Inequalities (1) express the fact that either node u can be selected or it can be assigned to another node. Inequalities (2) indicate that if a node u is assigned to a node v then this last node should be selected. The set of integer vectors that satisfy (1)–(4) corresponds to a *transitive packing* as defined in Müller and Schulz (2002).

Let $P(G)$ be the polytope defined by (1)–(4), and let $LP(G)$ be the convex hull of $P(G) \cap \{0, 1\}^{|V|+|A|}$. Clearly

$$LP(G) \subseteq P(G).$$

Here we characterize the graphs G for which $LP(G) = P(G)$. More precisely, we show that $LP(G) = P(G)$ if and only if G does not contain an odd cycle. We also give a polynomial algorithm to recognize the graphs in this class.

The *Uncapacitated Facility Location Problem* (UFLP) is a variation where V is partitioned into V_1 and V_2. The set V_1 corresponds to the customers, and the set V_2

corresponds to the potential facilities. Each customer in V_1 should be assigned to an opened facility in V_2. This is obtained by considering $A \subseteq V_1 \times V_2$, fixing to zero the variables y for the nodes in V_1 and setting into equation the inequalities (1) for the nodes in V_1. More precisely, the linear programming relaxation for this case is

$$\min \sum c(u, v)x(u, v) + \sum d(v)y(v)$$

$$\sum_{(u,v)\in A} x(u, v) = 1 \quad \forall u \in V_1, \tag{5}$$

$$x(u, v) \leq y(v) \quad \forall (u, v) \in A, \tag{6}$$

$$0 \leq y(v) \leq 1 \quad \forall v \in V_2, \tag{7}$$

$$x(u, v) \geq 0 \quad \forall (u, v) \in A. \tag{8}$$

Here we also characterize the cases for which (5)–(8) defines an integral polytope.

We omit the proofs of several technical lemmas, the full details appear in Baïou and Barahona (2006). The facets of the uncapacitated facility location polytope have been studied in Guignard (1980), Cornuejols and Thizy (1982), Cho et al. (1983a, 1983b) and Cánovas et al. (2002). In Baïou and Barahona (2005) we gave a description of $LP(G)$ for Y-free graphs. The UFLP has also been studied from the point of view of approximation algorithms in Shmoys (1997), Chudak and Shmoys (2003), Sviridenko (2002), Byrka and Aardal (2007) and others. Other references on this problem are Cornuejols et al. (1976), Mirchandani and Francis (1990). The relationship between location polytopes and the stable set polytope has been studied in Cornuejols and Thizy (1982), Cho et al. (1983a, 1983b), De Simone and Mannino (1996), and others. It would be interesting to know if our results also have an equivalent in terms of stable set polytopes, but so far we have not found the right transformation.

1.3.1 Decomposition

In this subsection we consider a graph $G = (V, A)$ that decomposes into two graphs $G_1 = (V_1, A_1)$ and $G_2 = (V_2, A_2)$, with $V = V_1 \cup V_2$, $V_1 \cap V_2 = \{u\}$, $A = A_1 \cup A_2$, $A_1 \cap A_2 = \emptyset$. We define G_1' that is obtained from G_1 after replacing u by u'. We also define G_2', obtained from G_2 after replacing u by u''. The theorem below shows that we have to concentrate on two-connected graphs.

Theorem 3.1. *Suppose that the system*

$$Az' \leq b, \tag{9}$$

$$z'\left(\delta_{G_1'}^+(u')\right) + z'(u') \leq 1 \tag{10}$$

describes $LP(G_1')$. Suppose that (9) contains the inequalities (1)–(4) except for (10). Similarly suppose that

$$Cz'' \leq d, \tag{11}$$

$$z''\left(\delta_{G_2'}^+(u'')\right) + z''(u'') \leq 1 \tag{12}$$

describes $LP(G_2')$. Also (11) contains the inequalities (1)–(4) except for (12). Then the system below describes an integer polytope.

$$Az' \leq b, \tag{13}$$

$$Cz'' \leq d, \tag{14}$$

$$z'\left(\delta_{G_1'}^+(u')\right) + z''\left(\delta_{G_2'}^+(u'')\right) + z'(u') \leq 1, \tag{15}$$

$$z'(u') = z''(u''). \tag{16}$$

We have the following corollary.

Corollary 3.2. *The polytope $LP(G)$ is defined by the system (13)–(16) after identifying the variables $z'(u')$ and $z''(u'')$.*

This last corollary shows that if $LP(G_1')$ and $LP(G_2')$ are defined by (1)–(4), then $LP(G)$ is also defined by (1)–(4). Thus we have to concentrate on graphs that are two-connected. A result analogous to Theorem 3.1, for the stable set polytope, has been given in Chvátal (1975).

1.3.2 Graph Transformations

First we plan to prove that if G has no odd cycle then $LP(G) = P(G)$. The proof consists of assuming that $\bar{z}$ is a fractional extreme point of $P(G)$ and arriving at a contradiction. Below we give several assumptions that can be made about $\bar{z}$ and G, they will be used in the next subsection. The proofs of the lemmas below consist of modifying the graph and the vector $\bar{z}$ so that we obtain a new extreme point associated with a new graph satisfying the assumptions below.

Lemma 3.3. *We can assume that G consists of only one connected component.*

Lemma 3.4. *If $0 < \bar{z}(u,v) < \bar{z}(v)$, we can assume that v is a pendent node with $|\delta^-(v)| = 1$ and $\bar{z}(v) = 1$.*

Lemma 3.5. *We can assume that $0 < \bar{z}(u,v) < 1$ for all $(u,v) \in A$.*

Lemma 3.6. *We can assume that G is either two-connected or it consists of a single arc.*

If the graph G consists of a single arc it is fairly easy to see that $LP(G) = P(G)$, so now we have to deal with the two-connected components. This is treated in the next subsection.

1.3.3 Treating Two-Connected Graphs

In this subsection we assume that the graph G is two-connected and it has no odd cycle. Let $\bar{z}$ be a fractional extreme point of $P(G)$, we are going to assign labels l to the nodes and arcs and define $z'(u,v) = \bar{z}(u,v) + l(u,v)\epsilon$, $z'(u) = \bar{z}(u) + l(u)\epsilon$, $\epsilon > 0$, for each arc (u,v) and each node u. We shall see that every constraint that is satisfied with equality by $\bar{z}$ is also satisfied with equality by z'. This is the required contradiction.

Given a path $P = v_0, a_0, \ldots, a_{p-1}, v_p$. Assume that the label of a_0, $l(a_0)$ has the value 1 or -1. We define the *labeling procedure* as follows.

For $i = 1$ to $p - 1$ do

- If v_i is the head of a_{i-1} and it is the tail of a_i then $l(v_i) = l(a_{i-1})$, $l(a_i) = -l(a_{i-1})$.
- If v_i is the head of a_{i-1} and it is the head of a_i then $l(v_i) = l(a_{i-1})$, $l(a_i) = l(a_{i-1})$.
- If v_i is the tail of a_{i-1} and it is the head of a_i then $l(v_i) = -l(a_{i-1})$, $l(a_i) = -l(a_{i-1})$.
- If v_i is the tail of a_{i-1} and it is the tail of a_i then $l(v_i) = 0$, $l(a_i) = -l(a_{i-1})$.

Notice that the labels of v_0 and v_p were not defined.

We have to study several cases as follows.

Case 1. G contains a directed cycle $C = v_0, a_0, \ldots, a_{p-1}, v_p$. Assume that the head of a_0 is v_1, set $l(v_0) = -1$, $l(a_0) = 1$ and extend the labels as above.

Case 2. G contains a cycle $C = v_0, a_0, \ldots, a_{p-1}, v_p$ and $\dot{C} \neq \emptyset$. Assume $v_0 \in \dot{C}$. Set $l(v_0) = 0$, $l(a_0) = 1$ and extend the labels.

The lemma below is needed to show that for v_0, the constraints that were satisfied with equality by $\bar{z}$ remain satisfied with equality.

Lemma 3.7. *After labeling as in Cases 1 and 2 we have $l(a_{p-1}) = -l(a_0)$.*

Notice that after the first cycle has been labeled as in Cases 1 or 2, the properties below hold, we shall see that these properties hold throughout the entire labeling procedure.

Property 1 *If a node has a nonzero label, then it is the tail of at most one labeled arc.*

Property 2 *If a node has a zero label, then it is the tail of exactly two labeled arcs.*

Once a cycle C has been labeled as in Cases 1 or 2, we have to extend the labeling as follows.

Case 3. Suppose that $l(v_0) \neq 0$ for $v_0 \in C$ (v_0 is the head of a labeled arc), and there is a path $P = v_0, a_0, v_1, a_1, \ldots, a_{p-1}, v_p$ in G such that:

- v_0 is the head of a_0,
- $v_p \in C$,
- $\{v_1, \ldots, v_{p-1}\}$ is disjoint from C.

We set $l(a_0) = l(v_0)$ and extend the labels. Case 3 is needed so that any inequality (2) associated with v_0 that is satisfied with equality, remains satisfied with equality.

We have to see that the label $l(a_{p-1})$ is such that constraints associated with v_p that were satisfied with equality remain satisfied with equality. This is discussed in the next lemma.

Lemma 3.8. *If v_p is the head of a_{p-1} then $l(a_{p-1}) = l(v_p)$. If v_p is the tail of a_{p-1} then $l(a_{p-1}) = -l(v_p)$.*

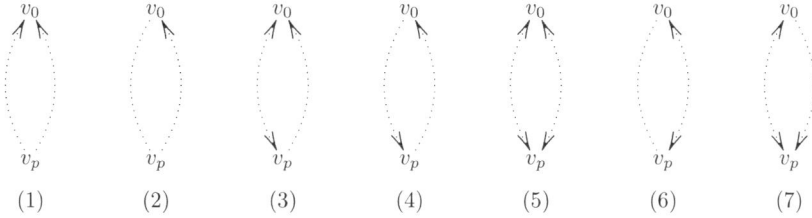

Fig. 1.1. Possible paths in C between v_0 and v_p. It is shown whether v_0 and v_p are the head or the tail of the arcs in C incident to them

Proof (cf. Baïou and Barahona 2006*).* Notice that $v_0 \notin \dot{C}$, in Fig. 1.1 we represent the possible configurations for the paths in C between v_0 and v_p. In this figure we show whether v_0 and v_p are the head or the tail of the arcs in C incident to them. These two paths are denoted by P_1 and P_2.

Consider configuration (1), these two paths should have different parity. When adding the path P, an odd cycle is created with either P_1 or P_2. So configuration (1) will not occur. The same happens with configuration (2).

Now we discuss configuration (3). These two paths should have the same parity. If v_p is the tail of a_{p-1} then P would create an odd cycle with either P_1 or P_2. If v_p is the head of a_{p-1} then P should have the same parity as P_1 and P_2. Then $l(a_{p-1}) = l(v_p)$.

The study of configuration (4) is similar. The two paths should have the same parity. If v_p is the tail of $a_{p\ 1}$ then P would create an odd cycle with either P_1 or P_2. If v_p is the head of a_{p-1} then P should have the same parity as P_1 and P_2, and $l(a_{p-1}) = l(v_p)$.

For configuration (5) again the two paths should have the same parity. If v_p is the head of a_{p-1} then P should have the same parity as P_1 and P_2, and $l(a_{p-1}) = l(v_p)$. If v_p is the tail of a_{p-1} then P should have the same parity as P_1 and P_2, and $l(a_{p-1}) = -l(v_p)$.

Also in configuration (6) the paths P_1 and P_2 should have the same parity. If v_p is the tail of a_{p-1} then P would form an odd cycle with either P_1 or P_2. If v_p is the head of a_{p-1} then P should have the same parity as P_1 and P_2, and $l(a_{p-1}) = l(v_p)$.

In configuration (7) also the two paths should have the same parity. If v_p is the head of a_{p-1} then P should have the same parity as P_1 and P_2, and $l(a_{p-1}) = l(v_p)$. If v_p is the tail of a_{p-1} then P should have the same parity as P_1 and P_2, and $l(a_{p-1}) = -l(v_p)$. $\square$

Based on this the labels are extended successively. Denote by G_l the subgraph defined by the labeled arcs. This is a two-connected graph, so for any two nodes v_0 and v_p it contains a cycle going through these two nodes. Thus we can check if Case 3 applies and extend the labels adding each time a path to the graph G_l. The two lemmas below show that Properties 1 and 2 remain satisfied.

Lemma 3.9. *Let v_p be a node with $l(v_p) \neq 0$. If v_p is the tail of an arc in G_l, then in Case 3 it cannot be the tail of a_{p-1}. Thus Property 1 remains satisfied.*

Lemma 3.10. *Let v_p be a node with $l(v_p) = 0$, thus v_p is the tail of exactly two arcs in G_l. Then in Case 3 it cannot be the tail of a_{p-1}. Therefore Property 2 remains satisfied.*

Once Case 3 has been exhausted we might have some nodes in G_l that are not pendent in G and that are only the head of labeled arcs. For such nodes we have to ensure that inequalities (1) that were satisfied as equality remain satisfied as equality. This is treated in the following.

Case 4. Suppose that v_0 is only the head of labeled arcs, ($l(v_0) \neq 0$), v_0 is not pendent. We have that $\delta^+(v_0) \neq \emptyset$ thus there is a cycle C in G_l and there is a path $P = v_0, a_0, v_1, a_1, \ldots, a_{p-1}, v_p$ in G such that:

- $v_0 \in C$ is the tail of a_0,
- $v_p \in C$,
- $\{v_1, \ldots, v_{p-1}\}$ is disjoint from G_l.

We set $l(a_0) = -l(v_0)$ and extend the labels. We have to see that the label $l(a_{p-1})$ is such that constraints associated with v_p, that were satisfied with equality, remain satisfied with equality. This is discussed below.

Lemma 3.11. *In Case 4 we have that v_p is the tail of a_{p-1} and $l(a_{p-1}) = -l(v_p)$. Also Properties 1 and 2 continue to hold.*

To summarize, the labeling algorithm consists of the following steps.

- Step 1. Identify a cycle C in G and treat it as in Cases 1 or 2. Set $G_l = C$.
- Step 2. For as long as needed label as in Case 3. Each time add to G_l the new set of labeled nodes and arcs.
- Step 3. If needed, label as in Case 4. Each time add to G_l the new set of labeled nodes and arcs. If some new labels have been assigned in this step go to Step 2, otherwise stop.

At this point we can discuss the properties of the labeling procedure. The labels are such that any inequality (2) that was satisfied with equality by $\bar{z}$ is also satisfied with equality by z'. To see that inequalities (1) that were tight remain tight, we need two observations about G_l:

- Any node that has a nonzero label is the tail of exactly one labeled arc having the opposite label.
- If u is a node with $l(u) = 0$, then there are exactly two labeled arcs having opposite labels and whose tail is u.

Finally we give the label "0" to all nodes and arcs that are unlabeled, this completes the definition of z'. Lemma 3.5 shows that inequalities (4) will not be violated. The fact that nodes v with $\bar{z}(v) = 0$ or $\bar{z}(v) = 1$ receive a zero label, shows that inequalities (3) will not be violated. Any constraint that is satisfied with equality by $\bar{z}$ is also satisfied with equality by z', this contradicts the assumption that $\bar{z}$ is an extreme point. We can state the main result of this subsection.

Theorem 3.12. *If the graph G is two-connected and has no odd cycle then LP(G) =*
P(G).

This implies the following.

Theorem 3.13. *If G is a graph with no odd cycle, then LP(G) = P(G).*

Theorem 3.14. *For graphs with no odd cycle, the uncapacitated facility location*
problem is polynomially solvable.

1.3.4 Odd Cycles

In this subsection we study the effect of odd cycles in $P(G)$. Let C be an odd cycle.
We can define a fractional vector $(\bar{x}, \bar{y}) \in P(G)$ as follows:

$$\bar{y}(u) = 0 \quad \text{for all nodes } u \in \dot{C}, \tag{17}$$

$$\bar{y}(u) = 1/2 \quad \text{for all nodes } u \in C \setminus \dot{C}, \tag{18}$$

$$\bar{x}(a) = 1/2 \quad \text{for } a \in A(C), \tag{19}$$

$$\bar{y}(v) = 0 \quad \text{for all other nodes } v \notin C, \tag{20}$$

$$\bar{x}(a) = 0 \quad \text{for all other arcs.} \tag{21}$$

Below we show a family of inequalities that separate the vectors defined above
from $LP(G)$. We call them *odd cycle* inequalities.

Lemma 3.15. *The following inequalities are valid for LP(G).*

$$\sum_{a \in A(C)} x(a) - \sum_{v \in \hat{C}} y(v) \leq \frac{|\tilde{C}| + |\hat{C}| - 1}{2} \tag{22}$$

for every odd cycle C.

These inequalities are {0, 1/2}-Chvatal–Gomory cuts, using the terminology of
Caprara and Fischetti (1996). A separation algorithm can be obtained from the results
of Caprara and Fischetti (1996). In Baïou and Barahona (2006) we gave an alternative
separation algorithm.

Now we can present the following result.

Theorem 3.16. *Let G be a directed graph, then LP(G) = P(G) if and only if G*
does not contain an odd cycle.

Proof (cf. Baïou and Barahona 2006). If G contains and odd cycle C, then we can
define a vector $(\bar{x}, \bar{y}) \in P(G)$ as in (17)–(21). We have

$$\sum_{a \in A(C)} \bar{x}(a) - \sum_{v \in \hat{C}} \bar{y}(v) = \frac{|\tilde{C}| + |\hat{C}|}{2}.$$

Lemma 3.15 shows that $\bar{z} \notin LP(G)$.

Then the theorem follows from Theorem 3.13.   □

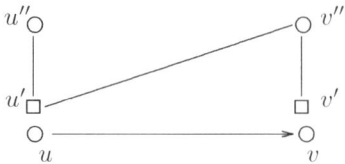

Fig. 1.2.

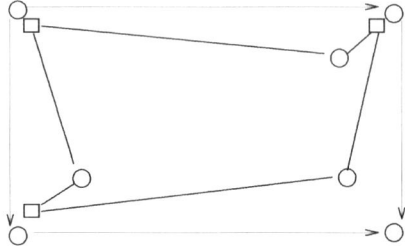

Fig. 1.3. An odd cycle in G and the corresponding cycle in H. The nodes of H close to a node $u \in G$ correspond to u' or u''

1.3.5 Detecting Odd Cycles

Now we study how to recognize the graphs G for which $LP(G) = P(G)$. We start with a graph G and several transformations are needed.

The first transformation consists of building an undirected graph $H = (N, E)$. For every node $u \in G$ we have the nodes u' and u'' in N, and the edge $u'u'' \in E$. For every arc $(u, v) \in G$ we have an edge $u'v'' \in E$. See Fig. 1.2.

Consider a cycle C in G, we build a cycle C_H in H as follows.

- If (u, v) and (u, w) are in C, then the edges $u'v''$ and $u'w''$ are taken.
- If (u, v) and (w, v) are in C, then the edges $u'v''$ and $v''w'$ are taken.
- If (u, v) and (v, w) are in C, then the edges $u'v''$, $v''v'$, and $v'w''$ are taken.

On the other hand, a cycle in H corresponds to a cycle in G. Thus there is a one to one correspondence among cycles of G and cycles of H. Moreover, if the cycle in H has cardinality $2q$, then $q = |\dot{C}| + |\tilde{C}|$, where C is the corresponding cycle in G. Therefore an odd cycle in G corresponds to a cycle in H of cardinality $2(2p + 1)$ for some positive integer p. See Fig. 1.3.

In other words, finding an odd cycle in G reduces to finding a cycle of cardinality $2(2p + 1)$, for some positive integer p, in the bipartite graph H.

For this question, a linear time algorithm was given in Yannakakis (1985), a simple $O(|V||A|^2)$ has been given in Conforti and Rao (1987).

1.3.6 Uncapacitated Facility Location

Now we assume that V is partitioned into V_1 and V_2, $A \subseteq V_1 \times V_2$, and we deal with the system

$$\sum_{(u,v)\in A} x(u, v) = 1 \quad \forall u \in V_1, \tag{23}$$

$$x(u, v) \leq y(v) \quad \forall (u, v) \in A, \tag{24}$$

$$0 \leq y(v) \leq 1 \quad \forall v \in V_2, \tag{25}$$

$$x(u, v) \geq 0 \quad \forall (u, v) \in A. \tag{26}$$

If the variables x and y are constrained to be integer, then we have the uncapacitated facility location problem (UFLP). We denote by $\Pi(G)$ the polytope defined by (23)–(26). Notice that $\Pi(G)$ is a face of $P(G)$. Let $\bar{V}_1$ be the set of nodes $u \in V_1$ with $|\delta^+(u)| = 1$. Let $\bar{V}_2$ be the set of nodes in V_2 that are adjacent to a node in $\bar{V}_1$. It is clear that the variables associated with nodes in $\bar{V}_2$ should be fixed, i.e., $y(v) = 1$ for all $v \in \bar{V}_2$. Let us denote by $\bar{G}$ the subgraph induced by $V \setminus \bar{V}_2$. In this section we prove that $\Pi(G)$ is an integer polytope if and only if $\bar{G}$ has no odd cycle.

Let us first assume that $\bar{G}$ has no odd cycle. As before, we suppose that $\bar{z}$ is a fractional extreme point of $\Pi(G)$. The analogues of Lemmas 3.3, 3.4 and 3.5 apply here. Thus we can assume that we deal with a connected component G'. Lemma 3.4 implies that any node in $\bar{V}_2$ is not in a cycle of G'. Therefore G' has no odd cycle and $P(G')$ is an integer polytope. Since $\Pi(G')$ is a face of $P(G')$, we have a contradiction.

Now let C be an odd cycle of $\bar{G}$. We can define a fractional vector as follows:

$$\bar{y}(v) = 1/2 \quad \text{for all nodes } v \in V_2 \cap V(C),$$
$$\bar{x}(a) = 1/2 \quad \text{for } a \in A(C),$$
$$\bar{y}(v) = 1 \quad \text{for all nodes } v \in V_2 \setminus V(C).$$

For every node $u \in V_1 \setminus V(C)$, we look for an arc $(u, v) \in \delta^+(u)$. If $\bar{y}(v) = 1$ we set $\bar{x}(u, v) = 1$. If $\bar{y}(v) = 1/2$, then there is another arc $(u, w) \in \delta^+(u)$ such that $\bar{y}(w) = 1/2$ or $\bar{y}(w) = 1$. We set $\bar{x}(u, v) = \bar{x}(u, w) = 1/2$. Finally we set $\bar{x}(a) = 0$ for each remaining arc a. This vector satisfies (23)–(26), but it violates the inequality (22) associated with C. This shows that in this case (23)–(26) does not define an integer polytope. Thus we can state our main results.

Theorem 3.17. *The system (23)–(26) defines an integral polytope if and only if $\bar{G}$ has no odd cycle.*

Theorem 3.18. *The UFLP is polynomially solvable for graphs G such that $\bar{G}$ has no odd cycle.*

This class of graphs can be recognized in polynomial time as described in Sect. 1.3.5.

1.4 The p-Median Problem

The p-median problem is closely related to the uncapacitated facility location problem. Here we need to select a specific number of centers. Formally, let $G = (V, A)$

be a directed graph, not necessarily connected. We assume that G is simple, i.e., between any two nodes u and v there is at most one arc directed from u to v. Also for each arc $(u, v) \in A$ and node $v \in V$ there is an associated cost $c(u, v)$ and $w(v)$, respectively. The *p-median problem* (*p*MP) consists of selecting p nodes, usually called *centers*, and then assign each non-selected node to a selected node. The goal is to select p nodes that minimize the sum of the costs of the selected nodes plus the sum of the costs yield by the assignment of the non-selected nodes. This problem has several applications such as location of bank accounts (Cornuejols et al. 1976), placement of web proxies in a computer network (Vigneron et al. 2000), semistructured data bases (Toumani 2002; Nestorov et al. 1998).

The following define an integer linear programming formulation for the *p*MP:

$$\min \sum_{(u,v)\in A} c(u, v)x(u, v) + \sum_{v\in V} d(v)y(v) \tag{27}$$

$$\sum_{v\in V} y(v) = p, \tag{28}$$

$$\sum_{v:(u,v)\in A} x(u, v) + y(u) = 1 \quad \forall u \in V, \tag{29}$$

$$x(u, v) \leq y(v) \quad \forall (u, v) \in A, \tag{30}$$

$$0 \leq y(v) \leq 1 \quad \forall v \in V, \tag{31}$$

$$x(u, v) \geq 0 \quad \forall (u, v) \in A. \tag{32}$$

Denote by $P_p(G)$ the polytope defined by (28)–(32), this gives a linear programming relaxation of the *p*MP. Let $pMP(G)$ be the convex hull of $P_p(G) \cap \{0, 1\}^{|A|+|V|}$.

The facets of $pMP(G)$ have been studied in Avella and Sassano (2001) and de Farias (2001). In Avella and Sassano (2001), new facets have been presented using a reduction to the stable set problem in the *intersection graph of G*. The intersection graph of G is defined as follows: its nodes are the arcs of G and there is an edge between two nodes (u, v) and (w, t) if $u = w$ or $v = w$. If we associate the cost $c(u, v)$ with each node (u, v) of the intersection graph, then the p-median problem in G, when the cost associated with the nodes of G is zero, is equivalent to find a stable set with minimum weight of cardinality $|V| - p$ in the intersection graph of G. In de Farias (2001), other class of facets have been presented in the class of bipartite graphs.

In this section we characterize all directed graphs such that $P_p(G) = pMP(G)$. To state our main result we need some definitions.

In Fig. 1.4, we show four directed graphs and for each of them a fractional extreme point of $P_p(G)$. The numbers near the nodes correspond to the variables y, all the arcs variables are equal to $\frac{1}{2}$.

Definition 4.1. *A simple cycle C is called a Y-cycle if for every $v \in \hat{C}$ there is an arc $(v, \bar{v})$, where $\bar{v}$ is in $V \setminus \dot{C}$.*

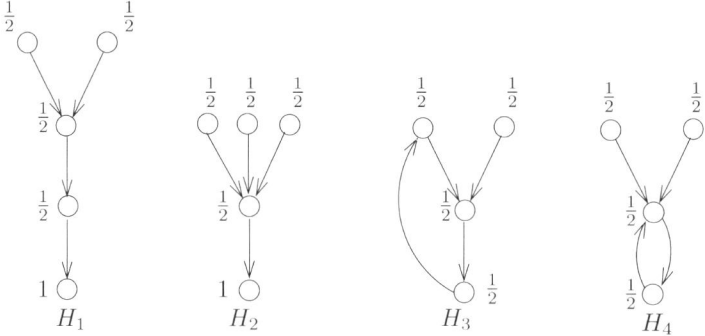

Fig. 1.4. Fractional extreme points of $P_p(G)$

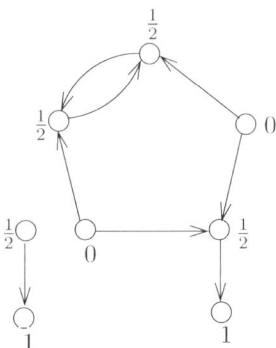

Fig. 1.5. An odd Y-cycle with an arc outside the cycle

In Fig. 1.5 we show a fractional extreme point of $P_p(G)$ different from those given in Fig. 1.4. It consists of an odd Y-cycle with an arc having both of its endnodes outside the cycle. The values reported near each node represent the node variables, the arc variables are all equal to $\frac{1}{2}$. These values form a fractional extreme point of $P_p(G)$, with $p = 4$.

The theorem below is the main result of this section. It shows that the configurations in Figs. 1.4 and 1.5 are the only configurations that should be forbidden in order to have an integral polytope.

Theorem 4.2. *Let $G = (V, A)$ be a directed graph, then $P_p(G)$ is integral if and only if*

(i) *it does not contain as a subgraph any of the graphs H_1, H_2, H_3 or H_4 of Fig. 1.4, and*

(ii) *it does not contain an odd Y-cycle C and an arc (u, v) with neither u nor v in $V(C)$.*

The proof of this theorem consists of three parts presented in Sects. 1.4.2, 1.4.3 and 1.4.4. The last two parts are the subject of two papers, see Baïou and Barahona

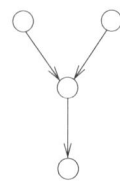

Fig. 1.6. The graph Y

(2007a, 2007b), each requires more than twenty pages. For these reasons, here we only present an overview of the proof.

In the first part of the proof, Sect. 1.4.2, we show that $P_p(G)$ is integral in Y-free graphs with no odd directed cycles. A *Y-free graph* is an oriented graph that does not contain as a subgraph the graph Y of Fig. 1.6. This class of graphs has been introduced in (Baïou and Barahona 2005).

In the second part, Sect. 1.4.3, we prove Theorem 4.2 when restricted to oriented graphs. This proof uses an induction on the number of subgraphs Y. The last part is devoted to the proof of Theorem 4.2 in general directed graphs and uses the result in oriented graphs as starting point. We will only present the sufficiency proof. The necessity proof is illustrated in Figs. 1.4 and 1.5. The fractional extreme points given in these figures can be easily extended to any graph that does not satisfy conditions (i) and (ii) of Theorem 4.2. Thus the graphs we consider do not contain as a subgraph any of the graphs H_1, H_2, H_3 or H_4 of Fig. 1.4.

1.4.1 Preliminaries

Let $G = (V, A)$ be a directed graph. Let $l : V \cup A \to \{0, -1, 1\}$ be a labeling function that associates to each node and arc of G a label 0, -1 or 1.

A vector $(x, y) \in P_p(G)$ will be denoted by z, i.e., $z(u) = y(u)$ for all $u \in V$ and $z(u, v) = x(u, v)$ for all $(u, v) \in A$. Given a vector z and a labeling function l, we define a new vector z_l from z as follows: $z_l(u) = z(u) + l(u)\epsilon$, for all $u \in V$, and $z_l(u, v) = z(u, v) + l(u, v)\epsilon$, for all $(u, v) \in A$, where ϵ is a sufficiently small positive scalar.

The Labeling Procedure for Even Cycles

Let $C = v_0, a_0, v_1, a_1, \dots, a_{p-1}, v_p$ be an even cycle, not necessarily a Y-cycle.

- If C is a directed cycle, assume that v_0 is the tail of a_0, then set $l(v_0) \leftarrow 1$; $l(a_0) \leftarrow -1$. Otherwise, assume $v_0 \in \dot{C}$ and set $l(v_0) \leftarrow 0$; $l(a_0) \leftarrow 1$.
- Extend the labels as in Sect. 1.3.3.

Remark 4.3. If C is a directed even cycle, then $l(a_{p-1}) = l(v_0)$ and $\sum l(v_i) = 0$.

This remark is easy to see. The second property is given in the following lemma and it concerns non-directed cycles.

Lemma 4.4. *If C is a non-directed even cycle, then $l(a_{p-1}) = -l(a_0)$ and $\sum l(v_i) = 0$.*

We are going to deal with a vector z that is a fractional extreme point of $P_p(G)$.

Recall that the graph G we consider in Sect. 1.4.2 is Y-free and with no odd directed cycles and the graph G in Sects. 1.4.3 and 1.4.4 do not contain as a subgraph any of the graphs H_1, H_2, H_3 or H_4 of Fig. 1.4. In these graphs the following two lemmas hold:

Lemma 4.5. *We may assume that $z(u, v) > 0$ for all $(u, v) \in A$.*

Proof (cf. Baïou and Barahona 2007a). Let G' be the graph obtained after removing all arcs (u, v) with $\bar{z}(u, v) = 0$. The graph G' has the same properties as G. Let z' be the restriction of $\bar{z}$ on G'. Then z' is a fractional extreme point of $P_p(G')$.   □

Lemma 4.6. *We may assume that $|\delta^-(v)| \leq 1$ for every pendent node v in G.*

Proof (cf. Baïou and Barahona 2007a). If v is a pendent node in G and $\delta^-(v) = \{(u_1, v), \ldots, (u_k, v)\}$, we can split v into k pendent nodes $\{v_1, \ldots, v_k\}$ and replace every arc (u_i, v) with (u_i, v_i). Then we define z' such that $z'(u_i, v_i) = z(u_i, v)$, $z'(v_i) = 1$, for all i, and $z'(u) = z(u)$, $z'(u, w) = z(u, w)$ for every other node and arc. Let G' be this new graph. The graph G' has the same properties as G. Moreover, it is easy to check that z' is a fractional extreme point of $P_{p+k-1}(G')$.   □

1.4.2 Y-Free Graphs

In Baïou and Barahona (2005), we characterized the fractional extreme points of $P_p(G)$ for Y-free graphs. Then we showed that by adding the family of odd cycle inequalities associated with each directed odd cycle in G we obtain an integral polytope. An alternate proof of this result based on matching theory is given in Stauffer (2007).

To prove our main result we do not need the description of $pMP(G)$ in Y-free graphs. We need its description in a smaller class described by those Y-free graphs with no odd directed cycle. In this restricted class of graphs $P_p(G)$ is integral, this is a directed consequence of Theorem 14 in Baïou and Barahona (2005). Below we give a proof based on the matching polytope in bipartite graphs, which is along the same lines of the proof given in Stauffer (2007).

Theorem 4.7. *If $G = (V, A)$ is a Y-free graph with no odd directed cycle, then for any p the polytope $P_p(G)$ is integral.*

Proof. Let $G = (V, A)$ be a Y-free graph with no odd directed cycle. Assume the contrary, and let $z = (x, y)$ be an extreme fractional point of $P_p(G)$.

Using Fourier-Motzkin elimination, we obtain the following system of linear inequalities, that defines the projection of $P_p(G)$ onto the arc variables space; call it $Q_p(G)$.

$$\sum_{(u,v)\in A} x(u,v) = |V| - p, \tag{33}$$

$$x(w,u) + \sum_{v:(u,v)\in A} x(u,v) \le 1 \quad \forall (w,u) \in A, \tag{34}$$

$$x(u,v) \ge 0 \quad \forall (u,v) \in A. \tag{35}$$

Remark that by Lemma 4.6 and the fact that G is a Y-free graph, we have that $|\delta^-(v)| \le 1$ for all $v \in V$. Hence if we omit the orientation of the arcs in G we obtain a undirected graph $I(G) = (V, E)$, and inequalities (34) and (35) are equivalent to

$$x(\delta_{I(G)}(v)) \le 1 \quad \forall v \in V, \tag{36}$$

$$x(e) \ge 0 \quad \forall e \in E. \tag{37}$$

Combining Lemma 4.6 and the fact that G does not contain an odd directed cycle, we obtain that $I(G)$ is a bipartite graph and hence the polytope defined by inequalities (36) and (37) is the matching polytope of a bipartite graph, so it is integral. Now by adding the equality $\sum_{e\in E} x(e) = |V| - p$ to the linear system defined by (36) and (37) the resulting polytope still integral, this is a well known property of the matching polytope, see for instance Lawler (1976). This proves that $Q_p(G)$ is integral.

To finish the proof of our theorem it suffices to see that if $z = (x, y)$ is an extreme point of $P_p(G)$, then x is an extreme point of $Q_p(G)$, which is easy to verify. □

1.4.3 Oriented Graphs

Let $G = (V, A)$ be an oriented graph that satisfies conditions (i) and (ii) of Theorem 4.2. First we study the case when G has no odd Y-cycle, and in the second case we assume that G has an odd Y-cycle.

G Does not Contain an Odd Y-Cycle

Let $t \in V$. The node t is called a Y-node in $G = (V, A)$ if there are three different nodes u_1, u_2, w in V such that (u_1, t), (u_2, t) and (t, w) belong to A. Denote by Y_G the set of Y-nodes in G.

The proof is done by induction on the number of Y-nodes. If $|Y_G| = 0$ then, the graph is Y-free with no odd directed cycle, it follows from Theorem 4.7 in Sect. 1.4.2 that $P_p(G)$ is integral. Assume that $P_p(G')$ is integral for any positive integer p and for any oriented graph G', with $|Y_{G'}| < |Y_G|$, that satisfies condition (i) and does not contain an odd Y-cycle. Now we suppose that $z = (x, y)$ is a fractional extreme point of $P_p(G)$ and we plan to obtain a contradiction. The next lemma we need is as follows.

Lemma 4.8. *G does not contain a cycle.*

Proof (sketch, cf. Baïou and Barahona 2007a*).* The proof of this lemma is a direct application of the labeling procedure of Sect. 1.4.1. We assume that there is a cycle, using Lemma 4.6, we can derive an even Y-cycle C, and we assign labels to the nodes and arcs of C following the labeling procedure of Sect. 1.4.1. Extend the labels as follows: for each node $v \in \hat{C}$, choose an arc $(v, \bar{v})$, $\bar{v} \notin V(C)$, and assign the label $-l(v)$ to it. Assign a zero label to all remaining nodes and arcs. In the last step, using Lemma 4.5 we show that any constraint that is satisfied with equality by z is also satisfied with equality by z_l. This contradicts the fact that z is an extreme point of $P_p(G)$. □

The graph G must contain at least one Y-node t with its incident arcs (u_1, t), (u_2, t), (t, w). Using Lemma 4.8 we can prove that V can be partitioned into W_1 and W_2 so that $\{u_1, t, w\} \subseteq W_1$ and $u_2 \in W_2$, and that the only arc in G between W_1 and W_2 is (u_2, t).

Next we show that $z(t) = \frac{1}{2}$. We have that $Q(G)$, the polytope defined by (29)–(32), is a face of the polytope $P(G)$ defined by (1)–(4)) studied in Sect. 1.3. And by Theorem 3.13, we know that $P(G)$ is integral when G does not contain an odd cycle, which is the case here. Thus $Q(G)$ is also integral. The polytope $P_p(G)$ is obtained from $Q(G)$ by adding exactly one equation. A simple polyhedral fact is that if $Q(G)$ is integral, then the values of z are in $\{0, 1, \alpha, 1 - \alpha\}$, for some number $\alpha \in [0, 1]$. But since $z(t) = \frac{1}{2}$ we have that all fractional values of z are equal to $\frac{1}{2}$.

Define $p_1 = \sum_{v \in W_1} z(v)$ and $p_2 = \sum_{v \in W_2} z(v)$, so $p = p_1 + p_2$. We distinguish two cases: p_1 and p_2 are integer; and they are not.

If the numbers p_1 and p_2 are integer, we define the graphs G^1 and G^2 as follows. Let $A(W_1)$ and $A(W_2)$ be the set of arcs in G having both endnodes in W_1 and W_2, respectively. Let $G^1 = (W_1, A(W_1))$ and $G^2 = (W_2 \cup \{t', v', w'\}, A(W_2) \cup \{(u_2, t'), (t', v'), (v', w')\})$.

Let z_1 be the restriction of z to G^1. Clearly $z_1 \in P_{p_1}(G^1)$. Define z_2 as follows, $z_2(u_2, t') = z(u_2, t) = \frac{1}{2}$, $z_2(t') = \frac{1}{2}$, $z_2(t', v') = \frac{1}{2}$, $z_2(v') = \frac{1}{2}$, $z_2(v', w') = \frac{1}{2}$, $z_2(w') = 1$ and $z_2(u) = z(u)$, $z_2(u, v) = z(u, v)$ for all other nodes and arcs of G^2. We have that $z_2 \in P_{p_2+2}(G^2)$.

Both graphs G^1 and G^2 satisfy condition (i) of Theorem 4.2 and do not contain an odd Y-cycle. Moreover, $|Y_{G^1}| < |Y_G|$ and $|Y_{G^2}| < |Y_G|$. Since z_1 and z_2 are both fractional, the induction hypothesis implies that they are not extreme points of $P_{p_1}(G^1)$ and $P_{p_2+2}(G^2)$, respectively. Thus there must exist a 0-1 vector $z_1' \in P_{p_1}(G^1)$ with $z_1'(t) = 0$ so that the same constraints that are tight for z_1 are also tight for z_1'. Also there must exist a 0-1 vector $z_2' \in P_{p_2+2}(G^2)$ with $z_2'(t') = 0$ such that the same constraints that are tight for z_2 are also tight for z_2'. Now by combining z_1' and z_2' one can define a solution $z' \in P_p(G)$ that satisfies as equality each constraint that is satisfied as equality by z.

In the case where the numbers p_1 and p_2 are not integer, we use the same idea as above but applied for new graphs G^1 and G^2, where $G^1 = (W_1 \cup \{u_1'\}, (A(W_1) \setminus \{(u_1, t)\}) \cup \{(u_1, u_1'), (u_1', t)\})$ and $G^2 = (W_2 \cup \{t', w'\}, A(W_2) \cup \{(u_2, t'), (t', w')\})$.

Notice that $\sum_{v \in W_1} z(v) = p_1 = \alpha + \frac{1}{2}$ and $\sum_{v \in W_2} z(v) = p_2 = \beta - \frac{1}{2}$, where α and β are integers and $\alpha + \beta = p$. Thus the number of nodes to be selected in G^1 (resp. G^2) is $\alpha + 1$ (resp. $\beta + 1$). This concludes the proof.

G Contains an Odd Y-Cycle

We assume that G satisfies conditions (i) and (ii) of Theorem 4.2 and contains an odd Y-cycle. We also assume that z is a fractional extreme point of $P_p(G)$. The first lemma we need is the following.

Lemma 4.9. *The graph G contains exactly one odd Y-cycle.*

Let C be the unique odd Y-cycle in G. We showed in Baïou and Barahona (2007a) that in this case G has the following special structure. The node set V is partitioned into three subsets, $V(C)$, V' and V''. Each node in V' has exactly one arc incident to it, this arc is directed into a node in $\dot{C}$. Also, each node in V'' has exactly one arc incident to it, this arc is directed away from a node in $V(C)$. Each node in $V(C)$ is adjacent to at most one node in $V' \cup V''$. We denote by A' (resp. A'') the set of arcs incident to the nodes in V' (resp. V''). These arcs together with $A(C)$ define the arc-set of G.

Lemma 4.10. *We may assume that $z(u, v) = z(v)$ for each arc in $A(C)$.*

Proof (cf. Baïou and Barahona 2007a). If we have an arc $(u, v) \in A(C)$ with $z(u, v) < z(v)$, then we remove this arc and add a new arc (u, v') and we assign the value $z(u, v)$ to the arc (u, v') and 1 to the node v'. From Lemma 4.9, this new graph does not contain an odd Y-cycle and the associated solution is fractional extreme point. But from Sect. 1.4.3, this is impossible. □

Next we concentrate on $Q(G)$, the polytope defined by (29)–(32). Notice that since $P_p(G)$ is obtained from $Q(G)$ by adding one equation, then an extreme point of $P_p(G)$ is either an extreme point of $Q(G)$ or a convex combination of two extreme points of $Q(G)$. We omit the proof of the following lemma.

Lemma 4.11. *If z is a fractional extreme point of $Q(G)$ with $z(u, v) = z(v)$ for each $(u, v) \in A(C)$, then $z(u, v) = \frac{1}{2}$ for each arc $(u, v) \in A(C)$, $z(v) = \frac{1}{2}$ for each node $v \in \hat{C} \cup \tilde{C}$ and $z(v) = 0$ for each node $v \in \dot{C}$.*

From this lemma and the definition of G we obtain the following corollary.

Corollary 4.12. *z cannot be an extreme point of $Q(G)$.*

Proof (cf. Baïou and Barahona 2007a). Assume that z is an extreme point of $Q(G)$. By definition we have $z(v) = 1$ for each node $v \in V''$. From Lemma 4.11, $z(v) = 0$ if $v \in \dot{C}$, so by the definition of V' we have $z(v) = 1$ for each $v \in V'$. Again from Lemma 4.11 we have $z(v) = \frac{1}{2}$ if $v \in \tilde{C} \cup \hat{C}$. Hence $\sum_{v \in V} z(v) = |V'| + |V''| + \frac{|\tilde{C}| + |\hat{C}|}{2}$ but $|\tilde{C}| + |\hat{C}|$ is odd, so $\sum_{v \in V} z(v)$ is not an integer, a contradiction. □

The corollary above implies that the extreme point z of $P_p(G)$ is a convex combination of two extreme points of $Q(G)$, they are $\tilde{z}$ and $\hat{z}$. Thus $z = \alpha\tilde{z} + (1 - \alpha)\hat{z}$, with $0 < \alpha < 1$.

Denote by A_1'' the arcs in A'' that are incident to a node in $\dot{C} \cup \tilde{C}$. Also let $\dot{C}^+$ be the set of nodes $v \in \dot{C}$ with $z(v) > 0$. Using the structure of the graph and the fact that z is a fractional extreme point of $P_p(G)$, the proof reduces to the following four cases: (1) $A' = \{(u, v)\}$, $A_1'' = \emptyset$, $\dot{C}^+ = \{v\}$; (2) $A_1'' = \{(u, v)\}$, $A' = \emptyset$, $\dot{C}^+ = \emptyset$; (3) $\dot{C}^+ = \{v\}$, $A' \cup A_1'' = \emptyset$; (4) $\dot{C}^+ = \emptyset$, $A' \cup A_1'' = \emptyset$.

In each of theses cases, we show that z cannot be an extreme point of $P_p(G)$ if both $\tilde{z}$ and $\hat{z}$ are fractional or both are integral. It remains the case when one is integral and the other is fractional. Notice that if $\tilde{z}$ or $\hat{z}$ is fractional, then it satisfies the conditions of Lemma 4.11, this follows from Lemma 4.10 where $z(u, v) = z(v)$ if $(u, v) \in A(C)$. Using this together with Lemma 4.11 the contradiction we obtain is that $\sum_{v \in V} z(v) = q + \frac{\alpha}{2}$, where q is an integer.

1.4.4 General Directed Graphs

Let us redefine a Y-cycle in this context, that is in the graphs that do not contain any of the graphs of Fig. 1.4 as a subgraph. With this we can distinguish the nodes in $\hat{C}$ that do not satisfy Definition 4.1, which is useful in the proof of Lemma 4.15.

Definition 4.13. *A simple cycle C is called a Y-cycle if for every $v \in \hat{C}$ at least one of the following hold*:

(i) *there exists an arc $(v, \bar{v}) \notin A(C)$, $\bar{v} \notin V(C)$, or*
(ii) *there exists an arc $(v, \bar{v}) \notin A(C)$, $\bar{v} \in \tilde{C}$ and $\bar{v}$ is one of the two neighbors of v in C.*

For a simple cycle C, denote by $\hat{C}_{(i)}$ the set of nodes in $\hat{C}$ that satisfy condition (i) of the above definition. Notice that we may have nodes in $\hat{C}$ that satisfy both (i) and (ii).

We study two cases as follows.

G Does not Contain an Odd Y-Cycle

We assume that $G = (V, A)$ is a directed graph that does not contain any of the graphs H_1, H_2, H_3 or H_4 of Fig. 1.4 as a subgraph. Also we assume that G does not contain an odd Y-cycle.

Assume that z is a fractional extreme point of $P_p(G)$. The first step in the proof is to show the following lemma, that is an analogue of Lemma 4.8. Its proof is harder than the proof of Lemma 4.8 and requires new definitions and notions. This proof illustrates one of the main differences between the oriented and the directed case.

Definition 4.14. *Let C be a Y-cycle in a directed graph $G = (V, A)$. A node $v \in V(C)$ is called a blocking node if one of the following hold*:

(i) $v \in \tilde{C}$, $(v, u) \in A(C)$, $(u, v) \in A \setminus A(C)$ and $u \in \tilde{C}$, or
(ii) $v \in \hat{C}$, $(u, v) \in A(C)$, $(w, v) \in A(C)$, $(v, u) \in A \setminus A(C)$, $(v, w) \in A \setminus A(C)$
   and both u and w are in $\tilde{C}$.

Lemma 4.15. *If $z(u, v) = z(v)$, for all (u, v) with v not a pendent node, then G does not contain a cycle of size at least three.*

Proof (cf. Baïou and Barahona 2007b). Assume the contrary. Suppose that G admits such a cycle. The first step is to derive an even Y-cycle. Let $C' = v_0, a_0, v_1, a_1, \ldots,$ a_{p-1}, v_p, be a simple cycle with $p \geq 3$. Suppose that C' is not a Y-cycle. Then we can show that there is a node $v_i \in \hat{C}'$ with $\delta^+(v_i) = \{(v_i, v_{i-1}), (v_i, v_{i+1})\}$, where v_{i-1} and v_{i+1} are the two neighbors of v_i in C' and they belong to $\dot{C}$. Now it suffices to define C from C', recursively, following the procedure below:

*   Step 1. $A(C) \leftarrow A(C')$, $V(C) \leftarrow V(C')$, $C \leftarrow C'$.
*   Step 2. If there exist $v_i \in \hat{C}$, a node not satisfying Definition 4.13(i) and (ii), go to Step 3. Otherwise stop, C is a Y-cycle.
*   Step 3. $A(C) \leftarrow (A(C) \setminus \{(v_{i-1}, v_i), (v_{i+1}, v_i)\}) \cup \{(v_i, v_{i-1}), (v_i, v_{i+1})\}$. C is the new cycle defined by $A(C)$. Go to Step 2.

Each Step 3 decreases by one the number of nodes in $\hat{C}$. Thus the procedure must end with a Y-cycle C.

The next step is to apply the labeling procedure to an even Y-cycle. Call this labeling l. We extend l to all other nodes and arcs in order to get a solution z_l that satisfies as equality each constraint satisfied as equality by z. The extension of l is possible only when the Y-cycle does not contain a blocking node. We show that if we choose a Y-cycle C with $|C_{(i)}|$ maximum, then this cycle does not contain a blocking node. This completes the proof of this lemma. $\square$

The lemma above is used to prove the following:

Lemma 4.16. *We cannot have $z(u, v) = z(v)$, for all (u, v) with v not a pendent node.*

Proof (cf. Baïou and Barahona 2007b). Denote by $Pair(G)$ the set of pair of nodes $\{u, v\}$ such that both arcs (u, v) and (v, u) belong to A.

The proof is by induction on $|Pair(G)|$. If $|Pair(G)| = 0$ then G is an oriented graph that satisfies conditions (i) and (ii) of Theorem 4.2. Thus from Sect. 1.4.3, $P_p(G)$ has no fractional extreme point so the lemma is true. Suppose now that $|Pair(G)| = m + 1$, for $m \geq 0$.

Suppose that $z(u, v) = z(v)$ for each arc (u, v) with v not a pendent node. Notice that Lemma 4.15 applies, so G does not contain a cycle. Let (u, v) and (v, u) be two arcs in G. Denote by $G(u, v)$ the graph obtained from G by removing the arc (u, v) and adding a new arc (u, t), where t is a new pendent node. Define $\tilde{z} \in P_{\tilde{p}}(G(u, v))$, $\tilde{p} = p + 1$, to be $\tilde{z}(u, t) = z(u, v)$, $\tilde{z}(t) = 1$ and $\tilde{z}(r) = z(r)$, $\tilde{z}(r, s) = \tilde{z}(r, s)$ for every other node and arc.

The graph $G(u, v)$ is directed with no multiple arcs and satisfies condition (i) of Theorem 4.2. Since G does not contain a cycle, we have that $G(u, v)$ has no odd Y-cycle. Moreover $|Pair(G(u, v))| \leq m$, hence the induction hypothesis applies for $G(u, v)$. We have that $\bar{z}$ is a fractional vector in $P_{\tilde{p}}(G(u, v))$ with $\bar{z}(u, v) = \bar{z}(v)$ for each arc (u, v), with v not pendent. By the induction hypothesis $\bar{z}$ is not an extreme point. Thus, there exists a set of extreme points of $P_{\tilde{p}}(G(u, v))$, $z^1, \ldots, z^k$, where each constraint that is tight for $\bar{z}$ is also tight for each of $z^1, \ldots, z^k$, and $\bar{z}$ is a convex combination of $z^1, \ldots, z^k$. We can show that all these extreme points are in 0-1.

Let z^1, with $z^1(v, u) = 1$. Define $z'' \in P_p(G)$ as follows: $z''(u, v) = z^1(u, t)$ and $z''(r, s) = z^1(r, s)$, $z''(r) = z^1(r)$, for all other nodes and arcs. All constraints that are tight for z are also tight for z''. To see this, it suffices to remark that $z''(v) = z^1(v) = 0$ and $z''(u, v) = z^1(u, t) = 0$. This contradicts the fact that z is an extreme point of $P_p(G)$.   □

Let v a node in G. We call v a *knot* if $\delta^-(v) = \{(u, v), (w, v)\}$, $u \neq w$ and both (v, u) and (v, w) belong to $\delta^+(v)$.

Suppose that G does not contain a knot. From Lemma 4.16 we may assume that there is an arc (u, v) with $z(u, v) < z(v)$ and v is not a pendent node. We may assume that G' the graph obtained from G by removing (u, v) and adding a new pendent node v' and the arc (u, v') contains an odd Y-cycle C. Otherwise, instead of considering G with z, we consider G' with z', $z'(u, v') = z(u, v)$, $z'(v') = 1$, and $z'(s, t) = z(s, t)$, $z'(r) = z(r)$ for all other arcs and nodes.

Since G contains no knot, this implies that $\delta_G^+(u) = \{(u, v)\}$ and $\delta_G^-(u) = \{(s, u), (v, u)\}$, where s and v are the nodes that are adjacent to u in C. Remark that v must be in $\dot{C}$, otherwise C is also an odd Y-cycle in G, which is not possible.

We have that $\delta^-(v) = \{(u, v)\}$. In fact, since $v \in \dot{C}$ we must have an arc (v, w) in C. Because G has no knot this implies that the arc (w, v) cannot exist. So suppose (w', v) is an arc of G with $w' \neq w$, $w' \neq u$. Since w is in C, it is not a pendent node and hence G does not satisfies condition (i) of Theorem 4.2.

We must have $z(v, u) = z(u)$. Otherwise, we can construct the pair G' and z' as above. But in this case, one can check that (29) with respect to v is violated.

This permit us to apply an induction on the number of knots in G to finish the proof.

G Contains an Odd Y-Cycle

Let C be an odd Y-cycle in G. Assume that z is a fractional extreme point of $P_p(G)$.

Lemma 4.17. *The node set of any cycle of size at least three in G coincides with $V(C)$.*

Lemma 4.18. *Let $G = (V, A)$ be a directed graph and (u, v) and (v, u) two arcs in A. If $P_p(G)$ admits a fractional extreme point $\bar{z}$ with $\bar{z}(v, u) > 0$, then $P_{\tilde{p}}(G(u, v)) \neq \tilde{p}MP(G(u, v))$, where $\tilde{p} = p + 1$. The graph $G(u, v)$ was defined in the last subsection.*

Proof (cf. Baïou and Barahona 2007b*).* Let $\bar{z}$ be a fractional extreme point of $P_p(G)$ with $\bar{z}(v, u) > 0$. Suppose that $P_{\bar{p}}(G(u, v)) = \tilde{p}MP(G(u, v))$. Define $\tilde{z} \in P_{\bar{p}}(G(u, v))$ to be $\tilde{z}(u, t) = \bar{z}(u, v)$, $\tilde{z}(t) = 1$ and $\tilde{z}(r) = \bar{z}(r)$, $\tilde{z}(r, s) = \bar{z}(r, s)$ for all other nodes and arcs. The solution $\tilde{z}$ is fractional, so $\tilde{z}$ is not an extreme point of $P_{\bar{p}}(G(u, v))$. Since $P_{\bar{p}}(G(u, v))$ is integral, there is a 0-1 vector $z^* \in P_{\bar{p}}(G(u, v))$ with $z^*(v, u) = 1$, so that the same constraints that are tight for $\tilde{z}$ are also tight for z^*. From z^* define $z'' \in P_p(G)$ as follows: $z''(u, v) = z^*(u, t)$ and $z''(r) = z^*(r)$, $z''(r, s) = z^*(r, s)$, for all other nodes and arcs. All constraints that are tight for $\tilde{z}$ are also tight for z''. To see this, it suffices to remark that $z''(v) = z^*(v) = 0$ and $z''(u, v) = z^*(u, t) = 0$. This contradicts the fact that $\bar{z}$ is an extreme point of $P_p(G)$. □

The proof of Theorem 4.2 in this case is by induction on $|Pair(G)|$, the number of pairs of nodes $\{u, v\}$ with both (u, v) and (v, u) in A. If $|Pair(G)| = 0$ then G is an oriented graph that satisfies conditions (i) and (ii) of Theorem 4.2. Hence the result follows from Sect. 1.4.3.

Let (u, v) and (v, u) be two arcs in A. Lemma 4.5 implies $z(v, u) > 0$, so Lemma 4.18 applies and implies that

$$P_{\bar{p}}(G(u, v)) \neq \tilde{p}MP(G(u, v)). \tag{38}$$

Using Lemma 4.17 and the definition of G, we can see that $G(u, v)$ satisfies conditions (i) and (ii) of Theorem 4.2. Since $|Pair(G(u, v))| < |Pair(G)|$, we can apply the induction hypothesis so $P_{\bar{p}}(G(u, v)) = \tilde{p}MP(G(u, v))$. This contradicts (38). Thus the proof of our main result is complete.

1.5 Recognizing the Graphs Defined in Theorem 4.2

In this section we show how to decide if a graph satisfies conditions (i) and (ii) of Theorem 4.2. Clearly condition (i) can be tested in polynomial time. Thus we assume that we have a graph satisfying condition (i), then we split all pendent nodes as in Lemma 4.6, then we pick an arc (u, v), we remove u and v, and look for an odd Y-cycle in the new graph. We repeat this for every arc. It remains to show how to find an odd Y-cycle.

In Sect. 1.3.5 we gave a procedure that finds an odd cycle if there is any. We remind the reader that a cycle C is odd if $|V(C)| + |\hat{C}|$ is odd. Since an odd cycle is not necessarily a Y-cycle, we are going to modify the graph so that an odd cycle in the new graph gives an odd Y-cycle in the original graph. The main difficulty resides in how to deal with nodes that satisfy condition (ii) of Definition 4.13. Such a node should appear in a pair $\{(u, v), (v, u)\}$. Instead of working with such a pair we are going to work with a maximal bidirected path $P = v_1, \ldots, v_q$, this is a path where the arcs (v_{i+1}, v_i) and (v_i, v_{i+1}) belong to G, for $i = 1, \ldots, q - 1$. Notice that if the graph contains a bidirected cycle (when $v_1 = v_q$), then it is easy to derive an odd Y-cycle. So in what follows we assume that there is no bidirected cycle. The transformation is based on the following two remarks.

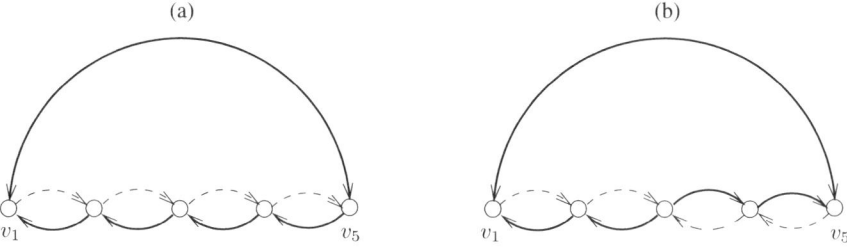

Fig. 1.7. Case 1, $q \geq 5$. In *bold* the Y-cycle C. In *dashed line* the other arcs of P

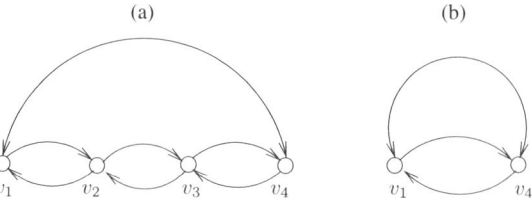

Fig. 1.8. Case 1, $q = 4$: (**a**) before transformation. (**b**) after transformation

Remark 5.1. There is at most one arc (u, v_1), $u \notin P$, and at most one arc (v, v_q), $v \notin P$. Otherwise the graph H_4 is present.

Remark 5.2. If the arc (u, v_1) is in A, $u \notin P$, and there is an arc (v_1, w) also in A, $w \notin P$, then w is a pendent node. Otherwise we obtain one of the graphs in Fig. 1.4.

Let C be a Y-cycle that goes through P. We have three cases to study.

Case 1. $\delta^-(P) = \{(u, v_1), (v, v_q)\}$. In this case C contains all nodes in P and also the arcs (u, v_1) and (v, v_q). Since C contains all nodes from P, the only variable that can change the parity of C is the parity of $|\hat{C} \cap P|$.

Notice that if $q \geq 5$ and if there is a Y-cycle going through P then we can always change the parity of it if needed. In fact, we can always join the nodes v_1 and v_q using arcs of P in such a way that $|\hat{C} \cap P| = 1$ as shown in Fig. 1.7(a), or $|\hat{C} \cap P| = 2$ as shown in Fig. 1.7(b). It follows that if there is a cycle C' going through P then there is a cycle C of the same parity, whose nodes in $|\hat{C} \cap P|$ satisfy Definition 4.13(ii).

It remains to analyze the cases when $q \leq 4$. The only cases when a transformation is required, are the following two:

- $q = 4$ and neither v_1 nor v_4 is adjacent to a pendent node. In this case we should have $|\hat{C} \cap P| = 1$. To impose that when looking for an odd cycle, we replace P by a bidirected path with two nodes. See Fig. 1.8.
  Let P' the new bidirected path. Any cycle C' with $|\hat{C}' \cap P'| = 1$ can be extended to a cycle C with $|\hat{C} \cap P| = 1$ and where the node in $\hat{C} \cap P$ satisfies Definition 4.13(ii).

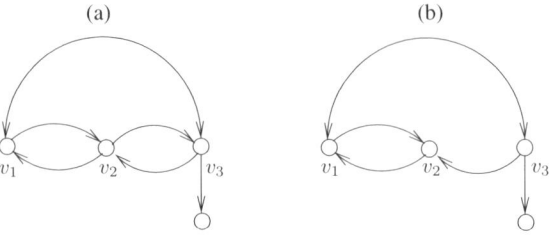

Fig. 1.9. Case 1, $q = 3$: (**a**) before transformation. (**b**) after transformation

- $q = 3$ and at most one of v_1 or v_3 is adjacent to a pendent node. Also here we have $|\hat{C} \cap P| = 1$. To impose that when looking for an odd cycle, we remove the arc (v_2, v_3).
  In Fig. 1.9, we supposed that v_3 is adjacent to a pendent node and v_1 is not.

The two remaining cases below follow the same philosophy as above.

Case 2. $\delta^-(P) = \{(u, v_1)\}$. In this case C contains (u, v_1), all the nodes in P and one arc (v_q, v), $v \notin P$. Here we have two cases to analyze.

- $q \geq 3$ or $q = 2$ and v_1 is adjacent to a pendent node. If $|\hat{C} \cap P|$ is even, we can assume that $|\hat{C} \cap P| = 0$. If $|\hat{C} \cap P|$ is odd, we can assume that $|\hat{C} \cap P| = 1$. Here no transformation is needed.
- $q = 2$ and v_1 is not adjacent to a pendent node. Here we should have $|\hat{C} \cap P| = 0$. To impose that when looking for an odd cycle, we remove (v_2, v_1).

Case 3. $\delta^-(P) = \emptyset$. In this case C contains an arc (v_1, u), $u \notin P$, all nodes in P, and an arc (v_q, v), $v \notin P$. Again we have two cases to analyze.

- $q \neq 3$ or $q = 3$ and v_2 is adjacent to a pendent node. If $|\hat{C} \cap P|$ is even, we can assume that $|\hat{C} \cap P| = 0$. If $|\hat{C} \cap P|$ is odd, we can assume that $|\hat{C} \cap P| = 1$. Here no transformation is needed.
- $q = 3$ and v_2 is not adjacent to a pendent node. Here we should have $|\hat{C} \cap P| = 0$. To impose that when looking for an odd cycle, we remove (v_1, v_2) and (v_3, v_2).

After preprocessing the graph as in Cases 1, 2, and 3, we look for an odd cycle; if there is one, it gives an odd Y-cycle in the original graph.

1.6 Related Polyhedra

In many applications, the underlying graph associated with the p-median problem may be a bipartite (oriented) or an undirected graph. Also, when the cost function is positive, then the problem always reduces to an instance in a bipartite graph. In this section we will consider the polyhedra associated with these two special cases. We will show how Theorem 4.2 is used to study the integrality of the linear relaxation of the p-median problem in bipartite and undirected graphs.

1.6.1 Bipartite Graphs

This is the standard case when V is partitioned into V_1 and V_2 and $A \subseteq V_1 \times V_2$. The customers are the nodes in V_1 and the potential locations are the nodes in V_2. Here we deal with the system

$$\sum_{v \in V_2} y(v) = p, \tag{39}$$

$$\sum_{(u,v) \in A} x(u, v) = 1 \quad \forall u \in V_1, \tag{40}$$

$$x(u, v) \le y(v) \quad \forall (u, v) \in A, \tag{41}$$

$$y(v) \ge 0 \quad \forall v \in V_2, \tag{42}$$

$$y(v) \le 1 \quad \forall v \in V_2, \tag{43}$$

$$x(u, v) \ge 0 \quad \forall (u, v) \in A. \tag{44}$$

Let $\Pi_p(G)$ be the polytope defined by (39)–(44), in this section we characterize the bipartite graphs for which $\Pi_p(G)$ is an integral polytope.

Let us recall some notations introduced in Sect. 1.3.6 when dealing with bipartite graphs. Let $\bar{V}_1$ be the set of nodes $u \in V_1$ with $|\delta^+(u)| = 1$. Let $\bar{V}_2$ be the set of nodes in V_2 that are adjacent to a node in $\bar{V}_1$. It is clear that the variables associated with nodes in $\bar{V}_2$ should be fixed, i.e., $y(v) = 1$ for all $v \in \bar{V}_2$. Let $\bar{G}$ be the graph induced by $V \setminus \bar{V}_2$.

Let H be a graph with node set $\{u_1, u_2, u_3, v_1, v_2, v_3, v_4\}$ and arc set

$$\{(u_1, v_1), (u_2, v_2), (u_3, v_3), (u_1, v_4), (u_2, v_4), (u_3, v_4)\}.$$

If the graph $\bar{G}$ contains H as a subgraph then we can construct a fractional extreme point as follows: Assign the value $\frac{1}{2}$ to each arc in H and to each node v_i, $i = 1, \ldots, 4$, set to zero all other node and arc variable of G. If $\bar{G}$ contains an odd cycle and one extra node in $V_2 \setminus \bar{V}_2$, we can also construct a fractional extreme point. Now we prove that these are only configurations that should be forbidden in order to have an integral polytope.

Theorem 6.1. *The polytope $\Pi_p(G)$ is integral if and only if*

(i) *$\bar{G}$ does not contain the graph H as a subgraph, and*
(ii) *$\bar{G}$ does not contain an odd cycle C and one extra node in $V_2 \setminus \bar{V}_2$.*

So let G be a graph such that $\bar{G}$ does not contain these two configurations. We assume that z is a fractional extreme point of $\Pi_p(G)$. As in Sect. 1.4.1, we can assume that $z(u, v) > 0$ for every arc $(u, v) \in A$.

Lemma 6.2. *We can assume that $z(u, v) = z(v)$ for each arc (u, v) such that $v \in V_2 \setminus \bar{V}_2$.*

Proof (cf. Baïou and Barahona 2007b*).* Suppose that $z(u, v) < z(v)$ for an arc (u, v) and $v \in V_2 \setminus \bar{V}_2$. We can add the nodes u', v', the arcs (u', v'), (u, v') and remove the arc (u, v). Then define $z'(u', v') = z(v') = 1$, $z'(u, v') = z(u, v)$, and $z'(s, t) = z(s, t)$, $z'(w) = z(w)$, for all other nodes and arcs. Let G' be the new graph. Then z' is an extreme point of $\Pi_{p+1}(G')$. The graph G' satisfies the hypothesis of Theorem 6.1.
□

The proof of Theorem 6.1 is divided into the following three cases:

(1) $\bar{G}$ does not contain an odd cycle nor the graph H.
(2) $\bar{G}$ does not contain H and contains an odd cycle C that includes all nodes in $V_2 \setminus \bar{V}_2$, and $|V_2 \setminus \bar{V}_2| \geq 5$.
(3) $\bar{G}$ does not contain H and contains an odd cycle C that includes all nodes in $V_2 \setminus \bar{V}_2$, and $|V_2 \setminus \bar{V}_2| = 3$.

We treat these three cases below.

$\bar{G}$ Does not Contain an Odd Cycle nor the Graph H

Lemma 6.3. *For all $u \in V_1$ we have $|\delta^+(u)| \leq 2$.*

Proof (cf. Baïou and Barahona 2007b*).* Since $\bar{G}$ has no odd cycle, the polytope defined by (40)–(44) is integral, this is Theorem 3.17 in Sect. 1.3.6. Thus z is a convex combination of two integral vectors satisfying (40)–(44). Therefore $|\delta^+(u)| \leq 2$.   □

Now we build an auxiliary undirected graph G' whose node-set is $V_2 \setminus \bar{V}_2$. For each node $u \in V_1$ such that $\delta^+(u) = \{(u, s), (u, t)\}$, $\{s, t\} \subseteq V_2 \setminus \bar{V}_2$, we have an edge in G' between s and t. This could create parallel edges. Notice that any node v in G' is adjacent to at most two other nodes. If v was adjacent to three other nodes, we would have the sub-graph H in $\bar{G}$.

Lemma 6.2 implies that if $z(v) = 1$ for $v \in V_2 \setminus \bar{V}_2$, then v is not adjacent to any other node in G'. A node $v \in V_2 \setminus \bar{V}_2$ is called fractional if $0 < z(v) < 1$. So G' consists of a set of isolated nodes, and a set of cycles and paths. We have to study the four cases below.

- If G' contains a cycle, it should be even, because $\bar{G}$ has no odd cycle. For a cycle in G' we can label the nodes with $+1$ and -1 so that adjacent nodes in the cycle have opposite labels. This labeling translates into a labeling in G as follows: If s and t have the labels $+1$ and -1 respectively, and the arcs (u, s) and (u, t) are in G, then (u, s) receives the label $+1$ and (u, t) receives the label -1. If s has the label $l(s)$ and the arcs (u, s) and (u, t) are in G with $t \in \bar{V}_2$, then (u, s) receives the label $l(s)$ and (u, t) receives the label $-l(s)$. All other nodes and arcs receive the label 0. This defines a new vector that satisfies with equality the same constraints that z satisfies with equality.

- If there is a path with an even number of fractional nodes we label them as before. This translates into a labeling in G as follows. If s and t have the labels $+1$ and -1 respectively, and the arcs (u, s) and (u, t) are in G, then (u, s) receives the label $+1$ and (u, t) receives the label -1. If s has the label $l(s)$ and the arcs (u, s) and (u, t) are in G with $t \in \bar{V}_2$, then (u, s) receives the label $l(s)$ and (u, t) receives the label $-l(s)$. All other nodes and arcs receive the label 0. This defines a new vector that satisfies with equality the same constraints that z satisfies with equality.

- If G' has two paths with an odd number of fractional nodes then again we can label the fractional nodes in these two paths and proceed as before.

- It remains the case where G' contains just one path with an odd number of fractional nodes. Let $v_1, \ldots, v_{2q+1}$ be the ordered sequence of nodes in this path. We should have $z(v_i) = \alpha$ if i is odd, and $z(v_i) = 1 - \alpha$ if i is even, with $0 < \alpha < 1$. This implies $\sum_{v \in V_2} z(v) = r + \alpha$ where r is an integer. We have then a contradiction.

$\bar{G}$ Does not Contain H and Contains an Odd Cycle C that Includes All Nodes in $V_2 \setminus \bar{V}_2$, and $|V_2 \setminus \bar{V}_2| \geq 5$

Here we use several transformations to obtain a new graph $\tilde{G}$ that satisfies conditions (i) and (ii) of Theorem 4.2, and we use the fact that $P_p(\tilde{G})$ is an integral polytope.

Lemma 6.4. *Let $u, v \in V(C)$, then there is no arc $(u, v) \notin A(C)$.*

Proof (cf. Baïou and Barahona 2007b). If such an arc exists, then the graph H would be present. $\square$

Lemma 6.5. *A node $u \in (V_1 \setminus \bar{V}_1)$ cannot be adjacent to more than one node in $\bar{V}_2$.*

Proof (cf. Baïou and Barahona 2007b). Suppose that the arcs (u, v_1) and (u, v_2) exist with v_1 and v_2 in $\bar{V}_2$. We can add and subtract ϵ to $z(u, v_1)$ and $z(u, v_2)$ to obtain a new vector that satisfies with equality the same constraints that z does. $\square$

Lemma 6.6. *We can assume that $(V_1 \setminus \bar{V}_1) \setminus V(C) = \emptyset$*

Proof (cf. Baïou and Barahona 2007b). Consider a node $u \in (V_1 \setminus \bar{V}_1) \setminus V(C)$ and suppose that the arcs (u, v_1) and (u, v_2) exist, with $v_1, v_2 \in V(C)$. If both paths in C between v_1 and v_2 contain another node in V_2, then there is an odd cycle in $\tilde{G}$ and an extra node in $V_2 \setminus \bar{V}_2$. Then we can assume that there is a node $w \in V(C)$ and $(w, v_1), (w, v_2) \in A(C)$. If there is another node $v_3 \in V(C)$ such that the arc (u, v_3) exists, then the graph H is present, this is because $|V_2 \setminus \bar{V}_2| \geq 5$. Thus u cannot be adjacent to any other node in $V(C)$. Lemma 6.2 implies

$$z(u, v_1) = z(w, v_1), \tag{45}$$

$$z(u, v_2) = z(w, v_2). \tag{46}$$

Then we remove the node u and study the vector z' that is the restriction of z to $G \setminus u$. If there is another vector z'' that satisfies with equality the same constraints that z' does, we can extend z'' using (45) and (46), to obtain a vector that satisfies with equality the same constraints that z does.

If there is a node $u \in (V_1 \setminus \bar{V}_1) \setminus V(C)$ that is adjacent to exactly one node $v \in V(C)$, then u is adjacent also to a node $w \in \bar{V}_2$. It follows from Lemma 6.5 that the node in $\bar{V}_2$ is unique. Lemma 6.2 implies

$$z(u, v) = z(v), \qquad (47)$$

and we also have

$$z(u, v) + z(u, w) = 1. \qquad (48)$$

Then we remove the node u and study the vector z' that is the restriction of z to $G \setminus u$. If there is another vector z'' that satisfies with equality the same constraints that z' does, we can extend z'' using (47) and (48), to obtain a vector that satisfies with equality the same constraints that z does.

The resulting graph does not contain H and contains the odd cycle C.   □

Now consider a node $u \in \bar{V}_1$ that is adjacent to $v \in \bar{V}_2$. We should have $z(u, v) = 1$ and $z(v) = 1$. We remove u from the graph and keep v with $z(v) = 1$.

Finally we add slack variables to the inequalities (43) for each node in $V_2 \setminus \bar{V}_2$. For that we add a node v' and the arc (v, v'), for each node $v \in V_2 \setminus \bar{V}_2$. Then we add the constraints

$$z(v) + z(v, v') = 1,$$
$$z(v, v') \leq z(v'),$$
$$z(v') = 1,$$
$$z(v, v') \geq 0.$$

Let $\tilde{G}$ be this new graph, and $\tilde{p} = p + |V_2 \setminus \bar{V}_2|$. It follows from Lemmas 6.4, 6.5 and 6.6 that $\tilde{G}$ is a graph satisfying conditions (i) and (ii) of Theorem 4.2. Here we have a face of $P_{\tilde{p}}(\tilde{G})$; because $z(v) = 0$ for all $v \in V_1$. Since $P_{\tilde{p}}(\tilde{G})$ is an integral polytope, there is an integral vector $\tilde{z}$ that satisfies with equality the same constraints that z does. From $\tilde{z} \in P_{\tilde{p}}(\tilde{G})$ one can easily derive $\tilde{z}' \in P_p(G)$ that satisfies with equality the same constrains that $z \in P_p(G)$ satisfies with equality.

$\bar{G}$ Does not Contain H and Contains an Odd Cycle C that Includes All Nodes in $V_2 \setminus \bar{V}_2$, and $|V_2 \setminus \bar{V}_2| = 3$

Let $p' = p - |\bar{V}_2|$. If $p' = 3$, we should have $z(v) = 1$ for all $v \in V_2$. Then it is easy to see that we have an integral polytope. So we assume that $p' \leq 2$. Let $V_2 \setminus \bar{V}_2 = \{v_1, v_2, v_3\}$.

Consider first $p' = 2$. If z is fractional, then at most one variable $z(v_i)$ can take the value one, so assume that

$$z(v_1) = 1,$$
$$1 > z(v_2) > 0,$$
$$1 > z(v_3) = 1 - z(v_2) > 0.$$

We give the label $l(v_2) = +1$ to v_2, the label $l(v_3) = -1$ to v_3, and $l(v) = 0$ for every other node in V_2. Then for each arc (u, v) with $z(u, v) = z(v)$, we give it the label $l(u, v) = l(v)$. If there is a node $u \in V_1$ that has only one arc (u, v) incident to it that is labeled, pick another arc (u, w) with $z(u, w) > 0$ and give it the label $l(u, w) = -l(u, v)$. For all the other arcs give the label 0. These labels define a new vector that satisfies with equation the same constraints that z does.

Now suppose that

$$1 > z(v_1) > 0,$$
$$1 > z(v_2) > 0,$$
$$1 > z(v_3) > 0.$$

Then for every node $u \in V_1$ there is at most one arc (u, v) such that $z(u, v) = z(v)$. Otherwise there is a node $w \in V_2 \setminus \bar{V}_2$ with $z(w) = 1$. Let us define a new vector z' as follows. Start with $z' = 0$. Set $z'(v_1) = z'(v_2) = 1$, $z'(v_3) = 0$, and $z'(v) = 1$ for all $v \in \bar{V}_2$. Then for each arc (u, v_1) with $z(u, v_1) = z(v_1)$ set $z'(u, v_1) = 1$. Also for each arc (u, v_2) with $z(u, v_2) = z(v_2)$ set $z'(u, v_2) = 1$. For each node u with $\sum_{(u,v) \in \delta^+(u)} z'(u, v) = 0$, pick an arc (u, v) with $v \neq v_3$ and set $z'(u, v) = 1$. This new vector satisfies with equality all the constraints that z does.

Finally suppose $p' = 1$ and

$$z(v_1) > 0,$$
$$z(v_2) > 0,$$
$$z(v_3) > 0.$$

We define a new vector z' as below. We set $z'(v_1) = 1$, $z'(v_2) = z'(v_3) = 0$, and $z'(v) = 1$ for $v \in \bar{V}_2$. For each node $u \in V_1$, if the arc (u, v_1) exists, we set $z(u, v_1) = 1$; otherwise there is a node $v \in \bar{V}_2$ such that the arc (u, v) exists, we set $z'(u, v) = 1$. We set $z'(s, t) = 0$ for every other arc. Every constraint that is satisfied with equality by z is also satisfied with equality by z'.

1.6.2 Undirected Graphs

For a undirected graph $G = (V, E)$ we denote by $\overset{\leftrightarrow}{G} = (V, A)$ the directed graph obtained from G by replacing each edge $uv \in E$ by two arcs (u, v) and (v, u).

Theorem 6.7. *Let G be a connected undirected graph. Then $P_p(\overset{\leftrightarrow}{G})$ is integral for all p if and only if G is a path or a cycle.*

Proof (cf. Baïou and Barahona 2007b). If G is a path or a cycle, then $\overset{\leftrightarrow}{G}$ satisfies conditions (i) and (ii) of Theorem 4.2 and so $P_p(\overset{\leftrightarrow}{G})$ is integral.

Suppose G is not a path nor a cycle. Then G contains a node of degree at least 3. Thus $\overset{\leftrightarrow}{G}$ contains H_4 as a subgraph. Again Theorem 4.2 implies that $P_p(\overset{\leftrightarrow}{G})$ is not integral for all p. □

Acknowledgements

We are grateful to the referees for several comments that helped us to improve the presentation.

References

Avella, P., Sassano, A.: On the p-median polytope. Math. Program. **89**, 395–411 (2001)

Baïou, M., Barahona, F.: On the p-median polytope of Y-free graphs Discrete Optim. **5**(2), 205–219 (2008)

Baïou, M., Barahona, F.: On the integrality of the uncapacitated facility location polytope. Technical Report RC24101, IBM Watson Research Center (2006)

Baïou, M., Barahona, F.: On the linear relaxation p-median problem I: oriented graphs. Technical Report, IBM Watson Research Center (2007a)

Baïou, M., Barahona, F.: On the linear relaxation p-median problem II: directed graphs. Technical report, IBM Watson Research Center (2007b)

Byrka, J., Aardal, K.: The approximation gap for the metric facility location problem is not yet closed. Oper. Res. Lett. **35**, 379–384 (2007)

Cánovas, L., Landete, M., Marín, A.: On the facets of the simple plant location packing polytope. Discrete Appl. Math. **124**, 27–53 (2002). Workshop on Discrete Optimization (Piscataway, NJ, 1999)

Caprara, A., Fischetti, M.: $\{0, \frac{1}{2}\}$-Chvátal–Gomory cuts. Math. Program. **74**, 221–235 (1996)

Cho, D.C., Johnson, E.L., Padberg, M., Rao, M.R.: On the uncapacitated plant location problem. I. Valid inequalities and facets. Math. Oper. Res. **8**, 579–589 (1983a)

Cho, D.C., Padberg, M.W., Rao, M.R.: On the uncapacitated plant location problem. II. Facets and lifting theorems. Math. Oper. Res. **8**, 590–612 (1983b)

Chudak, F.A., Shmoys, D.B.: Improved approximation algorithms for the uncapacitated facility location problem. SIAM J. Comput. **33**, 1–25 (2003)

Chvátal, V.: On certain polytopes associated with graphs. J. Comb. Theory, Ser. B **18**, 138–154 (1975)

Conforti, M., Rao, M.R.: Structural properties and recognition of restricted and strongly unimodular matrices. Math. Program. **38**, 17–27 (1987)

Cornuejols, G., Fisher, M.L., Nemhauser, G.L.: Location of bank accounts to optimize float: an analytic study of exact and approximate algorithms. Manage. Sci. **23**, 789–810 (1976/77)

Cornuejols, G., Thizy, J.-M.: Some facets of the simple plant location polytope. Math. Program. **23**, 50–74 (1982)

de Farias, I.R. Jr.: A family of facets for the uncapacitated p-median polytope. Oper. Res. Lett. **28**, 161–167 (2001)

De Simone, C., Mannino, C.: Easy instances of the plant location problem. Technical Report R. 427, IASI, CNR (1996)

Guignard, M.: Fractional vertices, cuts and facets of the simple plant location problem, Math. Program. Stud. (1980), pp. 150–162. Combinatorial optimization

Lawler, E.L.: Combinatorial Optimization: Networks and Matroids. Holt, Rinehart & Winston, New York (1976)

Mirchandani, P.B., Francis, R.L. (eds.): Discrete Location Theory, Wiley-Interscience Series in Discrete Mathematics and Optimization. Wiley, New York (1990)

Müller, R., Schulz, A.S.: Transitive packing: a unifying concept in combinatorial optimization. SIAM J. Optim. **13**, 335–367 (2002) (electronic)

Nestorov, S., Abiteboul, S., Motwani, R.: Extracting schema from semistructured data. In: SIGMOD '98: Proceedings of the 1998 ACM SIGMOD International Conference on Management of Data, pp. 295–306. ACM Press, New York (1998)

Shmoys, D.B., Tardos, É., Aardal, K.: Approximation algorithms for facility location problems (extended abstract). In: 29th ACM Symposium on Theory of Computing, pp. 265–274 (1997)

Stauffer, G.: The p-median polytope of Y-free graphs: an application of the matching theory. Technical Report (2007). To appear in Oper. Res. Lett.

Sviridenko, M.: An improved approximation algorithm for the metric uncapacitated facility location problem. In: Integer Programming and Combinatorial Optimization. Lecture Notes in Computer Science, vol. 2337, pp. 240–257. Springer, Berlin (2002)

Toumani, F.: Personal communication (2002)

Vigneron, A., Gao, L., Golin, M.J., Italiano, G.F., Li, B.: An algorithm for finding a k-median in a directed tree. Inf. Process. Lett. **74**, 81–88 (2000)

Yannakakis, M.: On a class of totally unimodular matrices. Math. Oper. Res. **10**, 280–304 (1985)

Facet Generating Techniques

Sylvia Boyd and William R. Pulleyblank

Summary. A major goal of polyhedral combinatorics is to find classes of essential, i.e. facet inducing, inequalities which describe a combinatorially defined polyhedron P. In general this is a difficult task. We consider the case in which we have knowledge of facets for a face F of P, and present a general technique for exploiting the close relationship between such polyhedra in order to obtain facets for P from facets of F. We demonstrate the usefulness of this technique by applying it in three instances where this relationship holds, namely the linear ordering polytope and the acyclic subgraph polytope, the asymmetric travelling salesman polytope and the monotone asymmetric travelling salesman polytope, and the symmetric travelling salesman polytope and the two-connected spanning subgraph polytope. In the last case we obtain a class of facet inducing inequalities for the two-connected spanning subgraph polytope by our procedure. This technique has also been applied by Boyd and Hao (1993) to show a class of inequalities is facet inducing for the two edge connected spanning subgraph polytope, by Leung and Lee (1994) to show a class of inequalities is facet inducing for the acyclic subgraph polytope and by Bauer (1997) to show several classes of inequalities are facet inducing for the circuit polytope.

The above technique requires that we demonstrate validity of an inequality. We discuss the problem of proving an inequality is valid for the integer hull of a polyhedron, and show that this problem is in NP for classes of polyhedra having bounded Chvátal–Gomory rank. This has the following consequence. Suppose we have an integer programming formulation of the members of an NP-complete class of problems with the property that we can, in polynomial time, show the validity of our defining inequalities. Then there will be problems in the class for which a linear system sufficient to define the integer hull will necessarily contain an inequality of arbitrarily large Chvátal–Gomory rank unless NP = co-NP.

Research partially supported by grants from N.S.E.R.C. of Canada.

2.1 Introduction and Notation

Polyhedra defined by combinatorial structures are often closely, and sometimes non-trivially, related. In some cases, one polyhedron is a face of another. For example, the Travelling Salesman Polytope is a face of the Two Connected Polytope, as well as a face of the Monotone Travelling Salesman Polytope and the Graphical Travelling Salesman Polytope.

Given a polyhedron P of interest, a major goal of polyhedral combinatorics is to find classes of essential, i.e. facet inducing inequalities which describe P. In general, finding such classes of inequalities is a difficult task, as is proving that these inequalities are facet inducing. Thus if we know of any polyhedra related to P for which facet inducing inequalities are known, it is desirable to exploit the connections between P and these related polyhedra in order to obtain facet inducing inequalities for P.

In this paper, we consider the situation in which a polyhedron P has a face F for which facet inducing inequalities are known. Often in the past, these faces F have been studied separately without exploiting their relationship to P. As a consequence, classes of inequalities have been shown to be facet inducing with separate arguments for P and F. Although this was originally the case for the Travelling Salesman Polytope (Q^n) and the Monotone Travelling Salesman Polytope ($\tilde{Q}^n$), Grötschel and Pulleyblank (1986) showed that a single argument was sufficient, in that once an inequality was shown to be facet inducing for Q^n and valid for $\tilde{Q}^n$, then either it was already facet inducing for $\tilde{Q}^n$, or else it could be efficiently transformed into an equivalent inequality which was facet inducing for $\tilde{Q}^n$.

Our goal here is to generalize this procedure. In Sect. 2.2, we consider the situation in which we have an inequality $ax \leq a_0$, which is valid for P, and is known to be facet inducing for a nonempty face F of P. We define the notion of an independent direction set, and show how such a direction set can be used to show that $ax \leq a_0$ is also facet inducing for P. We also show how this technique simplifies when P is the so called monotone completion of F. In Sect. 2.3, we demonstrate the usefulness of the technique described in Sect. 2.2 by discussing three applications of it. The first of these applications relates facets of the Linear Ordering Polytope to facets of the Acyclic Subgraph Polytope, and the second relates facets of the Asymmetric Travelling Salesman Polytope to facets of the Monotone Asymmetric Travelling Salesman Polytope. For both of these applications, we have that P is the monotone completion of F. We also discuss a third more complicated application which uses the more general technique from Sect. 2.2, in which we describe a class of facet inducing inequalities for the Two Connected Polytope (see Boyd and Pulleyblank 1993). Note that the general technique from Sect. 2.2 was also used successfully in Boyd and Hao (1993) to obtain other classes of facet inducing inequalities for the Two Connected Polytope and by Leung and Lee (1994) to show a class of inequalities is facet inducing for the acyclic subgraph polytope. It was also used by Bauer (1997) to obtain several classes of facet inducing inequalities for the circuit polytope using known classes of facets of the Travelling Salesman Polytope.

Given a facet inducing inequality $ax \le a_0$ for a face F of a polyhedron P which is valid for P, there exists many equivalent forms of it (with respect to F), only some of which are facet inducing for P. If $ax \le a_0$ is already in the correct form, we can go ahead and use the technique from Sect. 2.2 to prove it is facet inducing for P. If it is not, we must find a way to "pivot" it into a correct equivalent form. In Sect. 2.4, we discuss briefly how this "pivot" operation can be performed in general, and describe how it can be carried out efficiently in the case that P is the monotone completion of F. We also briefly discuss the converse problem, i.e. under what conditions an inequality inducing a facet of P also induces a facet of a face F of P.

Note that the methods discussed in Sects. 2.2 and 2.4 require proving an inequality is valid for a polyhedron. In Sect. 2.5 we discuss how to obtain short proofs of validity of an inequality.

The remainder of this section is devoted to definitions and notation.

For any finite set E we let $\mathbb{R}^E$ denote the set of all real vectors indexed by E. For any $J \subseteq E$ and $x \in \mathbb{R}^E$, we let $x(J)$ denote $(\sum x_j : j \in J)$. For any subset F of E, the incidence vector of F is the vector $x \in \mathbb{R}^E$ defined by

$$x_e = \begin{cases} 1 & \text{if } e \in F, \\ 0 & \text{otherwise.} \end{cases}$$

Given a matrix $A \in \mathbb{R}^{L \times E}$ and subset $J \subseteq E$, we let A_J represent the $(|L| \times |J|)$-submatrix of A consisting of those columns of A indexed by J. We abbreviate $A_{\{j\}}$ by A_j. The linear column rank of A we denote by $r_l(A)$.

Given a graph $G = (V, E)$ and any $S \subseteq V$, we let $\delta(S)$ denote the set of edges with exactly one end in S. We abbreviate $\delta(\{v\})$ by $\delta(v)$. We let $\gamma(S)$ denote the set of edges with both ends in S, and let $G[S] = (S, \gamma(S))$ denote the subgraph induced by S. For any $J \subseteq E$, we let $G(J)$ denote the subgraph induced by J. That is, the node set of $G(J)$ consists of all nodes incident with edges in J, and the edge set is J. If X and Y are two distinct subsets of V, then we let $[X : Y]$ denote the set of edges with one end in X and the other end in Y.

Given a directed graph $D = (V, E)$ and any $S \subseteq V$, we let $\overrightarrow{\delta}(S)$ denote the set of arcs for which only the tail is in S, and we let $\overleftarrow{\delta}(S)$ denote the set of arcs for which only the head is in S. If X and Y are two distinct subsets of V, then we let $(X : Y)$ denote the set of arcs with tail in X and head in Y. We let $\gamma(S)$ denote the set of arcs with both their head and tail in S.

We assume the reader is familiar with the basic definitions and concepts of polyhedral combinatorics, and here only summarize our notation and specialized definitions. We refer to Nemhauser and Wolsey (1988) or Pulleyblank (1989) for the necessary background.

For any polyhedron $P \subseteq \mathbb{R}^E$, we let dim(P) represent the dimension of P. For any finite $X \subseteq \mathbb{R}^E$, we denote the convex hull of X by conv(X) and the cone of X by cone(X).

Given a valid inequality $ax \le a_0$ for a polyhedron P, we let $P_a^=$ represent the face of P induced by $ax \le a_0$, i.e.,

$$P_a^= = \{\bar{x} \in P : a\bar{x} = a_0\}.$$

Given a linear system defining a polyhedron P, the set of constraints which are satisfied with equality by all $x \in P$ is called an *equation system for P*. For any face F of a polyhedron $P \subseteq E$, we call $A^F x = b^F$ a *minimal equation system for F with respect to P* if it consists of a minimal equation system for F excluding any equations satisfied by all $x \in P$. Note that $\dim(F) = \dim(P) - r_l(A^F)$, since for any polyhedron P, $\dim(P) = |E| - r_l(A^P)$, where $A^P x = b^P$ is a minimal equation system for P.

Let P be a polyhedron with equation system $Ax = b$ and let $ax \leq a_0$, $\bar{a}x \leq \bar{a}_0$ be valid inequalities. Let $F = \{x \in P : ax = a_0\}$ and $\bar{F} = \{x \in P : \bar{a}x = \bar{a}_0\}$. We say that $ax \leq a_0$ and $\bar{a}x \leq \bar{a}_0$ are *equivalent* if $F = \bar{F}$. Note that if $ax \leq a_0$ and $\bar{a}x \leq \bar{a}_0$ are facet inducing for P, they are equivalent if and only if there exists $\gamma > 0$ and a vector λ such that $a = \gamma\bar{a} + \lambda A$ and $a_0 = \gamma\bar{a}_0 + \lambda b$.

2.2 Extending Affine Bases Using Independent Direction Sets

Let F be a nonempty face of a polyhedron $P \subseteq \mathbb{R}^E$, and let $ax \leq a_0$ be a facet inducing inequality for F which is valid for P. We call a set $X \subseteq P_a^=$ an *extension set* for $F_a^=$ if $Y \cup X$ is affinely independent for an affine basis Y of $F_a^=$. Note that the choice of Y is independent of X; i.e. $X \cup Y$ will be affinely independent for any affine basis Y of $F_a^=$.

In order to show $ax \leq a_0$ is also facet inducing for P, it suffices to show there exists an extension set for $F_a^=$ of cardinality $\dim(P) - \dim(F)$. In this section we describe how to create such an extension set by using a set of carefully chosen direction vectors, each of which provides a way of moving from a point in $F_a^=$ to a new point in $P_a^=$. More formally, we define a set $D \subseteq \mathbb{R}^E$ to be an *independent direction set* for $F_a^=$ if the following conditions are satisfied:

(D1) For every $d \in D$, there exists $\hat{x}^d \in F_a^=$ such that $x^d := \hat{x}^d + d \in P$.
(D2) For every $d \in D$, $ad = 0$.
(D3) For some minimal equation system $A^F x = b^F$ for F with respect to P, $\{A^F d : d \in D\}$ is linearly independent.

For x^d as defined in (D1), we call $X^D := \{x^d : d \in D\}$ a *set generated by D*. We remark that here we are taking exactly one x^d for each $d \in D$, although there may be many choices for it. We also remark that in (D3), if the condition is true for some minimal equation system, it will in fact be true for any minimal equation system for F with respect to P.

The following theorem shows that there exists a strong relationship between independent direction sets and extension sets.

Theorem 2.1. *Let F be a nonempty proper face of a polyhedron P, and let $ax \leq a_0$ be a facet inducing inequality for F which is valid for P.*

(i) *If D is an independent direction set for $F_a^=$, then the set X^D generated by D is an extension set for $F_a^=$.*

(ii) *If X is an extension set for $F_a^=$ and $\hat{x}$ is any member of $F_a^=$, then $D^X := \{d^x : d^x = x - \hat{x},\ x \in X\}$ is an independent direction set for $F_a^=$.*

Proof. First we prove (i). Consider any $x^d \in X^D$. By (D1), $x^d = \hat{x}^d + d \in P$ for some $\hat{x}^d \in F_a^=$ and $d \in D$. Thus $ax^d = a\hat{x}^d + ad = a_0 + ad = a_0$ by (D2), and we have $X^D \subseteq P_a^=$.

To complete the proof of (i), we now show that $Y \cup X^D$ is affinely independent for any affine basis Y of $F_a^=$. Let $\alpha \in \mathbb{R}^{Y \cup D}$ be such that

$$\sum(\alpha_y : y \in Y) + \sum(\alpha_d : d \in D) = 0, \tag{1}$$

$$\sum(\alpha_y y : y \in Y) + \sum(\alpha_d x^d : d \in D) = \mathbf{0}. \tag{2}$$

By (D3), there exists a minimal equation system $A^F x = b^F$ for F with respect to P such that $\{A^F d : d \in D\}$ is linearly independent. Substituting $x^d = \hat{x}^d + d$ into (2) and multiplying through by A^F gives

$$\left(\sum(\alpha_y : y \in Y) + \sum(\alpha_d : d \in D)\right) b^F + \sum(\alpha_d(A^F d) : d \in D) = \mathbf{0}.$$

Using (1), it then follows that

$$\sum(\alpha_d(A^F d) : d \in D) = \mathbf{0},$$

which implies $\alpha_d = 0$ for all $d \in D$, since $\{A^F d : d \in D\}$ is linearly independent. It then follows from (1) and (2) that $\alpha_y = 0$ for all $y \in Y$, since Y is affinely independent. Thus $\alpha = \mathbf{0}$, and $Y \cup X^D$ is affinely independent as required.

Now we prove (ii). Clearly D^X satisfies properties (D1) and (D2). It is left to show that $\{A^F d : d \in D^X\}$ forms a linearly independent set, where $A^F x = b^F$ is a minimal equation system for F with respect to P.

Let Y be an affine basis of $F_a^=$. Since $F_a^=$ is a facet of F, there exists $x^F \in F \backslash F_a^=$ such that $\{x^F\} \cup Y$ is an affine basis of F. Clearly we have $x^F \notin P_a^=$ and $X \cup Y \subseteq P_a^=$, thus $\{x^F\} \cup X \cup Y$ is affinely independent. Complete $\{x^F\} \cup X \cup Y$ to an affine basis of P using a set W, and let $D = D^X \cup D^W$, where $D^W = \{d^w : d^w = w - \hat{x},\ w \in W\}$. Since $\dim(P) = \dim(F) + r_l(A^F)$, it follows that $|D| = r_l(A^F)$, and thus the matrix B whose columns consist of the vectors $\{A^F d : d \in D\}$ is a square matrix. We complete the proof of (ii) by showing B is nonsingular.

Let λ be a vector such that $\lambda B = \mathbf{0}$. Then

$$\lambda A^F d^x = 0 \quad \text{for all } d^x \in D,\ x \in X \cup W.$$

Substituting $d^x = x - \hat{x}$ in the above, we obtain,

$$(\lambda A^F)x = \lambda b^F \quad \text{for all } x \in X \cup W.$$

Since $\{x^F\} \cup Y \subseteq F$, it follows that $(\lambda A^F)x = \lambda b^F$ for $\{x^F\} \cup Y$ and thus

$$\lambda A^F x = \lambda b^F \quad \text{for all } x \in P, \tag{3}$$

since $\{x^F\} \cup Y \cup X \cup W$ is an affine basis for P.

Recall that $A^F x = b^F$ is a minimal equation system for F with respect to P. Thus there exists a minimal equation system $A^P x = b^P$ for P such that $A^F x = b^F$, $A^P x = b^P$ forms a minimal equation system for F. By (3), the equation $\lambda A^F x = \lambda b^F$ must be a linear combination of the equations $A^P x = b^P$; i.e. there exists a vector μ such that $\mu A^P = \lambda A^F$ and $\mu b^P = \lambda b^F$. But since the equations $A^P x = b^P$ and $A^F x = b^F$ are linearly independent, we must have $\lambda = 0$ and $\mu = 0$. $\square$

Note that Theorem 2.1 shows that an extension set and independent direction set for $F_a^=$ are equivalent in the sense that one can always be obtained from the other.

Using Theorem 2.1, we obtain the following corollary which relates the facets of a face of a polyhedron P to the facets of P. This corollary is used extensively in the applications described in Sect. 2.3.

Corollary 2.2. *Let $ax \le a_0$ be an inequality which is valid for a polyhedron $P \subseteq \mathbb{R}^E$ and facet inducing for a nonempty face F of P. Let $A^F x = b^F$ be a minimal equation system for F with respect to P. If there exists an independent direction set for $F_a^=$ of size $r_l(A^F)$, then $ax \le a_0$ is also facet inducing for P.*

Proof. Let Y be any affine basis of $F_a^=$. Since $|Y| = \dim(F)$ and $\dim(P) = \dim(F) + r_l(A^F)$, it follows from Theorem 2.1 that Y plus the set generated by the independent direction set forms a set of $\dim(P)$ affinely independent points in $P_a^=$. The result follows. $\square$

Another useful corollary obtained from Theorem 2.1 is the following.

Corollary 2.3. *Let $ax \le a_0$ be an inequality which is valid for a polyhedron $P \subseteq \mathbb{R}^E$ and facet inducing for a nonempty face F of P. If D is a maximal independent direction set for $F_a^=$, then $\dim(P_a^=) = \dim(F_a^=) + |D|$.*

Proof. Let X be the set generated by D and let Y be an affine basis of $F_a^=$. By Theorem 2.1, $X \cup Y$ is an affine basis of $P_a^=$. Thus $\dim(P_a^=) = (|Y| - 1) + |X| = \dim(F_a^=) + |D|$. $\square$

We conclude this section by discussing how the above technique simplifies in the case that polyhedron P is the monotone completion of F.

Given a polyhedron $Q \subseteq \mathbb{R}^E$ which lies in the nonnegative orthant of $\mathbb{R}^E$, the *monotone completion*, or *submissive*, $\tilde{Q}$ of Q is defined by

$$\tilde{Q} = \{x \in \mathbb{R}^E : 0 \le x \le x' \text{ for some } x' \in Q\}.$$

In the case of many combinatorial optimization polyhedra, Q can be expressed in the form $Q = \{x : x \ge 0, Ax = b, \tilde{A}x \le \tilde{b}\}$ where $A, \tilde{A}, b, \tilde{b} \ge 0$. In this case Q is a face of $\tilde{Q}$. Moreover, in most cases of interest, Q is a proper face of $\tilde{Q}$. Also, unless there exists $e \in E$ such that $x_e = 0$ for all $x \in Q$, $\tilde{Q}$ is full dimensional. We will assume that Q is a nonempty proper face of the full dimensional monotone completion $\tilde{Q}$ for the rest of this section.

Let $I \in \mathbb{R}^{E \times E}$ represent the $|E| \times |E|$ identity matrix. The following lemma is very useful in applying the previous method of this section to a polyhedron Q and its monotone completion, as well as for other applications.

Lemma 2.4. *Let F be a nonempty proper face of a polyhedron $P \subseteq \mathbb{R}^E$, let $ax \leq a_0$ be a facet inducing inequality for F which is valid for P, and let $A^F x = b^F$ be a minimal equation system for F with respect to P. Suppose there exists $S \subseteq E$ such that the following hold:*

(i) $a_e = 0$ for all $e \in S$,
(ii) $\{A_e^F : e \in S\}$ is a linearly independent set,
(iii) for every $e \in S$ there exists $\hat{x}^e \in F_a^=$ and scalar $\lambda^e \neq 0$ such that $x^e := \hat{x}^e + \lambda^e I_e \in P$.

Then $X^S := \{x^e : e \in S\}$ is an extension set for $F_a^=$.

Proof. Let $D = \{\lambda^e I_e : e \in S\}$. Clearly by (iii), D satisfies the independent direction set property (D1), by (i) D satisfies (D2), and by (ii) D satisfies (D3). Thus D is an independent direction set for $F_a^=$. It then follows by Theorem 2.1 that the set X^S generated by D is an extension set for $F_a^=$, as required. $\square$

Given a polyhedron Q and its monotone completion $\tilde{Q}$, let $ax \leq a_0$ be facet inducing for Q, and let $A^Q x = b^Q$ be a minimal equation system for Q. We let $E^0(a) := \{j \in E : a_j = 0\}$ ($E \setminus E^0(a)$ is usually called the *support* of a), and we say $ax \leq a_0$ is *support reduced* if a subset of $E^0(a)$ indexes a column basis of A^Q. Using Lemma 2.4, we obtain the following specialization of Corollary 2.2 which relates facets of a polyhedron to facets of its monotone completion.

Theorem 2.5. *Let Q be a polyhedron lying in the nonnegative orthant of $\mathbb{R}^E$ and let $\tilde{Q}$ be its monotone completion. Let $ax \leq a_0$, $a \geq 0$, be a facet inducing inequality for Q which is support reduced and does not induce the trivial facet $x_e \geq 0$ for Q. Then $ax \leq a_0$ is facet inducing for $\tilde{Q}$.*

Proof. Since $a \geq 0$, $ax \leq a_0$ is valid for $\tilde{Q}$. Also, since $ax \leq a_0$ does not induce the trivial facet, it follows that for every $e \in E^0(a)$ there exists $\hat{x}^e \in Q_a^=$ such that $\hat{x}_e^e > 0$. Thus $x^e := \hat{x}^e - \hat{x}_e^e I_e \in \tilde{Q}$. Since $ax \leq a_0$ is support reduced, it then follows by Lemma 2.4 that there exists an extension set for $Q_a^=$ of cardinality $r_l(A^Q)$, where $A^Q x = b^Q$ is a minimal equation system for Q. Thus it follows that $ax \leq a_0$ is facet inducing for $\tilde{Q}$. $\square$

2.3 Applications

In this section we discuss three applications of the preceding technique. The first two relate facets of polyhedra to facets of their so called monotone completions. In these applications we use the corresponding specialized form of the technique in Sect. 2.2. The third application is a more complicated example which relates facets of the Symmetric Travelling Salesman Polytope to facets of the Two Connected Polytope. In this application we use the general form of the method in Sect. 2.2 to prove that a class of inequalities is facet inducing for the Two Connected Polytope (see Boyd and Pulleyblank 1993).

Our first application deals with the Linear Ordering Polytope and its monotone completion, the Acyclic Subgraph Polytope. Let $D_n = (N_n, A_n)$ denote the complete directed graph on n nodes. A *tournament* in D_n is a subdigraph $G = (N, T)$ of D_n such that for every two nodes $u, v \in N$ there exists exactly one arc in T with endnodes u and v. Given a positive integer n and a vector $c \in \mathbb{R}^{A_n}$ of real arc weights, the *Linear Ordering Problem* is to find an acyclic tournament $D = (N, T)$ in D_n such that $c(T)$ is maximized. The *Linear Ordering Polytope* is denoted by P_{LO}^n, and defined by $P_{LO}^n = \text{conv}\{x \in \mathbb{R}^{A_n} : x$ is the incidence vector of an arc set of an acyclic tournament in $D_n\}$. The monotone completion of P_{LO}^n is the so called *Acyclic Subgraph Polytope*, which is denoted by P_{AC}^n, and defined by $P_{AC}^n = \text{conv}\{x \in \mathbb{R}^{A_n} : x$ is the incidence vector of an arc set of an acyclic subdigraph of $D_n\}$.

The Linear Ordering Polytope P_{LO}^n has been studied in Grötschel et al. (1982a), and the Acyclic Subgraph Polytope P_{AC}^N has been studied in Grötschel et al. (1982b). In these papers it is shown that P_{AC}^n can be defined as the convex hull of all $x \in \mathbb{R}^{A_n}$ such that x is integral and satisfies the following linear constraints:

$$x_{ij} \geq 0, \quad 1 \leq i, j \leq n, \tag{4}$$

$$x_{ij} \leq 1, \quad 1 \leq i, j \leq n, \tag{5}$$

$$x_{ij} + x_{ji} \leq 1, \quad \text{for all } i, j \in N_n, i \neq j, \tag{6}$$

$$x(C) \leq |C| - 1 \quad \text{for all dicycle arc sets } C \subseteq A_n, |C| \geq 3. \tag{7}$$

It is also shown that P_{LO}^n is the face of P_{AC}^n obtained by taking the convex hull of all integral x satisfying the above constraints (4)–(7) which also satisfy constraints (6) with equality.

In Grötschel et al. (1982a), the following four classes of inequalities are shown to be facet inducing for P_{LO}^n, as well as nonequivalent:

$$x_{ij} \geq 0, \quad 1 \leq i, j \leq n, \tag{8}$$

$$x_{ij} + x_{jk} + x_{ki} \leq 2, \quad 1 \leq i < j < k \leq n, \tag{9}$$

$$x(A) \leq k^2 - k + 1 \quad \text{for all simple } k\text{-fences } A \subseteq A_n, \ k \geq 4, \tag{10}$$

$$x(M) \leq 3k - \frac{k+1}{2} \quad \text{for all simple Möbius ladders } M \subseteq A_n,$$

$$k \geq 3. \tag{11}$$

(We refer the reader to Grötschel et al. 1982a for the definition of simple k-fences and simple Möbius ladders.) Using Theorem 2.5, we easily show below that all three nontrivial classes of inequalities (9)–(11) are also facet inducing for P_{AC}^n. Note that these inequalities are among those shown to be facet inducing for P_{AC}^n in Grötschel et al. (1982b) by more technical methods.

In Grötschel et al. (1982a) it is shown that $\dim(P_{LO}^n) = \binom{n}{2}$, and that

$$x_{ij} + x_{ji} = 1 \quad \text{for all } i, j \in N_n, i \neq j \tag{12}$$

is a minimal equation system for P_{LO}^n. Letting $Bx = b$ represent the system (12), then clearly for any $S \subset A_n$, $\{B_e : e \in S\}$ contains a column basis of B if and only if $A_n \setminus S$ contains at most one of the arcs ij and ji for all $i, j \in N_n$. Thus given an inequality $ax \leq a_0$, it is support reduced if $A_n \setminus E^0(a)$ contains at most one of the arcs ij and ji for all $i, j \in N_n$. This is easily verified for all inequalities (9)–(11). Since inequalities (9)–(11) are not equivalent to the trivial inequalities (8), it follows by Theorem 2.5 that all of these inequalities are facet inducing for P_{AC}^n.

Our second application deals with the Asymmetric Travelling Salesman Polytope and its monotone completion. Given the complete directed graph $D_n = (N_n, A_n)$ on n nodes with weighted arcs, the *Asymmetric Travelling Salesman Problem* is to find a minimum weight directed Hamiltonian cycle in D_n. Letting $\mathcal{H}_n$ represent the set of all directed Hamiltonian cycles in D_n, the *Asymmetric Travelling Salesman Polytope*, denoted by P^n, is defined by

$$P^n = \text{conv}\{x : x \text{ is the incidence vector of some } H \in \mathcal{H}_n\}.$$

The polytope P^n can also be defined as the convex hull of all $x \in \mathbb{R}^{A_n}$ such that x is integral and satisfies the following linear constraints:

$$x_e \geq 0, \quad \text{for all } e \in A_n, \tag{13}$$

$$x(\overrightarrow{\delta}(v)) = 1 \quad \text{for all } v \in N_n, \tag{14}$$

$$x(\overleftarrow{\delta}(v)) = 1 \quad \text{for all } v \in N_n, \tag{15}$$

$$x(\gamma(W)) \leq |W| - 1 \quad \text{for all } W \subseteq A_n, \ 2 \leq |W| \leq n - 1. \tag{16}$$

We denote the monotone completion of P^n, the *Monotone Asymmetric Travelling Salesman Polytope*, by $\tilde{P}^n$. Note that $\tilde{P}^n$ can be described as the convex hull of all integral $x \in \mathbb{R}^{A_n}$ satisfying the linear constraints (13)–(16) for P^n with the constraints (14) and (15) relaxed to less-than-or-equal-to inequalities, so it is easily seen that P^n is a face of $\tilde{P}^n$. A survey of known polyhedral results for P^n and $\tilde{P}^n$ can be found in Grötschel and Padberg (1985) and Balas and Fischetti (2007).

It is known (see Grötschel and Padberg 1985) that $\dim(P^n) = |A_n| - 2|N_n| + 1$, and that the following set of equations with any one removed forms a minimal equation system for P^n:

$$\begin{aligned} x(\overrightarrow{\delta}(i)) = 1 \quad \text{for all } i \in N_n, \\ x(\overleftarrow{\delta}(i)) = 1 \quad \text{for all } i \in N_n. \end{aligned} \tag{17}$$

Let $Bx = b$ represent (17), and let $B'x = b'$ represent (17) with any one removed. Note that B represents the node-edge incidence matrix of a bipartite graph $G = (V_1 \cup V_2, E)$, where E is in one-to-one correspondence with A_n, and each of V_1 and V_2 are in one-to-one correspondence with N_n. It is easily verified that $J \subseteq E$ indexes a set of linearly independent columns in B' if and only if $G(J)$ contains no cycle. For any $J \subseteq E$, let J' represent the arcs in A_n which correspond to J. Call a cycle C in D_n an *alternating cycle* if every pair of consecutive arcs are opposite in direction (i.e. every node in C has in-degree 2 or out-degree 2). Since $G(J)$ contains no cycle if and only if $D_n(J')$ contains no alternating cycle, we have the following.

Lemma 3.1. *Let $ax \leq a_0$, $a \geq 0$, be an inequality which is facet inducing for P^n. Then $ax \leq a_0$ is support reduced if there exists $J \subseteq E^0(a)$ such that $|J| = 2|N_n|-1$, and $D_n(J)$ contains no alternating cycle.*

The following lemma helps us to recognize when a facet inducing inequality for P^n induces a trivial inequality. Recall that for distinct subsets X and Y of N_n, $(X : Y)$ denotes the set of arcs with tail in X and head in Y.

Lemma 3.2. *Let $ax \leq a_0$, $a \geq \mathbf{0}$, be a facet inducing inequality for P^n which induces the trivial facet $-x_k \leq 0$ for some $k \in A_n$. Then*

(a) *either $E^0(a)$ or $E^0(a)\backslash\{k\} \subseteq \overrightarrow{\delta}(v) \cup \overleftarrow{\delta}(v)$ for some $v \in N_n$; or*

(b) *either $E^0(a)$ or $E^0(a)\backslash\{k\} = \gamma(Z) \cup (X : Y) \cup (X : Z) \cup (Z : Y)$ for three mutually disjoint node sets $X, Y, Z \subseteq N_n$. (Note that any of X, Y and Z may be empty.)*

Proof. Recall $Bx = b$ described earlier is an equation system for P^n. If $ax \leq a_0$ induces the trivial facet $-x_k \leq 0$ then there exists $\gamma > 0$ and vector λ indexed by the equations in $Bx = b$ such that $a_j = \lambda B_j$ for $j \in A_n\backslash\{k\}$ and $a_k = \lambda B_k - \gamma$. Let $\lambda_{\overrightarrow{u}}$ (resp. $\lambda_{\overleftarrow{u}}$) represent the components of λ corresponding to the equation $x(\overrightarrow{\delta}(u)) = 1$ (resp. $x(\overleftarrow{\delta}(u)) = 1$).

Suppose $\lambda \geq 0$. Let $X = \{u \in N_n : \lambda_{\overrightarrow{u}} = 0, \lambda_{\overleftarrow{u}} > 0\}$, $Y = \{u \in N_n : \lambda_{\overleftarrow{u}} = 0, \lambda_{\overrightarrow{u}} > 0\}$, and $Z = \{u \in N_n : \lambda_{\overrightarrow{u}} = \lambda_{\overleftarrow{u}} = 0\}$, and let $S = \gamma(Z) \cup (X : Y) \cup (X : Z) \cup (Z : Y)$. For all $j \in A_n\backslash\{k\}$, $a_j = 0$ if and only if $j \in S$ since $a_j = \lambda B_j$. For arc k, $a_k = \lambda B_k - \gamma$. If $k \in S$, then $a_k < 0$ since $\lambda B_k = 0$ and $\gamma > 0$, contradicting the fact that $a \geq \mathbf{0}$. Thus $k \notin S$, and may or may not belong to $E^0(a)$. Hence either $E^0(a)$ or $E^0(a)\backslash\{k\} = S$, which is (b) above.

Now suppose some component of λ is negative; say $\lambda_{\overrightarrow{u}} < 0$. Since $a \geq \mathbf{0}$, this implies

$$\lambda_{\overleftarrow{v}} \geq -\lambda_{\overrightarrow{u}} \quad \text{for all } v \in N_n\backslash\{u\}. \tag{18}$$

Since for all $j \in A_n\backslash\{k\}$ we have $a_j = \lambda B_j$, the only arcs for which $a_j = 0$ are either k, or else arcs pq for which $\lambda_{\overleftarrow{q}} = -\lambda_{\overrightarrow{p}}$. If $\lambda_{\overleftarrow{q}}$ has value $\alpha > 0$ for all such arcs pq, then let $X = \{i \in N_n : -\lambda_{\overrightarrow{i}} = \alpha, \lambda_{\overleftarrow{i}} \neq \alpha\}$, $Y = \{i \in N_n : -\lambda_{\overrightarrow{i}} \neq \alpha, \lambda_{\overleftarrow{i}} = \alpha\}$, and $Z = \{i \in N_n : -\lambda_{\overrightarrow{i}} = \lambda_{\overleftarrow{i}} = \alpha\}$, and let $S = \gamma(Z) \cup (X : Y) \cup (X : Z) \cup (Z : Y)$. Then either $E^0(a)$ or $E^0(a)\backslash\{k\} = S$, which is (b) above.

Now suppose there exists arcs pq and rs such that $\lambda_{\overleftarrow{q}} = -\lambda_{\overrightarrow{p}} = \alpha$, and $\lambda_{\overleftarrow{s}} = -\lambda_{\overrightarrow{r}} = \beta$, and $\alpha \neq \beta$. By (18), this can only occur if $p = u$ for all arcs pq such that $\lambda_{\overleftarrow{q}} = -\lambda_{\overrightarrow{p}} = \alpha$, and $s = u$ for all arcs rs such that $\lambda_{\overleftarrow{s}} = -\lambda_{\overrightarrow{r}} = \beta$. The only other possible $j \in A_n$ with $a_j = 0$ is k. This gives (a) above. $\square$

Combining Lemma 3.1, Lemma 3.2, and Theorem 2.5, we have the following:

Theorem 3.3. *Let $ax \leq a_0$, $a \geq \mathbf{0}$, be a facet inducing inequality for P^n not of the form (a) or (b) of Lemma 3.2. If there exists $J \subseteq E^0(a)$ such that $|J| = 2|N_n| - 1$, and $D_n(J)$ contains no alternating cycle, then $ax \leq a_0$ is facet inducing for $\tilde{P}^n$.*

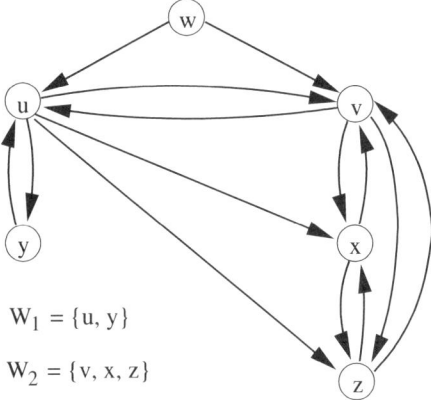

Fig. 2.1. Positive coefficients in a C3-inequality

Using Theorem 3.3, all of the facet inducing inequalities for P^n listed in Grötschel and Padberg (1985) can easily be shown to be facet inducing for $\tilde{P}^n$ (note these have previously been shown to be facet inducing for $\tilde{P}^n$ in Grötschel (1977) by more technical, direct methods). As an example, we illustrate this for the so called C3-inequalities.

Let u, v, w be three distinct nodes in N_n, and let W_1 and W_2 be subsets of N_n such that (a) $W_1 \cap W_2 = \emptyset$, (b) $W_1 \cap \{u, v, w\} = u$, (c) $W_2 \cap \{u, v, w\} = v$, (d) $|W_j| \geq 2$, $j = 1, 2$. Then

$$x(\gamma(W_1)) + x(\gamma(W_2)) + \sum_{j \in W_2} x_{uj} + x_{vu} + x_{wu} + x_{wv} \leq |W_1| + |W_2| - 1$$

is called a *C3-inequality*. Figure 2.1 shows the arcs having positive coefficients in a C3-inequality. The C3-inequalities are known to be facet inducing for P^n if $|W_1| + |W_2| = n - 1$ (see Grötschel and Padberg 1985).

In order to show the C3-inequalities $ax \leq a_0$ are facet inducing for $\tilde{P}^n$, we first find a set J of $2|N_n| - 1$ arcs in $E^0(a)$ such that $D_n(J)$ contains no alternating cycle. Since $|W_j| \geq 2$ for $j = 1, 2$, there exist nodes $y \in W_1 \setminus \{u\}$ and $z \in W_2 \setminus \{v\}$. Let

$$J = \{iw : i \in N_n \setminus \{w\}\} \cup \{wi : i \in N_n \setminus \{u, v, w\}\} \cup \{yz, zy, yv\}.$$

Then J satisfies the required properties.

Next we need to show that any C3-inequality is not of the form (a) or (b) of Lemma 3.2. Since $|W_j| \geq 2$ for $j = 1, 2$ for $ax \leq a_0$, there exist nodes $y \in W_1 \setminus \{u\}$ and $z \in W_2 \setminus \{v\}$. Since $\gamma(\{w, y, z\}) \subseteq E^0(a)$, we have $ax \leq a_0$ is not of form (a). Furthermore, if it is of form (b), then $\{w, y, z\} \in Z$. Since $\{yv, vy\} \subseteq E^0(a)$, we must have $v \in Z$ as well. However $\{vz, zv\} \subseteq A_n \setminus E^0(a)$, and thus $ax \leq a_0$ is not of form (b). It follows from Theorem 3.3 that the C3-inequalities are facet inducing for $\tilde{P}^n$.

Our third and final application deals with the Two Connected Polytope. Given the complete graph $K_n = (V, E)$ on n nodes and a vector $c \in \mathbb{R}^E$ of edge costs, the *Two Edge* (resp. *Node*) *Connected Problem* is to find a minimum weight two-edge (resp. node) connected spanning subgraph of K_n. By taking the convex hull of the incidence vectors of all two edge (resp. node) connected spanning subgraphs of K_n, we obtain the associated polytopes, namely the *Two Edge Connected Polytope* Q_{2E}^n, and the *Two Node Connected Polytope* Q_{2N}^n. Alternatively, Q_{2N}^n can be described as the convex hull of all integral $x \in \mathbb{R}^E$ satisfying the following linear constraints:

$$x_e \geq 0 \quad \text{for all } e \in E, \tag{19}$$

$$x_e \leq 1 \quad \text{for all } e \in E, \tag{20}$$

$$x(\delta(v)) \geq 2 \quad \text{for all } v \in V, \tag{21}$$

$$x(\delta(S)) \geq 2 \quad \text{for all } S \subset V, \emptyset \neq S \neq V, \tag{22}$$

$$x([W : V \backslash (W \cup \{z\})]) \geq 1 \quad \text{for any } z \in V \text{ and } W \subseteq V \backslash \{z\}. \tag{23}$$

Similarly, Q_{2E}^n can be described as the convex hull of all integral $x \in \mathbb{R}^E$ satisfying linear constraints (19)–(22).

In a more general form, both Q_{2E}^n and Q_{2N}^n have been studied in Grötschel and Monma (1990) and in Grötschel et al. (1992), and Q_{2E}^n has been studied in Mahjoub (1994) and Boyd and Hao (1993). It is known that both polytopes are full dimensional, and that they are closely related to the Symmetric Travelling Salesman Polytope Q^n, which is the convex hull of all incidence vectors of Hamilton cycles of K_n. The polytope Q^n can also be described as all x in Q_{2E}^n or in Q_{2N}^n that satisfy the linear inequalities (21) with equality. Thus Q^n is a face of Q_{2E}^n and Q_{2N}^n, which implies that for every facet inducing inequality for Q^n, there exists at least one equivalent inequality (with respect to Q^n) which is facet inducing for Q_{2E}^n, and similarly for Q_{2N}^n.

Using the results of Sect. 2.2, we define a class of facet inducing inequalities for Q_{2N}^n which are equivalent to the simple comb constraints of Q^n, and show a subclass of these are also facet inducing for Q_{2E}^n.

First we require some well known results related to Q^n. The *degree constraints* for Q^n are the constraints

$$x(\delta(v)) = 2 \quad \text{for all } v \in V.$$

For notational convenience, we denote the degree constraints by $Ax = 2$, where A is the node-edge incidence matrix of K_n.

Theorem 3.4 (Grötschel and Padberg 1979a). *The degree constraints form a minimal equation system for Q^n.*

A consequence of Theorem 3.4 is the following.

Theorem 3.5. *The dimension of Q^n equals $|E| - n$.*

Finally, we have the following well-known and easily verified result.

Theorem 3.6. *Let A be the node-edge incidence matrix for the complete graph $K_n = (V, E)$, and let $J \subseteq E$. The J indexes a set of linearly independent columns of A if and only if each component of the graph (V, J) contains no even cycle and at most one odd cycle.*

We are now ready to define the class of facet inducing inequalities for Q_{2N}^n. It is an adaptation of the class of *comb inequalities* for Q^n. A *comb* consists of a *handle* $H \subseteq V$ and mutually disjoint *teeth* $T_1, T_2, \ldots, T_k \subseteq V$, ($k \geq 3$ and odd) such that

$$T_j \cap H \neq \emptyset \neq T_j \backslash H, \quad 1 \leq j \leq k.$$

The associated *comb inequality*

$$x(\gamma(H)) + \sum_{i=1}^{k} x(\gamma(T_i)) \leq |H| + \sum_{i=1}^{k}(|T_i| - 1) - \frac{k+1}{2}$$

is facet inducing for Q^n (see Grötschel and Padberg 1979a, 1979b).

A comb is called *simple* if $|T_i \cap H| = 1$ for $i = 1, 2, \ldots, k$, and the associated inequality is called a *simple comb inequality*. If in addition we have $|T_i \backslash H| = 1$, i.e. each tooth consists of 2 nodes, the comb is called a *2-matching comb* and the corresponding inequality a *2-matching inequality*.

If we negate a simple comb inequality and add one half times the degree constraint for each $v \in V$, we obtain the *complemented simple comb inequality $ax \geq a_0$* which is defined by

$$a_e = \begin{cases} 0 & \text{for } e \in \gamma(H) \text{ or } e \in \gamma(T_i), \ i = 1, 2, \ldots, k, \\ 1 & \text{otherwise} \end{cases}$$

and

$$a_0 = |\mathring{V}| + \frac{k+1}{2},$$

where $\mathring{V} \subset V$ is the set of nodes not contained in H or any $T_i, i = 1, 2, \ldots, k$. Note that the edges whose coefficients have value 1 and the edges whose coefficients have value 0 in $ax \geq a_0$ are exactly reversed in the corresponding simple comb constraint. Also note that $ax \geq a_0$ is equivalent to a simple comb inequality with respect to Q^n, and thus the complemented simple comb constraints are facet inducing for Q^n.

Theorem 3.7. *The complemented simple comb constraints are valid for Q_{2N}^n.*

Proof. Let $ax \geq a_0$ represent a complemented simple comb inequality. Clearly, from inequalities (21) and (19) for Q_{2N}^n the following inequalities are valid for Q_{2N}^n:

$$x(\delta(v)) \geq 2 \quad \text{for all } v \in \mathring{V}, \tag{24}$$

$$x_e \geq 0 \quad \text{for all } e \in \delta(H) \Big\backslash \bigcup_{i=1}^{k} \gamma(T_i). \tag{25}$$

Also, if we take $z = T_i \cap H$ in inequality (23) for Q_{2N}^n we see that the following inequalities are valid for Q_{2N}^n:

$$x([T_i \setminus H : V \setminus T_i]) \geq 1 \quad \text{for } i = 1, 2, \ldots, k. \tag{26}$$

Summing the inequalities (24), (25), and (26) and dividing by two we obtain

$$ax \geq |\mathring{V}| + \frac{k}{2}.$$

Since for every vertex of Q_{2N}^n the left-hand side of the above inequality is an integer, we can round up the right-hand side to the next integer. Thus $ax \geq a_0$ is valid for Q_{2N}^n, as claimed. $\square$

Theorem 3.8. *The complemented simple comb constraints are facet inducing for* Q_{2N}^n.

Proof. Let $ax \geq a_0$ represent a complemented simple comb constraint with handle H and teeth $T_1, T_2, \ldots, T_k$. By Theorem 3.7, $ax \geq a_0$ is valid for Q_{2N}^n. Thus it suffices to show there exists $|E|$ affinely independent points in Q_{2N}^n satisfying $ax = a_0$. We generate a large portion of such points using an easily found independent direction set.

For each tooth T_i, $i = 1, 2, \ldots, k$, define v_i to be the single node in $H \cap T_i$, and define $\mathring{H}$ to be the set of nodes in H which are not in any tooth. Let C be a Hamiltonian cycle of $K_n[\{v_i : i = 1, 2, \ldots, k\}]$, let P_0 be a Hamiltonian path of $K_n[\{v_1\} \cup \mathring{H}]$, let P_i, $i = 1, 2, \ldots, k$, be a Hamiltonian path of $K_n[T_i]$ which starts at node v_i and ends at some node u_i, and let

$$B = C \cup \{P_i, i = 0, 1, 2, \ldots, k\}.$$

We show the set B satisfies the conditions of Lemma 3.1, i.e. X^B forms an extension set.

Since $a_e = 0$ for all $e \in B$, the set B satisfies condition (i) of Lemma 2.4. Let $Ax = 2$ represent the degree constraints for Q^n, which form a minimal equation system for Q^n by Theorem 3.4. By Theorem 3.6, the set $\{A_e : e \in B\}$ is linearly independent, and thus B satisfies condition (ii). To see that B also satisfies condition (iii) of Lemma 2.4, consider any $e \in B$. It is known that the simple comb inequality is not equivalent (with respect to Q^n) to the upper bound inequality $x_e \leq 1$ (see Grötschel and Padberg 1985). Thus for every $e \in B$, there exists a Hamiltonian cycle H of K_n which does not use edge e, and such that its incidence vector x^H satisfies $ax^H = a_0$. Adding edge e to H yields a two-node connected spanning subgraph of K_n. Let x^e be the edge incidence vector. Since $ax^H = a_0$ and $a_e = 0$, we have $ax^e = a_0$ as required.

Let $X = \{x^e : e \in B\}$ and let Y be an affine basis of $\{x \in Q^n : ax = a_0\}$. Let $F = \{x \in Q_{2N}^n : ax = a_0\}$. By Lemma 2.4, $X' = X \cup Y$ is an affinely independent set in F. Since $|X| = n - |\mathring{V}|$, and $|Y| = \dim(Q^n) = |E| - n$ by Theorem 3.5, we have $|X'| = |E| - |\mathring{V}|$. If $|\mathring{V}| = 0$, then X' is an affine basis of F, as required. Otherwise, we must find a set $\bar{X} \in F$ of size $|\mathring{V}|$ such that $\bar{X} \cup X'$ is affinely independent.

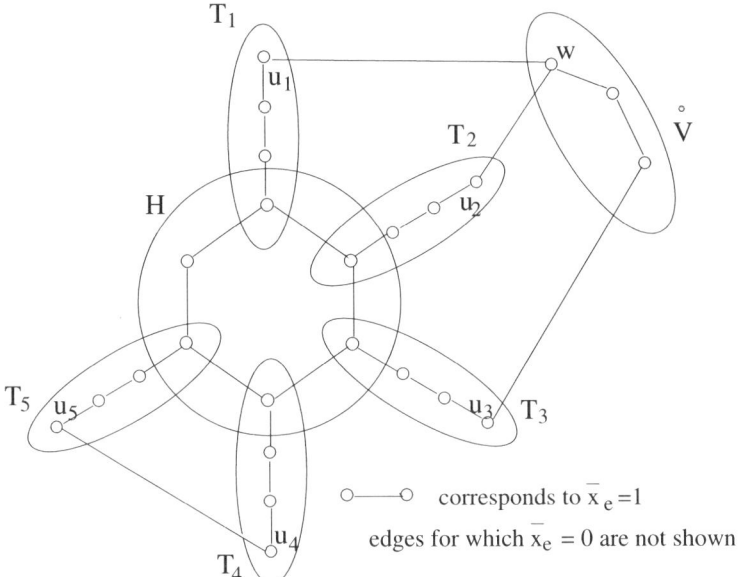

Fig. 2.2. An example of $\bar{x}$

We find $\bar{X}$ as follows. Every $x \in X'$ satisfies $x(\delta(w)) = 2$ for all $w \in \mathring{V}$. For each $w \in \mathring{V}$ we will construct a point $\bar{x}$ in F which satisfies $\bar{x}(\delta(v)) = 2$ for all $v \in \mathring{V}\backslash\{w\}$ and $\bar{x}(\delta(w)) = 3$. We will let $\bar{X}$ be the set of all these points. Then no such $\bar{x}$ can be expressed as an affine combination of other members of $X' \cup \bar{X}$, since any affine combination x will satisfy $x(\delta(w)) = 2$.

Let paths P_i and nodes $u_i \in T_i \backslash H$ for $i = 1, 2, \ldots, k$ be as previously defined. To construct $\bar{x}$, find a Hamiltonian path $P = w, \ldots, u_3$ of $K_n[\mathring{V} \cup \{u_3\}]$, a Hamiltonian cycle S of $K_n[H]$, and a perfect matching $M \subseteq E$ of $\{u_i : i = 4, 5, \ldots, k\}$. Then $\bar{x}$ is defined by

$$\bar{x}_e = \begin{cases} 1 & \text{for } e \in S \cup P \cup M \cup \{P_i : i = 1, 2, \ldots, k\} \text{ or } e = u_i w, \; i = 1, 2 \\ 0 & \text{otherwise.} \end{cases}$$

(See Fig. 2.2.)

Clearly $\bar{x}$ is the incidence vector of a two-node connected graph. Moreover,

$$a\bar{x} = |P| + |M| + 2$$
$$= |\mathring{V}| + \frac{k - 3}{2} + 2$$
$$= |\mathring{V}| + \frac{k + 1}{2}$$
$$= a_0.$$

Since $\bar{x}(\delta(w)) = 3$ but $\bar{x}(\delta(v)) = 2$ for all $v \in \mathring{V}\backslash\{w\}$, we are done. $\square$

In general, the complemented simple comb constraints are not valid for Q_{2E}^n. In fact, they are only valid for Q_{2E}^n when they originate from 2-matching combs. We call this subset of the complemented simple comb constraints the *complemented 2-matching constraints*, and show below that they are facet inducing for Q_{2E}^n. These inequalities were independently shown facet inducing for Q_{2N}^n and Q_{2E}^n (in a different form) in Grötschel et al. (1992) as a subset of a class of facets called the lifted 2-cover inequalities, and are also a subclass of a set of constraints shown to be facet inducing for Q_{2E}^n in Boyd and Hao (1993). Note that, to the best of our knowledge, the rest of the complemented simple comb constraints belong to no known class of facet inducing inequalities for Q_{2N}^n.

Theorem 3.9. *The complemented 2-matching constraints are valid for Q_{2E}^n.*

Proof. Let $ax \geq a_0$ represent a complemented 2-matching inequality. Clearly from the linear constraints (19)–(22) for Q_{2E}^n the following inequalities are valid for Q_{2E}^n:

$$x(\delta(v)) \geq 2 \quad \text{for all } v \in V \setminus H,$$

$$x_e \geq 0 \quad \text{for all } e \in \delta(H) \setminus \bigcup_{i=1}^k \gamma(T_i),$$

$$-x_e \geq -1 \quad \text{for all } e \in \bigcup_{i=1}^k \gamma(T_i).$$

Summing the above inequalities and dividing by two we obtain

$$ax \geq |\mathring{V}| + \frac{k}{2}.$$

Since for every vertex of Q_{2E}^n the left-hand side of the above inequality is an integer, we can round up the right-hand side to the next integer. Thus $ax \geq a_0$ is valid for Q_{2E}^n, as required.   □

Theorem 3.10. *The complemented 2-matching constraints are facet inducing for Q_{2E}^n.*

Proof. Let $ax \geq a_0$ represent a complemented 2-matching constraint. Since $\{x \in Q_{2N}^n : ax = a_0\} \subset \{x \in Q_{2E}^n : ax = a_0\}$, the result follows directly from Theorems 3.8 and 3.9.   □

2.4 Facets of Polyhedra and Facets of Faces

Given a facet inducing inequality $ax \leq a_0$ for a face F of a polyhedron $P \subseteq \mathbb{R}^E$ which is valid for P, there exists many equivalent inequalities $\bar{a}x \leq \bar{a}_0$ with respect to F, some of which are facet inducing for P. In some combinatorial cases, $ax \leq a_0$ will already satisfy this property. If not, an independent direction set can be used to determine whether this is the case, and if not, find an equivalent inequality (with respect to F) for which this holds. We describe an iterative procedure for this called *Dimension Augmentation*. It first finds a maximal independent direction set D for $F_a^=$. By Corollary 2.3, $\dim(P_a^=) = \dim(F_a^=) + |D|$. If $|D| = r_l(A^F)$ for some minimal equation system $A^F x = b^F$ for F with respect to P, then $ax \leq a_0$ is facet inducing for P by Corollary 2.2. If $|D| < r_l(A^F)$, we find an equivalent inequality $a'x \leq a_0'$ (with respect to F) and a set of directions D' for $F_{a'}^= (= F_a^=)$ such that $D \subsetneq D'$. Thus by Corollary 2.3, $\dim(P_{a'}^=) > \dim(P_a^=)$, i.e. our new inequality induces a face of P of higher dimension. Hence after performing this process at most $r_l(A^F)$ times we will obtain an inequality $a'x \leq a_0'$ which is facet inducing for P. For a more detailed description of the general Dimension Augmentation procedure, we refer the reader to Boyd and Pulleyblank (1993).

Note that the Dimension Augmentation process is not easily implemented in the general case, as some of the steps involve solving complex problems. However, for some special cases this process simplifies greatly. For example, in Bauer (1997), Bauer shows how this process can be adapted and simplified in the case of the circuit polytope, which is the convex hull of all circuits of K_n and for which the Travellings Salesman Polytope is a face. She then applies this process to successfully obtain many classes of facet inducing inequalities for the circuit polytope from known classes of facet inducing inequalities of the Travelling Salesman Polytope.

The Dimension Augmentation procedure can also be simplified and performed efficiently in the case in which $\tilde{Q}$ is the monotone completion of a polyhedron Q, which lies in the nonnegative orthant of $\mathbb{R}^E$. Given an inequality $ax \leq a_0, a \geq \mathbf{0}$, which is facet inducing for Q, the following specialized form of the Dimension Augmentation process can be used to obtain an equivalent inequality which is support reduced, and thus facet inducing for $\tilde{Q}$ by Theorem 2.5. (Note that this process was used by Grötschel and Pulleyblank (1986) for the Symmetric Travelling Salesman Polytope.) Let $A^Q x = b^Q$ be a minimal equation system for Q, and suppose no subset of $E^0(a)$ indexes a column basis of A^Q. It follows that the rows of $A^Q_{E^0(a)}$ are linearly dependent, and hence there exists a vector $\lambda \neq \mathbf{0}$ such that $\lambda A^Q_{E^0(a)} = \mathbf{0}$. Clearly $\lambda A^Q \neq \mathbf{0}$ since the rows of A^Q are linearly independent. Hence by subtracting an appropriate multiple of λA^Q from a and λb^Q from a_0 we can obtain an inequality $\bar{a}x \leq \bar{a}_0$ which induces the same facet of Q, and satisfies $\bar{a} \geq \mathbf{0}$ and $E^0(a) \subsetneq E^0(\bar{a})$. More specifically, let $\bar{a} = a - \alpha(\lambda A^Q)$ and let $\bar{a}_0 - \alpha\lambda b^Q$, where

$$\alpha = \min\left(\frac{a_e}{\lambda A_e^Q} : e \in E, \lambda A_e^Q \neq 0 \right).$$

Clearly we only have to repeat this process at most $|E|$ times before obtaining an equivalent inequality $\hat{a}x \leq \hat{a}_0$ such that $\hat{a}x \leq \hat{a}_0$ is support reduced.

As an aside, note that the monotone completion $\tilde{Q}$ of a polyhedron $Q \subseteq \mathbb{R}^E$ can be described as $(Q - C) \cap C$, where C is the nonnegative orthant; i.e. $\tilde{Q}$ is Q extended by the negative of cone C and then intersected with C. Both Theorem 2.5 and the process of support reducing an inequality can be generalized to include the case where C is an arbitrary cone (see Boyd 1986).

Given a face F of a polyhedron P, thus far we have shown how knowledge of facet inducing inequalities for F can be used to find facet inducing inequalities for P. As it is often technically much simpler to obtain results about facets for a full dimensional polyhedron than one of lower dimension, it would be nice to also have a converse result, i.e. know under what conditions an inequality inducing a facet of P also induces a facet of a face F of P. Although we know of no reasonable general result of this type, such conditions were found by Naddef and Rinaldi (1993) for the Graphical Travelling Salesman Polytope and the Symmetric Travelling Salesman Polytope. In this case the two polyhedra have the nice property that there is a one-to-one correspondence between their facets. Unfortunately this is not the case for general P and F. It is true that every facet inducing inequality $ax \leq a_0$ of F has an equivalent form which is facet inducing for P, however several of these equivalent forms may induce different facets of P, and P may have other facets which correspond to no facet of F.

The case of matchings in nonbipartite graphs provides an example of this. For a graph G, we let $M(G)$ denote the *matching polytope of G*, the convex hull of the incidence vectors of all matchings of G. We let $M^=(G)$ denote the *perfect matching polytope of G*, the convex hull of the incidence vectors of the perfect matchings of G. Then $M^=(G)$ is a face of $M(G)$. Pulleyblank and Edmonds (1974) described a complete characterization of the minimal set of inequalities necessary to define $M(G)$. Edmonds et al. (1982) provided a characterization of a minimal defining system for $M^=(G)$. This latter characterization is much more complex and required the creation of a new decomposition of a graph into so called *bricks*, three connected bicritical subgraphs.

2.5 Proving Validity

The procedures we describe in Sects. 2.2 and 2.4 require that we determine the validity of an inequality. This can be done with the rounding procedure we used in Sect. 2.3 to prove the validity of constraints for Q_{2N}^n and Q_{2E}^n.

With no loss of generality, we can restrict our attention to polyhedra defined by inequalities. Let $P = \{x \in \mathbb{R}^E : Ax \leq b\}$ be a polyhedron and let P_I be the convex hull of the integral members of P. It follows from Farkas' Lemma (see Pulleyblank 1989 or Nemhauser and Wolsey 1988) that an inequality $ax \leq a_0$ is valid for P if and only if there exists $\lambda \geq 0$ such that $a = \lambda A$ and $a_0 \geq \lambda b$. In general we require a more complicated argument to show that an inequality is valid for P_I. Such a procedure was proposed by Chvátal (1973), based on earlier work of Gomory (see Gomory 1960, 1963).

Suppose we have an inequality $ax \leq a_0$ defined by $a = \lambda A$ and $a_0 = \lambda b$, where $\lambda \geq 0$ and a is integral valued. Then every integral vector in P, and hence every member x of P_I will satisfy $ax \leq \lfloor a_0 \rfloor$, where $\lfloor a_0 \rfloor$ is the largest integer no greater than a_0. Such an inequality $(\lambda A)x \leq \lfloor \lambda b \rfloor$, for $\lambda \geq 0$ and λA integral is called a *first-order Chvátal–Gomory cut*, or simply an *elementary cut*.

The following is equivalent to the fact that we can restrict our attention to basic solutions when solving linear programs.

Proposition 5.1. *Suppose $ax \leq a_0$ is valid for nonempty $P = \{x \in \mathbb{R}^E : Ax \leq b\}$. Then there exists a subsystem $\bar{A}x \leq \bar{b}$ of $Ax \leq b$, consisting of at most $|E|$ inequalities, and $\lambda \geq 0$ such that $a = \lambda \bar{A}$, $a_0 \geq \lambda \bar{b}$. Moreover λ is rational and the largest number of digits required to represent any component λ_i as p/q, with p, q integers, is polynomial in $|E|$ and the number of digits in the largest magnitude number appearing in A, a. Let $\tilde{A}$ denote the magnitude of this largest number.*

Proof. Consider the linear program: maximize ax subject to $Ax \leq b$. Since $ax \leq a_0$ is valid, and $P \neq \emptyset$, there exists a basic optimum solution x^B having objective value at most a_0. A basic optimum solution to the dual linear program will be a vector λ with at most $|E|$ nonzero components, satisfying $\lambda A = a$, $\lambda b \leq a_0$ and $\lambda \geq 0$. This is the linear combination we require. By Cramer's rule, each component λ_i is the quotient of the determinants of two $|E| \times |E|$ submatrices of A, with the row a adjoined. Each of these determinants is at most $\tilde{A}^{|E|} \cdot |E|!$. The number of digits in each determinant is at most $\lfloor \log(\tilde{A}^{|E|} \cdot |E|!) \rfloor + 1 \leq |E| \log(\tilde{A}) + |E| \log|E| + 1$, giving the result. □

This implies that when doing the Chvátal–Gomory procedure, it is sufficient to consider subsystems consisting of at most $|E|$ inequalities.

The following corollary is used in Chvátal (1973).

Corollary 5.2. *Let $ax \leq \alpha$ be an elementary cut for $P = \{x \in \mathbb{R}^E : Ax \leq b\}$. Then this inequality, or a stronger one, can be obtained as an elementary cut from a subsystem consisting of at most $|E|$ inequalities.*

Proof. Let $ax \leq a_0$ be a valid inequality for P for which a_0 is minimized. By Proposition 5.1, this inequality is a nonnegative linear combination of at most $|E|$ inequalities from $Ax \leq b$. Since $ax \leq \alpha$ is elementary, we must have $\lfloor a_0 \rfloor \leq \alpha$, giving the result. □

Let $Cl(P)$ denote the set of all member of P which satisfy all elementary cuts. Then $Cl(P)$ is a polyhedron (Schrijver 1980). Clearly $P_I \subseteq Cl(P)$, but in general these polyhedra are not equal. Let $Cl^0(P) = P$ and, for $k \geq 1$, let $Cl^k(P) = Cl(Cl^{k-1}(P))$. Chvátal (1973) (for bounded polyhedra) and Schrijver (1980) (for general polyhedra) showed that for any polyhedron P, there exists some finite $k \geq 0$ such that $P_I = Cl^k(P)$. This therefore provides a framework for proving the validity of an inequality. Let $ax \leq a_0$ be any inequality valid for P_I. The *rank* of this inequality is the smallest value of k such that the inequality is valid for $Cl^k(P)$. We can prove the validity of $ax \leq a_0$ by expressing it, or a strengthening of it, as a nonnegative linear combination of inequalities obtained by at most k iterations of the Chvátal–Gomory procedure, where k is the rank of the inequality.

Suppose $ax \leq \alpha$ is valid for $Cl(P)$ with a, α integer. It is not necessarily true that $ax \leq \alpha$ can be deduced as an elementary cut from $Ax \leq b$. It may be a linear combination of elementary cuts which gives a smaller right hand side than any elementary cut with coefficients a.

We now show that the Chvátal–Gomory procedure yields a proof of the validity of an inequality of polynomial length, for fixed rank k. We assume that A and b are integral and let L be the largest number of digits required to represent any integer in A or b. Recall that $P = \{x \in \mathbb{R}^E : Ax \leq b\}$.

Theorem 5.3. *Every facet of $Cl(P)$ is induced by an inequality $ax \leq a_0$, with a, a_0 integral, such that a_0 and all coefficients in a are expressible in a number of digits polynomial in L and $|E|$.*

Proof. Let S be the set of all inequalities $ax \leq a_0$ where $a = \lambda A$, $a_0 = \lfloor \lambda b \rfloor$, such that a is integral, $0 \leq \lambda \leq 1$ and at most $|E|$ components of λ are nonzero. Note that in this case a_0 and each element of a is expressible with at most $L + \log |E| + 1$ digits. Now let $\bar{a}x \leq \bar{a}_0$ be any inequality which induces a facet of $Cl(P)$, where $\bar{a}, \bar{a}_0$ are integral valued. Then we have $\bar{a} = \mu A$ and $\bar{a}_0 = \lfloor \mu b \rfloor$ for some $\mu \geq 0$, since every other valid inequality is a linear combination of elementary cuts. By Corollary 5.2, we can assume that at most $|E|$ components of μ are nonzero.

Let $\langle x \rangle := x - \lfloor x \rfloor$ denote the fractional part of x, for any real number x. Then

$$\bar{a} = \mu A = \lfloor \mu \rfloor A + \langle \mu \rangle A$$

$$\bar{a}_0 = \lfloor \mu b \rfloor = \lfloor \mu \rfloor b + \lfloor \langle \mu \rangle b \rfloor.$$

Since at most $|E|$ components of μ are nonzero, the fractional part $\langle \mu \rangle$ has at most $|E|$ nonzero components, so since $0 \leq \langle \mu \rangle < 1$, the inequality $\langle \mu \rangle Ax \leq \lfloor \langle \mu \rangle b \rfloor$ is in S. Suppose that $\lfloor \mu \rfloor \neq 0$. Let $\lfloor \mu \rfloor_i \neq 0$. If we have $A_i x = b_i$ for every $x \in Cl(P)$, then we can replace μ_i with $\langle \mu_i \rangle$ and get the same facet. If not, then $A_i x \leq b_i$ must induce a proper face of $Cl(P)$, and hence the same face of $Cl(P)$ as does $\bar{a}x \leq \bar{a}_0$.

It thus follows that the system S plus $Ax \leq b$ provides a complete description of $Cl(P)$, and each of these inequalities has small coefficients as claimed. $\square$

Note that since $|S|$ in the preceding proof is finite, this proves that $Cl(P)$ is defined by a finite set of inequalities, that is, $Cl(P)$ is a polytope.

Let $P = \{x \in \mathbb{R}^E : Ax \leq b\}$. The *rank* of P_I, *with respect to* P, is the smallest k such that $P_I = Cl^k(P)$. Since many different polyhedra can have the same integer hull, we can only define rank with respect to a starting polyhedron.

Consider the decision problem: Is the inequality $ax \leq a_0$ valid for P_I? This is equivalent to the optimization problem: maximize ax for $x \in P_I$. If the answer is "No", then we can prove this by exhibiting an integer $\hat{x}$ satisfying $A\hat{x} \leq b$ for which $a\hat{x} > a_0$. (We ignore the work it could take to find such an $\hat{x}$.) If the answer is "Yes", we could show this by giving a derivation of the inequality from the starting system $Ax \leq b$, using the Chvátal–Gomory procedure. That is, we could construct a rooted derivation tree wherein each node corresponds to an inequality. With the exception of the root, the inequality corresponding to each node is obtained by the Chvátal–Gomory procedure from the inequalities corresponding to its children. the inequality $ax \leq a_0$ is associated with the root, and is either obtained using the Chvátal–Gomory procedure or is obtained as a nonnegative linear combination of the inequalities corresponding to its children.

It follows from Proposition 5.1 and Corollary 5.2 that each node of the tree need have at most $|E|$ children. Therefore the total number of nodes in the tree is at most $1 + |E| + |E|^2 + \cdots + |E|^{k+1} = (|E|^{k+2} - 1)/(|E| - 1)$, if $ax \leq a_0$ had rank k with respect to $Ax \leq b$. Therefore, it follows from Theorem 5.3 and Proposition 5.1, that the total length of the proof of validity grows polynomially with L, $|E|$ and the length of the longest number in a, a_0, but exponentially with k. Thus we have the following:

Theorem 5.4. *Let $P = \{x \in \mathbb{R}^E : Ax \leq b\}$ and suppose P_I has rank k, for fixed k. Then every valid inequality $ax \leq a_0$ for P_I has a derivation whose length grows polynomially in $|E|$ and the length of the largest magnitude number in A, b, a and a_0.*

Corollary 5.5. *Suppose that some NP-complete problem can be formulated as the decision problem: Does there exist integer $\hat{x} \in P = \{x \in \mathbb{R}^E : Ax \leq b\}$? Suppose we can in polynomial time verify that an inequality $ax \leq a_0$ belongs to $Ax \leq b$. Then if there exists a fixed integer k such that $P_I = Cl^k(P)$, for all instances of the problem, then $NP = coNP$.*

A consequence of this is that if we are considering an integer programming formulation of an NP-hard problem, with the property that the validity of the inequalities is polynomially verifiable, then unless $NP = coNP$, there will exist instances for which the Chvátal–Gomory rank of the integer hull is arbitrarily large.

A similar result, showing that cutting plane proofs of validity are of polynomial length appears in Cook et al. (1986).

The classical integer programming problem is to maximize cx for $x \in P_I$ where P_I is the set of integer points contained in P. The linear programming problem of maximizing cx for $x \in P$ is a relaxation of this integer programming problem. A stronger relaxation is to maximize cx for $x \in Cl(P)$. However Eisenbrand (1999) showed that the problem of determining whether there exists an elementary cut that separates a vector $\hat{x}$ from P is NP-complete. From this it follows that the problem of maximizing cx for $x \in Cl(P)$ is already NP-hard. Fischetti and Lodi (2007) showed, however, that the problem of separation over $Cl(P)$ could be formulated as a mixed integer programming problem (MIP) and this could be solved through appropriate usage of a general-purpose commercial MIP solver. They then used this to optimally solve a number of instances of integer programming problems, including a previously unsolved problem from MIPLIB 2003 (Achterberg et al. 2003).

References

Achterberg, T., Koch, T., Martin, A.: MIPLIB 2003. http://miplib.zib.de/ (2003)

Balas, E., Fischetti, M.: Polyhedral theory for the asymmetric traveling salesman problem. In: Gutin, G., Punnen, A. (eds.) The Traveling Salesman Problem and Its Variations. Springer, New York (2007)

Bauer, P.: The circuit polytope: Facets. Math. Oper. Res. **22**, 110–145 (1997)

Boyd, S.: The subtour polytope of the travelling salesman problem. Ph.D. Thesis, University of Waterloo, Waterloo, Canada (1986)

Boyd, S., Hao, T.: An integer polytope related to the design of survivable communication networks. SIAM J. Discrete Math. **6**, 612–630 (1993)

Boyd, S., Pulleyblank, W.R.: Facet generating techniques. Technical Report TR-91-31, Dept. of Computer Science, University of Ottawa (1993)

Chvátal, V.: Edmonds polytopes and a hierarchy of combinatorial problems. Discrete Math. **4**, 305–337 (1973)

Cook, W., Gerards, A.M.H., Schrijver, A., Tardos, E.: Sensitivity theorems in integer linear programming. Math. Program. **34**, 251–264 (1986)

Edmonds, J., Lovász, L., Pulleyblank, W.R.: Brick decompositions and the matching rank of graphs. Combinatorica **2**, 247–274 (1982)

Eisenbrand, F.: On the membership problem for the elementary closure of a polyhedron. Combinatorica **19**, 297–300 (1999)

Fischetti, M., Lodi, A.: Optimizing over the first Chvátal closure. Math. Program. B **110**(1), 3–20 (2007)

Gomory, R.: Solving linear programming problems in integers. In: Bellman, R.E., Hall, M. Jr. (eds.) Combinatorial Analysis. Am. Math. Soc., Providence (1960)

Gomory, R.: An algorithm for integer solutions to linear programs. In: Graves, R., Wolfe, P. (eds.) Recent Advances in Mathematical Programming. McGraw-Hill, New York (1963)

Grötschel, M.: Polyedrische Charakterisierungen kombinatorischer Optimierungs-probleme. Hain, Meisenheim am Glan (1977)

Grötschel, M., Monma, C.L.: Integer polyhedra arising from certain network design problems with connectivity constraints. SIAM J. Discrete Math. **3**, 502–523 (1990)

Grötschel, M., Padberg, M.: On the symmetric travelling salesman problem I: Inequalities. Math. Program. **16**, 265–280 (1979a)

Grötschel, M., Padberg, M.: On the symmetric travelling salesman problem II: Lifting theorems and facets. Math. Program. **16**, 281–302 (1979b)

Grötschel, M., Padberg, M.: Polyhedral theory. In: Lawler, E.L., et al. (eds.) The Travelling Salesman Problem. Wiley, New York (1985)

Grötschel, M., Pulleyblank, W.R.: Clique tree inequalities and the symmetric travelling salesman problem. Math. Oper. Res. **11**, 537–569 (1986)

Grötschel, M., Jünger, M., Reinelt, G.: Facets of the linear ordering polytope. Math. Program. **33**, 43–60 (1982a)

Grötschel, M., Jünger, M., Reinelt, G.: On the acyclic subgraph polytope. Math. Program. **33**, 1–27 (1982b)

Grötschel, M., Monma, C.L., Stoer, M.: Facets for polyhedra arising in the design of communication networks with low-connectivity constraints. SIAM J. Optim. **2**, 474–504 (1992)

Leung, J., Lee, J.: More facets from fences for linear ordering and acyclic subgraph polytopes. Discrete Appl. Math. **50**, 185–200 (1994)

Mahjoub, A.R.: Two-edge connected spanning subgraphs and polyhedra. Math. Program. **64**, 199–208 (1994)

Naddef, D., Rinaldi, G.: The graphical relaxation: A new framework for the travelling salesman polytope. Math. Program. **58**, 53–88 (1993)

Nemhauser, G.L., Wolsey, L.A.: Integer and Combinatorial Optimization. Wiley-Interscience, New York (1988)

Padberg M.W., Rao, M.R.: Odd minimum cut-sets and b-matchings. Math. Oper. Res. **7**, 67–80 (1982)

Pulleyblank, W.R.: Polyhedral combinatorics. In: Nemhauser, G.L., et al. (eds.) Handbooks in Operational Research and Management Science, vol. 1. Elsevier Science/North-Holland, Amsterdam (1989)

Pulleyblank, W.R., Edmonds, J.: Facets of 1-matching polyhedra. In: Berge, C., Ray-Chaudhuri, D.K. (eds.) Hypergraph Seminar. Lecture Notes in Mathematics, vol. 411, pp. 214–242. Springer, Berlin (1974)

Schrijver, A.: On cutting planes. Ann. Discrete Math. **9**, 291–296 (1980)

3

Antimatroids, Betweenness, Convexity

Vašek Chvátal

Summary. Two classical examples of antimatroids arise from double shellings of partially ordered sets and from simplicial shellings of triangulated graphs. The corresponding convex geometries have Carathéodory number two and admit a natural description in terms of the ternary relation of betweenness in the underlying structure. We characterize a nested pair of classes of betweenness which generate convex geometries of Carathéodory number two. The corresponding antimatroids include all antimatroids arising from double shellings of partially oredred sets and all antimatroids arising from simplicial shellings of triangulated graphs.

3.1 Introduction

Korte and Lovász (1981, 1984) founded the theory of *greedoids*. These combinatorial structures characterize a class of optimization problems that can be solved by greedy algorithms. In particular, greedoids generalize *matroids*, introduced earlier by Whitney (1935). *Antimatroids*, introduced by Dilworth (1940) as particular examples of semimodular lattices, make up another class of greedoids.

Antimatroids are related to abstract convexity; let us explain how. Kay and Womble (1971) defined a *convexity space* on a ground set E as a tuple $(E, \mathcal{N})$, where $\mathcal{N}$ is a collection of subsets of E such that $\emptyset \in \mathcal{N}$, $E \in \mathcal{N}$, and $\mathcal{N}$ is closed under intersections. Members of $\mathcal{N}$ are called *convex sets*. The *convex hull* of a subset X of E is defined as the intersection of all convex supersets of X and is denoted by $\tau_{\mathcal{N}}(X)$. Independently of each other, Edelman (1980) and Jamison-Waldner (1980) initiated the study of convexity spaces $(E, \mathcal{N})$ with the *anti-exchange property*

if $X \subseteq E$ and y, z are distinct points outside $\tau_{\mathcal{N}}(X)$,
then at most one of $y \in \tau_{\mathcal{N}}(X \cup \{z\})$ and $z \in \tau_{\mathcal{N}}(X \cup \{y\})$ holds true.

This research was funded by the Canada Research Chairs Program and by the Natural Sciences and Engineering Research Council of Canada.

Jointly, Edelman and Jamison (1985) proposed to call such convexity spaces *convex geometries*. An antimatroid is a tuple $(E, \mathcal{F})$ such that $(E, \{E - X: X \in \mathcal{F}\})$ is a convex geometry.

In the present paper, we deal exclusively with finite ground sets. Our starting point are two examples of antimatroids. One of these arises from double shelling of a poset (Example 2.4 in Chap. III of the monograph Korte et al. 1991) and the other from simplicial shelling of a triangulated graph (Example 2.7 in Chapter III of Korte et al. 1991). Let us describe them in terms of convex geometries.

In the first example, given a partially ordered set $(E, \preceq)$, we let $\mathcal{N}$ consist of all subsets K of E such that

$$a, c \in K, \; a \prec b \prec c \quad \Longrightarrow \quad b \in K;$$

the resulting tuple $(E, \mathcal{N})$ is a convex geometry. In the second example, given an undirected graph with a vertex-set E, we let $\mathcal{N}$ consist of all subsets K of E such that

$$a, c \in K, \; b \text{ is an interior vertex of a chordless path from } a \text{ to } c \quad \Longrightarrow \quad b \in K.$$

The resulting tuple $(E, \mathcal{N})$ is a convexity space, but not necessarily a convex geometry: for instance, take a chordless cycle through four vertices as the graph and consider the convex set X consisting of two adjacent vertices in this graph. Nevertheless, if the graph is *triangulated*, meaning that it contains no chordless cycle through four or more vertices, then $(E, \mathcal{N})$ is a convex geometry. To elucidate this point, we appeal to a characterization of convex geometries that involves the notion of an *extreme point* of a convex set K, defined as a point b of K such that $K - \{b\}$ is convex.

Nine equivalent characterizations of convex geometries are stated in Jamison-Waldner (1980); equivalence of the following four is proved in Korte et al. (1991), Chapter III, Theorem 1.1.

Fact 1.1. *For every convexity space $(E, \mathcal{N})$, the following four propositions are logically equivalent:*

(G1) $(E, \mathcal{N})$ *has the anti-exchange property.*
(G2) *If $X \in \mathcal{N}$ and $X \neq E$, then $X \cup \{y\} \in \mathcal{N}$ for some y in $E - X$.*
(G3) *Every set in $\mathcal{N}$ is the convex hull of its extreme points.*
(G4) *Every subset X of E contains a unique minimal subset Y such that $\tau_{\mathcal{N}}(Y) = \tau_{\mathcal{N}}(X)$.*

To see that our second example has property (G3), we invoke a theorem of Dirac (1961). There, a vertex is called *simplicial* if its neighbours are pairwise adjacent.

Fact 1.2. *For every finite undirected graph G, the following three propositions are logically equivalent:*

- *G is triangulated.*
- *Every minimal cutset in G is a clique.*
- *Every induced subgraph of G either includes two nonadjacent simplicial vertices or is complete.*

A corollary of this theorem (stated and proved in Farber and Jamison 1986 as Theorem 3.2) asserts that

every non-simplicial vertex in a triangulated graph
lies on a chordless path between two simplicial vertices;

since the extreme points of E in the second example are precisely the simplicial vertices of G, it follows that every point of E lies in the convex hull of at most two extreme points of E.

Each of these two examples of convex geometries is constructed through the intermediary of betweenness in the underlying structure. In the first example, we may say that b lies between a and c if, and only if, $a \prec b \prec c$ or $c \prec b \prec a$; in the second example, we may say that b lies between a and c if, and only if, b is an interior vertex of a chordless path from a to c; in either case, a subset K of E is convex if and only if

$$a, c \in K, \ b \text{ lies between } a \text{ and } c \ \implies \ b \in K.$$

This construction generalizes: as in Calder (1971), every ternary relation $\mathcal{B}$ on a finite ground set E defines a convexity space $(E, \mathcal{N}_{\mathcal{B}})$ by

$$\mathcal{N}_{\mathcal{B}} = \{K \subseteq E : a, c \in K, \ (a, b, c) \in \mathcal{B} \ \Rightarrow \ b \in K\}.$$

Our objective is to characterize a nested pair of classes of ternary relations $\mathcal{B}$ on finite ground sets E such that the corresponding classes of convexity spaces $(E, \mathcal{N}_{\mathcal{B}})$ consist exclusively of convex geometries, include all convex geometries that arise from double shelling of a poset, and include all convex geometries that arise from simplicial shelling of a triangulated graph.

3.2 Results

Note that, for every ternary relation $\mathcal{B}$ on a finite ground set, every set K in $\mathcal{N}_{\mathcal{B}}$, and every point b of K,

$$K - \{b\} \in \mathcal{N}_{\mathcal{B}} \ \iff \ \text{there are no points } a, c \text{ of } K \text{ such that } (a, b, c) \in \mathcal{B}.$$

This observation allows us to extend the definition of extreme points: given an arbitrary, not necessarily convex, subset X of the ground set and given an arbitrary point b of X, we shall say that b is an *extreme point of X* if, and only if, there are no points a, c of X such that $(a, b, c) \in \mathcal{B}$. The set of all extreme points of X will be denoted by $\mathrm{ex}_{\mathcal{B}}(X)$.

In addition, note that $\mathcal{N}_{\mathcal{B}}$ does not change if $\mathcal{B}$ is made symmetric by including (c, b, a) in $\mathcal{B}$ whenever (a, b, c) is in $\mathcal{B}$; it does not change either if all triples (b, b, b), (b, b, c), (a, b, b) are removed from $\mathcal{B}$. Note also that $\mathcal{N}_{\mathcal{B}}$ includes all singletons $\{a\}$ with $a \in E$ (Kay and Womble 1971 designate such convexity spaces T_1) if and only if $\mathcal{B}$ includes no triple (a, b, a) with $b \neq a$. We will restrict our attention to ternary relations $\mathcal{B}$ such that

$$(a, b, c) \in \mathcal{B} \ \implies \ (c, b, a) \subset \mathcal{B} \text{ and } a, b, c \text{ are pairwise distinct;}$$

any such $\mathcal{B}$ will be called a *strict betweenness*.

Theorem 2.1. *For every strict betweenness $\mathcal{B}$ on a finite ground set E, the following two propositions are logically equivalent:*

(i) *For all subsets X of E and all x_1, x_2, x_3 in X such that $(x_1, x_2, x_3) \in \mathcal{B}$, there are $\overline{x}_1, \overline{x}_3$ in $\mathrm{ex}_\mathcal{B}(X)$ such that $(\overline{x}_1, x_2, \overline{x}_3) \in \mathcal{B}$.*
(ii) *$(a, b, c_2), (c_1, c_2, c_3) \in \mathcal{B} \Rightarrow (a, b, c_1) \in \mathcal{B}$ or $(a, b, c_3) \in \mathcal{B}$ or $(c_1, b, c_3) \in \mathcal{B}$.*

Following Kay and Womble (1971), a convexity space is said to have *Carathéodory number d* if, and only if, d is the smallest positive integer with the following property:

if a point lies in the convex hull of a set X,
then it lies in the convex hull of a subset X' of X such that $|X'| \leq d$.

Theorem 2.1 characterizes a class of ternary relations $\mathcal{B}$ on finite ground sets E such that the corresponding class of convexity spaces $(E, \mathcal{N}_\mathcal{B})$ consists exclusively of convex geometries with Carathéodory number at most 2. However, it does not characterize all such relations: for instance, if $E = \{a, b, c_2, c_3\}$ and

$$\mathcal{B} = \{(a, b, c_2), (c_2, b, a), (a, c_2, c_3), ((c_3, c_2, a)\},$$

then $(E, \mathcal{N}_\mathcal{B})$ is a convex geometry with Carathéodory number 2 and yet $\mathcal{B}$ does not satisfy the conditions of Theorem 2.1.

Strict order betweenness (Lihová 2000) in a partially ordered set $(E, \preceq)$ is defined by

$$\mathcal{B} = \{(a, b, c) \in E^3 : a \prec b \prec c \text{ or } c \prec b \prec a\};$$

monophonic (Jamison-Waldner 1982), or *minimal path* (Duchet 1988), betweenness in an undirected graph with a vertex-set E is defined by

$$\mathcal{B} = \{(a, b, c) \in E^3 : b \text{ is an interior vertex of a chordless path from } a \text{ to } c\}.$$

Strict order betweenness satisfies condition (ii) of Theorem 2.1: it is a straightforward exercise to verify that it satisfies the stronger condition

$$(a, b, c_2), (c_1, c_2, c_3) \in \mathcal{B} \quad \Longrightarrow \quad (a, b, c_1) \in \mathcal{B} \text{ or } (a, b, c_3) \in \mathcal{B}.$$

We shall prove that monophonic betweenness in triangulated graphs, too, satisfies this stronger condition.

Theorem 2.2. *Let G be a finite triangulated graph and let a, b, c_1, c_2, c_3 be vertices of G. If*

b is an interior vertex of a chordless path between a and c_2 and
c_2 is an interior vertex of a chordless path between c_1 and c_3,

then

b is an interior vertex of a chordless path between a and c_1, or else
b is an interior vertex of a chordless path between a and c_3.

Theorem 2.3. *For every strict betweenness $\mathcal{B}$ on a finite ground set E, the following two propositions are logically equivalent:*

(i) *For all subsets X of E and all x_1, x_2, x_3 in X such that $(x_1, x_2, x_3) \in \mathcal{B}$, there is an $\overline{x}_3$ in $\mathrm{ex}_\mathcal{B}(X)$ such that $(x_1, x_2, \overline{x}_3) \in \mathcal{B}$.*

(ii) *$(a, b, c_2), (c_1, c_2, c_3) \in \mathcal{B} \Rightarrow (a, b, c_1) \in \mathcal{B}$ or $(a, b, c_3) \in \mathcal{B}$.*

3.3 Proofs

Our proof of Theorem 2.1 parallels a proof of the theorem of Dietrich (1987) that characterizes antimatroids in terms of circuits (see also Theorem 3.9 in Chap. III of Korte et al. 1991). It begins with a pair of auxiliary results.

Lemma 3.1. *Let $\mathcal{B}$ be a strict betweenness on a finite ground set E. If*

$$(a, b, c_2), (c_1, c_2, c_3) \in \mathcal{B} \implies (a, b, c_1) \in \mathcal{B} \text{ or } (a, b, c_3) \in \mathcal{B} \text{ or } (c_1, b, c_3) \in \mathcal{B},$$

then, with $\mathcal{N} = \mathcal{N}_\mathcal{B}$,

$$\tau_\mathcal{N}(X) = X \cup \{b: \text{ there are } a, c \text{ in } X \text{ with } (a, b, c) \in \mathcal{B}\}$$

for all subsets X of E.

Proof. Write $X' = \{b: \text{ there are } a, c \text{ in } X \text{ with } (a, b, c) \in \mathcal{B}\}$. Since $\tau_\mathcal{N}(X)$ is a convex superset of X, it is a superset of $X \cup X'$; our task reduces to proving that $X \cup X'$ is convex. For this purpose, consider an arbitrary b in E such that $(a, b, c) \in \mathcal{B}$ for some a, c in $X \cup X'$: we are going to prove that $b \in X \cup X'$.

CASE 1: $a, c \in X$.

In this case, $b \in X'$ by definition of X'.

CASE 2: $a \in X, c \in X'$.

By definition of X', there are c_1, c_3 in X with $(c_1, c, c_3) \in \mathcal{B}$; the hypothesis of the lemma with $c_2 = c$ guarantees that $(c_1, b, c_3) \in \mathcal{B}$ or $(a, b, c_1) \in \mathcal{B}$ or $(a, b, c_3) \in \mathcal{B}$. But then we are back in Case 1.

CASE 3: $a, c \in X'$.

As in Case 2, we find c_1, c_3 in X such that $(c_1, b, c_3) \in \mathcal{B}$ or $(a, b, c_1) \in \mathcal{B}$ or $(a, b, c_3) \in \mathcal{B}$. If $(c_1, b, c_3) \in \mathcal{B}$, then we are back in Case 1; if $(a, b, c_1) \in \mathcal{B}$ or $(a, b, c_3) \in \mathcal{B}$, then we are back in Case 2. $\square$

Lemma 3.2. *Let $\mathcal{B}$ be a strict betweenness on a finite ground set E. If*

$$(a, b, c_2), (c_1, c_2, c_3) \in \mathcal{B} \implies (a, b, c_1) \in \mathcal{B} \text{ or } (a, b, c_3) \in \mathcal{B} \text{ or } (c_1, b, c_3) \in \mathcal{B},$$

then $(E, \mathcal{N}_\mathcal{B})$ is a convex geometry.

Proof. Write $\mathcal{N} = \mathcal{N}_\mathcal{B}$. We will show that the convexity space $(E, \mathcal{N})$ has the anti-exchange property. For this purpose, assume the contrary: there are a subset X of E and distinct points y, z outside $\tau_\mathcal{N}(X)$ such that $y \in \tau_\mathcal{N}(X \cup \{z\})$ and $z \in \tau_\mathcal{N}(X \cup \{y\})$. Since $y \in \tau_\mathcal{N}(X \cup \{z\}) - \tau_\mathcal{N}(X)$ and $y \neq z$, Lemma 3.1 guarantees that $(x_y, y, z) \in \mathcal{B}$ for some x_y in X; similarly, $(x_z, z, y) \in \mathcal{B}$ for some x_z in X. But then the hypothesis with $a = x_y, b = y, c_1 = y, c_2 = z, c_3 = x_z$ implies that $(x_y, y, x_z) \in \mathcal{B}$, and so $y \in \tau_\mathcal{N}(X)$, a contradiction. $\square$

Proof of Theorem 2.1. To see that (i) implies (ii), set $X = \{a, b, c_1, c_2, c_3\}$ and $x_1 = a$, $x_2 = b$, $x_3 = c_2$ in (i). To show that (ii) implies (i), consider an arbitrary subset X of E and arbitrary x_1, x_2, x_3 in X such that $(x_1, x_2, x_3) \in \mathcal{B}$; let K denote the convex hull of X. Since $x_1, x_2, x_3 \in K$, we have $x_2 \notin \mathrm{ex}_{\mathcal{B}}(K)$, but Lemma 3.2 guarantees that x_2 belongs to the convex hull of $\mathrm{ex}_{\mathcal{B}}(K)$; now Lemma 3.1 (with $\mathrm{ex}_{\mathcal{B}}(K)$ in place of X) provides $\overline{x}_1, \overline{x}_3$ in $\mathrm{ex}_{\mathcal{B}}(K)$ such that $(\overline{x}_1, x_2, \overline{x}_3) \in \mathcal{B}$. Finally, Lemma 3.1 shows that $\mathrm{ex}_{\mathcal{B}}(K) \subseteq X$, and so $\overline{x}_1, \overline{x}_3 \in \mathrm{ex}_{\mathcal{B}}(X)$. $\square$

Proof of Theorem 2.2. Let P_1 denote the chordless path from a to c_2 that passes through b and let P_2 denote the chordless path from c_1 to c_3 that passes through c_2. Proceeding along P_1 from a to c_2, we label the vertices consecutively as $v_1, v_2, \ldots,$ v_m, so that

$$a = v_1, \quad b = v_s \quad \text{for some } s \text{ such that } 2 \le s \le m - 1, \quad c_2 = v_m;$$

proceeding along P_2 from c_1 to c_3, we label the vertices consecutively as $w_1, w_3,$ $\ldots, w_n$, so that

$$c_1 = w_1, \quad c_2 = w_t \quad \text{for some } t \text{ such that } 2 \le t \le n - 1, \quad c_3 = w_n.$$

We claim that

($\star$) none of $v_1, v_2, \ldots, v_{m-2}$ has a neighbour w_i with $i < t$ or
  none of $v_1, v_2, \ldots, v_{m-2}$ has a neighbour w_j with $j > t$.

To justify this claim, assume the contrary: there are edges $v_k w_i$ and $v_\ell w_j$ with $k, \ell \le m-2$ and $i < t, j > t$. Choose them so that $|k-\ell|$ is minimized (we may have $k = \ell$) and, subject to this constraint, i is maximized and j is minimized; let P denote the segment of P_1 that stretches between v_k and v_ℓ. Now v_k is the only vertex on P that has a neighbour in $\{w_1, w_2, \ldots, w_{t-1}\}$ and v_ℓ is the only vertex on P that has a neighbour in $\{w_{t+1}, w_{t+2}, \ldots, w_n\}$; as $w_t = v_m$ and $k, \ell \le m - 2$, no vertex on P is adjacent to w_t or identical with w_t. It follows that the paths P and $w_i w_{i+1} \ldots w_j$ are vertex-disjoint and that their union induces a chordless cycle through at least four vertices; this contradiction completes the proof of ($\star$).

After flipping P_2 if necessary, ($\star$) lets us assume that none of $v_1, v_2, \ldots, v_{s-1}$ has a neighbour w_i with $i < t$. Since the walk $v_s v_{s+1} \ldots v_m w_{t-1} \ldots w_2 w_1$ connects b to c_1, some subset of its vertices induces a chordless path P from b to c_1; now b is an interior vertex of the chordless path $v_1 v_2 \ldots v_{s-1} P$ between a and c_1. $\square$

Proof of Theorem 2.3. To see that (i) implies (ii), set $X = \{a, b, c_1, c_2, c_3\}$ and $x_1 = a$, $x_2 = b$, $x_3 = c_2$ in (i). To show that (ii) implies (i), we shall use induction on $|X|$. If $|X| \le 2$, then the conclusion is vacuously true. For the induction step, consider arbitrary x_1, x_2, x_3 in X such that $(x_1, x_2, x_3) \in \mathcal{B}$. Setting

$$Z = \{z \in X \colon (x_1, x_2, z) \in \mathcal{B}\},$$

we shall proceed to prove that $Z \cap \mathrm{ex}_{\mathcal{B}}(X) \ne \emptyset$.

First we claim that, with $\mathcal{N}$ a shorthand for $\mathcal{N}_\mathcal{B}$ as usual,

- $\tau_\mathcal{N}(X - Z) \cap Z = \emptyset$.

To justify this claim, assume the contrary: there is a triple (c_1, c_2, c_3) in $\mathcal{B}$ such that $c_1 \in X - Z, c_2 \in Z, c_3 \in X - Z$. But then proposition (ii) is contradicted by $a = x_1$, $b = x_2$.

Next, let us write $z' \prec z''$ if and only if $z', z'' \in Z$ and there exists a y in $X - Z$ such that $(y, z', z'') \in \mathcal{B}$; note that $z' \prec z'' \Rightarrow z' \neq z''$. We claim that

- $\prec$ is antisymmetric.

To justify this claim, assume the contrary: there are z_1, z_2 in Z with $z_1 \prec z_2$, $z_2 \prec z_1$. By definition, there are y_1, y_2 in $X - Z$ such that $(y_1, z_1, z_2) \in \mathcal{B}$ and $(y_2, z_2, z_1) \in \mathcal{B}$. But then proposition (ii) is contradicted by $a = y_1, b = z_1, c_1 = y_2$, $c_2 = z_2, c_3 = z_1$: since $\tau_\mathcal{N}(X - Z) \cap Z = \emptyset$, we have $(y_1, z_1, y_2) \notin \mathcal{B}$.

In addition, we claim that

- $\prec$ is transitive.

To justify this claim, consider any z_1, z_2, z_3 in Z such that $z_1 \prec z_2$ and $z_2 \prec z_3$. By definition, there are y_1, y_2 in $X - Z$ such that $(y_1, z_1, z_2), (y_2, z_2, z_3) \in \mathcal{B}$; as $\tau_\mathcal{N}(X - Z) \cap Z = \emptyset$ guarantees that $(y_1, z_1, y_2) \notin \mathcal{B}$, proposition (ii) with $a = y_1$, $b = z_1, c_1 = y_2, c_2 = z_2, c_3 = z_3$ implies $(y_1, z_1, z_3) \in \mathcal{B}$, and so $z_1 \prec z_3$.

Our set Z is nonempty (it includes x_3) and it is partially ordered by $\prec$. Let Z_{max} denote the set of its maximal elements. We claim that

- $Z_{max} \cap ex_\mathcal{B}(Z) \subseteq ex_\mathcal{B}(X)$.

To justify this claim, assume the contrary: there are a, z_2, c in X such that $z_2 \in Z_{max} \cap ex_\mathcal{B}(Z)$ and $(a, z_2, c) \in \mathcal{B}$. Since $\tau_\mathcal{N}(X - Z) \cap Z = \emptyset$, at least one of a, c belongs to Z; since $z_2 \in ex_\mathcal{B}(Z)$, at most one of a, c belongs to Z; now symmetry allows us to assume that $a \in X - Z$ and $c \in Z$. But then $z_2 \prec c$, contradicting the assumption that $z_2 \in Z_{max}$.

We shall complete the proof by showing that

- $Z_{max} \cap ex_\mathcal{B}(Z) \neq \emptyset$.

For this purpose, we rely on the induction hypothesis; note that $|Z| < |X|$ as Z includes neither x_1 nor x_2.

CASE 1: $Z_{max} \neq Z$.

In this case, let z be any maximal element of $Z - Z_{max}$. Since $z \notin Z_{max}$, there are a y in $X - Z$ and a z_2 in Z_{max} such that $(y, z, z_2) \in \mathcal{B}$. If $z_2 \in ex_\mathcal{B}(Z)$, then we are done; else there are elements z_1, z_3 of Z such that $(z_1, z_2, z_3) \in \mathcal{B}$. Now the induction hypothesis applied to Z and (z_1, z_2, z_3) yields a $\bar{z}_3$ in $ex_\mathcal{B}(Z)$ such that $(z_1, z_2, \bar{z}_3) \in \mathcal{B}$; next, the induction hypothesis applied to Z and $(\bar{z}_3, z_2, z_1)$ yields a $\bar{z}_1$ in $ex_\mathcal{B}(Z)$ such that $(\bar{z}_3, z_2, \bar{z}_1) \in \mathcal{B}$. Proposition (ii) with $a = y, b = z$, $c_1 = \bar{z}_1, c_2 = z_2, c_3 = \bar{z}_3$ guarantees that a subscript i in $\{1, 3\}$ satisfies $z \prec \bar{z}_i$; now maximality of z implies $\bar{z}_i \in Z_{max}$.

CASE 2: $Z_{max} = Z$.

In this case, our task reduces to proving that $\mathrm{ex}_{\mathcal{B}}(Z) \neq \emptyset$. We may assume that $\mathrm{ex}_{\mathcal{B}}(Z) \neq Z$ (else we are done), and so $\mathcal{B}$ includes a triple (z_1, z_2, z_3) such that z_1, z_2, z_3 are elements of Z. But then the induction hypothesis applied to Z and (z_1, z_2, z_3) yields a $\bar{z}_3$ in $\mathrm{ex}_{\mathcal{B}}(Z)$. $\square$

References

Calder, J.R.: Some elementary properties of interval convexities. J. Lond. Math. Soc. **3**(2), 422–428 (1971)

Dietrich, B.L.: A circuit set characterization of antimatroids. J. Comb. Theory Ser. B **43**, 314–321 (1987)

Dilworth, R.P.: Lattices with unique irreducible decompositions. Ann. Math. **41**(2), 771–777 (1940)

Dirac, G.A.: On rigid circuit graphs. Abh. Math. Sem. Univ. Hamburg **25**, 71–76 (1961)

Duchet, P.: Convex sets in graphs. II. Minimal path convexity. J. Comb. Theory, Ser. B **44**, 307–316 (1988)

Edelman, P.H.: Meet-distributive lattices and the anti-exchange closure. Algebra Univers. **10**, 290–299 (1980)

Edelman, P.H., Jamison, R.E.: The theory of convex geometries. Geom. Dedic. **19**, 247–270 (1985)

Farber, M., Jamison, R.E.: Convexity in graphs and hypergraphs. SIAM J. Algebr. Discrete Methods **7**, 433–444 (1986)

Jamison-Waldner, R.E.: Copoints in antimatroids. Congr. Numer. **29**, 535–544 (1980)

Jamison-Waldner, R.E.: A perspective on abstract convexity: classifying alignments by varieties. In: Convexity and Related Combinatorial Geometry, Norman, Okla, 1980. Lecture Notes in Pure and Applied Mathematics, vol. 76, pp. 113–150. Dekker, New York (1982)

Kay, D.C., Womble, E.W.: Axiomatic convexity theory and relationships between the Carathéodory, Helly, and Radon numbers. Pac. J. Math. **38**, 471–485 (1971)

Korte, B., Lovász, L.: Mathematical structures underlying greedy algorithms. In: Fundamentals of Computation Theory, Szeged, 1981. Lecture Notes in Computer Science, vol. 117, pp. 205–209. Springer, Berlin (1981)

Korte, B., Lovász, L.: Greedoids—a structural framework for the greedy algorithm. In: Progress in Combinatorial Optimization, Waterloo, Ont., 1982, pp. 221–243. Academic Press, Toronto (1984)

Korte, B., Lovász, L., Schrader, R.: Greedoids. Springer, Berlin (1991)

Lihová, J.: Strict order-betweennesses. Acta Univ. M. Belii Ser. Math. **8**, 27–33 (2000). http://matematika.fpv.umb.sk/papers8.ps

Whitney, H.: On the abstract properties of linear dependence. Am. J. Math. **57**, 509–533 (1935)

4

Euler Complexes

Jack Edmonds

Summary. We present a class of instances of the existence of a second object of a specified type, in fact, of an even number of objects of a specified type, which generalizes the existence of an equilibrium for bimatrix games. The proof is an abstract generalization of the Lemke–Howson algorithm for finding an equilibrium of a bimatrix game.

A *d-oik*, $C = (V, F)$, short for *d-dimensional Euler complex*, $d \geq 1$, is a finite set V of elements called the vertices of C and a family F of $d + 1$ element subsets of V, called the *rooms* of C, such that every d element subset of V is in an even number of the rooms.

A *wall of a room* means a set obtained by deleting one vertex of the room—and so any wall of a room in an oik is the wall of a positive even number of rooms of the oik.

Example 1. A d-dimensional simplicial pseudo-manifold is a d-oik where every d element subset of vertices is in exactly zero or two rooms, i.e., in a simplicial pseudo-manifold any wall is the wall of exactly two rooms. An important special case of simplicial pseudo-manifold is a triangulation of a compact manifold such as a sphere.

Example 2. Let $Ax = b, x \geq 0$, be a tableau as in the simplex method, whose solution-set is bounded and whose basic variables in each basic feasible solution are all non-zero (non-degenerate). Let V be the column-set of A. Let the rooms be the subsets S of columns such that $V - S$ is a feasible basis of the tableau. This is an $(n - r - 1)$-oik where n is the number of columns of A and r is the rank (the number of rows) of A. In fact, it is a triangulation of an $(n - r - 1)$-dimensional sphere—in particular, it is combinatorially the boundary of a 'simplicial polytope'.

Example 3. Let the n members of set V be colored with r colors. Let the rooms be the subsets S of V such that $V - S$ contains exactly one vertex of each color. This is an $(n \quad r \quad 1)$-oik. In fact it is the oik of Example 2 where each column of A is all zeroes except for one positive entry.

Example 4. An Euler graph, that is a graph such that each of its vertices is in an even number of its edges (the rooms), is a 1-oik.

Example 5. For any connected Euler graph G with n vertices ($n \geq 3$), we have an $(n-2)$-oik (V, K) where V is the set of edges of G and the rooms are the edge-sets of the spanning trees of G.

Example 6. For any connected bipartite graph G with m edges and n vertices we have an $(m-n)$-oik where V is the edge-set of G, and the rooms are the edge-complements of spanning trees of G.

Example 7 (Generalizing Examples 5 and 6). Where M is an Euler binary matroid, that is, a binary matroid of rank r such that each cocircuit, in fact each cocycle, is even, we have an $(r-1)$-oik, where V is the set of elements of the matroid, and the rooms are the bases of the matroid.

(A *binary matroid M* is given by a 0-1 matrix, A, mod 2. The elements of M are the columns. The bases of M are the linearly independent sets of columns. The cocycles are the supports of the row vectors generated by the rows of A. The cocircuits are the minimal cocycles. Matroid M is *Euler* when each row of A has an even number of ones. See, e.g., Korte and Vygen 2008; Schrijver 2003.)

Let $M = [(V, F_i) : i = 1, \ldots, h]$ be an indexed collection of oiks (which we call an *oik-family*) all on the same vertex-set V.

The oiks of M are not necessarily of the same dimension. Of course, all of them may be the same oik.

A *room-family*, $R = [R_i : i = 1, \ldots, h]$, for oik-family M, is where, for each i, R_i is a room of oik i (i.e., a member of F_i). A *room-partition R* for M means a room-family whose rooms partition V, i.e., each vertex is in exactly one room of R.

Theorem 1. *Given an oik-family M and a room-partition R for M, there exists another different room-partition for M. In fact, for any oik-family M, there is an even number of room-partitions.*

Proof. Choose a vertex, say w, to be special. A *w-skew room-family for oik-family M* means a room-family, $R = [R_i : i = 1, \ldots, h]$, for M such that w is not in any of the rooms R_i, some vertex v is in exactly two of the R_i, and every other vertex is in exactly one of the R_i.

Consider the so-called exchange-graph X, determined by M and w, where the nodes of X are all the room-partitions for M and all the w-skew room-families for M. Two nodes of X are joined by an edge of X if each is obtained from the other by replacing one room by another. It is easy to see that the odd-degree nodes of X are all the room-partitions for M, and all the even-degree nodes of X are the w-skew room-families for M. Hence there is an even number of room-partitions for M.   □

'Exchange algorithm': An algorithm for getting from one room-partition for M to another is to walk along a path in X, not repeating any edge of X, from one to another. Where each oik of the oik-family M is a simplicial pseudo-manifold, X consists of disjoint simple paths and simple cycles, and so the algorithm is uniquely determined by M and w.

Where oik-family M consists of two oiks of the kind in Example 2, the exchange algorithm is the Lemke–Howson algorithm for finding a Nash equilibrium of a 2-person game. Savani and von Stengel (2006) show that the number of steps in the Lemke–Howson algorithm can grow exponentially relative to the size of the two tableaus of the game.

It is not known whether there is a polytime algorithm for finding a Nash equilibrium of a 2-person game. Chen and Deng (2006) (see also Papadimitriou 1994) proved a deep completeness result which is regarded as some evidence that there might not be a polytime algorithm.

Suppose each oik of M is given by an explicit list of its rooms, each oik perhaps a simplicial pseudo-manifold, perhaps a 2-dimensional sphere. Is some path of the exchange graph not well-bounded by the number of rooms?

How about the exchange algorithm when each oik of M is a 1-oik? If each oik of M is the same 1-oik then the well-known, non-trivial, non-bipartite matching algorithm (Korte and Vygen 2008; Schrijver 2003) can be used to find, if there is one, a first and a second room-partition.

How about the exchange algorithm where each oik of M is an Euler binary matroid? For an oik-family like that, the well-known, non-trivial, 'matroid partition' algorithm (Korte and Vygen 2008; Schrijver 2003) can be used to find, if there is one, a first and a second room-partition.

Example 8. A pure $(d + 1)$-complex, $C = (V; F)$, means simply a finite set, V, and a family, F, of $d + 2$ element subsets. The boundary, $bd(C) = (V; bd(F))$, of any pure $(d + 1)$-complex, C, means the pure d-complex where $bd(F)$ is the family of those $d + 1$ element subsets of V which are subsets of an odd number of members of F.

For any pure $(d + 1)$-complex, C, its boundary, $bd(C)$, is a d-oik.

This is more-or-less the first theorem of simplicial homology theory. By recalling the meaning of d-oik, it is saying that for any pure $(d + 1)$-complex, C, every d element subset, H, of V is a subset of an even number of $(d + 1)$ element sets which are subsets of an odd number of the $(d + 2)$ element members of F. It can be proved graph theoretically by observing that, for any d element subset, H, of V, the following graph, G, has an even number of odd degree vertices: The vertices of G are the $(d + 1)$ element subsets of V which contain H. Two of these $(d + 1)$ element vertices are joined by an edge in G when their union is a $(d + 2)$ element member of F. Clearly a vertex of G is a subset of an odd number members of F, and hence is a member of $bd(F)$, when it is an odd degree vertex of G.

What can we say about $bd(F)$, besides Theorem 1, when F is the set of bases of a matroid?

In Cameron and Edmonds (1999), different exchange graphs were studied. In Cameron (2001), it was shown that Thomason's (1978) exchange graph algorithm for finding a second Hamiltonian circuit in a cubic graph is exponential relative to the size of the given graph.

References

Cameron, K.: Thomason's algorithm for finding a second Hamiltonian circuit through a given edge in a cubic graph is exponential on Krawczyk's graphs. Discrete Math. **235**, 69–77 (2001)

Cameron, K., Edmonds, J.: Some graphic uses of an even number of odd nodes. Ann. Inst. Fourier **49**, 815–827 (1999)

Chen, X., Deng, X.: Settling the complexity of two-player Nash equilibrium, In: Proc. of FOCS 2006, pp. 261–272 (2006)

Korte, B., Vygen, J.: Combinatorial Optimization: Theory and Algorithms, 4th edn. Algorithms and Combinatorics, vol. 21. Springer, Berlin (2008)

Papadimitriou, C.H.: On the complexity of the parity argument and other inefficient proofs of existence. J. Comput. Syst. Sci. **48**(3), 498–532 (1994)

Savani, R., von Stengel, B.: Hard-to-solve bimatrix games. Econometrica **74**(2), 397–429 (2006)

Schrijver, A.: Combinatorial Optimization: Polyhedra and Efficiency. Algorithms and Combinatorics, vol. 24. Springer, Berlin (2003)

Thomason, A.G.: Hamiltonian cycles and uniquely edge colourable graphs. Ann. Discrete Math. **3**, 259–268 (1978)

Strongly Polynomial Algorithm for the Intersection of a Line with a Polymatroid

Jean Fonlupt and Alexandre Skoda

Summary. We present a new algorithm for the problem of determining the intersection of a half-line $\Delta_u = \{x \in \mathbb{R}^N \mid x = \lambda u$ for $\lambda \geq 0\}$ with a polymatroid. We then propose a second algorithm which generalizes the first algorithm and solves a parametric linear program. We prove that these two algorithms are strongly polynomial and that their running time is $O(n^8 + \gamma n^7)$ where γ is the time for an oracle call. The second algorithm gives a polynomial algorithm to solve the submodular function minimization problem and to compute simultaneously the strength of a network with complexity bound $O(n^8 + \gamma n^7)$.

5.1 Introduction

Let f be a *set function* on a finite set $N = \{1, \ldots, n\}$. f is *submodular* if

$$f(S \cap T) + f(S \cup T) \leq f(S) + f(T)$$

for all subsets S, T of N. If f is also *normalized* ($f(\emptyset) = 0$) and *monotone* ($f(U) \leq f(T)$ whenever $U \subseteq T$), the *polymatroid associated with f* is defined by:

$$P(f) = \{x \in \mathbb{R}^N \mid x \geq 0, \ x(T) \leq f(T) \text{ for each } T \subseteq N\}.$$

Let u be a vector of $\mathbb{R}^N$ which describes the direction of the half-line $\Delta_u = \{x \in \mathbb{R}^N \mid x = \lambda u$ for $\lambda \geq 0\}$. $P(f) \cap \Delta_u$ is a closed interval $[0, \lambda_{\max}]u$ where $\lambda_{\max}$ is the value of the linear program:

$$\max \{\lambda \mid x \in P(f) \cap \Delta_u\} \quad (\mathcal{P})$$

This problem will be called the *Intersection Problem*.

This research was supported by a grant of "France Telecom R&D Sophia Antipolis" during three years.

We propose a strongly polynomial combinatorial algorithm for solving $\mathcal{P}$. This algorithm called the *Intersection Algorithm* runs in time $O(n^8 + \gamma n^7)$ where γ is the time of an oracle call.

The Intersection Problem has interesting applications: consider the following parametric linear program:

$$z(\lambda) = \max \left\{ \sum_{i \in N} x_i \mid x \in P(f),\ x \leq \lambda u \right\} \quad (\mathcal{P}_\lambda)$$

Standard results in parametric linear programming show that the function $\lambda \longrightarrow z(\lambda)$ is non-decreasing, concave and piecewise linear.

We prove that we need to solve at most n times the Intersection Problem to obtain a complete description of $z(\lambda)$. This will be the object of the *Parametric Intersection Algorithm* which is a simple variant of the Intersection Algorithm. This second algorithm runs also in time $O(n^8 + \gamma n^7)$. As the solution of the parametric linear program for a certain choice of the parameter λ solves the *submodular function minimization* problem and for another choice of this parameter provides the value of the *strength of a network* if f is the rank function of a graphic matroid of a graph, we obtain a strongly polynomial algorithm with complexity bound $O(n^8 + \gamma n^7)$ for solving simultaneously these two problems.

Polymatroids were introduced by Edmonds (1970). Examples of submodular functions are described in Fleischer (2000), Fujishige (2005) and Korte and Vygen (2008), Schrijver (2003). The Intersection Problem can be solved in strongly polynomial time using the ellipsoid method (Grötschel et al. 1981). Schrijver (2000) and Iwata et al. (2001) independently proposed combinatorial strongly polynomial algorithms for submodular function minimization. Vygen (2003) showed that the running time of Schrijver's original algorithm was $O(n^8 + \gamma n^7)$ which was also the improvement given by Fleischer and Iwata (2000). Iwata (2003) improved this bound to $O((n^7 + \gamma n^6) \log n)$. Orlin (2007) developed a strongly polynomial algorithm in $O(n^6 + \gamma n^5)$ which is the best complexity bound known so far.

All these approaches depend on the decomposition of feasible points of the polymatroid as convex combination of extreme points of this polytope; this idea was originated by Cunningham (1984, 1985b). Our own approach is based on the polymatroid itself and not on the base polyhedron associated to the polymatroid as do the above-cited authors; this is a major difference. Note however that an n-dimensional polymatroid can be seen as a projection of an $(n + 1)$-dimensional base polyhedron (see Fujishige (2005)) but this observation can not be easily exploited from an algorithmic point of view.

As a generalization of lexicographically optimal flows of Megiddo (1974), Fujishige (1980) introduced the concept of lexicographically optimal base of a polymatroid. Problem $\mathcal{P}_\lambda$ is known to be equivalent to the lexicographically optimal base problem (Fujishige 2005) but we will not use this result.

Cunningham (1985a) in his study of the strength of a network had an approach based on the forest polytope of a graph. Cheng and Cunningham (1994) studied, for the problem of the strength of a network, a parametric linear program which can be seen as a specific case of our parametric linear program.

Nagano (2004) developed a strongly polynomial algorithm for the intersection of an extended polymatroid and a straight line, based on a direct application of the submodular function minimization problem. Although our algorithm is only designed for the intersection with a polymatroid, it provides a new algorithm for the submodular function minimization problem. Recently Nagano (2007), extending Orlin's (2007) algorithm proposed an algorithm in $O(n^6 + \gamma n^5)$ to solve the parametric problem $\mathcal{P}_\lambda$. Our algorithm is not as good as Nagano's in time complexity but we hope that the interest of our approach is justified by the method we employ: we work directly on the polymatroid itself and we do not use known submodular function minimization algorithms.

Many of the results presented here appear in the Ph.D thesis of Skoda (2007).

The content of the paper can be summarized as follows. Section 5.2 recalls basic definitions and properties of polymatroids and present the general framework of the Intersection Algorithm. Section 5.3 describes the Intersection Algorithm and evaluates the time complexity of this algorithm. Section 5.4 describes the Parametric Intersection Algorithm and proves the strongly polynomiality of this algorithm; it is also shown in Sect. 5.4 why this algorithm simultaneously solve the submodular minimization problem and finds the strength of a network. Many of the results in Sect. 5.4 are standard (see for instance Fujishige 2005) but we include them with short proofs for completeness. Section 5.5 summarizes the results of the previous sections and proposes some new problems.

5.2 Preliminaries

1. If $\emptyset \subseteq T \subseteq N$, define $f_{N \setminus T}$ by

$$f_{N \setminus T}(A) = f(A) \quad \text{for all } \emptyset \subseteq A \subseteq N \setminus T.$$

$f_{N \setminus T}$ is normalized, monotone and submodular.

The polymatroid $P(f_{N \setminus T})$ is embedded in the vector space $\mathbb{R}^{N \setminus T}$; moreover

$$P(f_{N \setminus T}) = P(f) \cap \{x \in \mathbb{R}^N \mid x_i = 0 \text{ for } i \in T\}.$$

2. We can treat and eliminate the following cases:

If $u \notin \mathbb{R}^N_+$, $\lambda_{\max} = 0$.

If there exists $i \in N$ with $u_i = 0$, then $P(f) \cap \Delta_u = P(f_{N \setminus \{i\}}) \cap \Delta_u$ and we can replace N by $N \setminus \{i\}$.

Finally, assume that there exists $i \in N$ with $f(\{i\}) = 0$; for any feasible point $x \in P(f)$, $x_i = 0$ since $0 \leq x_i \leq f(\{i\})$ and $[0, \lambda_{\max}]u = P(f_{N \setminus \{i\}}) \cap \Delta_u$.

If $u_i > 0$, $\lambda_{\max} = 0$.

If $u_i = 0$, we can replace N by $N \setminus \{i\}$.

So, we will assume from now on that $u > 0$ and that $f(\{i\}) > 0$ for all $i \in N$.

3. By order we mean in this paper a subset of N, called the support of the order, and a total order on this support. Elements in the support will be called ordered elements and elements not in the support will be called singletons.

$\{\mathcal{O}^0, \mathcal{O}^1, \ldots, \mathcal{O}^k, \ldots, \mathcal{O}^{\bar{k}}\}$ is the set of distinct orders, $\mathcal{K} = \{0, 1, \ldots, k, \ldots, \bar{k}\}$ is the set of upper indices listing the distinct orders.

The order relation of $\mathcal{O}^k$ will be denoted $\prec_k$. We will assume that, in the description of $\mathcal{O}^k$, the elements of the support S^k are sorted in increasing order with respect to the order relation $\prec_k$:

$$\mathcal{O}^k = [i_1, \ldots, i_l, \ldots, i_m].$$

It was proved by Edmonds (1970) that extreme points of $P(f)$ can be generated by the greedy algorithm: given an order $\mathcal{O}^k = [i_1, \ldots, i_m]$ with support S^k, the point $a^k \in \mathbb{R}^N$ whose components are:

$$a^k_{i_1} = f(\{i_1\}),$$
$$a^k_{i_l} = f(\{i_1, \ldots, i_l\}) - f(\{i_1, \ldots, i_{l-1}\}) \quad \text{for } l = 2, \ldots, m,$$
$$a^k_i = 0 \quad \text{for all } i \in N \setminus S^k,$$

is an extreme point of $P(f)$. a^k is the extreme point generated by $\mathcal{O}^k$. Edmonds proved also that any extreme point of $P(f)$ is generated by an order. Note also that

$$a^k(\{i_1, \ldots, i_l\}) = f(\{i_1, \ldots, i_l\}) \quad \text{for } l = 1, \ldots, m. \tag{1}$$

Two distinct orders may generate the same extreme point. Bixby et al. (1985) gave a complete characterization of extreme points of $P(f)$. In particular they proved that the intersection of all the orders which generate the same extreme point is a minimal partial order and two minimal distinct partial orders generate two distinct extreme points. However we will not use this result here.

4. We will associate to a matroid order $\mathcal{O}^k$ the following oriented graph $G(k)$:

- The node-set of $G(k)$ is $\{0\} \cup N$ where 0 is a new element; from now on we will use the term element rather than node.
- The edge-set of $G(k)$ denoted $E(k)$ is defined as follows:
  - $e = (0, i)_k \in E(k)$ if and only if i is a singleton of $\mathcal{O}^k$,
  - $e = (i, j)_k \in E(k)$ if and only if i and j are ordered elements of $\mathcal{O}^k$ and $i \prec_k j$.

If $K \subseteq \mathcal{K}$ we define the graph $G(K)$ as:

$$G(K) = \left(\{0\} \cup N, \bigcup_{k \in K} E(k)\right).$$

Note that $G(K)$ may have (many) parallel edges: if $i \prec_{k_1} j$ and $i \prec_{k_2} j$ for $k_1, k_2 \in K$, then $(i, j)_{k_1}$ and $(i, j)_{k_2}$ are parallel edges.

5. We will now define two operations on the order $\mathcal{O}^k = [i_1, \ldots, i_m]$.

If j is a singleton of $\mathcal{O}^k$, $\mathcal{O}^{k'}$ is obtained by inserting j after the last element i_m of $\mathcal{O}^k$ and $\mathcal{O}^{k'}$ will be called the order associated to the edge $e = (0, j)_k$.

If $i \prec_k j$, $\mathcal{O}^{k'}$ is obtained by moving j just before i in $\mathcal{O}^k$ and $\mathcal{O}^{k'}$ will be called the order associated to the edge $e = (i, j)_k$. In both cases, we say that $\mathcal{O}^{k'}$ is an adjacent order to $\mathcal{O}^k$.

Let us set $b = a^{k'} - a^k$. We recall (see for instance Korte and Vygen 2008 or (Schrijver 2003)) and prove for completeness the following result:

Lemma 2.1. *If $\mathcal{O}^{k'}$ is associated to $e = (0, j)_k$, $b_j \geq 0$; $b_i = 0$ for $i \in N \setminus \{j\}$. If $\mathcal{O}^{k'}$ is associated to $e = (i, j)_k$, $b_j \geq 0$; $b_l \leq 0$ for $i \prec_k l \prec_k j$ or $l = i$, $b_l = 0$ otherwise; moreover $\sum_{i=1}^{n} b_i = 0$.*

Proof. Set $T_s = \{r \mid r \in N, r \prec_k s\}$ for $s \in N$ and let S^k be the support of $\mathcal{O}^k$.

If $\mathcal{O}^{k'}$ is associated to $e = (0, j)_k$, $b_i = 0$ for $i \in N \setminus \{j\}$; as f is monotone, $b_j = a_j^{k'} = f(S^k \cup \{j\}) - f(S^k) \geq 0$.

If $\mathcal{O}^{k'}$ is associated to $e = (i, j)_k$, as $T_i \subset T_j$ and f is submodular,

$$b_j = a_j^{k'} - a_j^k = f(T_i \cup \{j\}) - f(T_i) - (f(T_j \cup \{j\}) - f(T_j)) \geq 0;$$

if $i \prec_k l \prec_k j$ or $l = i$,

$$b_l = a_l^{k'} - a_l^k = f(T_l \cup \{j, l\}) - f(T_l \cup \{j\}) - (f(T_l \cup \{l\}) - f(T_l)) \leq 0;$$

$\mathcal{O}^k$ and $\mathcal{O}^{k'}$ have the same support S^k and $\sum_{i=1}^{n} a_i^k = \sum_{i=1}^{n} a_i^{k'} = f(S^k)$; thus $\sum_{i=1}^{n} b_i = 0$. □

6. We will express points of the half-line Δ_u as convex combinations of extreme points of $P(f)$. So, let $K \subseteq \mathcal{K}$; we say that K is a feasible set if the system:

$$\mathcal{L}(K) \quad \begin{cases} \alpha_k \geq 0 \quad \text{for } k \in K, \\ \lambda u = \sum_{k \in K} \alpha_k a^k, \\ \sum_{k \in K} \alpha_k = 1 \end{cases}$$

has a feasible solution. If K is feasible and the vectors $a^k : k \in K$ are linearly independent, $|K| \leq n$ and the solution $(\alpha_k, k \in K; \lambda)$ of $\mathcal{L}(K)$ is unique. If this solution is strictly positive, K is called a positive basis. λ will be called the value of the basis. If K is a feasible set, K contains a positive basis (this is Caratheodory's theorem).

A set $\{b^1, \ldots, b^n\}$ of vectors of $\mathbb{R}^n$ is called triangular if:

- $b_i^k = 0$ whenever $1 \leq k < i \leq n$,
- $b_i^k \leq 0$ whenever $1 \leq i < k \leq n$,
- $b_k^k > 0$ for $k = 1, \ldots, n$.

If $\{b^1, \ldots, b^n\}$ is triangular, the $n \times n$ matrix B induced by this set of column-vectors is upper triangular and non-singular since $b_k^k > 0$ for $k = 1, \ldots, n$. Thus the linear system $\mu_1 b^1 + \cdots + \mu_n b^n = u > 0$ has a unique solution $\mu = (\mu_1, \ldots, \mu_n)$ and by standard results in linear algebra the inverse of B is non-negative and $\mu > 0$.

This solution can be computed in $O(n^2)$ time.

Let K be a positive basis and let K_1 be a subset of K. Assume that there exists $K' \subseteq \mathcal{K}$ and a triangular set $\{b^1, \ldots, b^n\}$ with $b^r = a^{k'_r} - a^{k_r}$ where $k'_r \in K'$ and $k_r \in K_1$ for $r = 1, \ldots, n$.

We can now establish the following easy but fundamental lemma:

Lemma 2.2. *There exists a positive basis contained in* $(K \cup K') \setminus \{k_1\}$ *for some* $k_1 \in K_1$.

Proof. As $\{b^1, \ldots, b^n\}$ is a triangulated set, there exist n positive numbers μ_r, $r = 1, \ldots, n$ such that

$$u = \sum_{r=1}^n \mu_r a^{k'_r} - \sum_{r=1}^n \mu_r a^{k_r}.$$

So, there exists non-negative numbers v_k, $k \in K_1 \cup K'$ with:

$$\begin{cases} u = \sum_{k \in K'} v_k a^k - \sum_{k \in K_1} v_k a^k, \\ \sum_{k \in K'} v_k - \sum_{k \in K_1} v_k = 0. \end{cases}$$

Let $(\alpha_k, k \in K; \lambda)$ be the solution of $\mathcal{L}(K)$. For any $\lambda' > 0$,

$$\begin{cases} (\lambda + \lambda')u = \sum_{k \in K} \alpha_k a^k + \lambda'(\sum_{k \in K'} v_k a^k - \sum_{k \in K_1} v_k a^k), \\ \sum_{k \in K} \alpha_k + \lambda'(\sum_{k \in K'} v_k - \sum_{k \in K_1} v_k) = 1. \end{cases}$$

Set $\lambda' = \min_{k \in K_1} \frac{\alpha_k}{v_k}$ (if $v_k = 0$, $\frac{\alpha_k}{v_k} = +\infty$) and let $k_1 \in K_1$ be an upper index which realizes this minimum. The set $(K \cup K') \setminus \{k_1\}$ is feasible and contains a positive basis. □

$\mathcal{L}((K \cup K') \setminus \{k_1\})$ has $n + 1$ rows and at most $2n$ columns. By standard results in linear programming, we can find the new feasible basis in a time $O(n^3)$ (see for instance Korte and Vygen 2008, Chap. 4, Exercise 5).

7.   Our initial positive basis will be $K = \{1, \ldots, n\}$ where $\mathcal{O}^k = [k]$ for $k = 1, \ldots, n$. Thus $a_i^k = 0$ if $i \neq k$ and $a_k^k = f(\{k\})$; the solution associated to this basis is:

$$\alpha_k := \frac{\lambda u_k}{f(\{k\})}, \quad k = 1, \ldots, n; \; \lambda := \frac{1}{\sum_{k=1}^n \frac{u_k}{f(\{k\})}}.$$

5.3 The Intersection Algorithm

Let $(\alpha_k$ for $k \in K; \lambda)$ be the solution associated to a positive basis K. $\lambda u(T) \leq f(T)$ for any $T \subseteq N$; equality holds for λ_{max} and $T_{min} \subseteq N$ with

$$\frac{f(T_{min})}{u(T_{min})} = \min \left(\frac{f(T)}{u(T)} \mid \emptyset \subset T \subseteq N \right).$$

Optimality occurs when the following condition holds:

Proposition 3.1. *If* 0 *is not a root of* $G(K)$, *let* T *be the set of elements of* N *not reachable from* 0; *then* $\lambda_{\max} = \lambda$ *and* $T_{\min} = T$.

Proof. If 0 is not a root of $G(K)$, let T be the set of elements of N not reachable from 0. Let $\mathcal{O}^k = [i_1, \ldots, i_m]$ for some $k \in K$.

If $(0, i)_k$ is an edge of $G(K)$, i is a singleton of $\mathcal{O}^k$ and $i \notin T$; so $T \subseteq \{i_1, \ldots, i_m\}$. If $r \prec_k l$ and $l \in T$, then $r \in T$. This implies that $T = \{i_1, i_2, \ldots, i_l\}$ for some $1 \le l \le m$; equations (1) imply that $a^k(T) = f(T)$.

$$\lambda u(T) = \sum_{k \in K} \alpha_k a^k(T) = \left(\sum_{k \in K} \alpha_k \right) f(T) = f(T)$$

and our statement is proved.   □

So we will study now the case where 0 is a root of $G(K)$. As the set K will be updated at each iteration of the algorithm, we can always assume that $K = \{1, 2, \ldots, |K|\}$.

Let $d_1(i)$ denote the distance from 0 to i for each $i \in N$. $d_1(i)$ will be the first parameter assigned to i. We say that an element $j \in N$ is in bad position on $\mathcal{O}^k$ if either j is a singleton of $\mathcal{O}^k$ or if there exists an element i with $i \prec_k j$ and $d_1(i) < d_1(j)$. The number of elements in bad position on $\mathcal{O}^k$ will be noted $\alpha(k)$ and we will set: $K(j) = \{k \in K \mid j$ is in bad position on $\mathcal{O}^k\}$.

Our second parameter $d_2(j)$ is defined by: $d_2(j) = \max_{k \in K(j)} \alpha(k)$. We say that $k \in K(j)$ is critical for j if $d_2(j) = \alpha(k)$ and we say that k is a critical upper index of K if k is critical for at least one element of N. K_1 will denote the set of critical indices.

Finally the third parameter $d_3(j)$ assigned to j is: $d_3(j) = k_j = \min(k \mid k$ is critical for $j)$.

We will also associate to K three parameters:

$$\Delta_1(K) = \sum_{j=1}^{n} d_1(j); \qquad \Delta_2(K) = \sum_{j=1}^{n} d_2(j); \qquad \Delta_3(K) = |K_1|.$$

Before describing our algorithm, we need to implement a special procedure which renames the elements of N according to the following rules:

- $d_1(i) < d_1(j) \Longrightarrow i < j$ (rule 1)
- $d_1(i) = d_1(j), d_2(i) > d_2(j) \Longrightarrow i < j$ (rule 2)
- $d_1(i) = d_1(j), d_2(i) = d_2(j), d_3(i) < d_3(j) \Longrightarrow i < j$ (rule 3)
- $d_1(i) = d_1(j), d_2(i) = d_2(j), d_3(i) = d_3(j) = k, i \prec_k j \Longrightarrow i < j$ (rule 4)

Note that there exists a unique way to rename the elements of N if we apply these four rules (except for singletons which can be renamed in many ways since rule 4 does not apply to singletons).

Let $j \in N$; if $d_1(j) = 1$, j is a singleton for $\mathcal{O}^{k_j}$ and we will associate to j the edge $e_j = (0, j)_{k_j}$. If $d_1(j) > 1$, j is in bad position on $\mathcal{O}^{k_j}$ for $k_j \in K$; there exists an element $i \in N$ with $i \prec_{k_j} j$ and $d_1(i) < d_1(j)$ and we can choose for i the smallest predecessor of j on $\mathcal{O}^{k_j}$ which satisfies this property; we will associate to j the edge $e_j = (i, j)_{k_j}$.

Let $\mathcal{O}^{k_j'}$ be the adjacent order induced by e_j for $j \in N$ and consider the set of vectors $\{b^1, \ldots, b^n\}$ with $b^j = a^{k'_j} - a^{k_j}$ for $j = 1, \ldots, n$.

The following result holds:

Lemma 3.2. *Either one of the vectors $b^1, \ldots, b^n$ is the null vector, or the set $\{b^1, \ldots, b^n\}$ is triangulated.*

Proof. Assume that $b^j \neq 0$ for $j = 1, \ldots, n$.

If $e_j = (0, j)_{k_j}$, then $b_j^j \geq 0$ and $b_i^j = 0$ for $i \neq j$ by Lemma 2.1. As $b^j \neq 0$, $b_j^j > 0$.

If $e_j = (i, j)_{k_j}$, then $b_j^j \geq 0$ but $\sum_{l=1}^{n} b_l^j = 0$ by Lemma 2.1 and $b_l^j \leq 0$ for $l \neq j$ by Lemma 2.1, therefore $b_j^j > 0$. As $d_1(i) < d_1(j)$, $i < j$ by rule 1.

Let $l \in N$ distinct from i and j. If l does not satisfy the relation $i \prec_{k_j} l \prec_{k_j} j$, then $b_l^j = 0$; if $i \prec_{k_j} l \prec_{k_j} j$, then $b_l^j \leq 0$ by Lemma 2.1.

But $d_1(l) = d_1(j) - 1$ or $d_1(l) = d_1(j)$.

- If $d_1(l) = d_1(j) - 1$, then $l < j$ (rule 1).
- If $d_1(l) = d_1(j)$, then l is in bad position on $\mathcal{O}^{k_j}$; $d_2(l) \geq \alpha(k_j) = d_2(j)$.
  - If $d_2(l) > d_2(j)$, then $l < j$ (rule 2);
  - If $d_2(l) = d_2(j)$, then k_j is critical for l and $k_l \leq k_j$. If $k_l < k_j$, then $l < j$ (rule 3). If $k_l = k_j$, $l \prec_{k_j} j$, then $l < j$ (rule 4).

So, for all possible situations, we have $b_l^j \leq 0$ for $1 \leq l < j$, $b_j^j > 0$ and $b_l^j = 0$ for $j < l \leq n$; this proves our statement. □

We propose now a strongly polynomial algorithm to compute $\lambda_{\max}$:

Intersection Algorithm

Input: A normalized, monotone, submodular function f; a direction $u \in \mathbb{R}^N$.
Output: $\lambda_{\max}$, $T_{\min}$.

step 1. Take for initial positive basis the set $K := \{1, \ldots, n\}$ and compute the initial solution $(\alpha_1, \ldots, \alpha_n, \lambda)$ associated to this basis.

step 2. Represent $G(K)$ by the list of the set of out-neighbors of each node of $G(K)$.

step 3. If 0 is not a root, let T be the set of elements not reachable from 0, $\lambda_{\max} = \lambda$, $T_{\min} = T$;
return $\lambda_{\max}$, $T_{\min}$; **then stop**.
else compute the distance $d_1(j)$ from 0 to each element $j \in N$.

step 4. Compute for each $k \in K$ the number $\alpha(k)$ and for each element $j \in N$ the parameters $d_2(j)$ and $d_3(j) = k_j$; compute the numbers $\Delta_1(K)$, $\Delta_2(K)$, $\Delta_3(K)$.

step 5. Implement the relabeling procedure on the set N using rules 1, 2, 3, 4.

step 6. Find the edges $e_j := (i, j)_{k_j}$ and the vectors $b^j := a^{k'_j} - a^{k_j}$ for $j \in N$.

step 7. If one of this vector say $b^j = a^{k'_j} - a^{k_j}$ is the null vector, set $K' := K \cup \{k'_j\}$,
$K := K' \setminus \{k_j\}$ and **go to step 2**.

else set $K' := \{k'_1, \dots, k'_n\}$ and K_1 be the set of critical indices of K; apply the
procedure described in Lemma 2.2. Let $K_2 \subseteq (K \cup K') \setminus \{k_j\}$ for some $k_j \in K_1$
be the positive basis obtained after application of this procedure; $K := K_2$, **go
to step 2**.

It remains to analyze the complexity of this algorithm. First, we study the running
time of each step of an iteration. The most critical step is **step 2** where we have to
represent the graph $G(K)$ which may have $O(n^3)$ edges.

Let L be the $(n+1) \times (n+1)$ adjacency matrix of $G(K)$ where $L(i, j) = 1$ if j is
an out-neighbor of i and $L(i, j) = 0$ otherwise. We start to build the adjacency matrix L_k for each graph $G(\{k\})$ ($k \in K$): first we read the sequence $\mathcal{O}^k = [i_1, \dots, i_m]$
in $O(n \log n)$ time. Then we fill the coefficients $L_k(0, j)$ for $1 \le j \le n$ in $O(n)$
time (these coefficients are associated to singletons). To get the rest of the adjacency matrix L_k we start from the $m \times m$ adjacency matrix associated to the ordered set: $[1, \dots, m]$. This matrix is upper triangular with coefficients equal to 1
above the diagonal and 0 otherwise and can be obtained in $O(n^2)$ time. Permuting
the rows and the columns of this matrix requires $O(n^2)$ elementary operations. The
permutation which produces the matrix L_k is obtained by sorting the elements of
$\mathcal{O}^k = [i_1, \dots, i_m]$ by increasing value of the indices in time $O(n \log n)$. So the
global time is $O(n^2)$. All the matrices L_k are obtained in $O(n^3)$ time since $|K| \le n$.

Now, $L(i, j) = 1$ if, for some $k \in K$, $L_k(i, j) = 1$ and $L(i, j) = 0$ otherwise;
each coefficient $L(i, j)$ is computed in $O(n)$ time. Thus the total running time for
obtaining L is $O(n^3)$.

Using the adjacency matrix for $G(K)$, the running time of **step 3** is $O(n^2)$.

In **step 4** each $\alpha(k)$ is computed in $O(n^2)$ and all the $\alpha(k)$ in $O(n^3)$. All parameters $d_2(j)$ and $d_3(j)$ for $j \in N$ can be computed in $O(n^2)$ time. The three numbers
Δ_1, Δ_2, Δ_3 are obtained in linear time. So, the total amount of work for **step 4** is
$O(n^3)$.

The relabeling procedure of **step 5** can be implemented in $O(n^2)$ time: we first
have to sort the elements of N by increasing distance, then by decreasing value of
the second parameter $d_2(j)$, then by increasing value of the third parameter k_j and
finally by describing the list of the ordered elements of $\mathcal{O}^{k_j}$.

In **step 6** we have n^2 oracle calls and a time $O(\gamma n^2)$ if γ is the time for an oracle
call. We compute each vector $a^{k'_j}$ in $O(n^2)$ time. Thus, the total amount of work to
implement all the parts of **step 6** is $O(n^3 + \gamma n^2)$.

We already noticed in Lemma 2.2 that **step 7** can be implemented in $O(n^3)$ time.

As a consequence, the time needed to perform all the steps of the current iteration
is $O(n^3 + \gamma n^2)$.

We need now to find the maximum number of iterations of the algorithm. Let
K and $\overline{K}$ be the positive bases in two successive iterations of the algorithm. The
analysis of the complexity of the algorithm is based on the three following claims:

Claim. $\Delta_1(\overline{K}) \geq \Delta_1(K)$.

Proof. Let $e_j = (i, j)_{k_j}$ be the edge associated to $j \in N$ in **step 6**; $\mathcal{O}^{k'_j}$ is the adjacent order of $\mathcal{O}^{k_j}$. Let $(l, l')_{k'_j}$ be an edge of $G(k'_j)$.

If $(l, l')_{k_j}$ is also an edge of $G(k_j)$, $d_1(l') - d_1(l) \leq 1$; (i) if $(l, l')_{k_j}$ is not an edge of $G(k_j)$, there are two possible cases: either j is a singleton of $\mathcal{O}^{k_j}$ and $l' = j$; as $d_1(j) = 1, d_1(l') - d_1(l) \leq 1$; or (ii) j is an ordered element of $\mathcal{O}^{k_j}, l = j$ and $l' = i$ or $i \prec_{k_j} l' \prec_{k_j} j$; but $d_1(i) = d_1(j) - 1 \leq d_1(l') \leq d_1(j)$ and $d_1(l') - d_1(l) \leq 1$.

So, the relation $d_1(l') - d_1(l) \leq 1$ always hold. This shows that the distances in $G(K)$ and in $G(K \cup \{k'_j\})$ are equal. $G(K \cup K')$ is obtained by adding successively to $G(K)$ all the edge-sets of the graphs $G(k'_j)$ for all $k'_j \in K'$. Thus the distances are the same in $G(K \cup K')$ and in $G(K)$. But $G(\overline{K})$ is obtained from $G(K \cup K')$ by deletion of a subset of edges. Hence the distances can only increase in $G(\overline{K})$ and $\Delta_1(\overline{K}) \geq \Delta_1(K)$.   □

Claim. If $\Delta_1(\overline{K}) = \Delta_1(K)$, then $\Delta_2(\overline{K}) \leq \Delta_2(K)$.

Proof. Let $e_j = (i, j)_{k_j}$ be the edge of the previous claim. As the distances are equal in $G(K)$ and $G(\overline{K})$ j is in bad position on $\mathcal{O}^{k_j}$ but is not in bad position on $\mathcal{O}^{k'_j}$; this is obvious if j is a singleton; this is also true if j is not a singleton by our choice of i; indeed recall that i is the first element such that $d_1(i) < d_1(j)$ and $i \prec_{k_j} j$ when we describe $\mathcal{O}^{k_j} = [i_1, \ldots, i_m]$ from i_1 to i_m. Elements distinct from j are in bad position on $\mathcal{O}^{k'_j}$ if and only if they are in bad position on $\mathcal{O}^{k_j}$. Thus $\alpha(k'_j) = \alpha(k_j) - 1$.

For an element l in bad position on $\mathcal{O}^{k_j}$ and $\mathcal{O}^{k'_j}$, we have $d_2(l) \geq \alpha(k_j) > \alpha(k'_j)$. No upper index $k \in K'$ can be critical in the set $K \cup K'$; the parameters $d_2(l)$ remain unchanged when we replace $G(K)$ by $G(K \cup K')$. But $G(\overline{K})$ is obtained from $G(K \cup K')$ by deletion of a subset of upper indices. Therefore, either $d_2(l)$ remains unchanged or $d_2(l)$ strictly decreases. This proves the claim.   □

Claim. If $\Delta_1(\overline{K}) = \Delta_1(K)$ and $\Delta_2(\overline{K}) = \Delta_2(K)$, then $\Delta_3(\overline{K}) < \Delta_3(K)$.

Proof. Let us continue the proof of the preceding claim. If the parameters $d_2(j)$ are unchanged and l is in bad position on $\mathcal{O}^{k'_j}$, $d_2(l) \geq \alpha(k_j) > \alpha(k'_j)$. Hence k'_j is not critical for l and k'_j cannot be a critical index of $\overline{K}$. The set of critical indices of $\overline{K}$ is included in K_1; but there exists at least a critical index k_j of K_1 which does not belong to $\overline{K}$; this implies that $\Delta_3(\overline{K}) < \Delta_3(K)$.   □

We can state now our main result:

Theorem 3.3. *The running time of the Intersection Algorithm is* $O(n^8 + \gamma n^7)$, *where* γ *is the time for an oracle call.*

Proof. Define a lexicographic order among the positive bases enumerated during the algorithm. We say that $K \prec \overline{K}$ if $\Delta_1(\overline{K}) > \Delta_1(K)$ or $\Delta_1(\overline{K}) = \Delta_1(K), \Delta_2(\overline{K}) <$

$\Delta_2(K)$ or $\Delta_1(\overline{K}) = \Delta_1(K)$, $\Delta_2(\overline{K}) = \Delta_2(K)$, $\Delta_3(\overline{K}) < \Delta_3(K)$. By the preceding claims these bases are totally ordered with the lexicographic order; hence the triplets $(\Delta_1(K), \Delta_2(K), \Delta_3(K))$ are all different.

But the value of a distance cannot exceed n and $\Delta_1(K) \leq n^2$; the value of each second parameter cannot exceed n and $\Delta_2(K) \leq n^2$; finally $\Delta_3(K) \leq n$. so the total number of iterations cannot exceed n^5. The total running time is $O(n^8 + \gamma n^7)$.   □

Note that this is precisely the running time of the original algorithms for submodular function minimization. It may be possible to improve this complexity bound by, for instance, updating informations from an iteration to the next iteration. It may also be possible that the number of iterations is not greater than n^4 but we did not investigate these questions.

5.4 The Parametric Intersection Algorithm

5.4.1 A Parametric Linear Program

We consider in this section the following parametric linear program:

$$z(\lambda) = \max\left\{\sum_{i \in N} x_i \mid x \in P(f), \ x \leq \lambda u\right\} \quad (\mathcal{P}_\lambda)$$

H_λ will be the polytope: $\{x \in \mathbb{R}^N; 0 \leq x \leq \lambda u\}$.

A subset $T \subseteq N$ such that $z(\lambda) = \lambda u(N \setminus T) + f(T)$ for some $\lambda \geq 0$ will be called λ-tight.

The following two propositions are standard results. We give short proofs for completeness. For a generalization of Proposition 4.1, see Theorem 7.15 of Fujishige (2005). The property given in Proposition 4.2 corresponds to Theorem 4.6 in Megiddo (1974) and is also used in Fujishige's algorithm (Fujishige 1980, 2005) for finding the lexicographically optimal base of a polymatroid with respect to a weight vector.

Proposition 4.1. *There exists $\lambda_0 = 0 < \lambda_1 < \cdots < \lambda_s < \lambda_{s+1} = +\infty$ with $s \leq n$ and a unique increasing family of sets $T_0 = \emptyset \subset T_1 \subset \cdots \subset T_s = N$ such that:*

$$z(\lambda) = \lambda u(N \setminus T_r) + f(T_r) \quad \text{for } \lambda_r \leq \lambda \leq \lambda_{r+1}, \ r = 0, \ldots, s.$$

Moreover T_r is the unique λ-tight set for $\lambda_r < \lambda < \lambda_{r+1}$, $r = 0, \ldots, s$.

Proof. We will use the following theorem of Edmonds (1970) (this theorem gives the rank formula of a polymatroid):

$$z(\lambda) = \min(f(T) + \lambda u(N \setminus T) \mid T \subseteq N).$$

The function $\lambda \in \mathbb{R}^+ \longrightarrow z(\lambda)$ which is the minimum of a finite set of non-decreasing affine functions is a piecewise-linear non-decreasing concave function. There exists $\lambda_0 = 0 < \lambda_1 < \cdots < \lambda_s < \lambda_{s+1} = +\infty$ and a family of sets $T_0, T_1, \ldots, T_s$ such that:

$$z(\lambda) = \lambda u(N \setminus T_r) + f(T_r) \quad \text{for } \lambda_r \leq \lambda \leq \lambda_{r+1}, \ r = 0, \dots, s.$$

For $0 \leq \lambda \leq \lambda_{\max}$, $P(f) \cap H_\lambda = H_\lambda$; $x = \lambda u$ is the solution of $(\mathcal{P}_\lambda)$ and $z(\lambda) = \lambda u(N)$. For $\lambda > \lambda_{\max}$, $\lambda u \notin P(f)$ and $\lambda u(N) > z(\lambda)$. This implies that $\lambda_0 = 0$, $\lambda_1 = \lambda_{\max}$, $T_0 = \emptyset$.

For λ large enough, $H_\lambda \cap P(f) = P(f)$ and $z(\lambda) = f(N)$; so $z(\lambda) = f(N)$ for $\lambda \geq \lambda_s$ and $T_s = N$.

Assume that there exists two distinct λ-tight sets A, B for some $0 < \lambda < +\infty$. By the submodularity of f, $A \cap B$ and $A \cup B$ are also λ-tight sets; hence we can assume that A (resp. B) is the unique λ-tight set of minimum (resp. maximum) cardinality and that $A \subset B$.

Therefore $u(N \setminus B) < u(N \setminus A)$ since $u > 0$; by the concavity and the continuity of $z(\lambda)$, there exists $\varepsilon > 0$ such that $z(\lambda') = f(A) + \lambda' u(N \setminus A)$ if $\lambda - \varepsilon \leq \lambda' \leq \lambda$ and $z(\lambda') = f(B) + \lambda' u(N \setminus B)$ if $\lambda \leq \lambda' \leq \lambda + \varepsilon$. Hence $z(\lambda)$ is not a linear function in any neighborhood of λ; thus $\lambda \in \{\lambda_1, \dots, \lambda_s\}$.

If $\lambda = \lambda_r$ for $1 \leq r \leq s$, $T_{r-1} = A \subset B = T_r$.

Finally $r \leq n$ since the number of sets T_r cannot exceed n. This finishes the proof. $\square$

Let x^r be the solution of $(\mathcal{P}_{\lambda_r})$ and define a new direction:

$$u^r \quad \begin{cases} u_i^r = 0 & \text{if } i \in T_r, \\ u_i^r = u_i & \text{if } i \in N \setminus T_r. \end{cases}$$

Consider the following half-line:

$$\Delta_{u^r} = \{x \in \mathbb{R}^N \mid x = x^r + (\lambda - \lambda_r) u^r, \ \lambda \geq \lambda_r\}.$$

We want to find now the extremities of the closed interval $P(f) \cap \Delta_{u^r}$.

Proposition 4.2. $\lambda_{r+1} = \max\{\lambda \mid x \in P(f), \ x = x^r + (\lambda - \lambda_r) u^r, \ \lambda \geq \lambda_r\}$.

Proof. If $x = x^r + (\lambda - \lambda_r) u^r$, $x(N) = x^r(N) + (\lambda - \lambda_r) u(N \setminus T_r)$. But $x^r(N) = z(\lambda_r) = f(T_r) + \lambda_r u(N \setminus T_r)$. Therefore $x(N) = f(T_r) + \lambda u(N \setminus T_r)$.

If x is the solution of the linear program defined in the statement of Proposition 4.2, there exists a tight set $S \subseteq N$ for x which contains at least one element $i \in S \setminus T_r$. But $S \cup T_r$ which strictly contains T_r is also tight for x by the submodularity of f and $x(N) = f(S \cup T_r) + \lambda u(N \setminus (S \cup T_r))$.

As $x \in H_\lambda$, x is solution of $\mathcal{P}_\lambda$. But we know by Proposition 4.1 that T_r is the largest λ_r-tight set and that T_r is the unique λ-tight set in the interval $\lambda_r < \lambda < \lambda_{r+1}$. Thus the set $S \cup T_r$ can exist only if $\lambda = \lambda_{r+1}$. $\square$

5.4.2 The Parametric Intersection Algorithm

Some of the ideas of our Parametric Intersection Algorithm appear in Fujishige (2005). We will study the case when the Intersection Algorithm returns $\lambda_1 = \lambda_{\max}$ and the set $T_{\min}$. The case λ_r for $r > 1$ is similar and will not be treated. $x^1 = \lambda_1 u$ is the solution of $\mathcal{P}_{\lambda_1}$; $(\alpha_k^1, k \in K_1; \lambda_1)$ is the solution of $\mathcal{L}(K_1)$ where K_1 is the last positive basis returned by the Intersection Algorithm:

$$\mathcal{L}(K_1) \quad \begin{cases} \alpha_k \geq 0 \quad \text{for } k \in K_1, \\ \lambda u = \sum_{k \in K_1} \alpha_k a^k, \\ \sum_{k \in K_1} \alpha_k = 1 \end{cases}$$

and T_1 is the set of nodes of $G(K_1)$ not reachable from 0. Instead of stopping as in the Intersection Algorithm, we will use these informations in order to find the second value λ_2. By Proposition 4.2, λ_2 is the value of the linear program:

$$\lambda_2 = \max\{\lambda \mid x \in P(f), \ x = \lambda_1 u + (\lambda - \lambda_1)u^1, \ \lambda \geq \lambda_1\}$$

where u^1 is the new direction defined as in Proposition 4.2. So the linear system associated to any set $K \subseteq \mathcal{K}$ is now:

$$\mathcal{L}^1(K) \quad \begin{cases} \alpha_k \geq 0 \quad \text{for } k \in K, \\ \lambda_1 u + (\lambda - \lambda_1)u^1 = \sum_{k \in K} \alpha_k a^k, \\ \sum_{k \in K} \alpha_k = 1. \end{cases}$$

Since λ_1 is fixed, the variables of $\mathcal{L}^1(K)$ are $(\alpha_k, k \in K; \lambda)$ and $(\alpha_k^1, k \in K; \lambda = \lambda_1)$ is the unique solution of $\mathcal{L}^1(K_1)$ and K_1 becomes our initial basis. T_1 is the set of elements of $G(K)$ not reachable from 0 and $N_1 = N \setminus T_1$ is the set of elements reachable from 0; we will set $|T_1| = l \geq 1$. If $T_1 = N$, $\lambda_1 = \lambda_s$ and we stop. So, we will assume that $l < n$. $d_1(i)$ is the distance from 0 to i for $i \in N_1$. We observed in the proof of Proposition 3.1 that the elements of T_1 are the first elements of the sequence describing $\mathcal{O}^k$ for each $k \in K_1$: if

$$\mathcal{O}^k = [i_1, \ldots, i_l, i_{l+1}, \ldots, i_m],$$

$T_1 = \{i_1, \ldots, i_l\}$. To maintain this property throughout the next iterations of the algorithm, we assign to each element i of T_1 a first parameter $d_1(i)$ equal to n; if $j \in N$ is a predecessor of i for $\mathcal{O}^k$, $j \in T_1$ and $d_1(i) = d_1(j)$; therefore i is not in bad position for $\mathcal{O}^k$. Thus, if $i \in N$ is in bad position for $\mathcal{O}^k$, $d_1(i) < n$; if an adjacent order $\mathcal{O}^{k'}$ of $\mathcal{O}^k$ moves i in the next step of the algorithm, we know that in the description of $\mathcal{O}^{k'}$ i will be always situated after all the elements of T_1.

The numbers $d_2(i)$ for $i \in N_1$ are computed as in the Intersection Algorithm. Note however that no upper index k is critical for $i \in T_1$, so we will set $d_2(i) = 0$ for $i \in T_1$. The third parameter $d_3(i)$ for $i \in N_1$ is computed as in the Intersection Algorithm. The third parameter $d_3(i)$ plays no role for elements in T_1 and will not be computed.

We can assume that $T_1 = \{n - l + 1, \ldots, n\}$. The first rule of the relabeling procedure insures that the elements of T_1 will never be relabeled. Next, we will find the edges $e_j = (i, j)_{k_j}$ and the vectors $b^j = a^{k'_j} - a^{k_j}$ as in step 6. However these edges exist only for $j \in N_1$ since no index k is critical for $j \in T_1$. Thus there exists $|N_1|$ edges e_j and $|N_1|$ vectors b^j. The elements of T_1 remain in the first positions in any new order $\mathcal{O}^{k'_j}$ introduced in step 6. Since $b_i^j = 0$ for $i \in T_1$ and $j = 1, \ldots, n-l$ the linear system $\mu_1 b^1 + \cdots + \mu_{n-l} b^{n-l} = u^1$ has a non-negative solution. We can terminate step 7 as in the Intersection Algorithm.

Parametric Intersection Algorithm

Input: A normalized, monotone, submodular function f; a direction $u \in \mathbb{R}^N$.
Output: $s \geq 1$; $\lambda_0, \ldots, \lambda_s$; $T_0, \ldots, T_s$.

step 1. Take for initial positive basis $K := \{1, \ldots, n\}$ and compute the initial solution $(\alpha_1, \ldots, \alpha_n, \lambda)$ associated to this basis. Set $r = 0$, $\lambda_0 = 0$, $T_0 = \emptyset$.
step 2. Represent $G(K)$ by the list of the set of out-neighbors of each node of $G(K)$.
step 3. Let T be the set of elements not reachable from 0; compute the distances
$d_1(j)$ from 0 to each element $j \in N \setminus T$;
set $d_1(j) := n$ and $d_2(j) := 0$ for $j \in T$.
If $T = T_r$ **go to step 4**;
else $T_r \subset T$; set $r := r + 1$, $T_r := T$, $\lambda_r := \lambda$, $u_i := 0$ for $i \in T$.
If $T = N$ **stop**;
else **go to step 4.**
step 4. Compute for each $k \in K$ $\alpha(k)$ and for each element $j \in N \setminus T$ the parameters
$d_2(j)$ and k_j. Compute the numbers $\Delta_1(K)$, $\Delta_2(K)$, $\Delta_3(K)$.
step 5. Implement the relabeling procedure on the set $N \setminus T$ with rules 1, 2, 3, 4.
step 6. Find the edges $e_j := (i, j)_{k_j}$ and the vectors $b^j := a^{k'_j} - a^{k_j}$ for $j \in N \setminus T$.
step 7. If one of this vector say $b^j = a^{k'_j} - a^{k_j}$ is the null vector, set $K' := K \cup \{k'_j\}$,
$K := K' \setminus \{k_j\}$ and **go to step 2.**
else set $K' := \{k'_1, \ldots, k'_n\}$ and let K_1 be the set of critical indices of K; apply the procedure described in Lemma 2.2. Let $K_2 \subseteq (K \cup K') \setminus \{k_j\}$ for
some $k_j \in K_1$ be the positive basis obtained after application of this procedure; $K := K_2$ and **go to step 2**.

At the first iteration of the algorithm, all the elements of N are singletons and
$K_1 = K$. Thus, the triplet associated to the initial basis is $(n, (n - 1) \times n, n)$. At
the final iteration no element is reachable from 0 and the triplet is $(n^2, 0, 0)$. As in
the proof of the Intersection Algorithm, all the bases are distinct and the number of
iterations cannot exceed n^5; the complexity of the Parametric Intersection Algorithm
is $O(n^8 + \gamma n^7)$.
Let us give now two applications of this algorithm.

5.4.3 Strength of a Polymatroid

Consider the last discontinuity λ_s of $z(\lambda)$:

$$z(\lambda) = f(N) = f(N \setminus T_{s-1}) + \lambda_s u(T_{s-1}), \qquad \lambda_s = \frac{f(N) - f(N \setminus T_{s-1})}{u(T_{s-1})}.$$

If $(T, N \setminus T)$ is a partition of N with $f(N) > f(N \setminus T)$,

$$f(N) = f(N \setminus T_{s-1}) + \lambda_s u(T_{s-1}) \leq f(N \setminus T) + \lambda_s u(T)$$

and it is easy to see that:

$$\lambda_s = \frac{f(N) - f(N \setminus T_{s-1})}{u(T_{s-1})} \geq \frac{f(N) - f(N \setminus T)}{u(T)}.$$

If we set

$$\sigma(f, u) = \min\left\{\frac{u(T)}{f(N) - f(N \setminus T)} \text{ for all } T \subset N; \text{ such that } f(N) > f(N \setminus T)\right\},$$

then $\sigma(f, u) = \frac{1}{\lambda_s}$.

By analogy with graphs, we call $\sigma(f, u)$ the strength of the polymatroid $P(f)$. If f is the rank function of a graphic matroid, $\sigma(f, u)$ is precisely the strength of a network.

5.4.4 Minimization of a Submodular Function

Let $\overline{f}$ be a submodular set function defined on N. $\overline{f}$ may not be monotone or normalized. The submodular function minimization problem consists in finding a subset of N which realizes the minimum of $\overline{f}$. We will prove that this subset can be returned by the Parametric Intersection Algorithm for a well chosen value of the parameter λ.

Let us first make $(n + 2)$ oracle calls: $\overline{f}(i)$ for $i \in N$, $\overline{f}(N)$ and $\overline{f}(\emptyset)$. Set $\delta = \max_{i \in N} \overline{f}(i)$ and $\delta_0 = \min(0, \overline{f}(N))$.

Lemma 4.3. *Let* $\lambda = 2(n\delta - \delta_0) + 1$. *Define* f *by* $f(T) = \overline{f}(T) + \lambda|T| - \overline{f}(\emptyset)$, f *is normalized, monotone and submodular.*

Proof. Clearly f is normalized and submodular.

Let $T \subseteq N$; by the submodularity of $\overline{f}$, $\overline{f}(T) \leq \sum_{i \in N} \overline{f}(\{i\}) \leq n\delta \leq n\delta - \delta_0$. As $\delta_0 \leq \overline{f}(N) \leq n\delta$, $\lambda > 0$. $\overline{f}(T) \geq \overline{f}(N) - \overline{f}(N \setminus T) \geq \delta_0 - n\delta$. If S and T are two subsets of N, $-\lambda \leq \overline{f}(S) - \overline{f}(T) \leq \lambda$.

Assume that $S \subset T$: $f(T) - f(S) \geq \overline{f}(T) - \overline{f}(S) + \lambda \geq 0$ and f is monotone. $\square$

Let u be the vector with all components equal to 1. If $S \subseteq N$ is λ-tight,

$$z(\lambda) = f(S) + \lambda|N \setminus S| = \min(f(T) + \lambda|N \setminus T| \text{ for all } T \subseteq N),$$

$\overline{f}(S) + \lambda|S| - \overline{f}(\emptyset) + \lambda|N \setminus S| = \min(\overline{f}(T) + \lambda|T| - \overline{f}(\emptyset) + \lambda|N \setminus T|$ for all $T \subseteq N$). This proves that $\overline{f}(S) = \min(\overline{f}(T)| T \subseteq N)$.

We just proved the following result:

Theorem 4.4. *The Parametric Intersection Algorithm solves the submodular function minimization problem and simultaneously computes the strength of the polymatroid in* $O(n^8 + \gamma n^7)$ *time.*

5.5 Conclusion

We were motivated in the study of the intersection of a polymatroid and a half-line by problems in the Telecommunication industry, suggested by Jérôme Galtier and

Alexandre Laugier, where the strength of a network is a useful parameter; by choosing randomly the direction u according to a uniform distribution law it was possible to show that the average number of iterations for the Intersection Algorithm is $O(n)$; in a forthcoming paper, we will prove that the theoretical average complexity of a very simple algorithm for solving the Intersection Problem is $O(n^5)$, confirming the good practical bound observed by numerical tests.

The complexity of the Intersection Problem of a polytope with a straight line is trivial if the inequalities describing the polytope are explicitly given as input, by enumeration of these inequalities; if the polytope is given by its extreme points, the existence of a combinatorial algorithm to solve the Intersection Problem is a difficult open problem since it is possible to show that the existence of such an algorithm would answer positively the following question: does there exist a combinatorial, "simplex-like" algorithm for linear programming?

Finally the remaining case is when the polytope and the straight line are embedded in a vector space $\mathbb{R}^n$ with the number of inequalities describing the polytope not polynomial in n. The question is to find a strongly polynomial algorithm in n for solving the Intersection Problem. Of course a condition should be satisfied: the optimization problem on this polytope has to be solvable in polynomial time thanks to the ellipsoid method for instance. So we may hope that there exists strongly polynomial combinatorial algorithms when we replace the polymatroid by the matching polytope, the branching polytope, or the stable set polytope of a perfect graph. All these questions seem to be opened.

References

Bixby, R.E., Cunningham, W.H., Topkis, D.M.: The partial order of a polymatroid extreme point. Math. Oper. Res. **10**(3), 367–378 (1985)

Cheng, E., Cunningham, W.H.: A faster algorithm for computing the strength of a network. Inf. Process. Lett. **49**, 209–212 (1994)

Cunningham, W.H.: Testing membership in matroid polyhedra. J. Comb. Theory, Ser. B **36**, 161–188 (1984)

Cunningham, W.H.: Optimal attack and reinforcement of a network. J. Assoc. Comput. Mach. **32**(3), 549–561 (1985a)

Cunningham, W.H.: On submodular function minimization. Combinatorica **5**, 185–192 (1985b)

Edmonds, J.: Submodular functions, matroids and certain polyhedra. In: Combinatorial Structures and Their Applications, Proceedings of the Calgary International Conference, pp. 69–87. Gordon and Breach, New York (1970)

Fleischer, L.: Recent progress in submodular function minimization. Optima **64**, 1–11 (2000)

Fleischer, L., Iwata, S.: Improved algorithms for submodular function minimization and submodular flow. In: STOC '00: Proceedings of the Thirty-Second Annual ACM Symposium on Theory of Computing, pp. 107–116 (2000)

Fujishige, S.: Lexicographically optimal base of a polymatroid with respect to a weight vector. Math. Oper. Res. **5**, 186–196 (1980)

Fujishige, S.: Submodular Functions and Optimization. Annals of Discrete Mathematics, vol. 58. Elsevier, Amsterdam (2005)

Grötschel, M., Lovász, L., Schrijver, A.: The ellipsoid method and its consequences in combinatorial optimization. Combinatorica **1**, 169–197 (1981)

Iwata, S.: A faster scaling algorithm for minimizing submodular functions. SIAM J. Comput. **32**, 833–840 (2003)

Iwata, S., Fleischer, L., Fujishige, S.: A combinatorial strongly polynomial time algorithm for minimizing submodular functions. J. Assoc. Comput. Mach. **48**, 761–777 (2001)

Korte, B., Vygen, J.: Combinatorial Optimization: Theory and Algorithms, 4th edn. Algorithms and Combinatorics, vol. 21. Springer, Berlin (2008)

Megiddo, N.: Optimal flows in networks with multiple sources and sinks. Math. Program. **7**, 97–107 (1974)

Nagano, K.: A strongly polynomial algorithm for line search in submodular polyhedra. Mathematical engineering technical report, Department of Mathematical Informatics, The University of Tokyo (2004)

Nagano, K.: A faster parametric submodular function minimization algorithm and applications. Mathematical engineering technical report, Department of Mathematical Informatics, The University of Tokyo (2007)

Orlin, J.B.: A faster strongly polynomial time algorithm for submodular function minimization. In: Proc of IPCO, pp. 240–251 (2007)

Schrijver, A.: A combinatorial algorithm minimizing submodular functions in strongly polynomial time. J. Comb. Theory, Ser. B **80**(2), 346–355 (2000)

Schrijver, A.: Combinatorial Optimization: Polyhedra and Efficiency. Springer, Berlin (2003)

Skoda, A.: Force d'un graphe, multicoupes et fonctions sous-modulaires: aspects structurels et algorithmiques. PhD thesis, Université Pierre et Marie Curie, Paris 6 (2007)

Vygen, J.: A note on Schrijver's submodular function minimization algorithm. J. Comb. Theory, Ser. B **88**(2), 399–402 (2003)

6

A Survey on Covering Supermodular Functions

András Frank and Tamás Király

Summary. In this survey we present recent advances on problems that can be described as the construction of graphs or hypergraphs that cover certain set functions with supermodular or related properties. These include a wide range of network design and connectivity augmentation and orientation problems, as well as some results on colourings and matchings.

In the first part of the paper we survey results that follow from the totally dual integral (TDI) property of various systems defined by supermodular-type set functions. One of the aims of the survey is to emphasize the importance of relaxing the supermodularity property to include a wider range of set functions. We show how these relaxations lead to a unified understanding of different types of applications.

The second part is devoted to results that, according to our current knowledge, cannot be explained using total dual integrality. We would like to demonstrate that an extensive theory independent of total dual integrality has been developed in the last 15 years, centered around various connectivity augmentation problems.

Our survey concentrates on the theoretical foundations, and does not include every detail on applications, since the majority of these applications are described in detail in another survey paper by the first author (Frank 2006). The comprehensive book "Combinatorial Optimization: Polyhedra and Efficiency" by Schrijver (2003) is also a rich resource of results related to submodular functions.

It should be noted that sub- and supermodularity have several applications in areas not discussed in this paper. In particular, we should mention the book "Submodular Functions and Optimization" by Fujishige (2005) and the book "Discrete Convex Analysis" by Murota (2003). The former explains the foundations of the theory of submodular functions and describes the methods of submodular analysis, while the latter presents a unified framework for nonlinear discrete optimization by extending submodular function theory using ideas from continuous optimization. Our survey focuses on topics not discussed in detail in those books.

Supported by the Hungarian National Foundation for Scientific Research, OTKA K60802. The authors are members of the MTA-ELTE Egerváry Research Group.

6.1 Introduction

The role of matroids in combinatorial optimization is well-known. Basic results like the greedy algorithm, the matroid intersection and partition theorems are discussed in standard textbooks (Cook et al. 1998; Korte and Vygen 2008). More general notions like polymatroids, submodular flows and other submodular frameworks, by uniting polyhedral techniques with powerful properties of submodularity, have been successful in solving graph problems related to paths, flows, trees, and cuts. Submodular flows proved to be particularly versatile as they unify apparently unrelated areas like graph orientation, (poly)matroid intersection, network flows or the famous Lucchesi–Younger theorem. A unifying feature of all these frameworks is that the describing linear system is totally dual integral. It is this property that makes tractable the weighted versions of the corresponding optimization problems, and indeed there is a rich literature of polynomial time algorithms.

Another general trend concerning submodular functions attempts to capture the special role played by parity considerations. The notion of odd components showed up first in Tutte's characterization of perfectly matchable graphs and was later extended by results of Mader on packing paths. A general theory, called matroid parity, has been founded by Lovász. Although in some special cases (e.g. non-bipartite matchings) the weighted version of these problems are tractable, no TDI-ness result is known for general matroid parity (not even for linear matroids).

These frameworks are traditionally formulated for submodular functions but one could speak of supermodular functions as well since the role of submodular and supermodular functions are symmetric. Therefore we use sometimes the term *semimodular* to refer to a function which is either submodular or supermodular.

There is a third direction of semimodular optimization. Interestingly, at about the same time as submodular flows were introduced by Edmonds and Giles (1977), Eswaran and Tarjan (1976) found a min-max theorem (and algorithm) on the minimum number of additional edges needed to make an initial digraph strongly connected. Though the proof is not difficult at all, this result is set apart by the fact that the minimum cost version is obviously NP-complete, so it cannot fit into the submodular flow (or any other TDI) framework. Another development that proved later crucial was the introduction of the splitting off operation by Lovász. These two results became the starting points of supermodular optimization, which mainly focuses on obtaining minimum size solutions for (di)graph and hypergraph problems where minimizing an arbitrary cost function is NP-hard. An interesting phenomenon in several problems of this type is that results about edge-connectivity and node-connectivity augmentation can be derived from abstract results about supermodular functions. This first appeared in Frank (1994) where edge-connectivity augmentation of digraphs was generalized to the covering of crossing supermodular functions by directed edges, and several results in the same vein followed.

The main objective of this survey is to present recent developments in this branch of supermodular optimization. We concentrate on the theoretical aspects, and show applications only as illustrations. A detailed account of possible applications concerning graph and hypergraph connectivity can be found in Frank (2006).

Though these results cannot be explained using the framework of submodular polyhedra and total dual integrality, there is a strong interplay between the two areas. Therefore we present a summary of polyhedral results and their relation to hypergraph problems in Sect. 6.2. We also include some results that are at the borderline of the two areas.

We step across the border in Sect. 6.3 and describe a series of problems where good characterizations exist but do not come from a polyhedral description. Most of these results are obtained using some variant of the splitting-off method, although some of them require quite different techniques. We include only a few short proofs which allow the reader to take a glimpse at the methods used.

Some explanation is due on the title of the paper. There are many possible ways to define a "covering" of a set function. For example, set functions may be covered by elements of the ground set (e.g. the generators of matroids), or by vectors (as in the case of contra-polymatroids). One may also consider set functions covered by colourings, for which a good example is the supermodular colouring theorem of Schrijver (1985).

The present survey deals with the covering of set functions by graphs, hypergraphs and directed hypergraphs. In the following we define precisely what is meant by this.

6.1.1 Notation for Hypergraphs

Let V be a finite ground set. A *hyperedge* is a non-empty subset of V; it is called an *r-hyperedge* if its size is r. A *hypergraph* $H = (V, \mathcal{E})$ consists of a family $\mathcal{E}$ of hyperedges on the ground set V. We allow the same subset to appear multiple times in $\mathcal{E}$. The *rank* of a hypergraph is the size of its largest hyperedge. In this paper a hypergraph of rank at most r will be called an *r-hypergraph* for short. A hypergraph is *uniform* if every hyperedge has the same size; it is *nearly uniform* if the difference between the sizes of two hyperedges is at most one.

For a hypergraph $H = (V, \mathcal{E})$ and a node set $X \subseteq V$ we introduce the following notation:

$$\Delta_H(X) = \{e \in \mathcal{E} : e \cap X \neq \emptyset, e - X \neq \emptyset\},$$
$$d_H(X) = |\Delta_H(X)|,$$
$$i_H(X) = |\{e \in \mathcal{E} : e \subseteq X\}|,$$
$$e_H(X) = |\{e \in \mathcal{E} : e \cap X \neq \emptyset\}|.$$

For a partition $\mathcal{P}$ of V, let

$$e_H(\mathcal{P}) = |\mathcal{E}| - \sum_{X \in \mathcal{P}} i_H(X). \tag{1}$$

If $x : \mathcal{E} \to \mathbb{R}$ is a function on the hyperedge set, then we use the notation

$$d_x(Z) = \sum_{e \in \Delta_H(Z)} x(e) \quad \text{for } Z \subseteq V.$$

The set functions i_x and e_x are defined analogously. We say that a hypergraph $H = (V, \mathcal{E})$ *covers* a set function $p : 2^V \rightarrow \mathbb{R} \cup \{-\infty\}$ if $d_H(X) \geq p(X)$ for every $X \subseteq V$. A function $x : \mathcal{E} \rightarrow \mathbb{R}$ covers p if $d_x(Z) \geq p(Z)$ for every $Z \subseteq V$.

6.1.2 Notation for Directed Hypergraphs

A *hyperarc* is a pair (e, h_e) where e is a hyperedge and $h_e \in e$ is called the *head node*. With a slight abuse of notation, we will use e to denote both the hyperarc and the associated hyperedge. If $|e| = r$ then the hyperarc is called an *r-hyperarc*.

A hyperarc e *enters* a node-set X if $h_e \in X$ and $e \not\subseteq X$. A hyperarc e *leaves* a node-set X if $h_e \notin X$ and $e \cap X \neq \emptyset$.

A *directed hypergraph* $D = (V, \mathcal{A})$ consists of a node set V and a family of hyperarcs $\mathcal{A}$. We allow $\mathcal{A}$ to contain a hyperarc multiple times. The *rank* of a directed hypergraph is the size of its largest hyperarc. A directed hypergraph of rank at most r will be called a *directed r-hypergraph*.

We use the following notation for directed hypergraphs:

$$\Delta_D^-(X) = \{e \in \mathcal{A} : e \text{ enters } X\},$$
$$\Delta_D^+(X) = \{e \in \mathcal{A} : e \text{ leaves } X\},$$
$$\varrho_D(X) = |\Delta_D^-(X)|,$$
$$\delta_D(X) = |\Delta_D^+(X)|.$$

Analogously to undirected hypergraphs, a directed hypergraph $D = (V, \mathcal{A})$ is said to *cover* a set function $p : 2^V \rightarrow \mathbb{R} \cup \{-\infty\}$ if $\varrho_D(X) \geq p(X)$ for every $X \subseteq V$. If we have a function $x : \mathcal{A} \rightarrow \mathbb{R}$ on the hyperarc set, then we define

$$\varrho_x(Z) = \sum_{e \in \Delta_D^-(Z)} x(e) \quad \text{for } Z \subseteq V.$$

We say that x covers p if $\varrho_x(Z) \geq p(Z)$ for every $Z \subseteq V$.

Given an (undirected) hypergraph $H = (V, \mathcal{E})$, an *orientation* of H is a directed hypergraph obtained by assigning a head node to each hyperedge in $\mathcal{E}$.

6.2 Semimodular Frameworks with TDI Describing Systems

6.2.1 G-polymatroids

Let S be a finite ground set. A large part of the survey deals with the properties of set functions of type $f : 2^S \rightarrow \mathbb{R} \cup \{-\infty, +\infty\}$. Unless explicitly stated otherwise, we always assume that $f(\emptyset) = 0$.

Given a function $m : S \rightarrow \mathbb{R}$, we can define a set function by $m(X) = \sum_{v \in X} m(v)$ for $X \subseteq S$ (we denote this associated set function by the same letter as the original function). Set functions obtained this way are called *modular*.

A set function $b : 2^S \to \mathbb{R} \cup \{\infty\}$ is called *submodular* if the submodular inequality

$$b(X) + b(Y) \geq b(X \cap Y) + b(X \cup Y) \tag{2}$$

holds for every pair $X, Y \subseteq S$ of subsets. A non-decreasing submodular function is called a *polymatroid function*. A set function $p : 2^S \to \mathbb{R} \cup \{-\infty\}$ is *supermodular* if $-p$ is submodular, i.e. the supermodular inequality

$$p(X) + p(Y) \leq p(X \cap Y) + p(X \cup Y) \tag{3}$$

holds for every pair $X, Y \subseteq S$ of subsets. It is an easy observation that a finite-valued supermodular function is non-negative if and only if it is non-decreasing.

A pair (p, b) of set functions is called *paramodular* if p is supermodular, b is submodular, and they are *compliant* in the sense that the *cross-inequality*

$$b(X) - p(Y) \geq b(X - Y) - p(Y - X) \tag{4}$$

holds for every pair of subsets $X, Y \subseteq S$. Paramodular pairs appear naturally in many contexts. For example, the following is true.

Proposition 2.1. *If r is the rank function of a matroid and r' is its co-rank function, then (r', r) is a paramodular pair.*

We can state this a bit more generally. For a set function $f : 2^S \to \mathbb{R}\cup\{-\infty, +\infty\}$ with $f(S)$ finite, we define a kind of complementary set function $\overline{f}$ on the same ground set by $\overline{f}(X) = f(S) - f(S - X)$. If r is a matroid rank function, then $\overline{r}$ is the co-rank function. The above proposition can be generalized as follows:

Proposition 2.2. *If f is a supermodular function, then $\overline{f}$ is submodular, and $(f, \overline{f})$ is a paramodular pair.*

As an example, consider a hypergraph $H = (V, \mathcal{E})$. The set function $f(X) := i_H(X)$ is supermodular, and $\overline{f}$ is the set function $e_H(X)$. So (i_H, e_H) is a paramodular pair.

For set functions $p : 2^S \to \mathbb{R} \cup \{-\infty\}$ and $b : 2^S \to \mathbb{R} \cup \{\infty\}$ we define the following polyhedra:

$$P(b) := \{x \in \mathbb{R}^S : x \geq 0, x(Z) \leq b(Z) \text{ for every } Z \subseteq S\},$$

$$S(b) := \{x \in \mathbb{R}^S : x(Z) \leq b(Z) \text{ for every } Z \subseteq S\},$$

$$B(b) := \{x \in \mathbb{R}^S : x(Z) \leq b(Z) \text{ for every } Z \subseteq S, \ x(S) = b(S)\},$$

$$C(p) := \{x \in \mathbb{R}^S : x \geq 0, x(Z) \geq p(Z) \text{ for every } Z \subseteq S\},$$

$$S'(p) := \{x \in \mathbb{R}^S : x(Z) \geq p(Z) \text{ for every } Z \subseteq S\},$$

$$B'(p) := \{x \in \mathbb{R}^S : x(Z) \geq p(Z) \text{ for every } Z \subseteq S, \ x(S) = p(S)\},$$

$$Q(p, b) := \{x \in \mathbb{R}^S : p(Z) \leq x(Z) \leq b(Z) \text{ for every } Z \subseteq S\}.$$

If b is a polymatroid function, then $P(b)$ is called a *polymatroid*, and b is its *border function*. For a submodular function b, the polyhedron $S(b)$ is the *submodular polyhedron* of b and $B(b)$ is the *base polyhedron* of b. If p is a non-negative supermodular function, then $C(p)$ is called a *contra-polymatroid*, and p its border function. If p, b is a paramodular pair then $Q(p.b)$ is called a *generalized polymatroid* or *g-polymatroid* for short, and p and b are respectively the *lower* and *upper* *border functions*.

Proposition 2.3. *If p is supermodular, then $B'(p) = B(\overline{p})$, so $B'(p)$ is the base polyhedron of the submodular function $\overline{p}$.*

For technical reasons, we extend the above classes of polyhedra to include the empty polyhedron. However, the following is true.

Lemma 2.4. *Polymatroids, contra-polymatroids, base polyhedra and g-polymatroids defined by border functions are never empty. Moreover, their defining border function (or pair of border functions) is unique, i.e. different border functions give different polyhedra.*

A crucial property of these polyhedra is that if the border functions are integer-valued then the polyhedron is automatically integer. The theoretical background is that the linear systems defining these polyhedra are totally dual integral.

Theorem 2.5 (Frank 1981; Frank and Tardos 1988). *A g-polymatroid is integer if and only if its upper and lower border functions are integer.*

As an example of g-polymatroids which are not base polyhedra, consider a hypergraph $H = (V, \mathcal{E})$, and a subset of nodes $U \subseteq V$. Let the polyhedron $P \subseteq \mathbb{R}^U$ be the convex hull of the in-degree vectors on U of all possible orientations of the hypergraph H.

Proposition 2.6. *The polyhedron P is a g-polymatroid. The lower border function of P is i_H restricted to U, and the upper border function of P is e_H restricted to U.*

Note that if $U = V$, then P is a base polyhedron.

Operations on g-Polymatroids

An important and extremely useful property of the class of g-polymatroids is that it is closed under several natural operations. Moreover, the subclass of integer g-polymatroids is also closed under these operations, with some obvious restrictions. Here we present a brief summary of these properties.

Let $f : S \rightarrow \mathbb{Z} \cup \{-\infty\}$ and $g : S \rightarrow \mathbb{Z} \cup \{\infty\}$ be functions with $f \leq g$. The polyhedron $T(f, g) := \{x \in \mathbb{R}^S : f \leq x \leq g\}$ is called a *box*. For two numbers $\alpha \leq \beta$, the polyhedron $K(\alpha, \beta) = \{x \in \mathbb{R}^S : \alpha \leq x(S) \leq \beta\}$ is called a *plank*.

Proposition 2.7. *Any box or plank is a g-polymatroid.*

We may define several operations on polyhedra, or specifically on g-polymatroids. The operations *direct sum*, *translation*, *reflection* and *face* have the usual meaning. The *projection* of a g-polymatroid along a subset $T \subsetneq S$ is a polyhedron in $\mathbb{R}^{S-T}$ obtained by removing the components corresponding to T from each vector in the g-polymatroid. The *sum* of g-polymatroids is the Minkowski sum.

Given a surjective function $\varphi : S \to S'$ and a vector $x \in \mathbb{R}^S$, we define a vector $x_\varphi \in \mathbb{R}^{S'}$ by $x_\varphi(s') = x(\varphi^{-1}(s'))$ ($s' \in S'$). The *aggregate* of a polyhedron $P \in \mathbb{R}^S$ with respect to φ is $P_\varphi = \{x_\varphi : x \in P\}$. We also define the *aggregate* of a set function $f : 2^S \to \mathbb{R} \cup \{-\infty, +\infty\}$ by $f_\varphi(X) = f(\varphi^{-1}(X))$.

Theorem 2.8. *The class of g-polymatroids is closed under the following operations*:

- *direct sum*
- *translation*
- *reflection through a point*
- *projection along a subset*
- *aggregate*
- *sum*
- *face*
- *intersection with a box*
- *intersection with a plank*

In all of these cases, if the original g-polymatroid is integer and the operation involves integer numbers, then the resulting g-polymatroid is integer too.

The upper and lower border functions of sums and aggregates can be explicitly given.

Theorem 2.9. *Let (p, b) be a paramodular pair on S, and let $\varphi : S \to S'$ be a surjective function. Then (p_φ, b_φ) is a paramodular pair on S', and $Q(p, b)_\varphi = Q(p_\varphi, b_\varphi)$. If p and b are integral, then for each integer vector x in $Q(p, b)_\varphi$ there is an integer vector $y \in Q(p, b)$ such that $y_\varphi = x$.*

Theorem 2.10. *Let $(p_1, b_1), \ldots, (p_k, b_k)$ be paramodular pairs. The sum of the g-polymatroids $Q(p_1, b_1), \ldots, Q(p_k, b_k)$ is the g-polymatroid given by $Q(\sum_{i=1}^k p_i, \sum_{i=1}^k b_i)$. If all functions p_i and b_i are integral, then each integer vector x in $Q(\sum_{i=1}^k p_i, \sum_{i=1}^k b_i)$ can be written as $x = x^1 + \cdots + x^k$, where x^i is an integer vector in $Q(p_i, b_i)$.*

Base polyhedra are special cases of g-polymatroids, so projections of base polyhedra are g-polymatroids. The following theorem states that every g-polymatroid arises this way.

Theorem 2.11. *A g-polymatroid $Q(p, b)$ defined by a paramodular pair (p, b) arises as the projection of a base polyhedron along one element, namely, $Q(p, b)$ is the projection of $B(b^*)$ along s^* where*

$$b^*(X) = \begin{cases} b(X) & \text{if } X \subseteq S, \\ -p(S - X) & \text{if } s^* \in X. \end{cases} \tag{5}$$

The intersection of a g-polymatroid with a box can of course be empty. Luckily, the characterization of non-emptiness is quite simple and elegant, and it implies the *linking property* that we will use several times in the rest of the paper.

Theorem 2.12. *Let (p, b) be a paramodular pair, and $f : S \to \mathbb{Z} \cup \{-\infty\}$, $g : S \to \mathbb{Z} \cup \{\infty\}$ such that $f \leq g$. Then the g-polymatroid $Q(p, b) \cap T(f, g)$ is non-empty if and only if the following two conditions hold:*

$$f(X) \leq b(X) \quad \text{for every } X \subseteq S, \tag{6}$$
$$g(X) \geq p(X) \quad \text{for every } X \subseteq S. \tag{7}$$

Corollary 2.13 (Linking property). *Let p, b, f, g be as in Theorem 2.12. If there is an element x of $Q(p, b)$ with $x \geq f$ and there is an element y of $Q(p, b)$ with $y \leq g$, then there is an element z of $Q(p, b)$ with $f \leq z \leq g$. In addition, if p, b, f, g are all integer-valued, then z can be integer too.*

It is worth formulating Theorem 2.12 for the special case of base-polyhedra.

Theorem 2.14. *For a submodular function b for which $b(S)$ is finite, the intersection $B(b) \cap T(f, g)$ is non-empty if and only if*

$$f(Y) \leq b(Y) \quad \text{for every } Y \subseteq S \quad \text{and} \tag{8}$$
$$g(Y) \geq \overline{b}(Y) \quad \text{for every } Y \subseteq S. \tag{9}$$

A stronger theorem characterizing non-emptiness of the intersection of a g-polymatroid with a box and a plank can also be derived, leading to the *strong linking property*.

Theorem 2.15. *Let (p, b) be a paramodular pair, and $f : S \to \mathbb{Z} \cup \{-\infty\}$, $g : S \to \mathbb{Z} \cup \{\infty\}$ such that $f \leq g$. In addition, let $\alpha \leq \beta$ be two numbers. Then the g-polymatroid $Q(p, b) \cap T(f, g) \cap K(\alpha, \beta)$ is non-empty if and only if the following four conditions hold:*

$$f(X) \leq \min\{b(X), \beta - p(S - X)\} \quad \text{for every } X \subseteq S, \tag{10}$$
$$g(X) \geq \max\{p(X), \alpha - b(S - X)\} \quad \text{for every } X \subseteq S. \tag{11}$$

Proof. Let $M = Q(p, b) \cap T(f, g) \cap K(\alpha, \beta)$. If $x \in M$, then $f(X) \leq x(X) \leq b(X)$ and $f(X) \leq x(X) = x(S) - x(S - X) \leq \beta - p(S - X)$, from which the necessity of (10) follows. Furthermore, $g(X) \geq x(X) \geq p(X)$ and $g(X) \geq x(X) = x(S) - x(S - X) \geq \alpha - b(S - X)$, that is, (11) is also necessary.

To prove the sufficiency, we invoke Theorem 2.11 stating that $Q(p, b)$ arises as the projection of a base-polyhedron $B(b^*)$ where b^* is defined on $S^* = S + s^*$ by (5). Let f^* denote the extension of the f to S^* where $f^*(s^*) := -\beta$, and let g^* be the extension of g to S^* where $g^*(s^*) := -\alpha$.

We claim that f^* and b^* meet (8). Indeed, for $Y \subseteq S$ (10) implies $f^*(Y) = f(Y) \leq b(Y) = b^*(Y)$ while in case $s^* \in Y$ we have $f^*(Y) = f(X) - \beta \leq -p(S - X) = b^*(Y)$ for $X := Y - s^*$. Similarly, g^* and b^* meet (9), since for

$Y \subseteq S$ (11) implies $g^*(Y) = g(Y) \geq p(Y) = -b^*(S^* - Y) = \overline{b^*}(Y)$, while in case $s^* \in Y$ we have $g^*(Y) = g(X) - \alpha \geq -b(S - X) = -b^*(S - X) = \overline{b^*}(Y)$ for $X := Y - s^*$. Therefore Theorem 2.14 implies that there is an element $x^* \in B(b^*)$ and the restriction of x^* to S is in M by the construction.   □

Corollary 2.16 (Strong linking property). *Let $p, b, f, g, \alpha, \beta$ be as in Theorem 2.15. If there is an element x of $Q(p, b)$ with $x \geq f$ and $x(S) \leq \beta$, and there is an element y of $Q(p, b)$ with $y \leq g$ and $y(S) \geq \alpha$, then there is an element z of $Q(p, b)$ with $f \leq z \leq g$ and $\alpha \leq z(S) \leq \beta$. In addition, if $p, b, f, g, \alpha, \beta$ are all integer-valued, then z can be integer too.*

Relaxations of Submodularity and Supermodularity

The results of the previous paragraphs raise the question whether there is a common reason behind the fact that g-polymatroids are closed under all these different operations. The answer is yes: the deeper reason is that g-polymatroids may be defined using set functions with weaker properties. This fact is very important in applications, where set functions with only these weaker properties show up regularly.

Two subsets X and Y of the ground set S are called *intersecting* if $X \cap Y \neq \emptyset$, $X - Y \neq \emptyset$, and $Y - X \neq \emptyset$. If in addition $S - (X \cup Y) \neq \emptyset$, then they are called *crossing*. A set function b is called *intersecting (crossing) submodular* if the submodular inequality (2) holds whenever X and Y are intersecting (crossing). The definitions are analogous for intersecting and crossing supermodular functions.

Let b be a set function on the ground set S. A subset $X \subseteq S$ is called *b-separable from below* (or *separable* for short) if X can be partitioned into at least two non-empty disjoint subsets $X_1, \ldots, X_t$ for which $\sum_{i=1}^t b(X_i) \leq b(X)$. A set function b is *near submodular* if the submodular inequality holds for all intersecting pairs of non-separable subsets. A set function p is *near supermodular* if $-p$ is near submodular. A pair of set functions (p, b) is a *near paramodular pair* if p is near supermodular, b is near submodular, and the cross-inequality (4) holds for all intersecting pairs of subsets (X, Y) where X is not b-separable from below and Y is not p-separable from above.

Theorem 2.17. *A polyhedron $Q = Q(p, b)$ defined by a near paramodular pair (p, b) is a g-polymatroid. If p and b are integral, then Q is an integer polyhedron.*

Theorem 2.18. *A g-polymatroid $Q = Q(p, b)$ defined by a near paramodular pair (p, b) is non-empty if and only if*

$$p\left(\bigcup_{Z \in \mathcal{F}} Z\right) \leq \sum_{Z \in \mathcal{F}} b(Z) \quad and \quad \sum_{Z \in \mathcal{F}} p(Z) \leq b\left(\bigcup_{Z \in \mathcal{F}} Z\right) \tag{12}$$

for every sub-partition $\mathcal{F}$ of S.

As a special case, we may obtain Fujishige's theorem on the non emptiness of base polyhedra.

Theorem 2.19 (Fujishige 1984). *Let b be a crossing submodular function with $b(S) = 0$. Then $B(b)$ is a base polyhedron, and it is non-empty if and only if*

$$\sum_{Z \in \mathcal{P}} b(Z) \geq 0 \quad and \quad \sum_{Z \in \mathcal{P}} b(S - Z) \geq 0$$

for every partition $\mathcal{P}$ of S.

Proof. Let $s \in S$ be an arbitrary element, and let us define two new set functions:

$$b'(X) := \begin{cases} b(X) & \text{if } s \notin X, \\ 0 & \text{if } X = S, \\ \infty & \text{otherwise,} \end{cases}$$

$$p'(X) := \begin{cases} -b(S - X) & \text{if } s \notin X, \\ 0 & \text{if } X = S, \\ -\infty & \text{otherwise.} \end{cases}$$

It is easy to see that b' is intersecting submodular, p' is intersecting supermodular, and the cross-inequality holds for intersecting pairs. Furthermore, one can check that $B(b) = Q(p', b')$. By Theorem 2.18, we get exactly the non-emptiness conditions of the theorem. If $Q(p', b')$ is non-empty, then it has a lower border function p^* with $p^*(S) = 0$ and an upper border function b^* with $b^*(S) = 0$. It follows that $p^* \geq \overline{b^*}$ and $b^* \leq \overline{p^*}$, hence both inequalities must hold with equality, therefore $B(b) = B(b^*)$ is a base polyhedron. $\square$

Let $S(b)$ be a non-empty submodular polyhedron defined by a near submodular function b (this is a submodular polyhedron because of Theorem 2.17). We know that $S(b)$ is defined by a unique submodular border function b'. Similarly, if we consider the supermodular polyhedron $S'(p)$ for a near supermodular function p, it has a uniquely defined supermodular border function. The question arises whether we can define these two set functions in terms of p and b. This leads us to the notions of upper and lower truncation.

Given a near submodular function b, we define its *lower truncation* by

$$b^{\vee}(X) = \min \left\{ \sum_{Z \in \mathcal{P}} b(Z) : \mathcal{P} \text{ is a partition of } X \right\}.$$

Analogously, the *upper truncation* of a near supermodular function p is defined by

$$p^{\wedge}(X) = \max \left\{ \sum_{Z \in \mathcal{P}} p(Z) : \mathcal{P} \text{ is a partition of } X \right\}.$$

Theorem 2.20. *The lower truncation $b^{\vee}$ of a near submodular function is submodular, and the upper truncation $p^{\wedge}$ of a near supermodular function p is supermodular. Furthermore, $S(b) = S(b^{\vee})$ and $S'(p) = S'(p^{\wedge})$.*

We note that while $Q(p, b)$ is a g-polymatroid for a near paramodular pair (p, b), its lower and upper border functions are not necessarily $p^{\wedge}$ and $b^{\vee}$.

Orientations Covering Crossing Supermodular Functions

The first application we show concerns the orientation of hypergraphs. We assume that each hyperedge has size at least two. The following lemma establishes the link between orientations of a hypergraph and base polyhedra.

Lemma 2.21 (Orientation Lemma). *Given a hypergraph $H = (V, \mathcal{E})$ and an in-degree specification $m : V \to \mathbb{Z}_+$, there is an orientation D of H such that $\varrho_D(v) = m(v)$ for every $v \in V$ if and only if $m(V) = |\mathcal{E}|$ and $m(Z) \geq i_H(Z)$ for every $Z \subseteq V$.*

This lemma, which is relatively easy to prove, implies a rather strong result on orientations covering a crossing supermodular function.

Theorem 2.22 (Frank, Király and Király 2003a). *Let $H = (V, \mathcal{E})$ be a hypergraph, and p a non-negative crossing supermodular function on the node set with $p(V) = 0$. The in-degree vectors of orientations of H that cover p are the integer vectors in the base polyhedron $B'(p + i_H)$.*

Since $p + i_H$ is crossing supermodular, we can use Theorem 2.19 to derive a necessary and sufficient condition for the existence of an orientation covering p.

Theorem 2.23 (Frank, Király and Király 2003a). *Let $H = (V, \mathcal{E})$ be a hypergraph, and p a non-negative crossing supermodular function on the node set. There is an orientation of H covering p if and only if the following two conditions hold for every partition $\mathcal{P}$ of V:*

$$e_H(\mathcal{P}) \geq \sum_{Z \in \mathcal{P}} p(Z), \tag{13}$$

$$\sum_{e \in \mathcal{E}}(|\{Z \in \mathcal{P} : e \cap Z \neq \emptyset\}| - 1) \geq \sum_{Z \in \mathcal{P}} p(V - Z). \tag{14}$$

One can go even further by considering the intersection with a box.

Theorem 2.24. *Let $H = (V, \mathcal{E})$ be a hypergraph, p a non-negative crossing supermodular function on V, and $f : V \to \mathbb{Z}_+$, $g : V \to \mathbb{Z}_+$ lower and upper bounds such that $f \leq g$. There is an orientation of H covering p for which $f(v) \leq \varrho(v) \leq g(v)$ for every $v \in V$ if and only if the following two conditions hold for every sub-partition $\mathcal{F}$ of V:*

$$|\mathcal{E}| - \sum_{Z \in \mathcal{F}} i_H(Z) \geq f\left(V - \bigcup_{Z \in \mathcal{F}} Z\right) + \sum_{Z \in \mathcal{F}} p(Z), \tag{15}$$

$$\sum_{e \in \mathcal{E}}(|\{Z \in \mathcal{F} : e \cap Z \neq \emptyset\}| - 1) \geq \sum_{Z \in \mathcal{F}} p(V - Z) - g\left(V - \bigcup_{Z \in \mathcal{F}} Z\right). \tag{16}$$

Corollary 2.25 (Linking property for orientations). *Let* $H = (V, \mathcal{E})$ *be a hypergraph,* p *a non-negative crossing supermodular function on* V, *and* $f : V \to \mathbb{Z}_+$, $g : V \to \mathbb{Z}_+$ *lower and upper bounds such that* $f \leq g$. *If there is an orientation of* H *covering* p *for which* $\varrho(v) \geq f(v)$ *for every* $v \in V$, *and there is one for which* $\varrho(v) \leq g(v)$ *for every* $v \in V$, *then there is an orientation of* H *covering* p *for which* $f(v) \leq \varrho(v) \leq g(v)$ *for every* $v \in V$.

Packing Arborescences with Free Roots

One of the fundamental theorems in graph connectivity is Edmonds' disjoint arborescence theorem (Edmonds 1973). Here we state it in an equivalent form that allows multiple prescribed roots.

Theorem 2.26. *Let* $D = (V, A)$ *be a directed graph, and* $m : V \to \mathbb{Z}_+$ *an integer vector for which* $m(V) = k$. *There are* k *edge-disjoint spanning arborescences in* D *such that each* $v \in V$ *is the root of* $m(v)$ *arborescences if and only if*

$$\varrho_D(Z) \geq k - m(Z) \quad \text{for every } \emptyset \neq Z \subseteq V.$$

By combining this theorem with the theory of submodular polyhedra, we can derive a generalization that allows lower and upper bounds on the number of roots at each node.

Theorem 2.27 (Frank 1978; Cai 1983). *Let* $D = (V, A)$ *be a directed graph,* $f : V \to \mathbb{Z}_+$ *and* $g : V \to \mathbb{Z}_+$ *lower and upper bounds for which* $f \leq g$. *There exist* k *edge-disjoint spanning arborescences in* D *such that each* $v \in V$ *is the root of at least* $f(v)$ *and at most* $g(v)$ *arborescences if and only if*

$$f(V) \leq k, \tag{17}$$
$$g(Z) \geq k - \varrho_D(Z) \quad \text{for every } \emptyset \neq Z \subseteq V, \tag{18}$$

$$\sum_{Z \in \mathcal{F}} \varrho_D(Z) \geq k(|\mathcal{F}| - 1) + f\left(V - \bigcup_{Z \in \mathcal{F}} Z\right) \quad \text{for every subpartition } \mathcal{F} \text{ of } V. \tag{19}$$

The proof uses the fact that the set function $p(X) := k - \varrho(X)$ ($\emptyset \neq X \subseteq V$) is intersecting supermodular, so we can consider the g-polymatroid $C(p) \cap T(f, g)$ and characterize its non-emptiness using Theorem 2.12.

Corollary 2.28. *A digraph* $D = (V, A)$ *has* k *edge-disjoint spanning arborescences if and only if*

$$\sum_{Z \in \mathcal{F}} \varrho_D(Z) \geq k(|\mathcal{F}| - 1)$$

for every subpartition $\mathcal{F}$ *of* V.

As in the case of orientations, the presence of a g-polymatroid implies a kind of linking property.

Corollary 2.29 (Linking property for arborescences). *If a digraph $D = (V, A)$ has k edge-disjoint spanning arborescences such that each node $v \in V$ is the root of at least $f(v)$ of them, and it has k edge-disjoint spanning arborescences so that each node $v \in V$ is the root of at most $g(v)$ of them (where $f(v) \leq g(v)$), then D has k edge-disjoint spanning arborescences where the number of arborescences rooted at v is between $f(v)$ and $g(v)$.*

6.2.2 Intersection of Two g-Polymatroids

While the intersection of two g-polymatroids is not necessarily a g-polymatroid, it still has very nice properties. A central result of combinatorial optimization is Edmonds' (poly)matroid intersection theorem (Edmonds 1970), which can be extended to g-polymatroids.

Theorem 2.30. *Let (p_1, b_1) and (p_2, b_2) be paramodular pairs. The linear system $\{x \in \mathbb{R}^S : \max\{p_1(Z), p_2(Z)\} \leq x(Z) \leq \min\{b_1(Z), b_2(Z)\} \, \forall Z \subseteq V\}$ is totally dual integral. If p_1, b_1, p_2, b_2 are all integral, then $Q(p_1, b_1) \cap Q(p_2, b_2)$ is an integer polyhedron.*

Theorem 2.31. *Let (p_1, b_1) and (p_2, b_2) be paramodular pairs. The polyhedron $P = Q(p_1, b_1) \cap Q(p_2, b_2)$ is non-empty if and only if*

$$p_1 \leq b_2 \quad and \quad p_2 \leq b_1.$$

A corollary is the following theorem, which is a kind of discrete analogue of the theorem that a convex and a concave function can be separated by a linear function.

Corollary 2.32 (Discrete separation theorem). *Let $p : 2^S \to \mathbb{R} \cup \{-\infty\}$ be a supermodular function and $b : 2^S \to \mathbb{R} \cup \{\infty\}$ be a submodular function such that $p \leq b$. Then there is a modular function m that satisfies $p \leq m \leq b$. In addition, if p and b are integral, then m can be chosen to be integral too.*

To show a simple application, we may consider the problem of finding an orientation of a graph that covers two crossing supermodular set functions h_1 and h_2 simultaneously. To get the intersection of two g-polymatroids, we have to assume that one of them is non-negative. We state the result only in the case when the set functions are symmetric, otherwise the necessary and sufficient condition is much more complicated.

Theorem 2.33. *Let h_1 and h_2 be symmetric crossing supermodular set functions, and let us assume that h_1 is non-negative. A graph $G = (V, E)$ has an orientation covering $\max\{h_1, h_2\}$ if and only if $d_G(X) \geq 2 \max\{h_1(X), h_2(X)\}$ for every $X \subseteq V$.*

Supermodular Colourings

A more involved application of g-polymatroid intersection is the supermodular colouring theorem of Schrijver (1985). Here we state it in a slightly generalized form that involves skew supermodular functions. A set function $p : 2^S \to \mathbb{R} \cup \{-\infty\}$ is *skew supermodular* if for any two subsets $X \subseteq S$ and $Y \subseteq S$ at least one of the following two inequalities holds:

$$p(X) + p(Y) \le p(X \cap Y) + p(X \cup Y),$$
$$p(X) + p(Y) \le p(X - Y) + p(Y - X).$$

If p is a skew supermodular function, and we raise $p(\{v\})$ to 0 on all elements where it is negative, we get a near supermodular function. This is the basis of the following result.

Theorem 2.34. *If p is a skew supermodular function, then $C(p)$ is a contra-polymatroid.*

A *k-colouring* of a ground set S is a partition $S_1, \ldots, S_k$ of S, where we allow some members of the partition to be empty. The sets S_i are called the *colour classes*. Given a set function $p : 2^S \to \mathbb{Z}$, we say that a k-colouring $\{S_1, \ldots, S_k\}$ *dominates* p if $|\{i : S_i \cap Z \ne \emptyset\}| \ge p(Z)$ for every $Z \subseteq S$.

Lemma 2.35. *Let p be a skew supermodular set function that satisfies $p(X) \le \min\{k, |X|\}$ for every $X \subseteq S$. Then the polyhedron*

$$\{x \in [0, 1]^S : x(Z) \ge 1 \text{ if } p(Z) = k, \ x(Z) \le |Z| - p(Z) + 1 \text{ for every } Z \subseteq S\} \tag{20}$$

is an integer g-polymatroid.

Corollary 2.36. *The polyhedron defined by (20) is the convex hull of the possible colour classes of k-colourings that dominate p.*

A k-colouring of S is called *equitable* if the size of each colour class is $\lfloor |S|/k \rfloor$ or $\lceil |S|/k \rceil$. The following skew supermodular colouring theorem is a slight extension of the supermodular colouring theorem of Schrijver, and it can be proved by combining the g-polymatroid intersection theorem with Corollary 2.36.

Theorem 2.37. *Let k be a positive integer, p_1 and p_2 two skew supermodular functions such that $\max\{p_1(X), p_2(X)\} \le \min\{k, |X|\}$ for every $X \subseteq S$. Then there is a k-colouring of S that dominates $\max\{p_1, p_2\}$. Moreover, the colouring may be chosen to be equitable.*

Interestingly, this theorem can be used to prove a result on hypergraphs covering symmetric skew supermodular functions. The key to the connection between the two topics is the following lemma.

Lemma 2.38 (Bernáth and Király 2008). *Let p be a symmetric skew supermodular function such that $\max\{p(Z) : Z \subseteq S\} = k$. If $\{S_1, \ldots, S_k\}$ is a k-colouring that dominates p, then the hypergraph with edge set $\{S_1, \ldots, S_k\}$ covers p.*

This means that with the aid of the skew supermodular colouring theorem we can find a hypergraph covering two symmetric skew supermodular functions simultaneously, provided that their maximal value is the same.

Corollary 2.39. *Let* $p_1 : 2^V \to \mathbb{Z} \cup \{-\infty\}$ *and* $p_2 : 2^V \to \mathbb{Z} \cup \{-\infty\}$ *be two symmetric skew supermodular set functions such that* $\max\{p_1(X) : X \subseteq V\} = \max\{p_2(X) : X \subseteq V\} = k$, *and let* $m : V \to \mathbb{Z}_+$ *be a degree prescription such that* $m(v) \le k$ *for any* $v \in V$ *and*

$$\sum_{v \in X} m(v) \ge \max\{p_1(X), p_2(X)\} \quad \text{for every } X \subseteq V.$$

Then there exists a nearly uniform hypergraph H *of exactly* k *hyperedges such that* $d_H(v) = m(v)$ *for every* $v \in V$ *and* H *covers both* p_1 *and* p_2.

Corollary 2.40. *Let* $p_1 : 2^V \to \mathbb{Z} \cup \{-\infty\}$ *and* $p_2 : 2^V \to \mathbb{Z} \cup \{-\infty\}$ *be two symmetric skew supermodular set functions such that* $\max\{p_1(X) : X \subseteq V\} = \max\{p_2(X) : X \subseteq V\} = k$. *Given the right oracles for* p_1 *and* p_2, *it is possible to find in polynomial time a hypergraph of minimum total size that covers both* p_1 *and* p_2. *In addition, the hypergraph can be chosen to be nearly uniform.*

As an application let us consider the simultaneous local edge-connectivity augmentation of two hypergraphs. For a hypergraph $H = (V, \mathcal{E})$ and nodes $u, v \in V$, $\lambda_H(u, v)$ denotes the size of the minimum cut that separates u and v. Given two hypergraphs H_1, H_2 on the same ground set V, two symmetric requirement functions $r_1, r_2 : V \times V \to \mathbb{Z}_+$, we want to find a hypergraph H of minimum total size such that $\lambda_{H_i + H}(u, v) \ge r_i(u, v)$ for any $u, v \in V$ and $i \in \{1, 2\}$. Corollary 2.40 implies that if we assume that $\max\{r_1(u, v) - \lambda_{H_1}(u, v) : u, v \in V\} = \max\{r_2(u, v) - \lambda_{H_2}(u, v) : u, v \in V\}$, then we can solve the problem in polynomial time. Furthermore, we can even achieve that the hyperedges in H are of almost equal size. It should be mentioned that without assuming the equality of the maximum deficiencies the problem is NP-hard.

6.2.3 Submodular Flows

In this section we briefly describe submodular flows and present some applications related to hypergraph orientations that cover supermodular functions. Since we concentrate on supermodularity as opposed to submodularity, we will define submodular flows in terms of supermodular functions. Due to the symmetry involved (by Proposition 2.3, base polyhedra may be defined by supermodular as well as submodular functions) the definition is equivalent to the original one.

Let $D = (V, A)$ be a directed graph, $f : A \to \mathbb{Z} \cup \{-\infty\}$, $g : A \to \mathbb{Z} \cup \{+\infty\}$ lower and upper bounds for which $f \le g$. Let $p : 2^V \to \mathbb{R} \cup \{-\infty\}$ be a crossing supermodular function with $p(V) = 0$. A function $x : A \to \mathbb{R}$ is called a *submodular flow* if

$$\lambda_x(Z) := \varrho_x(Z) - \delta_x(Z) \ge p(Z) \quad \text{for every } Z \subseteq V. \tag{21}$$

Equivalently, the vector $\lambda_x \in \mathbb{R}^V$ must be in the base polyhedron $B'(p)$. A submodular flow x is *feasible* if $f \le x \le g$. The set $Q(f, g; p)$ of feasible submodular flows is called a *submodular flow polyhedron*. The fundamental result of Edmonds and Giles (1977) is that the natural linear system describing the submodular flow polyhedron is TDI. The characterization of non-emptiness is also simple if p is supermodular.

Theorem 2.41 (Frank 1982). *Let $D = (V, A)$ be a digraph, p a supermodular function for which $p(V) = 0$, and $f \le g$ lower and upper bounds on the edges. There exists a feasible submodular flow if and only if*

$$\varrho_g(Z) - \delta_f(Z) \ge p(Z) \quad \text{for every } Z \subseteq V. \tag{22}$$

If p, f, g are integral, then the submodular flow can be chosen to be integer too.

If the set function is only crossing supermodular, then the condition of the existence of a feasible submodular flow is much more complicated. To describe it, we have to introduce the notion of a tree-composition of a subset of V.

Let Z be a subset of the ground set V. A tree-composition of Z is given using a partition $\{Z_1, \ldots, Z_k\}$ of Z, a partition $\{U_1, \ldots, U_l\}$ of $V - Z$, and a directed tree T with node set $\{z_1, \ldots, z_k\} \cup \{u_1, \ldots, u_l\}$ where every edge of the tree is of type $u_i z_j$. The associated *tree-composition* is a family of subsets of V with one set corresponding to each edge of T. To obtain the set corresponding to an edge $u_i z_j$, we take the union of all partition members that correspond to nodes reachable from z_j in the underlying undirected graph of $T - \{u_i z_j\}$.

By convention, a tree-composition of V is either a partition or a co-partition of V.

Theorem 2.42 (Frank 1993). *Let $D = (V, A)$ be a digraph, p a crossing supermodular function for which $p(V) = 0$, and $f \le g$ lower and upper bounds on the edges. There exists a feasible submodular flow if and only if*

$$\varrho_g(X) - \delta_f(X) \ge \sum_{Z \in \mathcal{F}} p(Z) \tag{23}$$

for every subset $X \subseteq V$ and every tree-composition $\mathcal{F}$ of X. If p, f, g are integral, then the flow can be chosen to be integer too.

Submodular flow polyhedra are closely related to the intersection of two g-polymatroids.

Theorem 2.43 (Frank and Tardos 1988). *A polyhedron P is a submodular flow polyhedron if and only if it is a projection of the intersection of two g-polymatroids.*

Applications

Our first application comes from the simple observation that if $\delta_D(X) = 0$ whenever $p(X)$ is finite, then the submodular flow problem corresponds to the covering of p by a vector $x : A \to \mathbb{R}$ with $f \le x \le g$.

Theorem 2.44. *Let $D = (V, A)$ be a directed graph, let $p : 2^V \to \mathbb{R} \cup \{-\infty\}$ be crossing supermodular set function such that $\delta_D(X) = 0$ whenever $p(X)$ is finite and $p(V) = 0$. Let furthermore $g : A \to \mathbb{R}_+$ be an upper bound function. Then the linear system*

$$\{x \in \mathbb{R}^A : 0 \le x \le g, \ \varrho_x(Z) \ge p(Z) \text{ for every } Z \subseteq V\}$$

is totally dual integral. If p and g are integral, then the existence of a solution implies the existence of an integer solution.

Corollary 2.45. *Let $D = (V, A)$ be a directed graph whose underlying undirected graph is connected, and let $\mathcal{C}$ be a crossing family of directed cuts. The minimum number of edges in A that cover every cut in $\mathcal{C}$ equals the maximum number of edge-disjoint dicuts in $\mathcal{C}$.*

The next application concerns the orientation of hypergraphs. A *mixed hypergraph* is a triple $M = (V; \mathcal{E}, \mathcal{A})$, where $\mathcal{E}$ is a set of hyperedges and $\mathcal{A}$ is a set of hyperarcs. We can extend Theorem 2.24 to mixed hypergraphs using submodular flows.

Theorem 2.46. *Let $M = (V; \mathcal{E}, \mathcal{A})$ be a mixed hypergraph, and $p : 2^V \to \mathbb{Z} \cup \{-\infty\}$ a crossing supermodular set function. Suppose that a cost is assigned to each possible orientation of every hyperedge in $\mathcal{E}$. Then the problem of finding a minimum cost orientation of M covering p can be formulated as a submodular flow problem, solvable in polynomial time.*

6.2.4 Covering Intersecting Supermodular Functions

We start with a special case of Theorem 2.54 which was proved in Frank (1979).

Theorem 2.47. *Let $D = (V, A)$ be a digraph, p an intersecting supermodular function on V, and $g : A \to \mathbb{Z}_+$ a function on the edges for which $\varrho_g(X) \ge p(X)$ for every $X \subseteq V$. Then the linear system*

$$\{x \in \mathbb{R}^A : 0 \le x \le g, \ \varrho_x(Z) \ge p(Z) \text{ for every } Z \subseteq V\}$$

is totally dual integral.

Corollary 2.48. *The minimum number of (not necessarily distinct) edges of D covering an intersecting supermodular function p is equal to the maximum of $\sum_{X \in \mathcal{F}} p(X)$ over laminar families $\mathcal{F}$ which are independent in the sense that no edge of D enters more than one member of $\mathcal{F}$.*

This implies the Lucchesi–Younger theorem on dicut covers (Lucchesi and Younger 1978) if we consider the set function

$$p(X) = \begin{cases} \sigma(X) & \text{if } \emptyset \ne X \subseteq V \text{ and } \delta_D(X) = 0, \\ 0 & \text{otherwise} \end{cases}$$

where $\sigma(X)$ denotes the number of undirected components of $V - X$.

Corollary 2.49 (Lucchesi–Younger). *In a digraph the minimum size of a dicut cover equals the maximum number of disjoint dicuts.*

We now show a generalization where upper bounds are given on the in-degrees of the nodes. In a digraph $D = (V, A)$ the *entrance* $\Gamma^-(X)$ of a set $X \subseteq V$ is the set

$$\{v \in X : \exists uv \in A, \ u \in V - X\}.$$

For a function $g : V \to \mathbb{Z}_+$ let $\beta_g(X) := \sum_{v \in \Gamma^-(X)} g(v)$. It can be shown that the function β_g is submodular, which is the basis of the following result, proved in Frank and Tardos (1989).

Theorem 2.50. *Let $D = (V, A)$ be a digraph, $g : V \to \mathbb{Z}_+$ a function on the nodes, and p an intersecting supermodular function. There is an integer-valued function $x : A \to \mathbb{Z}_+$ for which $\varrho_x(Z) \geq p(Z)$ for every $Z \subseteq V$ and $\varrho_x(v) \leq g(v)$ for every $v \in V$ if and only if*

$$p(X) \leq \beta_g(X) \tag{24}$$

for every $X \subseteq V$.

It should be noted that if we have bounds on the out-degrees of the nodes instead of the in-degrees, then the problem becomes NP-complete, since the Hamiltonian path problem can be formulated this way.

Applications

By combining Theorem 2.50 with Edmonds' disjoint arborescence theorem (Theorem 2.26) it is possible to prove the following result of Vidyasankar (1978).

Theorem 2.51 (Vidyasankar 1978). *Let s be a node of in-degree 0 in the digraph $D = (V, A)$. The edges of D can be covered by k arborescences of root s if and only if $\varrho(v) \leq k$ for every $v \in V - s$ and*

$$k - \varrho(X) \leq \sum_{v \in \Gamma^-(X)} (k - \varrho(v)) \tag{25}$$

for every $X \subseteq V - s$.

Another application of Theorem 2.50 is that we can add in-degree bounds in the Lucchesi–Younger theorem. A *branching* is an acyclic digraph where every node has in-degree at most one.

Theorem 2.52. *Let $D = (V, A)$ be a digraph. There is a branching whose edges cover every directed cut if and only if the deletion of any nonempty subset $X \subset V$ results in at most $|X|$ components with in-degree 0.*

The proof uses the fact that the function $\sigma(X)$ defined on the kernels (which denotes the number of components of $D - X$ in the undirected sense) is intersecting supermodular, so Theorem 2.50 can be applied with this and $g \equiv 1$.

The next application is a common generalization of Rado's theorem on matroids (Rado 1942) and a theorem of Lovász on superadditive intersecting supermodular functions (Lovász 1970). For an edge set F and a node set $X \subseteq V$, let $\Gamma_F(X)$ denote the set of nodes in $V - X$ that are joined by an edge in F to a node in X.

Theorem 2.53. *Let $G = (S, T; E)$ be a simple bipartite graph, $p : 2^S \to \mathbb{Z} \cup \{-\infty\}$ an intersecting supermodular function, and $g : S \to \mathbb{Z}_+$ an upper bound function on the nodes. Let furthermore $M = (T, r)$ be a matroid of rank function r on the ground set T. There exists an edge set $F \subseteq E$ for which*

$$r(\Gamma_F(X)) \geq p(X) \quad \text{for every } X \subseteq S \quad \text{and} \tag{26}$$

$$d_F(v) \leq g(v) \quad \text{for every } v \in S \tag{27}$$

if and only if

$$p(X) \leq r(\Gamma_E(Y)) + g(X - Y) \quad \text{for every } Y \subseteq X \subseteq S. \tag{28}$$

6.2.5 Oriented Subgraphs of Mixed Hypergraphs

It was observed by Khanna et al. (2005) (for graphs, extended to hypergraphs in Frank et al. 2003a) that if p is intersecting supermodular, then the mixed (hyper)graph orientation problem remains tractable if we allow not only the orientation of the undirected hyperedges, but also their deletion, or their replacement by multiple hyperarcs. The next theorem formulates this result in the most general setting, and then we present two special cases.

Theorem 2.54 (Khanna, Naor and Shepherd 2005; Frank, Király and Király 2003a). *Let $D = (V, \mathcal{A})$ be a directed hypergraph for which $\mathcal{A}$ can be partitioned into hyperarc sets $\mathcal{A}_1, \ldots, \mathcal{A}_k$ where the underlying hyperedge of the hyperarcs in $\mathcal{A}_i$ is the same hyperedge e_i. Let $p : 2^V \to \mathbb{Z} \cup \{-\infty\}$ be an intersecting supermodular set function. Let furthermore $f : \mathcal{A} \to \mathbb{Z} \cup \{-\infty\}$, $g : \mathcal{A} \to \mathbb{Z} \cup \{+\infty\}$ be lower and upper bounds for which $f \leq g$, and let $l_i \leq u_i$ $(i = 1, \ldots, k)$. The linear system*

$$\{x \in \mathbb{R}^{\mathcal{A}} : f \leq x \leq g, \; l_i \leq x(\mathcal{A}_i) \leq u_i, \; \varrho_x(Z) \geq p(Z) \; \forall Z \subseteq V\}$$

is totally dual integral. The optimization problem over this polyhedron can be solved using submodular flows.

An *oriented sub-hypergraph* of a mixed hypergraph M is a sub-hypergraph of a directed hypergraph obtained from M by orienting the hyperedges in $\mathcal{E}$.

Corollary 2.55. *Let $M = (V; \mathcal{E}, \mathcal{A})$ be a mixed hypergraph, and $p : 2^V \to \mathbb{Z} \cup \{-\infty\}$ an intersecting supermodular set function. Suppose that a cost is assigned to each hyperarc in $\mathcal{A}$, and to each possible orientation of every hyperedge in $\mathcal{E}$. Then the problem of finding a minimum cost oriented sub-hypergraph of M covering p can be formulated as a submodular flow problem, solvable in polynomial time.*

The next result considers only graphs. A *partial orientation* of a mixed graph $M = (V; E, A)$ is a mixed graph obtained by orienting some edges in E. If a weight is assigned to both orientations of every edge in E, then the weight of a partial orientation is the sum of the weights of the oriented edges (so the un-oriented edges do not count).

Corollary 2.56. *Let $M = (V; E, A)$ be a mixed graph, and $p : 2^V \to \mathbb{Z} \cup \{-\infty\}$ an intersecting supermodular set function. Suppose that M covers p, and a weight is assigned to both orientations of every edge in E. Then the problem of finding a maximum weight partial orientation of M covering p can be formulated as a submodular flow problem, solvable in polynomial time.*

For crossing supermodular functions the above problem is NP-hard, even if the aim is to maximize the number of oriented edges.

6.2.6 Covering Intersecting Bi-set Functions

In this last section discussing TDI problems, we introduce a concept that will have an important role in the second part of the survey. The idea is to consider bi-set functions instead of set functions.

Let V be a ground set. A *bi-set* is a pair (X_O, X_I) where $X_I \subseteq X_O \subseteq V$. The set X_O is called the *outer member* of the bi-set, while X_I is called the *inner member*. A bi-set is called *trivial* if $X_O = V$ or $X_I = \emptyset$.

We say that a directed edge uv *enters* the bi-set $X = (X_O, X_I)$ if $u \in V - X_O$ and $v \in X_I$. Given a digraph $D = (V, A)$, $\varrho_D(X)$ denotes the number of edges entering the bi-set X. Two bi-sets are *independent* if their inner members are disjoint or their outer members are co-disjoint, i.e. if no edge can enter both. A family $\mathcal{I}$ of bi-sets is independent if its members are pairwise independent.

The *intersection* of two bi-sets $X = (X_O, X_I)$ and $Y = (Y_O, Y_I)$ is defined by $X \cap Y = (X_O \cap Y_O, X_I \cap Y_I)$. Analogously, the *union* of the two bi-sets is $X \cup Y = (X_O \cup Y_O, X_I \cup Y_I)$. We can consider the supermodular inequality for a bi-set function p:

$$p(X) + p(Y) \le p(X \cap Y) + p(X \cup Y). \tag{29}$$

We say that a bi-set function p is *intersecting supermodular* if the supermodular inequality (29) holds whenever $X \cap Y$ is non-trivial. The bi-set function p is *crossing supermodular* if the supermodular inequality (29) holds whenever $X \cap Y$ and $X \cup Y$ are non-trivial.

It turned out that Theorem 2.47 can be extended to bi-set functions.

Theorem 2.57 (Frank 2008). *Let $D = (V, A)$ be a digraph and let $g : A \to \mathbb{Z}_+$ be a capacity function. Let p be an intersecting supermodular bi-set function for which $\varrho_g(X) \ge p(X)$ for every bi-set X. Then the linear system*

$$\{x \in \mathbb{R}^A : 0 \le x \le g, \; \varrho_x(Z) \ge p(Z) \text{ for every bi-set } Z\}$$

is totally dual integral. The polyhedron described by the system is a submodular flow polyhedron.

Let us return for a moment to set functions. Let S be a ground set and $T \subseteq S$ a subset. A set function $p : 2^S \to \mathbb{Z} \cup \{-\infty\}$ is called *T-intersecting supermodular* if the supermodular inequality holds for any pair of subsets X, Y for which $X \cap Y \cap T \neq \emptyset$.

Given a T-intersecting supermodular function p, we can define an intersecting supermodular bi-set function by

$$p'(X_O, X_I) = \begin{cases} p(X_O) & \text{if } X_O \cap T \neq \emptyset \text{ and } X_I = X_O \cap T, \\ -\infty & \text{otherwise.} \end{cases}$$

Using this and Theorem 2.57 we can derive a theorem on covering T-intersecting supermodular functions.

Theorem 2.58. *Let $D = (V, A)$ be a digraph, let $g : A \to \mathbb{Z}_+$ be a capacity function, and let T be a subset of the nodes that contains the head of each edge in A. Let p be a T-intersecting supermodular function for which $\varrho_g(X) \geq p(X)$ for every set X. Then the linear system*

$$\{x \in \mathbb{R}^A : 0 \leq x \leq g, \ \varrho_x(Z) \geq p(Z) \text{ for every } Z \subseteq V\}$$

is totally dual integral. The polyhedron described by the system is a submodular flow polyhedron.

Corollary 2.59. *Let $D = (V, A)$ be a digraph, and $r \in V$ a specified root node. Then the convex hull of the characteristic vectors of rooted k-connected subgraphs of D is a submodular flow polyhedron.*

6.3 Beyond Total Dual Integrality

In the second part of the survey, we discuss problems involving graphs and hypergraphs that do not fit into the TDI frameworks described in the previous sections. As it was mentioned in the Introduction, the origins of this area can be traced back to the paper of Eswaran and Tarjan on making a graph 2-edge-connected (Eswaran and Tarjan 1976), and the splitting-off theorem of Lovász (1979). In fact, the majority of the results presented here are some kind of generalization of the result of Eswaran and Tarjan, and the proofs rely on various extensions of the splitting-off method.

One characteristic that sets these problems apart from those discussed previously is that the minimum cost versions are hard to solve. Still, we will show that for some special types of cost functions the problems are tractable. The reason behind this phenomenon is that while the feasible graphs or hypergraphs cannot be described by polyhedral methods, their degree vectors often can be.

While the problems described in this part have some common features like the nice description of the feasible in-degree vectors, there is no unifying framework similar to total dual integrality for the problems in the previous part. One may hope that such a framework will be discovered in the future.

6.3.1 Covering Crossing Supermodular Bi-set Functions by Digraphs

Recall the definition of bi-sets in Sect. 6.2.6, where we have shown that the problem of covering an intersecting supermodular bi-set function with a directed subgraph of minimum cost is solvable. This is not true for crossing supermodular bi-set functions; however, it was proved in Frank and Jordán (1995) that it is still possible to find a minimum number of edges that cover the function.

Theorem 3.1 (Frank and Jordán 1995). *Let p be a crossing supermodular bi-set function that has value 0 on all trivial bi-sets. Then the minimum number of directed edges covering p equals to the maximum of $\sum_{X \in \mathcal{I}} p(X)$ on all possible independent families $\mathcal{I}$.*

Furthermore, while it is not possible to give a polyhedral description of the covering edge-sets, the in-degree vectors of these edge-sets form a familiar polyhedron.

Theorem 3.2. *Let p be a crossing supermodular bi-set function that has value 0 on all trivial bi-sets. The possible in-degree vectors of digraphs that cover p are the integer vectors in an integer contra-polymatroid.*

Another important property is that the in-degrees and the out-degrees can be chosen independently, which implies a linking theorem.

Theorem 3.3. *Let p be a crossing supermodular bi-set function that has value 0 on all trivial bi-sets, and let $g_{in} : V \to \mathbb{Z}_+$ and $g_{out} : V \to \mathbb{Z}_+$ be upper bounds of the in-degrees and out-degrees, respectively. If p can be covered by at most γ edges, it can be covered by a digraph whose in-degrees are bounded by g_{in}, and it can be covered by a digraph whose out-degrees are bounded by g_{out}, then p can be covered by a digraph that satisfies all these conditions simultaneously.*

Applications

First, let us consider the node-connectivity augmentation of directed graphs.

Lemma 3.4. *A digraph $D = (V, A)$ is k-connected if and only if $\varrho_D(X) + |X_O - X_I| \geq k$ for every non-trivial bi-set X.*

Let $D = (V, A)$ be a digraph, and let us define the bi-set function

$$p_D(X) = \begin{cases} k - \varrho_D(X) - |X_O - X_I| & \text{if } X \text{ is non-trivial,} \\ 0 & \text{otherwise.} \end{cases}$$

This bi-set function is crossing supermodular, which gives us the following theorem.

Theorem 3.5. *A directed graph* $D = (V, A)$ *can be made k-connected by adding at most γ new edges if and only if*

$$\sum_{X \in \mathcal{I}} p_D(X) \leq \gamma$$

holds for every family $\mathcal{I}$ of pairwise independent bi-sets.

The second application comes from a completely different area: matching theory. Given a bipartite graph $G = (S, T; E)$, a *t-matching* is a subgraph with maximum degree at most t. It is $K_{t,t}$-*free* if it does not contain $K_{t,t}$ as a subgraph. As a generalization of the square-free 2-matching problem, we would like to find a $K_{t,t}$-free t-matching with a maximum number of edges.

It is rather surprising that one can describe this problem as the covering of a bi-set function. The key idea is that the covering will contain the edges which are *not* in the t-matching.

Theorem 3.6 (Frank 2003). *The maximum number of edges in a $K_{t,t}$-free t-matching of a bipartite graph $G = (S, T; E)$ is equal to*

$$\min_{Z \subseteq V} (t|Z| + i_G(V - Z) - c_t(Z)),$$

where $c_t(Z)$ denotes the number of components of $G - Z$ that are isomorphic to $K_{t,t}$.

A different generalization of the square-free 2-matching problem can also be deduced from Theorem 3.1. A *bi-clique* in a bipartite graph $G = (S, T; E)$ is a subgraph that is a $K_{i,j}$ for some $i, j \geq 1$. The *size* of the bi-clique is $i + j$.

Theorem 3.7 (Frank 2003). *In a bipartite graph $G = (S, T; E)$ the maximum number of edges in a subgraph not containing bi-cliques of size more than t equals*

$$\min\left\{|E| - \sum_{B \in \mathcal{B}}(|V(B)| - t) : \mathcal{B} \text{ is a family of edge-disjoint bi-cliques}\right\}.$$

The last application is the theorem of Győri on subpaths of a directed path (Győri 1984). Let P be a directed path, and let $\mathcal{P}$ be a family of subpaths of P. We say that a family of subpaths $\mathcal{G}$ is a *generator* of $\mathcal{P}$ if every member of $\mathcal{P}$ is the union of some members of $\mathcal{G}$. Let P_1 and P_2 be subpaths in $\mathcal{P}$, and let e_1 be an edge of P_1 and e_2 be an edge of P_2. The two path-edge pairs (P_1, e_1) and (P_2, e_2) are called *independent* if either there is no node that precedes e_1 in P_1 and precedes e_2 in P_2, or there is no node that follows e_1 in P_1 and follows e_2 in P_2.

Theorem 3.8 (Győri 1984). *Let P be a directed path, and let $\mathcal{P}$ be a family of subpaths of P. The minimum cardinality of a generator of $\mathcal{P}$ equals the maximum number of pairwise independent path-edge pairs in $\mathcal{P}$.*

This theorem follows from Theorem 3.1; in fact, that theorem implies that the characterization is also true for subpaths of a directed cycle, which does not follow from the original proof of Győri.

Theorem 3.8 has a surprising application in combinatorial geometry. Let us call a bounded region in the plane *angular* if its boundary is composed of a finite number of horizontal and vertical segments, and let us call it *vertically convex* if every vertical line intersects it in a segment.

Corollary 3.9. *Let R be an angular, vertically convex region of the plane. The minimum number of angular rectangles whose union is R equals the maximum number of points of R such that no two of them can be covered by a single angular rectangle in R.*

6.3.2 Covering Two Supermodular Bi-set Functions

If our bi-set functions are not only crossing supermodular, but supermodular, then it is possible to simultaneously cover two functions by a minimum number of directed edges. More precisely, the following has been proved in Bérczi and Frank (2008).

Theorem 3.10. *Let p_1 and p_2 be two supermodular bi-set functions on a ground set V. The minimum number of directed edges that cover both p_1 and p_2 equals*

$$\max\{p_1(X) + p_2(Y) : X \text{ and } Y \text{ are independent bi-sets}\}.$$

Note that the analogue of Theorem 2.57 is not true: the linear system

$$\{x \in \mathbb{R}^A : 0 \le x \le g, \ \varrho_x(Z) \ge \max\{p_1(Z), p_2(Z)\} \text{ for every bi-set } Z\}$$

is not necessarily TDI. Theorem 3.2 does not carry over to this case either: the in-degrees of covering digraphs do not necessarily form the integer vectors of a contra-polymatroid.

6.3.3 Bi-set Extension of Edmonds' Disjoint Arborescence Theorem

In this section we show an abstract generalization of Edmonds' disjoint arborescence theorem that involves families of bi-sets (a weaker generalization involving families of sets was proved by Szegő 2001). Let V be a ground set and let $\mathcal{F}_1, \ldots, \mathcal{F}_k$ be intersecting bi-set families that satisfy the following *mixed intersection property*:

If $X \in \mathcal{F}_i$ and $Y \in \mathcal{F}_j$ are intersecting bi-sets, then $X \cap Y \in \mathcal{F}_i \cap \mathcal{F}_j.$ (30)

We say that a set A of directed edges *covers* a family of bi-sets if for every member of the family there is an edge that enters both the inner and outer sets of that member. The following theorem is proved in Bérczi and Frank (2008).

Theorem 3.11. *Let* $D = (V, A)$ *be a digraph, and let* $\mathcal{F}_1, \ldots, \mathcal{F}_k$ *be intersecting bi-set families that satisfy* (30). *The edge set* A *can be partitioned into* k *parts* $F_1, \ldots, F_k$ *such that* F_i *covers* $\mathcal{F}_i$ *if and only if*

$$\varrho_D(X) \geq |\{i : X \in \mathcal{F}_i\}| \quad \text{for every bi-set } X.$$

This implies a nice result of Kamiyama et al. (2008) where instead of spanning arborescences one is interested in arborescences that contain each node reachable from their prescribed root.

Theorem 3.12. *Let* $D = (V, A)$ *be a directed graph and let* $R = \{r_1, \ldots, r_k\}$ *be a set of* k *distinct roots. Let* V_i *denote the set of nodes reachable from* r_i *in* D. *There exist edge-disjoint arborescences* $A_i \subseteq A$ $(i = 1, \ldots, k)$ *such that* A_i *is an* r_i-*arborescence spanning* V_i *if and only if*

$$\varrho_D(X) \geq |\{i : r_i \notin X, V_i \cap X \neq \emptyset\}| \quad \text{for every } X \subseteq V.$$

We give only an indication how this follows from Theorem 3.11. Call two nodes u and v in V *equivalent* if they are not separated by any V_i. A *subatom* is a subset of equivalent nodes. Let us define the bi-set families $\mathcal{F}_i$ $(i = 1, \ldots, k)$ as follows. For each non-empty subatom $X \subseteq V_i - r_i$ and each subset $Y \subseteq V - V_i$, let the bi-set $(X \cup Y, X)$ be a member of $\mathcal{F}_i$. We can apply Theorem 3.11 to this bi-set family to obtain Theorem 3.12.

6.3.4 Covering Crossing Supermodular Functions by Directed Hypergraphs

We return to problems involving set functions, and consider their covering by hyperarcs. First we describe an abstract form of the splitting-off operation. Let $p : 2^V \to \mathbb{Z} \cup \{-\infty\}$ be a crossing supermodular set function. Let furthermore $g_i : V \to \mathbb{Z}_+$ be an indegree-bound and $g_o : V \to \mathbb{Z}_+$ an outdegree-bound for which $g_i(V) \leq g_o(V)$. Suppose that $g_i(X) \geq p(X)$ and $g_o(V - X) \geq p(X)$ for every $X \subseteq V$. We define the *splitting-off operation* as follows. A hyperarc e can be *split off* from (p, g_i, g_o) if $g_i(h_e) > 0$ and $g_o(v) > 0$ for every $v \in e - h_e$. For such a hyperarc let

$$g_i^e(v) := \begin{cases} g_i(v) - 1 & \text{if } v = h_e, \\ g_i(v) & \text{otherwise,} \end{cases}$$

$$g_o^e(v) := \begin{cases} g_o(v) - 1 & \text{if } v \in e - h_e, \\ g_o(v) & \text{otherwise,} \end{cases}$$

$$p^e(X) := \begin{cases} p(X) - 1 & \text{if } e \text{ enters } X, \\ p(X) & \text{otherwise.} \end{cases}$$

The splitting-off operation is *feasible* if $g_i^e(X) \geq p^e(X)$ and $g_o^e(V - X) \geq p^e(X)$ for every $X \subseteq V$. The operation is called a *feasible r-splitting* if e is an r-hyperarc. It can be verified that p^e is crossing supermodular.

The following theorem from Király and Makai (2007) describes conditions when a feasible splitting-off is available (the special case corresponding to k-edge-connectivity augmentation was proved in Berg et al. (2003)).

Theorem 3.13. *Let p be a crossing supermodular function, $g_i : V \to \mathbb{Z}_+$ and $g_o : V \to \mathbb{Z}_+$ degree bounds such that $g_i(V) \leq g_o(V) \leq rg_i(V)$ for some integer r, and*

$$g_i(X) \geq p(X) \quad \text{for every } X \subseteq V, \tag{31}$$
$$g_o(V - X) \geq p(X) \quad \text{for every } X \subseteq V. \tag{32}$$

Let $u \in V$ be such that $g_i(u) > 0$. Then there is a hyperarc e with $h_e = u$ and $|e| \leq r + 1$ that can be feasibly split off.

The splitting-off operation has the following nice property:

Lemma 3.14. *If for some $r_1 > r > r_2$ there is a feasible r_1-spitting and a feasible r_2-splitting with head u, then there is a feasible r-splitting with head u.*

Using this, we can prove a theorem on the existence of a degree-bounded directed hypergraph covering a crossing supermodular set function.

Theorem 3.15 (Király and Makai 2007). *Let $p : 2^V \to \mathbb{Z} \cup \{-\infty\}$ be a crossing supermodular set function, $g_i : V \to \mathbb{Z}_+$ and $g_o : V \to \mathbb{Z}_+$ degree bounds such that $g_i(V) \leq g_o(V) \leq rg_i(V)$ for some positive integer r, and*

$$g_i(X) \geq p(X) \quad \text{for every } X \subseteq V, \tag{33}$$
$$g_o(V - X) \geq p(X) \quad \text{for every } X \subseteq V. \tag{34}$$

Then there is a directed $(r+1)$-hypergraph D such that $\delta_D(v) \leq g_o(v)$ and $\varrho_D(v) \leq g_i(v)$ for every $v \in V$, and

$$\varrho_D(X) \geq p(X) \quad \text{for every } X \subseteq V.$$

Proof. We can assume that $r = \lceil g_o(V)/g_i(V) \rceil$. According to Theorem 3.13 we can obtain a directed hypergraph D^* by successive feasible splitting-off operations such that $\delta_{D^*}(v) \leq g_o(v)$, $\varrho_{D^*}(v) \leq g_i(v)$ for every $v \in V$, and $\varrho_{D^*}(X) \geq p(X)$ for every $X \subseteq V$. Suppose that D^* contains a hyperarc e of size $r_1 > r + 1$. This means that there is a feasible r_1-splitting with head h_e for some $r_1 > r + 1$. However, we know by Theorem 3.13 that there is also a feasible r_2-splitting with head h_e for some $r_2 \leq r + 1$ since $g_o(V) \leq rg_i(V)$.

So Lemma 3.14 implies that there is an $(r + 1)$-hyperarc e that can be feasibly split off. Since $g_i^e(V) \leq g_o^e(V) \leq rg_i^e(V)$, we can continue the splitting-off process, until we obtain a directed $(r + 1)$-hypergraph that covers p. □

Theorem 3.15 implies that both the feasible in-degree bounds and the feasible out-degree bounds can be described as the integer points of a contra-polymatroid. This means that we can efficiently find directed hypergraphs covering a crossing supermodular set function p with

- minimum total size,
- minimum number of hyperedges, all of bounded size,
- given total size and given number of hyperedges.

Theorem 3.16. *Let p be a crossing supermodular function, let $r \geq 2$ be an integer, and let γ denote the minimum number of r-hyperarcs that cover p. Let p^γ be the set function defined by*

$$p^\gamma(X) = \begin{cases} \gamma & \text{if } X = V, \\ p(X) & \text{otherwise.} \end{cases}$$

Then the in-degree vectors of directed r-hypergraphs that cover p are exactly the integer vectors in the contra-polymatroid $C(p^\gamma)$.

Since a feasible in-degree vector and a feasible out-degree vector can be chosen independently, we can obtain a linking result.

Corollary 3.17. *Let p be a crossing supermodular set function, and let $g_i : V \to \mathbb{Z}_+$ and $g_o : V \to \mathbb{Z}_+$ be upper bounds of the in-degrees and out-degrees, respectively. If p can be covered by at most γ r-hyperarcs, it can be covered by a directed r-hypergraph whose in-degrees are bounded by g_i, and it can be covered by a directed r-hypergraph whose out-degrees are bounded by g_o, then p can be covered by a directed r-hypergraph that satisfies all these conditions simultaneously.*

6.3.5 Combined Augmentation and Orientation

We have seen that for a crossing supermodular function both the problem of covering it with a minimum number of hyperarcs and the problem of orienting a hypergraph to cover it can be solved. In this section we present a combination of these two problems. Let p be a crossing supermodular function on the ground set V. Suppose we have a hypergraph that cannot be oriented to cover it; what is the minimum number (or total size) of hyperedges that have to be added in order to have a good orientation?

We first show a characterization for the degree-bounded version of the problem, taken from Frank and Király (2003) and Király and Makai (2007).

Theorem 3.18. *Let $H_0 = (V, \mathcal{E}_0)$ be a hypergraph, $p : 2^V \to \mathbb{Z}_+$ a monotone decreasing or symmetric non-negative crossing supermodular set function, $g : V \to \mathbb{Z}_+$ a degree bound function and $0 \leq \gamma \leq g(V)/2$ an integer. There exists a hypergraph H with γ hyperedges satisfying the degree bounds such that $H_0 + H$ has an orientation covering p if and only if the following hold for every partition $\mathcal{P}$ of V:*

$$\gamma \geq \sum_{Z \in \mathcal{P}} p(Z) - e_{H_0}(\mathcal{P}), \tag{35}$$

$$\min_{X \in \mathcal{P}} g(V - X) \geq \sum_{Z \in \mathcal{P}} p(Z) - e_{H_0}(\mathcal{P}). \tag{36}$$

In addition, the rank of H can be bounded by $\lceil g(V)/\gamma \rceil$.

Proof. We only show the main ideas behind the proof. Suppose that (35) and (36) are satisfied. We extend the hypergraph H_0 by adding a new node z and for every $v \in V$ adding $g(v)$ parallel edges between v and z. It can be shown using Theorem 2.24 that this hypergraph has an orientation D^* so that $\min\{\varrho_{D^*}(X), \varrho_{D^*}(X + z)\} \geq p(X)$ for every $X \subseteq V$, and $\varrho_{D^*}(V) = \gamma$ (the property that p is monotone decreasing or symmetric implies that the conditions of Theorem 2.24 are satisfied).

For $v \in V$ let $g_i(v)$ be the number of new edges oriented towards v, and let $g_o(v)$ be the number of new edges oriented away from v. Let D_0 denote the orientation of H_0 induced by D^*. The construction implies that $g_i(v) + g_o(v) = g(v)$ for every $v \in V$, $g_i(V) = \gamma$, $g_i(X) \geq p(X) - \varrho_{D_0}(X)$ for every $X \subseteq V$ and $g_o(V - X) \geq p(X) - \varrho_{D_0}(X)$ for every $X \subseteq V$. Hence we can use Theorem 3.15 to obtain a directed hypergraph D of γ hyperarcs such that

- $\varrho_D(X) \geq p(X) - \varrho_{D_0}(X)$ for every $X \subseteq V$,
- $\varrho_D(v) \leq g_i(v)$ for every $v \in V$,
- $\delta_D(v) \leq g_o(v)$ for every $v \in V$.

Let H be the underlying undirected hypergraph of D. Then H has γ hyperedges, it satisfies the degree bound g, and $H_0 + H$ has an orientation that covers p. This means that H satisfies the conditions of the theorem.   □

As in the previous sections, the characterization of the degree bounds that allow a good augmentation helps to deduce a characterization of the minimum number (or minimum total size) of hyperedges needed. Again, this follows from the fact that the feasible degree-specifications are the integer points of a contra-polymatroid (but in this case the proof is more involved). To formulate the theorem, we have to extend the definition of $e_H(\mathcal{F})$ to any family of sets: given a hypergraph $H = (V, \mathcal{E})$ and a family $\mathcal{F}$ of node sets let

$$e_H(\mathcal{F}) = \sum_{e \in \mathcal{E}} \max_{u \in e} |\{X \in \mathcal{F} : u \in X, e \nsubseteq X\}|.$$

Theorem 3.19. *Let $H_0 = (V, \mathcal{E}_0)$ be a hypergraph, $p : 2^V \to \mathbb{Z}_+$ a monotone decreasing or symmetric non-negative crossing supermodular set function, $\sigma \geq 0$ and $0 \leq \gamma \leq \sigma/2$ integers. There exists a hypergraph H with γ hyperedges of total size at most σ such that $H_0 + H$ has an orientation covering p if and only if the following hold:*

$$\gamma \geq \sum_{Z \in \mathcal{P}} p(Z) - e_{H_0}(\mathcal{P}) \quad \text{for every partition } \mathcal{P},$$

$$\sigma \geq \sum_{Z \in \mathcal{F}} p(Z) - e_{H_0}(\mathcal{F})$$

whenever $\mathcal{F} = \mathcal{F}_1 \cup \mathcal{F}_2$ where $\mathcal{F}_1$ is a partition of some $X \subseteq V$ and $\mathcal{F}_2$ is obtained by complementing the members of a partition of X that is coarser than $\mathcal{F}_1$. In addition, the rank of H can be bounded by $\lceil \sigma/\gamma \rceil$.

Application

A hypergraph H is *k-partition-connected* if for every t, one has to delete at least kt hyperedges to dismantle it into $t + 1$ components. Equivalently, $e_H(\mathcal{P}) \geq k(|\mathcal{P}| - 1)$ for every partition $\mathcal{P}$.

For non-negative integers k and l, we say that a hypergraph is *(k, l)-partition-connected* if $e_H(\mathcal{P}) \geq k(|\mathcal{P}| - 1) + l$ for every nontrivial partition $\mathcal{P}$. Clearly, $(k, 0)$-partition-connectivity is equivalent to k-partition-connectivity, and $(0, l)$-partition-connectivity is equivalent to l-edge-connectivity.

As an application of Theorem 3.19 we solve (k, l)-partition-connectivity augmentation problems if $k \geq l$. The following characterization, proved in Frank et al. (2003a), is at the heart of our approach. A directed hypergraph $D = (V, A)$ is called *(k, l)-edge-connected* from root $r \in V$ if $\varrho_D(X) \geq k$ and $\delta_D(X) \geq l$ for every $\emptyset \neq X \subseteq V - r$.

Theorem 3.20. *Let $k \geq l$ be non-negative integers. A hypergraph is (k, l)-partition-connected if and only if it has a (k, l)-edge-connected orientation (from any root).*

Theorems 3.19 and 3.20 imply the following on (k, l)-partition-connectivity augmentation if $k \geq l$:

Corollary 3.21. *Let $H_0 = (V, \mathcal{E}_0)$ be a hypergraph, $\sigma \geq 0$, $0 \leq \gamma \leq \sigma/2$, and $k \geq l$ non-negative integers. There is a hypergraph H with γ hyperedges of total size at most σ such that $H_0 + H$ is (k, l)-partition-connected if and only if the following two conditions are met:*

(i) *$\gamma \geq (|\mathcal{P}| - 1)k + l - e_{H_0}(\mathcal{P})$ for every nontrivial partition $\mathcal{P}$,*
(ii) *$\sigma \geq |\mathcal{F}_1|k + |\mathcal{F}_2|l - e_{H_0}(\mathcal{F}_1 + \mathcal{F}_2)$ whenever $\mathcal{F}_1$ is a partition of some $X \subseteq V$ and $\mathcal{F}_2$ is obtained by complementing the members of a partition of X that is coarser than $\mathcal{F}_1$.*

In addition, the rank of H can be bounded by $\lceil \sigma/\gamma \rceil$.

6.3.6 Covering Symmetric Crossing Supermodular Functions by Uniform Hypergraphs

In the previous section we showed how to solve (k, l)-partition-connectivity augmentation when $k \geq l$. In this section we deal with the case when $k = 0$, i.e. l-edge-connectivity augmentation. Very little is known about (k, l)-partition-connectivity augmentation of hypergraphs when $0 < k < l$ (note that this case is irrelevant for graphs, since there it is equivalent to $(k + l)$-edge-connectivity).

The generalization considered in this section is the covering of symmetric crossing supermodular functions. Let $p : 2^V \to \mathbb{Z} \cup \{-\infty\}$ be a symmetric crossing supermodular set function. Let $r \geq 2$ be an integer, and $g : V \to \mathbb{Z}_+$ a degree bound function such that r divides $g(V)$. First we consider the problem of finding an r-hypergraph of $g(V)/r$ hyperedges satisfying the degree bound that covers the set function p.

We call a partition $\mathcal{P} = \{V_1, \ldots, V_l\}$ *p-full* if $l > r$ and $p(\bigcup_{i \in I} V_i) > 0$ for every $\emptyset \neq I \subsetneq \{1, \ldots, l\}$. Suppose that we have an r-hypergraph H of $g(V)/r$ hyperedges that covers p, and let $\mathcal{P}$ be a p-full partition. Then the hypergraph obtained from H by contracting the members of $\mathcal{P}$ must be connected. Since a connected hypergraph of rank r on a ground set of size $|\mathcal{P}|$ must have at least $(|\mathcal{P}| - 1)/(r - 1)$ hyperedges, we have $(|\mathcal{P}| - 1)/(r - 1) \leq g(V)/r$. This motivates the following definition.

A p-full partition $\mathcal{P}$ is called a *deficient partition* if $(|\mathcal{P}| - 1)/(r - 1) > g(V)/r$.

Theorem 3.22 (Király 2004). *Let $p : 2^V \to \mathbb{Z} \cup \{-\infty\}$ be a symmetric crossing supermodular set function, $r \geq 2$ an integer, and $g : V \to \mathbb{Z}_+$ a degree bound function such that $r \mid g(V)$. There is an r-hypergraph H of $g(V)/r$ hyperedges covering p such that $d_H(v) \leq g(v)$ for every $v \in V$ if and only if the following hold:*

$$g(X) \geq p(X) \quad \text{for every } X \subseteq V,$$

$$\frac{g(V)}{r} \geq p(X) \quad \text{for every } X \subseteq V,$$

there are no deficient partitions.

Since the last two conditions do not depend on the individual values of g, but only on $g(V)$, the characterization of the degree-bounded problem can be used in the usual way to prove a min-max theorem on the corresponding minimum cardinality problem.

Theorem 3.23. *Let $p : 2^V \to \mathbb{Z} \cup \{-\infty\}$ be a symmetric crossing supermodular set function, and $r \geq 2$ an integer. There is an r-hypergraph with γ hyperedges that covers p if and only if the following hold:*

$$r\gamma \geq \sum_{X \in \mathcal{P}} p(X) \quad \text{for every partition } \mathcal{P},$$

$$\gamma \geq p(X) \quad \text{for every } X \subseteq V,$$

$$\gamma \geq \frac{l - 1}{r - 1} \quad \text{if there is a p-full partition with l members.}$$

We also obtain a contra-polymatroid containing feasible degree vectors, which gives rise to a kind of linking property.

Theorem 3.24. *Let γ be the minimum number of r-hyperedges that cover p. Let p^γ be the set function defined by*

$$p^\gamma(X) = \begin{cases} r\gamma & \text{if } X = V, \\ p(X) & \text{otherwise.} \end{cases}$$

Then the integer vectors in the contra-polymatroid $C(p^\gamma)$ are degree vectors of r-hypergraphs that cover p.

Corollary 3.25. *Let p be a symmetric crossing supermodular set function, let $g : V \to \mathbb{Z}_+$ be an upper degree bound. If p can be covered by at most $g(V)/r$ r-hyperedges, and it can be covered by an r-hypergraph whose degrees are bounded by g, then p can be covered by an r-hypergraph that satisfies both conditions simultaneously.*

The k-edge-connectivity augmentation problem of an initial hypergraph by uniform hyperedges is a special case of Theorem 3.23. For a hypergraph $H = (V, \mathcal{E})$ let $c(H)$ denote the number of components of H.

Corollary 3.26. *Let $H_0 = (V_0, \mathcal{E}_0)$ be a hypergraph, and $r \geq 2$ an integer. There is an r-uniform hypergraph H with γ hyperedges such that $H_0 + H$ is k-edge-connected if and only if the following hold:*

$$r\gamma \geq |\mathcal{F}|k - \sum_{X \in \mathcal{F}} d_{H_0}(X) \quad \text{for every subpartition } \mathcal{F},$$

$$\gamma \geq k - d_{H_0}(X) \quad \text{for every } X \subseteq V,$$

$$(r - 1)\gamma \geq c(H_0 - \mathcal{E}_0') - 1 \quad \text{for every } \mathcal{E}_0' \subseteq \mathcal{E}_0 \text{ for which } |\mathcal{E}_0'| = k - 1.$$

6.3.7 Parity-Constrained Orientations Covering Non-negative Intersecting Supermodular Functions

If parity constraints are added to problems studied in this survey, most of them become difficult. For example, we do not know how to decide whether a graph has a strongly connected orientation where all in-degrees are even. There are cases though when the parity-constrained problems are tractable, and the present section describes a fairly general such case.

A set function $p : 2^V \to \mathbb{Z}$ is *monotone decreasing* if $p(X) \geq p(Y)$ whenever $\emptyset \neq X \subseteq Y$. It can be proved that an intersecting supermodular function p with $p(V) = 0$ is monotone decreasing if and only if it is non-negative.

Let $H = (V, \mathcal{E})$ be a hypergraph, $T \subseteq V$ a fixed set, and $p : 2^V \to \mathbb{Z}$ a set function such that $p(V) = 0$. An orientation of H is called (p, T)-*feasible* if it covers p and the in-degree of $v \in V$ is odd if and only if $v \in T$. A set $X \subseteq V$ is called *even* if $|X \cap T| + i_H(X) + p(X)$ is even; X is called *odd* if $|X \cap T| + i_H(X) + p(X)$ is odd. Clearly, $\varrho_D(X) \geq p(X) + 1$ must hold for an odd set X in a (p, T)-feasible orientation of H. This motivates the definition of the following set function:

$$p^T(X) := \begin{cases} p(X) & \text{if } X \text{ is even,} \\ p(X) + 1 & \text{if } X \text{ is odd.} \end{cases} \tag{37}$$

Note that p^T depends on H too. The definition implies that

$$p^T(X) \equiv |X \cap T| + i_H(X) \pmod 2 \tag{38}$$

for every $X \subseteq V$. Given a partition $\mathcal{P}$, the value

$$\mu_T(\mathcal{P}) := \sum_{Z \in \mathcal{P}} p^T(Z) - e_H(\mathcal{P})$$

is called the *deficiency* of $\mathcal{P}$, which depends also on H and p. It can be checked that the deficiency of every partition has the same parity.

It is easy to see that if an orientation D of H is (p, U)-feasible for some $U \subseteq V$, then

$$|T \Delta U| \geq \max\{\mu_T(\mathcal{P}) : \mathcal{P} \text{ is a partition}\}.$$

It turns out that if p is non-negative intersecting supermodular, and there exists an orientation covering p, then equality can be attained, i.e. we can characterize the minimum number of nodes with wrong in-degree parity in an orientation covering p.

Theorem 3.27 (Király and Szabó to appear). *Let $H = (V, \mathcal{E})$ be a hypergraph, $T \subseteq V$ a fixed set, and $p : 2^V \to \mathbb{Z}_+$ an intersecting supermodular and non-negative set function for which $p(V) = 0$. Suppose that H has an orientation covering p. Then there exists a set $U \subseteq V$ such that*

$$|T \Delta U| = \max\{\mu_T(\mathcal{P}) : \mathcal{P} \text{ is a partition}\} \tag{39}$$

and H has a (p, U)-feasible orientation.

Recently Makai et al. (2007) proved the above theorem using a general result on matroid parity and gave a polynomial time combinatorial algorithm.

Applications

One can deduce from Theorem 3.27 the Berge–Tutte formula on the size of a maximum matching of a graph. Let $G = (V, E)$ be an undirected graph. Let $\nu(G)$ denote the maximum size of a matching in G; then the Berge–Tutte formula can be written in the following form.

Theorem 3.28.

$$|V| - 2\nu(G) = \max\{odd_G(W) - |W| : W \subseteq V\},$$

where $odd_G(W)$ denotes the number of components of $G - W$ with an odd number of nodes.

To obtain this from Theorem 3.27, we subdivide each edge $e \in E$ by a new node u_e, resulting in the graph $G' = (V', E')$. For $v \in V$ let $p(\{v\}) = \deg_G(v) - 1$, and let $p(X) = 0$ for all other sets $X \subseteq V'$. Define $T \subseteq V'$ to consist of those nodes $v \in V$ for which $\deg_G(v) - 1$ is odd.

If we have an orientation of G' covering p that has a minimal number of nodes with wrong in-degree parity, then we may assume that $\varrho(u_e) \neq 1$ for every $e \in E$, and so the edges for which $\varrho(u_e) = 2$ form a matching. It can be shown that the size of this matching is exactly the value given by the Berge–Tutte formula.

Another application of Theorem 3.27 comes from the fact that it characterizes which graphs have a rooted k-edge-connected orientation where every in-degree is odd. This characterization was proved earlier in Frank et al. (2001). The special case when $k = 1$, proved by Nebeský (1981), surprisingly leads to the characterization of upper embeddable graphs.

6.3.8 Eulerian Splitting-off

In the previous section we saw that the addition of parity constraints can make the problem significantly more difficult. In this section we show a way in which parity can actually make things easier.

In his early seminal paper on splitting-off in Eulerian graphs (Lovász 1976), Lovász proved three theorems showing how the splitting operation may preserve certain connectivity properties of the graph. Here we look at some other problems where a similar approach may work, based partly on earlier results by Bertsimas and Teo (1997).

In this section we consider only graphs, and we allow parallel edges and loops. Let $p : 2^V \rightarrow \mathbb{Z}_+$ be a non-negative set function on the ground set V and let $m : V \rightarrow \mathbb{Z}_+$ be a function on V. We consider the problem of finding a graph $G = (V, E)$ for which $d_G(v) = m(v)$ for every $v \in V$ and $d_G(X) \geq p(X)$ for every $X \subseteq V$. We assume in the rest of the section that $p(\emptyset) = 0$ and the set function p is symmetric.

An obvious necessary condition for the existence of G is $m(X) \geq p(X)$ for every $X \subseteq V$; this motivates the introduction of the *excess function* m_p:

$$m_p(X) := m(X) - p(X) \quad (X \subseteq V).$$

In addition to the requirement that $m_p(X) \geq 0$ for every $X \subseteq V$, it is also necessary for $m(V)$ (or equivalently $m_p(V)$) to be even. Our strategy is to make an additional, seemingly very strong requirement on m_p: it should also be even for every $X \subseteq V$ for which $p(X) > 0$. In exchange, we can relax the requirements on the set function p.

A set function p is called *semi-skew-supermodular* if for any 3 sets X_1, X_2, X_3 with $p(X_i) > 0$ ($i = 1, 2, 3$) at least one of the following four possibilities holds:

- $p(X_i) + p(X_j) \leq p(X_i \cap X_j) + p(X_i \cup X_j)$ for some $i \neq j$,
- $p(X_i) + p(X_j) \leq p(X_i - X_j) + p(X_j - X_i)$ for some $i \neq j$,
- $p(X_1) + p(X_2) + p(X_3) \leq p(X_1 \cap X_2 \cap X_3) + p(X_1 - (X_2 \cup X_3)) + p(X_2 - (X_1 \cup X_3)) + p(X_3 - (X_1 \cup X_2))$,
- $p(X_1) + p(X_2) + p(X_3) \leq p(X_1 \cup X_2 \cup X_3) + p((X_2 \cap X_3) - X_1) + p((X_1 \cap X_3) - X_2) + p((X_1 \cap X_2) - X_3)$.

Obviously every skew supermodular set function is semi-skew-supermodular. In addition, the following is implied by the definition.

- If p_1 and p_2 are skew supermodular set functions, then $\max\{p_1, p_2, 0\}$ is a semi-skew-supermodular set function.

- If p is semi-skew-supermodular and $G - (V, E)$ is a graph, then $\max\{p - d_G, 0\}$ is semi-skew-supermodular.
- If p is a non-symmetric semi-skew-supermodular set function, then $p'(X) := \max\{p(X), p(V - X)\}$ is also semi-skew-supermodular, hence the assumption of symmetry is not restrictive.

The next theorem (proved in Király 2007b) states that if p is semi-skew-super-modular and m_p is even-valued then the non-negativity of m_p is sufficient for the existence of a degree-specified graph that covers p. A similar result for a slightly more restricted class of set functions appeared in Bertsimas and Teo (1997).

Theorem 3.29. *Let* $p : 2^V \to \mathbb{Z}_+$ *be a symmetric and semi-skew-supermodular set function. Let* $m : V \to \mathbb{Z}_+$ *be a degree specification with the properties that* $m(V)$ *is even and* $m_p(X)$ *is non-negative and even-valued if* $p(X) > 0$. *Then there exists a graph* G *such that* $d_G(v) = m(v)$ *for every* $v \in V$ *and* $d_G(X) \geq p(X)$ *for every* $X \subseteq V$.

We present four applications of Theorem 3.29. We start with the parsimonious property that was the original motivation for the result of Bertsimas and Teo. The second application concerns the covering of graphs by edge-disjoint forests, and it is a slight extension of a result in Frank and Király (2003). The third one offers a simple proof for some known results on edge-disjoint paths. The fourth application is a proof of a theorem of Karzanov and Lomonosov (1978) on multiflows.

The Parsimonious Property

Let $G = (V, E)$ be an undirected graph with a cost function $c : E \to \mathbb{Z}_+$ on the edges, and let $p : 2^V \to \mathbb{Z}_+$ be a symmetric set function. Consider the following linear program:

$$
\begin{aligned}
\min \; & cx \\
& x(e) \geq 0 \quad \text{for every } e \in E, \\
& d_x(Z) \geq p(Z) \quad \text{for every } Z \subseteq V.
\end{aligned}
\tag{40}
$$

We say that a node $v \in V$ has the *parsimonious property* if the linear system (40) has an optimal solution x^* with $d_{x^*}(v) = p(v)$. This property has several structural consequences, and it is useful in the analysis of approximation algorithms (see e.g. Goemans and Bertsimas 1993). The following theorem is a generalization of the result in (Bertsimas and Teo 1997) where a property stronger than subadditivity was required. A node $v \in V$ is called *subadditive* if $p(X) + p(v) \geq p(X + v)$ for every $X \subseteq V - v$.

Theorem 3.30. *If* G *is the complete graph, c satisfies the triangle inequality, and* p *is semi-skew-supermodular, then the linear system (40) has an optimal solution* x^* *with the following property*:

$$
d_{x^*}(v) = p(v) \quad \text{for every subadditive node } v.
$$

Augmentation of k-Forest-Coverable Graphs

The following theorem of Nash-Williams (1964) characterizes graphs that can be covered by k forests.

Theorem 3.31. *The edge-set of a graph $G = (V, E)$ can be covered by k forests if and only if $i_G(X) \leq k(|X| - 1)$ for every non-empty subset X of V.*

Given a graph $G = (V, E)$ that can be covered by k forests and weights on the edges of the complete graph on V, we may want to find an edge set F of maximum weight for which the graph $G' = (V, E + F)$ can still be covered by k forests. This is an easy problem in the sense that it can be solved by finding the maximum weight independent set in a matroid. In fact, the following, more general problem can be solved using the weighted matroid intersection algorithm:

Given two graphs $G_1 = (V, E_1)$ and $G_2 = (V, E_2)$ that can both be covered by k forests and weights on the edges of the complete graph on V, find an edge set F of maximum weight such that the graphs $G'_1 = (V, E_1 + F)$ and $G'_2 = (V, E_2 + F)$ can still be covered by k forests.

What happens if we also want to prescribe the number of new edges incident to each node? The weighted problem cited above becomes NP-complete, even for $E_1 = E_2 = \emptyset$ and $k = 1$. However, the non-weighted degree-prescribed problem can be solved, and there is a simple necessary and sufficient condition for the existence of F. This extends the result in Frank and Király (2003) that dealt with the case $E_1 = E_2$. An interesting point about this application of Theorem 3.29 is that parity is not part of the definition of the problem; an even-valued excess function appears only implicitly.

Theorem 3.32. *Let $G_1 = (V, E_1)$ and $G_2 = (V, E_2)$ be two graphs that can be covered by k forests, and let $m : V \to \mathbb{Z}_+$ be a degree specification with $m(V)$ even. There exists an edge set F for which $d_F(v) = m(v)$ for every $v \in V$ and both $G'_1 = (V, E_1 + F)$ and $G'_2 = (V, E_2 + F)$ can be covered by k forests if and only if*

$$\max\left\{ m(X) - \frac{m(V)}{2}, 0 \right\} \leq k(|X| - 1) - \max\{i_{G_1}(X), i_{G_2}(X)\}$$
for every $\emptyset \neq X \subseteq V$.

Edge-Disjoint Paths: The Eulerian Case

Let $G = (V, E)$ and $H = (V, F)$ be undirected graphs on the same ground set. Our goal is to find a family $\{P_f : f \in F\}$ of edge-disjoint paths in G such that for every $f \in F$ the end-nodes of the path P_f are the end-nodes of f.

We say that *the cut condition holds* if $d_G(X) \geq d_H(X)$ for every $X \subseteq V$. The cut condition is clearly necessary for the existence of the required edge-disjoint paths. It is known that if H is a double star, a K_4 or a C_5, possibly with multiple parallel edges, and $G + H$ is Eulerian, then the cut condition is also sufficient (see e.g. Schrijver 1991).

Theorem 3.33. *Let $G = (V, E)$ and $H = (V, F)$ be undirected graphs such that*

- *H is a double star, a K_4 or a C_5, possibly with multiple parallel edges,*
- *$G + H$ is Eulerian,*
- *$d_G(X) \geq d_H(X)$ for every $X \subseteq V$.*

Then there exists a family $\{P_f : f \in F\}$ of edge-disjoint paths in G such that for every $f \in F$ the end-nodes of the path P_f are the end-nodes of f.

This theorem can be proved quite easily using Theorem 3.29. The main observation is that in the cases mentioned in the theorem the function d_H is semi-skew-supermodular. This essentially means that we can either split off edge-pairs or remove edges that appear both in G and in H until we get an obviously solvable configuration.

Multiflows in Inner Eulerian Graphs

Let $G = (V, E)$ be an undirected graph and $T \subseteq V$ a set of terminal nodes. We say that the pair (G, T) is *inner Eulerian* if $d_G(v)$ is even for every $v \in V - T$. A *T-path* is a path with both end-nodes in T. If $\mathcal{P}$ is a family of T-paths and $Z \subseteq T$, then $d_{\mathcal{P}}(Z)$ denotes the number of paths in $\mathcal{P}$ that have exactly one end-node in Z. Let

$$\lambda_G(Z) := \min\{d_G(X) : X \subseteq V, \ X \cap T = Z\}.$$

If $\mathcal{P}$ is a family of edge-disjoint T-paths, then obviously $d_{\mathcal{P}}(Z) \leq \lambda_G(Z)$ for every $Z \subseteq T$. Generalizing a result of Lovász (1976), Karzanov and Lomonosov (1978) proved that equality can be attained on any given family $\mathcal{L}$ of subsets of T that is *3-cross-free*: it has no three members that are pairwise crossing on the ground set T.

Theorem 3.34 (Karzanov and Lomonosov 1978). *Let (G, T) be inner Eulerian and let $\mathcal{L}$ be a 3-cross-free family of subsets of T. Then there is a family $\mathcal{P}$ of edge-disjoint T-paths for which $d_{\mathcal{P}}(Z) = \lambda_G(Z)$ for every $Z \in \mathcal{L}$.*

This theorem also follows easily from Theorem 3.29, by splitting off edge-pairs from nodes of $V - T$.

6.3.9 Merging Hyperedges

This final section is related to the connectivity augmentation of hypergraphs, but the problem is formulated somewhat differently. The starting point is the same: we have a hypergraph $H = (V, \mathcal{E})$ that does not cover a given set function p, and we want to modify it to cover p. But instead of adding new hyperedges, we are allowed only to merge existing hyperedges.

By *merging* two disjoint hyperedges of H we mean the operation of replacing them in H by their union. "Merging some hyperedges of H" means repeating this operation a few times. Let us define the set function

$$b_H(X) := |\{e \in \mathcal{E} : e \cap X \neq \emptyset\}|.$$

Clearly, if there is a node-set for which $b_H(X) < p(X)$, then it is impossible to obtain a hypergraph covering p by merging hyperedges. It is easy to see that b_H is fully submodular and

$$b_H(X) + b_H(Y) \geq b_H(X - Y) + b_H(Y - X) + |\{e \in \mathcal{E} : \emptyset \neq e \cap Y \subseteq X \cap Y\}|.$$

The following theorem from Király (2007a) states that for symmetric skew supermodular functions the condition $b_H(X) \geq p(X)$ is sufficient. We include the whole proof since it is fairly straightforward.

Theorem 3.35 (Király 2007a). *Let $H = (V, \mathcal{E})$ be a hypergraph, and let $p : 2^V \to \mathbb{Z} \cup \{-\infty\}$ be a symmetric skew supermodular set function for which*

$$b_H(X) \geq p(X) \quad \text{for every } X \subseteq V. \tag{41}$$

Then by merging some hyperedges of H we can obtain a hypergraph $H_* = (V, \mathcal{E}_*)$ such that

$$d_{H_*}(X) \geq p(X) \quad \text{for every } X \subseteq V. \tag{42}$$

Proof. We use induction on the number of hyperedges of H (it is clearly true if $\mathcal{E} = \emptyset$). A set $X \subseteq V$ is called *tight* if $b_H(X) = p(X)$. By the properties of b_H and p, if X and Y are tight, then either $X \cap Y$ and $X \cup Y$ are tight, or $X - Y$ and $Y - X$ are tight. Furthermore, if X and Y are tight and there is a hyperedge e such that $\emptyset \neq e \cap Y \subseteq X \cap Y$, then $X \cap Y$ and $X \cup Y$ are tight.

Let e_0 be an arbitrary hyperedge of H. If there is no tight set X such that $e_0 \subseteq X$, then let $H' := H - e_0$ and

$$p'(X) := \begin{cases} p(X) - 1 & \text{if } e_0 \cap X \neq \emptyset \text{ and } e_0 \cap (V - X) \neq \emptyset, \\ p(X) & \text{otherwise.} \end{cases}$$

The set function p' is symmetric and skew supermodular, and $b_{H'}(X) \geq p'(X)$ for every $X \subseteq V$, so by induction there is a hypergraph H'_*, obtained by merging some hyperedges of H', such that $d_{H'_*}(X) \geq p'(X)$ for every $X \subseteq V$. It follows that $H_* := H'_* + e_0$ covers p. We can thus assume that there is a tight set X_0 such that $e_0 \subseteq X_0$; we also assume that X_0 is a maximal tight set with that property.

Suppose that there is no hyperedge $e \in \mathcal{E}$ such that $e \cap X_0 = \emptyset$. Then $p(V - X_0) = p(X_0) = b_H(X_0) > b_H(V - X_0)$ since $e_0 \subseteq X_0$, contradicting (41). Thus there is a hyperedge $e_1 \in \mathcal{E}$ such that $e_1 \cap X_0 = \emptyset$. Consider the hypergraph $H' := (V, \mathcal{E} - \{e_0, e_1\} + (e_0 \cup e_1))$, i.e. the hypergraph obtained by merging e_0 and e_1. If $b_{H'}(X) < p(X)$ for some $X \subseteq V$, then $e_0 \cap X \neq \emptyset$, $e_1 \cap X \neq \emptyset$, and X was tight. Since $\emptyset \neq e_0 \cap X \subseteq X_0 \cap X$, $X_0 \cup X$ is also tight, which contradicts the maximality of X_0 since $X - X_0 \supseteq e_1 \cap X \neq \emptyset$.

We proved that H' and p satisfy (41), so by induction there is a hypergraph H_* obtained by merging some hyperedges of H' (hence obtained by merging some hyperedges of H) that satisfies (42). □

Acknowledgement

This paper is dedicated to Bernhard Korte on the occasion of his seventieth birthday. Many of the results overviewed in this work were obtained or initiated during the years the first author spent in Bonn at the institute that has been the product of Bernhard's vision and lifelong effort. His way of providing an exceptional research environment was not only enjoyed but it also served as a model to form the Egerváry Research Group (EGRES), a small team of young researchers in Budapest.

References

Bérczi, K., Frank, A.: Variations for Lovász' submodular ideas. In: Grötschel, M., Katona, G.O.H. (eds.) Building Bridges Between Mathematics and Computer Science. Bolyai Society Mathematical Studies, vol. 19, pp. 137–164. Springer, Berlin (2008)

Berg, A., Jackson, B., Jordán, T.: Edge splitting and connectivity augmentation in directed hypergraphs. Discrete Math. **273**, 71–84 (2003)

Bernáth, A., Király, T.: Covering symmetric skew-supermodular functions with hyperedges. EGRES Technical Report No. 2008-05 (2008)

Bertsimas, D., Teo, C.: The parsimonious property of cut covering problems and its applications. Oper. Res. Lett. **21**, 123–132 (1997)

Cai, M.-c.: Arc-disjoint arborescences of digraphs. J. Graph Theory **7**, 235–240 (1983)

Cook, W.J., Cunningham, W.H., Pulleyblank, W.R., Schrijver, A.: Combinatorial Optimization. Wiley, New York (1998)

Edmonds, J.: Submodular functions, matroids, and certain polyhedra. In: Guy, R., Hanani, H., Sauer, N., Schönheim, J. (eds.) Combinatorial Structures and Their Applications, pp. 69–87. Gordon and Breach, New York (1970)

Edmonds, J.: Edge-disjoint branchings. In: Rustin, B. (ed.) Combinatorial Algorithms, pp. 91–96. Academic Press, New York (1973)

Edmonds, J., Giles, R.: A min-max relation for submodular functions on graphs. Ann. Discrete Math. **1**, 185–204 (1977)

Eswaran, K.P., Tarjan, R.E.: Augmentation problems. SIAM J. Comput. **5**(4), 653–665 (1976)

Frank, A.: On disjoint trees and arborescences. In: Algebraic Methods in Graph Theory. Colloquia Mathematica Societatis Janos Bolyai, vol. 25, pp. 159–169. North-Holland, Amsterdam (1978)

Frank, A.: Kernel systems of directed graphs. Acta Sci. Math. **41**, 63–76 (1979)

Frank, A.: On the orientation of graphs. J. Comb. Theory, Ser. B **28**(3), 251–261 (1980)

Frank, A.: Generalized polymatroids. In: Finite and Infinite Sets, Eger, 1981. Colloquia Mathematica Societatis Janos Bolyai, vol. 37, pp. 285–294. North-Holland, Amsterdam (1981)

Frank, A.: An algorithm for submodular functions on graphs. Ann. Discrete Math. **16**, 97–120 (1982)

Frank, A.: Augmenting graphs to meet edge-connectivity requirements. SIAM J. Discrete Math. **5**(1), 22–53 (1992)

Frank, A.: Applications of submodular functions. In: Walker, K. (ed.) Surveys in Combinatorics. London Mathematical Society Lecture Note Series, vol. 187, pp. 85–136. Cambridge University Press, Cambridge (1993)

Frank, A.: Connectivity augmentation problems in network design. In: Birge, J.R., Murty, K.G. (eds.) Mathematical Programming: State of the Art, pp. 34–63. The University of Michigan Press, Ann Arbor (1994)

Frank, A.: Orientations of graphs and submodular flows. Congr. Numer. **113**, 111–142 (1996)

Frank, A.: Restricted t-matchings in bipartite graphs. Discrete Appl. Math. **131**, 337–346 (2003)

Frank, A.: Edge-connection of graphs, digraphs, and hypergraphs. In: More Sets, Graphs and Numbers. Bolyai Society Mathematical Studies, vol. 15, pp. 93–141. Springer, Berlin (2006)

Frank, A.: Rooted k-connections in digraphs. Discrete Appl. Math. (2008), to appear

Frank, A., Jordán, T.: Minimal edge-coverings of pairs of sets. J. Comb. Theory **65**(1), 73–110 (1995)

Frank, A., Király, T.: Combined connectivity augmentation and orientation problems. Discrete Appl. Math. **131**, 401–419 (2003)

Frank, A., Tardos, É.: Generalized polymatroids and submodular flows. Math. Program., Ser. B **42**, 489–563 (1988)

Frank, A., Tardos, É.: An application of submodular flows. Linear Algebra Appl. **114/115**, 329–348 (1989)

Frank, A., Karzanov, A.V., Sebő, A.: On integer multiflow maximization. SIAM J. Discrete Math. **10**(1), 158–170 (1997)

Frank, A., Jordán, T., Szigeti, Z.: An orientation theorem with parity conditions. Discrete Appl. Math. **115**, 37–45 (2001)

Frank, A., Király, T., Király, Z.: On the orientation of graphs and hypergraphs. Discrete Appl. Math. **131**, 385–400 (2003a)

Frank, A., Király, T., Kriesell, M.: On decomposing a hypergraph into k connected subhypergraphs. Discrete Appl. Math. **131**, 373–383 (2003b)

Fujishige, S.: Structures of polyhedra determined by submodular functions on crossing families. Math. Program. **29**, 125–141 (1984)

Fujishige, S.: Submodular Functions and Optimization, 2nd edn., Ann. Discrete Math., vol. 58, Elsevier, Amsterdam (2005)

Goemans, M.X., Bertsimas, D.: Survivable networks, linear programming relaxations and the parsimonious property. Math. Program. **60**, 145–166 (1993)

Győri, E.: A minmax theorem on intervals. J. Comb. Theory, Ser. B **37**, 1–9 (1984)

Kamiyama, N., Katoh, N., Takizawa, A.: Arc-disjoint in-trees in directed graphs. In: Proc. Nineteenth Annual ACM-SIAM Symposium on Discrete Algorithms (SODA 2008), pp. 518–526 (2008)

Karzanov, A.V., Lomonosov, M.V.: Systems of flows in undirected networks. In: Larychev, O.I. (ed.) Mathematical Programming, vol. 1, pp. 59–66. Institute for System Studies, Moscow (1978) (in Russian)

Khanna, S., Naor, J., Shepherd, F.B.: Directed network design with orientation constraints. SIAM J. Discrete Math. **19**(1), 245–257 (2005)

Király, T.: Covering symmetric supermodular functions by uniform hypergraphs. J. Comb. Theory, Ser. B **91**, 185–200 (2004)

Király, T.: Merging hyperedges to meet edge-connectivity requirements. EGRES Technical Report No. 2005-08. Hungarian version: Mat. Lapok **13**(1), 28–31 (2007a)

Király, T.: Applications of Eulerian splitting-off. EGRES Technical Report No. 2007-01. In: Proceedings of Hungarian–Japanese Symposium on Discrete Mathematics and its Applications, pp. 298–307 (2007b)

Király, T., Makai, M.: A note on hypergraph connectivity augmentation. EGRES Technical Report No. 2002-11. Hungarian version: Mat. Lapok **13**(1), pp. 32–39 (2007)

Király, T., Szabó, J.: A note on parity constrained orientations. Combinatorica, to appear

Korte, B., Vygen, J.: Combinatorial Optimization: Theory and Algorithms, 4th edn. Algorithms and Combinatorics, vol. 21, Springer, Berlin (2008)

Lorea, M.: Hypergraphes et matroides. Cahiers Centre Etud. Rech. Oper. **17**, 289–291 (1975)

Lovász, L.: A generalization of Kőnig's theorem. Acta Math. Acad. Sci. Hungar. **21**, 443–446 (1970)

Lovász, L.: On some connectivity properties of Eulerian graphs. Acta Math. Acad. Sci. Hungar. **28**, 129–138 (1976)

Lovász, L.: Combinatorial Problems and Exercises. North-Holland, Amsterdam (1979)

Lucchesi, C.L., Younger, D.H.: A minmax relation for directed graphs. J. Lond. Math. Soc. **17**(2), 369–374 (1978)

Makai, M., Szabó, J.: The parity problem of polymatroids without double circuits. Combinatorica, to appear

Makai, M., Pap, Gy., Szabó, J.: Matching problems in polymatroids without double circuits. In: Fischetti, M., Williamson, D.P. (eds.) Proc. IPCO XII. Lecture Notes in Computer Science, vol. 4513, pp. 167–181. Springer, Berlin (2007)

Murota, K.: Discrete Convex Analysis. Monographs on Discrete Mathematics and Applications, vol. 10, SIAM, Philadelphia (2003)

Nash-Williams, C.St.J.A.: Decomposition of finite graphs into forests. J. Lond. Math. Soc. **39**, 12 (1964)

Nebeský, L.: A new characterization of the maximum genus of a graph. Czechoslov. Math. J. **31**(106), 604–613 (1981)

Rado, R.: A theorem on independence relations. Q.J. Math. **13**(2), 83–89 (1942)

Schrijver, A.: Total dual integrality from directed graphs, crossing families and sub- and supermodular functions. In: Pulleyblank, W.R. (ed.) Progress in Combinatorial Optimization, pp. 315–362. Academic Press, New York (1984)

Schrijver, A.: Supermodular colourings. In: Matroid Theory, Szeged, 1982. Colloquia Mathematica Societatis János Bolyai, vol. 40, pp. 327–343. North-Holland, Amsterdam (1985)

Schrijver, A.: Short proofs on multicommodity flows and cuts. J. Comb. Theory, Ser. B **53**, 32–39 (1991)

Schrijver, A.: Combinatorial Optimization: Polyhedra and Efficiency. Algorithms and Combinatorics, vol. 24, Springer, Berlin (2003)

Szegő, L.: A note on covering intersecting set-systems by digraphs. Discrete Math. **234**(1–3), 187–189 (2001)

Szigeti, Z.: Hypergraph connectivity augmentation. Math. Program., Ser. B **84**(3), 519–527 (1999)

Vidyasankar, K.: Covering the edge-set of a directed graph with trees. Discrete Math. **24**, 79–85 (1978)

7

Theory of Principal Partitions Revisited

Satoru Fujishige

Summary. The theory of principal partitions of discrete systems such as graphs, matrices, matroids, and submodular systems have been developed since 1967. In the early stage of the developments during 1967–75 the principal partition was considered as a decomposition of a discrete system into its components together with a partially ordered structure of the set of the components. It then turned out that such a decomposition with a partial order on it arises from the submodularity structure pertinent to the system and it has been realized that the principal partitions are closely related to resource allocation problems with submodular structures, which are kind of dual problems.

The aim of this paper is to give an overview of the developments in the theory of principal partitions and some recent extensions with special emphasis on its relation to associated resource allocation problems in order to make it better known to researchers in combinatorial optimization.

7.1 Introduction

The concept of principal partition originated from Kishi and Kajitani's pioneering work Kishi and Kajitani (1968), which is concerned with the tri-partition of a graph determined by a maximally distant pair of spanning trees of the graph. Since then the theory of principal partitions has been extended from graphs (Ozawa 1974) to matrices (Iri 1968, 1969), matroids (Bruno and Weinberg 1971; Narayanan 1974; Tomizawa 1976), and submodular systems (Fujishige 1980a, 1980b; Iri 1979, 1984; Nakamura 1988; Nakamura and Iri 1981; Tomizawa and Fujishige 1982).

In the early stage of the developments around 1967–75 the principal partition was considered as a decomposition of a discrete system into its components together with a partially ordered structure of the set of the components. It then turned out that such a decomposition and the associated partial order (poset) come from the submodularity structure pertinent to the system and that the principal partition is closely related to resource allocation problems with submodular constraints.

The decomposition and its associated poset structure arise from minimization of a submodular function underlying the discrete system under consideration. We have a min-max theorem that characterizes the submodular function minimization, and we can relate optimal solutions of the dual maximization problem to a resource allocation problem with submodular constraints.

It should be noted that research developments closely related to principal partitions have independently been made for parametric optimization problems with special emphasis put on monotonicity of optimal solutions, by Topkis et al. (see, e.g., Brumelle et al. 2005; Granot and Veinott 1985; Milgrom and Shannon 1994; Topkis 1978, 1998).

The aim of this paper is to give an overview of the developments in the theory of principal partitions and some recent extensions to make it better known to and fully understood by researchers in combinatorial optimization.

The present paper is organized as follows. Section 7.2 gives basics of submodular functions that lay the foundations of principal partitions. We make a historical overview of principal partitions in Sect. 7.3 and some recent extensions in Sect. 7.4. Section 7.5 describes some applications of principal partitions and related topics.

7.2 Fundamentals of Submodular Functions and Associated Polyhedra

In this section we describe basic properties and facts in the theory of submodular functions, which will play a fundamental rôle in the developments of the theory of principal partitions (see also Fujishige 2005).

7.2.1 Posets, Distributive Lattices, and Submodular Functions

Let E be a finite nonempty set and $\mathcal{D}$ be a collection of subsets of E such that for every $X, Y \in \mathcal{D}$ we have $X \cup Y, X \cap Y \in \mathcal{D}$. Then $\mathcal{D}$ is a *distributive lattice* (or a *ring family*) with set union and intersection as the lattice operations, join and meet.

Let $\preceq$ be a *partial order* on set E, i.e., $\preceq$ is a binary relation on E such that (i) (reflexive) $e \preceq e$ for all $e \in E$, (ii) (antisymmetric) $e \preceq e'$ and $e' \preceq e$ imply $e = e'$ for all $e, e' \in E$, and (iii) (transitive) $e \preceq e'$ and $e' \preceq e''$ imply $e \preceq e''$ for all $e, e', e'' \in E$. The pair $(E, \preceq)$ is called a *partially ordered set* (or a *poset* for short). A subset I of E is called an *order-ideal* (or an *ideal*) of poset $(E, \preceq)$ if $e \preceq e' \in I$ implies $e \in I$ for all $e, e' \in E$.

Theorem 2.1 (Birkhoff). *Let $\mathcal{D}$ be a set of subsets of a finite set E with $\emptyset, E \in \mathcal{D}$. Then $\mathcal{D}$ is a distributive lattice with set union and intersection as the lattice operations if and only if there exists a poset $(\Pi(E), \preceq)$ on a partition $\Pi(E)$ of E such that $\mathcal{D}$ is expressed as follows:*

$$X \in \mathcal{D} \quad \Longleftrightarrow \quad \text{there exists an ideal } \mathcal{J} \text{ of } (\Pi(E), \preceq) \text{ such that } X = \bigcup_{F \in \mathcal{J}} F.$$

We denote the poset $(\Pi(E), \preceq)$ appearing in Theorem 2.1 by $\mathcal{P}(\mathcal{D})$. Conversely, for any poset $\mathcal{P}$ on a partition of E there uniquely exists a distributive lattice $\mathcal{D} \subseteq 2^E$ with set union and intersection as the lattice operations such that $\emptyset, E \in \mathcal{D}$, and $\mathcal{P} = \mathcal{P}(\mathcal{D})$. We denote such a distributive lattice by $\mathcal{D}(\mathcal{P})$.

Remark 2.2. The original Birkhoff theorem (Birkhoff 1967) says that a finite lattice (not necessarily given as a set lattice) is a distributive lattice if and only if it is isomorphic to the set lattice of ideals of a finite poset. It is a crucial observation in principal partitions that a finite distributive lattice given as a set lattice induces a partition of the underlying set and a partial order on it, which conversely gives the distributive lattice as a set of ideals of the poset. This was explicitly mentioned by Iri (1979, 1984).

Let $\mathcal{D} \subseteq 2^E$ be a finite distributive lattice with set union and intersection as the lattice operations. Also suppose that $f : \mathcal{D} \to \mathbb{R}$ satisfies

$$f(X) + f(Y) \geq f(X \cup Y) + f(X \cap Y) \quad (X, Y \in \mathcal{D}). \tag{1}$$

Then, f is called a *submodular function* on $\mathcal{D}$. When f is a submodular function, $-f$ is called a *supermodular function*. A function that is simultaneously submodular and supermodular is called a *modular function*.

Lemma 2.3. *For any submodular function $f : \mathcal{D} \to \mathbb{R}$ define*

$$\mathcal{D}_{\min}(f) = \{X \in \mathcal{D} \mid f(X) = \min\{f(Z) \mid Z \in \mathcal{D}\}\}. \tag{2}$$

Then, the collection $\mathcal{D}_{\min}(f)$ of minimizers of f forms a distributive lattice with set union and intersection as the lattice operations.

Proof. For any $X, Y \in \mathcal{D}_{\min}(f)$ we have

$$f(X) + f(Y) \geq f(X \cup Y) + f(X \cap Y) \geq f(X) + f(Y) \tag{3}$$

where note that $f(X) = f(Y) \leq \min\{f(X \cup Y), f(X \cap Y)\}$. Hence $X \cup Y, X \cap Y \in \mathcal{D}_{\min}(f)$. $\square$

Combining Theorem 2.1 and Lemma 2.3, we observe that the collection of minimizers of submodular function f gives a partition of the underlying set E and a partial order on it. More precisely,

Theorem 2.4. *Let $\mathcal{D}_{\min}(f)$ be the collection of minimizers of a submodular function f as in Lemma 2.3. Let $E_{\min}$ be the minimum element of $\mathcal{D}_{\min}(f)$ and $E_{\max}$ the maximum element of $\mathcal{D}_{\min}(f)$. Then, E is partitioned into*

$$E_{\min}, \qquad F_i \quad (i \in I), \qquad E \setminus E_{\max} \tag{4}$$

on which we have a poset $\mathcal{P}_{\min}(f)$ such that $E_{\min}$ (resp. $E \setminus E_{\max}$) is the minimum (resp. maximum) element of $\mathcal{P}_{\min}(f)$ and the poset structure restricted to the partition of $E_{\max} \setminus E_{\min}$ is the one on the partition of $E_{\max} \setminus E_{\min}$ determined by the distributive lattice $\mathcal{D}_{\min}(f)/E_{\min} \equiv \{X \setminus E_{\min} \mid X \in \mathcal{D}_{\min}(f)\}$. Moreover, we have $X \in \mathcal{D}_{\min}(f)$ if and only if it is expressed as

$$X = E_{\min} \cup \left(\bigcup \{ F_i \mid i \in I, \; F_i \subset X \} \right) \tag{5}$$

where $\{ E_{\min} \} \cup \{ F_i \mid i \in I, \; F_i \subseteq X \}$ is an ideal of the poset $\mathcal{P}_{\min}(f)$.

Remark 2.5. Note that sets expressed by (5) correspond to ideals of a poset. We have a partial order on $\{ F_i \mid i \in I \}$ as in the Birkhoff theorem. Hence, adding to it $E_{\min}$ and $E \setminus E_{\max}$ as the minimum element and the maximum element of the poset, respectively, we get a partial order on the partition of the whole set E.

Remark 2.6. There is a large class of combinatorial optimization problems that have min-max relations expressed by submodular functions. For such problems we often encounter the problem of submodular function minimization that characterizes the minimization side of the min-max relation. Then, this naturally leads us to the decomposition of the discrete system under consideration, due to Theorem 2.4, i.e., we obtain a partition and a partial order on it derived from the submodular function minimization. This is the essence of principal partitions in the early stage of its developments. Related arguments for cut functions of networks were made in Picard and Queyranne (1980).

We call $\mathcal{D}$ a *simple* distributive lattice if the length of a maximal chain of $\mathcal{D}$ is equal to $|E|$, where note that all the maximal chains of $\mathcal{D}$ have the same length. For a simple distributive lattice $\mathcal{D}$ and its corresponding poset $\mathcal{P}(\mathcal{D}) = (\Pi(E), \preceq)$ the partition $\Pi(E)$ consists of singletons only. Hence we regard poset $\mathcal{P}(\mathcal{D})$ as a poset $\mathcal{P}(\mathcal{D}) = (E, \preceq)$ on E.

For a poset $\mathcal{P} = (E, \preceq)$ with $m = |E|$ a sequence or ordering $(e_1, e_2, \ldots, e_m)$ of elements of E is called a *linear extension* of $\mathcal{P} = (E, \preceq)$ if for all $i, j = 1, \ldots, m$, $e_i \prec e_j$ implies $i < j$. Every linear extension $(e_1, e_2, \ldots, e_m)$ of $(E, \preceq)$ determines a maximal chain $S_0 = \emptyset \subset S_1 \subset \cdots \subset S_m = E$ of $\mathcal{D}(\mathcal{P})$ by defining S_i as the set of the first i elements of the linear extension for each $i = 0, 1, \ldots, m$. Conversely, every maximal chain $S_0 = \emptyset \subset S_1 \subset \cdots \subset S_m = E$ of simple $\mathcal{D}$ determines a linear extension $(e_1, e_2, \ldots, e_m)$ of $\mathcal{P}(\mathcal{D}) = (E, \preceq)$ by defining $\{ e_i \} = S_i \setminus S_{i-1}$ for each $i = 1, \ldots, m$.

7.2.2 Submodular Functions and Associated Polyhedra

Let $f : \mathcal{D} \to \mathbb{R}$ be a submodular function on a distributive lattice $\mathcal{D} \subseteq 2^E$. Assume that $\emptyset, E \in \mathcal{D}$ and $f(\emptyset) = 0$. Then we call the pair $(\mathcal{D}, f)$ a *submodular system* on E. If $\mathcal{D}$ is simple, we call $(\mathcal{D}, f)$ a *simple* submodular system. Similarly we define a (simple) *supermodular system*.

We define two polyhedra associated with submodular system $(\mathcal{D}, f)$ as follows.

$$\mathrm{P}(f) = \{ x \mid x \in \mathbb{R}^E, \; \forall X \in \mathcal{D} : x(X) \leq f(X) \}, \tag{6}$$
$$\mathrm{B}(f) = \{ x \mid x \in \mathrm{P}(f), \; x(E) = f(E) \}, \tag{7}$$

where for any $X \subseteq E$ and $x \in \mathbb{R}^E$ we define $x(X) = \sum_{e \in X} x(e)$. We call $\mathrm{P}(f)$ and $\mathrm{B}(f)$, respectively, the *submodular polyhedron* and the *base polyhedron* associated with submodular system $(\mathcal{D}, f)$. Informally, submodular polyhedron $P(f)$ is the set of vectors $x \in \mathbb{R}^E$ 'smaller' than or equal to f. Base polyhedron $B(f)$ is the face of $P(f)$ determined by the hyperplane $x(E) = f(E)$ and is the set of all maximal vectors in submodular polyhedron $P(f)$, which is always nonempty. Define $f^\#(E \setminus X) = f(E) - f(X)$ for all $X \in \mathcal{D}$. We call $f^\#$ the *dual supermodular function* of f, and $(\overline{\mathcal{D}}, f^\#)$ the *dual supermodular system* of $(\mathcal{D}, f)$, where $\overline{\mathcal{D}} = \{E \setminus X \mid X \in \mathcal{D}\}$. Similarly, for any supermodular system $(\mathcal{D}, g)$ on E we define the *dual submodular function* $g^\#$ of g by $g^\#(E \setminus X) = g(E) - g(X)$ for all $X \in \mathcal{D}$.

For a supermodular system $(\mathcal{D}, g)$ we define the *supermodular polyhedron* $P(g)$ and the *base polyhedron* $B(g)$ by $P(g) = \{x \mid x \in \mathbb{R}^E, \forall X \in \mathcal{D} : x(X) \geq g(X)\}$ and $B(g) = \{x \mid x \in P(g), x(E) = g(E)\}$, respectively. Note that $B(g) = B(g^\#)$.

An element of $B(f)$ is called a *base* of submodular system $(\mathcal{D}, f)$. An extreme point of $B(f)$ is called an *extreme base*. An element of submodular polyhedron $P(f)$ is called a *subbase* of $(\mathcal{D}, f)$. The following theorem in the case when $\mathcal{D} = 2^E$ is due to Edmonds (1970) and Shapley (1971).

Theorem 2.7 (Edmonds, Shapley). *For a simple submodular system $(\mathcal{D}, f)$ let $(e_1, \ldots, e_m)$ be a linear extension of poset $\mathcal{P}(\mathcal{D}) = (E, \preceq)$, and let $S_i = \{e_1, \ldots, e_i\}$ for $i = 1, \ldots, m$ and $S_0 = \emptyset$. Define a vector $x \in \mathbb{R}^E$ by*

$$x(e_i) = f(S_i) - f(S_{i-1}) \quad (i = 1, \ldots, m). \tag{8}$$

Then x is an extreme base of submodular system $(\mathcal{D}, f)$.

Conversely, every extreme base of a simple submodular system $(\mathcal{D}, f)$ is generated in this way.

We say that the base x defined by (8) is the *extreme base corresponding to the linear ordering* $(e_1, \ldots, e_m)$. Similarly, for a simple supermodular system $(\mathcal{D}, g)$ on E the extreme base y corresponding to a linear ordering $(e_1, \ldots, e_m)$ (a linear extension of $\mathcal{P}(\mathcal{D})$) is given by $y(e_i) = g(S_i) - g(S_{i-1})$ $(i = 1, \ldots, m)$, where S_i is the set of the first i elements of the linear ordering.

Remark 2.8. Note that $S_i = \{e_1, \ldots, e_i\}$ for $i = 0, 1, \ldots, m$ in Theorem 2.7 is a maximal chain of $\mathcal{D}$. Any (not necessarily maximal) chain

$$\mathcal{C} : C_0 = \emptyset \subset C_1 \subset \cdots \subset C_k = E \tag{9}$$

of $\mathcal{D}$ determines a face $\mathbf{F}(\mathcal{C})$ of $B(f)$ by

$$\mathbf{F}(\mathcal{C}) = \{x \mid x \in B(f), \forall i = 1, \ldots, k : x(C_i) = f(C_i)\}, \tag{10}$$

which is nonempty. Every maximal chain containing $\mathcal{C}$ determines an extreme point of the face $\mathbf{F}(\mathcal{C})$. It should also be noted that the face $\mathbf{F}(\mathcal{C})$ is again a base polyhedron, which is a direct sum of bases of minors of submodular system $(\mathcal{D}, f)$ defined in the sequel.

Let $G(\mathcal{P}) = (E, A)$ be the graph with vertex set E and arc set A representing the Hasse diagram of $\mathcal{P} = (E, \preceq)$, where $(e, e') \in A$ if and only if $e' \prec e$ and there exists no element e'' such that $e' \prec e'' \prec e$ in $\mathcal{P}$. Any function $\varphi : A \to \mathbb{R}$ is a *flow* in $G(\mathcal{P})$. The *boundary* $\partial\varphi : E \to \mathbb{R}$ of flow φ is defined by

$$\partial\varphi(e) = \sum_{(e,e')\in A} \varphi(e, e') - \sum_{(e'',e)\in A} \varphi(e'', e) \quad (e \in E). \tag{11}$$

Theorem 2.9. *The characteristic cone* $\mathrm{Cone}(\mathrm{B}(f))$ *of base polyhedron* $\mathrm{B}(f)$ *associated with a simple submodular system* $(\mathcal{D}, f)$ *on* E *is given by*

$$\mathrm{Cone}(\mathrm{B}(f)) = \{\partial\varphi \mid \varphi : \text{a nonnegative flow in } G(\mathcal{P}(\mathcal{D}))\}. \tag{12}$$

Consider a submodular system $(\mathcal{D}, f)$ on E. For any $X \in \mathcal{D}$ the *reduction* or *restriction* of submodular system $(\mathcal{D}, f)$ by X is a submodular system $(\mathcal{D}^X, f^X)$ on X defined by

$$\mathcal{D}^X = \{Z \mid Z \in \mathcal{D}, \ Z \subseteq X\}, \tag{13}$$
$$f^X(Z) = f(Z) \quad (Z \in \mathcal{D}^X). \tag{14}$$

Also the *contraction* of $(\mathcal{D}, f)$ by $X \in \mathcal{D}$ is a submodular system $(\mathcal{D}_X, f_X)$ on $E \setminus X$ defined by

$$\mathcal{D}_X = \{Z \setminus X \mid Z \in \mathcal{D}, \ Z \supseteq X\}, \tag{15}$$
$$f_X(Z) = f(Z \cup X) - f(X) \quad (Z \in \mathcal{D}_X). \tag{16}$$

Note that $\mathcal{D}_\emptyset = \mathcal{D}$ and $\mathcal{D}^E = \mathcal{D}$.

For any $X, Y \in \mathcal{D}$ such that $X \subset Y$ define

$$\mathcal{D}_X^Y = (\mathcal{D}^Y)_X, \tag{17}$$
$$f_X^Y = (f^Y)_X. \tag{18}$$

Here note that $(\mathcal{D}^Y)_X = (\mathcal{D}_X)^{Y\setminus X}$ and $(f^Y)_X = (f_X)^{Y\setminus X}$. We call the submodular system $(\mathcal{D}_X^Y, f_X^Y)$ on $Y \setminus X$ a *minor* of $(\mathcal{D}, f)$.

Theorem 2.10. *Let* $\mathcal{C}$ *be a chain of* $\mathcal{D}$ *given by* (9) *and* $\mathbf{F}(\mathcal{C})$ *be the face of the base polyhedron* $\mathrm{B}(f)$ *determined by* (10). *Then,* $\mathbf{F}(\mathcal{C})$ *is expressed as*

$$\mathbf{F}(\mathcal{C}) = \bigoplus_{i=1}^{k} \mathrm{B}\big(f_{C_{i-1}}^{C_i}\big), \tag{19}$$

which is the direct sum of the base polyhedra associated with minors $(\mathcal{D}_{C_{i-1}}^{C_i}, f_{C_{i-1}}^{C_i})$ $(i = 1, \ldots, k)$ *of* $(\mathcal{D}, f)$.

We need some other definitions. Consider a submodular system $(\mathcal{D}, f)$ on E. For any $x \in \mathrm{P}(f)$ we call $X \in \mathcal{D}$ a *tight set* for x if $x(X) = f(X)$, and let $\mathcal{D}_f(x)$ denote the collection of all tight sets for x. Note that $\mathcal{D}_f(x) = \mathcal{D}_{\min}(f - x)$, so that it is closed with respect to set union and intersection.

For any $x \in P(f)$ define

$$\text{sat}(x) = \bigcup \{X \mid X \in \mathcal{D}_f(x)\}. \tag{20}$$

The saturated set, $\text{sat}(x)$, is the unique maximal element of the distributive lattice $\mathcal{D}_f(x)$ and can be expressed as

$$\text{sat}(x) = \{e \in E \mid \forall \alpha > 0 : x + \alpha \chi_e \notin P(f)\}. \tag{21}$$

Here χ_e is the unit vector in $\mathbb{R}^E$ with $\chi_e(e') = 1$ if $e' = e$ and $\chi_e(e') = 0$ if $e' \in E \setminus \{e\}$. We call sat : $P(f) \to 2^E$ the *saturation function*. Note that $\text{sat}(x)$ is the empty set if and only if x lies in the interior of $P(f)$.

Also define for any $x \in P(f)$ and any element $e \in \text{sat}(x)$

$$\text{dep}(x, e) = \bigcap \{X \mid e \in X \in \mathcal{D}_f(x)\}, \tag{22}$$

which can be rewritten as

$$\text{dep}(x, e) = \{e' \in E \mid \exists \alpha > 0 : x + \alpha(\chi_e - \chi_{e'}) \in P(f)\}, \tag{23}$$

and we also define $\text{dep}(x, e) = \emptyset$ if $e \notin \text{sat}(x)$. We call dep : $P(f) \times E \to 2^E$ the *dependence function*. Note that when $e \in \text{sat}(x)$, $\text{dep}(x, e)$ is the unique minimal element of the distributive lattice $\{X \mid e \in X \in \mathcal{D}_f(x)\}$.

7.3 An Overview of Principal Partitions

We make an overview of the developments in the theory of principal partitions.

7.3.1 Kishi and Kajitani's Tri-partition for Graphs

Suppose that we are given a connected graph $G = (V, E)$ with a vertex set V and an edge set E. We identify a spanning tree with its edge set. Let $\mathcal{T} \subseteq 2^E$ be the set of all the spanning trees of G. For any two spanning trees T_1 and T_2 in $\mathcal{T}$ we denote by $\text{dist}(T_1, T_2)$ the *distance* $|T_1 \setminus T_2|$ of T_1 and T_2. A pair of spanning trees T_1 and T_2 is called a *maximally distant pair of spanning trees* if it attains the maximum of the distance.

Kishi and Kajitani's principal partition (Kishi and Kajitani 1968) of graph $G = (V, E)$ is the ordered tri-partition of the edge set into (E^-, E^0, E^+) such that the following three hold.

($-$) For any $e \in E^-$ there exists a maximally distant pair of spanning trees T_1 and T_2 such that $e \notin T_1 \cup T_2$.

(0) For any maximally distant pair of spanning trees T_1 and T_2 we have a bi-partition of E^0 into $E^0 \cap T_1$ and $E^0 \cap T_2$, i.e., for any $e \in E^0$ we have either $e \in T_1$ or $e \in T_2$.

(+) For any $e \in E^+$ there exists a maximally distant pair of spanning trees T_1 and T_2 such that $e \in T_1 \cap T_2$.

Graph $G = (V, E)$ is decomposed into $G \cdot E^-$, $G \cdot (E^0 \cup E^-)/E^-$, and $G/(E^0 \cup E^-)$, where for any edge set $F \subseteq E$, $G \cdot F$ is the restriction of G on F and G/F is the graph obtained by contraction of all the edges in F.

It can be shown that Kishi and Kajitani's tri-partition is characterized by the following theorem, which is a matroidal min-max theorem known earlier in graph theory (Nash-Williams 1961; Tutte 1961) and in matroid theory (Edmonds 1965a, 1965b; Edmonds and Fulkerson 1965).

Theorem 3.1. *For a connected graph $G = (V, E)$ with rank function $r_G : 2^E \to \mathbb{Z}_+$,*

$$\max\{|T_1 \cup T_2| \mid T_1, T_2 : \text{spanning trees of } G\}$$
$$= \min\{2r_G(X) + |E \setminus X| \mid X \subseteq E\}.$$

Theorem 3.2. *The set $\mathcal{D}_G$ of all the minimizers of the submodular function $f(X) = 2r_G(X) + |E \setminus X|$ in $X \in 2^E$ is closed with respect to set union and intersection and forms a distributive lattice. The unique minimal element of $\mathcal{D}_G$ is given by E^- and the unique maximal element of $\mathcal{D}_G$ by $E^- \cup E^0$ $(= E \setminus E^+)$, where (E^-, E^0, E^+) is the Kishi–Kajitani tri-partition of E for $G = (V, E)$.*

Ozawa (1974) generalized Kishi and Kajitani's principal partition of a graph to a pair of graphs, which is a special case of the principal partition of a pair of (poly-) matroids to be discussed in Sect. 7.3.5.

Remark 3.3. For an electrical network the *topological degree of freedom* is the minimum number of current and voltage variables whose values uniquely determine all current and voltage values of arcs through Kirchhoff's current and voltage laws. It was noticed that Kishi and Kajitani's principal tri-partition could be used to resolve the problem of determining the topological degree of freedom (see Iri 1968, 1983; Kishi and Kajitani 1968; Ohtsuki et al. 1968; Tsuchiya et al. 1967 and also Narayanan 1997, Chap. 14).

It should also be noted that Kishi and Kajitani's principal tri-partition gives a solution of *Shannon's switching game* (see Bruno and Weinberg 1971; Edmonds 1965b).

7.3.2 Iri's Maximum-Rank Minimum-Term-Rank Theorem for Pivotal Transforms of a Matrix

Iri (1968, 1969) considered a generalization of Kishi and Kajitani's framework for graphs to that for matrices and related the matroidal min-max theorem to what is called the maximum-rank minimum-term-rank theorem for pivotal transforms of a matrix. Moreover, he derived a finer poset structure on E^0 part, based on the Dulmage–Mendelsohn decomposition of bipartite graphs.

Suppose that we are given an $m \times n$ real matrix $M = [I_m | A]$, where I_m is the identity matrix of order m and A an $m \times (n - m)$ matrix. Let E be the index set of the columns of M. Then consider the matroid $\mathbf{M}$ on E represented by the matrix M defined by the linear independence among the column vectors of M.

For any base B of matroid $\mathbf{M}$ we can transform the original matrix $M = [I_m | A]$ so that the submatrix corresponding to the columns B becomes the identity matrix I_m by fundamental row operations. After an appropriate column permutation we obtain a new matrix $M(B) = [I_m | A(B)]$. We call $A(B)$ a *pivotal transform* of A. Define

$$\mathcal{A}(M) = \{A(B) \mid B : \text{a base of } \mathbf{M}\}. \tag{24}$$

For any matrix $C \in \mathcal{A}(M)$ consider the bipartite graph $G(C)$ corresponding to the nonzero elements of matrix C. The size of a maximum matching in the bipartite graph $G(C)$ is the *term rank* of C, which we denote by t-rank C.

Now we have

Theorem 3.4 (Iri).

$$\max\{\text{rank } C \mid C \in \mathcal{A}(M)\} = \min\{\text{t-rank } C \mid C \in \mathcal{A}(M)\}, \tag{25}$$

where the maximum and the minimum can be attained simultaneously by a matrix $C \in \mathcal{A}(M)$.

This theorem can be considered as a matrix variant, in terms of term rank, of the following matroidal min-max theorem about the union of matroids (Edmonds and Fulkerson 1965; Rado 1942). We denote by $r_{\mathbf{M}}$ the rank function of matroid $\mathbf{M}$.

Theorem 3.5. *For any matroid* $\mathbf{M}$ *with rank function* $r_{\mathbf{M}} : 2^E \to \mathbb{Z}_+$,

$$\max\{|B_1 \cup B_2| \mid B_1, B_2 : \text{bases of } \mathbf{M}\}$$
$$= \min\{2r_{\mathbf{M}}(X) + |E \setminus X| \mid X \subseteq E\}, \tag{26}$$

or equivalently,

$$\max\{|B_1 \setminus B_2| \mid B_1, B_2 : \text{bases of } \mathbf{M}\}$$
$$= \min\{r_{\mathbf{M}}(X) + r_{\mathbf{M}}^*(E \setminus X) \mid X \subseteq E\}, \tag{27}$$

where $r_{\mathbf{M}}^*$ *is the corank function of matroid* $\mathbf{M}$.

The left-hand side of (27) is equal to that of (25) when matroid $\mathbf{M}$ is represented by matrix M, but it is nontrivial to directly show the equality of the right-hand sides of (27) and (25). It is mentioned in Iri (1969) that D.R. Fulkerson noticed the matroidal structure of the result of Iri, which can be derived from Edmonds and Fulkerson (1965).

7.3.3 The Principal Partition of Matroids by Bruno and Weinberg, Tomizawa, and Narayanan

Bruno and Weinberg (1971) also noticed the matroidal structure of the result of Kishi and Kajitani. With any positive integer $k \geq 2$ as a parameter they considered the union of k copies of a given matroid. This leads us to the following min-max relation with integer parameter $k \geq 2$, known for unions of matroids (see Edmonds and Fulkerson 1965; Rado 1942).

Theorem 3.6. *For a matroid* **M** *on* E *with the base family* $\mathcal{B}$ *and the rank function* ρ,

$$\max\left\{\left|\bigcup_{i=1}^{k} B_i\right| \,\middle|\, B_i \in \mathcal{B}\right\} = \min\{k\rho(X) + |E \setminus X| \mid X \subseteq E\}. \tag{28}$$

For each positive integer k we have the distributive lattice $\mathcal{D}_k$ of the minimizers of a submodular function $f_k(X) = k\rho(X) + |E \setminus X|$ appearing in the right-hand side of (28). Denote by E_k^- and E_k^+ the minimum and the maximum element of $\mathcal{D}_k$, respectively. It follows from Theorem 2.4 that we have a partition of the underlying set E as in (4) and a poset structure on the partition of $E_k^+ \setminus E_k^-$ for each integer $k \geq 2$. Suppose that the collection of distinct $\mathcal{D}_k$ is given by $\mathcal{D}_{k_i}$ ($i = 1, \ldots, l$) with $k_1 < \cdots < k_l$.

Then we have the following theorem, which will be shown for more general setting later.

Theorem 3.7.

$$E_{k_1}^+ \supseteq E_{k_1}^- \supseteq E_{k_2}^+ \supseteq E_{k_2}^- \supseteq \cdots \supseteq E_{k_l}^+ \supseteq E_{k_l}^-. \tag{29}$$

Remark 3.8. For each $i = 1, \ldots, l$ we have a partition of the difference set $E_{k_i}^+ \setminus E_{k_i}^-$ and a poset on it determined by the distributive lattice $\mathcal{D}_{k_i}$.

Remark 3.9. If the difference set $E_{k_i}^+ \setminus E_{k_i}^-$ is nonempty, the minor of matroid **M** on $E_{k_i}^+ \setminus E_{k_i}^-$ with rank function $\rho_{E_{k_i}^-}^{E_{k_i}^+}$ has disjoint k_i bases that partition $E_{k_i}^+ \setminus E_{k_i}^-$.

Tomizawa (1976) and Narayanan (1974) independently generalized the decomposition scheme of Bruno and Weinberg by considering rational numbers instead of integers k. For a positive rational $\frac{l}{k}$ for positive integers l and k they find a minor that has k bases that uniformly cover each element of the underlying set l times. The Bruno–Weinberg decomposition corresponds to the case when $l = 1$.

The min-max theorem associated with the Tomizawa–Narayanan decomposition is given parametrically as follows. This will also be proved in a more general setting later.

Theorem 3.10. *For any positive integers k and l,*

$$\max\left\{\sum_{i=1}^{k}|I_i| \mid I_i \in \mathcal{I} \ (i = 1, \ldots, k), \ \forall e \in E : |\{i \mid i \in \{1, \ldots, k\}, \ e \in I_i\}| \leq l\right\}$$
$$= \min\{k\rho(X) + l|E \setminus X| \mid X \subseteq E\}, \tag{30}$$

where $\mathcal{I}$ is the family of the independent sets of matroid $\mathbf{M}$.

Note that when $l = 1$, Theorem 3.10 is reduced to Theorem 3.6.

For a nonnegative rational number $\lambda = \frac{l}{k}$ let $\mathcal{D}_\lambda$ be the distributive lattice formed by the minimizers of the submodular function $f_\lambda(X) = k\rho(X) + l|E \setminus X|$.

We call the value λ *critical* if $\mathcal{D}_\lambda$ contains more than one element. Because of the finiteness character we have a finite set of critical values, which are supposed to be given by $0 \leq \lambda_1 < \cdots < \lambda_p$. For each $i = 1, \ldots, p$ let $E_{\lambda_i}^-$ and $E_{\lambda_i}^+$ be the minimum and the maximum element of $\mathcal{D}_{\lambda_i}$, respectively.

Theorem 3.11.

$$E_{\lambda_1}^- \subset E_{\lambda_1}^+ = E_{\lambda_2}^- \subset E_{\lambda_2}^+ = E_{\lambda_3}^- \subset \cdots \subset E_{\lambda_{p-1}}^+ = E_{\lambda_p}^- \subset E_{\lambda_p}^+. \tag{31}$$

For each nonempty difference set $E_i^+ \setminus E_i^-$ we have a partition of it with a partial order associated with the distributive lattice $\mathcal{D}_{\lambda_i}$. Also note that the union of $\mathcal{D}_{\lambda_i}$ $(i = 1, \ldots, p)$ as a whole is again a distributive lattice, which determines the decomposition of matroid $\mathbf{M}$ and a poset structure on it. Each minor $\mathbf{M}_{E_i^-}^{E_i^+}$ of $\mathbf{M}$ on $E_i^+ \setminus E_i^-$, the restriction of $\mathbf{M}$ to E_i^+ followed by the contraction by E_i^-, with critical value $\lambda_i = l/k$ has k bases of the minor that cover uniformly l times every element of $E_i^+ \setminus E_i^-$. The decomposition given above is the finest one that has such a property. This is the principal partition of matroid $\mathbf{M}$ in the sense of Tomizawa and Narayanan (see also Narayanan and Vartak 1981).

7.3.4 A Polymatroidal Approach to the Principal Partition of Tomizawa and Narayanan: A Lexicographically Optimal Base

The author (Fujishige 1980a, 1980b) noticed that Tomizawa and Narayanan's principal partition was polymatroidal. Readers will see that a polymatroidal approach to the principal partition is quite natural and easy to understand. Also this can easily be extended to general submodular systems.

Let $\mathbf{P} = (E, \rho)$ be a polymatroid with a rank function $\rho : 2^E \to \mathbb{R}_+$ and let $w : E \to \mathbb{R}$ be a positive weight vector on E. Then we have the following min-max relation for polymatroids (Edmonds 1970).

Theorem 3.12 (Edmonds). *For any real parameter λ,*

$$\max\{x(E) \mid x \in \mathrm{P}(\rho), \ x \leq \lambda w\} = \min\{\rho(X) + \lambda w(E \setminus X) \mid X \subseteq E\}, \tag{32}$$

where $\mathrm{P}(\rho)$ is the submodular polyhedron associated with the rank function ρ.

It should be noted that when $\lambda \geq 0$, $P(\rho)$ in (32) can be replaced by the polymatroid polyhedron $P(\rho) \cap \mathbb{R}_+^E$ and that when $\lambda < 0$, the right-hand side of (32) has the unique minimizer $X = \emptyset$. Relation (32) in the form given above can more naturally be extended to submodular systems.

Lemma 3.13. *For any reals λ_1 and λ_2 with $\lambda_1 < \lambda_2$ there exist a maximizer $x = b_1$ of the left-hand side of (32) for $\lambda = \lambda_1$ and a maximizer $x = b_2$ for $\lambda = \lambda_2$ such that $b_1 \leq b_2$.*

Proof. Let b_1 be any maximizer for $\lambda = \lambda_1$. Since $\{x \mid x \in P(\rho),\ x \leq \lambda_2 w\}$ is a submodular polyhedron (the vector reduction of $P(\rho)$ by $\lambda_2 w$) and b_1 belongs to it, there exists a base b_2 of the reduction such that $b_1 \leq b_2$. Here, b_2 is a maximizer of the left-hand side of (32) for $\lambda = \lambda_2$. $\quad\square$

Because of this fact the following was observed in Fujishige (1980a).

Theorem 3.14. *For any given positive weight vector w there uniquely exists a base b^* of polymatroid (E, ρ) such that $b^* \wedge \lambda w$ is a maximizer of the left-hand side of (32) for each λ, where $b^* \wedge \lambda w = (\min\{b^*(e), \lambda w(e)\} \mid e \in E)$.*

The base b^* appearing in Theorem 3.14 is called the *universal base* for polymatroid (E, ρ) with weight vector w.

Remark 3.15. The universal base b^* can be defined geometrically as follows. We start with $b = \lambda w$ for a sufficiently small λ such that b lies in the interior of $P(\rho)$ (we can take any negative λ in the present case of polymatroid rank function ρ). Then increase λ until we reach the boundary of $P(\rho)$. Let $b_1 = \lambda_1 w$ be the boundary point of $P(\rho)$. Put S_1 as the maximum minimizer of the submodular function $\rho(X) - b_1(X)$ (note that $S_1 = \mathrm{sat}(b_1)$). Now fix the components $b(e)$ as $b_1(e)$ for $e \in S_1$, and if $S_1 \neq E$, increase the other components $b(e)$ ($e \in E \setminus S_1$) in proportion to $w(e)$ until we cannot increase them without leaving $P(\rho)$. Let b_2 be the new boundary point of $P(\rho)$, find the maximum minimizer $S_2(= \mathrm{sat}(b_2))$ of the submodular function $\rho(X) - b_2(X)$, and fix the components $b(e)$ as $b_2(e)$ for $e \in S_2$, where note that we have $S_1 \subset S_2$. Repeat this process until all the components of b are fixed, which determines a piecewise-linear path to the universal base b^*.

Let $\mathcal{D}(\rho, \lambda w)$ denote the set of all minimizers of the submodular function $\rho(X) - \lambda w(E \setminus X)$. In the same way as in the principal partition of Tomizawa and Narayanan we call the value λ *critical* if $\mathcal{D}(\rho, \lambda w)$ contains more than one element. We have a finite set of critical values $0 \leq \lambda_1 < \cdots < \lambda_p$ with $p \leq |E|$. For each $i = 1, \ldots, p$ let E_i^- and E_i^+ be the minimum and the maximum element of $\mathcal{D}(\rho, \lambda_i w)$, respectively.

Theorem 3.16.

$$E_{\lambda_1}^-(= \emptyset) \subset E_{\lambda_1}^+ = E_{\lambda_2}^- \subset \cdots \subset E_{\lambda_{p-1}}^+ = E_{\lambda_p}^- \subset E_{\lambda_p}^+ (= E). \tag{33}$$

Proof. For the universal base b^* let the distinct values of $b^*(e)/w(e)$ $(e \in E)$ be given by $\beta_1^* < \cdots < \beta_{q^*}^*$ and define

$$S_i^* = \{e \mid e \in E, \ b^*(e)/w(e) \leq \beta_i^*\} \quad (i = 1, \ldots, q^*). \tag{34}$$

Then we can show that $q^* = p$, $S_i^* = E_{\lambda_i}^+$ $(i = 1, \ldots, p)$, and $S_i^* = E_{\lambda_{i+1}}^-$ $(i = 0, \ldots, p - 1)$ where $S_0^* = \emptyset$ and $S_p^* = E$. $\square$

For any base $b \in B(\rho)$ let the distinct values of $b(e)/w(e)$ $(e \in E)$ be given by

$$\beta_1 < \cdots < \beta_q, \tag{35}$$

and define

$$S_i = \{e \mid e \in E, \ b(e)/w(e) \leq \beta_i\} \tag{36}$$

for each $i = 1, \ldots, q$.
Then we have

Theorem 3.17. *A base $b \in B(\rho)$ is the universal base of (E, ρ) for weight vector w if and only if the sets S_i $(i = 1, \ldots, q)$ defined by (35) and (36) are tight sets of b, i.e.,*

$$\rho(S_i) = b(S_i) \quad (i = 1, \ldots, q). \tag{37}$$

Note that $\lambda_i w(E_i^+ \setminus E_i^-) = \rho(E_i^+) - \rho(E_i^-)$ $(i = 1, \ldots, p)$. Hence the critical values for the principal partition of Tomizawa and Narayanan are rational, where $w(X) = |X|$ $(X \subseteq E)$ and ρ is a matroid rank function.

The universal base b^* can be characterized as a lexicographically optimal base of polymatroid (E, ρ) with weight vector w and as a base that minimizes a separable convex function. Both were discussed in Fujishige (1980a).

Given a positive weight vector $w \in \mathbb{R}^E$, for any vector $x \in \mathbb{R}^E$ define a sequence

$$T_w(x) = (x(e_1)/w(e_1), \ldots, x(e_m)/w(e_m)) \tag{38}$$

of ratios $x(e)/w(e)$ $(e \in E)$ such that

$$x(e_1)/w(e_1) \leq \cdots \leq x(e_m)/w(e_m), \tag{39}$$

where $E = \{e_1, \ldots, e_m\}$. A base $b \in B(\rho)$ is called a *lexicographically optimal base* with respect to the weight vector w if it lexicographically maximizes $T_w(x)$ among all the bases $x \in B(\rho)$. We can easily see that a lexicographically optimal base with respect to the weight vector w uniquely exists.

Theorem 3.18. *The lexicographically optimal base with respect to the weight vector w coincides with the universal base b^* for the same w.*

Proof. We can show that a base $\hat{b} \in B(\rho)$ is the lexicographically optimal base with respect to the weight vector w if and only if for all $e, e' \in E$ such that $\hat{b}(e)/w(e) < \hat{b}(e')/w(e')$ we have $e' \notin \text{dep}(\hat{b}, e)$. (Recall (22) and (23).) The latter condition is equivalent to (37) with (35) and (36). $\square$

We also have

Theorem 3.19. *Let* $x = \hat{b}$ *be an optimal solution of the following problem.*

$$Minimize \sum_{e \in E} \frac{x^2(e)}{w(e)} \quad subject\ to\ x \in B(\rho). \tag{40}$$

Then $\hat{b}$ *is the universal base* b^* *for* w.

Proof. We can also show that a base $\hat{b}$ is an optimal solution of (40) if and only if for all $e, e' \in E$ such that $\hat{b}(e)/w(e) < \hat{b}(e')/w(e')$ we have $e' \notin \text{dep}(\hat{b}, e)$. □

Fujishige (1980a) gave an $O(|E|\text{SFM})$ algorithm for finding a lexicographically optimal base with respect to weight w, where SFM denotes the complexity of submodular function minimization (see Iwata 2008; McCormick 2005 for submodular function minimization). When specialized to multi-terminal flows, this improved Megiddo's algorithms for lexicographically optimal multi-terminal flows (Megiddo 1974, 1977). Also, Gallo et al. (1989) devised a faster algorithm for finding a lexicographically optimal multi-terminal flow with weights, which requires running time of a single max-flow computation. More general separable convex function minimization problems over polymatroids and their incremental algorithms were considered by Federguen and Groenevelt (1986) and Groenevelt (1991) (see also Hochbaum and Hong 1995). An O(SFM) algorithm for finding a lexicographically optimal base with weights has been obtained by Fleischer and Iwata (2003) (see also related recent algorithms by Nagano 2007a, 2007b).

The results in this subsection do not depend on the monotonicity of the rank function ρ, so that we can easily extend the results to those for general submodular systems with positive weight vectors. (Just replace the polymatroid rank function ρ with the rank function f of any submodular system. For details see Fujishige 2005, Sects. 7 and 9.)

Getting rid of the monotonicity assumption on the rank function is very important and extends the applicability of the theory of principal partitions.

Remark 3.20. The concept of a lexicographically optimal base of a polymatroid was rediscovered in convex games by Dutta (1990), Dutta and Ray (1989) (see also Hokari 2002; Hokari and van Gellekom 2002), where the lexicographically optimal base is called the *egalitarian solution* of a convex game. Note that the core of a convex game is the same as the base polyhedron of a polymatroid (Shapley 1971).

Remark 3.21. Consider a submodular system $(\mathcal{D}, f)$ on E and a positive weight vector w. If we are given the universal base b^* (or the lexicographically optimal base) with respect to weight w, $b^* \wedge \mathbf{0} = (\min\{b^*(e), 0\} \mid e \in E)$ is a maximizer of

$$\max\{x(E) \mid x \in P(f),\ x \le \mathbf{0}\} = \min\{f(X) \mid X \in \mathcal{D}\} \tag{41}$$

due to a generalized version of Theorem 3.12. Moreover, the sets $A^- = \{e \mid e \in E,\ b^*(e) < 0\}$ and $A^0 = \{e \mid e \in E,\ b^*(e) \le 0\}$ are, respectively, the unique minimal minimizer and the unique maximal minimizer of f, which minimize the right-hand side of (41). Hence we can minimize a given submodular function by solving the minimum-norm-point problem (40). Here we may choose a uniform weight vector w such that $w(X) = |X|$ for all $X \subseteq E$ to get the Euclidean norm. Polynomial algorithms for submodular function minimization have been developed so far (Grötschel et al. 1988; Iwata 2002, 2003; Iwata et al. 2001; Orlin 2007; Schrijver 2000) (see also Iwata 2008; McCormick 2005), but it seems to be worth investigating to apply the minimum-norm-point algorithm of Wolfe (1976) to submodular function minimization (see Fujishige et al. 2006).

7.3.5 The Principal Partition of a Pair of Polymatroids of Iri and Nakamura

Let (E, ρ_i) $(i = 1, 2)$ be two polymatroids. Then we have the following min-max theorem parametrically.

Theorem 3.22 (Edmonds). *For any $\lambda \geq 0$ we have*

$$\max\{x(E) \mid x \in P(\rho_1) \cap P(\lambda \rho_2)\} = \min\{\rho_1(X) + \lambda \rho_2(E \setminus X) \mid X \subseteq E\}. \quad (42)$$

For the sake of simplicity we suppose that ρ_2 is strictly monotone increasing, i.e., all the extreme bases of (E, ρ_2) are positive vectors (or $B(\rho_2)$ is included in the interior of the nonnegative orthant $\mathbb{R}_+^E$).

Iri (1979, 1984), Nakamura (1988) and Nakamura and Iri (1981) developed the principal partition of a pair of polymatroids, based on Theorem 3.22. Define $\mathcal{D}(\rho_1, \lambda \rho_2)$ as the collection of minimizers of the submodular function $\rho_1(X) + \lambda \rho_2(E \setminus X)$ in X. Let E_λ^- and E_λ^+ be, respectively, the minimum and the maximum element of the distributive lattice $\mathcal{D}(\rho_1, \lambda \rho_2)$ for all $\lambda \geq 0$. We call λ a *critical value* if $\mathcal{D}(\rho_1, \lambda \rho_2)$ contains more than one element. It should be noted that when ρ_2 is a modular function represented by a positive vector $w \in \mathbb{R}^E$, Theorem 3.22 reduces to Theorem 3.12.

Theorem 3.23 (Iri, Nakamura). *For two critical values λ and λ' with $\lambda < \lambda'$ we have*

$$E_\lambda^- \subset E_\lambda^+ \subseteq E_{\lambda'}^- \subset E_{\lambda'}^+. \quad (43)$$

Moreover, for any $X \in \mathcal{D}(\rho_1, \lambda \rho_2)$ and $X' \in \mathcal{D}(\rho_1, \lambda' \rho_2)$ we have

$$X \cap X' \in \mathcal{D}(\rho_1, \lambda \rho_2), \qquad X \cup X' \in \mathcal{D}(\rho_1, \lambda' \rho_2). \quad (44)$$

Proof. For any $\lambda < \lambda'$ (not necessarily critical values) and for any $X \in \mathcal{D}(\rho_1, \lambda \rho_2)$ and $X' \in \mathcal{D}(\rho_1, \lambda' \rho_2)$ we have

$$\begin{aligned}
\rho_1(X') &+ \lambda' \rho_2(E \setminus X') + \rho_1(X) + \lambda \rho_2(E \setminus X) \\
&\geq \rho_1(X \cup X') + \lambda' \rho_2(E \setminus (X \cup X')) + \rho_1(X \cap X') + \lambda \rho_2(E \setminus (X \cap X')) \\
&\quad + (\lambda' - \lambda)(\rho_2(E \setminus (X \cap X')) - \rho_2(E \setminus X)) \\
&\geq \rho_1(X \cup X') + \lambda' \rho_2(E \setminus (X \cup X')) + \rho_1(X \cap X') + \lambda \rho_2(E \setminus (X \cap X')). \quad (45)
\end{aligned}$$

This implies (44) and hence

$$E_\lambda^- \subseteq E_{\lambda'}^-, \qquad E_\lambda^+ \subseteq E_{\lambda'}^+. \tag{46}$$

When λ is a critical value, for a sufficiently small $\epsilon > 0$ $\mathcal{D}(\rho_1, (\lambda + \epsilon)\rho_2)$ contains only one element E_λ^+ since $\rho_2(X) < \rho_2(Y)$ for all $X \subset Y \subseteq E$ by the assumption that B(ρ_2) lies in the interior of $\mathbb{R}_+^E$. This together with (46) implies (43). □

Note that (44) and (46) hold without the assumption that B(ρ_2) lies in the interior of the nonnegative orthant $\mathbb{R}_+^E$.

It follows from Theorem 3.23 that there exist at most $|E|$ critical values $\lambda_1 < \cdots < \lambda_p$ and that

$$\mathcal{D}_{\text{all}}(\rho_1, \rho_2) \equiv \bigcup_{i=1}^{p} \mathcal{D}(\rho_1, \lambda_i \rho_2) \tag{47}$$

forms a distributive lattice, which leads us to a decomposition of the pair of polymatroids (ρ_1, ρ_2) as follows (Iri 1979, 1984; Nakamura 1988; Nakamura and Iri 1981; Tomizawa and Fujishige 1982).

The whole distributive lattice (47) yields a chain

$$E_{\lambda_1}^-(= \emptyset) \subset E_{\lambda_1}^+ = E_{\lambda_2}^- \subset \cdots \subset E_{\lambda_{p-1}}^+ = E_{\lambda_p}^- \subset E_{\lambda_p}^+ (= E). \tag{48}$$

Then polymatroids $\mathbf{P}_k = (E, \rho_k)$ $(k = 1, 2)$ are decomposed into

$$\mathbf{P}_1 \cdot E_{\lambda_i}^+ / E_{\lambda_i}^- \quad (i = 1, \ldots, p), \tag{49}$$

$$\mathbf{P}_2 \cdot \overline{E_{\lambda_i}^-} / \overline{E_{\lambda_i}^+} \quad (i = 1, \ldots, p) \tag{50}$$

where for any $X \subseteq E$ we denote by $\overline{X}$ its complement $E \setminus X$, by $\mathbf{P}_k \cdot X$ the restriction of $\mathbf{P}_k$ to X, and by $\mathbf{P}_k / X$ the contraction of $\mathbf{P}_k$ by X. For any $\lambda \geq 0$ we denote $\lambda \mathbf{P}_k = (E, \lambda \rho_k)$.

Theorem 3.24 (Iri, Nakamura). *For the minors of polymatroids* $\mathbf{P}_k = (E, \rho_k)$ $(k = 1, 2)$ *in* (49) *and* (50) *the following holds.*

- *For each* $i = 1, \ldots, p$ *the pair of minors* $\mathbf{P}_1 \cdot E_{\lambda_i}^+ / E_{\lambda_i}^-$ *and* $\lambda_i \mathbf{P}_2 \cdot \overline{E_{\lambda_i}^-} / \overline{E_{\lambda_i}^+}$ *has a common base* (*denoted by* $b^{(i)}$ *for later use*).

Note that any maximal chain of (47) includes (48) as a subchain. Taking a maximal chain of (47) and considering minors of $\mathbf{P}_1$ and $\mathbf{P}_2$ determined by the chain as in the above theorem, we obtain the finest decomposition of the pair of $\mathbf{P}_1$ and $\mathbf{P}_2$ that has the similar intersection property as given in the theorem and is independent of the choice of a maximal chain of (47). This is the principal partition of a pair of polymatroids in the sense of Iri and Nakamura.

Using $b^{(i)}$ $(i = 1, \ldots, p)$ in Theorem 3.24, define $b_1 \in$ B(ρ_1) and $b_2 \in$ B(ρ_2) as follows. Define $b_1 = b^{(1)} \oplus \cdots \oplus b^{(p)}$. If $\lambda_1 \neq 0$, then define $b_2 = (1/\lambda_1)b^{(1)} \oplus \cdots \oplus (1/\lambda_p)b^{(p)}$; otherwise ($\lambda_1 = 0$), choose any base $b_2^{(1)}$ of $\mathbf{P}_2 / E_{\lambda_1}^+$ and define $b_2 = b_2^{(1)} \oplus (1/\lambda_2)b^{(2)} \oplus \cdots \oplus (1/\lambda_p)b^{(p)}$. Then for any $\lambda \geq 0$, $b_1 \wedge \lambda b_2$ is a maximum common subbase of $\mathbf{P}_1$ and $\lambda \mathbf{P}_2$. Hence,

Theorem 3.25 (Nakamura). *There exist a base b_1 of $\mathbf{P}_1$ and a base b_2 of $\mathbf{P}_2$ such that for any $\lambda \geq 0$ $b_1 \wedge \lambda b_2$ is a maximum common subbase of $\mathbf{P}_1$ and $\lambda \mathbf{P}_2$.*

This generalizes Theorem 3.14. The pair (b_1, b_2) is called a *universal pair of bases*, where note that such a pair is not necessarily unique (see also Murota 1988).

It is not difficult to generalize the principal partition of a pair of polymatroids to that of a submodular system and a polymatroid. The range of parameter λ can also be extended to negative values by defining $\lambda \rho$ for $\lambda < 0$ by

$$\lambda \rho(X) = \lambda \rho^{\#}(X) \quad (X \subseteq E) \tag{51}$$

(see Fujishige 2005; Tomizawa and Fujishige 1982).

We shall discuss a further generalization later in Sect. 7.4.

7.3.6 The Principal Structure of a Submodular System

A related decomposition slightly different from principal partitions was considered in Fujishige (1980b).

Let $(\mathcal{D}, f)$ be any submodular system on E. Then for any $e \in E$ define

$$\mathcal{D}_f(e) = \{X \mid e \in X \in \mathcal{D}, \ f(X) = \min\{f(Y) \mid e \in Y \in \mathcal{D}\}\}. \tag{52}$$

Note that $\mathcal{D}_f(e)$ is a distributive lattice with set union and intersection as the lattice operations. Denote by $D_f(e)$ the minimum element of $\mathcal{D}_f(e)$.

Now we have the following.

Theorem 3.26. *For any $e_1, e_2 \in E$ such that $e_2 \in D_f(e_1)$ we have*

$$D_f(e_2) \subseteq D_f(e_1). \tag{53}$$

Proof. Putting $F_i = D_f(e_i)$ for $i = 1, 2$, we have

$$f(F_1) \leq f(F_1 \cup F_2), \tag{54}$$

since $e_1 \in F_1 \cup F_2$. It follows from (54) and the submodularity of f that

$$f(F_2) \geq f(F_1 \cap F_2) + f(F_1 \cup F_2) - f(F_1)$$
$$\geq f(F_1 \cap F_2). \tag{55}$$

This implies (53) since $e_2 \in F_1 \cap F_2$, and hence $F_2 \subseteq F_1 \cap F_2$. □

Let $\mathcal{F}$ be the collection of $D_f(e)$ ($e \in E$). Then we see from this theorem that for any $F_1, F_2 \in \mathcal{F}$ we have $F_1 \cap F_2 = \bigcup_{e \in F_1 \cap F_2} D_f(e)$.

We can define a transitive binary relation $\to$ on E by

$$e_1 \to e_2 \quad \Longleftrightarrow \quad e_2 \in D_f(e_1). \tag{56}$$

The transitive binary relation $\to$ on E naturally defines a directed graph G_f with a vertex set E whose strongly connected components are complete directed graphs with selfloops at every vertex. Decomposing G_f into strongly connected components, we obtain a decomposition with a poset structure on it, which is called the *principal structure* of the submodular system $(\mathcal{D}, f)$.

Remark 3.27. For a submodular system $\mathbf{S} = (\mathcal{D}, f)$ on E the principal structure of submodular system $\mathbf{S}$ furnishes a further decomposition of $E \setminus D_f^{\max}$, where $D_f^{\max}$ is the maximum element of the set of minimizers of f.

Remark 3.28. The concepts of principal structure and principal partition have been effectively applied to systems analysis and examined in details in matrix and matroidal frameworks in (Iwata 1996; Iwata and Murota 1995, 1996; Murota 1987, 1990; Murota et al. 1987) (see Murota's book, Murota 2000).

7.4 Extensions

In the principal partitions viewed in Sect. 7.3 we have considered submodular functions with a parameter that appears linearly as follows. Vector **1** denotes the vector of all ones.

* $\rho(X) + \lambda w(E \setminus X)$ $(X \subseteq E)$,
  $\rho = r_G$, $w = \mathbf{1}$, $\lambda = \frac{1}{2}$ (Kishi and Kajitani)
  ρ: a matroid rank function, $w = \mathbf{1}$, $\lambda \geq 0$ (Tomizawa and Narayanan)
  ρ: a polymatroid rank function, a positive weight w, $\lambda \geq 0$ (Fujishige)
  extension to submodular systems, a positive weight w, $\lambda \in \mathbb{R}$ (Fujishige)
* $\rho_1(X) + \lambda \rho_2(E \setminus X)$ $(X \subseteq E)$,
  ρ_1, ρ_2: polymatroid rank functions, $\lambda \geq 0$ (Iri and Nakamura)
  extension to submodular systems, $\lambda \in \mathbb{R}$ (Fujishige and Tomizawa)

We shall examine how the linear form in the parameter can be extended to a nonlinear form in Sect. 7.4.1. We also examine possible extension of the domain 2^E or $\mathcal{D}$ to the integer lattice $\mathbb{Z}^E$ in Sect. 7.4.2.

7.4.1 Parameters Nonlinearly

The result of this section is based on joint work of Fujishige and Nagano (2008) (see also Nagano 2007b).

In the principal partition with a parameter λ described in Sect. 7.3 a kind of monotonicity of λw and $\lambda \rho_2$ plays a crucial rôle. The essence of the monotonicity is the strong map relation of submodular systems.

Consider two submodular systems $\mathbf{S}_i = (\mathcal{D}_i, f_i)$ $(i = 1, 2)$ on E. The ordered pair $(\mathbf{S}_1, \mathbf{S}_2)$ is called a *strong map* if for all $X \in \mathcal{D}_1$ and $Y \in \mathcal{D}_2$ such that $X \subseteq Y$ we have

$$f_1(Y) - f_1(X) \geq f_2(Y) - f_2(X), \tag{57}$$

where if $X \notin \mathcal{D}_2$ or $Y \notin \mathcal{D}_1$, we understand that (57) holds. Following the convention, we write $f_1 \to f_2$ if $(\mathbf{S}_1, \mathbf{S}_2)$ is a strong map. For two supermodular functions g_1 and g_2 we write $g_1 \to g_2$ if we have a strong map relation $g_2^{\#} \to g_1^{\#}$, where recall that $g_i^{\#}$ is the dual submodular function of g_i.

The strong map relation is the monotonicity that we need to extend the principal partition having a parameter linearly.

Consider parameterized submodular systems $(\mathcal{D}, f_\lambda)$ $(\lambda \in \mathbb{R})$ and supermodular systems $(\mathcal{D}, g_\lambda)$ $(\lambda \in \mathbb{R})$ such that for all λ and λ' with $\lambda < \lambda'$

$$f_\lambda \to f_{\lambda'}, \qquad g_\lambda \to g_{\lambda'}. \tag{58}$$

We assume that for each $X \in \mathcal{D}$ functions $f_\lambda(X)$ and $g_\lambda(X)$ in $\lambda \in \mathbb{R}$ are continuous.

Now we have the following min-max theorem due to Edmonds. For any $x \in \mathbb{R}^E$ define $x^- = (\min\{x(e), 0\} \mid e \in E)$.

Theorem 4.1.

$$\max\{(x - y)^-(E) \mid x \in \mathrm{B}(f_\lambda), \ y \in \mathrm{B}(g_\lambda)\} = \min\{f_\lambda(X) - g_\lambda(X) \mid X \in \mathcal{D}\}.$$

Define a parameterized submodular function $h_\lambda(X)$ in $X \in \mathcal{D}$ as

$$h_\lambda(X) = f_\lambda(X) - g_\lambda(X) \quad (X \in \mathcal{D}). \tag{59}$$

It should be noted that for any λ and λ' such that $\lambda < \lambda'$ we have a strong map relation

$$h_\lambda \to h_{\lambda'}. \tag{60}$$

For any λ let $\mathcal{D}(h_\lambda)$ be the set of minimizers of h_λ.

Theorem 4.2. *For any λ and λ' such that $\lambda < \lambda'$ and for any $X \in \mathcal{D}(h_\lambda)$ and $Y \in \mathcal{D}(h_{\lambda'})$ we have*

$$X \cap Y \in \mathcal{D}(h_\lambda), \qquad X \cup Y \in \mathcal{D}(h_{\lambda'}). \tag{61}$$

Proof. Under the assumption of the present theorem,

$$\begin{aligned}
h_{\lambda'}(X) + h_\lambda(Y) &= h_{\lambda'}(X) + h_{\lambda'}(Y) - h_{\lambda'}(Y) + h_\lambda(Y) \\
&\geq h_{\lambda'}(X \cup Y) + h_{\lambda'}(X \cap Y) - h_{\lambda'}(Y) + h_\lambda(Y) \\
&= h_{\lambda'}(X \cup Y) + h_\lambda(X \cap Y) \\
&\quad + h_\lambda(Y) - h_\lambda(X \cap Y) - h_{\lambda'}(Y) + h_{\lambda'}(X \cap Y) \\
&\geq h_{\lambda'}(X \cup Y) + h_\lambda(X \cap Y).
\end{aligned} \tag{62}$$

Hence we have (61). □

It follows that the union of $\mathcal{D}(h_\lambda)$ $(\lambda \in \mathbb{R})$ is again a distributive lattice, denoted by $\mathcal{D}_{\mathrm{all}}(h)$. For each $\lambda \in \mathbb{R}$ denote the maximum and the minimum element of $\mathcal{D}(h_\lambda)$ by S_λ^+ and S_λ^-, respectively. From Theorem 4.2 we have

Theorem 4.3. *For any λ and λ' such that $\lambda < \lambda'$,*

$$S_\lambda^- \subseteq S_{\lambda'}^-, \qquad S_\lambda^+ \subseteq S_{\lambda'}^+. \tag{63}$$

Hence there exist at most $|E| + 1$ distinct S_λ^+ ($\lambda \in \mathbb{R}$), which we suppose are given by

$$S_0 \subset S_1 \subset \cdots \subset S_p. \tag{64}$$

Because of the finiteness character and the continuity of $h_\lambda(X)$ in λ, for each λ we have $\mathcal{D}(h_\lambda) \supseteq \mathcal{D}(h_{\lambda+\epsilon})$ for a sufficiently small $\epsilon > 0$. Hence, from Theorem 4.3, $\mathbb{R}$ is divided into the intervals

$$\Lambda_0 = (-\infty, \lambda_1), \quad \Lambda_1 = [\lambda_1, \lambda_2), \quad \ldots, \quad \Lambda_p = [\lambda_p, +\infty) \tag{65}$$

such that for any $i = 0, 1, \ldots, p$ and any $\lambda \in \Lambda_i$ we have $S_\lambda^+ = S_i$. We call λ_i ($i = 1, \ldots, p$) *upper critical values*.

For simplicity we assume that

$$S_0 = \emptyset, \qquad S_p = E. \tag{66}$$

Lemma 4.4. *For any $i = 2, \ldots, p$ we have $S_{\lambda_{i-1}}^+ \in \mathcal{D}(h_{\lambda_i})$ and $\emptyset \in \mathcal{D}(h_{\lambda_1})$.*

Proof. It follows from Theorems 4.2 and 4.3 that for any $\lambda \in \mathbb{R}$ we have

$$S_{\lambda-\epsilon}^- = S_\lambda^-, \qquad S_{\lambda+\epsilon}^+ = S_\lambda^+ \tag{67}$$

for a sufficiently small $\epsilon > 0$. That is to say, S_λ^- is left-continuous in λ and S_λ^+ is right-continuous in λ.

For any $i = 2, \ldots, p$ and a sufficiently small $\epsilon > 0$ we have $S_{\lambda_i}^- \in \mathcal{D}(h_{\lambda_i - \epsilon})$ and $S_{\lambda_i - \epsilon}^+ = S_{\lambda_{i-1}}^+$. Hence,

$$S_{\lambda_i}^- \subseteq S_{\lambda_{i-1}}^+. \tag{68}$$

It follows from Theorem 4.2 and (68) that $S_{\lambda_{i-1}}^+ = S_{\lambda_{i-1}}^+ \cup S_{\lambda_i}^- \in \mathcal{D}(h_{\lambda_i})$.

Similarly we can show $(S_0 =) \emptyset \in \mathcal{D}(h_{\lambda_1})$. □

For each λ let $\mathbf{S}_\lambda$ be the submodular system $(\mathcal{D}, h_\lambda)$ on E and for each $i = 1, \ldots, p$ consider minors $\mathbf{S}_{\lambda_i} \cdot S_i / S_{i-1}$. Note that for each $i = 1, \ldots, p$

$$\mathbf{S}_{\lambda_i} \cdot S_i / S_{i-1} = \left(\mathcal{D}_{S_{i-1}}^{S_i}, h_{\lambda_i}{}_{S_{i-1}}^{S_i} \right) \tag{69}$$

is a submodular system on $S_i \setminus S_{i-1}$ with rank function $h_{\lambda_i}{}_{S_{i-1}}^{S_i}$.

We use $\mathbf{0}$ to denote a zero vector of appropriate dimension. Its dimension is determined by the context.

Lemma 4.5. *For each $i = 1, \ldots, p$ we have $\mathbf{0} \in \mathrm{B}(h_{\lambda_i}{}_{S_{i-1}}^{S_i})$.*

Proof. We see from Lemma 4.4 that $S_{i-1} \subset S_i$ is a chain of $\mathcal{D}(h_{\lambda_i})$ for each $i = 1, \ldots, p$. Hence $h_{\lambda_i}{}_{S_{i-1}}^{S_i}$ is nonnegative and $h_{\lambda_i}{}_{S_{i-1}}^{S_i}(S_i \setminus S_{i-1}) = 0$, which shows the present lemma. □

Now we assume that $\mathcal{D}$ is simple, i.e., $\mathcal{D}$ is the collection of (lower) order-ideals of a poset $\mathcal{P} = (E, \preceq)$ on E. Let $G(\mathcal{P})$ be the graph representing the Hasse diagram of poset $\mathcal{P}$. Recall that for any $x \in \mathbb{R}^E$ and $F \subseteq E$ we denote $x^F = (x(e) \mid e \in F)$.

Then,

Theorem 4.6. *There exist at most $|E|$ linear extensions of poset $\mathcal{P}$ identified with linear orderings σ_j ($j \in I$) of E, a nonnegative flow φ in $G(\mathcal{P})$, and coefficients $\mu_j > 0$ ($j \in I$) with $\sum_{j \in I} \mu_j = 1$ such that for all $\lambda \in \mathbb{R}$, defining a base b_λ of submodular system $\mathbf{S}_\lambda$ by*

$$b_\lambda = \sum_{j \in I} \mu_j b_\lambda^{\sigma_j} + \partial\varphi, \tag{70}$$

the base b_λ satisfies

$$(b_{\lambda_i})^{S_i \setminus S_{i-1}} = 0 \quad (i = 1, \ldots, p), \tag{71}$$

where for each $j \in I$ $b_\lambda^{\sigma_j}$ appearing in (70) is the extreme base of $\mathrm{B}(h_\lambda)$ corresponding to the linear ordering σ_j and $\partial\varphi$ is the boundary of flow φ in $G(\mathcal{P})$.

Proof. For each $i = 1, \ldots, p$ base $b_i \equiv 0 \in \mathrm{B}(h_{\lambda_i \, S_{i-1}}^{S_i})$ is expressed by a convex combination of at most $|S_i \setminus S_{i-1}|$ extreme bases $b_{\lambda_i}^{\sigma_{ij}}$ ($j \in I_i$) of $\mathrm{B}(h_{\lambda_i \, S_{i-1}}^{S_i})$ and a nonnegative flow φ_i in $G(\mathcal{P}) \cdot (S_i \setminus S_{i-1})$ as follows.

$$b_i = \sum_{j \in I_i} \mu_{ij} b_{\lambda_i}^{\sigma_{ij}} + \partial\varphi_i. \tag{72}$$

Hence we can have an expression (70) satisfying (71), where we need at most $|E|$ extreme bases of $\mathrm{B}(h_\lambda)$ since

$$|S_1| + |S_2 \setminus S_1| + \cdots + |S_p \setminus S_{p-1}| = |E|. \tag{73}$$

For, the expression (70) can be constructed by the following procedure. Put $I = \emptyset$.

1. For each $i = 1, \ldots, p$ choose an index $k_i \in I_i$.
2. Find $i^* \in \{1, \ldots, p\}$ such that $\mu_{i^* k_{i^*}} = \min\{\mu_{ik_i} \mid i = 1, \ldots, p\}$.
3. Put $I \leftarrow I \cup \{k_{i^*}\}$.
   Let $\bar{\sigma}_{k_{i^*}}$ be the concatenation of $\sigma_{k_1}, \ldots, \sigma_{k_p}$ and define $\bar{\mu}_{k_{i^*}} = \mu_{i^* k_{i^*}}$.
4. For each $i = 1, \ldots, p$
   put $\mu_{ik_i} \leftarrow \mu_{ik_i} - \bar{\mu}_{k_{i^*}}$ and
   if $\mu_{ik_i} = 0$, then $I_i \leftarrow I_i \setminus \{k_i\}$ and
   if $I_i \neq \emptyset$, then choose an index $k_i \in I_i$,
   else go to Step 5.
   Go to Step 2.
5. Return $\bar{\sigma}_j$, $\bar{\mu}_j$ ($j \in I$), and I.

(Here we assume that I_i ($i = 1, \ldots, p$) are disjoint.)

It should be noted that the linear ordering defined by the concatenation of $\sigma_{k_1}, \ldots, \sigma_{k_p}$ in Step 2 is a linear extension of $\mathcal{P}$, so that it gives an extreme base of $\mathrm{B}(h_\lambda)$. We can see that $|I| \leq |E|$, because of (73). Then we have

$$b_\lambda = \sum_{j \in I} \bar{\mu}_j b_\lambda^{\bar{\sigma}_j} + \partial\varphi, \tag{74}$$

where $\varphi = \bigoplus_{i-1}^{p} \varphi_i$. We can also show that b_λ defined by (74) satisfies

$$(b_{\lambda_i})^{S_i \setminus S_{i-1}} = \mathbf{0} \tag{75}$$

for all $i = 1, \ldots, p$. $\square$

Moreover, we have

Theorem 4.7. *For any* $\lambda \in \mathbb{R}$ *the base* $b_\lambda \in \mathrm{B}(h_\lambda)$ *in Theorem 4.6 satisfies*

$$b_\lambda^-(E) \left(= \sum \{b_\lambda(e) \mid e \in E, \ b_\lambda(e) < 0\} \right) = \max\{x^-(E) \mid x \in \mathrm{B}(h_\lambda)\}. \tag{76}$$

Proof. Consider any $i \in \{0, 1, \ldots, p\}$ and $\lambda \in \Lambda_i$. Then, since S_i is a minimizer of h_λ, it suffices to show that

$$b_\lambda(e) \leq 0 \quad (e \in S_i), \tag{77}$$
$$b_\lambda(e) \geq 0 \quad (e \in E \setminus S_i), \tag{78}$$

and S_i is a tight set for b_λ in $\mathrm{B}(h_\lambda)$.

Because of Theorem 4.6 and the strong map relation we have (77) and (78) for $\lambda \in \Lambda_i$, where note that for any λ' and λ'' with $\lambda' < \lambda''$ we have $b_{\lambda'}^{\sigma_i} \geq b_{\lambda''}^{\sigma_i}$. Moreover, by the definitions of σ_k $(k \in I)$ and φ we have

$$b_\lambda^{\sigma_k}(S_i) = h_\lambda(S_i) \quad (k \in I), \tag{79}$$
$$\partial\varphi(S_i) = 0. \tag{80}$$

It follows that S_i is a tight set for b_λ. $\square$

From Theorems 4.6 and 4.7 we have

Theorem 4.8. *There exist at most* $|E|$ *linear orderings* σ_i $(i \in I)$ *of* E, *coefficients* μ_i $(i \in I)$ *of convex combination, and nonnegative flows* $\bar{\varphi}$ *and* φ *in* $G(\mathcal{P})$ *such that for all* $\lambda \in \mathbb{R}$, *defining*

$$\bar{b}_\lambda = \sum_{i \in I} \mu_i \bar{b}_\lambda^{\sigma_i} + \partial\bar{\varphi}, \qquad \underline{b}_\lambda = \sum_{i \in I} \mu_i \underline{b}_\lambda^{\sigma_i} - \partial\underline{\varphi} \tag{81}$$

by extreme bases $\bar{b}_\lambda^{\sigma_i}$ *of* $\mathrm{B}(f_\lambda)$ *and* $\underline{b}_\lambda^{\sigma_i}$ *of* $\mathrm{B}(g_\lambda)$ *corresponding to linear orderings* σ_i $(i \in I)$, *we have*

$$(\bar{b}_\lambda - \underline{b}_\lambda)^-(E) = \max\{(x - y)^-(E) \mid x \in \mathrm{B}(f_\lambda), \ y \in \mathrm{B}(g_\lambda)\} \tag{82}$$

for all $\lambda \in \mathbb{R}$.

Moreover, we have

$$(\bar{b}_\lambda)^{S_i \setminus S_{i-1}} \in \mathrm{B}\left(f_{\lambda S_{i-1}}^{S_i}\right), \qquad (\underline{b}_\lambda)^{S_i \setminus S_{i-1}} \in \mathrm{B}\left(g_{\lambda S_{i-1}}^{S_i}\right) \tag{83}$$

for all $\lambda \in \mathbb{R}$ *and* $i = 1, \ldots, p$, *and*

$$(\bar{b}_{\lambda_i})^{S_i \setminus S_{i-1}} = (\underline{b}_{\lambda_i})^{S_i \setminus S_{i-1}} \tag{84}$$

for all $i = 1, \ldots, p$.

It should be noted that Theorem 4.8 generalizes Theorems 3.24 and 3.25.

Remark 4.9. Besides upper critical values we can also define *lower critical values* as follows. Recall that S_λ^- is the minimum element of $\mathcal{D}(h_\lambda)$. Since we have $\mathcal{D}(h_{\lambda-\epsilon}) \subseteq \mathcal{D}(h_\lambda)$ for each λ and a sufficient small $\epsilon > 0$, let $S_1' \subset S_2' \subset \cdots \subset S_q'$ be the distinct elements of S_λ^- ($\lambda \in \mathbb{R}$). Then, $\mathbb{R}$ is divided into the intervals

$$\Lambda_0' = (-\infty, \lambda_1'], \quad \Lambda_1' = (\lambda_1', \lambda_2'], \quad \ldots, \quad \Lambda_q' = (\lambda_q', +\infty) \tag{85}$$

such that for any $j = 0, 1, \ldots, q$ and any $\lambda \in \Lambda_j'$ we have $S_\lambda^- = S_j'$. We call each λ_j' a *lower critical value*. By means of lower critical values and the chain $S_1' \subset \cdots \subset S_q'$ we can develop similar arguments as made for the above-mentioned principal partitions.

For any $\lambda, \lambda' \in \mathbb{R}$ with $\lambda < \lambda'$, if h_λ and $h_{\lambda'}$ satisfy

$$h_\lambda(Y) - h_\lambda(X) > h_{\lambda'}(Y) - h_{\lambda'}(X) \tag{86}$$

for all $X, Y \in \mathcal{D}$ with $X \subset Y$, we call $(h_\lambda, h_{\lambda'})$ a *strict strong map* and write $h_\lambda \twoheadrightarrow h_{\lambda'}$.

Theorem 4.10. *If $h_\lambda \twoheadrightarrow h_{\lambda'}$ for all λ and λ' with $\lambda < \lambda'$, then the upper critical values coincide with the lower critical values and we have*

$$S_i^+ = S_{i+1}^- \quad (i = 0, \ldots, p - 1). \tag{87}$$

Proof. For any $\lambda < \lambda'$ and for any $X \in \mathcal{D}(h_\lambda)$ and $Y \in \mathcal{D}(h_{\lambda'})$,

$$
\begin{aligned}
h_\lambda(X) + h_{\lambda'}(Y) &\geq h_\lambda(X \cup Y) + h_\lambda(X \cap Y) - h_\lambda(Y) + h_{\lambda'}(Y) \\
&= h_\lambda(X \cap Y) + h_{\lambda'}(X \cup Y) \\
&\quad + h_\lambda(X \cup Y) - h_\lambda(Y) - h_{\lambda'}(X \cup Y) + h_{\lambda'}(Y) \\
&\geq h_\lambda(X \cap Y) + h_{\lambda'}(X \cup Y). \tag{88}
\end{aligned}
$$

If $Y \subset X \cup Y$, i.e., $X \setminus Y \neq \emptyset$, then the second inequality is strict since $h_\lambda \twoheadrightarrow h_{\lambda'}$, which is a contradiction. Hence $X \setminus Y = \emptyset$, i.e., $X \subseteq Y$. The present theorem follows from this fact. $\square$

For a related parametric submodular intersection problem see Iwata et al. (1997).

7.4.2 Extension to Discrete Convex Functions

The result of this section is based on joint work of Fujishige et al. (2008).

Let $f : \mathbb{Z}^E \to \mathbb{R} \cup \{+\infty\}$ be a function on the integer lattice $\mathbb{Z}^E$ such that its effective domain $\mathrm{dom}\, f \equiv \{x \in \mathbb{Z}^E \mid f(x) < +\infty\}$ is nonempty. We suppose the following.

(S) f is submodular on $\operatorname{dom} f$, i.e.,

$$f(x) + f(y) \geq f(x \vee y) + f(x \wedge y) \quad (x, y \in \operatorname{dom} f), \tag{89}$$

where $(x \vee y)(e) = \max\{x(e), y(e)\}$ and $(x \wedge y)(e) = \min\{x(e), y(e)\}$ for $e \in E$.

Given a positive vector $w : E \to \mathbb{R}$, consider an optimization problem with a parameter $\lambda \in \mathbb{R}$ as follows.

$$(P_\lambda): \quad \text{Minimize } f(x) - \lambda \langle w, x \rangle, \tag{90}$$

where $\langle w, x \rangle = \sum_{e \in E} w(e) x(e)$. It should be noted that Problem (P_λ) generalizes the minimization problem appearing in Theorem 3.12. (For any $z \in \mathbb{Z}^E$ define $\mathcal{D}_z = \{X \mid X \subseteq E, \ f(z + \chi_X) < +\infty\}$, where χ_X is the characteristic vector of X, and if $\mathcal{D}_z \neq \emptyset$, also define a set function $f_z(X) = f(z + \chi_X)$ $(X \in \mathcal{D}_z)$. Then f_z is a submodular set function on the distributive lattice $\mathcal{D}_z$.)

Define $\mathcal{Z}(\lambda)$ to be the set of minimizers of $f(x) - \lambda \langle w, x \rangle$ in $x \in \mathbb{Z}^E$. Then,

Theorem 4.11. *For any $\lambda, \lambda' \in \mathbb{R}$ such that $\lambda \leq \lambda'$ and for any $x \in \mathcal{Z}(\lambda)$ and $x' \in \mathcal{Z}(\lambda')$ we have*

$$x \vee x' \in \mathcal{Z}(\lambda'), \qquad x \wedge x' \in \mathcal{Z}(\lambda). \tag{91}$$

Moreover, if $\lambda < \lambda'$, then $x \leq x'$.

Proof. Under the assumption of the present theorem we have

$$\begin{aligned}
f(x) &- \lambda \langle w, x \rangle + f(x') - \lambda' \langle w, x' \rangle \\
&\geq f(x \vee x') - \lambda' \langle w, x \vee x' \rangle + f(x \wedge x') - \lambda \langle w, x \wedge x' \rangle \\
&\quad + \lambda' \langle w, x \vee x' \rangle + \lambda \langle w, x \wedge x' \rangle - \lambda' \langle w, x' \rangle - \lambda \langle w, x \rangle \\
&= f(x \vee x') - \lambda' \langle w, x \vee x' \rangle + f(x \wedge x') - \lambda \langle w, x \wedge x' \rangle \\
&\quad + (\lambda' - \lambda) \langle w, x \vee x' - x' \rangle \\
&\geq f(x \vee x') - \lambda' \langle w, x \vee x' \rangle + f(x \wedge x') - \lambda \langle w, x \wedge x' \rangle. \tag{92}
\end{aligned}$$

Hence (91) follows. Moreover, since the inequalities in (92) must be equalities, if $\lambda < \lambda'$, the last inequality (now equality) implies $x \vee x' = x'$, i.e., $x \leq x'$, where note that $w > 0$. $\square$

Remark 4.12. Theorem 4.11 is subsumed by a result of Topkis (1978, 1998) (see also Iri 1984). Monotonicity of optimal solutions of parametric optimization problems has been investigated in the literature such as (Brumelle et al. 2005; Milgrom and Shannon 1994; Topkis 1978). The theory of principal partitions has been developed independently of these results and deals primarily with the critical values and the decomposition of systems, while the monotonicity of primal and dual optimal solutions with respect to the parameter plays a crucial rôle in the principal partitions.

Denote by z_λ^+ and z_λ^-, respectively, the maximum and the minimum element of $\mathcal{Z}(\lambda)$. Define

$$\Lambda^* = \{\lambda \in \mathbb{R} \mid z_\lambda^+ \neq z_\lambda^-\}. \tag{93}$$

Each $\lambda \in \Lambda^*$ is called a *critical value*.

Theorem 4.13. *Consider any critical values $\lambda, \lambda' \in \Lambda^*$ with $\lambda < \lambda'$. Then we have either $\mathcal{Z}(\lambda) \cap \mathcal{Z}(\lambda') = \emptyset$ or $z_\lambda^+ = z_{\lambda'}^-$.*

Proof. If $\mathcal{Z}(\lambda) \cap \mathcal{Z}(\lambda')$ contains two distinct elements x and x', then this contradicts the monotonicity in the last statement of Theorem 4.11. Hence we have $|\mathcal{Z}(\lambda) \cap \mathcal{Z}(\lambda')| = 0$ or 1. If $|\mathcal{Z}(\lambda) \cap \mathcal{Z}(\lambda')| = 1$, the element of $\mathcal{Z}(\lambda) \cap \mathcal{Z}(\lambda')$ must be z_λ^+ that is equal to $z_{\lambda'}^-$, due to Theorem 4.11.   □

For any two critical values $\lambda, \lambda' \in \Lambda^*$ with $\lambda < \lambda'$ we say that λ' *covers* λ if there is no critical value λ'' satisfying $\lambda < \lambda'' < \lambda'$.

Theorem 4.14. *For any critical values $\lambda, \lambda' \in \Lambda^*$ such that λ' covers λ we have*

$$z_\lambda^+ = z_{\lambda'}^-. \tag{94}$$

Moreover,

$$z_{\lambda''}^+ = z_{\lambda''}^- = z_\lambda^+ \ (= z_{\lambda'}^-) \quad (\lambda < \lambda'' < \lambda'). \tag{95}$$

Proof. Because of the continuity in the parameter, for any λ'' and sufficiently small $\epsilon > 0$ we have

$$\mathcal{Z}(\lambda'' \pm \epsilon) \subseteq \mathcal{Z}(\lambda''). \tag{96}$$

It follows from (96), Theorem 4.13, and the definition of a critical value that we have (94) and (95).   □

Remark 4.15. We can consider more general parametric submodular functions corresponding to those treated in Sect. 7.4.1. For each $\lambda \in \mathbb{R}$ let h_λ be a submodular function on $\mathbb{Z}^E$ that satisfies the following.

*   For any λ and λ' with $\lambda < \lambda'$ and for any $x, y \in \mathbb{Z}^E$ with $x \leq y$ we have

$$h_\lambda(y) - h_\lambda(x) \geq h_{\lambda'}(y) - h_{\lambda'}(x). \tag{97}$$

Then we say that $(h_\lambda, h_{\lambda'})$ is a *strong map* and write $h_\lambda \to h_{\lambda'}$. The arguments in Sect. 7.4.1 can be adapted to such parametric submodular functions on $\mathbb{Z}^E$ (cf. Topkis 1978, 1998). If (97) holds with strict inequality for all $x, y \in \mathbb{Z}^E$ with $x \leq y$ and $x \neq y$, we say that $(h_\lambda, h_{\lambda'})$ is a *strict strong map* and write $h_\lambda \twoheadrightarrow h_{\lambda'}$. Theorems 4.13 and 4.14 hold for parametric submodular functions satisfying the strict strong map condition.

Remark 4.16. It should be noted that Theorems 4.11–4.14 hold for f satisfying the submodularity condition (S). However, the submodularity on $\mathbb{Z}^E$ alone is not enough to treat the structure of $\mathcal{Z}(\lambda)$ $(\lambda \in \Lambda^*)$ algorithmically. In order to resolve this situation we consider discrete convex functions called $L^\natural$-*convex functions* by Murota (2003a). $L^\natural$-convex functions on $\mathbb{Z}^E$ are submodular functions on $\mathbb{Z}^E$ with discrete convexity defined in the following.

Denote by Conv the convex hull operator in $\mathbb{R}^E$. For any $z \in \mathbb{Z}^E$ and any linear ordering σ of E define a simplex

$$\Delta_z^\sigma = \mathrm{Conv}(\{z + \chi_{S_i} \mid i = 1, \ldots, m, \ S_i \text{ is the set of the first } i \text{ elements of } \sigma\}). \tag{98}$$

The collection of all such simplices Δ_z^σ for all points $z \in \mathbb{Z}^E$ and linear orderings σ of E forms a simplicial division of $\mathbb{R}^E$, which is called the *Freudentahl simplicial division*. We also call each Δ_z^σ a *Freudentahl cell*.

In addition to the submodularity condition (S) suppose

(A1) $\mathrm{Conv}(\mathrm{dom}\, f) \cap \mathbb{Z}^E = \mathrm{dom}\, f$.

Informally, (A1) means that there is no hole in $\mathrm{dom}\, f$.

We further assume

(A2) The convex hull $\mathrm{Conv}(\mathrm{dom}\, f)$ of the effective domain of f is full-dimensional and is the union of some Freudentahl cells.

The assumption of the full dimensionality is not essential but we assume it here for simplicity. Under Assumptions (A1) and (A2) we can uniquely construct a piecewise linear extension $\hat{f}$ of f by means of the Freudentahl simplicial division as follows. For any $x \in \Delta_z^\sigma$ we have a unique expression of x as a convex combination of extreme points of the cell Δ_z^σ as

$$x = \sum_{i=1}^{m} \alpha_i (z + \chi_{S_i}), \tag{99}$$

where S_i is the set of the first i elements of σ. According to the expression (99) we define

$$\hat{f}(x) = \sum_{i=1}^{m} \alpha_i f(z + \chi_{S_i}). \tag{100}$$

For all x outside $\mathrm{Conv}(\mathrm{dom}\, f)$ we put $\hat{f}(x) = +\infty$. Note that $\hat{f}$ is well defined. It should also be noted that when $\mathrm{dom}\, f = \{\chi_X \mid X \subseteq E\}$, $\hat{f}$ is called the *Lovász extension* (Fujishige 2005; Lovász 1983).

We add one more, crucial assumption as follows.

(A3) The piecewise linear extension $\hat{f} : \mathbb{R}^E \to \mathbb{R} \cup \{+\infty\}$ of f by (100) is a convex function on $\mathbb{R}^E$.

Remark 4.17. A function $f : \mathbb{Z}^E \to \mathbb{R} \cup \{+\infty\}$ satisfying Conditions (A1), (A2), and (A3) is exactly an $L^\natural$-*convex function* on $\mathbb{Z}^E$ (with full-dimensional $\mathrm{dom}\, f$) of Murota (Fujishige and Murota 2000; Murota 1998, 2003a, 2008). The original definition of an $L^\natural$-convex function on $\mathbb{Z}^E$ is different, but see Fujishige (2005, Chap. VII) for the proof of their equivalence. Note that Conditions (A1), (A2), and (A3) imply submodularity (S). It should also be noted that a submodular function $f : \mathbb{Z}^E \to \mathbb{R} \cup \{+\infty\}$ satisfying Condition (A3) with its effective domain being a standard box $[z_1, z_2]$ between two integer vectors z_1 and z_2 was first considered by Favati and Tardella (1990) and was called a *submodular integrally convex function*.

Now, suppose that we are given a positive vector $w : E \to \mathbb{R}$, a real constant β, and an $L^{\natural}$-convex function $f : \mathbb{Z}^E \to \mathbb{R} \cup \{+\infty\}$. Let us consider the following optimization problem with a linear inequality constraint.

$$(P^{\circ}): \quad \text{Minimize } \hat{f}(x)$$
$$\text{subject to } \langle w, x \rangle \leq \beta, \tag{101}$$

where $\hat{f}$ is the piecewise linear extension of f defined by (100).

We can relate critical values for f to Problem (P°) as follows. Recall that $\mathcal{Z}(\lambda)$ is the collection of minimizers of $h_{\lambda}(x) = f(x) - \lambda \langle w, x \rangle$.

Theorem 4.18. *Suppose that for a parameter $\lambda^* < 0$ there exist $x, x' \in \mathcal{Z}(\lambda^*)$ such that*

$$\langle w, x \rangle \leq \beta, \qquad \langle w, x' \rangle \geq \beta. \tag{102}$$

Then a vector x^* lying on the line segment between x and x' and satisfying $\langle w, x^* \rangle = \beta$ is an optimal solution of Problem (P°).

Proof. For any feasible solution y of Problem (P°),

$$\begin{aligned}
\hat{f}(y) &\geq \hat{f}(y) + \lambda^*(\beta - \langle w, y \rangle) \\
&\geq \min\{f(z) + \lambda^*(\beta - \langle w, z \rangle) \mid z \in \text{dom} f\} \\
&= \hat{f}(x^*) + \lambda^*(\beta - \langle w, x^* \rangle) \\
&= \hat{f}(x^*), \tag{103}
\end{aligned}$$

where note that $f(x) - \lambda^* \langle w, x \rangle = f(x') - \lambda^* \langle w, x' \rangle = \hat{f}(x^*) - \lambda^* \langle w, x^* \rangle$ because of (A1)–(A3). Hence x^* is an optimal solution of (P°). $\square$

Remark 4.19. Since Problem (P°) is an ordinary convex program, if (P°) has an optimal solution x^*, then either it is a global minimizer of $\hat{f}$ or it is the one that satisfies the condition of Theorem 4.18. In the latter case it suffices to find a critical value λ^* such that for some $x^* \in \text{Conv}(\mathcal{Z}(\lambda^*))$ we have $\langle w, x^* \rangle = \beta$. The last condition can be rephrased as $\langle w, z_{\lambda^*}^+ \rangle \geq \beta$ and $\langle w, z_{\lambda^*}^- \rangle \leq \beta$.

When $\text{dom} f$ is bounded, we can apply Murota's weakly polynomial algorithm (Murota 2003b) for minimizing $L^{\natural}$-convex functions to find a vector in $\mathcal{Z}(\lambda)$ for each λ. We can perform a binary search to find an optimal critical value λ^* by making use of algorithms for the minimum ratio problem described in Sect. 7.5.1. This gives a weakly polynomial algorithm for Problem (P°) with rational data (see Fujishige et al. 2008).

We can also consider multiple inequality constraints as follows.

$$(P): \quad \text{Minimize } \hat{f}(x)$$
$$\text{subject to } \langle w_i, x \rangle \leq \beta_i \quad (i = 1, \ldots, k), \tag{104}$$

where w_i $(i = 1, \ldots, k)$ are positive vectors and β_i $(i = 1, \ldots, k)$ are real constants. This leads us to the following multiple-parameter submodular function.

$$h_\lambda(x) = \hat{f}(x) - \sum_{i=1}^{k} \lambda_i \langle w_i, x \rangle, \tag{105}$$

where $\lambda = (\lambda_i \mid i = 1, \ldots, k)$. The present problem can also be treated theoretically in a similar way (cf. Iri 1984; Nakamura 1988 and Fujishige 2005, Sect. 7) but algorithmically it becomes much more difficult.

Remark 4.20. We can consider a class $\mathcal{F}$ of discrete convex functions $f : \mathbb{Z}^E \to \mathbb{R} \cup \{+\infty\}$ as follows.

(i) f satisfies Conditions (A1), i.e., $\text{Conv}(\text{dom} f) \cap \mathbb{Z}^E = \text{dom} f$.
(ii) For any $x \in \text{dom} f$ there exists a vector $\hat{w} \in \mathbb{R}^E$ such that x is a minimizer of $f(z) - \langle \hat{w}, z \rangle$ $(z \in \mathbb{Z}^E)$.

Provided that we can perform the minimization of $f(z) - \lambda \langle w, z \rangle$ in $z \in \mathbb{Z}^E$ for any $\lambda \in \mathbb{R}$, we can solve Problem (P°) in a similar way as described in this section. A typical example of such a class of discrete convex functions other than $L^\natural$-convex functions is that of $M^\natural$-convex functions on $\mathbb{Z}^E$ of Murota and Shioura (1999) (see Fujishige et al. 2008).

7.5 Applications and Related Topics

We often encounter problems described by submodular functions with parameters, for which the theory of principal partitions furnishes a powerful tool.

7.5.1 The Minimum Ratio Problem

Suppose that we are given a submodular system $(\mathcal{D}, f)$ and a supermodular system $(\mathcal{D}, g)$ on E, where $f(X) \geq 0$ $(X \in \mathcal{D})$, $g(X) \geq 0$ $(X \in \mathcal{D})$, and there exists an $X \in \mathcal{D}$ such that $g(X) > 0$.

Consider the minimum ratio problem described as follows.

$$\text{Minimize } \frac{f(X)}{g(X)} \quad \text{subject to } X \in \mathcal{D}, \ g(X) > 0. \tag{106}$$

Define a submodular function h_λ on $\mathcal{D}$ with a real parameter λ by

$$h_\lambda(X) = f(X) - \lambda g(X) \quad (X \in \mathcal{D}). \tag{107}$$

Then we have

Theorem 5.1. *Let $\hat{\lambda}$ be the minimum value of the objective function of Problem* (106). *Then,*

$$\min\{h_\lambda(X) \mid X \in \mathcal{D}\} = 0 \quad (0 \leq \lambda \leq \hat{\lambda}), \tag{108}$$
$$\min\{h_\lambda(X) \mid X \in \mathcal{D}\} < 0 \quad (\hat{\lambda} < \lambda). \tag{109}$$

Moreover, the converse also holds.

Remark 5.2. It should be noted that Theorem 5.1 does not depend on the submodularity (supermodularity) of f (g) and holds for any set functions. However, if f (g) is submodular (supermodular), then Problem (106) has a close relationship with the principal partition.

Theorem 5.1 means that $\hat{\lambda}$ is a critical value for h_λ such that

$$\mathcal{D}(h_\lambda) = \{\emptyset\} \quad (0 \le \lambda < \hat{\lambda}), \qquad \mathcal{D}(h_\lambda) \ne \{\emptyset\} \quad (\hat{\lambda} \le \lambda). \tag{110}$$

Hence the minimum ratio problem for submodular and supermodular functions f and g is reduced to finding such a critical value $\hat{\lambda}$ and a set $X \in \mathcal{D}(h_{\hat{\lambda}})$ for $h_{\hat{\lambda}} = f - \hat{\lambda}g$.

The Network Attack Problem of Cunningham

Cunningham (1985) introduced a measure of network (anti-)vulnerability as follows. For a connected graph $G = (V, E)$ and a positive weight vector $w : E \to \mathbb{R}_+$ the *strength* of the weighted graph is defined by

$$\sigma(G, w) = \min\left\{ \frac{w(X)}{\kappa(X)} \;\middle|\; X \subseteq E, \; \kappa(X) > 0 \right\}, \tag{111}$$

where $\kappa(X)$ denotes the number of the connected components of the subgraph $G \cdot (E \setminus X)$ minus one. We can easily see that $\kappa : 2^E \to \mathbb{Z}_+$ is a supermodular function expressed in terms of the rank function r_G of G as

$$\kappa(X) = r_G(E) - r_G(E \setminus X) \; (= r_G^{\#}(X)) \quad (X \subseteq E). \tag{112}$$

Hence the problem of computing the strength of G relative to weight w is a special case of the minimum ratio problem described above. Letting $\hat{\lambda}$ be the largest critical value for $r_G - \lambda w$, we obtain

$$\sigma(G, w) = 1/\hat{\lambda}. \tag{113}$$

Also see (Baïou et al. 2000; Barahona and Kerivin 2004) for related topics on *partition inequalities*, which is also closely related to the *principal lattice of partitions* of Narayanan (1997) (see also Desai et al. 2003; Patkar and Narayanan 2003 for their applications). Note that for a given submodular function f the principal lattice of partitions for f is concerned with the Dilworth truncation of the submodular function $f - \lambda$ with a real parameter λ.

Maximum Density Subgraphs

For a graph $G = (V, E)$ define the density of G by

$$d(G) = \frac{|E|}{|V| - 1}. \tag{114}$$

A subgraph of G of maximum density is connected, so that the problem of finding a maximum-density subgraph $H = (W, F)$ of G is reduced to the following problem.

$$\text{Maximize } \frac{|F|}{r_G(F)} \quad \text{subject to } \emptyset \neq F \subseteq E, \tag{115}$$

which is equivalent to the minimum-ratio problem

$$\text{Minimize } \frac{r_G(F)}{|F|} \quad \text{subject to } \emptyset \neq F \subseteq E. \tag{116}$$

Hence the problem is reduced to finding the minimum critical value λ_1 for $r_G(X) - \lambda|X|$.

The concept of density of a graph is closely related to connectivity and reliability of networks, to which the principal partitions can be applied effectively.

7.5.2 Resource Allocation Problems

Since the canonical simplex

$$\Delta_\beta = \{x \mid x \in \mathbb{R}^E_+, \ x(E) = \beta\} \tag{117}$$

for $\beta > 0$ is a special case of a base polyhedron, base polyhedra naturally arise in resource allocation problems. Also the core of a convex game (Shapley 1971) is a base polyhedron, so that we often consider allocation problems over cores or base polyhedra.

Given a positive weight vector $w : E \to \mathbb{R}_+$, the weighted min-max resource allocation problem over the base polyhedron $B(f)$ associated with a submodular system $(\mathcal{D}, f)$ on E is described as

$$\text{Minimize } \max\{x(e)/w(e) \mid e \in E\} \quad \text{subject to } x \in B(f). \tag{118}$$

Also, the weighted max-min resource allocation problem is described as

$$\text{Maximize } \min\{x(e)/w(e) \mid e \in E\} \quad \text{subject to } x \in B(f). \tag{119}$$

Then we can show the following (see also Fujishige 1980a and Fujishige 2005, Chap. V for more general and detailed discussions).

Theorem 5.3. *Let b^* be the universal base (or the lexicographically optimal base) for submodular system $(\mathcal{D}, f)$ with weight w. Then $x = b^*$ is an optimal solution of both problems* (118) *and* (119).

Moreover, the minimum (resp. maximum) critical value for $f - \lambda w$ is equal to the optimal objective function value of the max-min (resp. min-max) resource allocation problem (119) *(resp.* (118)).

The following *equitable resource allocation* problem was considered by Jain and Vazirani (2002). Let $w_\lambda : E \to \mathbb{R}$ be a vector with a parameter $\lambda \in \mathbb{R}$. We assume that for each $e \in E$ the component $w_\lambda(e)$ of w_λ is increasing in λ. Then, for a submodular system $(2^E, f)$ we have dual problems characterized by the following min-max relation for any λ.

$$\max\{x(E) \mid x \in \mathrm{P}(f),\ x \le w_\lambda\} = \min\{f(X) + w_\lambda(E \setminus X) \mid X \subseteq E\}. \quad (120)$$

This can be seen as a special case of the min-max relation given in Theorem 4.1. Hence Theorem 4.8 implies that there exists a (unique) base $b^* \in \mathrm{B}(f)$ such that for all λ

$$(b^* - w_\lambda)^-(E) = \min\{f(X) - w_\lambda(X) \mid X \subseteq E\}. \quad (121)$$

This is also equivalent to

$$(b^* \wedge w_\lambda)(E) = \max\{x(E) \mid x \in \mathrm{P}(f),\ x \le w_\lambda\} \quad (122)$$

for all λ. The universal base b^* is the desired equitable allocation.

More general convex minimization problems over base polyhedra have recently been examined by Nagano (2007a), which shows the equivalence between the lexicographic optimal base problem and the submodular utility allocation market problem (Jain and Vazirani 2007). Separable nonquadratic convex function minimization over base polyhedra is also considered in Fujishige (2005, Chap. V).

7.6 Concluding Remarks

Combinatorial optimization problems characterized by submodular functions arise in a lot of applications such as graph and network optimizations, scheduling problems, queueing network problems, information-theoretic data analysis and communication networks, games and economic equilibrium problems, etc. (see, e.g., Frank 1993; Fujishige 2005; Fujishige and Tamura 2007; Jain and Vazirani 2007; Lehmann et al. 2006; Murota 2000; Narayanan 1997; Queyranne 1993; Shanthikumar and Yao 1992; Studený 2005; Topkis 1998; Yeung 2002). Such combinatorial optimization problems often lead us to submodular function minimization, where the theory of principal partitions can provide us with the powerful tool for extracting useful structural information about the problems under consideration.

The essence of the theory of principal partitions is given in the author's book (Fujishige 2005) but it is rather scattered through the book (see also Tomizawa and Fujishige 1982). The author hopes that the present article will help readers fully appreciate the usefulness of the theory of principal partitions.

Acknowledgements

I would like to express my sincere gratitude to Professor Bernhard Korte for supporting my research at his institute in Bonn in 1982–83 and 1993–94 for more than two years besides other several shorter visits, which very much helped me to pursue my life work on the theory of submodular functions including principal partitions discussed here. Concerning this article, I am very grateful to Kazuo Murota and Kiyohito Nagano for their careful reading of an earlier version of the manuscript and giving valuable comments, and to the anonymous referee for his careful reading of the manuscript, which improved the presentation. Thanks are also due to Kiyohito Nagano and Takumi Hayashi for the joint work mentioned in Sects. 7.4.1 and 7.4.2.

References

Baïou, M., Barahona, F., Mahjoub, A.R.: Separation of partition inequalities. Math. Oper. Res. **25**, 243–254 (2000)

Barahona, F., Kerivin, H.: Separation of partition inequalities with terminals. Discrete Optim. **1**, 129–140 (2004)

Birkhoff, G.: Lattice Theory. American Mathematical Colloquium Publications, vol. 25, 3rd edn. Am. Math. Soc., Providence (1967)

Brumelle, S., Granot, D., Liu, L.: Ordered optimal solutions and parametric minimum cut problems. Discrete Optim. **2**, 123–134 (2005)

Bruno, J., Weinberg, L.: The principal minors of a matroid. Linear Algebra Appl. **4**, 17–54 (1971)

Cunningham, W.H.: Optimal attack and reinforcement of a network. J. Assoc. Comput. Mach. **32**, 549–561 (1985)

Desai, M.P., Narayanan, H., Patkar, S.B.: The realization of finite state machines by decomposition and the principal lattice of partitions of a submodular function. Discrete Appl. Math. **131**, 299–310 (2003)

Dutta, B.: The egalitarian solution and reduced game properties in convex games. Int. J. Game Theory **19**, 153–169 (1990)

Dutta, B., Ray, D.: A concept of egalitarianism under participation constraints. Econometrica **57**, 615–635 (1989)

Edmonds, J.: Minimum partition of a matroid into independent subsets. J. Res. Natl. Bur. Stand. B Math. Sci. **69**, 67–72 (1965a)

Edmonds, J.: Lehman's switching game and a theorem of Tutte and Nash-Williams. J. Res. Natl. Bur. Stand. B Math. Sci. **69**, 73–77 (1965b)

Edmonds, J.: Matroid intersection. Ann. Discrete Math. **4**, 39–49 (1979)

Edmonds, J.: Submodular functions, matroids, and certain polyhedra. In: Guy, R., Hanani, H., Sauer, N., Schönheim, J. (eds.) Proceedings of the Calgary International Conference on Combinatorial Structures and Their Applications, pp. 69–87. Gordon and Breach, New York (1970). Also in: Combinatorial Optimization—Eureka, You Shrink! Jünger, M., Reinelt, G., Rinaldi, G. (eds.) Lecture Notes in Computer Science, vol. 2570, pp. 11–26. Springer, Berlin (2003)

Edmonds, J., Fulkerson, D.R.: Transversals and matroid partition. J. Res. Natl. Bur. Stand. B Math. Sci. **69**, 147–157 (1965)

Favati, P., Tardella, F.: Convexity in nonlinear integer programming. Ric. Oper. **53**, 3–44 (1990)

Federguen, A., Groenevelt, H.: The greedy procedure for resource allocation problems—necessary and sufficient conditions for optimality. Oper. Res. **34**, 909–918 (1986)

Fleischer, L., Iwata, S.: A push-relabel framework for submodular function minimization and applications to parametric optimization. Discrete Appl. Math. **131**, 311–322 (2003)

Frank, A.: Applications of submodular functions. In: Walker, K. (ed.) Surveys in Combinatorics. London Mathematical Society Lecture Note Series, vol. 187, pp. 85–136. Cambridge University Press, Cambridge (1993)

Fujishige, S.: Lexicographically optimal base of a polymatroid with respect to a weight vector. Math. Oper. Res. **5**, 186–196 (1980a)

Fujishige, S.: Principal structures of submodular systems. Discrete Appl. Math. **2**, 77–79 (1980b)

Fujishige, S.: Submodular Functions and Optimization, 2nd edn. Annals of Discrete Mathematics, vol. 58. Elsevier, Amsterdam (2005)

Fujishige, S., Murota, K.: Notes on L-/M-convex functions and the separation theorems. Math. Program. **88**, 129–146 (2000)

Fujishige, S., Nagano, K.: A structure theory for the parametric submodular intersection problem. Preprint Series RIMS-1629, RIMS, Kyoto University, March 2008

Fujishige, S., Tamura, A.: A two-sided discrete-concave market with possibly bounded side payments: an approach by discrete convex analysis. Math. Oper. Res. **32**, 136–155 (2007)

Fujishige, S., Hayashi, T., Isotani, S.: The minimum-norm-point algorithm applied to submodular function minimization and linear programming. RIMS preprints series No. 1571, Research Institute for Mathematical Sciences, Kyoto University, September 2006

Fujishige, S., Hayashi, T., Nagano, K.: Minimizing discrete convex functions with inequality constraints. Preprint Series RIMS-1622, RIMS, Kyoto University, March 2008

Gallo, G., Grigoriadis, M.D., Tarjan, R.E.: A fast parametric maximum flow algorithm and applications. SIAM J. Comput. **18**, 30–55 (1989)

Granot, F., Veinott, A.F. Jr.: Substitutes, complements and ripples in network flows. Math. Oper. Res. **10**, 471–497 (1985)

Groenevelt, H.: Two algorithms for maximizing a separable concave function over a polymatroid feasible region. Eur. J. Oper. Res. **54**, 227–236 (1991)

Grötschel, M., Lovász, L., Schrijver, A.: Geometric Algorithms and Combinatorial Optimization. Algorithms and Combinatorics, vol. 2. Springer, Berlin (1988)

Hochbaum, D.S., Hong, S.P.: About strongly polynomial time algorithms for quadratic optimization over submodular constraints. Math. Program. **69**, 269–309 (1995)

Hokari, T.: Monotone-path Dutta-Ray solution on convex games. Soc. Choice Welf. **19**, 825–844 (2002)

Hokari, T., van Gellekom, A.: Population monotonicity and consistency in convex games: Some logical relations. Int. J. Game Theory **31**, 593–607 (2002)

Iri, M.: A min-max theorem for the ranks and term-ranks of a class of matrices—an algebraic approach to the problem of the topological degrees of freedom of a network. Trans. Inst. Electron. Commun. Eng. Jpn., Sect. A **51**, 180–187 (1968) (in Japanese)

Iri, M.: The maximum-rank minimum-term-rank theorem for the pivotal transforms of a matrix. Linear Algebra Appl. **2**, 427–446 (1969)

Iri, M.: A review of recent work in Japan on principal partitions of matroids and their applications. Ann. New York Acad. Sci. **319**, 306–319 (1979)

Iri, M.: Applications of matroid theory. In: Bachem, A., Grötschel, M., Korte, B. (eds.) Mathematical Programming—The State of the Art, pp. 158–201. Springer, Berlin (1983)

Iri, M.: Structural theory for the combinatorial systems characterized by submodular functions. In: Pulleyblank, W.R. (ed.) Progress in Combinatorial Optimization, pp. 197–219. Academic Press, Toronto (1984)

Iwata, S.: Principal structure of submodular systems and Hitchcock-type independent flows. Combinatorica **15**, 515–532 (1996)

Iwata, S.: A fully combinatorial algorithm for submodular function minimization. J. Comb. Theory Ser. B **84**, 203–212 (2002)

Iwata, S.: A faster scaling algorithm for minimizing submodular functions. SIAM J. Comput. **32**, 833–840 (2003)

Iwata, S.: Submodular function minimization. Math. Program. Ser. B **112**, 45–64 (2008)

Iwata, S., Murota, K.: A theorem on the principal structure for independent matchings. Discrete Appl. Math. **61**, 229–244 (1995)

Iwata, S., Murota, K.: Horizontal principal structure of layered mixed matrices— Decomposition of discrete systems by design-variable selections. SIAM J. Discrete Math. **9**, 71–86 (1996)

Iwata, S., Murota, K., Shigeno, M.: A fast submodular intersection algorithm for strong map sequences. Math. Oper. Res. **22**, 803–813 (1997)

Iwata, S., Fleischer, L., Fujishige, S.: A combinatorial strongly polynomial algorithm for minimizing submodular functions. J. Assoc. Comput. Mach. **48**, 761–777 (2001)

Jain, K., Vazirani, V.V.: Equitable cost allocations via primal-dual-type algorithms. In: Proceedings of STOC'02 (May 19–21, 2002, Montreal, Quebec, Canada), pp. 313–321 (2002)

Jain, K., Vazirani, V.V.: Eisenberg-Gale markets: algorithms and structural properties. In: Proceedings of STOC'07 (June 11–13, 2007, San Diego, California, USA), pp. 364–373 (2007)

Kishi, G., Kajitani, Y.: Maximally distant trees in a linear graphs. Trans. Inst. Electron. Commun. Eng. Jpn., Sect. A **51**, 196–203 (1968) (in Japanese). See also: On maximally distinct trees. In: Proceedings of the Fifth Annual Allerton Conference on Circuit and System Theory, pp. 635–643 (1967)

Lehmann, B., Lehmann, D., Nisan, N.: Combinatorial auctions with decreasing marginal utilities. Games Econ. Behav. **55**, 270–296 (2006)

Lovász, L.: Submodular functions and convexity. In: Bachem, A., Grötschel, M., Korte, B. (eds.) Mathematical Programming—The State of the Art, pp. 235–257. Springer, Berlin (1983)

McCormick, S.T.: Submodular function minimization. In: Aardal, K., Nemhauser, G.L., Weismantel, R. (eds.) Discrete Optimization. Handbooks in Operations Research, vol. 12, pp. 321–391. Elsevier, Amsterdam (2005)

Megiddo, N.: Optimal flows in networks with multiple sources and sinks. Math. Program. **7**, 97–107 (1974)

Megiddo, N.: A good algorithm for lexicographically optimal flows in multi-terminal networks. Bull. Am. Math. Soc. **83**, 407–409 (1977)

Milgrom, P., Shannon, C.: Monotone comparative statics. Econometrica **62**, 157–180 (1994)

Murota, K.: Menger-decomposition of a graph and its application to the structural analysis of a large-scale system of equations. Discrete Appl. Math. **17**, 107–134 (1987)

Murota, K.: Note on the universal bases of a pair of polymatroids. J. Oper. Res. Soc. Japan **31**, 565–572 (1988)

Murota, K.: Principal structure of layered mixed matrix. Discrete Appl. Math. **27**, 221–234 (1990)

Murota, K.: Discrete convex analysis. Math. Program. **83**, 313–371 (1998)

Murota, K.: Matrices and Matroids for Systems Analysis. Algorithms and Combinatorics, vol. 20. Springer, Berlin (2000)

Murota, K.: Discrete Convex Analysis. SIAM Monographs on Discrete Mathematics and Applications, vol. 10. SIAM, Philadelphia (2003a)

Murota, K.: On steepest descent algorithms for discrete convex functions. SIAM J. Optim. **14**, 699–707 (2003b)

Murota, K.: Recent developments in discrete convex analysis. A chapter in this book (2008)

Murota, K., Shioura, A.: M-convex function on generalized polymatroid. Math. Oper. Res. **24**, 95–105 (1999)

Murota, K., Iri, M., Nakamura, M.: Combinatorial canonical form of layered mixed matrices and its application to block-triangularization of systems of equations. SIAM J. Algebr. Discrete Methods **8**, 123–149 (1987)

Nagano, K.: On convex minimization over base polytopes. In: Proceedings of the 12th IPCO. LNCS, vol. 4513, pp. 252–266. Springer, Berlin (2007a)

Nagano, K.: A faster parametric submodular function minimization algorithm and applications. Mathematical Engineering Technical Reports, METR 2007-43, Department of Mathematical Informatics, Graduate School of Information Science and Technology, University of Tokyo, July 2007b

Nakamura, M.: Structural theorems for submodular functions, polymatroids and polymatroid intersections. Graphs Combinatorics **4**, 257–284 (1988)

Nakamura, M., Iri, M.: A structural theory for submodular functions, polymatroids and polymatroid intersections. Research Memorandum RMI 81-06, Department of Mathematical Engineering and Instrumentation Physics, Faculty of Engineering, University of Tokyo, August 1981. See also (Iri 1984; Nakamura 1988)

Narayanan, H.: Theory of matroids and network analysis. Ph.D. Thesis, Department of Electrical Engineering, Indian Institute of Technology, Bombay, February 1974

Narayanan, H.: Submodular Functions and Electrical Networks. Annals of Discrete Mathematics, vol. 54. North-Holland, Amsterdam (1997)

Narayanan, H., Vartak, M.N.: An elementary approach to the principal partition of a matroid. Trans. Inst. Electron. Commun. Eng. Jpn., Sect. E **64**, 227–234 (1981)

Nash-Williams, C.St.J.A.: Edge-disjoint spanning trees of finite graphs. J. Lond. Math. Soc. **36**, 445–450 (1961)

Ohtsuki, T., Ishizaki, Y., Watanabe, H.: Network analysis and topological degrees of freedom. Trans. Inst. Electron. Commun. Eng. Jpn., Sect. A **51**, 238–245 (1968) (in Japanese)

Orlin, J.B.: A faster strongly polynomial time algorithm for submodular function minimization. In: Proceedings of the 12th IPCO. LNCS, vol. 4513, pp. 240–251. Springer, Berlin (2007). Math. Program. (to appear)

Ozawa, T.: Common trees and partition of two-graphs. Trans. Inst. Electron. Commun. Eng. Jpn., Sect. A **57**, 383–390 (1974) (in Japanese)

Patkar, S.B., Narayanan, H.: Improving graph partitions using submodular functions. Discrete Appl. Math. **131**, 535–553 (2003)

Picard, J.C., Queyranne, M.: On the structure of all minimum cuts in a network and applications. Math. Program. Study **13**, 8–16 (1980)

Queyranne, M.: Structure of a simple scheduling polyhedron. Math. Program. **58**, 263–285 (1993)

Rado, R.: A theorem on independence relations. Q. J. Math. **13**, 83–89 (1942)

Schrijver, A.: A combinatorial algorithm minimizing submodular functions in strongly polynomial time. J. Comb. Theory, Ser. B **80**, 346–355 (2000)

Shanthikumar, J.G., Yao, D.D.: Multiclass queueing systems: polymatroidal structure and optimal scheduling control. Oper. Res. **40**, S293–S299 (1992)

Shapley, L.S.: Cores of convex games. Int. J. Game Theory **1**, 11–26 (1971)

Studený, M.: Probabilistic Conditional Independence Structures. Springer, Berlin (2005)

Tomizawa, N.: Strongly irreducible matroids and principal partition of a matroid into strongly irreducible minors. Trans. Inst. Electron. Commun. Eng. Jpn., Sect. A **59**, 83–91 (1976) (in Japanese)

Tomizawa, N., Fujishige, S.: Historical survey of extensions of the concept of principal partition and their unifying generalization to hypermatroids. Systems Science Research Report No. 5, Department of Systems Science, Tokyo Institute of Technology, April 1982; also its abridgment appeared in Proceedings of the 1982 International Symposium on Circuits and Systems (Rome, May 10–12, 1982), pp. 142–145

Topkis, D.M.: Minimizing a submodular function on a lattice. Oper. Res. **26**, 305–321 (1978)

Topkis, D.M.: Supermodularity and Complementarity. Princeton University Press, Princeton (1998)

Tsuchiya, T., Ohtsuki, T., Ishizaki, Y., Watanabe, H., Kajitani, Y., Kishi, G.: Topological degrees of freedom of electrical networks. In: Proceedings of the Fifth Annual Allerton Conference on Circuit and System Theory (October 4–6, 1967), pp. 644–653

Tutte, W.T.: On the problem of decomposing a graph into n connected factors. J. Lond. Math. Soc. **36**, 221–230 (1961)

Wolfe, P.: Finding the nearest point in a polytope. Math. Program. **11**, 128–149 (1976)

Yeung, R.W.: A First Course in Information Theory. Springer, Berlin (2002)

Locally Dense Independent Sets in Regular Graphs of Large Girth—An Example of a New Approach

Frank Göring, Jochen Harant, Dieter Rautenbach, and Ingo Schiermeyer

Summary. We present an example for a new approach which seems applicable to every graph theoretical concept defined by local conditions and regular graphs of large girth. It combines a random outer procedure processing the graph in rounds with a virtually arbitrary algorithm solving local instances within each round and combines the local solutions to a global one. The local uniformity of the considered instances and the randomness of the outer procedure make the asymptotic analysis possible. Here we apply this approach to the simplest yet fundamental example of a locally defined graph theoretical concept: independent sets in graphs.

For an integer $d \geq 3$ let $\alpha(d)$ be the supremum over all α with the property that for every $\epsilon > 0$ there exists some $g(\epsilon)$ such that every d-regular graph of order n and girth at least $g(\epsilon)$ has an independent set of cardinality at least $(\alpha - \epsilon)n$.

Considerably extending the work of Lauer and Wormald (Large independent sets in regular graphs of large girth, J. Comb. Theory, Ser. B **97**, 999–1009, 2007) and improving results due to Shearer (A note on the independence number of triangle-free graphs, II, J. Comb. Theory, Ser. B **53**, 300–307, 1991) and Lauer and Wormald, we present the best known lower bounds for $\alpha(d)$ for all $d \geq 3$.

8.1 Introduction

The results we will present here can be regarded as an example of the successful implementation of a new approach. In principle this approach can be applied to every graph theoretical concept defined by local conditions and regular graphs of large girth. The basic idea is to combine locally good or even optimal solutions to a good global solution. Which local solutions are selected is decided by a random procedure working in rounds. Within every round of this outer random framework a virtually arbitrary algorithm solving the local problem can be executed. The key observation is that the local uniformity of regular graphs of large girth allows an asymptotic analysis of the evolution of the underlying graph. After selecting certain local solutions the corresponding parts of the graph have to be discarded during the following rounds. By the randomness of the outer procedure this evolution is essentially governed by two probabilities: the survival probability of a vertex and the conditional survival probability of a neighbour of a survivor vertex. The only technical condition which the algorithm solving the local problems has to satisfy is that its expected performance when applied to a finite part of the infinite regular tree whose vertices survived according to these two probabilities can be quantified. Most graph algorithms naturally satisfy this condition.

Here we apply this approach to the simplest yet fundamental example of a locally defined graph theoretical concept: independent sets in graphs. While the treatment becomes quite technical already for this simplest example the reader should note that much of the analysis especially of the outer random framework and the associated survival probabilities is in fact totally independent of the actual local problem solved by the inner algorithm.

We consider the independence number $\alpha(G)$ of finite, simple and undirected graphs $G = (V, E)$ which are d-regular for some $d \geq 3$ and have large girth.

For integers $d \geq 3$ and $g \geq 3$ let $\mathcal{G}(d, g)$ denote the class of all d-regular graphs of girth at least g and let

$$\alpha(d, g) := \sup\{\alpha \mid \alpha(G) \geq \alpha \cdot |V| \text{ for all } G = (V, E) \in \mathcal{G}(d, g)\}.$$

Clearly, $\alpha(d, g)$ is monotonic non-decreasing in g and bounded above by 1 and we can consider

$$\alpha(d) := \lim_{g \to \infty} \alpha(d, g).$$

Note that this definition implies that for every $\epsilon > 0$ there exists some $g(\epsilon)$ such that $\alpha(G) \geq (\alpha(d) - \epsilon) \cdot |V|$ for every graph $G = (V, E) \in \mathcal{G}(d, g(\epsilon))$.

Our aim is to prove lower bounds on $\alpha(d)$. While the first result on the independence number in regular graphs of large girth is due to Hopkins and Staton (1982) who proved $\alpha(3) \geq \frac{7}{18} \approx 0.3888$, for quite a long time the best known estimates of $\alpha(d)$ were due to Shearer (1991).

Theorem 1.1 (Shearer 1991). *If*

$$\beta_{\text{Shearer}}(d) := \begin{cases} \frac{125}{302} \approx 0.4139 & \text{for } d = 3, \\ \frac{1 + d(d-1)\beta_{\text{Shearer}}(d-1)}{d^2 + 1} & \text{for } d \geq 4, \end{cases}$$

then

$$\alpha(d) \geq \beta_{\text{Shearer}}(d)$$

for all $d \geq 3$.

Only very recently Lauer and Wormald (2007) improved Shearer's result for $d \geq 7$.

Theorem 1.2 (Lauer and Wormald 2007). *For all $d \geq 3$*

$$\alpha(d) \geq \beta_{\text{LauWo}}(d) := \frac{1 - (d-1)^{-2/(d-2)}}{2}.$$

From a very abstract viewpoint their approaches are actually similar. On the one hand Shearer constructs an independent set by carefully selecting vertices according to some degree dependent weight function, adding them to the independent set, deleting them together with their neighbours and iterating this process. On the other hand Lauer and Wormald construct an independent set by randomly selecting vertices, adding most of them to the independent set, deleting them together with their neighbours and iterating this process.

In order to get some intuition about how to improve these approaches it is instructive to see that a very simple argument allows to improve Shearer's bound on $\alpha(3)$.

Proposition 1.3. $\alpha(3) \geq 0.4144 > \beta_{\text{Shearer}}(3) \approx 0.4139$.

Proof. It follows from Theorem 4 in Shearer (1991) that for every $\epsilon > 0$ there is some $g(\epsilon)$ such that: *If $G = (V, E)$ is a graph of order n and girth at least $g(\epsilon)$, with n_2 vertices of degree 2 no two of which are adjacent and $n_3 = n - n_2$ vertices of degree 3, then*

$$\alpha(G) \geq \left(\frac{79}{151} - \epsilon\right)n_2 + \left(\frac{125}{302} - \epsilon\right)n_3. \tag{1}$$

For a cubic graph G of order n and sufficiently large girth $g(\epsilon)$ the 7-th power G^7 is $(3 + 3 \cdot 2 + 3 \cdot 2^2 + \cdots + 3 \cdot 2^6) = 381$-regular. Therefore, G^7 has an independent set I^7 with $|I^7| \geq n/(\Delta(G^7) + 1) = n/382$ where $\Delta(G^7)$ denotes the maximum degree of G^7 (cf. e.g. Caro 1979; Wei 1981).

Let H arise from G by deleting all vertices within distance at most 2 from I^7. Clearly, H has minimum degree 2 and no two vertices of degree 2 in H are adjacent. We construct an independent set of G by adding all $3|I^7|$ many vertices at distance 1 from a vertex in I^7 and by applying (1) to H. It follows that

$$\alpha(G) \geq 3|I^7| + \alpha(H)$$
$$\geq 3|I^7| + \left(\frac{79}{151} - \epsilon\right)12|I^7| + \left(\frac{125}{302} - \epsilon\right)(n - 22|I^7|)$$
$$\geq (0.4144 - \epsilon)n$$

which completes the proof. $\square$

The proof of Proposition 1.3 suggests that it is worthwhile to consider the iterative deletion not just of a vertex and its neighbours—which would induce a rooted tree of depth 1—but of rooted trees of larger depths. Locally this should allow us to pack the vertices of the independent set more densely which hopefully yields an overall improvement.

We follow exactly this intuition by generalizing the random procedure and its analysis using differential equations proposed by Lauer and Wormald (2007).

8.2 The Algorithm $TREE(k, l, p, f)$

In this section we describe a random procedure $TREE(k, l, p, f)$ which depends on two integers $k, l \geq 0$, a real value $p \in [0, 1]$ and a function f which maps rooted trees T with root r to independent subsets $f(T, r)$ of their vertex set. We assume that $|f(T, r)| = |f(T', r')|$ for isomorphic rooted trees, T rooted at r and T' rooted at r'.

The algorithm $TREE(k, l, p, f)$ will be applied to a graph $G = (V, E)$ of girth at least $2(l + 1)$. It executes k rounds and determines disjoint rooted subtrees of G of depth at most l. The value p will serve as a probability.

We need a little more notation to describe the algorithm. Let $u \in V$ be a vertex of the graph G. For an integer $i \geq 0$ let $N_G^i(u)$ and $N_G^{\leq i}(u)$ denote the sets of vertices of G within distance—measured with respect to G—exactly i from u and at most i from u, respectively, i.e.

$$N_G^i(u) = \{v \in V \mid \operatorname{dist}_G(v, u) = i\} \quad \text{and}$$

$$N_G^{\leq i}(u) = \{v \in V \mid \operatorname{dist}_G(v, u) \leq i\}.$$

Furthermore, let $B_G^i(u)$ denote the set of vertices $v \in N_G^{\leq i}(u)$ which are not adjacent to a vertex in $V \setminus N_G^{\leq i}(u)$, i.e.

$$B_G^i(u) = \left\{v \in V \mid N_G^{\leq 1}(v) \subseteq N_G^{\leq i}(u)\right\}.$$

For a set $U \subseteq V$ of vertices of G the subgraph of G induced by $V \setminus U$ is denoted by $G - U$ or $G[V \setminus U]$.

$TREE(k, l, p, f)$ proceeds as follows:

(1) Set $G_0 = (V_0, E_0) := G$, $Z_0 := \emptyset$ and $i := 0$.
(2) While $i < k$ select a subset X_i of V by assigning every vertex of G to X_i independently at random with probability p.
Set

$$G_{i+1} := (V_{i+1}, E_{i+1}) := G_i - \bigcup_{u \in X_i \cap V_i} N_{G_i}^{\leq l}(u), \tag{2}$$

$$X_i^* := \left\{v \in X_i \mid \operatorname{dist}_G(v, u) \geq 2l + 1 \; \forall u \in X_i \setminus \{v\}\right\}, \tag{3}$$

$$T_i(u) := G_i\left[B_{G_i}^l(u)\right], \tag{4}$$

$$\Delta Z_i := \bigcup_{u \in X_i^* \cap V_i} f(T_i(u), u), \tag{5}$$

$$Z_{i+1} := Z_i \cup \Delta Z_i, \tag{6}$$

$$i := i + 1.$$

(3) Output Z_k.

There are some subtleties we want to stress: The definition of G_{i+1} in (2) and ΔZ_i in (5) use neighbourhoods within the graph G_i while X_i^* is defined in (3) with respect to G. Furthermore, the construction of Z_{i+1} in (4), (5) and (6) does not influence the evolution of the G_i. By the girth condition, $T_i(v)$ is a tree and $f(T_i(v), v)$ is a well-defined subset of $B_{G_i}^l(v)$.

We first observe that $TREE(k, l, p, f)$ really produces an independent set of G.

Lemma 2.1. Z_k *is an independent set of* G.

Proof. For contradiction, we assume that $v, w \in Z_k$ with $vw \in E$.

Let $v \in \Delta Z_i$ and $w \in \Delta Z_j$ with, say, $i \leq j$. Let $v \in f(T_i(u), u)$ for some $u \in X_i^* \cap V_i$. Since, by the definition of X_i^* in (3), the set $N_{G_i}^{\leq l}(u) \cap N_{G_i}^{\leq l}(u')$ is empty for all distinct $u, u' \in X_i^*$ we obtain that $w \in V_i$ is a neighbour of v outside of $N_{G_i}^{\leq l}(u)$ which implies the contradiction $v \notin B_{G_i}^l(u)$. $\square$

8.3 The Analysis of *TREE*(k, l, p, f)

Throughout this section we will assume that $G = (V, E)$ is a d-regular graph for some $d \geq 3$ and sufficiently large girth. We consider the behaviour of $TREE(k, l, p, f)$ when applied to this graph. We will specify the necessary girth conditions which are all in terms of k and l more exactly whenever they are explicitly needed.

It is one of the key observations made by Lauer and Wormald (2007) that for a sufficiently large girth the probabilities which are suitable to describe the behaviour of their randomized algorithm can be well understood. The next lemma corresponds to Lemma 2 in Lauer and Wormald (2007).

Lemma 3.1. *Let* $k \geq 2$ *and* $0 \leq i \leq k$. *Let the girth of* G *be at least* $2(k+1)l + 2$ *and let* $u \in V$ *and* $vv' \in E$.

  (i) *The probabilities* $\mathbf{P}[u \in V_i]$, $\mathbf{P}[(v \in V_i) \wedge (v' \in V_i)]$ *and* $\mathbf{P}[u \in \Delta Z_i]$ *as well as the conditional expected value* $\mathbf{E}[|f(T_i(u), u)| \mid u \in X_i^* \cap V_i]$ *do not depend on the choice of the vertex* u *or the edge* vv'.

  (ii) *Conditional upon the event* $(v \in V_i)$, *the event* $(v' \in V_i)$ *depends only on the intersection of the sets* $X_0, X_1, \ldots, X_{i-1}$ *with* $N_{G-v}^{\leq il}(v')$.

Proof. It follows immediately, by induction on i, from the description of $TREE(k, l, p, f)$ that the events $(u \in V_i)$ and $(v \in V_i) \wedge (v' \in V_i)$ depend only on the intersection of the sets $X_0, X_1, \ldots, X_{i-1}$ with $N_G^{\leq il}(u)$ and $N_G^{\leq il}(v) \cup N_G^{\leq il}(v')$, respectively. Furthermore, the event $(u \in \Delta Z_i)$ depends only on the intersection of the sets $X_0, X_1, \ldots, X_{i-1}$ with $N_G^{\leq (i+1)l}(u)$ and the intersection of the set X_i with $N_G^{\leq 2l}(u)$. Finally, conditional upon the event $(u \in X_i^* \cap V_i)$, the cardinality of $f(T_i(u), u)$ depends only on the intersection of the sets $X_0, X_1, \ldots, X_{i-1}$ with $N_G^{\leq (i+1)l}(u)$.

Since, by the girth condition, the induced subgraphs

$$G\left[N_G^{\leq 2l}(u)\right], \quad G\left[N_G^{\leq (i+1)l}(u)\right] \quad \text{and} \quad G\left[N_G^{\leq il}(v) \cup N_G^{\leq il}(v')\right]$$

are isomorphic for all choices of the vertex u or the edge vv', we obtain (i). Similarly, (ii) follows immediately by induction on i. □

By Lemma 3.1, for $0 \leq i \leq k$ the following quantities

$$r_i := \mathbf{P}[u \in V_i],$$

$$w_i := \frac{\mathbf{P}[(v \in V_i) \wedge (v' \in V_i)]}{\mathbf{P}[v \in V_i]} = \mathbf{P}[(v \in V_i) \wedge (v' \in V_i) | v \in V_i],$$

$$f_l(w_i) := \mathbf{E}[|f(T_i(u), u)| \mid u \in X_i^* \cap V_i],$$

$$\Delta z_i := \mathbf{P}[u \in \Delta Z_i] \quad \text{and}$$

$$z_i := \mathbf{P}[u \in Z_k \setminus Z_i]$$

are the same for every vertex $u \in V$ and every edge $vv' \in E$.

Using Lemma 3.1, we can determine the following recursions for these probabilities.

Lemma 3.2. *Let the girth of G be at least $2(k + 1)l + 2$.*

(i) $r_0 = w_0 = 1$ *and* $z_{i+1} = z_i - \Delta z_i$ *for* $0 \leq i \leq k - 1$.
(ii) *For* $0 \leq i \leq k - 1$

$$r_{i+1} = r_i \left(1 - p \cdot \left(1 + \sum_{j=1}^{l} d(d-1)^{j-1} \cdot w_i^j \right) + O(p^2) \right),$$

$$w_{i+1} = w_i \left(1 - p \cdot \left(1 + \sum_{j=1}^{l} (d-2)(d-1)^{j-1} \cdot w_i^j \right) + O(p^2) \right) \quad \text{and}$$

$$\Delta z_i = f_l(w_i) \cdot r_i \cdot p \cdot \prod_{j=1}^{2l} (1 - p)^{d(d-1)^{j-1}}$$

where the constants implicit in the $O(\cdot)$-terms depend only on d and l.

Proof. (i) is immediate from the definitions and we proceed to the proof of (ii).

Let $u \in V$ be fixed. For $v \in N_G^{\leq l}(u)$ let P_v denote the vertex set of the unique path of length at most l from u to v. The event $(u \in V_{i+1})$ holds if and only if

$$(u \in V_i) \wedge (u \notin X_i) \wedge \bigwedge_{v:1 \leq \text{dist}_G(u,v) \leq l} \left((v \notin X_i) \vee \left((v \in X_i) \wedge (P_v \nsubseteq V_i) \right) \right).$$

Expanding this representation of the event $(u \in V_{i+1})$ to a disjunction of conjunctions, all events corresponding to the conjunctions are disjoint because they differ in

$$X_i \cap \{v : 1 \leq \text{dist}_G(u, v) \leq l\}.$$

Furthermore, all of those events for which two of the independent events $(v \in X_i)$ for some v with $1 \leq \text{dist}_G(u, v) \leq l$ hold, will contribute together only $\mathbf{P}[u \in V_i] \cdot O(p^2)$ to $\mathbf{P}[u \in V_{i+1}]$ where the constant implicit in the $O(\cdot)$-term depends only on d and l.

Therefore,

$$
\mathbf{P}[u \in V_{i+1}] = \mathbf{P}\left[(u \in V_i) \wedge (u \notin X_i) \wedge \bigwedge_{v:1 \leq \mathrm{dist}_G(u,v) \leq l} (v \notin X_i)\right]
$$

$$
+ \mathbf{P}[u \in V_i] \cdot O(p^2)
$$

$$
+ \sum_{v:1 \leq \mathrm{dist}_G(u,v) \leq l} \mathbf{P}\left[(u \in V_i) \wedge (v \in X_i) \wedge (P_v \nsubseteq V_i)\right.
$$

$$
\wedge \bigwedge_{v':(v' \neq v) \wedge (0 \leq \mathrm{dist}_G(u,v') \leq l)} \left.(v' \notin X_i)\right]
$$

$$
= \mathbf{P}[u \in V_i] \cdot (1-p)^{(1+\sum_{j=1}^{l} d(d-1)^{j-1})}
$$

$$
+ \sum_{v:1 \leq \mathrm{dist}_G(u,v) \leq l} \mathbf{P}[(u \in V_i) \wedge (P_v \nsubseteq V_i)] p (1-p)^{\sum_{j=1}^{l} d(d-1)^{j-1}}
$$

$$
+ \mathbf{P}[u \in V_i] \cdot O(p^2)
$$

$$
= \mathbf{P}[u \in V_i] \cdot \left(1 - \left(1 + \sum_{j=1}^{l} d(d-1)^{j-1}\right) p\right)
$$

$$
+ \sum_{v:1 \leq \mathrm{dist}_G(u,v) \leq l} \mathbf{P}[(u \in V_i) \wedge (P_v \nsubseteq V_i)] \cdot p + \mathbf{P}[u \in V_i] \cdot O(p^2).
$$

In order to evaluate $\mathbf{P}[(u \in V_i) \wedge (P_v \nsubseteq V_i)]$ let $u = u_0 u_1 u_2 \ldots u_j = v$ be the unique path from u to v for some $1 \leq j \leq l$.

By Lemma 3.1(i) and (ii), we have for $0 \leq \nu \leq j-1$

$$
\mathbf{P}[(u_0, u_1, \ldots, u_\nu \in V_i) \wedge (u_{\nu+1} \notin V_i)]
$$

$$
= \mathbf{P}[u_0 \in V_i] \cdot \mathbf{P}[u_1 \in V_i \mid u_0 \in V_i] \cdot \mathbf{P}[u_2 \in V_i \mid u_0, u_1 \in V_i] \cdot \ldots
$$

$$
\cdot \mathbf{P}[u_\nu \in V_i \mid u_0, u_1, \ldots, u_{\nu-1} \in V_i] \cdot \mathbf{P}[u_{\nu+1} \notin V_i \mid u_0, u_1, \ldots, u_\nu \in V_i]
$$

$$
= \mathbf{P}[u_0 \in V_i] \cdot \mathbf{P}[u_1 \in V_i \mid u_0 \in V_i] \cdot \mathbf{P}[u_2 \in V_i \mid u_1 \in V_i] \cdot \ldots
$$

$$
\cdot \mathbf{P}[u_\nu \in V_i \mid u_{\nu-1} \in V_i] \cdot (1 - \mathbf{P}[u_{\nu+1} \in V_i \mid u_\nu \in V_i])
$$

$$
= r_i w_i^\nu (1 - w_i)
$$

and we obtain

$$
\mathbf{P}[(u \in V_i) \wedge (P_v \nsubseteq V_i)]
$$

$$
= \mathbf{P}[(u_0 \in V_i) \wedge ((u_0 \notin V_i) \vee (u_1 \notin V_i) \vee \cdots \vee (u_j \notin V_i))]
$$

$$
= \mathbf{P}[(u_0 \in V_i) \wedge (u_1 \notin V_i)] + \mathbf{P}[(u_0, u_1 \in V_i) \wedge (u_2 \notin V_i)]
$$

$$
+ \mathbf{P}[(u_0, u_1, u_2 \in V_i) \wedge (u_3 \notin V_i)] + \cdots
$$

$$
+ \mathbf{P}[(u_0, u_1, \ldots, u_{j-1} \in V_i) \wedge (u_j \notin V_i)]
$$

$$= r_i(1 - w_i) + r_i w_i(1 - w_i) + \cdots + r_i w_i^{j-1}(1 - w_i)$$
$$= r_i(1 - w_i^j).$$

Putting everything together, we obtain

$$r_{i+1} = \mathbf{P}[u \in V_{i+1}]$$

$$= r_i \cdot \left(1 - \left(1 + \sum_{j=1}^{l} d(d-1)^{j-1}\right)p\right)$$

$$+ \sum_{j=1}^{l} d(d-1)^{j-1} r_i \cdot (1 - w_i^j) \cdot p + r_i \cdot O(p^2)$$

$$= r_i \cdot (1 - p) - \sum_{j=1}^{l} d(d-1)^{j-1} r_i \cdot w_i^j \cdot p + r_i \cdot O(p^2)$$

$$= r_i \left(1 - p \cdot \left(1 + \sum_{j=1}^{l} d(d-1)^{j-1} \cdot w_i^j\right) + O(p^2)\right).$$

By the same type of argument, it follows that for every edge $vv' \in E$

$$\mathbf{P}[(v \in V_{i+1}) \wedge (v' \in V_{i+1})]$$

$$= \mathbf{P}[(v \in V_i) \wedge (v' \in V_i)]\left(1 - p\left(2 + \sum_{j=1}^{l}(2d-2)(d-1)^{j-1}w_i^j\right) + O(p^2)\right).$$

Since $w_i = \frac{\mathbf{P}[(v \in V_i) \wedge (v' \in V_i)]}{r_i}$, the desired equation for w_i follows.

Finally, we consider Δz_i. By the definitions in (3) and (5) and Lemma 3.1, we have

$$\Delta z_i = \mathbf{P}[u \in \Delta Z_i]$$
$$= \frac{\mathbf{E}[|\Delta Z_i|]}{|V|}$$
$$= f_l(w_i) \cdot \mathbf{P}[u \in X_i^* \cap V_i] = f_l(w_i) \cdot \mathbf{P}[u \in V_i] \cdot \mathbf{P}[u \in X_i^*]$$
$$= f_l(w_i) \cdot r_i \cdot p \prod_{j=1}^{2l}(1 - p)^{d(d-1)^{j-1}}$$

which completes the proof. $\square$

Setting

$$a_i := \frac{z_i w_i}{r_i} = \mathbf{P}[u \in Z_k \setminus Z_i \mid u \in V_i] \cdot w_i \le w_i$$

for $0 \le i \le k$ and

$$\Delta r_i := r_{i+1} - r_i,$$

$$\Delta w_i := w_{i+1} - w_i \quad \text{and}$$
$$\Delta a_i := a_{i+1} - a_i$$

for $0 \le i \le k - 1$, we obtain the following.

Lemma 3.3. *For $0 \le i \le k - 1$*

$$\frac{\Delta a_i}{\Delta w_i} = \frac{f_l(w_i) - 2a_i \frac{((d-1)w_i)^l - 1}{(d-1)w_i - 1}}{1 + (d-2)w_i \frac{((d-1)w_i)^l - 1}{(d-1)w_i - 1}} + O(p)$$

where the constant implicit in the $O(\cdot)$-term depends only on d and l.

Proof. Note that, by definition, $\Delta z_i = z_i - z_{i+1}$. Immediately from the previous definitions it is straightforward to verify that

$$\frac{\Delta a_i}{\Delta w_i} = \frac{w_i}{\Delta w_i} \left(a_i \frac{\frac{\Delta w_i}{w_i} \frac{r_i}{r_{i+1}} - \frac{\Delta r_i}{r_{i+1}}}{w_i} - \frac{\Delta z_i}{r_i} \frac{w_{i+1}}{w_i} \frac{r_i}{r_{i+1}} \right).$$

By Lemma 3.2,

$$\frac{w_{i+1}}{w_i} = 1 - O(p) \quad \text{and}$$
$$\frac{r_i}{r_{i+1}} = 1 + O(p).$$

Furthermore,

$$\frac{\Delta w_i}{w_i} = -\left(1 + (d-2)w_i \sum_{j=0}^{l-1} (w_i(d-1))^j \right) p + O(p^2),$$

$$\frac{\Delta r_i}{r_i} = -\left(1 + dw_i \sum_{j=0}^{l-1} (w_i(d-1))^j \right) p + O(p^2) \quad \text{and}$$

$$\frac{\Delta z_i}{r_i} = f_l(w_i)p + O(p^2)$$

which implies that also

$$\frac{\Delta r_i}{r_{i+1}} = -\left(1 + dw_i \sum_{j=0}^{l-1} (w_i(d-1))^j \right) p + O(p^2).$$

Putting everything together, we obtain

$$\frac{\Delta a_i}{\Delta w_i} = \frac{f_l(w_i) - 2a_i \sum_{j=0}^{l-1} ((d-1)w_i)^j + O(p)}{1 + (d-2)w_i \sum_{j=0}^{l-1} ((d-1)w_i)^j + O(p)}.$$

Note that $f_l(w_i)$ is bounded from above by the order of a d-regular tree of radius l, i.e. it is bounded above in terms of d and l. Clearly,

$$\left(2a_i \sum_{j=0}^{l-1}((d-1)w_i)^j\right)$$

is bounded from above in terms of d and l while

$$\left(1 + (d-2)w_i \sum_{j=0}^{l-1}((d-1)w_i)^j\right)$$

is bounded from below by 1 and bounded from above in terms of d and l. Altogether this implies the stated equation for $\frac{\Delta a_i}{\Delta w_i}$. □

We proceed to our main result which extends Theorem 1 of Lauer and Wormald (2007).

Theorem 3.4. *Let $d \geq 3$ and $l \geq 0$. If $f_l(w)$ is continuous on $[0, 1]$, then*

$$\alpha(d) \geq b_{l,f}(1)$$

where $b_{l,f}$ is the solution of the linear differential equation

$$b'_{l,f}(w) = c_{l,f,0}(w) + c_{l,f,1}(w)b_{l,f}(w) \quad and \quad b_{l,f}(0) = 0 \tag{7}$$

with

$$c_{l,f,0}(w) = \frac{f_l(w)}{1 + (d-2)w\frac{((d-1)w)^l - 1}{(d-1)w - 1}} \quad and$$

$$c_{l,f,1}(w) = -\frac{2\frac{((d-1)w)^l - 1}{(d-1)w - 1}}{1 + (d-2)w\frac{((d-1)w)^l - 1}{(d-1)w - 1}}.$$

Proof. Note that by definition,

$$a_0 = \frac{z_0 w_0}{r_0} = z_0 = \mathbf{P}[u \in Z_k \setminus Z_0 \mid u \in V_0] = \mathbf{P}[u \in Z_k].$$

Therefore, $TREE(k, l, p, f)$ produces an independent set of $G = (V, E)$ of expected cardinality $a_0|V|$ and hence, by the first moment principle, $\alpha(d) \geq a_0$.

Whenever we use the $O(\cdot)$-notation, the implicit constants will be in terms of d and l.

Similarly as in Lauer and Wormald (2007), we will prove that for every $\epsilon > 0$ there is some $c_0 = c_0(\epsilon)$ and a function $p_0(c) > 0$ such that for $pk = c > c_0$ and $p < p_0(c)$ we have

$$a_0 \geq b_{l,f}(1) - O(\epsilon)$$

which clearly implies the desired result.

Let some $\epsilon > 0$ be fixed. By Lemma 3.2, we have

$$w_{i+1} \leq w_i\left(1 - p + O(p^2)\right)$$

for $0 \leq i < k$. Therefore, for sufficiently small p,

$$w_k \leq (1 - p/2)^k = \left((1 - p/2)^{\frac{1}{p}}\right)^c. \tag{8}$$

Thus for p small enough and $c \to \infty$ we have $w_k \to 0$. Furthermore, by Lemma 3.2,

$$\Delta w_i = O(p) \tag{9}$$

and hence $\Delta w_i \to 0$ as $p \to 0$ uniformly for every $0 \leq i < k$.

Since $f_l(w)$ is continuous, the function $c_0(w) := c_{l,f,0}(w)$ is continuous and bounded on the compact set $[0, 1]$. (Note that $f_l(w)$ is always bounded in terms of d and l as already noted in the proof of Lemma 3.3.) Furthermore, the function $c_1(w) := c_{l,f,1}(w)$ is Lipschitz continuous on $[0, 1]$. Hence the solution $b(w) := b_{l,f}(w)$ of (7) is also Lipschitz continuous on $[0, 1]$ where all bounds and Lipschitz constants are in terms of d and l.

Clearly, $a_k = 0$. Let $\tilde{b}(w)$ be the solution of the differential equation with modified initial condition

$$\tilde{b}'(w) = c_0(w) + c_1(w)\tilde{b}(w) \quad \text{and} \quad \tilde{b}(w_k) = 0.$$

By the mentioned continuity/Lipschitz continuity conditions, it follows from standard results (cf. Corollary 4 and Corollary 6 in §7 of Arnol'd 1992) that the solution of (7) depends continuously on the initial condition. Hence, by (8), for p small enough and c large enough, $\tilde{b}(w_k) = b(w_k) + O(\epsilon)$ which implies

$$\tilde{b}(1) = b(1) + O(\epsilon).$$

By Lemma 3.3, we have for $0 \leq i < k$ that

$$a_i = a_{i+1} - \Delta a_i = a_{i+1} - \left(c_0(w_i) + c_1(w_i)a_i + O(p)\right)\Delta w_i$$

which implies

$$\begin{aligned}
a_i &= \frac{a_{i+1} - (c_0(w_i) + O(p))\Delta w_i}{1 + c_1(w_i)\Delta w_i} \\
&= \frac{a_{i+1} - c_0(w_i)\Delta w_i}{1 + c_1(w_i)\Delta w_i} + O(p)\Delta w_i \tag{10}
\end{aligned}$$

for p small enough.

Similarly, the differential equation for $\tilde{b}$ together with the mean value theorem imply for $0 \leq i < k$ and some $w_{i+1} \leq \tilde{w}_i \leq w_i$ that

$$\tilde{b}(w_i) = \tilde{b}(w_{i+1}) - \left(c_0(\tilde{w}_i) + c_1(\tilde{w}_i)\tilde{b}(\tilde{w}_i)\right)\Delta w_i.$$

By (9) and the continuity of c_0, c_1 and b, this implies that for every $\delta > 0$ there is some $p_1(\delta)$ such that for $p < p_1(\delta)$

$$\tilde{b}(w_i) = \tilde{b}(w_{i+1}) - \big(c_0(w_i) + c_1(w_i)\tilde{b}(w_i) + O(\delta)\big)\Delta w_i$$

and thus

$$
\begin{aligned}
\tilde{b}(w_i) &= \frac{\tilde{b}(w_{i+1}) - (c_0(w_i) + O(\delta))\Delta w_i}{1 + c_1(w_i)\Delta w_i} \\
&= \frac{\tilde{b}(w_{i+1}) - c_0(w_i)\Delta w_i}{1 + c_1(w_i)\Delta w_i} + O(\delta)\Delta w_i
\end{aligned}
\tag{11}
$$

for p small enough.

In view of (10) and (11), we deduce

$$\tilde{b}(w_k) - a_k = 0 \quad \text{and}$$

$$\tilde{b}(w_i) - a_i = \frac{\tilde{b}(w_{i+1}) - a_{i+1}}{1 + c_1(w_i)\Delta w_i} + \big(O(p) + O(\delta)\big)\Delta w_i \quad \text{for } 0 \le i < k.$$

Since for p small enough

$$\frac{1}{1 + c_1(w_i)\Delta w_i} = 1 + O(\Delta w_i) = 1 + O(p),$$

we obtain, by induction,

$$\tilde{b}(w_0) - a_0 = \big(O(p) + O(\delta)\big)\sum_{i=0}^{k-1}\Delta w_i\big(1 + O(p)\big)^i.$$

We have

$$\big(1 + O(p)\big)^k = \big((1 + O(p))^{\frac{1}{p}}\big)^c$$

which is bounded in terms of c. Therefore, choosing δ small enough in terms of c and choosing p small enough in terms of c (and δ), we finally obtain

$$b(1) = a_0 + \big(b(1) - \tilde{b}(1)\big) + \big(\tilde{b}(1) - a_0\big) = a_0 + O(\epsilon)$$

and the proof is complete. $\square$

8.3.1 Some Instructive Choices for l

It is very instructive to consider the behaviour of $TREE(k, l, p, f)$ for $l \in \{0, 1\}$ and appropriate choices for f.

For $l = 0$ we have $N_{\bar{G}_i}^{\le 0}(v) = \{v\}$ and $X_i^* = X_i$ for all $0 \le i \le k - 1$. Furthermore, the set $B_{G_i}^l(v)$ contains the vertex v exactly if all neighbours of v in G are not contained in V_i and is empty otherwise.

Choosing $f(T, r) = V_T$ whenever the vertex set V_T of T satisfies $|V_T| \leq 1$ and $f(T, r) = \emptyset$ otherwise, $TREE(k, 0, p, f)$ produces an independent set by Lemma 2.1. Conditional upon the event $(u \in X_i^* \cap V_i)$, the expected value of $|f(T_i(u), u)|$ equals

$$f_0(w_i) = (1 - w_i)^d,$$

because, by Lemma 3.1(ii), each of the d neighbours of u are in V_i independently at random with probability w_i. The differential equation (7) simplifies to

$$b'_{0,f}(w) = f_0(w) \quad \text{and} \quad b_{0,f}(0) = 0$$

which has the solution $b_{0,f}(w) = \frac{1 - (1-w)^{d+1}}{1+d}$ and thus

$$b_{0,f}(1) = \frac{1}{1+d}.$$

Therefore, by Theorem 3.4, asymptotically $TREE(k, 0, p, f)$ produces an independent set of G which contains exactly the same fraction of the vertices, namely $\frac{1}{1+d}$, as guaranteed by the lower bound on the independence number of d-regular graphs proved by Caro (1979) and Wei (1981).

The reason for this is that for $pk \to \infty$ and $p \to 0$, $TREE(k, 0, p, f)$ essentially processes the vertices of G according to a random linear ordering $v_1, v_2, \ldots, v_n$ and adds an individual vertex v_i to the constructed independent set exactly if all neighbours of v_i are among $\{v_1 v_2, \ldots, v_{i-1}\}$. Applying this algorithm to a random linear ordering, the probability that an individual vertex v belongs to the constructed independent set equals exactly $1/(1+d)$, because this is the probability that v is the last vertex from $N_G^{\leq 1}(v)$ with respect to the linear ordering. In fact, this is exactly the argument used in Alon and Spencer (1992) to prove the bound due to Caro (1979) and Wei (1981).

For $l = 1$ the set $B_{G_i}^1(v)$ always contains the vertex v itself and we can choose $f(T, r) = \{r\}$ in order to obtain an independent set according to Lemma 2.1.

For this choice $TREE(k, 1, p, f)$ essentially coincides with the randomized **p**-*greedy algorithm* used by Lauer and Wormald (2007) for $\mathbf{p} = (p_1, p_2, \ldots, p_k) = (p, p, \ldots, p)$. Solving the differential equation (7) yields the same values for $b_{1,f}(1)$ as obtained by Lauer and Wormald. In the next section, we consider a choice of f which generalizes their **p**-greedy algorithm.

8.3.2 A Reasonable Choice for f

A reasonable choice for f is to select all vertices within some even distance from the root. Therefore, let $l = 2h + 1$ for some integer $h \geq 0$ and let

$$f_{\text{even}}(T, r) = \{u \in V_T \mid \text{dist}_T(r, u) \text{ is even}\}.$$

Note that $B_{G_i}^{2h+1}(u)$ contains all vertices of $N_{\bar{G}_i}^{\leq 2h+1}(u)$ within distance at most $2h$ from u.

Conditional upon the event $(u \in X_i^* \cap V_i)$, the probability for the event $(v \in B_{G_i}^{2h+1}(u))$ for some vertex v with $\text{dist}_G(u, v) = 2j$ for some $1 \leq j \leq h$ equals exactly the probability that all vertices of the unique path from u to v within $N_G^{\leq 2h+1}(u)$ lie in V_i. Using Lemma 3.1 in the same way as for the calculation of $\mathbf{P}[(u_0, u_1, \ldots, u_v \in V_i) \wedge (u_{v+1} \notin V_i)]$ in the proof of Lemma 3.2, this conditional probability equals w_i^{2j}.

By linearity of expectation, we deduce

$$(f_{\text{even}})_{2h+1}(w) = 1 + \sum_{j=1}^{h} d(d-1)^{2j-1} w^{2j}$$

$$= 1 + d(d-1)w^2 \sum_{j=0}^{h-1} ((d-1)w)^{2j}$$

$$= 1 + d(d-1)w^2 \frac{((d-1)w)^{2h} - 1}{((d-1)w)^2 - 1}.$$

Now the differential equation (7) reads as follows.

$$b'_{2h+1, f_{\text{even}}}(w) = \frac{\left(1 + d(d-1)w^2 \frac{((d-1)w)^{2h}-1}{((d-1)w)^2-1}\right) - 2b_{2h+1, f_{\text{even}}}(w) \frac{((d-1)w)^{2h+1}-1}{(d-1)w-1}}{1 + (d-2)w \frac{((d-1)w)^{2h+1}-1}{(d-1)w-1}},$$

$$b_{2h+1, f_{\text{even}}}(0) = 0.$$

(12)

The algorithm proposed by Lauer and Wormald (2007) corresponds to the choice $h = 0$ in which case (12) simplifies to

$$b'_{1, f_{\text{even}}}(w) = \frac{1 - 2b_{1, f_{\text{even}}}(w)}{1 + (d-2)w}.$$

The solution of this differential equation is

$$b_{1, f_{\text{even}}}(w) = \frac{1 - (1 + wd - 2w)^{-2/(d-2)}}{2}$$

which together with Theorem 3.4 immediately implies one of the main results from Lauer and Wormald (2007).

Corollary 3.5 (Lauer and Wormald 2007, cf. Theorem 1). *For every $d \geq 3$, we have*

$$\alpha(d) \geq \frac{1 - (d-1)^{-2/(d-2)}}{2}.$$

Next we consider the behaviour of $b_{2h+1, f_{\text{even}}}(w)$ for $h \to \infty$. Our analysis naturally splits into the two cases $(d-1)w < 1$ and $(d-1)w > 1$.

The intuitive reason for this is that for values of w_i with $(d-1)w_i < 1$ the sets $N_{G_i}^{\leq l}(u)$ typically contain no vertices far from u, while for values of w_i with $(d-1)w_i > 1$ the sets $N_{G_i}^{\leq l}(u)$ may contain vertices up to distance l from u, i.e. the trees induced by the sets $B_{G_i}^l(u)$ "*die out*" quickly for small values of the "*probability of survival*" w_i.

Considering the two intervals $\left[0, \frac{1}{d-1}\right)$ and $\left(\frac{1}{d-1}, 1\right]$ it follows from standard results (cf. Corollary 6 in §7 of Arnol'd 1992) that for $h \to \infty$ the solutions of (12) converge to the solutions of

$$b'_{\infty, f_{even}}(w) = \begin{cases} \dfrac{1 - \frac{d(d-1)w^2}{((d-1)w)^2-1} + \frac{2b_{\infty, f_{even}}(w)}{(d-1)w-1}}{1 - \frac{(d-2)w}{(d-1)w-1}} & \text{for } 0 \leq w < 1/(d-1), \\[4mm] \dfrac{\frac{d(d-1)w^2}{((d-1)w)^2-1} - 2b_{\infty, f_{even}}(w)\frac{(d-1)w}{(d-1)w-1}}{(d-2)w\frac{(d-1)w}{(d-1)w-1}} & \text{for } 1/(d-1) < w \leq 1 \end{cases}$$

$$(13)$$

$$= \begin{cases} \dfrac{(d-1)w^2+1}{((d-1)w+1)(1-w)} - \dfrac{2}{1-w}b_{\infty, f_{even}}(w) & \text{for } 0 \leq w < 1/(d-1), \\[4mm] \dfrac{d}{((d-1)w+1)(d-2)} - \dfrac{2}{(d-2)w}b_{\infty, f_{even}}(w) & \text{for } 1/(d-1) < w \leq 1, \end{cases}$$

$$b_{\infty, f_{even}}(0) = 0.$$

Corollary 3.6. *For every $d \geq 3$, we have $\alpha(d) \geq b_{\infty, f_{even}}(1)$.*

Solving (13) for $d = 3$ yields

$$b_{\infty, f_{even}}(w) = \begin{cases} \frac{w}{3} + \frac{w^2}{6} + \frac{2}{9}\ln\left(\frac{2w+1}{1-w}\right) & \text{for } 0 \leq w < 1/(d-1), \\[3mm] \frac{3}{4}\frac{w-1}{w} + \frac{1}{w^2}\left(\frac{23}{96} + \frac{3}{8}\ln(2w+1) - \frac{25}{72}\ln(2)\right) \\[2mm] \quad \text{for } 1/(d-1) < w \leq 1 \end{cases}$$

and hence in this case

$$b_{\infty, f_{even}}(1) = \frac{23}{96} + \frac{3}{8}\ln(3) - \frac{25}{72}\ln(2) \approx 0.4108.$$

Solving (13) exactly or numerically for further values of d leads to the values for $b_{\infty, f_{even}}(1)$ given in Table 8.1 at the end of the paper.

It is plausible to ask what would happen for the function f_{odd} which would chose all vertices at odd distance from the root. Since during the execution of *TREE(k, l, p, f)* the average degree of the remaining graphs drops to 0, there will eventually be many isolated vertices. The main weakness of f_{odd} is that these isolated vertices would not be included in the independent set. In terms of the corresponding differential equations this means that for $w(d-1) < 1$ the contribution to the independent set would be $w(d-1)$ times smaller for f_{odd} than for f_{even}.

8.3.3 An Optimal Choice for f

In this section we consider a function f_{opt} for which $f_{opt}(T, r)$ is a maximum independent set within the rooted tree T. For some tree T with root r the set $f_{opt}(T, r)$ is obtained by applying the following algorithm $\mathcal{A}_{\text{opt}}$: *Start with $f_{\text{opt}}(T, r) = \emptyset$. Iteratively add to $f_{\text{opt}}(T, r)$ all vertices at maximum distance from the root r within the current tree and delete them together with their parents.*

For this algorithm it follows immediately from the definition of $B^l_{G_i}(v)$ that

$$f_{\text{opt}}(T_i(v), v) = B^l_{G_i}(v) \cap f_{\text{opt}}\big(G_i\big[N^{\leq l+1}_{G_i}(v)\big], v\big)$$

for every $v \in X^*_i \cap V_i$.

By Lemma 3.1, for $0 \leq i < k$ and $-1 \leq j \leq l+1$ the probability

$$p_l(j, i) = \mathbf{P}\Big[v \in f_{\text{opt}}\big(N^{\leq l+1}_{G_i}(u), u\big)\Big|\big((u \in X^*_i \cap V_i) \wedge \big(v \in N^{l+1-j}_{G_i}(u)\big)\big)\Big]$$

does not depend on the choice of $u, v \in V$ with $\text{dist}_G(u, v) = l+1-j$, i.e. $p_l(j, i)$ is well-defined.

Furthermore, by Lemma 3.1(ii), the events

$$\Big(v \in f_{\text{opt}}\big(N^{\leq l+1}_{G_i}(u), u\big)\Big)\Big|\Big((u \in X^*_i \cap V_i) \wedge \big(v \in N^{l+1-j}_{G_i}(u)\big)\Big)$$

and

$$\Big(v' \in f_{\text{opt}}\big(N^{\leq l+1}_{G_i}(u), u\big)\Big)\Big|\Big((u \in X^*_i \cap V_i) \wedge \big(v' \in N^{l+1-j}_{G_i}(u)\big)\Big)$$

are independent for different $v, v' \in N^{l+1-j}_G(u)$.

Let $t_j(w)$ be defined recursively for integers $j \geq -1$ by

$$t_j(w) := \begin{cases} 0 & \text{for } j = -1, \\ (1 - wt_{j-1}(w))^{d-1} & \text{for } j \geq 0. \end{cases}$$

Obviously, by definition and the first step of the algorithm $\mathcal{A}_{\text{opt}}$,

$$t_{-1}(w_i) = p_l(-1, i) = 0 \quad \text{and}$$
$$t_0(w_i) = p_l(0, i) = 1.$$

For $j \geq 0$ the event

$$\Big(v \in f_{\text{opt}}\big(N^{\leq l+1}_{G_i}(u), u\big)\Big)\Big|\Big((u \in X^*_i \cap V_i) \wedge \big(v \in N^{l+1-j}_{G_i}(u)\big)\Big)$$

is equivalent to the event that none of the vertices $v' \in N^1_G(v) \cap N^{l+1-(j-1)}_G(u)$ is in the set $f_{\text{opt}}\big(N^{\leq l+1}_{G_i}(u), u\big)$, i.e. either they are not in V_i or they are in V_i but not in the set $f_{\text{opt}}\big(N^{\leq l+1}_{G_i}(u), u\big)$.

By Lemma 3.1(ii), conditional upon the event

$$\Big((u \in X^*_i \cap V_i) \wedge \big(v \in N^{l+1-j}_{G_i}(u)\big)\Big),$$

the probability that $v' \in N^1_G(v) \cap N^{l+1-(j-1)}_G(u)$ is not in V_i equals $(1 - w_i)$, and the probability that such a v' is in V_i but not in $f_{\text{opt}}\big(N^{\leq l+1}_{G_i}(u), u\big)$ equals $w_i(1 - p_l(j - 1, i))$.

Hence, the probability that such a v' is in not $f_{\mathrm{opt}}\big(N_{G_i}^{\leq l+1}(u), u\big)$ equals

$$(1 - w_i) + w_i\big(1 - p_l(j - 1, i)\big) = 1 - w_i\, p_l(j - 1, i)$$

and, by Lemma 3.1(ii),

$$p_l(j, i) = \big(1 - w_i\, p_l(j - 1, i)\big)^{|N_G^1(v) \cap N_G^{l+1-(j-1)}(u)|}.$$

For $0 \leq j \leq l$ the values $p_l(j, i)$ satisfy the same recursion as the $t_j(w_i)$ starting with the same value 0 at $j = -1$ and altogether we obtain

$$p_l(j, i) = \begin{cases} t_j(w_i) & \text{for } -1 \leq j \leq l, \\ (1 - w_i t_l(w_i))^d & \text{for } j = l + 1 \end{cases}$$

(for $j = l + 1$ remember that the root u has d possible children while all internal vertices have $d - 1$).

By linearity of expectation,

$$
\begin{aligned}
(f_{\mathrm{opt}})_l(w_i) &= \mathbf{E}\Big[|f_{\mathrm{opt}}(B_{G_i}^l(u), u)| \mid u \in X_i^* \cap V_i\Big] \\
&= \sum_{v \in N_G^{\leq l}(u)} \mathbf{P}\Big[v \in f_{\mathrm{opt}}(B_{G_i}^l(u), u) \mid u \in X_i^* \cap V_i\Big] \\
&= p_l(l + 1, i) + \sum_{j=1}^{l} d(d - 1)^{l-j} w_i^{l+1-j} \cdot p_l(j, i) \\
&= p_l(l + 1, i) + \sum_{j=1}^{l} w_i d((d - 1)w_i)^{l-j} \cdot p_l(j, i) \\
&= (1 - w_i t_l(w_i))^d + \sum_{j=1}^{l} w_i d((d - 1)w_i)^{l-j} \cdot t_j(w_i) \\
&= (1 - w_i t_l(w_i))^d + w_i d \sum_{j=0}^{l-1} ((d - 1)w_i)^j \cdot t_{l-j}(w_i). \quad (14)
\end{aligned}
$$

In the case $w(d - 1) < 1$ and $l \to \infty$ the limit behaviour of the recursion for $t_j(w)$ becomes important. The function

$$t \mapsto (1 - wt)^{d-1}$$

maps the unit interval $[0, 1]$ into itself and the absolute value of its derivative $(d - 1)w(1 - wt)^{d-2}$ is strictly smaller than 1 for $t \in [0, 1]$ and $w(d - 1) < 1$. Hence the recursion for $t_j(w)$ is a contractive map and converges to a unique fixed point $t(w)$ which solves the equation

$$t(w) = (1 - wt(w))^{d-1}.$$

Because for $w(d-1) < 1$ the factors preceding $t_j(w)$ in (14) decrease exponentially in j, we obtain

$$\lim_{l\to\infty} (f_{opt})_l(w) = (1 - wt(w))^d + t(w)wd \sum_{j=0}^{\infty} ((d-1)w)^j$$

$$= t(w)\left(1 - wt(w) + \frac{wd}{1-(d-1)w}\right)$$

and

$$c_{\infty, f_{opt}, 0}(w) = \lim_{l\to\infty} c_{l, f_{opt}, 0}(w) = \frac{t(w)}{1-w}(dw + (1 - wt(w))(1 - (d-1)w)).$$

Because in this case

$$c_{\infty, f_{opt}, 1}(w) = \lim_{l\to\infty} c_{l, f_{opt}, 1} = \frac{2}{w-1},$$

we are able to solve the differential equation for $l \to \infty$ in the interval $[0, \frac{1}{d-1})$ and obtain

$$b_{\infty, f_{opt}}(w) = (w-1)^2 \int_0^w \frac{c_{\infty, f_{opt}, 0}(w)}{(t-1)^2} \delta t$$

$$= (w-1)^2 \int_0^w \frac{t(w)(dw + (1 - wt(w))(1 - (d-1)w))}{(1-w)(t-1)^2} \delta t.$$

The integral is solvable at least for $d = 3$ in which case we obtain

$$b_{\infty, f_{opt}}(w) = \frac{1 + 6w - \sqrt{4w+1}}{(1 + \sqrt{4w+1})^2}.$$

The most interesting value in this case is at $\frac{1}{d-1} = \frac{1}{2}$

$$b_{\infty, f_{opt}}\left(\frac{1}{2}\right) = \frac{11}{2} - 6\sqrt{3} \approx 0.3038.$$

Clearly, $(f_{even})_\infty(w) \leq (f_{opt})_\infty(w)$ and we can use $(f_{even})_\infty(w)$ in order to determine a lower bound for $b_{\infty, f_{opt}}(1)$ in the case $d = 3$ by solving (13) on the interval $[\frac{1}{2}, 1]$ using as initial condition the value $b_{\infty, f_{opt}}(\frac{1}{2})$ at $w = 1/2$. We obtain for $w \in [\frac{1}{2}, 1]$ the following lower bound

$$b_{\infty, f_{opt}}(w) \geq \frac{25 - 12\sqrt{3} - 12w + 12w^2 + 6\ln(\frac{1}{2} + w)}{16w^2}.$$

Corollary 3.7. $\alpha(3) \geq b_{\infty, f_{opt}}(1) \geq \dfrac{25 - 12\sqrt{3} + 6\ln(\frac{3}{2})}{16} \approx 0.4155 > \beta_{Shearer}(3).$

In general the following observations are useful for estimating $\lim_{l \to \infty} c_{l, f_{opt}, 0}(w)$ in the case $w(d-1) > 1$.

Observation 1. Because the recursion for $t_j(w)$ is based upon a strictly monotonic decreasing function which contracts the unit interval, and starts with $t_{-1}(w) = 0$ we obtain

$$t_{2j}(w) > t_{2j+1}(w),$$

$$t_{2(j+1)}(w) < t_{2j}(w) \quad \text{and}$$

$$t_{2(j+1)+1}(w) > t_{2j+1}(w)$$

for $j \geq 0$.

Observation 2. Consider a modified algorithm $\mathcal{A}^-$ applied to a tree T with root r which behaves like $\mathcal{A}_{opt}$ up to some distance $j+1$ from r, chooses less vertices at distance j from r than $\mathcal{A}_{opt}$ and continues like $\mathcal{A}_{opt}$ for smaller distances to r.

For distances larger than j to r the output of $\mathcal{A}^-$ coincides with the output of $\mathcal{A}_{opt}$. For distances at most j to r the output of $\mathcal{A}^-$ coincides with the output of $\mathcal{A}_{opt}$ when applied to a proper subtree of the tree induced by the vertices at distance at most j to r.

Therefore, the set produced by $\mathcal{A}^-$ will contain at most as many vertices as the set produced by $\mathcal{A}_{opt}$.

Conversely, if $\mathcal{A}^+$ chooses more vertices at distance j from r than $\mathcal{A}_{opt}$—possibly neglecting independence—and behaves like $\mathcal{A}_{opt}$ otherwise, then the set produced by $\mathcal{A}^+$ will contain at least as many vertices as the set produced by $\mathcal{A}_{opt}$.

Iteratively applying these observations allows to derive lower and upper bounds on $(f_{opt})_l(w)$ for $(d-1)w > 1$:

We choose an integer j^*.

If for all $j \geq j^*$ we replace in (14) t_{2j-1} by t_{2j^*-1} and t_{2j} by t_{2j^*}, then Observation 1 and Observation 2 for $\mathcal{A}^-$ imply that we obtain a lower bound on $(f_{opt})_l(w)$.

If for all $j \geq j^*$ we replace in (14) t_{2j+1} by t_{2j^*+1} and t_{2j} by t_{2j^*}, then Observation 1 and Observation 2 for $\mathcal{A}^+$ imply that we obtain an upper bound on $(f_{opt})_l(w)$.

Therefore, we obtain

$$\frac{(f_{opt})_l(w)}{(w(d-1))^l} = \frac{(1 - wt_l(w))^d}{(w(d-1))^l} + wd \sum_{j=1}^{l} \frac{t_j(w)}{(w(d-1))^j}$$

$$\geq wd \left(\sum_{j=1}^{2j^*-2} \frac{t_j(w)}{(w(d-1))^j} + \sum_{j=j^*}^{\lfloor \frac{l}{2} \rfloor} \frac{t_{2j^*}(w)}{(w(d-1))^{2j}} \right.$$

$$\left. + \sum_{j=j^*}^{\lfloor \frac{l+1}{2} \rfloor} \frac{t_{2j^*-1}(w)}{(w(d-1))^{2j-1}} \right)$$

and

$$\frac{(f_{\mathrm{opt}})_l(w)}{(w(d-1))^l} \le \frac{1}{(w(d-1))^l} + wd\left(\sum_{j=1}^{2j^*-1} \frac{t_j(w)}{(w(d-1))^j} + \sum_{j=j^*}^{\lfloor \frac{l}{2} \rfloor} \frac{t_{2j^*}(w)}{(w(d-1))^{2j}}\right.$$

$$\left. + \sum_{j=j^*}^{\lfloor \frac{l-1}{2} \rfloor} \frac{t_{2j^*+1}(w)}{(w(d-1))^{2j+1}}\right).$$

Using these inequalities it is possible to derive lower and upper bounds for $c_{\infty, f_{\mathrm{opt}}, 0}(w)$.

$$c_{\infty, f_{\mathrm{opt}}, 0}(w)$$
$$= \frac{w(d-1)-1}{(d-2)w} \lim_{l\to\infty} \frac{(f_{\mathrm{opt}})_l(w)}{(w(d-1))^l}$$
$$\ge \frac{d(w(d-1)-1)}{d-2}\left(\sum_{j=1}^{2j^*-2} \frac{t_j(w)}{(w(d-1))^j} + \sum_{j=j^*}^{\infty} \frac{t_{2j^*}(w)+w(d-1)t_{2j^*-1}}{(w(d-1))^{2j}}\right)$$
$$= \frac{d}{d-2}\left(\frac{t_{2j^*}(w)+w(d-1)t_{2j^*-1}(w)}{(w(d-1)+1)(w(d-1))^{2j^*-2}} + \sum_{j=1}^{2j^*-2} \frac{(w(d-1)-1)t_j(w)}{(w(d-1))^j}\right)$$

and

$$c_{\infty, f_{\mathrm{opt}}, 0}(w)$$
$$\le \frac{d(w(d-1)-1)}{d-2}\left(\sum_{j=1}^{2j^*-1} \frac{t_j(w)}{(w(d-1))^j} + \sum_{j=j^*}^{\infty} \frac{w(d-1)t_{2j^*+1}(w)+t_{2j^*}(w)}{(w(d-1))^{2j+1}}\right)$$
$$= \frac{d}{d-2}\left(\frac{t_{2j^*}(w)+w(d-1)t_{2j^*+1}(w)}{(w(d-1)+1)(w(d-1))^{2j^*-1}} + \sum_{j=1}^{2j^*-1} \frac{(w(d-1)-1)t_j(w)}{(w(d-1))^j}\right).$$

Table 8.1.

d	$\max\{\beta_{\mathrm{Shearer}}(d), \beta_{\mathrm{LauWo}}(d)\}$	$b_{\infty, f_{\mathrm{even}}}(1)$	$b_{\infty, f_{\mathrm{opt}}}(1)$	$\gamma(d)$
3	0.4139	0.4109	0.4193	0.4554
4	0.3510	0.3580	0.3664	0.4136
5	0.3085	0.3201	0.3279	0.3816
6	0.2771	0.2911	0.2982	0.3580
7	0.2558	0.2678	0.2744	0.3357
8	0.2386	0.2487	0.2548	0.3165
9	0.2240	0.2326	0.2382	0.2999
10	0.2113	0.2188	0.2241	0.2852
20	0.1395	0.1424	0.1455	0.1973
50	0.0748	0.0756	0.0770	0.1108
100	0.0447	0.0450	0.0457	0.0679

Choosing j^* sufficiently large and numerically solving the corresponding two differential equations, we can obtain estimates for $b_{\infty, f_{\text{opt}}}(1)$ with any desired precision.

Corollary 3.8. *For $d \geq 3$ we have $\alpha(d) \geq b_{\infty, f_{\text{opt}}}(1)$.*

Table 8.1 summarizes the numerically obtained values for selected values of d. The entry $\gamma(d)$ is an upper bound on $\alpha(d)$ which is derived from the analysis of random d-regular graphs (Bollobás 1981; McKay 1987).

References

Alon, N., Spencer, J.: The Probabilistic Method. Wiley, New York (1992)

Arnol'd, V.I.: Ordinary Differential Equations. Springer Textbook. Springer, Berlin (1992), 334 pp. (Translated from the 3rd Russian edn.)

Bollobás, B.: The independence ratio of regular graphs. Proc. Am. Math. Soc. **83**, 433–436 (1981)

Caro, Y.: New results on the independence number. Technical Report, Tel-Aviv University (1979)

Hopkins, G., Staton, W.: Girth and independence ratio. Can. Math. Bull. **25**, 179–186 (1982)

Lauer, J., Wormald, N.: Large independent sets in regular graphs of large girth. J. Comb. Theory, Ser. B **97**, 999–1009 (2007)

McKay, B.D.: Independent sets in regular graphs of high girth. Ars Comb. **23A**, 179–185 (1987)

Shearer, J.B.: A note on the independence number of triangle-free graphs. II. J. Comb. Theory, Ser. B **53**, 300–307 (1991)

Wei, V.K.: A lower bound on the stability number of a simple graph. Bell Laboratories Technical Memorandum, 81-11217-9, Murray Hill, NJ (1981)

Wormald, N.: Differential equations for random processes and random graphs. Ann. Appl. Probab. **5**, 1217–1235 (1995)

9

Linear Time Approximation Algorithms for Degree Constrained Subgraph Problems

Stefan Hougardy

Summary. Many real-world problems require graphs of such large size that polynomial time algorithms are too costly as soon as their runtime is superlinear. Examples include problems in VLSI-design or problems in bioinformatics. For such problems the question arises: What is the best solution that can be obtained in linear time? We survey linear time approximation algorithms for some classical problems from combinatorial optimization, e.g. matchings and branchings.

9.1 Introduction

For many combinatorial optimization problems arising from real-world applications, efficient, i.e., polynomial time algorithms are known for computing an optimum solution. However, there exist several applications for which the input size can easily exceed 10^9. In such cases polynomial time algorithms with a runtime that is quadratic or even higher are much too slow. It is therefore desirable to have faster algorithms that not necessarily find an optimum solution.

An *approximation algorithm* for a combinatorial optimization problem is an algorithm that for any possible input returns some feasible solution. An approximation algorithm has an *approximation ratio* of c if for any input it returns a solution with value at least c times the value of an optimum solution (in this paper we will consider maximization problems only).

For most reasonable problems the lowest possible runtime for a deterministic algorithm is linear, as at least the whole input must be read. In this paper we are interested in approximation algorithms that achieve this linear runtime. Moreover, we are interested here only in approximation algorithms that achieve a constant approximation ratio. The reason for the latter requirement is that in practice solutions that are far away from an optimum solution are quite useless. Thus, even an approximation ratio of $1/2$ may be too bad for a given application. However, as the approximation ratio is a guarantee for the worst case, in practice approximation algorithms with constant approximation ratios usually deliver solutions that are very close to the optimum. For example the greedy algorithm for the MAXIMUM WEIGHT MATCHING PROBLEM has an approximation ratio of $1/2$ but its solutions are usually within 5% of the optimum solution (Drake and Hougardy 2003a).

Linear time approximation algorithms offer several benefits against exact algorithms.

1. The most obvious benefit is its runtime: Exact algorithms even if their runtime is polynomial may simply be too slow to be applicable.
2. Linear time approximation algorithms are usually simpler than their exact counterparts. This not only means that the algorithms are simpler, but also their proofs of correctness may be simpler.
3. The implementation of linear time approximation algorithms can be much simpler than for exact algorithms. This is the main reason why in many applications approximation algorithms are used, even though exact algorithms would be fast enough (see for example Avis 1978).
4. There is another major reason, why in many applications approximation algorithms are used for combinatorial optimization problems even though exact algorithms would be fast enough: If the algorithm is used as a subroutine in some heuristic that does not give any performance guarantee, it may be a waste of time to compute exact solutions. In some cases one might even observe the weird effect that an exact solution yields results that are inferior to approximate solutions.
5. Finally, approximation algorithms can avoid problems with floating point arithmetic. In Althaus and Mehlhorn (1998) such a problem is analyzed for the maximum flow problem using a preflow-push algorithm. These unexpected difficulties may always occur when using floating point arithmetic in exact algorithms.

Of course, an algorithm with linear runtime may turn out to be completely useless in practice. A famous such example is the algorithm of Bodlaender (1996) for determining the treewidth of a graph. More precisely, if k is fixed, then for a given graph G the algorithm of Bodlaender finds in linear time a tree decomposition of width at most k, or decides that the treewidth of G exceeds k. A huge constant is involved in the linear runtime of Bodlaenders algorithm. Therefore, this algorithm is not feasible in practice, even not for $k = 4$.

All linear time algorithms that we present in this paper have constants in their runtime that are small, which means that they are at least not larger than the constants involved in the runtime of exact algorithms for the problem.

In this survey we will cover linear time deterministic algorithms only. There are related subjects as for example nearly linear time algorithms, sublinear time algorithms, or linear time randomized algorithms which we will not discuss in this paper. In the following the notion linear time algorithm always means a deterministic linear time algorithm.

9.1.1 A Technique for Obtaining Linear Runtime

There exists one general approach for obtaining linear time approximation algorithms for combinatorial optimization problems. Several exact algorithms for these problems work in *phases*, where each phase has linear runtime. Examples are the matching algorithm of Micali and Vazirani (1980) which needs $O(\sqrt{n})$ phases, each phase can be accomplished in $O(m)$, or the algorithm of Ford and Fulkerson (1956) for computing a flow of maximum value, where in each phase one flow augmenting path is computed in $O(m)$.

A simple way to turn such algorithms into linear time approximation algorithms is to simply stop after executing a constant number of phases. We will call this approach the *phase bounding approach*. This approach clearly results in linear time approximation algorithms, however, it is not at all clear what approximation ratio it achieves. One can prove for example (see below) that the phase bounding approach applied to the matching algorithm of Micali and Vazirani (1980) achieves an approximation ratio, that can be arbitrarily close to 1. However, the phase bounding approach applied to Ford and Fulkerson's maximum flow algorithm cannot guarantee any constant approximation ratio larger than 0.

The phase bounding approach while it can produce linear time approximation algorithms with constant approximation ratio misses one of the advantages that we listed above, namely to be much simpler than exact algorithms. Nevertheless, in cases where no other approach is known to yield similar results we do at least know what approximation ratio can be achieved in linear time. Such a result should be considered as a stimulation to look for simpler algorithms that do not use the phase bounding approach.

9.2 Matchings

A *matching* M in a graph $G = (V, E)$ is a subset of the edges of G such that no two edges in M are incident to the same vertex. A *perfect matching* in G is a matching M such that each vertex of G is contained in an edge of M. Computing a (perfect) matching that is optimal with respect to certain side constraints is one of the fundamental problems in combinatorial optimization.

The first polynomial time matching algorithms date back to the papers of Kőnig and Egerváry from 1931 (Frank 2005) which resulted in the famous Hungarian method for bipartite graphs due to Kuhn (1955). A major breakthrough was Edmonds' polynomial time algorithm for the MAXIMUM WEIGHT MATCHING PROBLEM (Edmonds 1965). Edmonds introduced in his paper for the first time the notion of a 'good' algorithm which led to the definition of the class $\mathcal{P}$ of polynomially time solvable problems.

Given a matching M in a graph $G = (V, E)$, we say that a vertex x is *matched* if $\{x, y\} \in M$ for some $y \in V$ and *free* otherwise. An *M-augmenting path* is a simple path $P = v_0, v_1, \ldots, v_k$ such that the endpoints v_0 and v_k are free, $\{v_i, v_{i+1}\} \in E(G)$ and the edges of P are alternately in $E(G) \setminus M$ and M. The *length* of a path is defined as the number of edges contained in it. As the endpoints of an augmenting path are free this implies that an augmenting path has odd length. Given an augmenting path, one can augment the matching M by deleting from M the edges on the path that are in M, and adding all of the other edges on the path to M. This results in a matching with one more edge. A well known result that is the basis of many matching algorithms says that the absence of an augmenting path implies optimality of the current matching. This result was proved by Petersen (1891) and first formulated in the language of modern graph theory by Berge (1957).

Theorem 2.1 (Petersen 1891; Berge 1957). *A matching M has maximum size if and only if there exists no M-augmenting path.*

9.2.1 Maximum Cardinality Matchings

The maximum cardinality matching problem simply asks for a matching of maximum size in an unweighted undirected graph. As a special case it contains the question whether a given graph has a perfect matching.

CARDINALITY MATCHING PROBLEM	
Input:	An undirected graph G.
Output:	A maximum cardinality matching in G.

The fastest known deterministic algorithm for the CARDINALITY MATCHING PROBLEM is due to Goldberg and Karzanov (2004) and has runtime $O(\sqrt{n}m \cdot \log(n^2/m)/\log n)$ (see also Fremuth-Paeger and Jungnickel 2003). It makes use of graph compression techniques that have been developed by Feder and Motwani (1995). For dense graphs Mucha and Sankowski (2004) presented a faster randomized algorithm which has been simplified by Harvey (2006).

We will show that the phase bounding approach can be used to get a linear time approximation algorithm for the CARDINALITY MATCHING PROBLEM with approximation ratio arbitrarily close to 1. The next lemma is the key to prove such a result. It says that if a matching does not admit short augmenting paths then its cardinality must be close to the cardinality of a maximum cardinality matching. This result is due to Hopcroft and Karp (1973) but has been rediscovered several times, e.g., (Fischer et al. 1993) or (Hassin and Lahav 1994).

Lemma 2.2 (Hopcroft and Karp 1973). *If M is a matching in a graph G such that every M-augmenting path has length at least $2k - 1$ then*

$$|M| \geq \frac{k-1}{k} \cdot |M^*|,$$

where M^* denotes a maximum cardinality matching in G.

Proof. Suppose the matching M does not admit augmenting paths of length less than k. Consider the symmetric difference between M and M^*. It contains $|M^*| - |M|$

vertex disjoint M-augmenting paths. Since each of these paths contains at least $k-1$ edges of M, we have

$$|M| \geq \frac{k-1}{k} \cdot |M^*|. \qquad \square$$

For the phase bounding approach one can make use of the maximum cardinality matching algorithm due to Micali and Vazirani (1980), see also (Vazirani 1994; Blum 1990) and (Gabow and Tarjan 1991). Their algorithm constructs in each phase a maximal set of vertex disjoint augmenting paths. Each phase can be implemented in $O(m)$ time. The size of the shortest augmenting path strictly increases from a phase to the next. Thus, by applying Lemma 2.2 one gets the following result which was first observed by Gabow and Tarjan (1988).

Theorem 2.3 (Gabow and Tarjan 1988). *For every fixed $\epsilon > 0$ there exists an algorithm that computes a matching of size at least $(1 - \epsilon) \cdot |M^*|$ in linear time.*

The number of phases needed to obtain a maximum cardinality matching is $O(\sqrt{n})$. One can easily construct instances where this number of phases is needed. In practice it turns out that usually a much fewer number of phases suffices to find a maximum cardinality matching in a graph. This observation can be proved for certain graph instances rigorously. A first such result is due to Motwani (1994). It has been improved by Bast et al. (2006) who proved the following statement.

Theorem 2.4 (Bast et al. 2006). *In $G_{n, \frac{33}{n}}$ the algorithm of Micali and Vazirani terminates with high probability after $O(\log n)$ phases.*

Here, $G_{n, \frac{33}{n}}$ denotes a random graph on n vertices where each edge is in the graph with probability $\frac{33}{n}$.

Theorem 2.3 shows that one can find in linear time a matching whose cardinality is arbitrarily close to the cardinality of a maximum cardinality matching. However, as we used the phase bounding approach to get this result, the algorithm is not simpler than the exact one. Therefore we will consider here alternative approaches to compute large matchings in linear time.

One such approach is to simply compute a maximal matching. A matching M is called *maximal* if for all $e \in E(G) \setminus M$ the set $M \cup \{e\}$ is not a matching. For maximal matchings we get the following simple result in the special case $k = 2$ of Lemma 2.2.

Lemma 2.5. *If M is a maximal matching and M^* a maximum cardinality matching then $|M| \geq \frac{1}{2} \cdot |M^*|$.*

Maximal matchings can easily be computed in linear time: start with $M = \emptyset$ and for each $e \in E(G)$ add e to M if both endpoints of e are free. Therefore, Lemma 2.5 immediately implies a very simple linear time approximation algorithm for the CARDINALITY MATCHING PROBLEM with approximation ratio $1/2$. It is easily seen that this approximation ratio is tight (see Fig. 9.1).

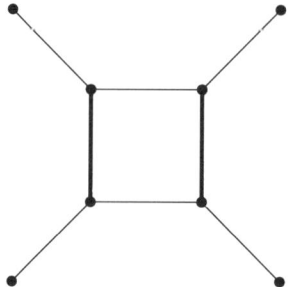

Fig. 9.1. An example where a maximum matching is twice as large as a maximal matching (*bold* edges)

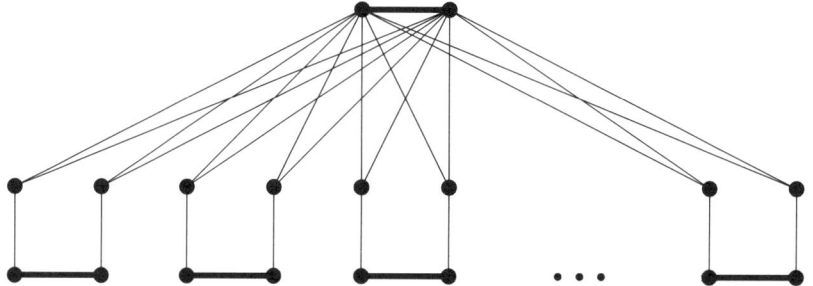

Fig. 9.2. An example where a maximum matching is asymptotically twice as large as a maximal matching (*bold* edges) that considers vertex degrees

Figure 9.1 suggest that it should be a good idea to consider the degrees when computing a maximal matching: First sort in linear time the vertices in increasing order by their degrees and then compute a maximal matching by choosing as the next edge one that is incident to a vertex of currently lowest degree. This approach is often used in practice (Karypis and Kumar 1998) and usually yields better results. But as Fig. 9.2 shows this approach does not improve the approximation ratio.

A better approximation ratio than 1/2 can be achieved by a simple algorithm that results from an algorithm of Drake Vinkemeier and Hougardy (2005) for the MAXIMUM WEIGHT MATCHING PROBLEM by specializing it to the unweighted case. The slightly complicated computation of 'good' augmentations that is needed in their algorithm becomes completely trivial in the unweighted case. This way one obtains a simple linear time approximation algorithm for the CARDINALITY MATCHING PROBLEM with an approximation ratio arbitrarily close to 2/3.

However, one can obtain even an approximation ratio of 4/5 in the unweighted case. For this we will make use of the following result of Hopcroft and Karp (1973).

Lemma 2.6 (Hopcroft and Karp 1973). *Let M be a matching in a graph G such that every M-augmenting path has length at least k. If P is a maximal set of vertex*

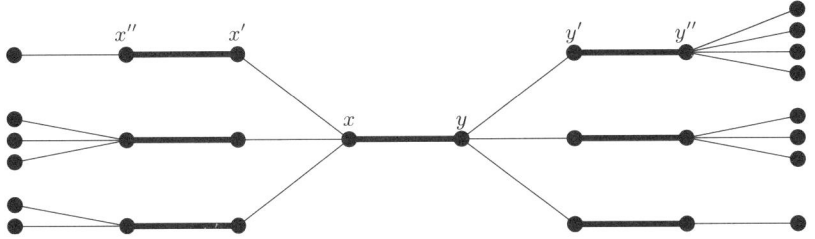

Fig. 9.3. How to find a maximal set of vertex disjoint augmenting paths of length 7 in linear time

disjoint M-augmenting paths of length k then after augmenting all paths in $\mathcal{P}$ the shortest augmenting path in the new matching has length at least $k + 2$.

A maximal set of vertex disjoint M-augmenting paths of length k can be found in linear time for small k.

Lemma 2.7. *Let M be a matching in a graph G. For $k \leq 7$ a maximal set of vertex disjoint M-augmenting paths of length k can be found in linear time.*

Proof. We prove the result here for $k = 7$ only. The proof shows that the statement also holds for smaller k. Let M be a matching in G such that each M-augmenting path has length at least 7. For each edge in M we check whether it is the middle edge of an M-augmenting path of length 7 in G.

Let $e = \{x, y\} \in M$ be such an edge. Then we scan all matched neighbors of y. For each such neighbor, say y', we scan its incident edge $\{y', y''\} \in M$. This can be done in time proportional to the degree of y. Similarly, we scan all possible neighbors x' and incident edges $\{x', x''\} \in M$ of x in time proportional to the degree of x (see Fig. 9.3).

Now we simply have to look for a vertex y'' that has a free neighbor y''' and a vertex x'' that has a free neighbor x'''. This clearly can be done in time proportional to the sum of the degrees of vertex x and y. However, we need to be a bit careful, as the vertices y', y'', y''' and x', x'', x''' need not to be distinct. Therefore we proceed as follows: If there is a vertex y'' that has at least two free neighbors and there is an edge $\{x', x''\}$ different from $\{y', y''\}$ such that x'' has a free neighbor then we have found an M-augmenting path of length 7 in time proportional to the sum of the degree of x and y. If all y'' have only one free neighbor, we check whether there are at least two different such neighbors. If so, we can find an M-augmenting path of length 7 as soon as there is a vertex x'' on the other side. If all vertices y'' are adjacent to the same free vertex we also can easily check whether an M-augmenting path of length seven can be found.

Therefore we can find an M-augmenting path of length 7 with the edge $\{x, y\}$ in the middle in time proportional to the sum of the degrees of x and y. Thus, a maximal vertex disjoint set of such paths can be found in linear time. $\square$

As a consequence of Lemma 2.6 and Lemma 2.7 we get a linear time algorithm with approximation ratio 4/5 for the CARDINALITY MATCHING PROBLEM.

Theorem 2.8. *If M^* is a maximum cardinality matching then a matching M with $|M| \geq \frac{4}{5} \cdot |M^*|$ can be computed in linear time.*

Better approximation algorithms can be obtained in the special case of bounded degree graphs. In this case one easily can check for all augmenting paths up to a fixed length k containing a given vertex x in constant time. More precisely: if all vertex degrees are bounded by a constant B, then there can exist at most B^k paths of length k containing a given vertex. Therefore, one gets a simple linear time approximation algorithm for bounded degree graphs with approximation ratio arbitrarily close to 1. However, as this approach involves the constant B^k in the runtime it is not useful in practice.

9.2.2 Maximum Weight Matchings

Given a graph $G = (V, E)$ and a weight function $c : E(G) \rightarrow \mathbb{R}$, the weight of a matching $M \subseteq E$ is defined as $w(M) := \sum_{e \in M} w(e)$. The MAXIMUM WEIGHT MATCHING PROBLEM now asks for a matching of maximum weight. The CARDINALITY MATCHING PROBLEM is a special case of this problem where all weights are equal.

MAXIMUM WEIGHT MATCHING PROBLEM

Input: An undirected graph G and edge weights $c : E(G) \rightarrow \mathbb{R}$.
Output: A maximum weight matching in G.

The fastest known algorithm for the MAXIMUM WEIGHT MATCHING PROB-LEM is due to Gabow (1990) and has runtime $O(nm + n^2 \log n)$. However, this algorithm involves rather complicated data structures that prevent it from being useful in practice. The fastest implementations known today for solving the MAXIMUM WEIGHT MATCHING PROBLEM are due to Cook and Rohe (1999) respectively Mehlhorn and Schäfer (2002). These algorithms have a worst case runtime of $O(n^3)$ respectively $O(nm \log n)$. Under the assumption that all edge weights are integers in the range $[1..N]$ Gabow and Tarjan (1991) presented an algorithm with runtime $O(\sqrt{n} \log n \alpha(m, n) m \log(Nn))$, where α is the inverse of Ackermann's function.

The greedy algorithm for the MAXIMUM WEIGHT MATCHING PROBLEM is a very simple approximation algorithm that achieves an approximation ratio of $1/2$ (Jenkyns 1976; Korte and Hausmann 1978; Avis 1978). This algorithm simply sorts the edges by decreasing weight and then computes a maximal matching using this ordering. The runtime of the greedy algorithm is $O(m \log n)$ as sorting the edges requires this amount of time. Surprisingly, for long time no linear time approximation algorithm for the MAXIMUM WEIGHT MATCHING PROBLEM was known that achieves a constant approximation ratio strictly larger than zero. The first such algorithm was presented by Preis (1999). The idea of this algorithm is that instead of using the heaviest edge in each step it is enough to consider a locally heaviest edge, i.e., an edge that is heavier than all adjacent edges. Using this approach it is easy to

---

PathGrowingAlgorithm $(G = (V, E), w : E \to \mathbb{R}_+)$

1    $M_1 := \emptyset, M_2 := \emptyset, i := 1$
2   **while** $E \neq \emptyset$ **do begin**
3   choose $x \in V$ of degree at least 1 arbitrarily
4   **while** x has a neighbor **do begin**
5   let $\{x, y\}$ be the heaviest edge incident to x
6   add $\{x, y\}$ to M_i
7            $i := 3 - i$
8   remove x from G
9            $x := y$
10   **end**
11   **end**
12   **return** $\max(w(M_1), w(M_2))$

Fig. 9.4. The Path Growing Algorithm for finding maximum weight matchings

see that the algorithm achieves an approximation ratio of $1/2$. However, it is quite complicated to prove that its runtime is indeed linear.

Drake and Hougardy (2003a) presented a much simpler linear time approximation algorithm for the MAXIMUM WEIGHT MATCHING PROBLEM with approximation ratio $1/2$. This algorithm is called the Path Growing Algorithm and is shown in Fig. 9.4. Its idea is to simultaneously compute two matchings and for the heavier of these two matchings the algorithm guarantees that its weight is at least half the weight of a maximum weight matching.

It is easily seen that the runtime of the Path Growing Algorithm is linear (Drake and Hougardy 2003a). The next result shows the correctness of the algorithm.

Theorem 2.9 (Drake and Hougardy 2003a). *The Path Growing Algorithm has a performance ratio of $1/2$.*

Proof. For the analysis of the performance ratio we will assign each edge of the graph to some vertex of the graph in the following way. Whenever a vertex is removed in line 8 of the algorithm all edges which are currently incident to that vertex x are assigned to x. This way each edge of G is assigned to exactly one vertex of G. Note that there might be vertices in G that have no edges assigned to them.

Now consider a maximum weight matching M in G. As M must not contain two incident edges, all edges of M are assigned to different vertices of G. In each step of the algorithm the heaviest edge that was currently incident to vertex x is chosen in line 5 of the algorithm and added to M_1 or M_2. Therefore the weight of $M_1 \cup M_2$ is at least as large as the weight of M. As

$$\max(w(M_1), w(M_2)) \geq \frac{1}{2} w(M_1 \cup M_2) \geq \frac{1}{2} w(M)$$

the weight returned by the Path Growing Algorithm is at least half the weight of the optimal solution.   □

It turns out that the greedy algorithm, the algorithm of Preis, and the Path Growing Algorithm of Drake and Hougardy are in practice much better than the approximation factor of $1/2$ suggests. In Drake and Hougardy (2003a) it is shown that the solution found by these algorithms is typically about 5% away from the weight of an optimum solution. For the Path Growing Algorithm one even can get a guarantee that the solution found by the algorithm is strictly larger than the approximation ratio of $1/2$. This is due to the fact that this algorithm computes two different matchings for which it guarantees that the sum of the weights of these two matchings is at least as large as the weight of an optimum solution. Therefore we have the following observation which allows to get a better guarantee how far the solution returned by the Path Growing Algorithm is away from an optimum solution.

Lemma 2.10. *The matching M returned by the Path Growing Algorithm has weight at least*

$$\frac{\max\{w(M_1), w(M_2)\}}{w(M_1) + w(M_2)} \cdot w(M^*),$$

where M^* is a maximum weight matching.

Note that we have $\frac{1}{2} \leq \frac{\max\{w(M_1), w(M_2)\}}{w(M_1) + w(M_2)} \leq 1$. This means that the Path Growing Algorithm might even return a matching M and a guarantee that this matching M is a maximum weight matching.

Drake Vinkemeier and Hougardy (2005) improved on the Path Growing Algorithm by presenting a linear time approximation algorithm for the MAXIMUM WEIGHT MATCHING PROBLEM that achieves an approximation ratio arbitrarily close to $2/3$. This approximation ratio is the best that is currently known to be achievable in linear time.

Theorem 2.11 (Drake Vinkemeier and Hougardy 2005). *For every fixed $\epsilon > 0$ there exists a linear time algorithm that computes a matching of weight at least $(2/3 - \epsilon) \cdot w(M^*)$ where M^* is a maximum weight matching.*

Pettie and Sanders (2004) improved on this result by presenting another linear time $2/3 - \epsilon$ approximation algorithm for the MAXIMUM WEIGHT MATCHING PROBLEM whose runtime has a better dependence on ϵ.

9.2.3 Minimum Weight Perfect Matchings

Computing a perfect matching of maximum or minimum weight is a problem that appears quite often in applications. The maximum and minimum perfect matching problem can easily be transformed into each other by just negating the edge weights.

MINIMUM WEIGHT PERFECT MATCHING PROBLEM

Input:	An undirected graph G and edge weights $c : E(G) \to \mathbb{R}$.
Output:	A minimum weight perfect matching in G or a proof that G has no perfect matching.

There is a simple reduction that allows to formulate the MAXIMUM WEIGHT MATCHING PROBLEM as a MAXIMUM WEIGHT PERFECT MATCHING PROBLEM. Take a copy G' of the input graph G and connect each vertex in G by an edge of weight 0 with its copy in G'. Then a maximum weight perfect matching in the new graph corresponds to a maximum weight matching in G (the weights differ exactly by a factor of 2).

This is a simple reduction that can be performed in linear time, unfortunately it does not preserve approximation ratios. However, we cannot expect to find any such reduction as every algorithm that finds an approximate solution to the MINIMUM WEIGHT PERFECT MATCHING PROBLEM must at least be able to decide whether the graph has a perfect matching. The fastest algorithm for doing so has runtime $O(\sqrt{n}m \log(n^2/m)/\log n)$ (Goldberg and Karzanov 2004). Therefore linear time approximation algorithms with constant approximation ratios do not exist for the MINIMUM WEIGHT PERFECT MATCHING PROBLEM unless one can decide in linear time whether a given graph has a perfect matching.

9.3 Degree Constrained Subgraphs

Let $G = (V, E)$ be an undirected graph and $b : V(G) \rightarrow \mathbb{N}$ a degree constraint for every vertex. A subgraph H of G is a *degree constrained subgraph* for G with respect to b, if the degree of each vertex x in H is at most $b(x)$. We consider here the problems of finding a degree constrained subgraph with the largest possible number of edges and a weighted version of this problem. A degree constrained subgraph is also known as *b-matching*. A subset $M \subset E$ is called a *b-matching* if for all $v \in V(G)$ the number of edges in M incident to v is at most $b(v)$. For the special case that $b(v) = 1$ for all $v \in V(G)$ the b-matching is a matching. Therefore, the degree constrained subgraph problems are generalizations of matching problems. As the connections to matching problems are quite strong, we prefer here the notion of b-matchings instead of degree constrained subgraphs.

CARDINALITY b-MATCHING PROBLEM
Input: An undirected graph G and $b : V(G) \rightarrow \mathbb{N}$.
Output: A maximum cardinality b-matching in G.

The CARDINALITY b-MATCHING PROBLEM can be reduced to the CARDINALITY MATCHING PROBLEM by the following reduction which is due to Shiloach (1981): Replace each vertex v of degree $d(v)$ by $d(v)$ copies, such that each edge incident to v uses another copy of v. Then add additional $b(v)$ vertices for each vertex v that are completely connected to the $d(v)$ copies of v. Figure 9.5 shows an example for this reduction.

Let G be an arbitrary graph with n vertices and m edges and let G' be the graph that results from this reduction. Then it is not difficult to prove that a matching in G' of size $\alpha + m$ corresponds to a b-matching in G of size α (Shiloach 1981). The graph

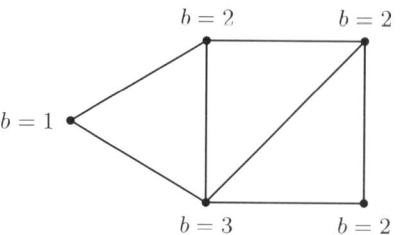
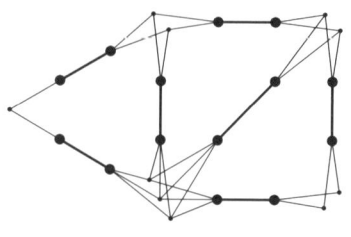

Fig. 9.5. An example illustrating the reduction of a b-matching problem to a matching problem

G' has $O(m)$ vertices and $O(Bm)$ edges, where $B = \max_{v \in V(G)}\{b(v)\}$. For constant B this reduction leads to a b-matching algorithm with runtime $O(m^{3/2})$ by using the algorithm of Micali and Vazirani (1980). For non-constant B the fastest known algorithm for the CARDINALITY b-MATCHING PROBLEM has runtime $O(nm \log n)$ and is due to Gabow (1983).

Unfortunately the above reduction does not preserve approximation ratios. Therefore, even for constant B we cannot use the constant factor approximation algorithms for the CARDINALITY MATCHING PROBLEM to obtain such algorithms for the CARDINALITY b-MATCHING PROBLEM. Instead, we directly have to adapt the approximation algorithms to the CARDINALITY b-MATCHING PROBLEM. A b-matching M in a graph $G = (V, E)$ is called *maximal* if $M \cup \{e\}$ is not a b-matching for all $e \in E \setminus M$. A maximal b-matching can easily be computed in linear time: Start with $M = \emptyset$ and for each $e \in E(G)$ add e to M if this does not violate the degree conditions. The following result shows that this already gives a linear time $1/2$-approximation algorithm for the CARDINALITY b-MATCHING PROBLEM.

Theorem 3.1. *The* CARDINALITY b-MATCHING PROBLEM *can be solved in linear time with approximation ratio* $\frac{1}{2}$.

Proof. As observed above a maximal b-matching can be computed in linear time. Thus we simply have to prove that if M^* is a maximum cardinality b-matching and M is a maximal b-matching then $|M| \geq \frac{1}{2} \cdot |M^*|$. By S we denote all vertices v such that there are exactly $b(v)$ edges of M incident to v. Then every edge in M^* must have at least one endpoint in S as otherwise M is not maximal. As for each vertex $v \in S$ there are at most $b(v)$ edges of M^* incident to v and each edge in M is incident to at most two vertices in S we have:

$$|M^*| \leq \sum_{v \in S} b(v) \leq 2 \cdot |M|$$

which proves the approximation ratio of $1/2$. □

We now also consider a weighted version of the b-matching problem.

MAXIMUM WEIGHT b-MATCHING PROBLEM

Input: An undirected graph G and edge weights $c : E(G) \to \mathbb{R}$.
Output: A maximum weight b-matching in G.

For the weighted version of the b-matching problem one can use the above mentioned reduction of Shiloach (1981) to get a MAXIMUM WEIGHT MATCHING PROBLEM. Again, faster algorithms are possible by adapting algorithms for the MAXIMUM WEIGHT MATCHING PROBLEM directly to the MAXIMUM WEIGHT b-MATCHING PROBLEM. This has been done by Gabow (1983) who obtained the currently fastest algorithm for the MAXIMUM WEIGHT b-MATCHING PROBLEM. His algorithm has runtime $O(m^2 \log n)$. It is easily seen that the Greedy algorithm for the MAXIMUM WEIGHT b-MATCHING PROBLEM achieves an approximation ratio of $1/2$ and has runtime $O(m \log n)$ (Avis 1978).

Linear time approximation algorithms for the MAXIMUM WEIGHT b-MATCHING PROBLEM are known only for the case that $b(x)$ is bounded by some constant for all $x \in V(G)$. In this case Mestre (2006) obtained the following result by adapting an algorithm of Drake and Hougardy (2003b) for the MAXIMUM WEIGHT MATCHING PROBLEM to the MAXIMUM WEIGHT b-MATCHING PROBLEM.

Theorem 3.2 (Mestre 2006). *If $b(x)$ is bounded by some constant for all $x \in V(G)$ then the* MAXIMUM WEIGHT b-MATCHING PROBLEM *can be solved in linear time with approximation ratio $\frac{1}{2}$.*

9.4 Branchings

A *branching* is a directed graph without circuits such that each vertex has indegree at most one. Computing branchings of maximum weight is one of the classical problems in combinatorial optimization.

MAXIMUM WEIGHT BRANCHING PROBLEM	
Input:	A directed graph G and edge weights $c : E(G) \to \mathbb{R}$.
Output:	A maximum weight branching in G.

Edmonds (1967) and independently Chu and Liu (1965) and Bock (1971) gave the first polynomial time algorithms for computing a maximum weight branching. Later Gabow et al. (1986) showed that Edmond's algorithm can be implemented in $O(m + n \log n)$ time. This is already quite close to a linear time algorithm but from a practical point of view one is still interested in simpler linear time approximation algorithms for the MAXIMUM WEIGHT BRANCHING PROBLEM.

One such algorithm is given by the following greedy approach: Start with an empty branching B and sort the edges by decreasing weight. Now check for each edge one by one whether its addition violates the degree condition. If not then add it to B. Now one can destroy each circuit that might appear by removing the lightest edge from each circuit. This can be done in linear time. As each circuit has at least two edges, the algorithm has an approximation ratio of $1/2$. The total runtime is linear, as sorting of the edges is actually not required: simply take the heaviest incoming edge for each vertex.

This result can be improved by applying the phase bounding approach to Edmonds' algorithm. This has been observed by Ziegler (2008). He obtained the following result.

Theorem 4.1 (Ziegler 2008). *For every $\epsilon > 0$ there exists a linear time approximation algorithm for the* MAXIMUM WEIGHT BRANCHING PROBLEM *that has an approximation ratio of* $1 - \epsilon$.

Acknowledgement

I am grateful to an anonymous referee for several useful comments.

References

Althaus, E., Mehlhorn, K.: Maximum network flow with floating point arithmetic. Inf. Process. Lett. **66**(3), 109–113 (1998)

Avis, D.: Two greedy heuristics for the weighted matching problem. Congr. Numer. **XXI**, 65–76 (1978). Proceedings of the Ninth Southeastern Conference on Combinatorics, Graph Theory, and Computing (1978)

Bast, H., Mehlhorn, K., Schäfer, G., Tamaki, H.: Matching algorithms are fast in sparse random graphs. Theory Comput. Syst. **39**(1), 3–14 (2006)

Berge, C.: Two theorems in graph theory. Proc. Natl. Acad. Sci. U.S.A. **43**(9), 842–844 (1957)

Blum, N.: A new approach to maximum matching in general graphs. In: Proc. of 17th ICALP (1990). Lecture Notes in Computer Science, vol. 443, pp. 586–597. Springer, Berlin (1990)

Bock, F.: An algorithm to construct a minimum directed spanning tree in a directed network. In: Avi Itzhak, B. (ed.) Developments in Operations Research, Proceedings of the Third Annual Israel Conference on Operations Research, July 1969, vol. 1, pp. 29–44. Gordon and Breach, New York (1971). Paper 1-2

Bodlaender, H.L.: A linear-time algorithm for finding tree-decompositions of small treewidth. SIAM J. Comput. **25**(6), 1305–1317 (1996)

Chu, Y.-J., Liu, T.-H.: On the shortest arborescence of a directed graph. Sci. Sin. **14**(10), 1396–1400 (1965)

Cook, W., Rohe, A.: Computing minimum-weight perfect matchings. INFORMS J. Comput. **11**(2), 138–148 (1999)

Drake, D.E., Hougardy, S.: Linear time local improvements for weighted matchings in graphs. In: Jansen, K. et al. (eds.) International Workshop on Experimental and Efficient Algorithms (WEA) 2003. Lecture Notes in Computer Science, vol. 2647, pp. 107–119. Springer, Berlin (2003a)

Drake, D.E., Hougardy, S.: A simple approximation algorithm for the weighted matching problem. Inf. Process. Lett. **85**(4), 211–213 (2003b)

Drake Vinkemeier, D.E., Hougardy, S.: A linear-time approximation algorithm for weighted matchings in graphs. ACM Trans. Algorithms **1**(1), 107–122 (2005)

Edmonds, J.: Paths, trees, and flowers. Can. J. Math. **17**(3), 449–467 (1965)

Edmonds, J.: Optimum branchings. J. Res. Natl. Bur. Stand., B Math. Math. Phys. **71**(4), 233–240 (1967)

Feder, T., Motwani, R.: Clique partitions, graph compression and speeding-up algorithms. J. Comput. Syst. Sci. **51**(2), 261–272 (1995)

Fischer, T., Goldberg, A.V., Haglin, D.J., Plotkin, S.: Approximating matchings in parallel. Inf. Process. Lett. **46**(3), 115–118 (1993)

Ford, L.R., Fulkerson, D.R.: Maximal flow through a network. Can. J. Math. **8**, 399–404 (1956)

Frank, A.: On Kuhn's Hungarian method—a tribute from Hungary. Nav. Res. Logist. **52**(1), 2–5 (2005)

Fremuth-Paeger, C., Jungnickel, D.: Balanced network flows. VIII. A revised theory of phase-ordered algorithms and the $O(\sqrt{n}m \log(n^2/m)/\log n)$ bound for the nonbipartite cardinality matching problem. Networks **41**(3), 137–142 (2003)

Gabow, H.N.: An efficient reduction technique for degree-constrained subgraph and bidirected network flow problems. In: STOC '83: Proceedings of the Fifteenth Annual ACM Symposium on Theory of Computing, pp. 448–456. ACM Press, New York (1983)

Gabow, H.N.: Data structures for weighted matching and nearest common ancestors with linking. In: SODA '90: Proceedings of the First Annual ACM–SIAM Symposium on Discrete Algorithms, pp. 434–443. SIAM, Philadelphia (1990)

Gabow, H.N., Tarjan, R.E.: Algorithms for two bottleneck optimization problems. J. Algorithms **9**(3), 411–417 (1988)

Gabow, H.N., Tarjan, R.E.: Faster scaling algorithms for general graph-matching problems. J. Assoc. Comput. Mach. **38**(4), 815–853 (1991)

Gabow, H.N., Galil, Z., Spencer, T., Tarjan, R.E.: Efficient algorithms for finding minimum spanning trees in undirected and directed graphs. Combinatorica **6**(2), 109–122 (1986)

Goldberg, A.V., Karzanov, A.V.: Maximum skew-symmetric flows and matchings. Math. Program. **100**(3), 537–568 (2004)

Harvey, N.J.A.: Algebraic structures and algorithms for matching and matroid problems. In: Proceedings of the 47th Annual IEEE Symposium on Foundations of Computer Science (FOCS'06), pp. 531–542. IEEE Computer Society, Washington (2006)

Hassin, R., Lahav (Haddad), S.: Maximizing the number of unused colors in the vertex coloring problem. Inf. Process. Lett. **52**(2), 87–90 (1994)

Hopcroft, J.E., Karp, R.M.: An $n^{5/2}$ algorithm for maximum matchings in bipartite graphs. SIAM J. Comput. **2**(4), 225–231 (1973)

Jenkyns, T.A.: The efficacy of the greedy algorithm. Congr. Numer. **17**, 341–350 (1976)

Karypis, G., Kumar, V.: Multilevel k-way partitioning scheme for irregular graphs. J. Parallel Distrib. Comput. **48**(1), 96–129 (1998)

Korte, B., Hausmann, D.: An analysis of the greedy heuristic for independence systems. Ann. Discrete Math. **2**, 65–74 (1978)

Kuhn, H.W.: The Hungarian method for the assignment problem. Nav. Res. Logist. Q. **2**, 83–97 (1955)

Mehlhorn, K., Schäfer, G.: Implementation of $O(nm \log n)$ weighted matchings in general graphs: the power of data structures. ACM J. Exp. Algorithmics **7**, 4 (2002)

Mestre, J.: Greedy in approximation algorithms. In: Azar, Y., Erlebach, T. (eds.) ESA 2006. Lecture Notes in Computer Science, vol. 4168, pp. 528–539. Springer, Berlin (2006)

Micali, S., Vazirani, V.V.: An $O(\sqrt{|v|} \cdot |E|)$ algorithm for finding maximum matching in general graphs. In: Proc. of 21st Annual Symposium on Foundations of Computer Science (21st FOCS, Syracuse, New York, 1980), pp. 17–27 (1980)

Motwani, R.: Average-case analysis of algorithms for matchings and related problems. J. Assoc. Comput. Mach. **41**(6), 1329–1356 (1994)

Mucha, M., Sankowski, P.: Maximum matchings via Gaussian elimination. In: Proceedings of the 45th Annual IEEE Symposium on Foundations of Computer Science (FOCS'04), pp. 248–255. IEEE Computer Society, Washington (2004)

Petersen, J.: Die Theorie der regulären Graphs. Acta Math. **15**(1), 193–220 (1891)

Pettie, S., Sanders, P.: A simpler linear time $2/3 - \varepsilon$ approximation for maximum weight matching. Inf. Process. Lett. **91**(6), 271–276 (2004)

Preis, R.: Linear time $\frac{1}{2}$-approximation algorithm for maximum weighted matching in general graphs. In: Meinel, C., Tison, S. (eds.) Symposium on Theoretical Aspects in Computer Science (STACS). Lecture Notes in Computer Science, vol. 1563, pp. 259–269. Springer, Berlin (1999)

Shiloach, Y.: Another look at the degree constrained subgraph problem. Inf. Process. Lett. **12**(2), 89–92 (1981)

Vazirani, V.V.: A theory of alternating paths and blossoms for proving correctness of the $O(\sqrt{V}E)$ general graph maximum matching algorithm. Combinatorica **14**(1), 71–109 (1994)

Ziegler, V.: Approximating optimum branchings in linear time. Technical report, Humboldt-Universität zu Berlin, Institut für Informatik (2008)

10

The Unbounded Knapsack Problem

T.C. Hu, Leo Landa, and Man-Tak Shing

Summary. This paper presents a survey of the unbounded knapsack problem. We focus on the techniques for obtaining the optimal solutions, particularly those using the periodic structure of the optimal solutions when the knapsack weight-carrying capacity b is sufficiently large. In addition to reviewing existing algorithms on the subject, the paper also includes two new algorithms, one for finding the onset of the optimal periodic solutions in time $O(nw_1)$, where w_1 is the weight of the best item, i.e. the item with the highest value-to-weight ratio, and a second one for finding the optimal solutions when the capacity b is below the critical value where the optimal periodic solution begins. The second algorithm has a worst-case time complexity of $O(nw_1v_1)$, where v_1 is the value of the best item.

10.1 Introduction

The unbounded knapsack problem (UKP) arises from the following hypothetical situation. Consider a hiker with a knapsack on a trip. He has to select items to fill the knapsack. Every ith type item has a value v_i and a weight w_i, and the knapsack has a weight-carry capacity b.

Mathematically, the problem is defined as follows:

$$\max\ V = \sum_{i=1}^{n} v_i x_i$$

$$\text{subject to}\quad \sum_{i=1}^{n} w_i x_i \le b \tag{1}$$

with v_i, w_i, b, x_i all non-negative integers.

The Knapsack problem is the simplest integer programming problem with a single constraint, and belongs to the class of NP-complete problems (Garey and Johnson 1979). The topic started with a sequence of papers by Gilmore and Gomory (1966). Since then, two books (Kellerer et al. 2004, Martello and Toth 1990a) and over three hundred technical papers have been devoted to the knapsack problems in the last forty years.

10.2 General Techniques for Finding Optimal Solutions

10.2.1 The Branch-and-Bound Solutions

The most straightforward way for solving the UKP is to use the branch and bound technique which has a worst-case time complexity of $O(2^n)$. The branch and bound algorithm can be speeded by successively pruning the search tree with revised lower and upper bounds. Any feasible solution gives a lower bound on the optimal value, and we can use linear programming relaxation to get an upper bound. Other concepts from duality can also be used.

The class of algorithms based on branch and bound can be further speeded up when dominance relation exists between the item types. Assuming that the ith and the kth items have values and weights satisfying the relation

$$v_i > v_k \quad \text{and} \quad w_i < w_k \tag{2}$$

then the kth item is never used in any optimal solution. We say that the ith item dominates the kth item. We can use relation (2) to reduce the input size of the knapsack problem. The simple dominance relation (2) has been extended by several researchers into more generalized dominance relations (e.g. the multiple dominance by Martello and Toth (1990b) and the threshold dominance by Andonov et al. (2000)). See (Kellerer et al. 2004) for details.

10.2.2 The Dynamic Programming Solutions

Due to the integer constraint on x_i, the first optimal knapsack algorithm based on dynamic programming was developed by Gilmore and Gomory (1966). Let $F_k(y)$ be the maximum value obtained if the knapsack has a weight-carrying capacity y and there are k kinds of items to choose from. We can systematically compute $F_k(y)$ for $k = 1, \ldots, n$ and $y = 1, \ldots, b$ as follows.

Clearly, the boundary conditions are

$$F_k(0) = 0, \quad \text{for all } k \quad \text{and}$$

$$F_1(y) = v_1 \left\lfloor \frac{y}{w_1} \right\rfloor.$$

In general,

$$F_k(y) = \max[F_{k-1}(y), \ v_k + F_k(y - w_k)]. \tag{3}$$

Based on (3), we can build a table of n rows and b columns, and solve the knapsack problem in $O(nb)$ time. Note that this algorithm runs in pseudo-polynomial time because any reasonable encoding of the knapsack problem requires only a polynomial in n and $\log_2 b$.

10.2.3 The Word RAM Algorithm

Since the input are given in binary digits, Pisinger developed a novel technique that uses the parallel binary bit operations in a computer word to speed up the dynamic programming algorithm for a special case of UKP where $v_i = w_i$ for $1 \leq i \leq n$ (Pisinger 2003).

Assume that the word size of the computer is $W = \Theta(\log b)$, which is large enough to hold the weight-carry capacity b. The Word RAM algorithm computes a table $g(k, y)$, $0 \leq k \leq n$, $0 \leq y \leq b$, where $g(k, y) = 1 \Leftrightarrow$ there is a multiset over the items $w_1, w_2, \ldots w_k$ with overall sum y. We have $g(0, 0) = 1$, $g(0, y) = 0$ for $0 \leq y \leq b$, and for $1 \leq k \leq n, 0 \leq y \leq b$,

$$g(k, y) = 1 \quad \Leftrightarrow \quad (g(k - 1, y) = 1 \text{ or } g(k, y - w_k) = 1).$$

By storing the W entries of the table g in each computer word and updating the entries in each word in parallel using the *binary shift*, *binary and*, and the *binary or* operations, the Word RAM algorithm computes $g(n, b)$ in time $O(b \log \log b + n b / \log b)$, which equals $O(n b / \log b)$ for sufficiently large n.

10.3 Fully-Polynomial-Time Approximation Algorithms

Since the unbounded knapsack problem is NP-complete, many researchers have developed approximation algorithms for solving the problem. Given an input instance I of the unbounded knapsack problem, let $V^*(I)$ be the value of the optimal packing and $V_A(I)$ be the value of the solution by an approximation algorithm A. We say that A is an ϵ-approximation algorithm if for every ϵ, $0 < \epsilon < 1$, and every input instance I, $V_A(I) \geq (1 - \epsilon)V^*(I)$. An ϵ-approximation algorithm is a fully-polynomial-time ϵ-approximation algorithm if it runs in time polynomial both in n and $\frac{1}{\epsilon}$.

The earliest fully-polynomial-time ϵ-approximation algorithm for UKP was given by Ibarra and Kim (1975). The Ibarra–Kim algorithm runs in time $O((3/\epsilon)^3 n)$ and space $O(n + (3/\epsilon)^3)$ and consists of the following major steps:

1. Given a positive real number ϵ, $0 < \epsilon < 1$, compute a cut-off threshold $\delta = \hat{V}(\epsilon/3)^2$, where (w_1, v_1) is the item type with the highest value-to-weight ratio and $\hat{V} = \lceil b/w_1 \rceil v_1$.
2. Separate the items into two classes, large value items with $v_i > \delta$ and small value items with $v_i \leq \delta$. Let the large value items be $(w_{L_1}, v_{L_1}), (w_{L_2}, v_{L_2}), \ldots,$ $(w_{L_\alpha}, v_{L_\alpha})$ and the small value items be $(w_{S_1}, v_{S_1}), (w_{S_2}, v_{S_2}), \ldots, (w_{S_\beta}, v_{S_\beta})$.
3. Create a multiset Q' by adding n_i copies of the large value item (w_{L_i}, v_{L_i}) to Q', $1 \leq i \leq \alpha$, where $n_i = \lfloor b/w_{L_i} \rfloor$, and replace the value v_{L_i} of every item (w_{L_i}, v_{L_i}), $1 \leq i \leq \alpha$, in Q' with the normalized value $\lfloor v_{L_i}/\delta \rfloor$.
4. Partition the interval $[0, \hat{V}]$ into $g = \lfloor \hat{V}/\delta \rfloor = \lfloor (3/\epsilon)^2 \rfloor$ sub-intervals. Use dynamic program to find at most one optimal packing in each sub-interval using the items with the modified values in Q', discarding all candidates with total value in the same interval except the one with the least total weight.

5. For each optimal packing k obtained in Step 4, let $W(k)$ be the total weight of the packing. Fill the remaining space $b - W(k)$ with multiple copies of a small value item (w_{S_i}, v_{S_i}), such that the total value

$$\left\lfloor \frac{b - W(k)}{w_{S_i}} \right\rfloor v_{S_i} = \max_{1 \le j \le \beta} \left\lfloor \frac{b - W(k)}{w_{S_j}} \right\rfloor v_{S_j}.$$

6. Return the packing with the largest combined value.

Lawler (1979) modified Ibarra–Kim's algorithm with a more efficient method of scaling, resulting in an ϵ-approximation algorithm that runs in $O(n + 1/\epsilon^3)$ time and $O(n + 1/\epsilon^2)$ space.

10.4 Optimal Periodic Structure

When there are no restrictions on the values of x_i and we need to calculate the optimal solutions for large values of b, Gilmore and Gomory discovered that the optimal solutions have a periodic structure when b is sufficiently large. For simplicity, we always assume

$$\frac{v_1}{w_1} \ge \max_j \frac{v_j}{w_j} \tag{4}$$

in the rest of the paper and, in the case of a tie in (4), we let $w_1 < w_j$, so that the first item shall be the best item. Note that the best item can be found in $O(n)$ time without sorting the items according to their value-to-weight ratios.

When the weight-carry capacity b exceeds a critical value b^{**}, the optimal solution for b equals to the optimal solution for $b - w_1$ plus a copy of the best item. In other words, we can first fill portion of the knapsack using the optimal solutions for a smaller weight-carry capacity and fill the rest of the knapsack with the best item only. The question is how large the critical value b^{**} is? Two simple upper bounds would be

$$b^{**} \le \sum_{j \ne 1} \mathrm{lcm}(w_1, w_j) \tag{5}$$

and

$$b^{**} \le \frac{w_1 \rho_1}{\rho_1 - \rho_2} \tag{6}$$

where

$$\rho_1 = \frac{v_1}{w_1} \quad \text{the highest value-to-weight ratio,}$$

$$\rho_2 = \frac{v_2}{w_2} \quad \text{the second highest value-to-weight ratio.}$$

To prove the above upper bounds, we need to show that there is always an optimal packing that contains at least one copy of the best item for each b greater than the upper bound. For condition (5), assume that none of the optimal packings for a boundary b that is greater than $\sum_{j \neq 1} \text{lcm}(w_1, w_j)$ contains a copy of the best item. Then there must be an item $j \neq 1$ such that $x_j \geq \text{lcm}(w_1, w_j)/w_j$ and we can obtain another packing for b by replacing $\text{lcm}(w_1, w_j)/w_j$ copies of w_j with $\text{lcm}(w_1, w_j)/w_1$ copies of w_1. Since the value of the new packing is equal to or greater than the original one, we have a contradiction.

For condition (6), we note that the maximum valve achievable in any packing b cannot exceed $\rho_2 b$ if the packing does not use the best item at all, and the optimal packing is always greater than or equal to $v_1 \lfloor \frac{b}{w_1} \rfloor$. We have

$$F(b) \geq v_1 \left\lfloor \frac{b}{w_1} \right\rfloor > v_1 \left(\frac{b}{w_1} - 1 \right) = \rho_1 (b - w_1) \geq \rho_2 b$$

if condition (6) holds, implying that the optimal packing cannot be accomplished without the best items.

For illustration purpose, we shall use the following numerical example throughout the rest of the paper.

$$\max \ V = 8x_1 + 3x_2 + 17x_3$$
$$\text{subject to} \quad W = 8x_1 + 5x_2 + 18x_3 \leq b \tag{7}$$

where b and x_i are non-negative integers.

To find the starting point of the periodic optimal structure, we shall partition the weight-carrying capacity b into w_1 classes (called *threads*), where

$\quad b \quad$ is in the thread 0 if $b(\text{mod } w_1) = 0$,
$\quad b \quad$ is in the thread 1 if $b(\text{mod } w_1) = 1$,
$\quad \quad \ldots$
$\quad b \quad$ is in the thread $(w_1 - 1)$ if $b(\text{mod } w_1) = (w_1 - 1)$.

We shall call the weight-carrying capacity as the *boundary* from now on, and present an algorithm to pinpoint the necessary and sufficient value of the boundary in each thread (called the *thread critical boundary*) where we start using the best item repeatedly. For brevity, we shall use

$\quad t(b) \quad$ to denote the thread of a boundary b,
$\quad b^*(i) \quad$ to denote the thread critical boundary of thread i,
$\quad b^k(i) \quad$ to denote the thread critical boundary of thread i using the first k types of items only.

If b happens to be a multiple of w_1, then b is in the thread 0 and the critical boundary of the thread 0 is zero. This is denoted by $b^*(0) = 0$. For our numerical example, we have

$$b^*(0) = 0, \quad b^*(1) = 1, \quad b^*(2) = 18, \quad b^*(3) = 19,$$
$$b^*(4) = 36, \quad b^*(5) = 5, \quad b^*(6) = 6, \quad b^*(7) = 23$$

and

$$b^{**} = \max_j b^*(j) = b^*(4) = 36. \tag{8}$$

The optimal solution for any boundary b larger than the critical value $b^{**} = 36$ will exhibit the periodic structure that uses one or more copies of the best items in our example.

10.4.1 Greenberg's Algorithm

Greenberg (1985) presented an algorithm for finding $b^*(t)$, the starting point of the periodic structure of the optimal solutions in each thread t. The algorithm defines, for each thread t, a function $H(t)$ where

$$H(t) = \min_{x \bmod w_1 = t} \left(\frac{x v_1}{w_1} - F_n(x) \right),$$

the smallest penalty among all boundaries in the thread t for not being able to fill up the knapsack with the best item only (i.e. the difference between the non-integer and the integer solution). Greenberg showed that $H(t)$ can be solved recursively using the relation, $H(0) = 0$, and

$$H(t) = \min_{2 \leq j \leq n} (d_j + H(t - w_j) \bmod w_1)$$

where $d_j = (v_1/w_1) \times w_j - v_j$, the penalty of using one copy of jth item to fill up a capacity of w_j.

Let $(t, \hat{b}, \hat{H}, i)$ denote a packing instance that falls on the thread t, has a total weight $\hat{b}$ and a penalty $\hat{H}$, and is obtained by adding an item w_i to another packing. Greenberg's algorithm works in the same spirit as Dijkstra's single source shortest path algorithm. It starts with the packing $(0, 0, 0, 1)$ and successively "grows" the partial packings in search of the $w_1 - 1$ packings that achieve $H(t)$ for the threads $2, 3, \ldots, w_1 - 1$. Let us illustrate Greenberg's algorithm with the numerical example (7).

Since w_1 is not relatively prime to w_3, we first add a slack variable x_4 with $w_4 = 1$ and $v_4 = 0$ to (7) and convert the constraint into an equation.

$$\max \ V = 8x_1 + 3x_2 + 17x_3 + 0x_4$$
$$\text{subject to} \quad W = 8x_1 + 5x_2 + 18x_3 + x_4 = b. \tag{9}$$

We then initialize $d_2 = 2$, $d_3 = 1$, $d_4 = 1$, $H(1) = 0$, $b(1) = 0$. For $2 \leq t \leq w_1 - 1$, we set $H(t) = \infty$ and $b(t) = 0$. Next, we set $H(5) = 2$, $b(5) = 5$, $H(2) = 1$, $b(2) = 18$, $H(1) = 1$, $b(1) = 1$ and insert the packing $(5, 5, 2, 2)$, $(2, 18, 1, 3)$, $(1, 1, 1, 4)$ into an empty first-in-first-out-queue.

While the queue is not empty, we remove a packing $(t, \hat{b}, \hat{H}, i)$ from the front of the queue. Since the queue may contain more than one partial packing that falls on the thread t as more and more partial packings are added to the queue while

Table 10.1. The packings generated successively by Greenberg's algorithm

t	0	5	2	1	2	7	4	6	3	2	4	2	6	3	0	5
$\hat{b}$	0	5	18	1	10	23	36	6	19	2	28	41	54	11	24	37
$\hat{H}$	0	2	1	1	2	3	2	3	2	2	5	4	3	5	4	3
i	1	2	3	4	2	2	3	2	3	4	2	2	3	2	2	3

reducing the penalty values of the corresponding threads, we need to make sure that the packing $(t, \hat{b}, \hat{H}, i)$ which we just remove from the queue is indeed the partial packing responsible for the smallest penalty for the thread t so far. We create $i - 1$ candidate packings by adding a copy of $w_2, \ldots, w_i$ to $\hat{b}$ respectively if $\hat{H} = H(t)$ and $\hat{b} = b(t)$, and discard the packing otherwise. For each candidate packing $(t', \hat{b}', \hat{H}', i')$, we set $H(t') = \hat{H}'$, $b(t') = \hat{b}'$ and insert $(t', \hat{b}', \hat{H}', i')$ into the queue if $\hat{H}' < H(t')$ or ($\hat{H}' = H(t')$ and $\hat{b}' < b(t')$).

In our numerical example, we will first remove $(5, 5, 2, 2)$ from the queue. Since $2 = H(5)$ and $5 = b(5)$, we create the candidate packing $(2, 10, 2, 2)$ by adding another copy of w_2 to $b(5)$. We compare the penalty of the candidate packing against $H(2)$, the best penalty obtained for thread 2 so far, and discard the candidate packing $(2, 10, 2, 2)$ since it cannot reduce $H(2)$.

Next, we remove the packing, $(2, 18, 1, 3)$ from the front of the queue. After confirming that $1 = H(2)$ and $18 = b(2)$, we create two candidate packings $(7, 23, 3, 2)$ and $(4, 36, 2, 3)$ by adding a copy of w_2 and w_3 to $b(2)$ respectively. Since $3 < H(7) = \infty$ and $2 < H(4) = \infty$, we set $H(7) = 3, b(7) = 23, H(4) = 2$, $b(4) = 36$, and insert both candidate packings into the queue.

We repeat the process and generate the packings shown in Table 10.1. The packings with shaded background are packings that have been examined and discarded by the algorithm because they do not reduce the penalty of their corresponding threads. The algorithm terminates when the queue is empty, and returns the $H(t)$ and $b^*(t) = b(t)$, $1 \leq t \leq w_1 - 1$.

Greenberg did not give any time complexity analysis of his algorithm in Greenberg (1985). However, one can easily conclude that his algorithm has a worst-case time complexity of $O(nw_1)$ since any optimal packing not containing w_1 items can have at most $w_1 - 1$ copies of w_i, $2 \leq i \leq n$.

10.4.2 Landa's Algorithm

Greenberg's algorithm has 3 drawbacks: (1) the penalty function $H(t)$ usually involves real numbers when $w_1 \neq v_1$; (2) it requires the scaling of the weights if the greatest common divisor of $w_1, w_2, \ldots, w_n > 1$; and (3) it requires the additional slack variable with unit weight if w_1 is not relatively prime to all other weights. The use of real numbers is undesirable because they do not always have finite length encodings and may introduce additional errors to the computation. Although one can use rational numbers instead of real numbers for $H(t)$, the computation for rational numbers is very inefficient.

Landa (2004) proposed a new algorithm for computing the critical thread boundaries. The major difference between Landa's algorithm and Greenberg's algorithm is that Landa focused on the *gain* for filling the otherwise unusable space (if we are limited to use integral copies of the best item only) while Greenberg focused on the *penalty*, i.e. the loss of not being able to fill up the knapsack completely with the best item due to integer constraint. One major advantage of using *gain* instead of *penalty* is that the gain computation involves only integers. Another advantage is that Landa's algorithm does not require any scaling of weights if the greatest common divisor of $w_1, w_2, \ldots, w_n > 1$, and does not require the addition of the unit-weight slack variable if w_1 is not relatively prime to all other weights. Moreover, the gain concept also enables us to obtain a very simple algorithm for finding the optimal solutions when the boundary b is less than $b^*(t(b))$.

In order to understand how Landa's algorithm works, we need to introduce the concept of *gain*, a measure of the efficiency of different packings for a boundary b.

Gain

The *gain* of a boundary b is the difference between the optimal value $V(b)$ and the value of the packing using the best items only. Mathematically, we define

$$g(b) = V(b) - v_1 \left\lfloor \frac{b}{w_1} \right\rfloor \tag{10}$$

Since we can always use the best-item-only solution as the default packing for any boundary b, $g(b) \geq 0$ and is an integer by its definition (10). Note that for a fixed b, the difference between $V(b)$ and $g(b)$ is a constant; i.e. larger $V(b)$ implies larger $g(b)$ and vice versa.

Let $b_b = b_s + w_1$. Then b_b and b_s reside in the same thread. If the optimal solution for b_b contains at least one copy of w_1, then its sub-solution for a knapsack of boundary b_s must also be optimal by the principle of optimality. To formalize this concept, we state it as Lemma 4.1.

Lemma 4.1. *If the optimal solution to the boundary $b_b = b_s + w_1$ equals to the optimal solution to the boundary b_s plus one copy of the best item, then the gains $g(b_b)$ and $g(b_s)$ are equal.*

Proof.

$$g(b_b) = V(b_b) - v_1 \left\lfloor \frac{b_b}{w_1} \right\rfloor = V(b_b) - v_1 \left\lfloor \frac{b_s + w_1}{w_1} \right\rfloor$$

$$= V(b_s) + v_1 - v_1 \left\lfloor \frac{b_s + w_1}{w_1} \right\rfloor$$

$$= V(b_s) - v_1 \left\lfloor \frac{b_s}{w_1} \right\rfloor = g(b_s). \quad \square$$

Corollary 4.2. *The gains are monotonically increasing in each thread.*

Proof. Let $V(b_s)$ and $g(b_s)$ be the values of the optimal packing for b_s and its gain. We can create a feasible solution for $b_b = b_s + w_1$ by using one copy of the best item. The optimal value of b_b, denoted by $V(b_b)$, must be greater than or equal to $V(b_s) + v_1$ since $b_s + w_1$ is a feasible solution. From the definition (10), we have $g(b_b) \geq g(b_s)$. □

From now on, if there is more than one optimal solution for a boundary b, we prefer the optimal solution with the largest x_1. Since we want to find the smallest boundary where the optimal solution becomes periodic in each thread, we shall partition the boundaries in each thread into two categories. The boundary $b_b = b_s + w_1$ is a *special* boundary if its gain $g(b_b)$ is larger than $g(b_s)$, the gain of the boundary b_s; the boundary is a *non-special* boundary if $g(b_b) = g(b_s)$. Note that the definitions of special and non-special boundaries are for boundaries in the same thread.

Let us build a table of eight columns for our numerical example (7), where each column corresponds to a thread, and write in each cell, the boundary and its gain. In Table 10.2, all special boundaries are marked with darker borders, and the special cell with the largest gain (i.e. thread critical boundary) in each column is marked with darker borders and shading. Since there is no cell above the first row, we define all cells in the first row as special boundaries.

Table 10.2. Threaded view of optimal gains

Threads							
0	1	2	3	4	5	6	7
$b = 0$ $g = 0$	$b = 1$ $g = 0$	$b = 2$ $g = 0$	$b = 3$ $g = 0$	$b = 4$ $g = 0$	$b = 5$ $g = 3$	$b = 6$ $g = 3$	$b = 7$ $g = 3$
$b = 8$ $g = 0$	$b = 9$ $g = 0$	$b = 10$ $g = 0$	$b = 11$ $g = 0$	$b = 12$ $g = 0$	$b = 13$ $g = 3$	$b = 14$ $g = 3$	$b = 15$ $g = 3$
$b - 16$ $g = 0$	$b = 17$ $g = 0$	$b = 18$ $g = 1$	$b = 19$ $g = 1$	$b = 20$ $g = 1$	$b = 21$ $g = 3$	$b = 22$ $g = 3$	$b = 23$ $g = 4$
$b = 24$ $g = 0$	$b = 25$ $g = 0$	$b = 26$ $g = 1$	$b = 27$ $g = 1$	$b = 28$ $g = 1$	$b = 29$ $g = 3$	$b = 30$ $g = 3$	$b = 31$ $g = 4$
$b = 32$ $g = 0$	$b = 33$ $g = 0$	$b = 34$ $g = 1$	$b = 35$ $g = 1$	$b = 36$ $g = 2$	$b = 37$ $g = 3$	$b = 38$ $g = 3$	$b = 39$ $g = 4$
$b = 40$ $g = 0$	$b = 41$ $g = 0$	$b = 42$ $g = 1$	$b = 43$ $g = 1$	$b = 44$ $g = 2$	$b = 45$ $g = 3$	$b = 46$ $g = 3$	$b = 47$ $g = 4$

Lemma 4.3. *The gain in any cell is strictly less than v_1.*

Proof. From the definition of gain in (10), we have

$$V(b) = g(b) + v_1 \left\lfloor \frac{b}{w_1} \right\rfloor.$$

If $g(b) = v_1$, then

$$V(b) = v_1 + v_1 \left\lfloor \frac{b}{w_1} \right\rfloor = v_1 \left\lfloor \frac{b + w_1}{w_1} \right\rfloor > \frac{v_1}{w_1} b,$$

a contradiction.   □

Lemma 4.4. *There exists a cell in each thread that has the maximum gain, and all boundaries larger than that cell have the same gain.*

Proof. We know that all gains are non-negative integers, and from Lemma 4.3, all gains are less than v_1. Since the gains are monotonically increasing in each thread from Corollary 4.2, the gain will stabilize somewhere in each thread.   □

Suppose we have an optimal solution to a boundary b_s with gain $g(b_s)$, and the optimal solution for the boundary $b_b = b_s + w_k$ (in a different thread) uses the same optimal solution for b_s plus one copy of the kth item. Then the gain $g(b_b)$ of the boundary b_b is related to $g(b_s)$ with the formula stated in Theorem 4.5.

Theorem 4.5. *If the optimal solution for the boundary $b_b = b_s + w_k$ in the thread $t(b_b)$ consists of the optimal solution for the boundary b_s in the thread $t(b_s)$ plus one copy of the kth item, then*

$$g(b_b) = g(b_s) + v_k - v_1 \left\lfloor \frac{t(b_s) + w_k}{w_1} \right\rfloor. \tag{11}$$

Proof. Since

$$g(b_b) = V(b_b) - v_1 \left\lfloor \frac{b_b}{w_1} \right\rfloor$$

$$= V(b_s) + v_k - v_1 \left\lfloor \frac{b_s + w_k}{w_1} \right\rfloor$$

and

$$b_s = b_s \,(\mathrm{mod}\ w_1) + \left\lfloor \frac{b_s}{w_1} \right\rfloor \times w_1 = t(b_s) + \left\lfloor \frac{b_s}{w_1} \right\rfloor \times w_1,$$

$$g(b_b) = V(b_s) + v_k - v_1 \left(\left\lfloor \frac{t(b_s) + w_k}{w_1} \right\rfloor + \left\lfloor \frac{b_s}{w_1} \right\rfloor \right)$$

$$= \left(V(b_s) - v_1 \left\lfloor \frac{b_s}{w_1} \right\rfloor \right) + v_k - v_1 \left\lfloor \frac{t(b_s) + w_k}{w_1} \right\rfloor$$

$$= g(b_s) + v_k - v_1 \left\lfloor \frac{t(b_s) + w_k}{w_1} \right\rfloor.   □$$

Note that the formula (11) does not contain the actual value of the optimal solutions $V(b_b)$ or $V(b_s)$. It relates the gains of two cells in two different threads $t(b_b)$ and $t(b_s)$ using only the values of $v_1, w_1, v_k, w_k, t(b_s)$ and $t(b_b)$.

Thread Critical Boundaries

In this section, we describe the method for obtaining the largest gain (and its corresponding boundary) in each thread. From the principle of optimality, the sub-solution of an optimal solution must be an optimal solution of the sub-boundary. The principle, however, does not say how to extend an optimal solution of a boundary to the optimal solution of a larger boundary. We shall show how to use the gain formula in Theorem 4.5 to discover the gain of a larger boundary from a smaller one. For brevity, we shall use

$g(b)$ to denote the gain of a boundary b,

$g_t^*(i)$ to denote the largest gain of the thread i, which is the gain of $b^*(i)$, the critical boundary in the thread i,

$g_t^k(i)$ to denote the gain of $b^k(i)$, the largest gain of the thread i using the first k types of items only.

Starting with only one kind of items with value v_1 and weight w_1, we have the thread critical boundary $b^1(i) = i$ and the gain $g_t^1(i) = 0$ for every $0 \le i \le w_1 - 1$.

Next, we consider packing the boundaries in each thread with the first and second types of items, and see if the addition of a second type of items with value v_2 and weight w_2 can increase the maximum gain in each thread. The newly introduced type is called the *challenge* type.

Assuming that we have obtained the correct value of $g_t^2(i)$, the largest gain in thread i using the first two types of items (e.g. $g_t^2(0) = g_t^1(0) = 0$), we can apply the gain formula to see if we can get a better gain for the thread $j = (i + w_2) \bmod w_1$ and set the gain $g_t^2(j)$ to

$$\max \left\{ g_t^1(j), \ g_t^2(i) + v_2 - v_1 \left\lfloor \frac{i + w_2}{w_1} \right\rfloor \right\}.$$

In general, to find $g_t^k(j)$ from $g_t^k(i)$, $j = (i + w_k) \bmod w_1$, we set $g_t^k(j)$ to

$$\max \left\{ g_t^{k-1}(j), \ g_t^k(i) + v_k - v_1 \left\lfloor \frac{i + w_k}{w_1} \right\rfloor \right\}. \tag{12}$$

Therefore, for each thread, we keep the boundary of the largest gain discovered so far. In case of tie, we will keep the smaller boundary for equal gain in each thread. We update the information as a new kind of items becomes available to get a larger gain for the thread. Let us illustrate the algorithm using the numerical example (7).

We start with Table 10.3 for the best item only solution, where $b^1(i) = b^*(i) = i$ and $g_t^1(i) = 0, 0 \le i \le 7$.

Next, we introduce the second item with $v_2 = 3$ and $w_2 = 5$. We start with the thread 0, and try to improve the gain of thread 5 using one copy of the second item.

Table 10.3. thread critical boundaries using the first item only

k	Threads							
	0	1	2	3	4	5	6	7
1	$b = 0$	$b = 1$	$b = 2$	$b = 3$	$b = 4$	$b = 5$	$b = 6$	$b = 7$
	$g = 0$	$g = 0$	$g = 0$	$g = 0$	$g = 0$	$g = 0$	$g = 0$	$g = 0$

Table 10.4. Thread critical boundaries using the item types 1 and 2

k	Threads							
	0	1	2	3	4	5	6	7
2	$b = 0$	$b = 1$	$b = 2$	$b = 3$	$b = 4$	$b = 5$	$b = 6$	$b = 7$
	$g = 0$	$g = 0$	$g = 0$	$g = 0$	$g = 0$	$g = 3$	$g = 3$	$g = 3$

Since $0 + 3 - 8\lfloor\frac{0+5}{8}\rfloor = 3 > 0$, we set $b^2(5) = 5$ and $g_t^2(5) = 3$. We then look at $b = 10$ in thread 2. We have $3 + 3 - 8\lfloor\frac{5+5}{8}\rfloor = -2 < 0$, we set $b^2(2) = b^1(2) = 2$ and $g_t^2(2) = g_t^1(2) = 0$.

From thread 2, we will visit the remaining threads in the order of 7, 4, 1, 6, 3 and use the formula (12) to obtain the thread critical boundaries and the gains shown in Table 10.4.

Table 10.4 contains the critical thread boundaries using the two types of items. The sequence of threads visited 0, 5, 2, 7, 4, 1, 6, 3 is called a *chain*. Because $\gcd(w_1, w_2) = 1$, we are able to visit all threads in one chain.

Now, let us introduce the third item with $v_3 = 17$ and $w_3 = 18$. Since $\gcd(w_1, w_3) = 2$, we can only visit 0, 2, 4, 6 in one chain if we start with thread 0. Instead of visiting all the threads in a single chain, we need two chains. The second chain involves the threads 1, 3, 5, 7. In general, we need $\gcd(w_1, w_k)$ chains to discover the $b^k(i)$ and $g_t^k(i)$, $0 \le i \le w_1 - 1$, and the more chains we have, the less number of threads we have in each chain. This is stated in Lemma 4.6.

Lemma 4.6. *Every boundary b has an optimal solution that uses at most $\frac{w_1}{\gcd(w_1,w_k)} - 1$ copies of the kth type items.*

Proof. Suppose an optimal solution of a boundary b has $\frac{w_1}{\gcd(w_1,w_k)}$ or more copies of the kth type items. Then we can obtain another packing by replacing $\frac{w_1}{\gcd(w_1,w_k)}$ copies of the kth type items which occupy a space of $\frac{w_1 w_k}{\gcd(w_1,w_k)}$ with $\frac{w_k}{\gcd(w_1,w_k)}$ copies of the best items, with a total value greater than or equal to the original one. □

To extend an optimal solution of a boundary b_s to the optimal solution of a larger boundary b_b, we can use the formula in Theorem 4.5, provided that the values in the cell $[b_s, g_s]$ are correct. When introducing the kth type items, we need to know the optimal gain $g_t^k(i)$ in order to compute the optimal gain $g_t^k(j)$ where $j = (i + w_k) \mod w_1$. We know that $[0, g_t^k(0) = 0]$ in the thread 0 is always a correct starting cell, and if $\gcd(w_1, w_s) = 1$, the chain would contain all cells in all threads. However, if $\gcd(w_1, w_s) \neq 1$, such as $w_1 = 8$ and $w_3 = 18$ in our example, it is not clear which cells should be the starting cells for the other chains that do not involve the thread 0.

From Lemma 4.6, we know that a cell can use at most $\frac{w_1}{\gcd(w_1,w_k)} - 1$ copies of the kth type items. In other words, there exists at least one cell in the chain which does not use any kth type item. Hence, for our numerical example, there must exist a thread $j \in \{1, 3, 5, 7\}$ such that

$$g_i^3(j) = g_i^2(j) \geq g_i^3(i) + 17 - 8\left\lfloor \frac{i+18}{8} \right\rfloor \geq g_i^2(i) + 17 - 8\left\lfloor \frac{i+18}{w_1} \right\rfloor.$$

Therefore, we can start with an arbitrarily thread, say thread 5 in the second chain, and loop once around the chain 7, 1, 3, 5 until we find the starting cell, then start with that cell and loop around the chain a second time, applying the formula (12) to compute the gains of the other threads along the chain. We call the procedure the *double loop* technique (Landa 2004), which will be used whenever $\gcd(w_1, w_k) \neq 1$. Table 10.5 shows the values of $b^3(t)$ and $g_i^3(t)$ of the threads 7, 1, 3, 5 after the first loop. Note that both thread 1 and 5 can serve as the starting cell for the chain.

We then start with the thread 5 and visit the threads in the second chain in the order of 7, 1, 3 one more time and obtain the thread critical boundaries and gains shown in Table 10.6.

Hence, we have

$$b^*(0) = 0, \quad b^*(1) = 1, \quad b^*(2) = 18, \quad b^*(3) = 19,$$
$$b^*(4) = 36, \quad b^*(5) = 5, \quad b^*(6) = 6, \quad b^*(7) = 23$$

and

$$b^{**} = \max_j b^*(j) = b^*(4) = 36. \tag{13}$$

Table 10.5. The $g_i^3(j)$ and $b^3(j)$ values after the first loop

i	j	$g_i^2(j)$	$g_i^3(i) + 17 - 8 \times \lfloor\frac{i+18}{8}\rfloor$	$g_i^3(j)$	$b^3(j)$	Remarks
5	7	3	4	4	23	
7	1	0	−3	0	1	Possible starting cell for the chain 1, 3, 5, 7
1	3	0	1	1	19	
3	5	3	3	3	5	Possible starting cell for the chain 1, 3, 5, 7

Table 10.6. Thread critical boundaries using the item types 1, 2 and 3

k	Threads							
	0	1	2	3	4	5	6	7
3	$b=0$	$b=1$	$b=18$	$b=19$	$b=36$	$b=5$	$b=6$	$b=23$
	$g=0$	$g=0$	$g=1$	$g=1$	$g=2$	$g=3$	$g=3$	$g=4$

High-Level Description of The Algorithm

The algorithm for finding the thread critical boundaries is outlined below.

1. Find the best item type, i.e. the item type that has the highest value-to-weight ratio, and rename it as the first item. (In case of tie, pick the one with the smallest weight.)
2. Create an array T with w_1 entries, and initialize the entry $T[i]$, $0 \le i \le w_1 - 1$, with the ordered pair $(b^1(i) = i, g_t^1(i) = 0)$.
3. For $k = 2, \ldots, n$, we introduce the challenge types one type at a time. Starting from the thread 0, we set $b^k(0) = 0$ and $g_t^k(0) = 0$, traverse the threads in the first chain in the order of $j = w_k \bmod w_1, 2w_k \bmod w_1, \ldots$, compute the $(b^k(j), g_t^k(j))$ using the formula (12) and update the entries in T accordingly. If $\gcd(w_1, w_k) > 1$, we will use the double loop technique to compute the $(b^k(j), g_t^k(j))$ for the threads in each of the remaining chains and update the entries in the array T accordingly.

Since Step 1 takes $O(n)$ time, Step 2 takes $O(w_1)$ time, and Step 3 takes $O(w_1)$ time for each new item type and there are $n - 1$ new item types, the above algorithm runs in $O(nw_1)$ time.

Now, given any $b \ge b^{**}$, we obtain $i = b \bmod w_1$ and return

$$V(b) = v_1 \frac{(b - b^*(i))}{w_1} + g_t^*(i) + v_1 \left\lfloor \frac{b^*(i)}{w_1} \right\rfloor.$$

Finding the Optimal Solutions for $b < b^{**}$

Note that the above algorithm does not give the optimal solutions for $b < b^{**}$. For that part, we need to modify the algorithm to keep track of all the special boundaries in each thread, instead of just the critical boundaries.

It follows from the Lemma 4.3 that there can be at most v_1 special boundaries in each thread, with gain values from 0 to $v_1 - 1$. Each gain value is either achievable in a thread, or it is not. Hence, we can construct a table with v_1 rows and w_1 columns to keep track of the gains that are achievable in each thread. And for each gain that is achievable in the thread, we shall remember the smallest boundary whose optimum packing exhibits this gain in the table.

When we introduce a new challenge type, say w_k, instead of just applying the formula (12) to the thread critical boundary involving the first $k - 1$ types of items in each thread, we apply the formula to all the packings corresponding to some gains achievable in each thread using the first $k - 1$ types of items. Let us illustrate the intuitive ideas of the algorithm with the numerical example (7).

We start with Table 10.7 and use $b = \infty$ to denote the gains that are not achievable in each thread so far. Note that only the gain 0 is achievable using the first type of item alone and the smallest boundary of achieving the gain 0 in thread j is an empty bin of capacity j.

We add the second item type with $v_2 = 3$ and $w_2 = 5$. We start from thread 0 and applying the formula (12) to the threads 5, 2, 7, 4, 1, 6, 3, resulting in the boundaries

Table 10.7. Achievable gains using the best item type only

Gain	Threads							
	0	1	2	3	4	5	6	7
0	$b = 0$	$b = 1$	$b = 2$	$b = 3$	$b = 4$	$b = 5$	$b = 6$	$b = 7$
1	$b = \infty$	$b = \infty$	$b = \infty$	$b = \infty$	$b = \infty$	$b = \infty$	$b = \infty$	$b = \infty$
2	$b = \infty$	$b = \infty$	$b = \infty$	$b = \infty$	$b = \infty$	$b = \infty$	$b = \infty$	$b = \infty$
3	$b = \infty$	$b = \infty$	$b = \infty$	$b = \infty$	$b = \infty$	$b = \infty$	$b = \infty$	$b = \infty$
4	$b = \infty$	$b = \infty$	$b = \infty$	$b = \infty$	$b = \infty$	$b = \infty$	$b = \infty$	$b = \infty$
5	$b = \infty$	$b = \infty$	$b = \infty$	$b = \infty$	$b = \infty$	$b = \infty$	$b = \infty$	$b = \infty$
6	$b = \infty$	$b = \infty$	$b = \infty$	$b = \infty$	$b = \infty$	$b = \infty$	$b = \infty$	$b = \infty$
7	$b = \infty$	$b = \infty$	$b = \infty$	$b = \infty$	$b = \infty$	$b = \infty$	$b = \infty$	$b = \infty$

Table 10.8. Achievable gains using the item types 1 and 2

Gain	Threads							
	0	1	2	3	4	5	6	7
0	$b = 0$	$b = 1$	$b = 2$	$b = 3$	$b = 4$	$b = 5$	$b = 6$	$b = 7$
1	$b = \infty$	$b = \infty$	$b = \infty$	$b = \infty$	$b = \infty$	$b = \infty$	$b = \infty$	$b = \infty$
2	$b = \infty$	$b = \infty$	$b = \infty$	$b = \infty$	$b = \infty$	$b = \infty$	$b = \infty$	$b = \infty$
3	$b = \infty$	$b = \infty$	$b = \infty$	$b = \infty$	$b = \infty$	$b = 5$	$b = 6$	$b = 7$
4	$b = \infty$	$b = \infty$	$b = \infty$	$b = \infty$	$b = \infty$	$b = \infty$	$b = \infty$	$b = \infty$
5	$b = \infty$	$b = \infty$	$b = \infty$	$b = \infty$	$b = \infty$	$b = \infty$	$b = \infty$	$b = \infty$
6	$b = \infty$	$b = \infty$	$b = \infty$	$b = \infty$	$b = \infty$	$b = \infty$	$b = \infty$	$b = \infty$
7	$b = \infty$	$b = \infty$	$b = \infty$	$b = \infty$	$b = \infty$	$b = \infty$	$b = \infty$	$b = \infty$

for the achievable gains shown in Table 10.8. Note that the boundaries $b = 5, 6$ and 7 are responsible for achieving the gains 0 and 3 for threads 5, 6 and 7 respectively, since we can either leave each of these bins empty, or fill it with a copy of the second item.

Next, we add the third item type with $v_3 = 17$ and $w_3 = 18$. Since $\gcd(8, 18) = 2$, we have two chains 0, 2, 4, 6 and 1, 3, 5, 7. For the first chain, we first apply the formula (12) to the boundary $b = 0$ in the thread 0 to obtain the new boundary $b = 18$ with a gain $= 1$ in the thread 2. Then we apply the formula (12) to both $b = 2$ and $b = 18$ in the thread 2 to obtain the new boundaries $b = 20$ with a gain $= 1$ and $b = 36$ with a gain $= 2$ in the thread 4. Next, we apply the formula (12) to the boundaries $b = 4, 20$ and 36 in the thread 4 to obtain the new boundaries $b = 22$ with a gain $= 1$ and $b = 38$ with a gain $= 2$ in the thread 6. Although the formula (12) also returns a gain of 3 for the new boundary $b = 54$, we reject the new boundary because we already have achieved the gain of 3 in the thread 6 using a smaller boundary ($b = 6$). For the second chain, we can start with thread 1, use the double loop technique and apply the formula (12) to the boundaries with achievable gains in the thread sequence 1, 3, 5, 7. The resultant special boundaries are shown in Table 10.9.

T.C. Hu et al.

Table 10.9. Achievable gains using the item types 1, 2 and 3

Gain	Threads							
	0	1	2	3	4	5	6	7
0	$b = 0$	$b = 1$	$b = 2$	$b = 3$	$b = 4$	$b = 5$	$b = 6$	$b = 7$
1	$b = \infty$	$b = \infty$	$b = 18$	$b = 19$	$b = 20$	$b = 21$	$b = 22$	$b = 23$
2	$b = \infty$	$b = \infty$	$b = \infty$	$b = \infty$	$b = 36$	$b = 37$	$b = 38$	$b = 39$
3	$b = \infty$	$b = \infty$	$b = \infty$	$b = \infty$	$b = \infty$	$b = 5$	$b = 6$	$b = 7$
4	$b = \infty$	$b = \infty$	$b = \infty$	$b = \infty$	$b = \infty$	$b = \infty$	$b = \infty$	$b = 23$
5	$b = \infty$	$b = \infty$	$b = \infty$	$b = \infty$	$b = \infty$	$b = \infty$	$b = \infty$	$b = \infty$
6	$b = \infty$	$b = \infty$	$b = \infty$	$b = \infty$	$b = \infty$	$b = \infty$	$b = \infty$	$b = \infty$
7	$b = \infty$	$b = \infty$	$b = \infty$	$b = \infty$	$b = \infty$	$b = \infty$	$b = \infty$	$b = \infty$

Table 10.10. Thread special boundaries using the item types 1, 2 and 3

Threads							
0	1	2	3	4	5	6	7
$b = 0$	$b = 1$	$b = 2$	$b = 3$	$b = 4$	$b = 5$	$b = 6$	$b = 7$
$g = 0$	$g = 0$	$g = 0$	$g = 0$	$g = 0$	$g = 3$	$g = 3$	$g = 3$
		$b = 18$	$b = 19$	$b = 20$			$b = 23$
		$g = 1$	$g = 1$	$g = 1$			$g = 4$
				$b = 36$			
				$g = 2$			

Table 10.9 records all the achievable gains and their corresponding boundaries in each thread using the three types of items. It includes many non-special boundaries. Hence, we need to sweep each column in Table 10.9 once from bottom to top, and eliminate any boundaries that are not smaller than all the boundaries below it in each column, resulting in Table 10.10.

Since we may need to apply the formula (12) to v_1 boundaries in each thread for each new item type, we can locate all the special boundaries in $O(nv_1w_1)$ time.

Now, given any instance of b where $b < b^{**}$, we can either find the optimal solution for b in time $O(nb)$ using dynamic programming directly, or we can pay the $O(nv_1w_1)$ pre-processing time to build the table, then compute $i = b \bmod w_1$, use binary search on the special boundaries in the thread i to locate the largest special boundary b_i that is less than or equal to b in the thread i in $O(\log v_1)$ time, and return

$$V(b) = v_1 \frac{(b - b_i)}{w_1} + g(b_i) + v_1 \left\lfloor \frac{b_i}{w_1} \right\rfloor.$$

Hence, given any b, we can compute the optimal solution in time $O(nw_1 + \min\{nv_1w_1, nb\})$.

10.5 Conclusion

This paper presents a survey of the unbounded knapsack problem. We focus on the techniques for obtaining the optimal solutions, particularly those using the periodic structure of the optimal solutions when the knapsack weight-carrying capacity b is sufficiently large.

As we study the periodic structure of the optimal solutions to the unbounded knapsack problem, we observed that we need *less* time to find the optimal solutions for large instances of b, and need *more* time to find the optimal solutions for the small instances of b. This non-intuitive behavior comes from the fact that smaller boundaries have more internal structure (based on special boundaries) than those larger than b^{**}, which only need to use the best item repeatedly. We believe that this phenomenon

large instances are easy, small instances are hard

may also appear in other NP-complete problems.

References

Andonov, R., Poirriez, V., Rajopadhye, S.: Unbounded knapsack problem: dynamic programming revisited. Eur. J. Oper. Res. **12**, 394–407 (2000)

Garey, M.R., Johnson, D.S.: Computer and Intractability. Freeman, San Francisco (1979)

Gilmore, P.C., Gomory, R.E.: The theory of computation of Knapsack functions. J. Oper. Res. Soc. Am. **14**(6), 1045–1074 (1966)

Greenberg, H.: An algorithm for the periodic solutions in the knapsack problem. J. Math. Anal. Appl. **111**, 327–331 (1985)

Hu, T.C.: Integer Programming and Network Flows. Addison-Wesley, Reading (1969)

Hu, T.C., Shing, M.T.: Combinatorial Algorithms, 2nd edn. Dover, Mineola (2002)

Ibarra, O.H., Kim, C.E.: Fast approximation algorithms for the knapsack and sum of subset problem. J. Assoc. Comput. Mach. **22**, 463–468 (1975)

Kellerer, H., Pferschy, U., Pisinger, D.: Knapsack Problems. Springer, Berlin (2004)

Landa, L.: Sage algorithms for Knapsack problem. Technical Report CS-2004-0794, UC San Diego, La Jolla, California (2004)

Lawler, E.L.: Fast approximation algorithms for knapsack problems. Math. Oper. Res. **4**, 339–356 (1979)

Martello, S., Toth, P.: Knapsack Problems: Algorithms and Computer Implementations. Wiley, New York (1990a)

Martello, S., Toth, P.: An exact algorithm for large unbounded knapsack problems. Oper. Res. Let. **9**, 15–20 (1990b)

Pisinger, D.: Dynamic programming on the word RAM. Algorithmica **35**, 128–145 (2003)

11

Recent Developments in Discrete Convex Analysis

Kazuo Murota

Summary. This paper describes recent developments in discrete convex analysis. Particular emphasis is laid on natural introduction of the classes of L-convex and M-convex functions in discrete and continuous variables. Expansion of the application areas is demonstrated by recent connections to submodular function maximization, finite metric space, eigenvalues of Hermitian matrices, discrete fixed point theorem, and matching games.

11.1 Introduction

This paper describes recent developments in discrete convex analysis. Particular emphasis is laid on natural introduction of the classes of L-convex and M-convex functions in discrete and continuous variables. Expansion of the application areas is demonstrated by recent connections to submodular function maximization, finite metric space, eigenvalues of Hermitian matrices, discrete fixed point theorem, and matching games.

Discrete convex analysis (Murota 1998b, 2001, 2003a) is aimed at establishing a general theoretical framework for solvable discrete optimization problems by means of a combination of the ideas in continuous optimization and combinatorial optimization. The framework of convex analysis is adapted to discrete settings and the mathematical results in matroid/submodular function theory are generalized. Viewed from the continuous side, it is a theory of convex functions $f : \mathbb{R}^n \to \mathbb{R}$ that have additional combinatorial properties. Viewed from the discrete side, it is a theory of discrete functions $f : \mathbb{Z}^n \to \mathbb{R}$ or $f : \mathbb{Z}^n \to \mathbb{Z}$ that enjoy certain nice properties comparable to convexity. Symbolically,

$$\text{Discrete Convex Analysis} = \text{Convex Analysis} + \text{Matroid Theory}.$$

The theory extends the direction set forth by J. Edmonds, A. Frank, S. Fujishige, and L. Lovász (Edmonds 1970; Frank 1982; Fujishige 1984; Lovász 1983); see also (Fujishige 2005, Chap. VII). The reader is referred to Rockafellar (1970) for convex analysis, Cook et al. (1998), Korte and Vygen (2008) and Schrijver (2003) for combinatorial optimization, Oxley (1992), Recski (1989) and White (1986) for matroid theory, and Fujishige (2005), Narayanan (1997) and Topkis (1998) for submodular function theory.

Two convexity concepts, called L-convexity and M-convexity, play primary roles. L-convex functions and M-convex functions are conjugate to each other through the (continuous or discrete) Legendre–Fenchel transformation. L-convex functions and M-convex functions generalize, respectively, the concepts of submodular set functions and base polyhedra. It is noted that "L" stands for "Lattice" and "M" for "Matroid."

The contents of this paper are as follows. The first part, Sects. 11.2 to 11.5, presents the fundamental facts with some new observations, whereas the second part, Sects. 11.6 to 11.10, deals with recent topics.

The set of all real numbers is denoted by $\mathbb{R}$, and $\overline{\mathbb{R}} = \mathbb{R} \cup \{+\infty\}$ and $\underline{\mathbb{R}} = \mathbb{R} \cup \{-\infty\}$. The set of all integers is denoted by $\mathbb{Z}$, and $\overline{\mathbb{Z}} = \mathbb{Z} \cup \{+\infty\}$ and $\underline{\mathbb{Z}} = \mathbb{Z} \cup \{-\infty\}$. Let $V = \{1, 2, \ldots, n\}$ for a positive integer n. The characteristic vector of $X \subseteq V$ is denoted by $\chi_X \in \{0, 1\}^n$. For $i \in V$, we write χ_i for $\chi_{\{i\}}$, which is the ith unit vector, and $\chi_0 = \mathbf{0}$ (zero vector).

11.2 Concepts of Discrete Convex Functions

The concepts of L-convex and M-convex functions can be obtained through discretization of two different characterizations of convex functions.

11.2.1 Ordinary Convex Functions

We start by recalling the definition of ordinary convex functions. A function $f : \mathbb{R}^n \to \overline{\mathbb{R}}$ is said to be *convex* if

$$\lambda f(x) + (1 - \lambda) f(y) \geq f(\lambda x + (1 - \lambda) y) \tag{1}$$

for all $x, y \in \mathbb{R}^n$ and for all λ with $0 \leq \lambda \leq 1$, where it is understood that the inequality is satisfied if $f(x)$ or $f(y)$ is equal to $+\infty$. A function $h : \mathbb{R}^n \to \underline{\mathbb{R}}$ is said to be *concave* if $-h$ is convex.

A set $S \subseteq \mathbb{R}^n$ is called *convex* if, for any $x, y \in S$ and $0 \leq \lambda \leq 1$, we have $\lambda x + (1-\lambda)y \in S$. The *indicator function* of a set S is a function $\delta_S : \mathbb{R}^n \to \{0, +\infty\}$ defined by

$$\delta_S(x) = \begin{cases} 0 & (x \in S), \\ +\infty & (x \notin S). \end{cases} \tag{2}$$

Then S is a convex set if and only if δ_S is a convex function.

For a function $f : \mathbb{R}^n \to \mathbb{R} \cup \{-\infty, +\infty\}$ in general, the set

$$\mathrm{dom}_{\mathbb{R}} f = \{x \in \mathbb{R}^n \mid f(x) \in \mathbb{R}\}$$

is called the *effective domain* of f. A point $x \in \mathbb{R}^n$ is said to be a *global minimum* of f if the inequality $f(x) \leq f(y)$ holds for every $y \in \mathbb{R}^n$. Point x is a *local minimum* if this inequality holds for every y in some neighborhood of x. The set of global minima (minimizers) is denoted as

$$\mathrm{argmin}_{\mathbb{R}} f = \{x \in \mathbb{R}^n \mid f(x) \leq f(y) \; (\forall y \in \mathbb{R}^n)\}.$$

Convex functions are tractable in optimization (or minimization) problems and this is mainly because of the following properties.

1. Local optimality (or minimality) guarantees global optimality.
2. Duality theorems such as min-max relation and separation hold.

Duality is a central issue in convex analysis, and is discussed in Sect. 11.5.

A *separable convex function* is a function $f : \mathbb{R}^n \to \overline{\mathbb{R}}$ that can be represented as

$$f(x) = \sum_{i=1}^{n} \varphi_i(x_i), \tag{3}$$

where $x = (x_i \mid i = 1, \ldots, n)$ and $\varphi_i : \mathbb{R} \to \overline{\mathbb{R}}$ $(i = 1, \ldots, n)$ are univariate convex functions.

11.2.2 Discrete Convex Functions

We now consider how convexity concept can (or should) be defined for functions in discrete variables. It would be natural to expect the following properties of any function $f : \mathbb{Z}^n \to \mathbb{R}$ that is qualified as a "discrete convex function."

1. Function f is extensible to a convex function on $\mathbb{R}^n$.
2. Local optimality (or minimality) guarantees global optimality.
3. Duality theorems such as min-max relation and separation hold.

Recall that $f : \mathbb{Z}^n \to \overline{\mathbb{R}}$ is said to be *convex-extensible* if there exists a convex function $\overline{f} : \mathbb{R}^n \to \overline{\mathbb{R}}$ such that $\overline{f}(x) = f(x)$ for all $x \in \mathbb{Z}^n$. It is widely understood that convex extensibility alone does not yield a fruitful theoretical framework, which fact motivates us to introduce L-convex and M-convex functions. In this section we focus on convex extensibility and local optimality while deferring duality issues to Sect. 11.5. The effective domain and the set of minimizers are denoted respectively as

$$\mathrm{dom}_\mathbb{Z} f = \{x \in \mathbb{Z}^n \mid f(x) \in \mathbb{R}\},$$
$$\mathrm{argmin}_\mathbb{Z} f = \{x \in \mathbb{Z}^n \mid f(x) \le f(y) \; (\forall y \in \mathbb{Z}^n)\}.$$

Univariate and Separable Convex Functions

The univariate case ($n = 1$) is simple and straightforward. We may regard a function $f : \mathbb{Z} \to \overline{\mathbb{R}}$ as a discrete convex function if

$$f(x - 1) + f(x + 1) \ge 2 f(x) \quad (\forall x \in \mathbb{Z}). \tag{4}$$

This is justified by the following facts.

Theorem 2.1. *A function $f : \mathbb{Z} \to \overline{\mathbb{R}}$ is convex-extensible if and only if it satisfies* (4).

Theorem 2.2. *For a function $f : \mathbb{Z} \to \overline{\mathbb{R}}$ satisfying* (4)*, a point $x \in \mathrm{dom}_\mathbb{Z} f$ is a global minimum if and only if it is a local minimum in the sense that*

$$f(x) \le \min\{f(x - 1), \; f(x + 1)\}.$$

Theorems 2.1 and 2.2 above can be extended in obvious ways to a *separable* (*discrete*) *convex function* $f : \mathbb{Z}^n \to \overline{\mathbb{R}}$, which is, by definition, representable in the form of (3) with univariate functions $\varphi_i : \mathbb{Z} \to \overline{\mathbb{R}}$ having property (4).

L-convex Functions

We explain the concept of L-convex functions (Murota 1998b) by featuring an equivalent variant thereof, called $L^\natural$-convex functions (Fujishige and Murota 2000) ("$L^\natural$" should be read "el natural").

We first observe that a convex function g on $\mathbb{R}^n$ satisfies

$$g(p) + g(q) \ge g\left(\frac{p + q}{2}\right) + g\left(\frac{p + q}{2}\right) \quad (p, q \in \mathbb{R}^n), \tag{5}$$

which is a special case of (1) with $\lambda = 1/2$. This property, called *midpoint convexity*, is known to be equivalent to convexity if g is a continuous function.

For a function $g : \mathbb{Z}^n \to \overline{\mathbb{R}}$ in discrete variables the above inequality does not always make sense, since the midpoint $\frac{p+q}{2}$ of two integer vectors p and q may not be integral. Instead we simulate (5) by

$$g(p) + g(q) \ge g\left(\left\lceil \frac{p + q}{2} \right\rceil\right) + g\left(\left\lfloor \frac{p + q}{2} \right\rfloor\right) \quad (p, q \in \mathbb{Z}^n), \tag{6}$$

where, for $z \in \mathbb{R}$ in general, $\lceil z \rceil$ denotes the smallest integer not smaller than z (rounding-up to the nearest integer) and $\lfloor z \rfloor$ the largest integer not larger than z (rounding-down to the nearest integer), and this operation is extended to a vector

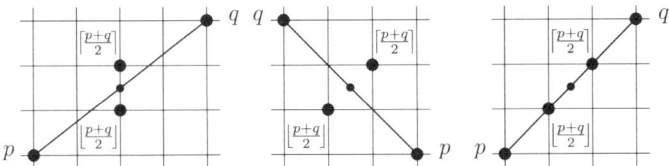

Fig. 11.1. Discrete midpoint convexity

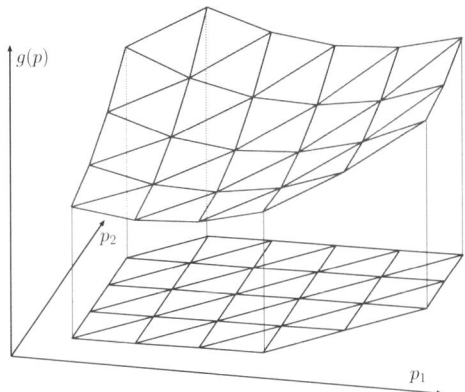

Fig. 11.2. An $L^\natural$-convex function ($n = 2$)

by componentwise applications, as illustrated in Fig. 11.1 in the case of $n = 2$. We refer to (6) as *discrete midpoint convexity*.

We say that a function $g : \mathbb{Z}^n \to \overline{\mathbb{R}}$ is $L^\natural$-*convex* if it satisfies discrete midpoint convexity (6). In the case of $n = 1$, $L^\natural$-convexity is equivalent to the condition (4). A concrete example of an $L^\natural$-convex function ($n = 2$) is shown in Fig. 11.2. Examples of $L^\natural$-convex functions are given in Sect. 11.4.1.

With this definition we can obtain the following desired statements in parallel with Theorems 2.1 and 2.2.

Theorem 2.3. *An $L^\natural$-convex function $g : \mathbb{Z}^n \to \overline{\mathbb{R}}$ is convex-extensible.*

Theorem 2.4. *For an $L^\natural$-convex function $g : \mathbb{Z}^n \to \overline{\mathbb{R}}$, a point $p \in \mathrm{dom}_{\mathbb{Z}} g$ is a global minimum if and only if it is a local minimum in the sense that*

$$g(p) \leq \min\{g(p - q), g(p + q)\} \quad (\forall q \in \{0, 1\}^n). \tag{7}$$

Although Theorem 2.4 affords a local criterion for global optimality of a point p, a straightforward verification of (7) requires $\mathrm{O}(2^n)$ function evaluations. The verification can be done in polynomial time as follows. We consider set functions ρ_p^+ and ρ_p^- defined by $\rho_p^\pm(Y) = g(p \pm \chi_Y) - g(p)$ for $Y \subseteq V$, both of which are submodular. Since (7) is equivalent to saying that both ρ_p^+ and ρ_p^- achieve the minimum at $Y = \emptyset$, this condition can be verified in polynomial time by submodular function minimization algorithms (Iwata 2007).

$L^\natural$-convexity is closely related with submodularity. For two vectors p and q, the vectors of componentwise maxima and minima are denoted respectively by $p \vee q$ and $p \wedge q$, that is,

$$(p \vee q)_i = \max(p_i, q_i), \qquad (p \wedge q)_i = \min(p_i, q_i).$$

A function $g : \mathbb{Z}^n \to \overline{\mathbb{R}}$ is called *submodular* if

$$g(p) + g(q) \geq g(p \vee q) + g(p \wedge q) \quad (p, q \in \mathbb{Z}^n), \tag{8}$$

and *translation submodular* if

$$g(p) + g(q) \geq g((p - \alpha\mathbf{1}) \vee q) + g(p \wedge (q + \alpha\mathbf{1})) \quad (\alpha \in \mathbb{Z}_+, \ p, q \in \mathbb{Z}^n), \tag{9}$$

where $\mathbf{1} = (1, 1, \ldots, 1)$ and $\mathbb{Z}_+$ denotes the set of nonnegative integers. The latter property characterizes $L^\natural$-convexity, as follows.

Theorem 2.5. *For a function $g : \mathbb{Z}^n \to \overline{\mathbb{R}}$, translation submodularity (9) is equivalent to discrete midpoint convexity (6).*

An *L-convex function* is defined as an $L^\natural$-convex function g that satisfies

$$g(p + \mathbf{1}) = g(p) + r \tag{10}$$

for some $r \in \mathbb{R}$ (which is independent of p). It is known that g is L-convex if and only if it satisfies (8) and (10); in fact this is the original definition of L-convexity. L-convex functions and $L^\natural$-convex functions are essentially the same, in that $L^\natural$-convex functions in n variables can be identified, up to the constant r in (10), with L-convex functions in $n + 1$ variables.

M-convex Functions

Just as L-convexity is defined through discretization of midpoint convexity, another kind of discrete convexity, called M-convexity (Murota 1996c, 1998b), can be defined through discretization of another convexity property. We feature an equivalent variant of M-convexity, called $M^\natural$-convexity (Murota and Shioura 1999) ("$M^\natural$" should be read "em natural").

We first observe that a convex function f on $\mathbb{R}^n$ satisfies the inequality

$$f(x) + f(y) \geq f(x - \alpha(x - y)) + f(y + \alpha(x - y)) \tag{11}$$

for every $\alpha \in \mathbb{R}$ with $0 \leq \alpha \leq 1$. This inequality follows from (1) for $\lambda = \alpha$ and $\lambda = 1 - \alpha$, whereas it implies (1) if f is a continuous function. The inequality (11) says that the sum of the function values evaluated at two points, x and y, does not increase if the two points approach each other by the same distance on the line segment connecting them (see Fig. 11.3). We refer to this property as *equidistance convexity*.

For a function $f : \mathbb{Z}^n \to \overline{\mathbb{R}}$ in discrete variables we simulate equidistance convexity (11) by moving a pair of points (x, y) to another pair (x', y') along the co-

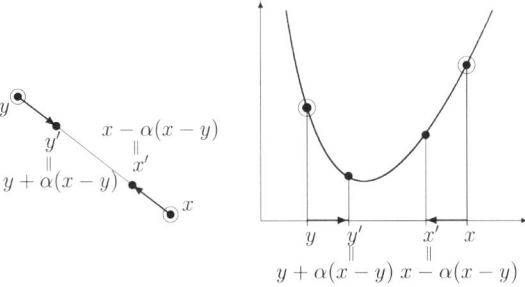

Fig. 11.3. Equidistance convexity

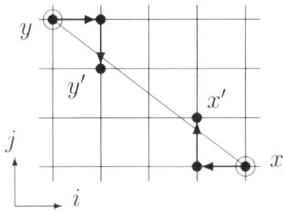

Fig. 11.4. Nearer pair in the definition of $M^\natural$-convex functions

ordinate axes rather than on the connecting line segment. To be more specific, we consider two kinds of possibilities

$$(x', y') = (x - \chi_i, y + \chi_i) \quad \text{or} \quad (x', y') = (x - \chi_i + \chi_j, y + \chi_i - \chi_j) \quad (12)$$

with indices i and j such that $x_i > y_i$ and $x_j < y_j$; see Fig. 11.4. For a vector $z \in \mathbb{R}^n$ in general, define the *positive* and *negative supports* of z as

$$\mathrm{supp}^+(z) = \{i \mid z_i > 0\}, \qquad \mathrm{supp}^-(z) = \{j \mid z_j < 0\}.$$

Then the expression (12) can be rewritten compactly as $(x', y') = (x - \chi_i + \chi_j, y + \chi_i - \chi_j)$ with $i \in \mathrm{supp}^+(x - y)$ and $j \in \mathrm{supp}^-(x - y) \cup \{0\}$, where χ_0 is defined to be the zero vector.

As a discrete analogue of equidistance convexity (11) we consider the following condition: For any $x, y \in \mathrm{dom}_\mathbb{Z} f$ and any $i \in \mathrm{supp}^+(x - y)$, there exists $j \in \mathrm{supp}^-(x - y) \cup \{0\}$ such that

$$f(x) + f(y) \geq f(x - \chi_i + \chi_j) + f(y + \chi_i - \chi_j), \quad (13)$$

which is referred to as the *exchange property*. A function $f : \mathbb{Z}^n \to \overline{\mathbb{R}}$ having this exchange property is called $M^\natural$-*convex*. In the case of $n = 1$, $M^\natural$-convexity is equivalent to the condition (4). A concrete example of an $M^\natural$-convex function $(n = 2)$ is shown in Fig. 11.5. Examples of $M^\natural$-convex functions are given in Sect. 11.4.2.

With this definition we can obtain the following desired statements comparable to Theorems 2.1 and 2.2.

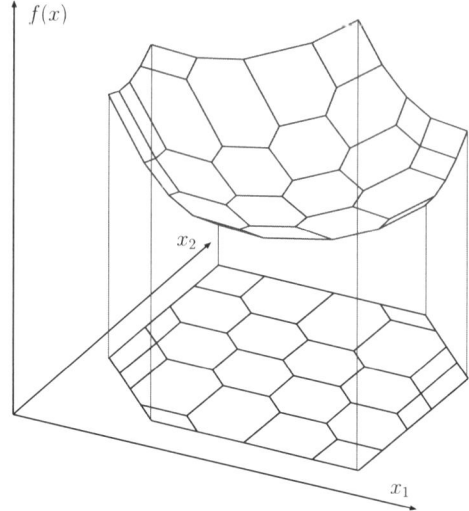

Fig. 11.5. An M$^\natural$-convex function ($n = 2$)

Theorem 2.6. *An M$^\natural$-convex function $f : \mathbb{Z}^n \to \overline{\mathbb{R}}$ is convex-extensible.*

Theorem 2.7. *For an M$^\natural$-convex function $f : \mathbb{Z}^n \to \overline{\mathbb{R}}$, a point $x \in \mathrm{dom}_{\mathbb{Z}} f$ is a global minimum if and only if it is a local minimum in the sense that*

$$f(x) \le f(x - \chi_i + \chi_j) \quad (\forall i, j \in \{0, 1, \ldots, n\}).$$

An *M-convex function* is defined as an M$^\natural$-convex function f that satisfies (13) with $j \in \mathrm{supp}^-(x - y)$. This is equivalent to saying that f is an M-convex function if and only if it is M$^\natural$-convex and $\mathrm{dom}_{\mathbb{Z}} f \subseteq \{x \in \mathbb{Z}^n \mid \sum_{i=1}^n x_i = r\}$ for some $r \in \mathbb{Z}$. M-convex functions and M$^\natural$-convex functions are essentially the same, in that M$^\natural$-convex functions in n variables can be obtained as projections of M-convex functions in $n + 1$ variables.

Classes of Discrete Convex Functions

We have thus defined L$^\natural$-convex functions and M$^\natural$-convex functions by discretization of midpoint convexity and equidistance convexity, respectively. The definitions are summarized in Fig. 11.6.

Figure 11.7 shows the classes of discrete convex functions we have introduced. L$^\natural$-convex functions contain L-convex functions as a special case. The same is true for M$^\natural$-convex and M-convex functions. By Theorems 2.3 and 2.6 both L$^\natural$-convex functions and M$^\natural$-convex functions are contained in the class of convex-extensible functions. It is known that the classes of L-convex functions and M-convex functions are disjoint, whereas the intersection of the classes of L$^\natural$-convex functions and M$^\natural$-convex functions is exactly the class of separable convex functions.

⟨Continuous variables⟩		⟨Discrete variables⟩
$f : \mathbb{R}^n \to \overline{\mathbb{R}}$		$f : \mathbb{Z}^n \to \overline{\mathbb{R}}$
midpoint convex	$\longrightarrow$	discrete midpoint convex ($\mathrm{L}^\natural$-convex)
$\updownarrow$	[discretization]	
(ordinary) convex		
$\updownarrow$	[discretization]	
equidistance convex	$\longrightarrow$	exchange property ($\mathrm{M}^\natural$-convex)

discrete midpoint convex: $f(x) + f(y) \geq f\left(\left\lceil \frac{x+y}{2} \right\rceil\right) + f\left(\left\lfloor \frac{x+y}{2} \right\rfloor\right)$

midpoint convex: $\qquad f(x) + f(y) \geq 2 f\left(\frac{x+y}{2}\right)$

(ordinary) convex: $\qquad \lambda f(x) + (1-\lambda) f(y) \geq f(\lambda x + (1-\lambda)y)$

equidistance convex: $\qquad f(x) + f(y) \geq f(x - \alpha(x-y)) + f(y + \alpha(x-y))$

exchange property: $\qquad f(x) + f(y) \geq \min[f(x - \chi_i) + f(y + \chi_i),$

$$\min_{x_j < y_j} \{f(x - \chi_i + \chi_j) + f(y + \chi_i - \chi_j)\}]$$

Fig. 11.6. Definitions of $\mathrm{L}^\natural$-convexity and $\mathrm{M}^\natural$-convexity by discretization

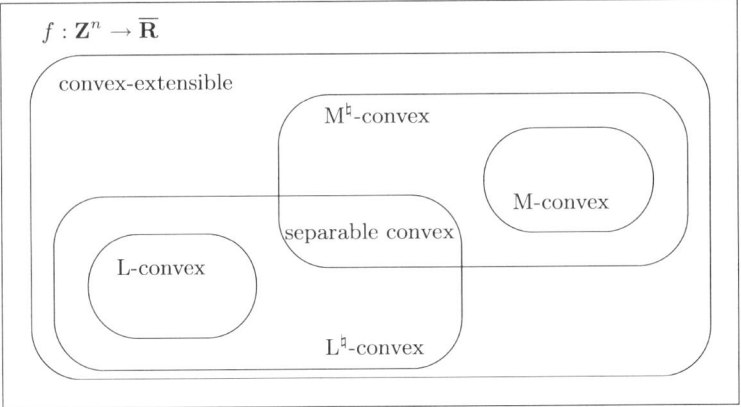

Fig. 11.7. Classes of discrete convex functions ($\mathrm{L}^\natural$-convex $\cap$ $\mathrm{M}^\natural$-convex = separable convex)

Discrete Convex Sets

In the continuous case the convexity of a set $S \subseteq \mathbb{R}^n$ can be characterized by that of its indicator function δ_S as

$$S \text{ is a convex set} \iff \delta_S \text{ is a convex function.}$$

We make use of this relation to define the concepts of discrete convex sets.

For a set $S \subseteq \mathbb{Z}^n$ the indicator function of S is a function $\delta_S : \mathbb{Z}^n \to \overline{\mathbb{R}}$ given by (2). $\mathrm{L}^\natural$-convex sets and $\mathrm{M}^\natural$-convex sets are defined as

$$S \text{ is an } \mathrm{L}^\natural\text{-convex set} \iff \delta_S \text{ is an } \mathrm{L}^\natural\text{-convex function,}$$
$$S \text{ is an } \mathrm{M}^\natural\text{-convex set} \iff \delta_S \text{ is an } \mathrm{M}^\natural\text{-convex function.}$$

Similarly for the definitions of L-convex and M-convex sets. We have $S = \overline{S} \cap \mathbb{Z}^n$ for an $L^\natural$-convex ($M^\natural$-convex, L-convex or M-convex) set S, where $\overline{S}$ denotes the convex hull of S.

For an $L^\natural$-convex function f, both $\mathrm{dom}_\mathbb{Z} f$ and $\mathrm{argmin}_\mathbb{Z} f$ are $L^\natural$-convex sets. This statement remains true when $L^\natural$-convexity is replaced by $M^\natural$-convexity, L-convexity or M-convexity.

11.2.3 Discrete Convex Functions in Continuous Variables

So far we have been concerned with the translation from "continuous" to "discrete." We have defined L-convex and M-convex functions by discretization of midpoint convexity and equidistance convexity, respectively. Although these two properties are both equivalent to (ordinary) convexity for continuous functions in continuous variables, their discrete versions have given rise to different concepts (cf. Fig. 11.6).

We are now interested in the reverse direction, from "discrete" to "continuous," to define the concepts of L-convex and M-convex functions in continuous variables (Murota and Shioura 2000, 2004a, 2004b). In so doing we intend to capture certain classes of convex functions with additional combinatorial structures. We refer to such functions as *discrete convex functions in continuous variables*. This may sound somewhat contradictory, but the adjective "discrete" indicates the discreteness in direction in the space $\mathbb{R}^n$ of continuous variables.

L-convex Functions

$L^\natural$-convex functions in discrete variables have been introduced in terms of a discretization of midpoint convexity. By Theorem 2.5, however, we can alternatively say that $L^\natural$-convex functions are those functions which satisfy translation submodularity (9).

This alternative definition enables us to introduce the concept of $L^\natural$-convex functions in continuous variables. That is, a convex function $g : \mathbb{R}^n \to \overline{\mathbb{R}}$ is defined to be *$L^\natural$-convex* if

$$g(p) + g(q) \geq g((p - \alpha\mathbf{1}) \vee q) + g(p \wedge (q + \alpha\mathbf{1})) \quad (\alpha \in \mathbb{R}_+, \ p, q \in \mathbb{R}^n), \ (14)$$

where $\mathbb{R}_+$ denotes the set of nonnegative reals. Examples of $L^\natural$-convex functions are given in Sect. 11.4.1.

$L^\natural$-convex functions constitute a subclass of convex functions that are equipped with certain combinatorial properties in addition to convexity. It is known (Murota and Shioura 2004b), for example, that a smooth function g is $L^\natural$-convex if and only if the Hessian matrix $H = (h_{ij} = \partial^2 g / \partial p_i \partial p_j)$ is a *diagonally dominant symmetric M-matrix*, i.e.,

$$h_{ij} \leq 0 \quad (i \neq j), \qquad \sum_{j=1}^{n} h_{ij} \geq 0 \quad (i = 1, \ldots, n) \qquad (15)$$

⟨Continuous variables⟩		⟨Discrete variables⟩
$g : \mathbb{R}^n \to \overline{\mathbb{R}}$		$g : \mathbb{Z}^n \to \overline{\mathbb{R}}$
(ordinary) convex		
$\Updownarrow$	[discretization]	
midpoint convex	$\longrightarrow$	discrete midpoint convex
		$\Updownarrow$
translation submodular	$\longleftarrow$	translation submodular
($L^\natural$-convex)	[prolongation]	($L^\natural$-convex)
(ordinary) convex:	$\lambda g(p) + (1-\lambda)g(q) \geq g(\lambda p + (1-\lambda)q)$	
midpoint convex:	$g(p) + g(q) \geq 2g\left(\frac{p+q}{2}\right)$	
discrete midpoint convex:	$g(p) + g(q) \geq g\left(\left\lceil \frac{p+q}{2}\right\rceil\right) + g\left(\left\lfloor \frac{p+q}{2}\right\rfloor\right)$	
translation submodular:	$g(p) + g(q) \geq g((p-\alpha\mathbf{1}) \vee q) + g(p \wedge (q+\alpha\mathbf{1}))$	

Fig. 11.8. Definitions of $L^\natural$-convexity by discretization and prolongation

at each point. This is a combinatorial property on top of positive semidefiniteness, which is familiar in operations research, mathematical economics, and numerical analysis. It may be said that $L^\natural$-convexity extends this well-known property to non-smooth functions.

An *L-convex function* in continuous variables is defined as an $L^\natural$-convex function $g : \mathbb{R}^n \to \overline{\mathbb{R}}$ that satisfies

$$g(p + \alpha\mathbf{1}) = g(p) + \alpha r \quad (\alpha \in \mathbb{R}, \ p \in \mathbb{R}^n) \tag{16}$$

for some $r \in \mathbb{R}$ (which is independent of p and α). L-convex functions and $L^\natural$-convex functions are essentially the same, in that $L^\natural$-convex functions in n variables can be identified, up to the constant r in (16), with L-convex functions in $n + 1$ variables.

The inequality (14) is a continuous version of the translation submodularity (9), in which we had $\alpha \in \mathbb{Z}_+$ and $p, q \in \mathbb{Z}^n$ instead of $\alpha \in \mathbb{R}_+$ and $p, q \in \mathbb{R}^n$. It may be said that (14) is obtained from (9) by *prolongation*, by which we mean a process converse to discretization. Figure 11.8 summarizes how we have defined $L^\natural$-convex functions in discrete and continuous variables. Note that prolongation of discrete midpoint convexity renders no novel concept, but reduces to midpoint convexity, which is (almost) equivalent to convexity.

M-convex Functions

$M^\natural$-convex functions in continuous variables can be defined by prolongation of the exchange property (13). We say that a convex function $f : \mathbb{R}^n \to \overline{\mathbb{R}}$ is $M^\natural$-*convex* if, for any $x, y \in \mathrm{dom}_{\mathbb{R}} f$ and any $i \in \mathrm{supp}^+(x - y)$, there exist $j \in \mathrm{supp}^-(x - y) \cup \{0\}$ and a positive real number α_0 such that

$$f(x) + f(y) \geq f(x - \alpha(\chi_i - \chi_j)) + f(y + \alpha(\chi_i - \chi_j)) \tag{17}$$

for all $\alpha \in \mathbb{R}$ with $0 \leq \alpha \leq \alpha_0$.

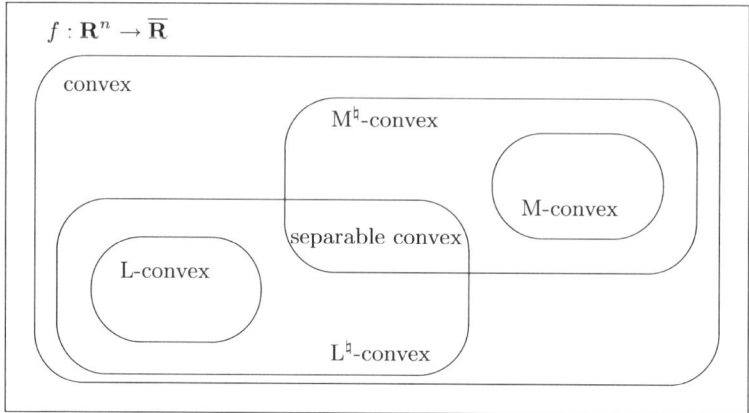

Fig. 11.9. Classes of convex functions (L$^\natural$-convex $\cap$ M$^\natural$-convex = separable convex)

M$^\natural$-convex functions in continuous variables constitute another subclass of convex functions, different from L$^\natural$-convex functions, that are equipped with another kind of combinatorial properties. See examples in Sect. 11.4.2.

An *M-convex function* in continuous variables is defined as an M$^\natural$-convex function $f : \mathbb{R}^n \to \overline{\mathbb{R}}$ that satisfies (17) with $j \in \mathrm{supp}^-(x - y)$. This is equivalent to saying that f is M-convex if and only if it is M$^\natural$-convex and $\mathrm{dom}_{\mathbb{R}} f \subseteq \{x \in \mathbb{R}^n \mid \sum_{i=1}^n x_i = r\}$ for some $r \in \mathbb{R}$. M-convex functions and M$^\natural$-convex functions are essentially the same, in that M$^\natural$-convex functions in n variables can be obtained as projections of M-convex functions in $n + 1$ variables.

Classes of Discrete Convex Functions in Continuous Variables

Figure 11.9 shows the classes of discrete convex functions in continuous variables. L$^\natural$-convex functions contain L-convex functions as a special case. The same is true for M$^\natural$-convex and M-convex functions. It is known that the classes of L-convex functions and M-convex functions are disjoint, whereas the intersection of the classes of L$^\natural$-convex functions and M$^\natural$-convex functions is exactly the class of separable convex functions.

Comparison of Fig. 11.9 with Fig. 11.7 shows the parallelism between the continuous and discrete cases.

11.3 Conjugacy

Conjugacy under the Legendre transformation is one of the most appealing facts in convex analysis. In discrete convex analysis, the discrete Legendre transformation gives a one-to-one correspondence between L-convex functions and M-convex functions.

11.3.1 Continuous Case

For a function $f : \mathbb{R}^n \to \overline{\mathbb{R}}$ (not necessarily convex) with $\mathrm{dom}_{\mathbb{R}} f \neq \emptyset$, the *convex conjugate* $f^{\bullet} : \mathbb{R}^n \to \overline{\mathbb{R}}$ is defined by

$$f^{\bullet}(p) = \sup\{\langle p, x \rangle - f(x) \mid x \in \mathbb{R}^n\} \quad (p \in \mathbb{R}^n), \tag{18}$$

where $\langle p, x \rangle = \sum_{i=1}^{n} p_i x_i$ is the inner product of $p = (p_i) \in \mathbb{R}^n$ and $x = (x_i) \in \mathbb{R}^n$. The function $f^{\bullet}$ is also referred to as the (convex) *Legendre(–Fenchel) transform* of f, and the mapping $f \mapsto f^{\bullet}$ as the (convex) *Legendre(–Fenchel) transformation*. Similarly to (18), the *concave conjugate* of $h : \mathbb{R}^n \to \underline{\mathbb{R}}$ is defined to be the function $h^{\circ} : \mathbb{R}^n \to \underline{\mathbb{R}}$ given by

$$h^{\circ}(p) = \inf\{\langle p, x \rangle - h(x) \mid x \in \mathbb{R}^n\} \quad (p \in \mathbb{R}^n). \tag{19}$$

Note that $h^{\circ}(p) = -(-h)^{\bullet}(-p)$.

The conjugacy theorem in convex analysis states that the Legendre transformation gives a one-to-one correspondence in the class of closed proper convex functions, where a convex function f is said to be *proper* if $\mathrm{dom}_{\mathbb{R}} f$ is nonempty, and *closed* if the epigraph $\{(x, y) \in \mathbb{R}^{n+1} \mid y \geq f(x)\}$ is a closed subset of $\mathbb{R}^{n+1}$. Notation $f^{\bullet\bullet}$ means $(f^{\bullet})^{\bullet}$.

Theorem 3.1. *The Legendre transformation* (18) *gives a symmetric one-to-one correspondence in the class of all closed proper convex functions. That is, for a closed proper convex function f, the conjugate function $f^{\bullet}$ is a closed proper convex function and $f^{\bullet\bullet} = f$.*

Addition of combinatorial ingredients to the above theorem yields the conjugacy between M-convex and L-convex functions.

Theorem 3.2 (Murota and Shioura 2004a). *The Legendre transformation* (18) *gives a one-to-one correspondence between the classes of all closed proper $M^{\natural}$-convex functions and $L^{\natural}$-convex functions. Similarly for M-convex and L-convex functions.*

The first statement above means that, for a closed proper $M^{\natural}$-convex function f, $f^{\bullet}$ is a closed proper $L^{\natural}$-convex function and $f^{\bullet\bullet} = f$, and that, for a closed proper $L^{\natural}$-convex function g, $g^{\bullet}$ is a closed proper $M^{\natural}$-convex function and $g^{\bullet\bullet} = g$. To express this one-to-one correspondence we have indicated $M^{\natural}$-convex functions and $L^{\natural}$-convex functions by congruent regions in Fig. 11.9. The second statement means similarly that, for a closed proper M-convex function f, $f^{\bullet}$ is a closed proper L-convex function and $f^{\bullet\bullet} = f$, and that, for a closed proper L-convex function g, $g^{\bullet}$ is a closed proper M-convex function and $g^{\bullet\bullet} = g$. It is also noted that the conjugate of a separable convex function is another separable convex function.

The L/M-conjugacy is also valid for polyhedral convex functions.

Theorem 3.3 (Murota and Shioura 2000). *The Legendre transformation* (18) *gives a one-to-one correspondence between the classes of all polyhedral $M^{\natural}$-convex functions and $L^{\natural}$-convex functions. Similarly for M-convex and L-convex functions.*

11.3.2 Discrete Case

We turn to functions defined on integer points. For functions $f : \mathbb{Z}^n \to \overline{\mathbb{R}}$ and $h : \mathbb{Z}^n \to \underline{\mathbb{R}}$ with $\mathrm{dom}_{\mathbb{Z}} f \neq \emptyset$ and $\mathrm{dom}_{\mathbb{Z}} h \neq \emptyset$, discrete versions of the Legendre transformations are defined by

$$f^{\bullet}(p) = \sup\{\langle p, x \rangle - f(x) \mid x \in \mathbb{Z}^n\} \quad (p \in \mathbb{R}^n), \tag{20}$$

$$h^{\circ}(p) = \inf\{\langle p, x \rangle - h(x) \mid x \in \mathbb{Z}^n\} \quad (p \in \mathbb{R}^n). \tag{21}$$

We call (20) and (21), respectively, *convex* and *concave discrete Legendre(–Fenchel) transformations*. The functions $f^{\bullet} : \mathbb{R}^n \to \overline{\mathbb{R}}$ and $h^{\circ} : \mathbb{R}^n \to \underline{\mathbb{R}}$ are called the *convex conjugate* of f and the *concave conjugate* of h, respectively.

Theorem 3.4. *For an $M^{\natural}$-convex function $f : \mathbb{Z}^n \to \overline{\mathbb{R}}$, the conjugate function $f^{\bullet} : \mathbb{R}^n \to \overline{\mathbb{R}}$ is a (locally polyhedral) $L^{\natural}$-convex function. For an $L^{\natural}$-convex function $g : \mathbb{Z}^n \to \overline{\mathbb{R}}$, the conjugate function $g^{\bullet} : \mathbb{R}^n \to \overline{\mathbb{R}}$ is a (locally polyhedral) $M^{\natural}$-convex function. Similarly for M-convex and L-convex functions.*

For an integer-valued function f, $f^{\bullet}(p)$ is integer for an integer vector p. Hence (20) with $p \in \mathbb{Z}^n$ defines a transformation of $f : \mathbb{Z}^n \to \overline{\mathbb{Z}}$ to $f^{\bullet} : \mathbb{Z}^n \to \overline{\mathbb{Z}}$; we refer to (20) with $p \in \mathbb{Z}^n$ as $(20)_{\mathbb{Z}}$.

The conjugacy theorem for discrete M-convex and L-convex functions reads as follows.

Theorem 3.5 (Murota 1998b). *The discrete Legendre transformation $(20)_{\mathbb{Z}}$ gives a one-to-one correspondence between the classes of all integer-valued $M^{\natural}$-convex functions and $L^{\natural}$-convex functions in discrete variables. Similarly for M-convex and L-convex functions.*

It should be clear that the first statement above means that, for an integer-valued $M^{\natural}$-convex function $f : \mathbb{Z}^n \to \overline{\mathbb{Z}}$, the function $f^{\bullet}$ in $(20)_{\mathbb{Z}}$ is an integer-valued $L^{\natural}$-convex function and $f^{\bullet\bullet} = f$, where $f^{\bullet\bullet}$ is a short-hand notation for $(f^{\bullet})^{\bullet}$ using the discrete Legendre transformation $(20)_{\mathbb{Z}}$, and similarly when f is $L^{\natural}$-convex.

11.4 Examples

11.4.1 L-convex Functions

Some examples of $L^{\natural}$- and L-convex functions are given in this section. The following basic facts are noted.

1. The effective domain of an $L^{\natural}$-convex function is an $L^{\natural}$-convex set.
2. An $L^{\natural}$-convex function remains to be $L^{\natural}$-convex when its effective domain is restricted to any $L^{\natural}$-convex set.
3. A sum of $L^{\natural}$-convex functions is $L^{\natural}$-convex.

Similar statements are true when "$L^\natural$-convex" is replaced by "L-convex" in the above.

We first consider functions in discrete variable $p = (p_1, \ldots, p_n) \in \mathbb{Z}^n$.

- **Linear function:** A linear (or affine) function

$$g(p) = \alpha + \langle p, x \rangle \tag{22}$$

with $x \in \mathbb{R}^n$ and $\alpha \in \mathbb{R}$ is L-convex (and hence $L^\natural$-convex).

- **Quadratic function:** A quadratic function

$$g(p) = \sum_{i=1}^{n} \sum_{j=1}^{n} a_{ij} p_i p_j \tag{23}$$

with $a_{ij} = a_{ji} \in \mathbb{R}$ $(i, j = 1, \ldots, n)$ is $L^\natural$-convex if and only if

$$a_{ij} \leq 0 \quad (i \neq j), \qquad \sum_{j=1}^{n} a_{ij} \geq 0 \quad (i = 1, \ldots, n). \tag{24}$$

It is L-convex if and only if

$$a_{ij} \leq 0 \quad (i \neq j), \qquad \sum_{j=1}^{n} a_{ij} = 0 \quad (i = 1, \ldots, n). \tag{25}$$

- **Separable convex function:** For univariate convex functions ψ_i $(i = 1, \ldots, n)$ and ψ_{ij} $(i, j = 1, \ldots, n; i \neq j)$,

$$g(p) = \sum_{i=1}^{n} \psi_i(p_i) + \sum_{i \neq j} \psi_{ij}(p_i - p_j) \tag{26}$$

is an $L^\natural$-convex function. This is L-convex if $\psi_i = 0$ for $i = 1, \ldots, n$.

- **Maximum-component function:** For any $\tau_0, \tau_1, \ldots, \tau_n \in \mathbb{R}$,

$$g(p) = \max\{\tau_0, \ p_1 + \tau_1, \ p_2 + \tau_2, \ \ldots, \ p_n + \tau_n\} \tag{27}$$

is an $L^\natural$-convex function. This is L-convex if τ_0 does not exist (i.e., $\tau_0 = -\infty$). Hence

$$g(p) = \max\{p_1, p_2, \ldots, p_n\} - \min\{p_1, p_2, \ldots, p_n\} \tag{28}$$

is an L-convex function. Furthermore, if ψ is a nondecreasing univariate convex function,

$$g(p) = \psi\left(\max_{1 \leq i \leq n} \{p_i + \tau_i\}\right) \tag{29}$$

is an $L^\natural$-convex function. It is also mentioned that, if $g_0(p, t)$ is $L^\natural$-convex in $(p, t) \in \mathbb{Z}^n \times \mathbb{Z}$ and nondecreasing in t, then the *max-aggregation* $g : \mathbb{Z}^n \times \mathbb{Z}^m \to \overline{\mathbb{R}}$ defined by

$$g(p, q) = g_0(p, \max(q_1, \ldots, q_m)) \quad (p \in \mathbb{Z}^n, q \in \mathbb{Z}^m) \tag{30}$$

is $L^\natural$-convex in (p, q), whereas g is L-convex if g_0 is L-convex.

- **Submodular set function:** A submodular set function $\rho : 2^V \to \overline{\mathbb{R}}$ can be identified with an $L^{\natural}$-convex function g under the correspondence $g(\chi_X) = \rho(X)$ for $X \subseteq V$, where $\mathrm{dom}_{\mathbb{Z}} g \subseteq \{0, 1\}^n$.
- **Multimodular function:** A function $h : \mathbb{Z}^n \to \overline{\mathbb{R}}$ is *multimodular* if and only if it can be represented as

$$h(p) = g(p_1, p_1 + p_2, \ldots, p_1 + \cdots + p_n)$$

for some $L^{\natural}$-convex function g; see (Altman et al. 2000, 2003; Hajek 1985; Murota 2005).

The constructions above work for functions in continuous variable $p \in \mathbb{R}^n$. That is, the functions $g : \mathbb{R}^n \to \overline{\mathbb{R}}$ defined by the expressions (22) to (30) are $L^{\natural}$- or L-convex functions, if all the variables are understood as real numbers or vectors. It is noteworthy that quadratic $L^{\natural}$-convex functions are exactly the same as the (finite dimensional case of) *Dirichlet forms* used in probability theory (Fukushima et al. 1994). The energy consumed in a nonlinear electrical network, when expressed as a function in terminal voltages, is an $L^{\natural}$-convex function (Murota 2003a, Sect. 2.2).

11.4.2 M-convex Functions

Some examples of $M^{\natural}$- and M-convex functions are given in this section. The following basic facts are noted.

1. The effective domain of an $M^{\natural}$-convex function is an $M^{\natural}$-convex set.
2. An $M^{\natural}$-convex function does not necessarily remain $M^{\natural}$-convex when its effective domain is restricted to an $M^{\natural}$-convex set.
3. A sum of $M^{\natural}$-convex functions is not necessarily $M^{\natural}$-convex.
4. The *infimal convolution* of $M^{\natural}$-convex functions f_1 and f_2, defined as

$$(f_1 \,\square\, f_2)(x) = \inf\{f_1(x_1) + f_2(x_2) \mid x = x_1 + x_2\}, \tag{31}$$

is $M^{\natural}$-convex if $f_1 \,\square\, f_2$ does not take $-\infty$, where $x_1, x_2 \in \mathbb{Z}^n$ in the discrete case and $x_1, x_2 \in \mathbb{R}^n$ in the continuous case.

Similar statements are true when "$M^{\natural}$-convex" is replaced by "M-convex" in the above.

We first consider functions in discrete variable $x = (x_1, \ldots, x_n) \in \mathbb{Z}^n$.

- **Linear function:** A linear (or affine) function

$$f(x) = \alpha + \langle p, x \rangle \tag{32}$$

with $p \in \mathbb{R}^n$ and $\alpha \in \mathbb{R}$ is $M^{\natural}$-convex. It is M-convex if $\mathrm{dom}_{\mathbb{Z}} f$ is an M-convex set.
- **Quadratic function:** A quadratic function

$$f(x) = \sum_{i=1}^{n} \sum_{j=1}^{n} a_{ij} x_i x_j \tag{33}$$

with $a_{ij} = a_{ji} \in \mathbb{R}$ $(i, j = 1, \ldots, n)$ is $M^\natural$-convex if and only if $a_{ij} \geq 0$ for all (i, j) and

$$a_{ij} \geq \min(a_{ik}, a_{jk}) \quad \text{if } \{i, j\} \cap \{k\} = \emptyset, \tag{34}$$

where $\mathrm{dom}_{\mathbb{Z}} f = \mathbb{Z}^n$. A function f of (33), with $\mathrm{dom}_{\mathbb{Z}} f = \{x \in \mathbb{Z}^n \mid \sum_{i=1}^n x_i = r\}$ for some $r \in \mathbb{Z}$, is M-convex if and only if

$$a_{ij} + a_{kl} \geq \min(a_{ik} + a_{jl}, a_{il} + a_{jk}) \quad \text{if } \{i, j\} \cap \{k, l\} = \emptyset. \tag{35}$$

- **Laminar convex function:** By a *laminar family* we mean a nonempty family $\mathcal{T}$ of subsets of V such that $X \cap Y = \emptyset$ or $X \subseteq Y$ or $X \supseteq Y$ for any $X, Y \in \mathcal{T}$. A function f is called *laminar convex* if it can be represented as

$$f(x) = \sum_{X \in \mathcal{T}} f_X(x(X)) \tag{36}$$

for a laminar family $\mathcal{T}$ and a family of univariate convex functions f_X indexed by $X \in \mathcal{T}$, where $x(X) = \sum_{i \in X} x_i$. A laminar convex function is $M^\natural$-convex. A separable convex function (3) is laminar convex and hence $M^\natural$-convex. It is known (Hirai and Murota 2004) that every quadratic $M^\natural$-convex function (in discrete variables) is laminar convex.
- **Minimum-value function:** Given a_i for $i \in V$ we define a set function $\mu : 2^V \to \overline{\mathbb{R}}$ as $\mu(X) = \min\{a_i \mid i \in X\}$ for nonempty $X \subseteq V$. By convention we put $\mu(\emptyset) = a_*$ by choosing $a_* \in \mathbb{R}$ such that $a_* \geq \max\{a_i \mid i \in V\}$. Then μ is $M^\natural$-convex when identified with a function $f : \mathbb{Z}^n \to \overline{\mathbb{R}}$ with $\mathrm{dom}_{\mathbb{Z}} f \subseteq \{0, 1\}^n$ by $f(\chi_X) = \mu(X)$ for $X \subseteq V$.
- **Bipartite matching:** Let $G = (V, W; E)$ be a bipartite graph with vertex set $V \cup W$ and edge set E, and suppose that each edge $e \in E$ is associated with weight $\gamma(e) \in \mathbb{R}$. For $X \subseteq V$ denote by $\Gamma(X)$ the minimum weight of a matching that matches with X, i.e.,

$$\Gamma(X) = \min\left\{ \sum_{e \in M} \gamma(e) \mid M \text{ is a matching, } V \cap \partial M = X \right\},$$

where $\Gamma(X) = +\infty$ if such M does not exist. Then Γ is $M^\natural$-convex when identified with a function $f : \mathbb{Z}^n \to \overline{\mathbb{R}}$ with $\mathrm{dom}_{\mathbb{Z}} f \subseteq \{0, 1\}^n$ by $f(\chi_X) = \Gamma(X)$ for $X \subseteq V$. This construction can be extended to the minimum convex-cost flow problem.
- **Stable marriage problem:** The payoff function of the stable marriage problem is $M^\natural$-concave; see (54) in Sect. 11.10.
- **Matroid:** Let $(V, \mathcal{B}, \mathcal{I}, \rho)$ be a matroid on V with base family $\mathcal{B}$, independent-set family $\mathcal{I}$ and rank function ρ. The characteristic vectors of bases $\{\chi_B \mid B \in \mathcal{B}\}$ form an M-convex set and those of independent sets $\{\chi_I \mid I \in \mathcal{I}\}$ form an $M^\natural$-convex set. The rank function $\rho : 2^V \to \mathbb{Z}$ is $M^\natural$-concave when identified with a function $f : \mathbb{Z}^n \to \overline{\mathbb{R}}$ with $\mathrm{dom}_{\mathbb{Z}} f = \{0, 1\}^n$ by $f(\chi_X) = \rho(X)$ for $X \subseteq V$; see Sect. 11.6.1. More generally, the *vector rank function* of an integral submodular system is $M^\natural$-concave (Fujishige 2005, p. 51).

- **Valuated matroid:** A *valuated matroid* $\omega : 2^V \to \mathbb{R}$ of Dress and Wenzel (1990, 1992) (see also Murota 2000, Chap. 5) can be identified with an $\mathrm{M}^\natural$-concave function f under the correspondence $f(\chi_X) = \omega(X)$ for $X \subseteq V$, where $\mathrm{dom}_{\mathbb{Z}} f \subseteq \{0, 1\}^n$. The tropical geometry (Speyer and Sturmfels 2004) is closely related with valuated matroids. For example, the tropical linear space (Speyer 2004) is essentially the same as the circuit valuation of matroids (Murota and Tamura 2001).

Next we turn to functions $f : \mathbb{R}^n \to \overline{\mathbb{R}}$ in continuous variable $x \in \mathbb{R}^n$. The infimal convolution (31) preserves $\mathrm{M}^\natural$-convexity when the infimum is taken over $x_1, x_2 \in \mathbb{R}^n$. Laminar convex functions (36) as well as linear functions (32) remain to be $\mathrm{M}^\natural$-convex when x is understood as a real vector. The energy consumed in a nonlinear electrical network, when expressed as a function in terminal currents, is an $\mathrm{M}^\natural$-convex function (Murota 2003a, Sect. 2.2).

A subtlety arises for quadratic functions. Condition (34), together with $a_{ij} \geq 0$ for all (i, j), is sufficient but not necessary for $f : \mathbb{R}^n \to \mathbb{R}$ of the form of (33) to be $\mathrm{M}^\natural$-convex. A necessary and sufficient condition in terms of the matrix $A = (a_{ij})$ is that, for any $\beta > 0$, $A + \beta I$ is nonsingular and $(A + \beta I)^{-1}$ satisfies (24). It is also mentioned that not every quadratic $\mathrm{M}^\natural$-convex function in real variables is laminar convex. As for M-convexity, condition (35) is sufficient but not necessary for f to be M-convex.

Thus the relation between discrete and continuous cases are not so simple in M-convexity as in L-convexity.

11.5 Separation and Fenchel Duality

11.5.1 Separation Theorem

The *duality* principle in convex analysis can be expressed in a number of different forms. One of the most appealing statements is in the form of the separation theorem, which asserts the existence of a separating affine function $y = \alpha^* + \langle p^*, x \rangle$ for a pair of convex and concave functions.

In the continuous case we have the following.

Theorem 5.1. *Let* $f : \mathbb{R}^n \to \overline{\mathbb{R}}$ *and* $h : \mathbb{R}^n \to \overline{\mathbb{R}}$ *be convex and concave functions, respectively* (*satisfying certain regularity conditions*). *If*

$$f(x) \geq h(x) \quad (\forall x \in \mathbb{R}^n),$$

there exist $\alpha^* \in \mathbb{R}$ *and* $p^* \in \mathbb{R}^n$ *such that*

$$f(x) \geq \alpha^* + \langle p^*, x \rangle \geq h(x) \quad (\forall x \in \mathbb{R}^n).$$

A *discrete separation theorem* means a statement like:

For any $f : \mathbb{Z}^n \to \overline{\mathbb{R}}$ and $h : \mathbb{Z}^n \to \underline{\mathbb{R}}$ belonging to certain classes of functions, if $f(x) \geq h(x)$ for all $x \in \mathbb{Z}^n$, then there exist $\alpha^* \in \mathbb{R}$ and $p^* \in \mathbb{R}^n$ such that

$$f(x) \geq \alpha^* + \langle p^*, x \rangle \geq h(x) \quad (\forall x \in \mathbb{Z}^n).$$

Moreover, if f and h are integer-valued, there exist integer-valued $\alpha^* \in \mathbb{Z}$ and $p^* \in \mathbb{Z}^n$.

Discrete separation theorems often capture deep combinatorial properties in spite of the apparent similarity to the separation theorem in convex analysis. In this connection we note the following facts (see Murota 2003a, Examples 1.5 and 1.6 for concrete examples), where $\overline{f}$ denotes the convex closure of f, $\overline{h}$ the concave closure of h, and $\not\Longrightarrow$ stands for "does not imply."

1. $f(x) \geq h(x)$ $(\forall x \in \mathbb{Z}^n)$ $\not\Longrightarrow$ $\overline{f}(x) \geq \overline{h}(x)$ $(\forall x \in \mathbb{R}^n)$.
2. $f(x) \geq h(x)$ $(\forall x \in \mathbb{Z}^n)$ $\not\Longrightarrow$ existence of $\alpha^* \in \mathbb{R}$ and $p^* \in \mathbb{R}^n$.
3. Existence of $\alpha^* \in \mathbb{R}$ and $p^* \in \mathbb{R}^n$ $\not\Longrightarrow$ existence of $\alpha^* \in \mathbb{Z}$ and $p^* \in \mathbb{Z}^n$.

The separation theorems for M-convex/M-concave functions and for L-convex/L-concave functions read as follows. It should be clear that $f^\bullet$ and h° are the convex and concave conjugate functions of f and h defined by (20) and (21), respectively.

Theorem 5.2 (M-separation theorem). *Let $f : \mathbb{Z}^n \to \overline{\mathbb{R}}$ be an $M^\natural$-convex function and $h : \mathbb{Z}^n \to \underline{\mathbb{R}}$ be an $M^\natural$-concave function such that $\mathrm{dom}_{\mathbb{Z}} f \cap \mathrm{dom}_{\mathbb{Z}} h \neq \emptyset$ or $\mathrm{dom}_{\mathbb{R}} f^\bullet \cap \mathrm{dom}_{\mathbb{R}} h^\circ \neq \emptyset$. If $f(x) \geq h(x)$ $(\forall x \in \mathbb{Z}^n)$, there exist $\alpha^* \in \mathbb{R}$ and $p^* \in \mathbb{R}^n$ such that*

$$f(x) \geq \alpha^* + \langle p^*, x \rangle \geq h(x) \quad (\forall x \in \mathbb{Z}^n).$$

Moreover, if f and h are integer-valued, there exist integer-valued $\alpha^* \in \mathbb{Z}$ and $p^* \in \mathbb{Z}^n$.

Theorem 5.3 (L-separation theorem). *Let $g : \mathbb{Z}^n \to \overline{\mathbb{R}}$ be an $L^\natural$-convex function and $k : \mathbb{Z}^n \to \underline{\mathbb{R}}$ be an $L^\natural$-concave function such that $\mathrm{dom}_{\mathbb{Z}} g \cap \mathrm{dom}_{\mathbb{Z}} k \neq \emptyset$ or $\mathrm{dom}_{\mathbb{R}} g^\bullet \cap \mathrm{dom}_{\mathbb{R}} k^\circ \neq \emptyset$. If $g(p) \geq k(p)$ $(\forall p \in \mathbb{Z}^n)$, there exist $\beta^* \in \mathbb{R}$ and $x^* \in \mathbb{R}^n$ such that*

$$g(p) \geq \beta^* + \langle p, x^* \rangle \geq k(p) \quad (\forall p \in \mathbb{Z}^n).$$

Moreover, if g and k are integer-valued, there exist integer-valued $\beta^* \in \mathbb{Z}$ and $x^* \in \mathbb{Z}^n$.

As an immediate corollary of the M-separation theorem we can obtain an optimality criterion for the problem of minimizing the sum of two M-convex functions, which we call the *M-convex intersection problem*. Note that the sum of M-convex functions is no longer M-convex and Theorem 2.7 does not apply.

Theorem 5.4 (M-convex intersection theorem). *For $M^\natural$-convex functions $f_1, f_2 :$ $\mathbb{Z}^n \to \overline{\mathbb{R}}$ and a point $x^* \in \mathrm{dom}_{\mathbb{Z}} f_1 \cap \mathrm{dom}_{\mathbb{Z}} f_2$ we have*

$$f_1(x^*) + f_2(x^*) \le f_1(x) + f_2(x) \quad (\forall x \in \mathbb{Z}^n)$$

if and only if there exists $p^* \in \mathbb{R}^n$ such that

$$(f_1 - p^*)(x^*) \le (f_1 - p^*)(x) \quad (\forall x \in \mathbb{Z}^n),$$
$$(f_2 + p^*)(x^*) \le (f_2 + p^*)(x) \quad (\forall x \in \mathbb{Z}^n).$$

These conditions are equivalent, respectively, to

$$(f_1 - p^*)(x^*) \le (f_1 - p^*)(x^* + \chi_i - \chi_j) \quad (\forall i, j \in \{0, 1, \dots, n\}),$$
$$(f_2 + p^*)(x^*) \le (f_2 + p^*)(x^* + \chi_i - \chi_j) \quad (\forall i, j \in \{0, 1, \dots, n\}),$$

and for such p^* we have

$$\mathrm{argmin}_{\mathbb{Z}}(f_1 + f_2) = \mathrm{argmin}_{\mathbb{Z}}(f_1 - p^*) \cap \mathrm{argmin}_{\mathbb{Z}}(f_2 + p^*).$$

Moreover, if f_1 and f_2 are integer-valued, we can choose integer-valued $p^* \in \mathbb{Z}^n$.

Frank's discrete separation theorem (Frank 1982) for submodular/supermodular set functions is a special case of the L-separation theorem. Frank's weight splitting theorem (Frank 1981) for the weighted matroid intersection problem is a special case of the M-convex intersection problem. The submodular flow problem can be generalized to the M-convex submodular flow problem (Murota 1999); see also (Iwata et al. 2005; Iwata and Shigeno 2003).

11.5.2 Fenchel Duality

Another expression of the duality principle is in the form of the Fenchel duality. This is a min-max relation between a pair of convex and concave functions and their conjugate functions. Such a min-max theorem is computationally useful in that it affords a certificate of optimality.

The Fenchel duality theorem in the continuous case reads as follows. Recall the notations $f^\bullet$ and h° in (18) and (19).

Theorem 5.5. *Let $f : \mathbb{R}^n \to \overline{\mathbb{R}}$ and $h : \mathbb{R}^n \to \underline{\mathbb{R}}$ be convex and concave functions, respectively (satisfying certain regularity conditions). Then*

$$\inf\{f(x) - h(x) \mid x \in \mathbb{R}^n\} = \sup\{h^\circ(p) - f^\bullet(p) \mid p \in \mathbb{R}^n\}.$$

We now turn to the discrete case. For any functions $f : \mathbb{Z}^n \to \overline{\mathbb{Z}}$ and $h : \mathbb{Z}^n \to \underline{\mathbb{Z}}$ we have a chain of inequalities:

$$
\begin{array}{ccc}
\inf\{f(x) - h(x) \mid x \in \mathbb{Z}^n\} & & \sup\{h^\circ(p) - f^\bullet(p) \mid p \in \mathbb{Z}^n\} \\
\mathrm{I}\vee & & \wedge\mathrm{I} \\
\inf\{\overline{f}(x) - \overline{h}(x) \mid x \in \mathbb{R}^n\} & \ge & \sup\{\overline{h}^\circ(p) - \overline{f}^\bullet(p) \mid p \in \mathbb{R}^n\}
\end{array} \tag{37}
$$

from the definitions (20) and (21) of conjugate functions $f^\bullet$ and h°, where $\overline{f}$ and $\overline{h}$ are convex and concave closures of f and h, respectively. It should be observed that

1. The second inequality in the middle of (37) is in fact an equality by the Fenchel duality theorem (Theorem 5.5) in convex analysis;
2. The first (left) inequality in (37) can be strict even when f is convex-extensible and h is concave-extensible, and similarly for the third (right) inequality. See Examples 5.6 and 5.7 below.

Example 5.6. For $f, h : \mathbb{Z}^2 \to \mathbb{Z}$ defined as

$$f(x_1, x_2) = |x_1 + x_2 - 1|, \qquad h(x_1, x_2) = 1 - |x_1 - x_2|$$

we have $\inf\{f - h\} = 0$, $\inf\{\overline{f} - \overline{h}\} = -1$. The discrete Legendre transforms are given by

$$f^\bullet(p_1, p_2) = \begin{cases} p_1 & ((p_1, p_2) \in S) \\ +\infty & \text{(otherwise)}, \end{cases} \qquad h^\circ(p_1, p_2) = \begin{cases} -1 & ((p_1, p_2) \in T) \\ -\infty & \text{(otherwise)} \end{cases}$$

with $S = \{(-1, -1), (0, 0), (1, 1)\}$ and $T = \{(-1, 1), (0, 0), (1, -1)\}$. Hence $\sup\{h^\circ - f^\bullet\} = h^\circ(0, 0) - f^\bullet(0, 0) = -1 - 0 = -1$. Then (37) reads as

$$\inf\{f - h\} > \inf\{\overline{f} - \overline{h}\} = \sup\{\overline{h}^\circ - \overline{f}^\bullet\} = \sup\{h^\circ - f^\bullet\}.$$
$$(0) \qquad\qquad (-1) \qquad\qquad (-1) \qquad\qquad (-1)$$

Example 5.7. For $f, h : \mathbb{Z}^2 \to \mathbb{Z}$ defined as

$$f(x_1, x_2) = \max(0, x_1 + x_2), \qquad h(x_1, x_2) = \min(x_1, x_2)$$

we have $\inf\{f - h\} = \inf\{\overline{f} - \overline{h}\} = 0$. The discrete Legendre transforms are given as $f^\bullet = \delta_S$ and $h^\circ = -\delta_T$ in terms of the indicator functions of $S = \{(0, 0), (1, 1)\}$ and $T = \{(1, 0), (0, 1)\}$. Since $S \cap T = \emptyset$, $h^\circ - f^\bullet$ is identically equal to $-\infty$, whereas $\sup\{\overline{h}^\circ - \overline{f}^\bullet\} = 0$ since $\overline{f}^\bullet = \delta_{\overline{S}}$, $\overline{h}^\circ = -\delta_{\overline{T}}$ and $\overline{S} \cap \overline{T} = \{(1/2, 1/2)\}$. Then (37) reads as

$$\inf\{f - h\} = \inf\{\overline{f} - \overline{h}\} = \sup\{\overline{h}^\circ - \overline{f}^\bullet\} > \sup\{h^\circ - f^\bullet\}.$$
$$(0) \qquad\qquad (0) \qquad\qquad (0) \qquad\qquad (-\infty)$$

From the observations above, we see that the essence of the following theorem is the assertion that the first and third inequalities in (37) are in fact equalities for $M^\natural$-convex/$M^\natural$-concave functions and $L^\natural$-convex/$L^\natural$-concave functions.

Theorem 5.8 (Fenchel-type duality theorem).

(1) *Let* $f : \mathbb{Z}^n \to \overline{\mathbb{Z}}$ *be an integer-valued $M^\natural$-convex function and* $h : \mathbb{Z}^n \to \underline{\mathbb{Z}}$ *be an integer-valued $M^\natural$-concave function such that* $\text{dom}_\mathbb{Z} f \cap \text{dom}_\mathbb{Z} h \neq \emptyset$ *or* $\text{dom}_\mathbb{Z} f^\bullet \cap \text{dom}_\mathbb{Z} h^\circ \neq \emptyset$. *Then we have*

$$\inf\{f(x) - h(x) \mid x \in \mathbb{Z}^n\} = \sup\{h^\circ(p) - f^\bullet(p) \mid p \in \mathbb{Z}^n\}. \qquad (38)$$

If this common value is finite, the infimum and the supremum are attained.

(2) *Let* $g : \mathbb{Z}^n \to \overline{\mathbb{Z}}$ *be an integer-valued* $L^\natural$*-convex function and* $k : \mathbb{Z}^n \to \overline{\mathbb{Z}}$ *be an integer-valued* $L^\natural$*-concave function such that* $\mathrm{dom}_{\mathbb{Z}} g \cap \mathrm{dom}_{\mathbb{Z}} k \neq \emptyset$ *or* $\mathrm{dom}_{\mathbb{Z}} g^\bullet \cap \mathrm{dom}_{\mathbb{Z}} k^\circ \neq \emptyset$. *Then we have*

$$\inf\{g(p) - k(p) \mid p \in \mathbb{Z}^n\} = \sup\{k^\circ(x) - g^\bullet(x) \mid x \in \mathbb{Z}^n\}. \qquad (39)$$

If this common value is finite, the infimum and the supremum are attained.

Edmonds' intersection theorem (Edmonds 1970) in the integral case is a special case of Theorem 5.8(1) above, and Fujishige's Fenchel-type duality theorem (Fujishige 1984) (see also Fujishige 2005, Sect. 6.1) for submodular set functions is a special case of Theorem 5.8 (2) above.

Whereas L-separation and M-separation theorems are parallel or conjugate in their statements, the Fenchel-type duality theorem is self-conjugate, in that the substitution of $f = g^\bullet$ and $h = k^\circ$ into (38) results in (39) by virtue of $g = g^{\bullet\bullet}$ and $k = k^{\circ\circ}$. With the knowledge of M-/L-conjugacy, these three duality theorems are almost equivalent to one another; once one of them is established, the other two theorems can be derived by relatively easy formal calculations.

11.6 Submodular Function Maximization

Maximization of a submodular set function is a difficult task in general. Many NP-hard problems can be reduced to this problem. Also known is that no polynomial algorithm exists in the ordinary oracle model (and this statement is independent of the P $\neq$ NP conjecture) (Jensen and Korte 1982; Lovász 1980, 1983). For approximate maximization under matroid constraints the performance bounds of greedy or ascent type algorithms were analyzed in Conforti and Cornuéjols (1984), Fisher et al. (1978), Nemhauser et al. (1978) and, recently, a pipage rounding algorithm has been designed for a subclass of submodular functions in Calinescu et al. (2007), which is extended in Vondrák (2008) to general submodular functions with the aid of randomization.

$M^\natural$-concave functions on $\{0, 1\}$-vectors form a subclass of submodular set functions that are algorithmically tractable for maximization. This is compatible with our general understanding that concave functions are easy to maximize, and explains why certain submodular functions treated in the literature are easier to maximize. To be specific, we have the following.

1. The greedy algorithm can be generalized for maximization of a single $M^\natural$-concave function.
2. The matroid intersection algorithm can be generalized for maximization of a sum of two $M^\natural$-concave functions.
3. The pipage rounding algorithm (Ageev and Sviridenko 2004) can be generalized for approximate maximization of a sum of nondecreasing $M^\natural$-concave functions under a matroid constraint.

Note that a sum of $M^\natural$-concave functions is not necessarily $M^\natural$-concave, though it is submodular. It is also mentioned that maximization of a sum of three $M^\natural$-concave functions is NP-hard, since it includes the three-matroid intersection problem as a special case.

11.6.1 $M^\natural$-concave Set Functions

Let us say that a set function $\rho : 2^V \to \mathbb{R}$ is $M^\natural$-*concave* if the function $h : \mathbb{Z}^n \to \mathbb{R}$ defined as $h(\chi_X) = \rho(X)$ for $X \subseteq V$ and $h(x) = -\infty$ for $x \notin \{0, 1\}^n$ is $M^\natural$-concave. In other words, ρ is $M^\natural$-concave if and only if, for any $X, Y \subseteq V$ and $i \in X \setminus Y$, we have $\rho(X) + \rho(Y) \le \rho(X \setminus \{i\}) + \rho(Y \cup \{i\})$ or $\rho(X) + \rho(Y) \le \rho((X \setminus \{i\}) \cup \{j\}) + \rho((Y \cup \{i\}) \setminus \{j\})$ for some $j \in Y \setminus X$. An $M^\natural$-concave set function is submodular (Murota 2003a, Theorem 6.19).

Not every submodular set function is $M^\natural$-concave. An example of a submodular function that is not $M^\natural$-concave is given by ρ on $V = \{1, 2, 3\}$ defined as $\rho(\emptyset) = 0$, $\rho(\{2, 3\}) = 2$, $\rho(\{1\}) = \rho(\{2\}) = \rho(\{3\}) = \rho(\{1, 2\}) = \rho(\{1, 3\}) = \rho(\{1, 2, 3\}) = 1$. The condition above fails for $X = \{2, 3\}$, $Y = \{1\}$ and $i = 2$.

A simple example of an $M^\natural$-concave set function is given by $\rho(X) = \varphi(|X|)$, where φ is a univariate concave function. This is a classical example of a submodular function (Edmonds 1970; Lovász 1983) that connects submodularity and concavity.

For a family of univariate concave functions $\{\varphi_A \mid A \in \mathcal{T}\}$ indexed by a family $\mathcal{T}$ of subsets of V, the function

$$\rho(X) = \sum_{A \in \mathcal{T}} \varphi_A(|A \cap X|) \quad (X \subseteq V)$$

is submodular. This function is $M^\natural$-concave if, in addition, $\mathcal{T}$ is a laminar family (i.e., $A, B \in \mathcal{T} \Rightarrow A \cap B = \emptyset$ or $A \subseteq B$ or $A \supseteq B$).

Given a set of real numbers a_i indexed by $i \in V$, the *maximum-value function*

$$\rho(X) = \max_{i \in X} a_i \quad (X \subseteq V)$$

is an $M^\natural$-concave function, where $\rho(\emptyset)$ is defined to be sufficiently small.

A *matroid rank function* is $M^\natural$-concave (Fujishige 2005, p. 51). Given a matroid on V in terms of the family $\mathcal{I}$ of independent sets, the rank function ρ is defined by

$$\rho(X) = \max\{|I| \mid I \in \mathcal{I}, I \subseteq X\} \quad (X \subseteq V),$$

which denotes the maximum size of an independent set contained in X. An interesting identity exists that indicates a kind of self-conjugacy of a matroid rank function. Let $g : \mathbb{Z}^n \to \overline{\mathbb{Z}}$ be such that $g(\chi_X) = \rho(X)$ for $X \subseteq V$ and $\mathrm{dom}_{\mathbb{Z}} g = \{0, 1\}^n$, and denote by $\rho^\bullet$ the discrete Legendre transform $g^\bullet$ of g defined by $(20)_{\mathbb{Z}}$ (i.e., (20) with $p \in \mathbb{Z}^n$). Then we have

$$\rho(X) = |X| - \rho^\bullet(\chi_X) \quad (X \subseteq V). \tag{40}$$

This can be shown as follows: $\rho^\bullet(\chi_X) = \max_Y\{|X \cap Y| - \rho(Y) \mid Y \subseteq V\} = \max_Y\{|X \cap Y| - \rho(Y) \mid X \subseteq Y \subseteq V\} = \max_Y\{|X| - \rho(Y) \mid X \subseteq Y \subseteq V\} = |X| - \rho(X)$; see also (Fujishige 2005, Lemma 6.2). Since ρ is submodular, g is $L^\natural$-convex, and hence $g^\bullet$ ($= \rho^\bullet$) is $M^\natural$-convex by conjugacy (Theorem 3.5). Then the expression (40) shows that ρ is $M^\natural$-concave.

A *weighted matroid rank function*, represented as

$$\rho(X) = \max\left\{\sum_{i \in I} w_i \mid I \in \mathcal{I}, \ I \subseteq X\right\} \quad (X \subseteq V) \tag{41}$$

with a nonnegative vector $w \in \mathbb{R}^n$, is also $M^\natural$-concave. This is a recent observation by Shioura (2008).

11.6.2 Greedy Algorithm

$M^\natural$-concave set functions admit the following local characterization of global maximum, an immediate corollary of Theorem 2.7.

Theorem 6.1. *For an $M^\natural$-concave set function $\rho : 2^V \to \mathbb{R}$ and a subset $X \subseteq V$, we have $\rho(X) \geq \rho(Y)$ ($\forall Y \subseteq V$) if and only if*

$$\rho(X) \geq \max_{i \in X, j \in V \setminus X}\{\rho((X \setminus \{i\}) \cup \{j\}), \ \rho(X \setminus \{i\}), \ \rho(X \cup \{j\})\}.$$

A natural greedy algorithm works for maximization of an $M^\natural$-concave set function ρ:

S0: Put $X := \emptyset$.
S1: Find $j \in V \setminus X$ that maximizes $\rho(X \cup \{j\})$.
S2: If $\rho(X) \geq \rho(X \cup \{j\})$, then stop ($X$ is a maximizer of ρ).
S3: Set $X := X \cup \{j\}$ and go to S1.

This algorithm may be regarded as a variant of the algorithm of Dress and Wenzel (1990) for valuated matroids, and the validity can be shown similarly.

11.6.3 Intersection Algorithm

Edmonds's matroid intersection/union algorithms show that we can efficiently maximize $\rho_1(X) + \rho_2(V \setminus X)$ and $\rho_1(X) + \rho_2(X) - |X|$ for two matroid rank functions ρ_1 and ρ_2. It should be clear that $\max_X\{\rho_1(X) + \rho_2(V \setminus X)\}$ is equal to the rank of the union of two matroids (V, ρ_1) and (V, ρ_2), and that $\max_X\{\rho_1(X) + \rho_2(X) - |X|\}$ is equal to the maximum size of a common independent set for matroid (V, ρ_1) and the dual of matroid (V, ρ_2). We note here that both $\rho_1(X) + \rho_2(V \setminus X)$ and $\rho_1(X) + (\rho_2(X) - |X|)$ are submodular functions that are represented as a sum of two $M^\natural$-concave functions.

Edmonds's intersection algorithm can be generalized for $M^\natural$-concave functions. A sum of two $M^\natural$-concave set functions can be maximized in polynomial time by means of a variant of the valuated matroid intersection algorithm (Murota 1996b); see also (Murota 1999, 2000, 2003a). It follows from the M-convex intersection theorem (Theorem 5.4) that, for two $M^\natural$-concave set functions ρ_1 and ρ_2, X maximizes $\rho_1(X) + \rho_2(X)$ if and only if there exists $p^* \in \mathbb{R}^n$ such that X maximizes both $\rho_1(X) + p^*(X)$ and $\rho_1(X) - p^*(X)$ at the same time, where $p^*(X) = \sum_{i \in X} p_i^*$.

11.6.4 Pipage Rounding Algorithm

Let ρ be a nondecreasing submodular set function on V and $(V, \mathcal{I})$ be a matroid on V with the family $\mathcal{I}$ of independent sets. We consider the problem of maximizing $\rho(X)$ subject to $X \in \mathcal{I}$. It is assumed that the function evaluation oracle for ρ and the membership oracle for $\mathcal{I}$ are available.

A recent paper of Calinescu et al. (2007) proposes a pipage rounding framework for approximate solution of this problem, showing that it works if the function ρ is represented as a sum of weighted matroid rank functions (41). Subsequently, it is pointed out by Shioura (2008) that this approach can be extended to the class of functions ρ represented as a sum of $M^\natural$-concave functions.

The framework of Calinescu et al. (2007) consists of three major steps.

1. Define a continuous relaxation: maximize $f(x)$ subject to $x \in P$, where P is the matroid polytope (convex hull of the characteristic vectors of independent sets) of $(V, \mathcal{I})$, and $f(x)$ is a nondecreasing concave function on P such that $f(\chi_X) = \rho(X)$ for all $X \subseteq V$.
2. Find an (approximately) optimal solution $x^* \in P$ of the continuous relaxation.
3. Round the fractional vector $x^* \in P$ to a $\{0, 1\}$-vector $\hat{x} \in P$ by applying the "pipage rounding scheme," and output the corresponding subset $\hat{X}$ (such that $\chi_{\hat{X}} = \hat{x}$) as an approximate solution to the original problem.

This algorithm, if computationally feasible at all, is guaranteed to output a $(1 - 1/e)$-approximate solution, where e denotes the base of natural logarithm.

In the case where $\rho = \sum_{k=1}^m \rho_k$ with nondecreasing $M^\natural$-concave set functions ρ_k, the above algorithm can be executed in polynomial time. As the concave extension f we may take the sum of the concave closures, say, $\bar{\rho}_k$ of ρ_k for $k = 1, \ldots, m$. The continuous relaxation can be solved by the ellipsoid method, which uses subgradients of $\bar{\rho}_k$. The subgradients of $\bar{\rho}_k$ can in turn be computed in polynomial time by exploiting the combinatorial structure of $M^\natural$-concave functions.

11.7 Finite Metric Space

Metrics are closely related to discrete convexity in several aspects. Distance functions satisfying triangle inequality are in one-to-one correspondence with positively homogeneous M-convex functions, and tree metrics are the same as valuated matroids of rank two. Furthermore, the Buneman construction and the Bandelt–Dress split decomposition can be derived as decompositions of polyhedral convex functions.

11.7.1 Positively Homogeneous M-convex Functions

Recall that $V = \{1, 2, \ldots, n\}$. By a *distance function* we mean a function $d : V \times V \to \overline{\mathbb{R}}$ such that $d(i, i) = 0$ for all $i \in V$, where d may take negative values and is not necessarily symmetric (i.e., $d(i, j) \neq d(j, i)$ in general). As usual, *triangle inequality* means the inequality:

$$d(i, j) + d(j, k) \geq d(i, k) \quad (\forall i, j, k \in V). \tag{42}$$

There exists a one-to-one correspondence between distance functions with triangle inequality and positively homogeneous (polyhedral) M-convex functions, as follows (see Murota 2003a, Sect. 6.12 for detail).

Theorem 7.1. *For a distance function $d : V \times V \to \overline{\mathbb{R}}$ satisfying triangle inequality, the function $f : \mathbb{R}^n \to \overline{\mathbb{R}}$ defined by*

$$f(x) = \inf_{\lambda} \left\{ \sum_{i,j \in V} \lambda_{ij} d(i, j) \;\middle|\; \sum_{i,j \in V} \lambda_{ij}(\chi_j - \chi_i) = x, \; \lambda_{ij} \geq 0 \; (i, j \in V) \right\} \tag{43}$$

is a positively homogeneous M-convex function, for which

$$d(i, j) = f(\chi_j - \chi_i) \quad (i, j \in V). \tag{44}$$

Conversely, for a positively homogeneous M-convex function f, the function d defined by (44) is a distance function with triangle inequality, for which (43) is true.

Figure 11.10 illustrates this correspondence when $V = \{1, 2, 3\}$; (a) shows the point set $\{\chi_j - \chi_i \mid i, j \in V\}$, (b) the function values of f, and (c) the corresponding positively homogeneous M-convex function f.

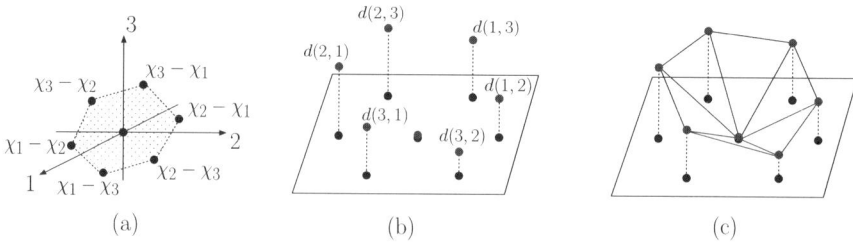

Fig. 11.10. Correspondence between distance functions and positively homogeneous M-convex functions ($n = 3$)

11.7.2 Tree Metrics and Buneman Construction

In the following we assume that d is a *metric*, which means that d is finite-valued $(d : V \times V \to \mathbb{R})$ and satisfies $d(i, i) = 0$ $(\forall i \in V)$, $d(i, j) = d(j, i) \geq 0$ $(\forall i, j \in V)$, and triangle inequality (42).

A *tree metric* means a metric that can be represented as the distance between vertices of a tree with nonnegative edge length. It is known that a metric d is a tree metric if and only if

$$d(i, j) + d(k, l) \leq \max\{d(i, k) + d(j, l), d(i, l) + d(j, k)\} \qquad (45)$$

for all distinct $i, j, k, l \in V$. This condition is called the *four-point condition*.

Remark 7.2. Consider the family $\mathcal{B} = \{\{i, j\} \mid 1 \leq i < j \leq n\}$ of unordered pairs of V. A function $d : V \times V \to \mathbb{R}$ with $d(i, j) = d(j, i) \geq 0$ and $d(i, i) = 0$ for all $i, j \in V$ can be identified with a function $\omega : \mathcal{B} \to \mathbb{R}$. Then d is a tree metric if and only if ω is a valuated matroid. Thus a tree metric is essentially equivalent to a valuated matroid on the uniform matroid of rank two.

Remark 7.3. A metric d is called an *ultrametric* if

$$d(i, j) \leq \max\{d(i, k), d(j, k)\} \qquad (46)$$

for all distinct $i, j, k \in V$. An ultrametric is a tree metric. For a tree metric d on V the function $\bar{d}$ defined by

$$\bar{d}(i, j) = d(i, j) - d(i, n) - d(j, n) \quad (i, j \in V \setminus \{n\}) \qquad (47)$$

satisfies (46), where $\bar{d}(i, j) \leq 0$.

The four point condition is closely related to M- or $M^\natural$-convexity of a quadratic function $f(x) = x^\top A x$ in $x \in \mathbb{Z}^n$. The condition (35) for M-convexity, $a_{ij} + a_{kl} \geq \min(a_{ik} + a_{jl}, a_{il} + a_{jk})$, is equivalent to the four point condition (45) for $d(i, j) = -a_{ij}$, and the condition (34) for $M^\natural$-convexity, $a_{ij} \geq \min(a_{ik}, a_{jk})$, is the same as (46). Note also that the substitution of $x_n = -(x_1 + \cdots + x_{n-1})$ into $f(x) = x^\top A x$ yields a quadratic function $\bar{f}(\bar{x}) = \bar{x}^\top \bar{A} \bar{x}$ in $\bar{x} = (x_1, \dots, x_{n-1})^\top$ with $\bar{a}_{ij} = a_{ij} - a_{in} - a_{jn} + a_{nn}$ $(i, j = 1, \dots, n-1)$. This is identical with (47) up to a constant term a_{nn}.

The Buneman construction decomposes a given metric d into a tree metric $\hat{d}$ and a residual d', as follows.

A partition of V into two nonempty sets is called a *split*. For a split $\sigma = \{A, B\}$, where $A \cap B = \emptyset$, $A \cup B = V$, $A \neq \emptyset$, $B \neq \emptyset$, we define *split metric* or *cut metric* $\Delta_\sigma : V \times V \to \mathbb{R}$ by

$$\Delta_\sigma(i, j) = \begin{cases} 1 & (|\{i, j\} \cap A| = |\{i, j\} \cap B| = 1), \\ 0 & (\{i, j\} \subseteq A \text{ or } \{i, j\} \subseteq B). \end{cases}$$

For a metric d and a split $\sigma = \{A, B\}$ the *Buneman index* is a real number defined as

$$\beta_\sigma(d) = \frac{1}{2} \min_{i,j \in A,\ k,l \in B} \{d(i,k) + d(j,l) - d(i,j) - d(k,l)\}.$$

With the notation $\mathcal{B}(d) = \{\sigma \mid \beta_\sigma(d) > 0\}$ we define $\hat{d} : V \times V \to \mathbb{R}$ as

$$\hat{d}(i,j) = \sum_{\sigma \in \mathcal{B}(d)} \beta_\sigma(d) \Delta_\sigma(i,j).$$

Then $\mathcal{B}(d)$ is *compatible* in the sense that for any two splits $\sigma_1 = \{A_1, B_1\}$, $\sigma_2 = \{A_2, B_2\}$ in $\mathcal{B}(d)$ at least one of $A_1 \cap A_2$, $A_1 \cap B_2$, $B_1 \cap A_2$, and $B_1 \cap B_2$ is empty. Accordingly, $\hat{d}$ is a tree metric with $\hat{d} \leq d$, where $\hat{d} = d$ if (and only if) d is a tree metric. Furthermore, $d' = d - \hat{d}$ is a metric such that $\beta_{\sigma'}(d') \leq 0$ for every split σ'. Note that we have obtained a decomposition of d in the form of

$$d = \sum_{\sigma \in \mathcal{B}(d)} \beta_\sigma(d) \Delta_\sigma + d'. \tag{48}$$

11.7.3 Discrete Convex Approach to Buneman Construction

The decomposition (48) of a metric d can be derived from a general decomposition method for polyhedral convex functions applied to the positively homogeneous M-convex function f that corresponds to d as in Theorem 7.1. The decomposition method for polyhedral convex functions, called *polyhedral split decomposition*, is as follows.

For a hyperplane H lying in $\mathbb{R}^n$ and a point $x \in \mathbb{R}^n$ let $l_H(x)$ denote half the distance between x and H. That is, $l_H(x) = |\langle a, x \rangle - b|/2$ if H is represented as $\langle a, x \rangle = b$, where $a \in \mathbb{R}^n$, $b \in \mathbb{R}$ with $\|a\|_2 = 1$. This function $l_H : \mathbb{R}^n \to \mathbb{R}$ is called the *split function* associated with H.

For a polyhedral convex function f and a hyperplane H let $c_H(f)$ be the maximum value of $t \in \mathbb{R}$ such that $f - t l_H$ is convex, i.e.,

$$c_H(f) = \sup\{t \in \mathbb{R} \mid f - t l_H \text{ is convex}\}.$$

With the notation

$$\mathcal{H}(f) = \{H : \text{hyperplane} \mid 0 < c_H(f) < +\infty\}$$

we obtain the following decomposition, called the *polyhedral split decomposition*.

Theorem 7.4 (Hirai 2006). *Any polyhedral convex function $f : \mathbb{R}^n \to \overline{\mathbb{R}}$ with* $\dim \mathrm{dom}_{\mathbb{R}} f = n$ *can be represented uniquely as*

$$f = \sum_{H \in \mathcal{H}(f)} c_H(f) l_H + f', \tag{49}$$

where f' is a polyhedral convex function such that $c_{H'}(f') \in \{0, +\infty\}$ for every hyperplane H'.

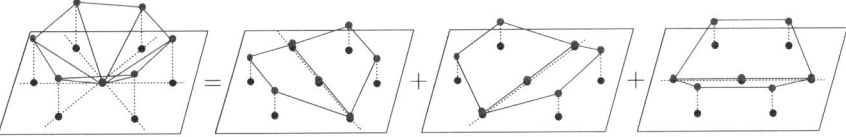

Fig. 11.11. Polyhedral split decomposition of the positively homogeneous M-convex function associated with a metric ($n = 3$)

Given a metric d we consider the polyhedral convex function f of (43) associated with d and apply the decomposition (49) to f with necessary modifications to adapt to the case of $\dim \mathrm{dom}_{\mathbb{R}} f = n - 1$; see Fig. 11.11, where $n = 3$. It turns out that each hyperplane in $\mathcal{H}(f)$ is represented as

$$H_\sigma = \{x \in \mathbb{R}^n \mid x(A) = x(B)\} \tag{50}$$

for a split $\sigma = \{A, B\}$. Moreover, the split function l_{H_σ} coincides essentially with the split metric Δ_σ in that

$$\Delta_\sigma(i, j) = \frac{1}{2}|x(A) - x(B)| = \sqrt{n}\, l_{H_\sigma}(x) \quad (x = \chi_j - \chi_i)$$

and the coefficient $c_{H_\sigma}(f)$ is given in terms of the Buneman index $\beta_\sigma(d)$ as

$$c_{H_\sigma}(f) = \sqrt{n}\, \max\{\beta_\sigma(d), 0\}.$$

Furthermore, the residual term f' turns out to be M-convex and it corresponds to a metric, which we denote as d'. Thus the decomposition (49) evaluated at $x = \chi_j - \chi_i$ ($i \neq j$) yields the decomposition (48) of d based on the Buneman index.

All the terms in the decomposition (49) for f associated with d are positively homogeneous M-convex functions. In other words, the sum of the positively homogeneous M-convex functions, l_{H_σ} and f', is another positively homogeneous M-convex function f. Compatibility of $\mathcal{B}(d)$ as a family of splits plays a crucial role here. Note that a sum of M-convex functions is not always M-convex.

11.7.4 Discrete Convex Approach to Split Decomposition

The split decomposition of Bandelt and Dress (1992) can also be derived through the polyhedral split decomposition.

For a metric d and a split $\sigma = \{A, B\}$ the *isolation index* is a real number defined as

$$\alpha_\sigma(d) = \frac{1}{2} \min_{i,j \in A,\ k,l \in B} \left\{ \max \left\{ \begin{array}{l} d(i, k) + d(j, l), \\ d(i, l) + d(j, k) \end{array} \right\} - d(i, j) - d(k, l) \right\}.$$

The *split decomposition* of d is defined as

$$d = \sum_{\sigma \in \mathcal{A}(d)} \alpha_\sigma(d)\Delta_\sigma + d'' \tag{51}$$

with $\mathcal{A}(d) = \{\sigma \mid \alpha_\sigma(d) > 0\}$. The "remainder term" d'' is a metric such that $\alpha_{\sigma'}(d'') \leq 0$ for every split σ', whereas the "main part" $\sum \alpha_\sigma(d) \Delta_\sigma$ admits a graphical representation (a generalization of tree representation).

Let f be the largest positively homogeneous convex function such that $f(\chi_i + \chi_j) = -d(i, j)$ for $i, j \in V$, which means, in particular, that $f(2\chi_i) = 0$ for $i \in V$. This function f is a polyhedral convex function, to which the decomposition (49) can be applied. It turns out that each hyperplane H_σ in $\mathcal{H}(f)$ appearing in this decomposition is represented as (50) for a split $\sigma = \{A, B\}$. Moreover, the split function l_{H_σ} coincides essentially with the split metric Δ_σ in that

$$\Delta_\sigma(i, j) = -\frac{1}{2}|x(A) - x(B)| + 1 = -\sqrt{n}\, l_{H_\sigma}(x) + 1 \quad (x = \chi_i + \chi_j)$$

and the coefficient $c_{H_\sigma}(f)$ is given in terms of the isolation index $\alpha_\sigma(d)$ as

$$c_{H_\sigma}(f) = \sqrt{n}\, \max\{\alpha_\sigma(d), 0\}.$$

Thus the polyhedral split decomposition (49) evaluated at $x = \chi_i + \chi_j$ $(i \neq j)$ yields the split decomposition (51) of d.

The reader is referred to Deza and Laurent (1997) and Semple and Steel (2003) for fundamental facts about metrics and phylogenetics, and to Dress et al. (1996) for a survey of T-theory. In particular the decomposition (48) based on the Buneman index is due to Buneman (1971). Discrete convex approach was initiated by Hirai (2006) for the split decomposition (51) of Bandelt and Dress (1992), whereas its application to the decomposition (48) based on the Buneman index is due to Koichi (2006).

11.8 Eigenvalue of Hermitian Matrices

An interesting connection exists between discrete concave functions in two variables and the range of eigenvalues of a sum of two Hermitian matrices with specified eigenvalues. For an $n \times n$ Hermitian matrix A we denote by $\lambda(A)$ the descending vector of eigenvalues of A, where a descending vector means a vector $\alpha = (\alpha_1, \ldots, \alpha_n)$ such that $\alpha_1 \geq \alpha_2 \geq \cdots \geq \alpha_n$.

Given two descending vectors $\alpha = (\alpha_1, \ldots, \alpha_n)$ and $\beta = (\beta_1, \ldots, \beta_n)$, we are concerned with the problem of determining the set

$$E(\alpha, \beta) = \{\gamma \in \mathbb{R}^n \mid \lambda(A) = \alpha, \lambda(B) = \beta, \lambda(A + B) = \gamma\},$$

which denotes the range of eigenvalues of $A + B$ when Hermitian matrices A and B vary subject to the constraint that $\lambda(A) = \alpha$ and $\lambda(B) = \beta$. This problem was first addressed by H. Weyl in 1912 and investigated intensively by A. Horn around 1960, who posed a conjecture that $E(\alpha, \beta)$ is a convex polyhedron described by the descending condition $\gamma_1 \geq \gamma_2 \geq \cdots \geq \gamma_n$, the trace condition $\sum_{k=1}^n \gamma_k = \sum_{i=1}^n \alpha_i + \sum_{j=1}^n \beta_j$ and a family of inequalities of the form

$$\sum_{k \in K} \gamma_k \leq \sum_{i \in I} \alpha_i + \sum_{j \in J} \beta_j,$$

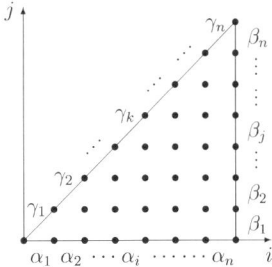

Fig. 11.12. Triangular region Δ

where (I, J, K) runs over a finite index set Π such that $|I| = |J| = |K|$ for (I, J, K) $\in \Pi$.

In the 1990's this problem received revived interest. With contributions by many researchers, in particular, by A. Klyachko, this problem has been settled in the affirmative. The range $E(\alpha, \beta)$ is now understood and described in terms of "puzzles" or "honeycombs." See (Danilov and Koshevoy 2003; Fulton 2000; Karzanov 2005; Klyachko 1998; Knutson and Tao 2001; Knutson et al. 2003) for details.

The connection to discrete concave functions is as follows. Consider an $L^\natural$-convex set

$$\Delta = \{(i, j) \in \mathbb{Z}^2 \mid 0 \le j \le i \le n\}$$

as depicted in Fig. 11.12. An $L^\natural$-concave function f on Δ determines three descending vectors α, β and γ from its boundary values as

$$
\begin{aligned}
\alpha_i &= f(i, 0) - f(i - 1, 0) & (i = 1, \ldots, n), \\
\beta_j &= f(n, j) - f(n, j - 1) & (j = 1, \ldots, n), \\
\gamma_k &= f(k, k) - f(k - 1, k - 1) & (k = 1, \ldots, n).
\end{aligned}
\tag{52}
$$

It then follows that $\sum_{k=1}^{n} \gamma_k = \sum_{i=1}^{n} \alpha_i + \sum_{j=1}^{n} \beta_j$. Conversely, given two descending vectors α and β, let $C(\alpha, \beta)$ be the set of γ such that (52) holds for some $L^\natural$-concave function $f : \Delta \to \mathbb{R}$, i.e.,

$$C(\alpha, \beta) = \{\gamma \in \mathbb{R}^n \mid \exists\, L^\natural\text{-concave } f : \Delta \to \mathbb{R} \text{ satisfying (52)}\}.$$

It is easy to see that $C(\alpha, \beta)$ is a polyhedral convex set, and moreover the following relationship is known.

Theorem 8.1 (Danilov and Koshevoy 2003). $E(\alpha, \beta) = C(\alpha, \beta)$.

A further problem has been posed by Danilov and Koshevoy (2003). Theorem 8.1 shows that for any (A, B) with $\lambda(A) = \alpha$ and $\lambda(B) = \beta$ there exists an $L^\natural$-concave function f that satisfies (52) for $\gamma = \lambda(A + B)$. How can we construct such f from (A, B)? It is conjectured in Danilov and Koshevoy (2003) that

$$f(i, j) = \max\{\text{tr}\,(AP + BQ) \mid \text{tr}\,P = i, \text{tr}\,Q = j, Q(I - P) = 0\},$$

where P and Q run over orthogonal projectors satisfying the specified conditions, is an $L^{\natural}$-concave function on Δ, and that every $L^{\natural}$-concave function on Δ can be represented in this form with some (A, B). Note that we have (52), since

$$f(i, 0) = \max\{\operatorname{tr}(AP) \mid \operatorname{tr} P = i\} = \alpha_1 + \cdots + \alpha_i,$$

$$f(n, j) = \max\{\operatorname{tr}(A + BQ) \mid \operatorname{tr} Q = j\} = (\alpha_1 + \cdots + \alpha_n) + \beta_1 + \cdots + \beta_j,$$

$$f(k, k) = \max\{\operatorname{tr}(AP + BQ) \mid \operatorname{tr} P = \operatorname{tr} Q = k, P = Q\} = \gamma_1 + \cdots + \gamma_k.$$

Some attempts have been made, but no answer has yet been obtained, as far as the present author knows.

11.9 Discrete Fixed Point Theorem

To motivate a discrete fixed point theorem we first take a glimpse at Kakutani's fixed point theorem. Then we explain how the conditions assumed in that theorem can be "discretized" to yield a discrete fixed point theorem.

Let S be a subset of $\mathbb{R}^n$ and F be a set-valued mapping (or a correspondence) from S to itself, which is denoted as $F : S \rightrightarrows S$ (or $F : S \to 2^S$). A point $x \in S$ satisfying $x \in F(x)$ is said to be a *fixed point* of F. Kakutani's fixed point theorem reads as follows.

Theorem 9.1. *A set-valued function $F : S \rightrightarrows S$ has a fixed point if*

(a) *S is a bounded closed convex subset of $\mathbb{R}^n$,*
(b) *For each $x \in S$, $F(x)$ is a nonempty closed convex set, and*
(c) *F is upper-semicontinuous.*

In a discrete fixed point theorem we are concerned with $F : S \rightrightarrows S$ where S is a subset of $\mathbb{Z}^n$. With reference to the three conditions in Theorem 9.1 above we proceed as follows to obtain a discrete fixed point theorem.

- Condition (a) assumes that the domain of definition S is nicely-shaped or well-behaved. In the discrete case we assume S to be "integrally convex."
- Condition (b) assumes that each value $F(x)$ is nicely-shaped or well-behaved. In the discrete case we assume that $F(x) = \overline{F(x)} \cap \mathbb{Z}^n$, where $\overline{F(x)}$ denotes the convex hull of $F(x)$.
- Condition (c) assumes that function F is continuous in some sense. In the discrete case we assume F to be "direction-preserving."

We will explain the key concepts, "integrally convex" and "direction-preserving", in turn.

The integral neighborhood of a point $y \in \mathbb{R}^n$ is defined to be

$$N(y) = \{z \in \mathbb{Z}^n \mid \|z - y\|_\infty < 1\},$$

where $\|\cdot\|_\infty$ means the maximum norm. A set $S \subseteq \mathbb{Z}^n$ is said to be *integrally convex* (Favati and Tardella 1990) if

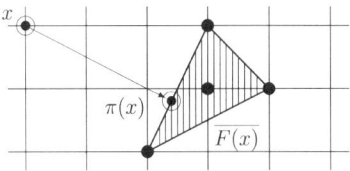

Fig. 11.13. Projection $\pi(x)$ with $\sigma(x) = \text{sign}(\pi(x) - x) = (+1, -1)$

$$y \in \bar{S} \implies y \in \overline{S \cap N(y)}$$

for any $y \in \mathbb{R}^n$. We have $S = \bar{S} \cap \mathbb{Z}^n$ for an integrally convex set S. It is known that $L^\natural$-convex sets and $M^\natural$-convex sets are integrally convex.

Given $F : S \rightarrow\rightarrow S$ and $x \in \mathbb{Z}^n$ we denote by $\pi(x)$ the projection of x to $\overline{F(x)}$. This means that $\pi(x)$ is the point of $\overline{F(x)}$ that is nearest to x with respect to the Euclidean norm (see Fig. 11.13). We also define the direction sign vector $\sigma(x) \in \{+1, 0, -1\}^n$ as

$$\sigma(x) = (\sigma_1(x), \ldots, \sigma_n(x)) = (\text{sign}(\pi_1(x) - x_1), \ldots, \text{sign}(\pi_n(x) - x_n)).$$

Then we say that F is *direction-preserving* if for all $x, z \in S$ with $\|x - z\|_\infty \leq 1$ we have

$$\sigma_i(x) > 0 \implies \sigma_i(z) \geq 0 \quad (i = 1, \ldots, n).$$

Note that this is equivalent to saying that $\sigma_i(x)\sigma_i(z) \neq -1$ for each $i = 1, \ldots, n$ if $x, z \in S$ and $\|x - z\|_\infty \leq 1$.

We are now ready to state the discrete fixed point theorem.

Theorem 9.2 (Iimura 2003; Iimura et al. 2005). *A set-valued function $F : S \rightarrow\rightarrow S$ has a fixed point if*

(a) *S is a nonempty finite integrally convex subset of $\mathbb{Z}^n$,*
(b) *For each $x \in S$, $F(x)$ is nonempty and $F(x) = \overline{F(x)} \cap \mathbb{Z}^n$, and*
(c) *F is direction-preserving.*

The proof of this theorem consists of three major steps.

1. We show that an integrally convex set S has a simplicial decomposition $\mathcal{T}$ such that for each $y \in \bar{S}$ all the vertices of $T(y)$ belong to $N(y)$, where $\bar{S}$ means the convex hull of S and $T(y)$ the smallest simplex in $\mathcal{T}$ that contains y.
2. We consider a piecewise linear extension $f : \bar{S} \to \bar{S}$ of π defined as

$$f(y) = \sum_{x \in V(y)} \lambda_x \pi(x) \quad \left(y = \sum_{x \in V(y)} \lambda_x x, \ \sum_{x \in V(y)} \lambda_x = 1, \ \lambda_x \geq 0 \right)$$

where $V(y) = T(y) \cap N(y)$. By Brouwer's fixed point theorem applied to f we obtain a fixed point $y \in \bar{S}$ of f (i.e., $y = f(y)$).

3. From the identity

$$\sum_{x \in V(y)} \lambda_x (\pi(x) - x) = \sum_{x \in V(y)} \lambda_x \pi(x) - \sum_{x \in V(y)} \lambda_x x = f(y) - y = \mathbf{0}$$

and the assumption of F being direction-preserving, we see that $\pi(x) - x = \mathbf{0}$ for some $x \in V(y)$, which is a fixed point of F.

The discrete fixed point theorem originates in Iimura (2003) with a subsequent rectification in Iimura et al. (2005). See Chen and Deng (2006) for a generalization and van der Laan et al. (2006) for an algorithm.

11.10 Stable Marriage and Assignment Game

Two-sided matching (Roth and Sotomayor 1990) affords a fairly general framework in game theory, including the stable matching of Gale and Shapley (1962) and the assignment model of Shapley and Shubik (1972) as special cases. An even more general framework has been proposed recently by Fujishige and Tamura (2007), in which the existence of an equilibrium is established on the basis of a novel duality-related property of $M^\natural$-concave functions.

Let P and Q be finite sets and put

$$E = P \times Q = \{(i, j) \mid i \in P, j \in Q\},$$

where we think of P as a set of workers and Q as a set of firms, respectively. We suppose that worker i works at firm j for x_{ij} units of time, gaining a salary s_{ij} per unit time. Then the *labor allocation* is represented by an integer vector

$$x = (x_{ij} \mid (i, j) \in E) \in \mathbb{Z}^E$$

and the salary by a real vector $s = (s_{ij} \mid (i, j) \in E) \in \mathbb{R}^E$. We are interested in the stability of a pair (x, s) in the sense to be made precise later.

For $i \in P$ and $j \in Q$ we put

$$E_{(i)} = \{i\} \times Q = \{(i, j) \mid j \in Q\}, \qquad E_{(j)} = P \times \{j\} = \{(i, j) \mid i \in P\},$$

and for a vector y on E we denote by $y_{(i)}$ and $y_{(j)}$ the restrictions of y to $E_{(i)}$ and $E_{(j)}$, respectively. For example, for the labor allocation x we obtain

$$x_{(i)} = (x_{ij} \mid j \in Q) \in \mathbb{Z}^{E_{(i)}}, \qquad x_{(j)} = (x_{ij} \mid i \in P) \in \mathbb{Z}^{E_{(j)}}$$

and this convention also applies to the salary vector s to yield $s_{(i)}$ and $s_{(j)}$.

It is supposed that for each $(i, j) \in E$ lower and upper bounds on the salary s_{ij} are given, denoted by $\underline{\pi}_{ij} \in \mathbb{R}$ and $\overline{\pi}_{ij} \in \overline{\mathbb{R}}$, where $\underline{\pi}_{ij} \leq \overline{\pi}_{ij}$. A salary s is called *feasible* if $\underline{\pi}_{ij} \leq s_{ij} \leq \overline{\pi}_{ij}$ for all $(i, j) \in E$. We put

$$\underline{\pi} = (\underline{\pi}_{ij} \mid (i, j) \in E) \in \underline{\mathbb{R}}^E, \qquad \overline{\pi} = (\overline{\pi}_{ij} \mid (i, j) \in E) \in \overline{\mathbb{R}}^E.$$

Each agent (worker or firm) $k \in P \cup Q$ evaluates his/her state $x_{(k)}$ of labor allocation in monetary terms through a function $f_k : \mathbb{Z}^{E(k)} \to \mathbb{R}$. Here the effective domain $\mathrm{dom}_\mathbb{Z} f_k = \{z \in \mathbb{Z}^{E(k)} \mid f_k(z) > -\infty\}$ is assumed to satisfy the following natural condition:

$$\mathrm{dom}_\mathbb{Z} f_k \text{ is bounded and hereditary, with unique minimal element } \mathbf{0}, \qquad (53)$$

where $\mathrm{dom}_\mathbb{Z} f_k$ being hereditary means that $\mathbf{0} \le z \le y \in \mathrm{dom}_\mathbb{Z} f_k$ implies $z \in \mathrm{dom}_\mathbb{Z} f_k$. In what follows we always assume that x is feasible in the sense that

$$\dot{x}_{(i)} \in \mathrm{dom}_\mathbb{Z} f_i \quad (i \in P), \qquad x_{(j)} \in \mathrm{dom}_\mathbb{Z} f_j \quad (j \in Q).$$

A pair (x, s) of feasible allocation x and feasible salary s is called an *outcome*.

Example 10.1. The *stable marriage problem* can be formulated as a special case of the present setting. Put $\underline{\pi} = \overline{\pi} = \mathbf{0}$ and define $f_i : \mathbb{Z}^{E(i)} \to \underline{\mathbb{R}}$ for $i \in P$ and $f_j : \mathbb{Z}^{E(j)} \to \mathbb{R}$ for $j \in Q$ as

$$f_i(y) = \begin{cases} a_{ij} & (y = \chi_j, j \in Q), \\ 0 & (y = \mathbf{0}), \\ -\infty & \text{(otherwise)}, \end{cases} \qquad f_j(z) = \begin{cases} b_{ij} & (z = \chi_i, i \in P), \\ 0 & (z = \mathbf{0}), \\ -\infty & \text{(otherwise)}, \end{cases} \qquad (54)$$

where the vector $(a_{ij} \mid j \in Q) \in \mathbb{R}^Q$ represents (or, is an encoding of) the preference of "man" $i \in P$ over "women" Q, and $(b_{ij} \mid i \in P) \in \mathbb{R}^P$ the preference of "woman" $j \in Q$ over "men" P. Then a matching X is stable if and only if $(x, s) = (\chi_X, \mathbf{0})$ is stable in the present model.

Example 10.2. The *assignment model* is a special case where $\underline{\pi} = (-\infty, \ldots, -\infty)$, $\overline{\pi} = (+\infty, \ldots, +\infty)$ and the functions f_i and f_j are of the form of (54) with some $a_{ij}, b_{ij} \in \mathbb{R}$ for all $i \in P, j \in Q$.

Given an outcome (x, s) the payoff of worker $i \in P$ is defined to be the sum of his/her evaluation of $x_{(i)}$ and the total income from firms:

$$f_i(x_{(i)}) + \sum_{j \in Q} s_{ij} x_{ij} \quad (=: (f_i + s_{(i)})(x_{(i)})).$$

Similarly, the payoff of firm $j \in Q$ is defined as

$$f_j(x_{(j)}) - \sum_{i \in P} s_{ij} x_{ij} \quad (=: (f_j - s_{(j)})(x_{(j)})).$$

Each agent ($i \in P$ or $j \in Q$) naturally wishes to maximize his/her payoff function.

A *market equilibrium* is defined as an outcome (x, s) that is stable under reasonable actions (i) by each worker i, (ii) by each firm j, and (iii) by each worker-firm pair (i, j). To be specific, we say that (x, s) is stable with respect to $i \in P$ if

$$(f_i + s_{(i)})(x_{(i)}) = \max\{(f_i + s_{(i)})(y) \mid y \le x_{(i)}\}. \qquad (55)$$

Similarly, (x, s) is said to be stable with respect to $j \in Q$ if

$$(f_j - s_{(j)})(x_{(j)}) = \max\{(f_j - s_{(j)})(z) \mid z \leq x_{(j)}\}. \tag{56}$$

In technical terms (x, s) is said to satisfy the *incentive constraint* if it satisfies (55) and (56).

The stability of (x, s) with respect to (i, j) is defined as follows. Suppose that worker i and firm j think of a change of their contract to a new salary $\alpha \in [\underline{\pi}_{ij}, \overline{\pi}_{ij}]_\mathbb{R}$ and a new working time of $\beta \in \mathbb{Z}_+$ units. Worker i will be happy with this contract if there exists $y \in \mathbb{Z}^{E_{(i)}}$ such that

$$y_j = \beta, \qquad y_k \leq x_{ik} \quad (k \in Q \setminus \{j\}), \tag{57}$$

$$(f_i + s_{(i)})(x_{(i)}) < (f_i + (s_{(i)}^{-j}, \alpha))(y), \tag{58}$$

where $(s_{(i)}^{-j}, \alpha)$ denotes the vector $s_{(i)}$ with its j-th component replaced by α. Note that y means the new labor allocation of worker i with an increased payoff given on the right-hand side of (58). Similarly, firm j is motivated to make the new contract if there exists $z \in \mathbb{Z}^{E_{(j)}}$ such that

$$z_i = \beta, \qquad z_k \leq x_{kj} \quad (k \in P \setminus \{i\}), \tag{59}$$

$$(f_j - s_{(j)})(x_{(j)}) < (f_j - (s_{(j)}^{-i}, \alpha))(z), \tag{60}$$

where $(s_{(j)}^{-i}, \alpha)$ is the vector $s_{(j)}$ with its i-th component replaced by α. Then we say that (x, s) is stable with respect to (i, j) if there exists no (α, β, y, z) that simultaneously satisfies (57), (58), (59) and (60).

We now define an outcome (x, s) to be *stable* if, for every $i \in P$, $j \in Q$, (x, s) is (i) stable with respect to i, (ii) stable with respect to j, and (iii) stable with respect to (i, j). This is our concept of market equilibrium.

A remarkable fact, Theorem 10.3 below, is that a market equilibrium exists if the functions f_k are M$^\natural$-concave. See (Murota 2003a, Sect. 11.3) for the relevance of M$^\natural$-concave functions for economic or game-theoretic problems; in particular, M$^\natural$-concave functions enjoy gross substitutes property, concave-extendibility. and submodularity. See (Topkis 1998) for the role of submodularity in this context.

Theorem 10.3 (Fujishige and Tamura 2007). *Assume that $\underline{\pi} \leq \overline{\pi}$ and, for each $k \in P \cup Q$, f_k is an M$^\natural$-concave function satisfying (53). Then a stable outcome $(x, s) \in \mathbb{Z}^E \times \mathbb{R}^E$ exists. Furthermore, we can take an integral $s \in \mathbb{Z}^E$ if $\underline{\pi} \in \underline{\mathbb{Z}}^E$, $\overline{\pi} \in \overline{\mathbb{Z}}^E$, and f_k is integer-valued for every $k \in P \cup Q$.*

The technical ingredients of the above theorem can be divided into the following two theorems. Note also that sufficiency part of Theorem 10.4 (which we need here) is independent of M$^\natural$-concavity.

Theorem 10.4 (Fujishige and Tamura 2007). *Under the same assumption as in Theorem 10.3 let x be a feasible allocation. Then (x, s) is a stable outcome for some s if and only if there exist $p \in \mathbb{R}^E$, $u = (u_{(i)} \mid i \in P) \in \overline{\mathbb{Z}}^E$ and $v = (v_{(j)} \mid j \in Q) \in \overline{\mathbb{Z}}^E$ such that*

$$x_{(i)} \in \mathrm{argmax}_{\mathbb{Z}}\{(f_i + p_{(i)})(y) \mid y \le u_{(i)}\}, \tag{61}$$

$$x_{(j)} \in \mathrm{argmax}_{\mathbb{Z}}\{(f_j - p_{(j)})(z) \mid z \le v_{(j)}\}, \tag{62}$$

$$\underline{\pi} \le p \le \overline{\pi}, \tag{63}$$

$$(i, j) \in E, u_{ij} < +\infty \implies p_{ij} = \underline{\pi}_{ij}, v_{ij} = +\infty, \tag{64}$$

$$(i, j) \in E, v_{ij} < +\infty \implies p_{ij} = \overline{\pi}_{ij}, u_{ij} = +\infty. \tag{65}$$

Moreover, (x, p) is a stable outcome for any (x, p, u, v) satisfying the above conditions.

Theorem 10.5 (Fujishige and Tamura 2007). *Under the same assumption as in Theorem 10.3 there exists (x, p, u, v) that satisfies (61)–(65). Furthermore, we can take an integral $p \in \mathbb{Z}^E$ if $\underline{\pi} \in \mathbb{Z}^E$, $\overline{\pi} \in \overline{\mathbb{Z}}^E$, and f_k is integer-valued for every $k \in P \cup Q$.*

It is worth while noting that the essence of Theorem 10.5 is an intersection-type theorem for a pair of $\mathrm{M}^\natural$-concave functions. Indeed it can be derived easily from Theorem 10.6 below applied to

$$f_P(x) = \sum_{i \in P} f_i(x_{(i)}), \qquad f_Q(x) = \sum_{j \in Q} f_j(x_{(j)}).$$

Theorem 10.6 (Fujishige and Tamura 2007). *Assume $\underline{\pi} \le \overline{\pi}$ for $\underline{\pi} \in \mathbb{R}^E$ and $\overline{\pi} \in \overline{\mathbb{R}}^E$, and let $f, g : \mathbb{Z}^E \to \mathbb{R}$ be $\mathrm{M}^\natural$-concave functions such that the effective domains are bounded and hereditary, with unique minimal element $\mathbf{0}$. Then there exist $x \in \mathrm{dom}_{\mathbb{Z}} f \cap \mathrm{dom}_{\mathbb{Z}} g$, $p \in \mathbb{R}^E$, $u \in \overline{\mathbb{Z}}^E$ and $v \in \overline{\mathbb{Z}}^E$ such that*

$$x \in \mathrm{argmax}_{\mathbb{Z}}\{(f + p)(y) \mid y \le u\},$$

$$x \in \mathrm{argmax}_{\mathbb{Z}}\{(g - p)(z) \mid z \le v\},$$

$$\underline{\pi} \le p \le \overline{\pi},$$

$$e \in E, u_e < +\infty \implies p_e = \underline{\pi}_e, v_e = +\infty,$$

$$e \in E, v_e < +\infty \implies p_e = \overline{\pi}_e, u_e = +\infty.$$

Furthermore, we can take an integral $p \in \mathbb{Z}^E$ if $\underline{\pi} \in \mathbb{Z}^E$, $\overline{\pi} \in \overline{\mathbb{Z}}^E$, and f and g are integer-valued.

The Fujishige–Tamura model contains recently proposed matching models such as (Eriksson and Karlander 2000; Fleiner 2001; Sotomayor 2002) special cases. In particular, the hybrid model of Eriksson and Karlander (2000), with flexible and rigid agents, is a special case where P and Q are partitioned as $P = P_\infty \cup P_0$ and $Q = Q_\infty \cup Q_0$, and $\underline{\pi}_{ij} = -\infty$, $\overline{\pi}_{ij} = +\infty$ for $(i, j) \in P_\infty \times Q_\infty$ and $\underline{\pi}_{ij} = \overline{\pi}_{ij} = 0$ for other (i, j). Fleiner's fixed point theorem approach (Fleiner 2003) seems to be independent of the Fujishige–Tamura model.

Concepts and results of discrete convex analysis are also useful for other problems of mathematical economics. For instance, Walrasian equilibria of indivisible markets are discussed in Murota (2003a, Chap. 11) and combinatorial auctions are treated in Lehmann et al. (2006). See Tamura (2004) for a survey.

Conclusion

Efficient algorithms are available for minimization of L-convex and M-convex functions (Murota 2003a, Chap. 10). The complexity analysis for the L-convex function minimization algorithm of Murota (2003b) is improved in Kolmogorov and Shioura (2007). As other recent papers we refer to Shioura (2003), Tamura (2005) for M-convex function minimization, and Iwata et al. (2005) for the submodular flow problem, or equivalently for the Fenchel duality. Most of the efficient algorithms employ scaling techniques based on proximity theorems; see (Iwata and Shigeno 2003; Moriguchi et al. 2002; Murota and Tamura 2004) for proximity theorems.

Discrete convex functions appear naturally in operations research. Multimodular functions, which are $L^\natural$-convex functions in disguise, are used in the analysis of queueing systems (or more generally, discrete event systems) (Altman et al. 2000, 2003; Hajek 1985; Murota 2005). In inventory theory Miller (1971) was a forerunner of discrete convexity in the 1970's and a recent paper of Zipkin (2008) sheds a new light on some classical results of Karlin, Scarf, and Morton.

A jump system (Bouchet and Cunningham 1995) is a generalization of a matroid, a delta-matroid and a base polyhedron of an integral polymatroid (or a submodular system). The concept of M-convex functions can be extended to functions on constant-parity jump systems (Murota 2006). For $x, y \in \mathbb{Z}^n$ we call $s \in \mathbb{Z}^n$ an (x, y)-increment if $s = \chi_i$ for some $i \in \mathrm{supp}^+(y - x)$ or $s = -\chi_i$ for some $i \in \mathrm{supp}^-(y - x)$. We call $f : \mathbb{Z}^n \to \overline{\mathbb{R}}$ an M-convex function (on a constant-parity jump system) if it satisfies the following exchange property: For any $x, y \in \mathrm{dom}_\mathbb{Z} f$ and any (x, y)-increment s, there exists an $(x + s, y)$-increment t such that

$$f(x) + f(y) \geq f(x + s + t) + f(y - s - t).$$

It then follows that $\mathrm{dom}_\mathbb{Z} f$ is a constant-parity jump system. Theorem 2.7 can be extended and operations such as infimal convolution can be generalized. See (Kobayashi and Murota 2007; Kobayashi et al. 2007; Kobayashi and Takazawa 2007; Shioura and Tanaka 2007).

Acknowledgements

My research of discrete convex analysis was started during my stay at Forschungsinstitut für Diskrete Mathematik, Universität Bonn, 1994–1995. In fact, the papers at the earlier stage were published as technical reports of the institute: (Murota 1996a) as No. 95837-OR (January 1995), (Murota 1996b) as No. 95838-OR (January 1995), (Murota 1998a) as No. 95839-OR (January 1995), (Murota 1999) as No. 95843-OR (March 1995), (Murota 1996c) as No. 95848-OR (June 1995). On this occasion I would like to express my deep gratitude to Professor Bernhard Korte for providing me with comfortable working environment.

I am indebted to Satoru Fujishige, Satoru Iwata, Shungo Koichi, Satoko Moriguchi, Akiyoshi Shioura, and Akihisa Tamura for helpful comments.

References

Ageev, A., Sviridenko, M.: Pipage rounding: A new method of constructing algorithms with proven performance guarantee. J. Comb. Optim. **8**, 307–328 (2004)

Altman, E., Gaujal, B., Hordijk, A.: Multimodularity, convexity, and optimization properties. Math. Oper. Res. **25**, 324–347 (2000)

Altman, E., Gaujal, B., Hordijk, A.: Discrete-Event Control of Stochastic Networks: Multimodularity and Regularity. Lecture Notes in Mathematics, vol. 1829. Springer, Heidelberg (2003)

Bandelt, H.-J., Dress, A.W.M.: A canonical decomposition theory for metrics on a finite set. Adv. Math. **92**, 47–105 (1992)

Bouchet, A., Cunningham, W.H.: Delta-matroids, jump systems, and bisubmodular polyhedra. SIAM J. Discrete Math. **8**, 17–32 (1995)

Buneman, P.: The recovery of trees from measures of dissimilarity. In: Hodson, R.F., Kendall, D.G., Tautu, P. (eds.) Mathematics in the Archaeological and Historical Sciences, pp. 387–395. Edinburgh University Press, Edinburgh (1971)

Calinescu, G., Chekuri, C., Pál, M., Vondrák, J.: Maximizing a submodular set function subject to a matroid constraint (extended abstract). In: Fischetti, M., Williamson, D.P. (eds.) Integer Programming and Combinatorial Optimization. Lecture Notes in Computer Science, vol. 4513, pp. 182–196. Springer, Berlin (2007)

Chen, X., Deng, X.: A simplicial approach for discrete fixed point theorems. In: Chen, D.Z., Lee, D.T. (eds.) Computing and Combinatorics. Lecture Notes in Computer Science, vol. 4112, pp. 3–12. Springer, Berlin (2006)

Conforti, M., Cornuéjols, G.: Submodular set functions, matroids and the greedy algorithm: tight worst-case bounds and some generalizations of the Rado–Edmonds theorem. Discrete Appl. Math. **7**, 251–274 (1984)

Cook, W.J., Cunningham, W.H., Pulleyblank, W.R., Schrijver, A.: Combinatorial Optimization. Wiley, New York (1998)

Danilov, V.I., Koshevoy, G.A.: Discrete convexity and Hermitian matrices. Proc. Steklov Inst. Math. **241**, 58–78 (2003)

Deza, M.M., Laurent, M.: Geometry of Cuts and Metrics. Springer, Berlin (1997)

Dress, A.W.M., Wenzel, W.: Valuated matroid: A new look at the greedy algorithm. Appl. Math. Lett. **3**, 33–35 (1990)

Dress, A.W.M., Wenzel, W.: Valuated matroids. Adv. Math. **93**, 214–250 (1992)

Dress, A.W.M., Moulton, V., Terhalle, W.: T-theory: an overview. Eur. J. Comb. **17**, 161–175 (1996)

Edmonds, J.: Submodular functions, matroids and certain polyhedra. In: Guy, R., Hanani, H., Sauer, N., Schönheim, J. (eds.) Combinatorial Structures and Their Applications, pp. 69–87. Gordon and Breach, New York (1970). Also in: Jünger, M., Reinelt, G., Rinaldi, G. (eds.) Combinatorial Optimization—Eureka, You Shrink! Lecture Notes in Computer Science, vol. 2570, pp. 11–26. Springer, Berlin (2003)

Eriksson, K., Karlander, J.: Stable matching in a common generalization of the marriage and assignment models. Discrete Math. **217**, 135–156 (2000)

Favati, P., Tardella, F.: Convexity in nonlinear integer programming. Ric. Oper. **53**, 3–44 (1990)

Fisher, M.L., Nemhauser, G.L., Wolsey, L.A.: An analysis of approximations for maximizing submodular set functions II. Math. Program. Study **8**, 73–87 (1978)

Fleiner, T.: A matroid generalization of the stable matching polytope. In: Gerards, B., Aardal, K. (eds.) Integer Programming and Combinatorial Optimization. Lecture Notes in Computer Science, vol. 2081, pp. 105–114. Springer, Berlin (2001)

Fleiner, T.: A fixed point approach to stable matchings and some applications. Math. Oper. Res. **28**, 103–126 (2003)

Frank, A.: A weighted matroid intersection algorithm. J. Algorithms **2**, 328–336 (1981)

Frank, A.: An algorithm for submodular functions on graphs. Ann. Discrete Math. **16**, 97–120 (1982)

Fujishige, S.: Theory of submodular programs: A Fenchel-type min-max theorem and subgradients of submodular functions. Math. Program. **29**, 142–155 (1984)

Fujishige, S.: Submodular Functions and Optimization, 2nd edn. Annals of Discrete Mathematics, vol. 58. Elsevier, Amsterdam (2005)

Fujishige, S., Murota, K.: Notes on L-/M-convex functions and the separation theorems. Math. Program. **88**, 129–146 (2000)

Fujishige, S., Tamura, A.: A two-sided discrete-concave market with possibly bounded side payments: An approach by discrete convex analysis. Math. Oper. Res. **32**, 136–155 (2007)

Fukushima, M., Oshima, Y., Takeda, M.: Dirichlet Forms and Symmetric Markov Processes. Walter de Gruyter, Berlin (1994)

Fulton, W.: Eigenvalues, invariant factors, highest weights, and Schubert calculus. Bull., New Ser., Am. Math. Soc. **37**, 209–249 (2000)

Gale, D., Shapley, L.S.: College admissions and stability of marriage. Am. Math. Mon. **69**, 9–15 (1962)

Hajek, B.: Extremal splittings of point processes. Math. Oper. Res. **10**, 543–556 (1985)

Hirai, H.: A geometric study of the split decomposition. Discrete Comput. Geom. **36**, 331–361 (2006)

Hirai, H., Murota, K.: M-convex functions and tree metrics. Jpn. J. Ind. Appl. Math. **21**, 391–403 (2004)

Iimura, T.: A discrete fixed point theorem and its applications. J. Math. Econ. **39**, 725–742 (2003)

Iimura, T., Murota, K., Tamura, A.: Discrete fixed point theorem reconsidered. J. Math. Econ. **41**, 1030–1036 (2005)

Iwata, S.: Submodular function minimization. Math. Program. Ser. B **112**, 45–64 (2007)

Iwata, S., Shigeno, M.: Conjugate scaling algorithm for Fenchel-type duality in discrete convex optimization. SIAM J. Optim. **13**, 204–211 (2003)

Iwata, S., Moriguchi, S., Murota, K.: A capacity scaling algorithm for M-convex submodular flow. Math. Program. **103**, 181–202 (2005)

Jensen, P.M., Korte, B.: Complexity of matroid property algorithms. SIAM J. Comput. **11**, 184–190 (1982)

Karzanov, A.V.: Concave cocirculations in a triangular grid. Linear Algebra Appl. **400**, 67–89 (2005)

Klyachko, A.A.: Stable bundles, representation theory and Hermitian operators. Sel. Math. **4**, 419–445 (1998)

Knutson, A., Tao, T.: Honeycombs and sums of Hermitian matrices. Not. Am. Math. Soc. **48**, 175–186 (2001)

Knutson, A., Tao, T., Woodward, C.: The honeycomb model of $GL_n(\mathbf{C})$ tensor products II: Puzzles determine facets of the Littlewood–Richardson cone. J. Am. Math. Soc. **17**, 19–48 (2003)

Kobayashi, Y., Murota, K.: Induction of M-convex functions by linking systems. Discrete Appl. Math. **155**, 1471–1480 (2007)

Kobayashi, Y., Takazawa, K.: Even factors, jump systems, and discrete convexity. METR 2007-36, Department of Mathematical Informatics, University of Tokyo (June 2007). J. Comb. Theory, Ser. B, to appear

Kobayashi, Y., Murota, K., Tanaka, K.: Operations on M-convex functions on jump systems. SIAM J. Discrete Math. **21**, 107–129 (2007)

Koichi, S.: The Buneman index via polyhedral split decomposition. METR 2006-57, Department of Mathematical Informatics, University of Tokyo (November 2006)

Kolmogorov, V., Shioura, A.: New algorithms for the dual of the convex cost network flow problem with application to computer vision. Preprint (2007)

Korte, B., Vygen, J.: Combinatorial Optimization: Theory and Algorithms, 4th edn. Springer, Berlin (2008)

Lehmann, B., Lehmann, D., Nisan, N.: Combinatorial auctions with decreasing marginal utilities. Games Econ. Behav. **55**, 270–296 (2006)

Lovász, L.: Matroid matching and some applications. J. Comb. Theory, Ser. B **28**, 208–236 (1980)

Lovász, L.: Submodular functions and convexity. In: Bachem, A., Grötschel, M., Korte, B. (eds.) Mathematical Programming—The State of the Art, pp. 235–257. Springer, Berlin (1983)

Miller, B.L.: On minimizing nonseparable functions defined on the integers with an inventory application. SIAM J. Appl. Math. **21**, 166–185 (1971)

Moriguchi, S., Murota, K., Shioura, A.: Scaling algorithms for M-convex function minimization. IEICE Trans. Fundam. Electron. Commun. Comput. Sci. **85-A**, 922–929 (2002)

Murota, K.: Valuated matroid intersection, I: optimality criteria. SIAM J. Discrete Math. **9**, 545–561 (1996a)

Murota, K.: Valuated matroid intersection, II: algorithms. SIAM J. Discrete Math. **9**, 562–576 (1996b)

Murota, K.: Convexity and Steinitz's exchange property. Adv. Math. **124**, 272–311 (1996c)

Murota, K.: Fenchel-type duality for matroid valuations. Math. Program. **82**, 357–375 (1998a)

Murota, K.: Discrete convex analysis. Math. Program. **83**, 313–371 (1998b)

Murota, K.: Submodular flow problem with a nonseparable cost function. Combinatorica **19**, 87–109 (1999)

Murota, K.: Matrices and Matroids for Systems Analysis. Springer, Berlin (2000)

Murota, K.: Discrete Convex Analysis—An Introduction. Kyoritsu Publishing Co., Tokyo (2001) (in Japanese)

Murota, K.: Discrete Convex Analysis. SIAM Monographs on Discrete Mathematics and Applications, vol. 10. SIAM, Philadelphia (2003a)

Murota, K.: On steepest descent algorithms for discrete convex functions. SIAM J. Optim. **14**, 699–707 (2003b)

Murota, K.: Note on multimodularity and L-convexity. Math. Oper. Res. **30**, 658–661 (2005)

Murota, K.: M-convex functions on jump systems: A general framework for minsquare graph factor problem. SIAM J. Discrete Math. **20**, 213–226 (2006)

Murota, K., Shioura, A.: M-convex function on generalized polymatroid. Math. Oper. Res. **24**, 95–105 (1999)

Murota, K., Shioura, A.: Extension of M-convexity and L-convexity to polyhedral convex functions. Adv. Appl. Math. **25**, 352–427 (2000)

Murota, K., Shioura, A.: Conjugacy relationship between M-convex and L-convex functions in continuous variables. Math. Program. **101**, 415–433 (2004a)

Murota, K., Shioura, A.: Fundamental properties of M-convex and L-convex functions in continuous variables. IEICE Trans. Fundam. Electron. Commun. Comput. Sci. **87-A**, 1042–1052 (2004b)

Murota, K., Tamura, A.: On circuit valuation of matroids. Adv. Appl. Math. **26**, 192–225 (2001)

Murota, K., Tamura, A.: Proximity theorems of discrete convex functions. Math. Program. **99**, 539–562 (2004)

Narayanan, H.: Submodular Functions and Electrical Networks. Annals of Discrete Mathematics, vol. 54. North-Holland, Amsterdam (1997)

Nemhauser, G.L., Wolsey, L.A., Fisher, M.L.: An analysis of approximations for maximizing submodular set functions I. Math. Program. **14**, 265–294 (1978)

Oxley, J.G.: Matroid Theory. Oxford University Press, Oxford (1992)

Recski, A.: Matroid Theory and Its Applications in Electric Network Theory and in Statics. Springer, Berlin (1989)

Rockafellar, R.T.: Convex Analysis. Princeton University Press, Princeton (1970)

Roth, A.E., Sotomayor, M.A.O.: Two-Sided Matching—A Study in Game-Theoretic Modeling and Analysis. Cambridge University Press, Cambridge (1990)

Schrijver, A.: Combinatorial Optimization—Polyhedra and Efficiency. Springer, Heidelberg (2003)

Semple, C., Steel, M.: Phylogenetics. Oxford University Press, Oxford (2003)

Shapley, L.S., Shubik, M.: The assignment game I: The core. Int. J. Game Theory **1**, 111–130 (1972)

Shioura, A.: Fast scaling algorithms for M-convex function minimization with application to the resource allocation problem. Discrete Appl. Math. **134**, 303–316 (2003)

Shioura, A.: On the pipage rounding algorithm for submodular function maximization: A view from discrete convex analysis. METR 2008-03, Department of Mathematical Informatics, University of Tokyo (January 2008)

Shioura, A., Tanaka, K.: Polynomial-time algorithms for linear and convex optimization on jump systems. SIAM J. Discrete Math. **21**, 504–522 (2007)

Sotomayor, M.: A labor market with heterogeneous firms and workers. International J. Game Theory **31**, 269–283 (2002)

Speyer, D.: Tropical linear spaces. arXiv:math.CO/0410455 (2004)

Speyer, D., Sturmfels, B.: The tropical Grassmannian. Adv. Geom. **4**, 389–411 (2004)

Sviridenko, M.: A note on maximizing a submodular set function subject to a knapsack constraint. Oper. Res. Lett. **32**, 41–43 (2004)

Tamura, A.: Applications of discrete convex analysis to mathematical economics. Publ. Res. Inst. Math. Sci. **40**, 1015–1037 (2004)

Tamura, A.: Coordinatewise domain scaling algorithm for M-convex function minimization. Math. Program. **102**, 339–354 (2005)

Topkis, D.M.: Supermodularity and Complementarity. Princeton University Press, Princeton (1998)

van der Laan, G., Talman, D., Yang, Z.: Solving discrete zero point problems. Math. Program. **108**, 127–134 (2006)

Vondrák, J.: Optimal approximation for the submodular welfare problem in the value oracle model. In: 40th ACM Symposium on Theory of Computing (May 2008)

White, N. (ed.): Theory of Matroids. Cambridge University Press, London (1986)

Zipkin, P.: On the structure of lost-sales inventory models. Oper. Res. (2008). doi:10.1287/opre.1070.0482

Multiflow Feasibility: An Annotated Tableau

Guyslain Naves and András Sebő

Summary. We provide a tableau of 189 entries and some annotations presenting the computational complexity of integer multiflow feasibility problems; 21 entries remain open. The tableau is followed by an introduction to the field, providing more problems, reproving some results with new insights, simple proofs, or slight sharpenings motivated by the tableau, paying particular attention to planar (di)graphs with terminals on the boundary. Last, the key-theorems and key-problems of the tableau are listed.

12.1 Introduction

Finding a set of (vertex- or edge-) disjoint paths in (directed or undirected) graphs between given pairs of terminals is one of the most ancestral and most studied themes of graph theory, with important applications such as routing problems of VLSI design (Korte et al. 1990). The scope of the methods and objectives is large and spread in time: Menger's theorems or more generally network flows are among the first consistent results of combinatorial optimization (Schrijver 2003), whereas finding edge- or vertex-disjoint paths between a given (fixed) number of terminal pairs in polynomial time is a deep pure graph theory result (Robertson and Seymour 1995). A multiflow is the packing of one of the simplest objects in graphs: paths. At the same time it is an integer point in a naturally defined polyhedral cone. The field has been developed in parallel with the tools of optimization, polyhedral combinatorics and graph theory. Some branches were and are still the subject of extensive studies both by the inner stimulus of the theory and the request of the applications.

In honor of Bernhard Korte's 70-th birthday and in memory of the determining impact of the Institut für Diskrete Mathematik, Ökonometrie und Operations Research and its successors on research in the field of Combinatorial Optimization, and in particular on results that have been achieved in the subject of routing, VLSI design, or simply, disjoint paths problems. The second author learnt the subject, proved and wrote down his first results in the subject during his stays in Bonn, had the opportunity to work with students having an excellent training by Professor Korte, and acknowledges gratitude to him for his multiple, generous contribution.
Supported by the Marie Curie Training Network "ADONET" of the European community.

Table 12.1.

G	|E(H)|	r	directed			directed acyclic			undirected		
			arc-disjoint		vertex-	arc-disjoint		vertex-	edge-disjoint		vertex-
			gen	Euler	disjoint	gen	Euler	disjoint	gen	Euler	disjoint
gen	arb	bin	NPC	NPC	NPC	NPC	NPC	NPC	NPC	NPC	NPC^4
		un	NPC	NPC		NPC	NPC		NPC	NPC	
	fix	bin	NPC	NPC	NPC	NPC	NPC	P^2	NPC	NPC	P^{12}
		un	NPC	NPC		NPC	NPC^{19}		NPC	NPC	
		fix	NPC	???^{3}		P	P		P	P	
	2	bin	NPC	P^{10}	NPC^2	NPC	P	P	NPC	P^{13}	P
		un	NPC	P		NPC^1	P		NPC^1	P	
		fix	NPC	P		P	P		P	P	
		2	NPC^2	P		P	P		P	P	
plan	arb	bin	NPC	NPC	NPC	NPC	NPC	NPC	NPC	NPC	NPC^7
		un	NPC	NPC		NPC^{19}	NPC^{20}		NPC^5	NPC	
	fix	bin	NPC	???	P^{14}	NPC	???	P	NPC	???	P^{15}
		un	NPC	???		NPC^{16}	???		NPC^{16}	???	
		fix	???	???		P	P		P	P	
	2	bin	NPC	P	P	NPC	P	P	NPC	P	P
		un	NPC^9	P		NPC^{11}	P		NPC^{11}	P	
		fix	???	P		P	P		P	P	
		2	???	P		P	P		P	P	
G+H plan	arb	bin	NPC	???	NPC	P^6	P	NPC^{19}	NPC	P^{18}	NPC^8
		un	NPC	???		P	P		NPC^8	P	
	fix	bin	NPC	???	P	P	P	P	P^{17}	P	P
		un	NPC	???		P	P		P	P	
		fix	???	???		P	P		P	P	
	2	bin	NPC	P	P	P	P	P	P	P	P
		un	NPC^{11}	P		P	P		P	P	
		fix	???	P		P	P		P	P	
		2	???	P		P	P		P	P	

[1] Even et al. (1976).

[2] Fortune et al. (1980). Moreover G acyclic and $|E(H)|$ fixed implies polynomiality.

[3] Polynomial for 3 demand edges (Ibaraki and Poljak 1991).

[4] Karp (1975).

[5] Kramer and Van Leeuwen (1984). See Sect. 12.4.1.

[6] Lucchesi and Younger (1978).

[7] Lynch (1975).

[8] Middendorf and Pfeiffer (1993). Even if maximum degree is 3.

[9] Müller (2006).

[10] Frank (1989), see also Nash-Williams' (2008) proof of Hu's theorem.

[11] Naves (2008).

[12] Robertson and Seymour (1995).

Nevertheless, while the variety of the possibilities is endless, some interesting questions may not even have been realized. It is even more frustrating that at the borderlines of existing theories there are forgotten problems that have no reason to be missing. The idea of making this tableau arose when the authors got confused in varying the defining parameters of problems: which are the combinations of the parameters that lead to polynomial solvable, NP-hard or unsolved problems. A careful focus on these showed that some of the interesting combinations have not yet been studied at all.

For some kind of disjoint paths problems there exist classifications, for instance in book of Schrijver (2003) or that of Korte and Vygen (2000), or in survey papers of the collection (Korte et al. 1990), like (Frank 1990).

A (integer) multiflow—first informally—is just a multiset of paths satisfying request and capacity constraints. The difference is not essential comparing to disjoint paths problems as far as assertions about them are concerned, however, there may be a difference in the algorithmic point of view: in multiflow problems there are numbers associated to edges or vertices, and in a solution—called a multiflow— a multiplicity is given with every path, and we want the algorithms to deal with the multiplicities in a clever way. From this viewpoint multiflows are points of a cone.

In this note we wish to focus merely on the *existence* of multiflows with particular attention to different natural special cases involving planarity, the number of demands, the way the capacities are given, and Eulericity. We restrict ourselves to *feasibility*, that is the existence of disjoint paths between *all* pairs of given terminal pairs. Another important direction is multiflow maximization (or maximum number of disjoint paths) that we do not treat here, since we would then have to cover yet other vast theories handled by quite different methods, and where approximation algorithms and APX-completeness should also be accounted. Exact methods concerning this subject, such as Mader's theorem are treated in the above mentioned books, and some other aspects like approximability are surveyed in the work of Bentz et al. (2005) and in the thesis work of Bentz (2006).

There are also many derivates of the problem. We had to be selective for keeping enough attention for the problems that occur in the most basic circle in the focus of our magnifying lens.

The main "product" of our work is the tableau above (Table 12.1). In the tableau we tried to concentrate on a small number of natural row (column) heads that can

---

[13] Rothschild and Whinston (1966). Polynomiality and the sufficiency of the cut condition extends to H being two stars, K_4 or C_5 (Lomonosov, 1985).

[14] Schrijver (1994).

[15] Schrijver (1990b). The condition is: the number of faces of demand is fixed.

[16] Schwärzler (2008). See Theorem 4.5 in Sect. 12.4.1.

[17] Sebő (1993).

[18] Seymour (1981). Also polynomial in graphs with no K_5 minor.

[19] Vygen (1995).

[20] See Theorem 4.8 in Sect. 12.4.2.

[21] Marx (2004).

be nontrivially matched by most columns (rows) so as to cover most of the relevant problems. More problems (like the Okamura–Seymour circle of questions) will be discussed in the text without charging the tableau.

In Sect. 12.2, we provide the first explanations concerning the tableau, and the most important notations. Section 12.3 is a short introduction to the basic methods concerning multiflows.

The traces of the unsolved problems of the tableau lead to the particular graphs treated in details in Sect. 12.4: planar (di)graphs. The undirected planar case (Sect. 12.4.1) seems to be almost the same as the acyclic planar case (Sect. 12.4.2), the arguments for one can be repeated for the other, but we do not see any formal reduction between the two sets of instances.

When we started our work, more than one third of these problems were open. While we were working, two fundamental problems have been solved, one of them stimulated by this tableau. Schwärzler (2008) result started the row, solving Problem 56 in Schrijver (2003): disjoint paths in planar graphs when all terminal pairs are on the boundary of the infinite face. This proof opened new hopes of reaching longstanding open problems and simplifying complicated proofs.

In Schwärzler's proof there are three natural classes of pairs of terminals, so it is not difficult to prove NP-hardness if the number of terminals is restricted to 3, and we will show the reduction below. With essential new ideas the first author has then shown (Naves 2008) that 3 can be decreased to 2, thus filling in new squares of the tableau, and solving a problem of Müller (2006) about planar graphs in general, and replacing Müller's quite involved proof for the directed version. We hope the tableau will provide similar stimulation for the 21 still unfilled squares.

To make this guided tour more pleasant, we occasionally provide some new viewpoints or variants of results, simple proofs or remarks on the way.

12.2 Basic Notation and Annotation

We hope the tableau is making clear the limits of different complexity behaviors (polynomiality and NP-completeness) and of the open cases. This also requires the realization of some connections. We introduce now the most important definitions, notations and conventions for a correct interpretation of the tableau.

Let $G = (V, E)$ be a graph, for the moment we allow G to be undirected or directed, and $n := |V|$. Let us call a function $c : E \to \mathbb{N}$ be a capacity function, and $H = (T, D), T \subseteq V$ a demand graph with a request function $r : D \to \mathbb{N}$. Then the *multiflow problem* is to find a multiset $\mathcal{C}$ of cycles in $G + H$ verifying the following condition:

–   for each cycle $C \in \mathcal{C}, |C \cap D| = 1$,
–   for each $d \in D$, there are exactly $r(d)$ cycles in $\mathcal{C}$ that contain d,
–   for each $e \in E$, there are at most $c(d)$ cycles in $\mathcal{C}$ that contain e.

The integrality of the multiplicities of cycles is supposed. In the rare cases when it is not, we will speak about *fractional multiflows*.

If r and c are both 1 everywhere we speak about *edge-disjoint* (or in digraphs *arc-disjoint*) paths problems.

By analogy, we could define, both in directed and undirected graphs, *vertex-capacitated multiflow problems*, *vertex-disjoint paths problems* by putting capacities and demands on vertices, and by repeating the three conditions above by replacing circuits C by their vertices, H simply by a vertex-set $D \subseteq V$, and E by $V \setminus D$.

If we still want to keep a demand graph H, we can, by putting a new vertex d_e in the middle of each edge (arc) $e = tu$ of H, and letting $r(d_e) := c(t) = c(u)$.

The choices for the rows and columns of the tableau are of course partly a matter of taste. However we tried to distinguish the different problems along some basic parameters that the community cares about:

The first three columns of the tableau concern

- restrictions of G and H: general, G planar or $G + H$ planar
- restrictions on the cardinality of $E(H)$: arbitrary, fix or 2
- restrictions on the size of r and the way it is given: "bin" means binary encoding, "un" unary encoding, that is, the size of the input is measured by the sum of the given numbers instead of the sum of their logarithms; "fix" means that $\sum_{e \in E(H)} r(e)$ is bounded.

Even though the restrictions never concern c directly, it is naturally affected: if r is unary, we can suppose without loss of generality that c is also. (The sum of c on all edges can be supposed to be at most n times the sum of r.)

The distinction between "bin" and "un" is the same as the usual distinction between pseudopolynomiality and strong NP-complete: for instance when H has two edges, unary encoding is equivalent to putting as many parallel copies of the edges in H as the demand, and similarly for the capacities; so the unary problem with $|E(H)| = 2$ is the same as the edge-disjoint paths problem with two parallel classes of demand edges, and is NP-complete. However, the "fix" version is polynomially solvable by Robertson and Seymour (1995).

The same holds for all edge-disjoint paths problems: multiflows with "unary" encoding are nothing more than edge-disjoint paths problems with maybe restricted H (like in the example) and several parallel demand edges.

The "bin" case could be essentially more difficult than the unary. Indeed, in a binary encoding we are not allowed to replace the capacities by parallel edges, since a polynomial algorithm must then work in time which is polynomial in the input size. In this case the input size is the sum of the *logarithms* of the capacities. Surprisingly, this does not drastically change the complexity of the problems: in our tableau the "bin" cases have exactly the same complexity as the "un" ones. An explanation of this lies probably in the classical Ford–Fulkerson theory of network flows: the paths through each demand edge obey the same rules as ordinary network flows, the difficult problem is to split the problem between the different terminal pairs.

Another kind of relation occurs between "fix" and "un" or "bin" if $G + H$ is planar and H has a bounded number of edges, that is, $|E(H)|$ is "fix": then r "fix" may again be settled by Robertson and Seymour (1995), but this does not solve the "un" or "bin" case, and turns out not to be the best solution for "fix" either. Indeed, "bin" can also be solved in polynomial time, by applying Lenstra's "cheaper" integer programming algorithm (Sebő 1993).

Thus in the edge-disjoint case "bin" and "un" can be thought of as being the same, and allowing an arbitrary number of parallel classes of demand edges; "fix" $|E(H)|$ restricts the number of parallel classes of demand edges, and "fix" in the r column the total number of demand edges. The latter of course implies the former.

The situation is somewhat more complicated in the vertex-disjoint case. For vertex-capacitated multiflows the unary case has to be distinguished from vertex-disjoint paths if G is restricted for instance to planar graphs. The replication of vertices (replacing the parallel edges of the reduction of multiflows to edge-disjoint paths), does not keep for instance the planarity of G.

Besides edge- or arc- and vertex-disjoint paths problems we also distinguish the same problems under the Eulerian condition:

We distinguish between $G + H$ $(r + c)$ Eulerian (Euler) or not (gen): if G, H are undirected, (G, H, r, c) is called *Eulerian* if for each $v \in V$,

$$\sum_{e \in \delta_G(v)} c(e) + \sum_{d \in \delta_H(v)} r(d) \text{ is even}$$

and if G, H are digraphs, then the Eulerian property means for each $v \in V$,

$$\sum_{e \in \delta_G^+(v)} c(e) + \sum_{d \in \delta_H^+(v)} r(d) = \sum_{e \in \delta_G^-(v)} c(e) + \sum_{d \in \delta_H^-(v)} r(d).$$

The four-tuple (G, H, r, c) will not necessarily be always explicitly mentioned—most of these parameters are fixed by the context.

In this paper the main focus is the edge-disjoint paths problem and multiflows. The columns concerning vertex-disjoint paths are present for comparison and all suppose that the request and capacity functions are both 1 everywhere, that is, we are looking only at vertex-disjoint paths problems, and none of the new problems that are raised:

Problem 2.1. Fill in additional columns of the tableau for vertex-capacitated problems, where the vertex requests and capacities are not supposed to be 1, but are encoded with a unary or binary encoding.

Note that the unary case cannot always be reduced to the vertex-disjoint paths problem in the same class of graphs.

Paths and cycles will always be simple, and the terms are used both in directed and undirected graphs. Our notations will be usual; $\delta(X)$ $(X \subseteq V)$ denotes the set of edges with exactly one endpoint if X, and X, $V \setminus X$ are the shores of this cut.

In several cases we will also have particular notes for the case when H has three edges. Another particular case of H is when it is a star: then the problem can be reduced to a flow problem, and thus the problems are polynomially solvable for any G.

12.3 Basic Facts

12.3.1 Well-Known Reductions

We recall some well-known reductions between the different cases that are fully exploited in the tableau.

The undirected case can be reduced to the directed one by replacing each edge by the gadget depicted in Fig. 12.1. Note that this reduction preserves the planarity of G and $G + H$, but does not preserve the Euler property, and the resulting graph is not acyclic.

The edge-disjoint case is reducible to the vertex-disjoint one by taking the line-graph. This operation does not keep the planarity of G. It works in the directed case as well with the appropriate definition of the line graph (stars of vertices become complete bipartite graphs by joining all the vertices corresponding to incoming edges to all those corresponding to outgoing edges).

In the edge-disjoint case, it is possible to reduce every graph with max-degree greater than 4 to a graph with degrees at most 4, by using the gadget of Fig. 12.2, which also keeps planarity. (The capacities must be 1.) In the particular case when $G + H$ is planar, it was remarked in Middendorf and Pfeiffer (1993) that in the un-capacitated case the maximum degree can be restricted to 3, thus the edge-disjoint paths problem is reducible to the vertex-disjoint paths problem. This allows to con-firm the negative complexity of some vertex-disjoint paths problems but one has to proceed carefully, since $|E(H)|$ increases.

Fig. 12.1. The undirected case is reducible to the directed case, using this gadget. Only one path can use these five arcs, either from left to right or from right to left

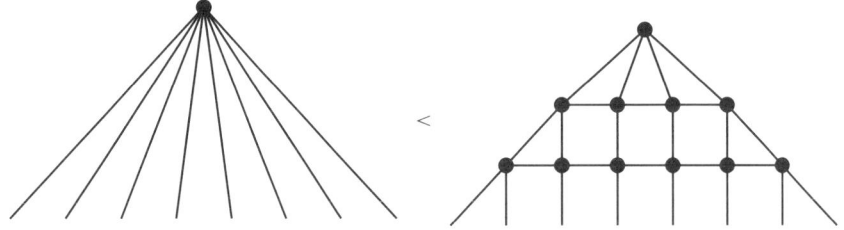

Fig. 12.2. In the edge-disjoint case, this gadget reduces the degrees of vertices to 4

The following lemma was proposed by Vygen (1995). It proves the equivalence between the acyclic arc-disjoint paths problem in Eulerian digraphs and the edge-disjoint paths problem (in Eulerian graphs).

Lemma 3.1 (Vygen 1995). *Let (G, H) be an instance of the arc-disjoint paths problem, assume $G + H$ is Eulerian and that G is acyclic. Let (G', H') be the instance of the edge-disjoint paths problem obtained by neglecting the orientation of G and H. Then there exists a solution for the arc-disjoint paths problem in (G, H) if and only if there exists a solution for the edge-disjoint paths problem (G', H').*

More exactly, it is proved that the solutions of these two problems can be transformed to one another by neglecting the orientation or conversely by orienting the edges depending on the orientation of G.

12.3.2 Conditions

A solution of the (fractional or integer) multiflow problem can be seen as the problem of deciding the existence of an (integer) point in a given particular polytope. Using an idea of Lomonosov (1985) we provide a compact formulation of a 'lifting' of this polytope, that is, we provide a polytope with a polynomial number of constraints in the input size whose projection is the multiflow polytope. The conditions for multiflow feasibility can be seen as valid inequalities for this polytope.

Let $G = (V, E)$, c, $H = (T, D)$, r be an instance of the multiflow problem. *Paths* and *cycles*, directed or undirected will always supposed to be simple, that is, contain each edge at most once. Let $\mathcal{C}$ be the set of the cycles of $G + H$ that contain exactly one edge from H. Then the solutions of the disjoint paths problem are in bijection with the integer solutions of the following linear program:

$$\sum_{d \in C, C \in \mathcal{C}} x_C = r(d) \quad (d \in D), \tag{1}$$

$$\sum_{e \in C, C \in \mathcal{C}} x_C \le c(e) \quad (e \in E), \tag{2}$$

$$x_C \ge 0 \quad\quad\quad (C \in \mathcal{C}). \tag{3}$$

Equations (1) and (2) define the (fractional) *multiflow polytope*. A multiflow is an integer point of this polytope.

The nonemptyness of this polytope can be characterized by Farkas's lemma:

Theorem 3.1 (Japanese theorem, Onaga and Kakusho 1971; Schrijver 2003). *The existence of a multiflow is equivalent to the distance criterion*

$$\text{For all } w : E \to \mathbb{R}_+, \quad \sum_{(u,v) \in D} r(u, v) \times d_{G,w}(u, v) \le \sum_{e \in E} w(e). \tag{4}$$

We can obtain an easy consequence by taking, for each cut C, the weight function $w : E \rightarrow \{0, 1\}$ defined by $w(e) = 1$ iff $e \in \delta_H(C)$. This gives the following condition called the cut condition:

$$\text{for all } C \subset V, \quad |\delta_G(C)| \geq |\delta_H(C)|. \tag{5}$$

A cut is called *tight* if equality holds in (5) for this cut. Another interesting necessary condition for the existence of integer multiflows when $G + H$ is not Eulerian is that the union of any number of tight cuts (as edge-sets) must not contain an odd cut. (The reason is that each edge of such an odd cut is used by a multiflow, that is, the disjoint circuits of a multiflow partition the cut. However, each class of this partition is even. Intersecting the shore of (any number of) cuts, if X is the intersection, $\delta(X)$ will be contained in the union of the cuts.)

Therefore if the intersected shores define all tight cuts, the intersection must define an even cut (if an integer multiflow exists). We do not know many papers where this is exploited; the nicest example is probably Frank (1990), which uses this condition for two tight cuts.

We show now that the condition (5) of the Japanese theorem can be handled as a linear program of polynomial size, and at the same time we show the polarity between metrics and multiflows.

A function $\mu : V \times V \rightarrow \mathbb{Z}_+$ is called a *metric* on V, if it satisfies the triangle inequality

$$\mu(x, y) + \mu(y, z) - \mu(x, z) \geq 0, \quad \text{for all } x, y, z \in V.$$

The integrality requirement is superfluous, we only suppose it for comfort. Let us denote $t(x, y, z) \in \{0, 1, -1\}^{V \times V}$ which takes the value 1 on (x, y) and (y, z), -1 on (x, z), and 0 on all the other ordered pairs. Denote T the matrix whose columns are the vectors of the form $t(x, y, z)$ for all ordered triples (x, y, z) $(x, y, z \in V)$. The metrics are then the solutions of the systems of inequalities $T \geq 0$. The following nice observation is due to Lomonosov (1985):

Let $P := (v_1, \ldots, v_k)$ be a path, and $v_P \in \{0, 1, -1\}^{V \times V}$, $v_P(x, y) = 1$ if $x = v_i, y = v_{i+1}$ for all $i = 1, \ldots, k - 1$, and $v_P(v_k v_1) = -1$. Then

$$v_P = \sum_{i=2}^{k-1} t(v_1, v_i, v_{i+1}),$$

and therefore for $c \in R^{V \times V}$, the solutions of the system of linear inequalities $Tx \leq c, x \geq 0$ are in one-to-one correspondence with the (fractional) multiflows in the graph $G = (V, \{e \in V \times V : c(e) > 0\})$, with capacity c, and demand graph H, $uv \in E(H)$ if and only if $c(uv) < 0$, and then $r(uv) := -c(uv)$. (For undirected multiflow problems we use only one of uv and vu.) Integer solutions of this system correspond to (integer) multiflows. Note that T has a polynomial number of entries, immediately implying polynomial solvability of fractional multiflow problems and the interested reader may find it useful to rewrite the Farkas' Lemma for this somewhat different system of inequalities.

12.4 Planar Graphs

In this section we state and sometimes improve or reprove results about the complexity of multiflows in planar graphs. The results concerning undirected graphs can often be translated to acyclic digraphs.

12.4.1 Undirected Graphs

This subsection updates the complexity of the planar edge-disjoint paths problem.

Two of the first important results of the subject are that of Lynch (1975) stating that the vertex-disjoint paths problem is NP-complete in planar graphs, and the sharpening of Kramer and Van Leeuwen (1984) to grid graphs. The latter result has the advantage of being easy to manipulate to prove NP-completeness of variants of the problem such as edge-disjoint paths problems: the authors themselves note that the problem remains NP-complete if common edges are still not allowed, but common vertices may occur provided the two paths "cross" in those. Raghavan (1986, Lemma 2.1) notes that the edge-disjoint problem is also NP-complete with "their reduction". This is right noting that the last part of Kramer and van Leeuwen's proof has to be—slightly and in a straightforward way—modified in order to get NP-completeness of the general planar edge-disjoint paths problem.

In planar routing problems the terminals are often on the boundary of the infinite face. We want to explore the complexity of problems satisfying this condition.

We start with a new proof of the classical Okamura–Seymour theorem providing a polynomial algorithm for planar Eulerian graphs with all terminals on the outer face. We continue by sketching Schwärzler's proof of the NP-completeness of the non-Eulerian case, and show a slight extension where in addition the number of demand edges can be restricted to three. Finally, we sketch the more involved new ideas that allowed the first author to achieve the last possible step and prove that the same holds for two demand edges.

Theorem 4.1 (Okamura and Seymour 1981). *Let $G = (V, E)$ be a planar graph and $H = (T, D)$, $T \subseteq V$ where the vertices of T are on the outer boundary of the embedding of G. Let $r : D \to \mathbb{N}$ and $c : E \to \mathbb{N}$ be weight functions, and suppose that $r + c$ is Eulerian. Then the cut condition is sufficient for the existence of a multiflow for (G, H).*

We first reformulate this theorem as a statement on metric packings, and provide a proof combining a technique of Schrijver for proving distance packing theorems (Schrijver 2003) with ideas in Lomonosov and Sebő (1993) for decomposing distance functions, and new ideas capturing the essence of Lins' theorem: in a critical situation saturated by a technique of Schrijver (2003), guided by the role of the "oppositeness relation" in Lomonosov and Sebő (1993)—but without using the related polyhedral statements—we decompose our graph into cuts. Schrijver applies the dual of this oppositeness relation to prove Lins' theorem in the context of an inductive proof. Despite these similarities, the use of the previous results remains implicit in the proof below, and our present proof is self-contained, fully combinatorial, and hopefully generalizable. Since it seems to provide some insight, we want to communicate it for possible future use.

The theorem is equivalent to a theorem on metric packings, see Corollary 74.2a in Schrijver (2003), proved there using the Okamura–Seymour theorem. Here we will prove this form directly. The advantage of this method may be to provide some insight of how the metrics guide the direction the (dual) paths take.

Let us call a circuit $C \subseteq E(G)$ *rigid*, if for any two, a and b of its vertices, one of the (a, b)-paths on the circuit is a geodesic in G. (The facial structure of the cone of metrics implies that the only way to write the distance function of a graph as the sum of metrics is using cuts intersecting rigid cycles with 0 or two opposite edges (Lomonosov and Sebő 1993). This statement did guide our proof without using it.)

We prove the following reformulation of the Okamura–Seymour theorem:

Theorem 4.2. *Let* $G = (V, E)$ *be a planar graph with only rigid faces, with all faces being 4-cycles except the infinite face, and where in addition any set of two successive edges of a face are together contained in a geodesic with both endpoints in the boundary* C *of the infinite face. Then the graph* (E, Ω) *on the edges of* G, *where* $\Omega := \{ef : e, f \in E, e \text{ is opposite to } f \text{ on some face}\}$ *is a graph that has* $|C|/2$ *components, where each component is a path joining two opposite edges of* C.

The conditions imply, of course, that G is bipartite. Before the proof let us sketch the reduction of the Okamura–Seymour theorem (Theorem 4.1) to this, which consists of simple and standard steps.

1. Reduce the Okamura–Seymour theorem to the case when the terminal pairs $D := \{s_1 t_1, \ldots, s_k t_k\}$ follow one another in the order $s_1, \ldots, s_k, t_1, \ldots, t_k$ on C, see Fig. 12.3. Reduce then to the 2-vertex-connected case without changing the order of the terminals.
2. Add a new vertex x_0, place it to the infinite face and join it with all the terminal vertices. Delete each vertex of degree 2, by merging its two edges.
3. Take the planar dual of the obtained graph.
4. Add the gadget of Fig. 12.4 to all faces that are not 4-cycles, until all faces are 4-cycles.
5. Identify the opposite vertices of 4-cycles if they are not contained on a geodesic with both endpoints in C.

It is easy to see that the cut condition implies that after applying these procedures the conditions of Theorem 4.2 are satisfied. The theorem then implies by dualization a set of edge-disjoint paths for the original problem, and the proof is algorithmic, straightforwardly providing a polynomial algorithm.

If P is a path and x, y are two of its vertices, $P(x, y)$ denotes the subpath of P from x to y.

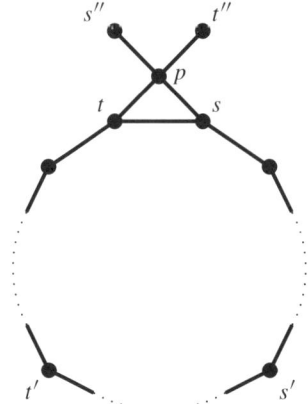

Fig. 12.3. In Okamura–Seymour's theorem, we can assume that terminals of each demand pair are diametrally opposed: add vertices p, s'' and t'' and the edges joining them as indicated on the figure, whenever s, t, t', s' occur in this "wrong" order along the boundary of the outer face. Replace the demand edges ss', tt' by $s's''$ and $t't''$

Fig. 12.4. Reducing the boundaries of faces to 4, without changing the distances, in poly-time

Proof.

Claim 1. The graph (E, Ω) is the disjoint union of cycles and of $|C|/2$ paths with both endpoints on C.

Indeed, in the graph (E, Ω) every edge of C is of degree 1, and any other $e \in E$ has degree 2.

Claim 2. For any cycle $D \subseteq E$ in G and $a, b \in V(D)$ such that both paths A and B between a and b on D are shortest paths in G, each component of (E, Ω) is a path that has one end in A, and another in B.

Indeed, if every edge of D incident to a is followed by a boundary edge of D, then D is a face of G, and the statement is evident. Otherwise there exist two edges e, f (Fig. 12.5) such that

(i) e and f are incident edges of a face—let their common point be p.
(ii) e is incident to a.
(iii) The interior of f is contained inside D, that is, in the open disk bounded by D.

By the condition there exists a geodesic path S containing e and f and with extremities on C. Then starting on S from a on e and then f and continuing, let q be the next

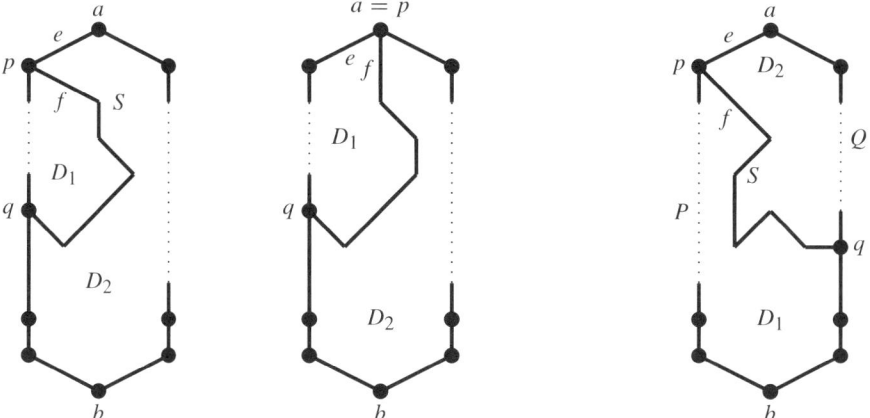

Fig. 12.5. Depending on the position of q, on the same (a, b)-path as p or not, we apply the induction hypothesis to D_1, p, q and D_2, a, b on the two first drawings (case 1), or D_1, p, b and D_2, a, q on the third one (case 2), for which the condition on the distances still holds

vertex of D (by (iii) and Jordan's theorem q exists) on S. The subpath $S(p, q)$ divides D into two cycles D_1 and D_2 that intersect in $S(p, q)$ (see Fig. 12.5). Since the subpath of a geodesic is also a geodesic, both $S(a, q)$ and $S(p, q)$ are geodesics.

Denote H, H_1, H_2 the subgraph of (E, Ω) induced by D, D_1, D_2 and the edges inside. Informally, these are the subgraphs describing oppositeness within D, D_1 or D_2. (In H the edges of $S(p, q)$ form a vertex-cut-set which, together with the two components of $H - S(p, q)$ induces the subgraphs H_1 and H_2.) Clearly, like in Claim 1, the components of H, H_1 and H_2 are paths, and there are respectively $|D|/2, |D_1|/2, |D_2|/2$ such paths.

Case 1: One of the two (a, b)-paths of D is disjoint from $S(p, q)$.

Then applying the induction hypothesis to D_1 and D_2 (see left drawings of Fig. 12.5) and merging the components of H_1 and H_2, we get the statement for D. (Each edge of $S(p, q)$ is the endpoint of a component in both of the graphs H_1 and H_2, and these pairs of components can be merged.)

Case 2: Both (a, b)-paths of D meet $S(p, q)$.

Then ap is the first edge of one of the two (a, b)-paths P of D, and q is on the other (a, b)-path Q (Fig. 12.5 right).

Both $S(a, q)$ and $Q(a, q)$ are geodesics, and the induction hypothesis can be applied to D_2 with these geodesics. By induction we get paths of (E, Ω) one of which, $S_{2,e}$ connects the edge $e = ap$ to an edge of $Q(a, q)$, and the others, $S_{2,h}$ ($h \in S(p, q)$) each connect edges of $S(p, q)$ to edges of $Q(a, q)$ (except the end of $S_{2,e}$ different from e). These are the components of H_2.

As a side-product $|S(a, q)| = |Q(a, q)|$, and therefore

$$|S(p, q) \cup Q(q, b)| = |Q(a, q)| - 1 + Q(q, b)| = |P(p, b)|,$$

whence $S(p, q) \cup Q(q, b)$ is also a geodesic and the induction hypothesis can be applied to D_1 as well, and with the geodesics $S(p, q) \cup Q(q, b)$ and $P(p, b)$. So by induction, the components of H_1 are the paths $S_{1,h}$ ($h \in S(p, q)$) connecting $S(p, q)$ to a subset of $P(p, b)$, and $S_{1,h}$ ($h \in Q(q, b)$) to another subset. Let $S := \{S_{1,h} \cup S_{2,h} : h \in S(p, q)\}$. Now clearly, the set

$$\{S_{2,e}\} \cup S \cup \{S_{1,h} : h \in Q(q, b)\}$$

is the set of components of H, and connects each edge of $P(a, b)$ to an edge of $Q(a, b)$, finishing the proof of Claim 2.

Now applying Claim 2 to $D := C$ and all the $|C|/2$ pairs of geodesics each of which (bi)partitions C, we get that each component of H connects two edges that do not lie in the same class of any of these bipartitions. It follows that the components of H join opposite edges of C. □

Frank proved that the problem is still polynomially solvable when only the inner vertices of G verify the Eulerian condition:

Theorem 4.3 (Frank). *Let* $G = (V, E)$ *be a planar graph and* $H = (T, D)$ *($T \subset V$), where the vertices of* T *are on the outer boundary of the embedding of* G; *let* $r : D \rightarrow \mathbb{N}$ *and* $c : E \rightarrow \mathbb{N}$ *be weight functions, and suppose that for each vertex* v *not contained in the outer boundary of* G, *$\sum_{e \in \delta(v)} c(e)$ is even. Then the edge-disjoint paths problem can be solved in polynomial time.*

However, the Euler property cannot be completely removed:

Theorem 4.4 (Schwärzler). *The edge-disjoint paths problem when* G *is planar and the terminals lie on the outer boundary of* G *is NP-complete.*

Schwärzler's gadget can be completed to reduce the number of demand edges to 3:

Theorem 4.5. *The multiflow problem when* G *is planar,* $|E(H)| = 3$ *and the terminals lie on the outer boundary of* G *is NP-complete.*

Proof (sketch). We sketch Schwärzler's proof, rearranged and completed by a reduction to three parallel classes of demand edges with a linear number of demands altogether. The reduction is from SATISFIABILITY. From a formula given in conjunctive normal form, a grid is built with as many columns and rows as there are clauses and variables respectively. There are two lines in each column and in each row, paths in the graph, but because of the placement of the terminals, these will not be paths in a solution, see Fig. 12.6. The extremities of the demand edges are labeled vertices and their primes.

In a solution two paths will join the two demand edges of each column, and one the demand edge of each row. The latter (horizontal) path of each row will be obliged to be one of the two horizontal lines of the row, and this choice corresponds to choosing a truth value for the corresponding variable: choosing the upper path means that TRUE is assigned to it, and the lower path means that FALSE is the assigned value.

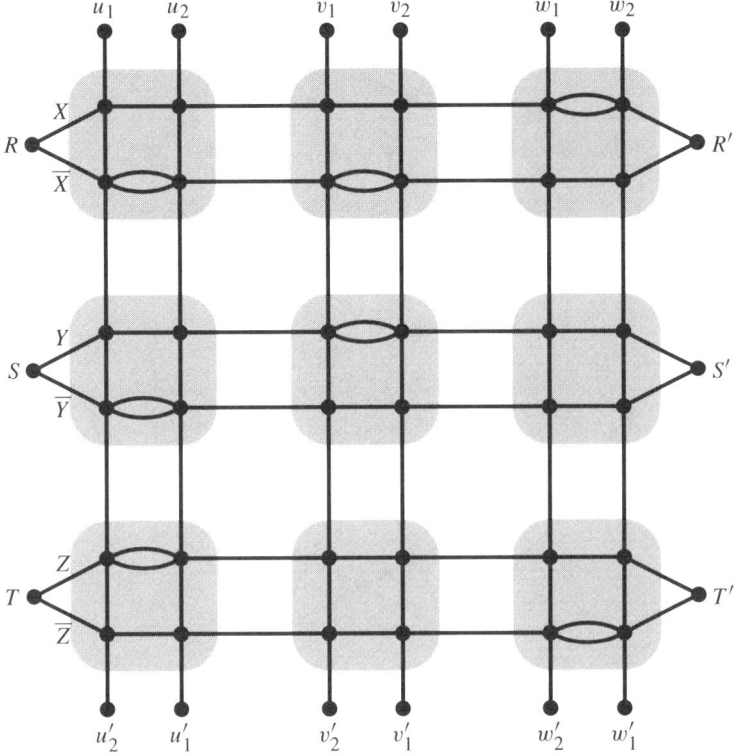

Fig. 12.6. An example of the reduction, from the formula $(X \vee Y \vee \neg Z) \wedge (X \vee \neg Y) \wedge (\neg X \vee Z)$. The drawn graph is the supply graph, the demand edges join vertices with their primes

The two paths of each column are forced by the order of their terminal vertices to exchange their lines. This exchange encodes the fact that each clause must be satisfied. Such an exchange is possible through two parallel horizontal "column-switch" edges. In each square of the grid the two parallel column-switch edges are placed in the upper or lower line or neither depending on whether a variable, its negation or neither are present in the corresponding clause.

By considering tight cuts, Schwärzler proves that the horizontal paths do not use any vertical edges. This is the way of forcing a horizontal path to stay in the same row and not to change lines, corresponding to a choice of truth value. Then, vertical paths can cross only through free column-switch parallel edges, making the choice of a true variable which is positive in the clause or a false one which is negated.

The number of parallel classes of demand edges can be reduced to 3 by introducing one parallel class for each "type" of demand edge: first introduce two new terminals and one demand edge for all the horizontal paths, this does not cause any difficulty; then construct the two parallel classes of demand edges for the vertical paths, one for paths switching from left to right, and one for those switching from right to left, with one demand edge per path, and a gadget making possible for these

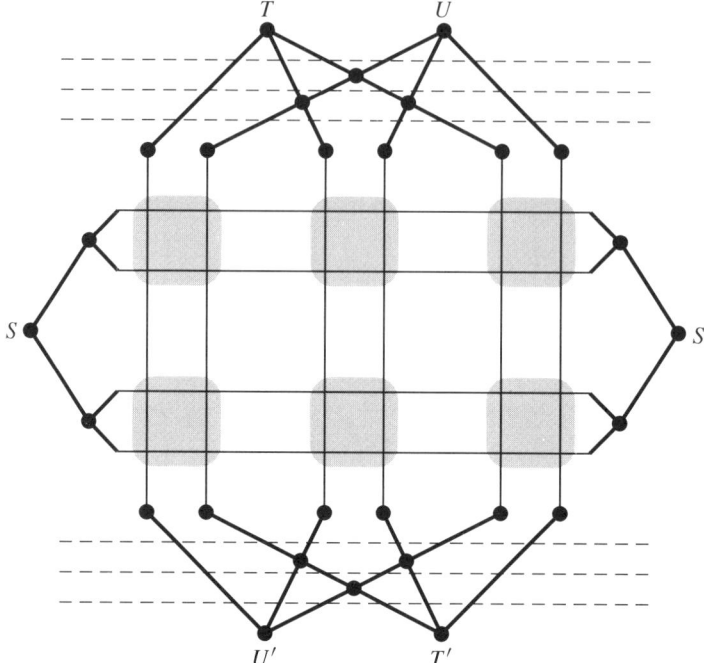

Fig. 12.7. Reduction of the demand graph to only three sets of parallel edges. The *central part* corresponds to the grid built before, *thick edges* are the new edges. The demands are equal to the number of variables for the demand edge (X, X'), and to the number of clauses for (U, U') and (V, V'). *Dashed lines* define tight cuts

paths to cross in a planar way (Fig. 12.7). The demand graph is reduced then to only three sets of parallel edges—one horizontal, and two vertical ones. To check that this operation does not change the problem, note:

The tight cuts represented by the dashed lines force all starting points of both lines of all columns to be contained in different vertical paths. It can be shown by induction from left to right that these paths are rooted like in the previous part of the proof. □

This result can be further strengthened to two demand edges (Naves 2008). Keeping Schwärzler's global idea of the reduction from 3-SAT, the details become much more complicated since the two classes of "vertical demand edges" are decreased to one with a tricky idea whose technical realization is also more complicated.

In Schwärzler's proof vertical paths usually do not switch columns (allowing then two horizontal paths per row to cross the column), columns are switched only in one row, where the corresponding variable is set to TRUE in the clause of the column. In this case the corresponding horizontal line is prevented from becoming a path in the multiflow.

The idea now is to do just the opposite in terms of switching columns: vertical paths *will usually switch columns, except in the row associated with the particular*

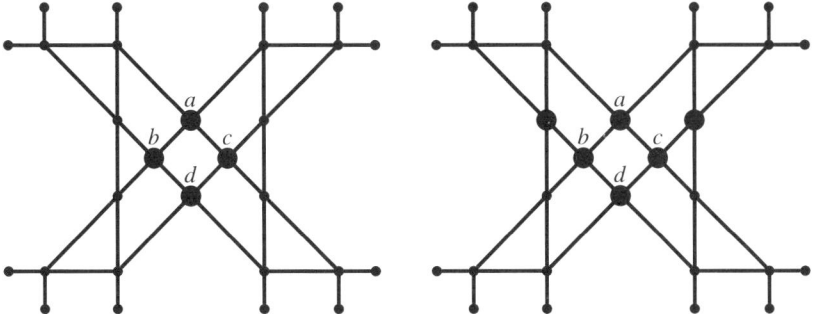

Fig. 12.8. The two gadgets used for the reduction to the case where there are only two demand edges. Two paths can cross in the bold vertices and nowhere else

TRUE valued variable of the column, when they don't switch. The number of rows can be supposed to be even, so in case of a feasible truth assignment there are an odd number of column-switches in every column!

This is realized this time by paths that run in pairs parallelly and never cross. The corresponding demand edges form one parallel class. Let us explain the main ideas of realizing this and the related technical problems of Naves (2008):

First, for convenience, the problem is generalized by forbidding crossing paths in a subset $W \subseteq V$ of vertices of degree 4. This is not really a generalization, since each $w \in W$ can be split into four vertices of degree 1, with one new vertex for each incident edge; then add a C_4 between the four new vertices in the cyclic order of the four edges in the planar embedding. Clearly, there is a bijection between the solutions of the edge-disjoint paths problem after the application of these gadgets and the solutions where paths are not crossing in the vertices of W in the original graph.

A combination of the two gadgets of Fig. 12.8 will be placed in the crossing of columns (associated with clauses) and rows (associated with variables).

These gadgets have the following properties:

The two parallel vertical paths we mentioned in our general description arrive either from the two left vertices in the upper left corner, or from the two vertices in the upper right corner. Either both stay in the same (left or right) side, or both change sides (column-switch) when they go through the gadget.

If one horizontal path goes *through the left gadget, then the two parallel vertical paths are obliged to switch sides, while in the right gadget they are allowed to stay on the same side.* If two horizontal paths go through the gadgets *both paths are obliged to switch sides.*

A combination of these gadgets placed in the same "grid" as before encodes a truth assignment satisfying the goal we have described. Again, the most difficult part of the proof is to ensure that the paths cannot deviate from their intended itineraries. There are indeed only two kinds of demand edges: vertical and horizontal, as required.

12.4.2 Acyclic Digraphs

Schwärzler gave also a directed acyclic version of his reduction (Schwärzler 2008):

Theorem 4.6 (Schwärzler). *The arc-disjoint paths problem is NP-complete, even if G is planar and acyclic, and all terminals lie on the outer boundary of G.*

The trick presented in Sect. 12.4.1 serves now again to reduce the number of terminals:

Theorem 4.7. *The arc-disjoint paths problem is NP-complete, even if G is planar and acyclic, $|E(H)| = 3$, and all terminals lie on the outer boundary of G.*

Both the arc-disjoint and the vertex-disjoint paths problems are polynomial-time solvable when the total number of demand is fixed. We show that the complexity of the vertex-disjoint version is again the same as the edge-disjoint versions when $|E(H)|$ is not bounded, both problems are NP-complete:

Theorem 4.8. *The vertex-disjoint paths problem is NP-complete in acyclic digraphs, even if G + H is planar.*

Proof. The proof is the directed acyclic version of Middendorf and Pfeiffer (1993) proof of their Theorem 1 establishing the NP-hardness of the edge-disjoint paths problem if $G + H$ is planar. (However, again, we cannot reduce the theorem to their result.)

We reduce PLANAR 3-SAT to the stated problem: let φ be a formula whose associated graph is planar, and suppose (without loss of generality) that each variable appears at most three times, exactly once negatively, and there is no clause with twice the same variable. Define the undirected bipartite graph (C, V, F) with the set of clauses C and the set V of variables as classes, and $F := \{xc : \text{variable } x \text{ appears in clause } c\}$ and subdivide each edge (x, c) into two edges by adding a new node v_{xc}.

Take now an arbitrary ordering of the set of variables, and define for each clause a gadget in the following way: choose z to be an arbitrary of the three variables of the clause, and then choose the notation x and y so that $x < y$. With this notation construct the gadget on the left of Fig. 12.9 upon the vertices v_{xc}, v_{yc}, v_{zc}, adding the other vertices of the figure anew for each clause. Finally delete the vertex representing c.

Now for each variable vertex x occurring in three clauses, let a and b be the clauses in which x occurs positively (in arbitrary order), and c the one in which it is negated, and put the gadget depicted in the right side of Fig. 12.9 upon the vertices v_{xa}, v_{xb}, v_{xc}. (If x occurs only twice, positively in a and negatively in c, we add the vertex v_{xb} artificially.) Let G_φ, H_φ denote the constructed graph and the constructed demand graph.

Then we have to prove that there exist arc-disjoint paths in (G_φ, H_φ) if and only if ϕ is satisfiable. The proof is similar to that of Middendorf and Pfeiffer (1993), let us sketch it:

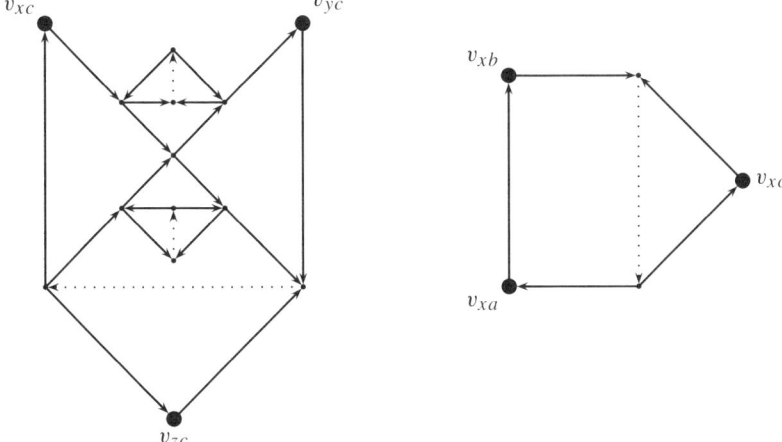

Fig. 12.9. The gadget for the clauses is on the *left*, for the variables it is on the *right*. *Dotted* edges are demand edges, *bold* vertices are those that subdivide the edges in the original graph

It is easy to see that the demand arc in a variable gadget is satisfied either by a path that contains v_{xc} corresponding to $x =$ TRUE, or a path that contains the arc $v_{xa}v_{xb}$, which corresponds to $x =$ FALSE.

The demands of a clause gadget can be satisfied if and only if at least one of the three bold vertices of the figure is not used by variable demands, encoding that the clause is satisfied by the variable assignment.

Finally we prove that the digraph is acyclic. Each gadget is acyclic, thus if there is a cycle, it uses at least two gadgets. Suppose for a contradiction that Q is a cycle. Then it intersects clause gadgets in (v_{xa}, v_{ya})-paths and variable gadgets in (v_{xa}, v_{xb})-paths. The cycle Q would then follow a sequence where the variable gadgets belong to variables forming an increasing sequence.   □

12.5 Key Assertions

In this section we state the assertions (theorems or problems) that provide (or would provide) most of the results of the tableau: the "minimal" NP-complete problems, and the "maximal" polynomial ones. Those that allowed filling in most of the tableau, also using the basic reductions of Sect. 12.3; and also those problems that were output by the tableau as missing. Some historical results cited in the footnotes of the tableau do not reappear here, because they are subsumed by more recent theorems that do reappear. By giving the main theorems in a "full-text" version, we try to provide the most precise formulation and thus a high credibility for the tableau.

We give the list without comment. We hope it will then be easy to switch between the tableau and this list hence and forth to see the facts and their reasons.

12.5.1 NP-completeness

Theorem 5.1 (Fortune et al. 1980). *The vertex-disjoint paths problem is NP-complete, even if $E(H) = 2$.*

Theorem 5.2 (Middendor and Pfeiffer 1993). *The edge-disjoint paths problem is NP-complete, even if $G + H$ is planar.*

Theorem 5.3 (Vygen 1995). *The multiflow problem is NP-complete, even if G is an acyclic digraph, $r + c$ is Eulerian and $|E(H)| = 3$.*

Note that under the same condition, supposing $r(h) = 1$, for all $h \in H$, the problem is solvable in polynomial-time but still non-trivial, see the nice algorithm of Ibaraki and Poljak (1991).

Theorem 4.8. *The vertex-disjoint paths problem is NP-complete in acyclic digraphs, even if $G + H$ is planar.*

Theorem 5.4 (Marx 2004). *The multiflow problem is NP-complete if G is planar (also if it is a grid) and $G + H$ is Eulerian, both in the undirected and directed acyclic case.*

Theorem 5.5 (Naves 2008). *The multiflow problem is NP-complete, even with one of the following restrictions:*

  (i) *G is a planar undirected graph, H has only two edges, both on the infinite face of G,*
 (ii) *G is a directed graph, $G + H$ is planar, H has only two terminals.*
(iii) *G is a directed acyclic digraph, H has only two edges, both on the infinite face of G.*

12.5.2 Polynomiality

Theorem 5.6 (Frank 1989). *The multiflow problem in Eulerian digraphs with $|E(H)| = 2$ is solvable in polynomial-time. The cut condition is sufficient for the existence of a solution.*

Theorem 5.7 (Lucchesi and Younger 1978). *The multiflow problem in directed acyclic graphs with $G + H$ planar is solvable in polynomial-time.*

Theorem 5.8 (Fortune 1980). *The vertex-disjoint paths problem in directed acyclic graphs with $|E(H)|$ bounded is solvable in polynomial-time.*

Theorem 5.9 (Seymour 1981). *The multiflow problem is solvable in polynomial time in undirected graphs, if $G + H$ is planar (or more generally if it does not have a K_5 minor) and $r + c$ is Eulerian. The cut condition is then necessary and sufficient for the existence of a solution.*

Theorem 5.10 (Lomonosov 1985). *The multiflow problem in Eulerian undirected graphs with $E(H)$ being the union of two stars, or K_4 or C_5, is solvable in polynomial-time. The cut condition is sufficient for the existence of a solution.*

Theorem 5.11 (Robertson and Seymour 1995). *The vertex-disjoint and edge-disjoint paths problems in undirected graphs with $r(E(H))$ bounded are solvable in polynomial-time.*

Theorem 5.12 (Schrijver 1993). *The vertex-disjoint paths problem in planar digraphs with $|E(H)|$ bounded is solvable in polynomial-time.*

Theorem 5.13 (Sebő 1993). *The integer multiflow problem in undirected graphs with $G + H$ planar and $|E(H)|$ bounded is solvable in polynomial-time.*

12.5.3 Relevant Open Problems

Last, we state 5 (in fact 7) of the 21 open problems that we find particularly nice or frustrating.

Problem 5.14 (Round-trip problem, Schrijver 2003, Problem 50). Is the problem of finding a connected Eulerian subgraph of a digraph, containing two pre-given vertices, polynomial-time solvable?

Problem 5.15. Let k be an integer. What is the complexity of routing k pairs of terminals in a Eulerian digraph if k is fixed? Is this problem easier if G is planar?

Ibaraki and Poljak (1991) found a polynomial-time algorithm for arbitrary graphs and $k = 3$. As far as we know, this is the only partial result about this question.

Problem 5.16. Is the integer multiflow problem solvable in polynomial-time when G is a Eulerian directed acyclic graph? And when the demand graph is fixed?

Problem 5.17. What is the complexity of the undirected multiflow problem if G is planar and $G + H$ (or more generally $r + c$) is Eulerian and $|E(H)|$ is fixed?

The complexity of this last problem is open already if $|E(H)| = 3$. One of the latest results (Marx 2004) establishes NP-completeness in both the undirected and directed case, if the number of edges of the demand graph is not fixed. For the directed case this settles Problem 5.15 if k is not fixed.

References

Bentz, C.: Résolution exacte et approchée de problèmes de multiflot entier et de multicoupe: algorithmes et complexité. Thèse de docteur en informatique, Conservatoire Nationale des Arts et Métiers (November 2006)

Bentz, C., Costa, M.-C., Létocart, L., Roupin, F.: Minimal multicut and maximal integer multiflow: A survey. Eur. J. Oper. Res. **162**, 55–69 (2005)

Even, S., Itai, A., Shamir, A.: On the complexity of timetable and multicommodity flow problems. SIAM J. Comput. **5**(4), 691–703 (1976)

Fortune, S., Hopcroft, J., Wyllie, J.: The directed subgraph homeomorphism problem. Theor. Comput. Sci. **10**, 111–121 (1980)

Frank, A.: On connectivity properties of Eulerian digraphs. In: Annals of Discrete Mathematics, vol. 41, pp. 179–194. North-Holland, Amsterdam (1989)

Frank, A.: Packing paths in planar graphs. Combinatorica **10**(4), 325–331 (1990)

Frank, A.: Packing paths, circuits and cuts—a survey. In: Korte, B., Lovász, L., Prömel, H.J., Schrijver, A. (eds.) Paths, Flows, and VLSI-Layout. Springer, Berlin (1990)

Ibaraki, T., Poljak, S.: Weak three-linking in Eulerian digraphs. SIAM J. Discrete Math. **4**, 84–98 (1991)

Karp, R.M.: On the complexity of combinatorial problems. Networks **5**, 45–68 (1975)

Korte, B., Vygen, J.: Combinatorial Optimization: Theory and Algorithms. Algorithms and Combinatorics, vol. 21. Springer, Berlin (2000)

Korte, B., Lovász, L., Prömel, H.-J., Schrijver, A. (eds.): Paths, Flows, and VLSI-Layout. Springer, Berlin (1990)

Kramer, M.R., Van Leeuwen, J.: The complexity of wire-routing and finding the minimum area layouts for arbitrary VLSI circuits. In: Preparata, F.P. (ed.) Advances in Computing Research 2: VLSI Theory, pp. 129–146. JAI Press, London (1984)

Lomonosov, M.: Combinatorial approaches to multiflow problems. Discrete Appl. Math. **11**, 1–94 (1985)

Lomonosov, M., Sebő, A.: On the geodesic structure of graphs: a polyhedral approach to metric decomposition. In: Rinaldi and Wolsey (eds.) Integer Programming and Combinatorial Optimization, Proceedings of the 3rd IPCO Conference, Erice, Italy, pp. 221–234 (1993)

Lovász, L.: Combinatorial Problems and Exercises. Akadémiai Kiadó, Budapest (1979)

Lucchesi, C.L., Younger, D.H.: A minimax theorem for directed graphs. J. Lond. Math. Soc. (2) **17**, 369–374 (1978)

Lynch, N.: The equivalence of theorem proving and the interconnexion problem. SIGDA News. **5**(3), 31–36 (1975)

Marx, D.: Eulerian disjoint paths problem in grid graphs is NP-complete. Discrete Appl. Math. **143**, 336–341 (2004)

Middendorf, M., Pfeiffer, F.: On the complexity of the disjoint paths problem. Combinatorica **13**(1), 97–107 (1993)

Müller, D.: On the complexity of the planar directed edge-disjoint paths problem. Math. Program. **105**(2–3), 275–288 (2006)

Nash-Williams, C.St.J.A.: Exercise 6.56 in Lovász (1979) and 71.1a in Schrijver (2003), p. 1254

Naves, G.: The hardness of routing two pairs on one face. In preparation (2008)

Okamura, H., Seymour, P.D.: Multicommodity flows in planar graphs. J. Comb. Theory, Ser. B **31**, 75–81 (1981)

Onaga, K., Kakusho, O.: On feasibility conditions of multicommodity flows in networks. IEEE Trans. Circuit Theory **18**, 425–429 (1971)

Raghavan, P., Ph.D. thesis. Report No. UCB/CSD 87/312, University of California, Berkeley (1986)

Robertson, N., Seymour, P.D.: Graph minors XIII. The disjoint paths problem. J. Comb. Theory, Ser. B **63**, 65–110 (1995)

Rothschild, B., Whinston, A.: Feasibility of two-commodity network flows. Oper. Res. **14**, 1121–1129 (1966)

Schrijver, A.: The Klein bottle and multicommodity flows. Combinatorica **9**, 375–384 (1989)

Schrijver, A.: Applications of polyhedral combinatorics to multicommodity flows and compact surfaces. In: Cook, W., Seymour, P. (eds.) Polyhedral Combinatorics. DIMACS Series in Discrete Mathematics and Theoretical Computer Science, vol. 1, pp. 119–137. Am. Math. Soc., Providence (1990a)

Schrijver, A.: Homotopic routing methods. In: Korte, B., Lovász, L., Prömel, H.J., Schrijver, A. (eds.) Paths, Flows, and VLSI-Layout. Springer, Berlin (1990b)

Schrijver, A.: Complexity of disjoint paths problems in planar graphs. In: Lengauer, T. (ed.) Algorithms—ESA '93. Lecture Notes in Computer Science, vol. 726. Springer, Berlin (1993)

Schrijver, A.: Finding k disjoint paths in a directed planar graph. SIAM J. Comput. **23**(4), 780–788 (1994)

Schrijver, A.: Combinatorial Optimization: Polyhedra and Efficiency. Springer, Berlin (2003), pp. 1–3

Schwärzler, W.: On the complexity of the planar edge-disjoint paths problem with terminals on the outer boundary. Combinatorica (2008, in press)

Sebő, A.: Integer plane multiflows with a fixed number of demands. J. Comb. Theory, Ser. B **59**, 163–171 (1993)

Seymour, P.D.: On odd cuts and planar multicommodity flows. Proc. Lond. Math. Soc. **42**, 178–192 (1981)

Vygen, J.: NP-completeness of some edge-disjoint paths problems. Discrete Appl. Math. **61**, 83–90 (1995)

Vygen, J.: Disjoint paths. Report No. 94816, Institute for Discrete Mathematics (revised in 1998)

Many Facets of Dualities

Jaroslav Nešetřil

Summary. In this paper we survey results related to homomorphism dualities for graphs, and more generally, for finite structures. This is related to some of the classical combinatorial problems, such as colorings of graphs and hypergraphs, and also to recently intensively studied Constraint Satisfaction Problems. On the other side dualities are related to the descriptive complexity and First Order definability as well as to universal graphs. And in yet another context they can be expressed as properties of the homomorphism order of structures. In the contemporary context homomorphism dualities are a complex area and it is our aim to describe some of the main ideas only. However we introduce the four conceptually different proofs of the existence of duals thus indicating the versatility of this notion. Particularly we describe setting of restricted dualities and the role of bounded expansion classes.

13.1 Introduction

Think of 3-colorability of a graph G. This is a well known hard (and a canonical NP-complete) problem. From the combinatorial point of view there is a standard way how to approach this problem (and monotone properties in general): we investigate minimal (usually called 4-critical) graphs without this property (i.e. which are not 3-colorable), we then denote by $\mathcal{F}$ the set (or language) of all such critical graphs and we define the set $Forb(\mathcal{F})$ of all structures which do not "contain" any $F \in \mathcal{F}$. Then the language $Forb(\mathcal{F})$ coincides with the language of 3-colorable graphs.

Unfortunately, in this case the set $\mathcal{F}$ is infinite and this seems to be a general phenomenon: in most "interesting" cases the set of minimal forbidden graphs seems to be infinite (and mostly even not "enough" structured). Yet in this paper we study exactly those cases which have finitely many obstructions. Such cases are called *finite homomorphism dualities*. And we shall see that nevertheless finitely many obstructions exists for a wide range of problems. However this holds not on the very

Part of this work was supported by ITI and DIMATIA of Charles University Prague under grant 1M0021620808 and by AEOLUS.

elementary level. (It has been proved already in Nešetřil and Pultr (1978) that, except in the single trivial case, no coloring problem for undirected graphs admits finitely many minimal obstructions.) We have to generalize and this may seem to be a leitmotiv of this paper. We have to generalize not for its own sake but in order to find a proper setting for concrete problems (like the 3-colorability, or, more generally, Constraint Satisfaction Problems).

There are three main ingredients in our approach:

1. The use of relational structures and their homomorphisms (i.e. we deal with the *category* of graphs and structures).
2. The use of existential statements in the form of lifts and shadows.
3. Restriction of the obstruction characterization to a particular class of structures (such as planar graphs, or proper minor closed classes, or classes with bounded expansion as defined by P. Ossona de Mendez and author, e.g. Nešetřil and Ossona de Mendez 2006c).

In this paper we survey the development in all 3 directions by giving concrete examples. Accordingly, the paper has 4 sections. In Sect. 13.2 we deal with dualities and their characterizations. In Sect. 13.3 we survey the results related to lifts and shadows, their descriptive complexity in the context of Constraint Satisfaction Problems (CSP). In Sect. 13.4 we relate dualities to classical model theoretic problems related to universal objects. In Sect. 13.5 we demonstrate the richness of Restricted Dualities and in the final section we summarize the main results together with some remarks and open problems.

In the whole paper we deal not only with graphs but also with relational structures. This language is essential for questions which will be considered in this paper. It can be briefly introduced as follows:

A type Δ is a sequence $(\delta_i; i \in I)$ of positive integers. A *relational structure* $\mathbf{A}$ of *type* Δ is a pair $(X, (R_i; i \in I))$ where X is a set and $R_i \in X^{\delta_i}$; thus R_i is a δ_i-ary relation on X. In this paper we shall always assume that X is a finite set (thus we consider finite relational structures only). Relational structures (of type Δ) will be denoted by capital letters $\mathbf{A}, \mathbf{B}, \mathbf{C}, \ldots$. A relational structure of type Δ is also called a Δ-*structure* (or just a *structure*). If $\mathbf{A} = (X, (R_i : i \in I))$ we also denote the base set X as $X(\mathbf{A})$ and the relation R_i by $R_i(\mathbf{A})$. We denote by $Rel(\Delta)$ the class of all Δ-type relational structures. The class $Rel(\Delta)$ will be considered as category endowed with all homomorphisms, which are just all relations preserving mappings. To be explicit, for relational structures $\mathbf{A}, \mathbf{B} \in Rel(\Delta)$ a mapping $f : X(\mathbf{A}) \longrightarrow X(\mathbf{B})$ is a *homomorphism* $\mathbf{A} \longrightarrow \mathbf{B}$ if for every relational symbol $R \in \Delta$ and for every tuple $(x_1, \ldots, x_t) \in R(\mathbf{A})$ holds $(f(x_1), \ldots, f(x_t)) \in R(\mathbf{B})$. The existence of such a homomorphism will be denoted by $\mathbf{A} \longrightarrow \mathbf{B}$ and its non-existence by $\mathbf{A} \nrightarrow \mathbf{B}$.

Natural examples of relational structures are abundant: graphs, hypergraphs, graphs with colored edges, ordered graphs, triples modelling betweeness, etc. A rich source and strong motivation is the theory of database queries, e.g. (Feder and Vardi 1999; Atserias et al. 2006; Gottlob et al. 2008; Kolaitis and Vardi 1998; Hell and Nešetřil 2004) and Constraint Satisfaction Problems (Feder and Vardi 1999; Hell and Nešetřil 2004; Lovász 1967; Kolaitis and Vardi 1998). Many of these questions are best formulated in the categorical language of graphs and their homomorphisms. For data systems this goes back to (Chandra and Merlin 1977), for combinatorial and algebraic setting see e.g. (Pultr and Trnková 1968; Lovász 1967; Hell and Nešetřil 2004). The notion of homomorphism now plays a role in many problems in areas as diverse as statistical physics, extremal theory, limit structure theorems, see e.g. (Borgs et al. 2006; Hell and Nešetřil 2004; Nešetřil and Ossona de Mendez 2008d).

13.2 Finite Dualities

What can be better than finitely many obstructions? Yes, sure, but is this a realistic goal which has some interesting instances? In the undirected graph case the answer is negative (Nešetřil and Pultr 1978). However the properties characterized by a finite set $\mathcal{F}$ are very interesting if we consider them for more complicated structures than undirected graphs only. Towards this end we define the notion of *finite (homomorphism) duality*.

Definition 2.1. *Let $\mathcal{F}$, $\mathcal{D}$ be finite sets of structures (in a fixed class $Rel(\Delta)$). We say that sets $\mathcal{F}$ and $\mathcal{D}$ establish* finite duality *if the following holds for every structure $\mathbf{A} \in Rel(\Delta)$:*

$$\mathbf{F} \nrightarrow \mathbf{A} \quad \text{for every } \mathbf{F} \in \mathcal{F} \quad \Longleftrightarrow \quad \mathbf{A} \longrightarrow \mathbf{D} \quad \text{for some } \mathbf{D} \in \mathcal{D}.$$

In this case we say that $(\mathcal{F}, \mathcal{D})$ is dual pair, *and that $\mathcal{D}$ is* dual set *of $\mathcal{F}$.*

The simplest non-trivial instance of dualities is for oriented graphs (again: for undirected graphs we have trivial examples only; see Nešetřil and Pultr 1978 where the term *homomorphism duality* was introduced) and it is usually expressed in terms of orientations of graphs. The connection between chromatic number and orientations is not new and goes back to Gallai (1968) and Roy (1967). These pioneering works provided a name ("Gallai–Roy" theorem) for this result although both of these papers are anticipated by Hasse (1964/1965) and Vitaver (1962) where the same thing is proved (in the more algebraic language). Note that another influential connection between orientations and chromatic number is given by Minty (1962) but that goes in a different direction (flows and matroids). For our purposes Gallai–Hasse–Roy–Vitaver theorem takes the following compact form:

Theorem 2.2. *For any directed graph G the following holds:*

$$P_k \nrightarrow G \quad \Longleftrightarrow \quad G \to T_k.$$

The undefined notions have the following meaning: P_k denotes the directed path of length k (i.e. with $k + 1$ vertices) and T_k denotes the transitive tournament with k vertices.

It may be seen easily that for undirected graph this has the following consequence.

Corollary 2.3. *For an undirected graph G the following statements are equivalent:*

1. *$\chi(G) \leq k$ (which is equivalent to $G \rightarrow K_k$);*
2. *There exists an orientation $\mathbf{G}$ of G such that $\mathbf{G} \rightarrow T_k$;*
3. *There exists an orientation $\mathbf{G}$ of G such that $P_k \nrightarrow \mathbf{G}$.*

This particular result was the starting point (Nešetřil and Pultr 1978) for the following result which characterizes homomorphisms dualities in general classes of relational structures (Nešetřil and Tardif 2000; Foniok et al. 2008). This characterization involves notion of relational tree and relational forest. These intuitive notions can be defined similarly as for graphs but perhaps the easiest way is to reduce them to graphs by means of the notion of *incidence graph* $G_\mathbf{A}$ *of a structure* $\mathbf{A} = (X, (R_i : i \in I))$. We include this construction for the sake of completeness. The vertex set of $G_\mathbf{A}$ is the set X together with all tuples of relations $R_i : i \in I$. ($G_\mathbf{A}$ is a bipartite graph, X is on the one side, tuples are on the other side of bipartition.) The edges will be formed by all incidencies between an element and a tuple (such as x and (x, y, z); $G_\mathbf{A}$ may have multiple edges such as between x and (x, x, y)). Now we can say that a structure $\mathbf{A}$ is a *relational tree* (*relational forest*, respectively) if the incidence graph $G_\mathbf{A}$ is a tree (forest, respectively). First we state the theorem for singleton sets only:

Theorem 2.4 (Singleton homomorphism dualities, Nešetřil and Tardif 2000).

1. *For every relational tree $\mathbf{T}$ there exists a structure $\mathbf{D_T}$ (called the dual of $\mathbf{T}$) such that the following holds (for every structure $\mathbf{A}$):*

$$\mathbf{T} \nrightarrow \mathbf{A} \quad \Longleftrightarrow \quad \mathbf{A} \rightarrow \mathbf{D_T}.$$

2. *Up to a homomorphism equivalence there are no other dual pairs (of singleton structures).*

This result was proved for oriented graphs in Komárek (1987) and in the full generality (and by different methods) in Nešetřil and Tardif (2000). The characterization of singleton dualities is the basis of the characterization of dual pairs of sets and of finite dualities:

Theorem 2.5 (Finite homomorphism dualities, Nešetřil and Tardif 2000; Foniok et al. 2008).

1. *For every finite set of relational forests $\mathcal{T}$ there exists a dual set of structures $\mathcal{D_T}$ such that the following holds (for every structure $\mathbf{A}$):*

$$\mathcal{T} \nrightarrow \mathbf{A} \quad \Longleftrightarrow \quad \mathbf{A} \rightarrow \mathcal{D_T}.$$

2. *Up to a homomorphism equivalence there are no other dual pairs (of finite sets of structures).*

(Here we of course write $\mathcal{T} \nrightarrow \mathbf{A}$ if $\mathbf{T} \nrightarrow \mathbf{A}$ for every $\mathbf{T} \in \mathcal{T}$. Similarly, we write $\mathbf{A} \rightarrow \mathcal{D_T}$ if $\mathbf{A} \rightarrow \mathbf{D}$ for some $\mathbf{D} \in \mathcal{D_T}$.)

Theorems 2.4, 2.5 are nontrivial in both directions and the existence of duals is established by various means. This will be briefly reviewed. Perhaps the variety of techniques will convince an interested reader that these are interesting objects.

13.2.1 Existence of Duals I—Explicit Construction

We review here the recent simple construction which appeared in Nešetřil and Tardif (2005, 2008a). For the notation simplicity we consider oriented graphs only (i.e. our type is $\Delta = (2)$).

Let $T = (V, E)$ be an oriented tree. We define its dual $D_T = (V', E')$ as follows: The set V' is the set of all mappings $f : V \longrightarrow V$ satisfying that $f(v)$ is adjacent to v for every $v \in V$ (such mappings are called *neighbourly*). For neighbourly mappings f, g we put $(f, g) \in E'$ if $(v, u) \neq (f(u), g(v))$ for every arc $(u, v) \in E$. The later condition simply means that no arc is jointly "flipped" by the pair (f, g).

Proof (a rough sketch). It is easy to see that in order to show that the graph D_T is indeed the dual of the tree T we have to prove two facts:

(i) $T \not\longrightarrow D_T$,
(ii) $T \not\longrightarrow G \Longrightarrow G \longrightarrow D_T$.

(i) is proved by a contradiction: The existence of a homomorphism $\phi : T \longrightarrow D_T$ implies the existence of an infinite walk on T defined as sequence

$$v_0, \ \phi(v_0)(v_0) = v_1, \ \phi(v_1)(v_1) = v_2, \ \ldots.$$

This walk has to eventually return. Thus there exists i such that $\phi(v_i)(v_i) = v_{i+1}, \phi(v_{i+1})(v_{i+1}) = v_i$ and either (v_i, v_{i+1}) or (v_{i+1}, v_i) is an arc of T. Assume e.g. that (v_i, v_{i+1}) is an arc. Then $(\phi(v_i), \phi(v_{i+1}))$ is an arc of D_T. However $(\phi(v_i)(v_i), \phi(v_{i+1})(v_{i+1})) = (v_{i+1}, v_i)$, which a contradiction (a flip).

(ii) is proved constructively and the homomorphism $\phi : G \longrightarrow D_T$ is given by the formula

$$\phi(x)(u) = v,$$

where $x \in V(G), u \in V$, v is adjacent to u and v determines a branch B of T at u for which there is no homomorphism $(B, u) \not\longrightarrow (G, x)$. (Such a branch obviously exists by the "freeness" of the tree T.) If there are more such branches then we take the first one in a depth first order.

Note that this construction of a dual uses $2^{n \log(n)}$ vertices. Up to logarithmic factor this is optimal as shown in Nešetřil and Tardif (2005, 2008a). Indeed, G. Kun and C. Tardif showed recently (Kun and Tardif 2008) that almost all oriented paths have exponentially large duals.

13.2.2 Existence of Duals II—Homomorphism Order

Let us consider the class $Rel(\Delta)$ of all finite relational structures of type Δ. The existence of the homomorphism (i.e. the simplification of the corresponding category) is a quasiorder which becomes a partial order if we factorize by the relation of the homomorphism equivalence. (Structures $\mathbf{A}, \mathbf{B}$ are said to be *homomorphism equivalent* if $\mathbf{A} \longrightarrow \mathbf{B}$ and $\mathbf{B} \longrightarrow \mathbf{A}$.) This partial order is called *homomorphism order* and it is denoted by $\mathcal{C}_\Delta$. It is important that the homomorphism equivalence may be described more effectively by the notion of core: A *core of structure* **A** is a minimal retract of **A**. Core of any finite structure is (up to an isomorphism) unique and thus we can speak about the core of **A**. Two structures are then homomorphism equivalent if and only if they have isomorphic cores. $\mathcal{C}_\Delta$ is the set of all non-isomorphic *core* structures ordered by the existence of a homomorphism, see (Hell and Nešetřil 2004) for an introduction to this area. The order of $\mathcal{C}_\Delta$ will be denoted by $\leq$ and its strict version by $<$.

The partial order $\mathcal{C}_\Delta$ has spectacular properties. It is not only a lattice, it is also Heyting poset (Nešetřil et al. 2007) and it has many interesting global and local properties. For example:

- $\mathcal{C}_\Delta$ is (countably) universal (Pultr and Trnková 1968; Hell and Nešetřil 2004),
- $\mathcal{C}_\Delta$ is dense for undirected graphs (Welzl 1982).

The later property leads to the following:

Definition 2.6 (Gaps and density). *Let* **A**, **B** *be structures in* $\mathcal{C}_\Delta$ *and let* **A** $<$ **B**. *If there is no* **C** $\in \mathcal{C}_\Delta$ *such that* **A** $<$ **C** $<$ **B** *then the pair* (**A**, **B**) *is called a* gap *in* $\mathcal{C}_\Delta$. *Density problem asks for a characterization of all gaps in* $\mathcal{C}_\Delta$.

Density problem is solved for all classes $\mathcal{C}_\Delta$. The characterization, given in Nešetřil and Tardif (2000) rests on a surprising connection of gaps and dualities.

Theorem 2.7. *Every gap* **A** $<$ **B** *with* **B** *connected, yields a duality* (**B**, **A**^**B**).

Theorem 2.8. *Every (singleton) duality* (**F**, **D**) *yields a gap* **F** $\times$ **D** $<$ **F**.

This one-to-one correspondence between (singleton) dualities and (connected) gaps was originally used to prove the existence of duals: We construct for every relational tree **F** its (up to the homomorphism equivalence unique) *predecessor* $\mathbf{P_T}$ such that $\mathbf{P_T} <$ **T** is a gap and thus $(\mathbf{P_T})^\mathbf{T}$ is dual of **T**. This connection was also used to prove the necessity in both Theorems 2.4 and 2.5, see (Nešetřil and Tardif 2000).

Proof (of Theorems 2.7, 2.8). Use categorial algebra (and the proofs can be generalized to Heyting posets, Nešetřil et al. 2007): If **A** $<$ **B** is a gap with **B** connected then for arbitrary structure **C** consider the object **A** $+$ (**B** $\times$ **C**). Clearly **A** $\leq$ **A** $+$ (**B** $\times$ **C**) $\leq$ **B** and thus either **B** $\longrightarrow$ (**B** $\times$ **C**) (by the connectivity of **B**) and thus also **B** $\longrightarrow$ **C**, or **B** $\not\longrightarrow$ **A** $+$ (**B** $\times$ **C**) and thus (**B** $\times$ **C**) $\longrightarrow$ **A**. But then, (using the definition of the power **A**^**B**), we have **C** $\longrightarrow$ **A**^**B**). This proves Theorem 2.7.

In Theorem 2.8 we want to prove that **F** $\times$ **D** $<$ **F** is a gap. We proceed as follows: If **C** satisfies **F** $\times$ **D** $\leq$ **C** $\leq$ **F** then either **F** $\longrightarrow$ **C** or not. In the later case **C** $\longrightarrow$ **D** (by duality) and thus also **C** $\longrightarrow$ **F** $\times$ **D**. □

It is quite remarkable that the algorithmically motivated concept of the dualities relates so closely to a concept of partial order theory. But duals seem to be natural objects. This is also indicated by the fact that there are two more constructions of duals and they have a different flavour. One of them will be introduced in the next section.

13.3 CSP as Duality

Let us return to our example of 3-colorability. Instead of a graph $G = (V, E)$ we consider the graph G together with three unary relations C_1, C_2, C_3 which *cover* the vertex set V; this structure will be denoted by G' and called a *lift* of G (thus G' has one binary and three unary relations). There are 3 *forbidden substructures*: For each $i = 1, 2, 3$ the single edge graph K_2 together with cover $C_i = \{1, 2\}$ and $C_j = \emptyset$ for $j \neq i$ form structure $\mathbf{F}'_i$ (where the signature of $\mathbf{F}'_i$ contains one binary and three unary relations). The language of all 3-colorable graphs is just the language $\Phi(Forb(\mathbf{F}'_1, \mathbf{F}'_2, \mathbf{F}'_3))$, where Φ is the forgetful functor which transforms G' to G. We then call G the *shadow* of G'.

Clearly this situation can be generalized and one of the main results of paper (Kun and Nešetřil 2008) states that every problem in NP is polynomially equivalent to the membership problem for a class $\Phi(Forb(\mathcal{F}'))$. Here $\mathcal{F}'$ is a finite set of (vertex or pair)-colored digraphs, $Forb(\mathcal{F}')$ is the class of all lifted graphs G' for which there is no homomorphism $F' \longrightarrow G'$ for an $F' \in \mathcal{F}'$. Thus $Forb(\mathcal{F}')$ is the class of all graphs G' with *forbidden* homomorphisms from $\mathcal{F}'$. More precisely this can be done as follows:

Let Γ denote a finite set we refer to as colors. A Γ-*colored structure* (shortly a *colored structure* is a structure together with either a coloring of its vertices or a coloring of all tuples (of arities from the type of the relational structure) of vertices by colors from Γ. Mostly it suffices to consider the coloring of vertices only. We denote colored relational structures by $\mathbf{A}', \mathbf{B}'$ etc. We call $\mathbf{A}'$ a *lift* of $\mathbf{A}$ and $\mathbf{A}$ is called the *shadow* of $\mathbf{A}'$. We also write $\Phi(\mathbf{A}') = \mathbf{A}$ and we think of Φ as a forgetful functor (which "forgets" the colors). (Note that both constructions of lifts and shadows are known in model theory and in this context they are called *extension* and *reduct*, see Hodges 1993. Our terminology is motivated by Computer Science applications of category theory; yet another approach is given in Madelaine 2003; Madelaine and Stewart 2005.) Thus (vertex-) colored structures can also be described as *monadic* lifts (monadic meaning that only vertices are colored, only unary relations are added). A homomorphism of relational structures preserves all the edges (arcs). A *homomorphism* of colored relational structures preserves the color of vertices (and colors of tuples), too. We call a mapping between two (colored) digraphs a *full homomorphism* if in addition the preimage of an edge is an edge. Full homomorphisms have very easy structure, as every full homomorphism which is onto is a retraction. The other special homomorphisms we will be interested in are *injective* homomorphisms.

Let $\mathcal{F}'$ be a finite set of colored relational structures. By $Forb(\mathcal{F}')$ we denote the set of all colored relational structures $\mathbf{A}'$ satisfying $\mathbf{F}' \not\longrightarrow \mathbf{A}'$ for every $\mathbf{F}' \in \mathcal{F}'$. (If we use injective or full homomorphisms this will be denoted by $Forb_{inj}(\mathcal{F}')$ or $Forb_{full}(\mathcal{F}')$, respectively.)

Contrary to our common sense, the left side of dualities (i.e. classes of form $Forb(\mathcal{F})$) is more powerful than the right side. We can prove that shadows of classes $\mathcal{F}'$ have computational power of the whole class NP. More precisely in Kun and Nešetřil (2007, 2008) we proved:

Theorem 3.1. *For every language $L \in NP$ there exist a finite set of colors Γ and a finite set of Γ-colored digraphs $\mathcal{F}'$, where we color the pairs of vertices, such that L is computationally equivalent to the membership problem for the set of all digraphs G for which there exists a Γ coloring graph G' of the pairs of vertices in G such that no $F' \in \mathcal{F}'$ is homomorphic to G'.*

Symbolically, $\Phi(Forb(\mathcal{F}'))$ is the class whose membership problem is polynomial equivalent to L.

Similar results hold also for classes $Forb_{inj}(\mathcal{F}')$ and $Forb_{full}(\mathcal{F}')$ (in these cases even with monadic lifts only!), see (Kun and Nešetřil 2007, 2008).

Let us consider the right side of dualities: The *Constraint Satisfaction Problem* corresponding to the relational structure **D** is the membership problem for the class of all structures defined by $\{\mathbf{B} : \mathbf{B} \longrightarrow \mathbf{D}\}$. Similarly, for a finite set of colored relational structures $\mathcal{D}'$ we denote by $CSP(\mathcal{D}')$ the class of all colored structures $\mathbf{A}'$ satisfying $\mathbf{A}' \longrightarrow \mathbf{D}'$ for some $\mathbf{D}' \in \mathcal{D}'$. ($CSP(\mathcal{D}')$ is a finite union of $CSP(\mathbf{D}')$ for $\mathbf{D}' \in \mathcal{D}'$. This is sometimes denoted by $\rightarrow \mathcal{D}'$.) If the classes $Forb(\mathcal{F}')$ and $CSP(\mathcal{D}')$ are equal then we get a finite homomorphism duality (for the lifted category) which we introduced earlier. Explicitly, in this notation, a finite duality means that the following equivalence holds for every (colored) relational structure $\mathbf{A}'$:

$$\forall \mathbf{F}' \in \mathcal{F}' \ \mathbf{F}' \not\longrightarrow \mathbf{A}' \quad \Longleftrightarrow \quad \exists \mathbf{D}' \in \mathcal{D}' \ \mathbf{A}' \longrightarrow \mathbf{D}'.$$

By Φ we denote the forgetful functor which associates to a Γ-colored relational structure the uncolored one, i.e. it forgets about the coloring. We will investigate classes of the form $\Phi(Forb(\mathcal{F}'))$. We call the pair $(\mathcal{F}', \mathcal{D})$ *shadow duality* if $\Phi(Forb(\mathcal{F}')) = CSP(\mathcal{D})$. An example of shadow duality is the language of 3-colorable graphs discussed in the introduction (or, as we shall see, any CSP problem in general).

We should add one remark. We of course do not only claim that every problem in NP can be polynomially *reduced* to a problem in any of these classes. This would only mean that each of these classes contains an NP-complete problem. What we claim is that these classes have the *computational power* of the whole NP class. More precisely, to each language L in NP there exists a language M in any of these three classes such that M is *polynomially equivalent* to L, i.e. there exist *polynomial reductions* of L to M and M to L. E.g. assuming P $\neq$ NP there is a language of form $\Phi(Forb(\mathcal{F}'))$ that is neither in P nor NP-complete, since there is such a language in NP by Ladner's celebrated result (Ladner 1975).

The expressive power of classes $\Phi(Forb(\mathcal{F}'))$ corresponds to many combinatorially studied problems and presents a combinatorial counterpart to the celebrated result of Fagin (1974) who expressed every NP problem in logical terms by means of an Existential Second Order formula. The proof of Theorem 3.1 uses refinements of Fagin's theorem due to Feder and Vardi (1999).

The fact that the membership problem for classes $\Phi(Forb(\mathcal{F}'))$ (and also their injective and full variants $\Phi(Forb_{inj}(\mathcal{F}'))$ and $\Phi(Forb_{full}(\mathcal{F}'))$, see Kun and Nešetřil 2008) have full computational power is pleasing from the combinatorial point of view as these classes cover well known examples of hard combinatorial problems:

Ramsey type problems (where as in Theorem 3.1 we consider edge colored graphs), colorings of bounded degree graphs (defined by an injectivity condition) and structural partitions (studied e.g. in Feder et al. (1999)). It follows that, in the full generality, one cannot expect dichotomies here. On the other side of the spectrum, Feder and Vardi have formulated the celebrated *Dichotomy conjecture* for all coloring problems (CSP).

The shadow dualities are related to the decision problems for classes $CSP(\mathbf{D})$.

The main result of (Kun and Nešetřil 2008, 2007) presents an easy characterization of those languages $\Phi(Forb(\mathcal{F}'))$ which are coloring problems (CSP):

Theorem 3.2.   *Consider the finite set of colors Γ and the language $\Phi(Forb(\mathcal{F}'))$ for a finite set $\mathcal{F}'$ of vertex Γ-colored relational structures.*

If no $\mathbf{F}' \in \mathcal{F}'$ contains a cycle then there is a finite set of relational structures $\mathcal{D}$ such that $\Phi(Forb(\mathcal{F}')) = CSP(\mathcal{D})$.

If one of the lifts $\mathbf{F}'$ in a minimal subfamily of $\mathcal{F}'$ contains a cycle in its core then the language $\Phi(Forb(\mathcal{F}'))$ is not a finite union of CSP languages.

This can be viewed as an extension of the duality characterization theorem for structures (Foniok et al. 2008). However the proof given in Kun and Nešetřil (2008) uses the Theorem 2.4, 2.5 together with the homomorphism properties of structures not containing short cycles (i.e. with a large girth). This is a combinatorial problem studied intensively since times of P. Erdős. The following result has proved to be repeatedly useful in various applications. It is often called the *Sparse Incomparability Lemma*:

Theorem 3.3. *Let k, ℓ be positive integers and let $\mathbf{A}$ be a structure. Then there exists a structure $\mathbf{B}$ with the following properties:*

1. *There exists a homomorphism $f : \mathbf{B} \longrightarrow \mathbf{A}$;*
2. *For every structure $\mathbf{C}$ with at most k points the following holds: there exists a homomorphism $\mathbf{A} \longrightarrow \mathbf{C}$ if and only if there exists a homomorphism $\mathbf{B} \longrightarrow \mathbf{C}$;*
3. *$\mathbf{B}$ has girth $\geq \ell$.*

This result was proved by probabilistic method in Nešetřil and Rödl (1989), Nešetřil and Zhu (2004), see also Hell and Nešetřil (2004). The polynomial time construction of $\mathbf{B}$ is possible, too: in the case of binary relations (digraphs) this was done in Matoušek and Nešetřil (2008) and for relational structures in Kun (2007).

On a higher level Theorem 3.2 may be interpreted as stability of dualities for finite structures. While shadows of the classes $Forb(\mathcal{F}')$ are computationally equivalent to the whole NP, the shadow dualities are not bringing anything new: these are just shadows of dualities. This is interesting also from the point of view of descriptive complexity as one can show that the coloring problems in the class MMSNP (see Feder and Vardi 1999) are just shadow finite dualities. This holds for graphs as well for relational structures, see (Kun and Nešetřil 2007, 2008) for details of these aspects of dualities.

13.3.1 Existence of Duals III—Deletion Method

Inspired by the previous connection of lifts and shadows and CSP we can construct the dual structures by monadic lifts. We only sketch the construction which in its spirit goes back to Komárek (1987) and it is implicit also in Feder and Vardi (1999) (I thank to V. Dalmau and J. Foniok who informed me about this). Allow me here to mention a bit of history: Pavel Komárek was my student in 80's and I directed his attention to dualities in the broad setting. I have been convinced that this is a good and elegant approach to *good characterizations* (in the sense of Edmonds) and from this point of view I also wrote Czech graph theory book (Nešetřil 1979). With A. Pultr we also wrote (Nešetřil and Pultr 1978) where we coined the term duality. We originally (Nešetřil and Pultr 1978) conjectured that Gallai–Hasse–Roy–Vitaver theorem is the only instance of duality for oriented graphs. Nearly 10 years later Komárek quickly found a new example and then infinitely many new examples which were reported in Komárek (1984) and Welzl (1982). This revolutionized the scene and we conjectured a converse: that any oriented tree leads to a duality. This has been proved by Komárek in his thesis (Komárek 1987). The proof has never been published and (unfortunately) Komárek himself did not pursue an academic career. In a different and general setting (and by different techniques) the theorem was proved in Nešetřil and Tardif (2000). This was the start of the theory covered here.

The deletion method essentially uses monadic lifts. Let us sketch it at least briefly now again on the case of oriented trees (type $\Delta = (2)$). Let F be a fixed (forbidden) tree. Let $(B_i, x_i) : i = 1, \ldots, t$ be the set consisting of all possible branches which appear in F. Thus every branch is determined by a vertex x_i and an edge e_i containing x_i. Let X consists from all subsets I of $\{1, 2, \ldots, t\}$ for which there is no homomorphism $F \longrightarrow B_I$ where B_I denotes the disjoint union of all $(B_i, x_i), i \in I$, with all roots x_i identified. X will be the vertex set of our dual graph D_F.

The edges of D_F will be defined in two steps: First, we consider all pairs (I, J). And then we delete a pair (I, J) if there exists $i \in I$ and $j \in J$ with an edge $e = (x_i, x_j) \in E(F)$ such that both branches (B_i, x_i) and (B_j, x_j) contain edge e.

Of course the language of both (Komárek 1987) and (Feder and Vardi 1999) is different and proofs and constructions more complicated So this is a good example of the use of lifts and shadows.

13.4 Universality and Existence of Duals

Homomorphism duality may be rephrased in yet another context. Consider a finite set of connected structures $\mathcal{F}$ and the class $Forb(\mathcal{F})$. Then the dual object $\mathbf{D}_{\mathcal{F}}$ is the maximum (or greatest) element of the class $Forb(\mathcal{F})$ in the homomorphism order $\mathcal{C}_\Delta$. Consequently Theorem 2.4 characterizes all the classes $Forb(F)$ which have maximum and Theorem 2.5 characterizes all classes of form $Forb(\mathcal{F})$ which are bounded by a finitely many maximal elements.

We can also say that $\mathbf{D}_{\mathcal{F}}$ is *hom-universal* object (Nešetřil and Rödl 1989) for the class $Forb(\mathcal{F})$. Hom-universal objects should be distinguished from embedding universality: Given a class $\mathcal{K}$ of countable structures, an object $\mathbf{U} \in \mathcal{K}$ is called *(embedding) universal* for $\mathcal{K}$ if for every object $\mathbf{A} \in \mathcal{K}$ there exists an embedding $\mathbf{A} \longrightarrow \mathbf{U}$).

The characterization of those classes $\mathcal{K}$ which have an universal object is a well known open problem in model theory which was studied intensively, see e.g. (Komjáth 1999; Komjáth and Mekler 1988; Cherlin and Shi 1996; Cameron 1998). The whole area was inspired by the negative results (see Hajnal and Pach 1981; Cherlin and Komjáth 1994): for example the class of graphs not containing C_l (= cycle of length l) fails to be universal for any $l > 3$. Until (Cherlin et al. 1999) in fact there were not many classes known with universal objects. The strongest results in the positive direction were obtained by Cherlin et al. (1999). Particularly, they proved the following

Theorem 4.1. *For every finite set $\mathcal{F}$ of finite connected graphs the class $Forb(\mathcal{F})$ has an embedding universal object.*

This result was extended to relational structures in Cherlin and Shi (2003). The proof of Theorem 4.1 given in Cherlin et al. (1999) is based on techniques of model theory and it is possible to say that no explicit universal object is constructed. Using lifts and shadows with J. Hubička we recently gave an alternative and more explicit combinatorial proof of Theorem 4.1 for structures (Hubička and Nešetřil 2008). Along the lines above we can get universal structure for the classes $Forb(\mathcal{F})$ in particularly easy way as the shadow $\mathbf{U}$ of the direct (Fraïssé) limit $\mathbf{U}$ of an explicitly defined lifted class $Forb(\mathcal{F}')$ (which is an amalgamation class), see (Hubička and Nešetřil 2008) for details.

13.4.1 Existence of Duals IV—Generic Duals

Clearly every universal object is also hom-universal. This however does not hold conversely (as shown by examples of classes with bounded chromatic numbers: K_4 is hom-universal for the class of planar graphs by virtue of the 4-color theorem). As we already know, of special interests are classes $Forb(\mathcal{F})$ which have finite hom-universal graph: these are just duals. The proof of Theorem 4.1 given in Hubička and Nešetřil (2008) gives more: In the case that $\mathcal{F}$ is a finite set of relational trees then the theorem is proved just by monadic lifts (similarly as in the above Construction III) and the resulting universal object $\mathbf{U}_{\mathcal{F}}$ has a finite retract $\mathbf{D}_{\mathcal{F}}$ which is consequently hom-universal. This implies yet another proof of the existence of duals.

Moreover, as the universal lifted structure $\mathbf{U}_{\mathcal{F}}$ may be chosen to be *generic* (meaning ultrahomogeneous and universal) then we see that we may think of duals as a retract of a generic object—*duals are generic*; see (Hubička and Nešetřil 2008).

13.5 Restricted Dualities

Finite dualities became much more abundant when we demand the validity of the duality formula just for all graphs from a given class $\mathcal{K}$. In such cases we speak about

$\mathcal{K}$-*restricted duality.* It has been proved in Nešetřil and Ossona de Mendez (2008c) that so called *Bounded Expansion* classes (which include both proper minor closed classes and classes of graphs with bounded degree) have a restricted duality for every choice of $\mathcal{F}$. As a consequence of this we can show that the shadow $\Phi(Forb(\mathcal{F}))$ of a vertex colored class of structures $Forb(\mathcal{F})$ is always the restriction of a CSP language when restricted to a bounded expansion class (this notion generalizes bounded degree and proper minor closed classes) (Nešetřil and Ossona de Mendez 2006c).

More explicitly, the following definition is the central definition of this section:

Definition 5.1. *A class of structures $\mathcal{K}$ admits* all restricted dualities *if, for any finite set of connected structures $\mathcal{F} = \{F_1, F_2, \ldots, F_t\}$, there exists a finite structure $\mathbf{D}_{\mathcal{F}}^{\mathcal{K}}$ such that $\mathbf{F}_i \not\longrightarrow \mathbf{D}_{\mathcal{F}}^{\mathcal{K}}$ for $i = 1, \ldots, t$ and for all $\mathbf{G} \in \mathcal{K}$,*

$$(\mathbf{F}_i \not\longrightarrow \mathbf{G}), \quad i = 1, 2, \ldots, t \quad \Longleftrightarrow \quad (\mathbf{G} \longrightarrow \mathbf{D}_{\mathcal{F}}^{\mathcal{K}}). \tag{1}$$

Any instance of (1) is called a restricted duality *(for the class $\mathcal{K}$).*

To motivate this definition let us consider the following example.

The Grötzsch's celebrated theorem (see e.g. Thomassen 1994) says that every triangle-free planar graph is 3-colorable. In the language of homomorphisms this says that for every triangle-free planar graph G there is a homomorphism of G into K_3. Using the partial order terminology, Grötzsch's theorem says that K_3 is an upper bound (in the homomorphism order) for the class $\mathcal{P}_3$ of all planar triangle-free graphs. The fact that $K_3 \notin \mathcal{P}_3$ suggests a natural question (first formulated in Nešetřil 1999): Is there yet a smaller bound? The answer, which may be viewed as a strengthening of Grötzsch's theorem, is positive: there exists a triangle free 3-colorable graph H such that $G \longrightarrow H$ for every graph $G \in \mathcal{P}_3$. This has been proved in Nešetřil and Ossona de Mendez (2006a, 2006b) in a stronger version for minor-closed classes.

One can view these results as restricted dualities (which hold in the class of planar graphs). Restricted duality results have since been generalized not only to proper minor closed classes of graphs and but also to other forbidden subgraphs, in fact to any finite set of connected graphs thus yielding all restricted dualities for the class of planar graphs. This then implies that Grötzsch's theorem can be strengthened by a sequence of even stronger bounds and that the supremum (in the homomorphism order) of the class of all triangle free planar graphs does not exist (Nešetřil and Ossona de Mendez 2005).

What is the proper setting for the restricted dualities? This is presently an open problem but the strongest result in this direction is the notion of a class with bounded expansion. Such a class may be defined in several (very) different ways, see (Nešetřil and Ossona de Mendez 2006c, 2008c, 2008d).

It is important that this seemingly elusive global property (having an upper bound) has a localized version by means of the densities of *shallow minors*. We can proceed as follows (Nešetřil and Ossona de Mendez 2008a):

The *maximum average degree* mad(G) of a graph G is the maximum over all subgraphs H of G of the average degree of H, that is mad(G) = $\max_{H \subseteq G} \frac{2|E(H)|}{|V(H)|}$. The *distance* $d(x, y)$ between two vertices x and y of a graph is the minimum length of a path linking x and y, or ∞ if x and y do not belong to same connected component. Also we denote by $G[A]$ the subgraph of G induced by a subset A of its vertices.

We introduce several notations:

- The *radius* $\rho(G)$ of a connected graph G is:

$$\rho(G) = \min_{r \in V(G)} \max_{x \in V(G)} d(r, x)$$

- A *center* of G is a vertex r such that $\max_{x \in V(G)} d(r, x) = \rho(G)$.

Definition 5.2. *Let G be a graph. A* ball *of G is a subset of vertices inducing a connected subgraph. The set of all the families of pairwise disjoint balls of G is noted $\mathfrak{B}(G)$.*

Let $\mathcal{P} = \{V_1, \ldots, V_p\}$ be a family of pairwise disjoint balls of G.

- *The* radius $\rho(\mathcal{P})$ *of $\mathcal{P}$ is $\rho(\mathcal{P}) = \max_{X \in \mathcal{P}} \rho(G[X])$.*
- *The* quotient $G/\mathcal{P}$ *of G by $\mathcal{P}$ is a graph with vertex set $\{1, \ldots, p\}$ and edge set $E(G/\mathcal{P}) = \{\{i, j\} : (V_i \times V_j) \cap E(G) \neq \emptyset \text{ or } V_i \cap V_j = \emptyset\}$.*
- *If $\rho(\mathcal{P}) \leq r$ then graph $G/\mathcal{P}$ is called* shallow minor at depth r *of graph G.*

The following invariants generalize maximum average degree:

Definition 5.3. *The* greatest reduced average density (grad) *of a graph G with rank r is*

$$\nabla_r(G) = \max \frac{|E(G/\mathcal{P})|}{|\mathcal{P}|},$$

where maximum is taken over all $\mathcal{P} \in \mathfrak{B}(G)$ satisfying $\rho(\mathcal{P}) \leq r$.

The following is our key definition:

Definition 5.4. *A* class of graphs $\mathcal{C}$ has bounded expansion *if there exists a function $f : \mathbb{N} \to \mathbb{N}$ such that for every graph $G \in \mathcal{C}$ and every r,*

$$\nabla_r(G) \leq f(r). \tag{2}$$

f is called the expansion function.

The definition of bounded expansion can be carried over to general structures by means of incidence graphs. Thus we may speak about classes of structures with bounded expansion.

The definition of bounded expansion is very robust: it may be alternatively defined by means of forbidden shallow subdivisions (Dvořák 2008), by means of special colorings of vertices (Zhu 2008; Nešetřil and Ossona de Mendez 2008d). The definition is preserved by most local operations (for example by doubling of vertices). Proper minor closed classes have bounded expansion with the constant expanding function. Graphs with all vertices bounded by d have exponential expansion function. Several geometrically defined graphs have polynomial expansion function. See (Nešetřil and Ossona de Mendez 2006c, 2008a, 2008b, 2008c, 2008d) for many more examples. Despite of this generality we have the following (Nešetřil and Ossona de Mendez 2008c):

Theorem 5.5. *Any class of structures with bounded expansion has all restricted dualities.*

13.6 Remarks

1. The existence of a homomorphism from an oriented path P to a graph G may be sometimes conveniently tested by means of matrix multiplication: Let $G = (V, E)$, $V = \{v_1, \ldots, v_n\}$ and let $A = (a_{ij})$ be the adjacency matrix (we put $a_{ij} = 1$ iff $(v_i, v_j) \in E$. For a path P with $k + 1$ vertices (and thus of k arcs) we consider the product B of matrices $B_1, B_2, \ldots, B_k$ where $B_i = A$ if the k-th arc of P is forward and $B_i = A^T$ if the k-th arc of P leads backwards. Then $b_{ij} = 1$ if and only if there exists a homomorphism from P to G mapping the first vertex of P to v_i and the last vertex to v_j. Thus $P \longrightarrow G$ if and only if the matrix $B \neq 0$.

This connection (which is already made in Vitaver 1962) can be used for an effective testing of large (recursively defined) paths, see (Nešetřil and Tardif 2008b). In this context it is fitting to note that the fastest algorithm for testing the existence of a homomorphism $G \longrightarrow H$ for a *fixed* G is based on the fast matrix multiplication (Nešetřil and Poljak 1985).

2. We proved that shadow dualities and lifted monadic dualities are in 1–1 correspondence. This abstract result has several consequences and streamlines some earlier results in descriptive complexity theory (related to MMSNP and CSP classes) (Madelaine 2003; Madelaine and Stewart 2005). The simplicity of this approach suggests some other problems. It is tempting to try to relate Ladner's diagonalization method (Ladner 1975) in this setting (as it was pioneered by Gács and Lovász (1977) for NP ∩ coNP in a similar context). The characterization of Lifted Dualities is beyond reach but particular cases are interesting as they generalize results of (Nešetřil and Tardif 2000; Foniok et al. 2008) and as the corresponding duals present polynomial instances of CSP.

But perhaps more importantly, our approach to the complexity subclasses of NP is based on lifts and shadows as a combination of algebra, combinatorics and logic. We believe that it has further applications and that it forms a useful paradigm.

3. Let us finish this paper by listing the characterization theorem for finite dualities. (We say that a class of structures $\mathcal{K}$ is *homomorphism closed* if $\mathbf{A} \in \mathcal{K}, \mathbf{A} \longrightarrow \mathbf{B}$ implies $\mathbf{B} \in \mathcal{K}$. We also denote by $\mathcal{F} \longrightarrow$ the class of all structures $\mathbf{A}$ for which there exists $\mathbf{F} \in \mathcal{F}$ such that $\mathbf{F} \longrightarrow \mathbf{A}$. The class $\mathcal{F} \longrightarrow$ is the complementary class of $Forb(\mathcal{F})$. By a combination of results (Atserias 2008; Rossman 2005; Nešetřil and Tardif 2000; Foniok et al. 2008) we obtain the following result.

Theorem 6.1. *Let $\mathcal{K}$ be a class of structures closed under homomorphisms. For $\mathcal{K}$ are the following statements equivalent:*

- *$\mathcal{K}$ is first order definable class;*
- *$\mathcal{K} = \mathcal{F} \longrightarrow$ for a finite set $\mathcal{F}$ of structures.*

It follows that any first order definable class $\mathcal{K} = CSP(\mathcal{D})$ is defined by a finite duality (and thus the corresponding set $\mathcal{F}$ is a set of finite relational trees).

By a combination of (Atserias et al. 2006; Nešetřil and Ossona de Mendez 2008c, 2008d) we have a surprisingly strong relativized version of this result:

Theorem 6.2. *Let $\mathcal{K}$ be a bounded expansion class. For a homomorphism closed subclass $\mathcal{L}$ of $\mathcal{K}$ are the following statements equivalent:*

- $\mathcal{L}$ *is first order definable in* $\mathcal{K}$;
- $\mathcal{L} = (\mathcal{F}) \longrightarrow$ *for a finite set* $\mathcal{F}$ *of structures;*
- $\mathcal{L}$ *is defined by a restricted finite duality.*

In Nešetřil and Ossona de Mendez (2008d, 2008e) we developed further the connections of classes of sparse graphs to logic and descriptive complexity.

Acknowledgements

The writing of this paper for the Bernhard Korte volume is bringing back some good memories of times at Nassestrasse and then Lennéstrasse, times when the topics discussed in this paper were still at their cradle.

References

Atserias, A.: On digraph coloring problems and treewidth duality. Eur. J. Comb. **29**(4), 796–820 (2008)

Atserias, A., Dawar, A., Kolaitis, Ph.G.: On preservation under homomorphisms and conjunctive queries. J. Assoc. Comput. Mach. **53**(2), 208–237 (2006)

Borgs, C., Chayes, J., Lovász, L., Sós, V.T., Vesztergombi, K.: Counting Graph Homomorphisms. Topics in Discrete Mathematics, pp. 315–371. Springer, Berlin (2006)

Cameron, P.J.: The random graph. In: Graham, R.L., Nešetřil, J. (eds.) The Mathematics of Paul Erdös, pp. 333–351. Springer, Berlin (1998)

Chandra, A.K., Merlin, P.M.: Optimal implementation of conjunctive queries in relational data bases. In: STOC'77, pp. 77–90. Am. Math. Soc., Providence (1977). Also: Springer, Berlin (1998), pp. 333–351

Cherlin, G., Komjáth, P.: There is no universal countable pentagon free graph. J. Graph Theory **18**, 337–341 (1994)

Cherlin, G., Shelah, S., Shi, N.: Universal graphs with forbidden subgraphs and algebraic closure. Adv. Appl. Math. **22**, 454–491 (1999)

Cherlin, G., Shi, N.: Graphs omitting a finite set of cycles. J. Graph Theory **21**, 351–355 (1996)

Cherlin, G., Shi, N.: Forbidden subgraphs and forbidden substructures. J. Symb. Logic **66**(3), 1342–1352 (2003)

Dvořák, Z.: On forbidden subdivision characterization of graph classes. Eur. J. Comb. (2008, to appear)

Fagin, R.: Generalized first-order spectra and polynomial-time recognizable sets. In: Karp, R. (ed.) Complexity of Computation. SIAM–AMS Proceedings, vol. 7, pp. 43–73 (1974)

Feder, T., Vardi, M.: The computational structure of monotone monadic SNP and constraint satisfaction: A study through Datalog and group theory. SIAM J. Comput. **28**(1), 57–104 (1999)

Feder, T., Hell, P., Klein, S., Motwani, R.: Complexity of graph partition problems. In: 31st Annual ACM STOC, pp. 464–472 (1999)

Foniok, J., Nešetřil, J., Tardif, C.: Generalized dualities and maximal finite antichains in the homomorphism order of relational structures. Eur. J. Comb. **29**(4), 881–899 (2008)

Gács, P., Lovász, L.: Some remarks on generalized spectra. Z. Math. Log. Grdl. **23**(6), 547–554 (1977)

Gallai, T.: On directed paths and circuits. In: Theory of Graphs, Proc. Colloq., Tihany, 1966, pp. 115–118. Academic Press, New York (1968)

Gottlob, G., Koch, C., Schulz, K.U.: Conjunctive queries over trees. Manuscript (2008)

Hajnal, A., Pach, J.: Monochromatic paths in infinite graphs. In: Finite and Infinite Sets. Coll. Math. Soc. J. Bolyai, vol. 37, pp. 359–369. Eger, Hungary (1981)

Hasse, M.: Zur algebraischen Begründung der Graphentheorie. I. Math. Nachr. **28**, 275–290 (1964/1965)

Hell, P., Nešetřil, J.: Graphs and Homomorphism. Oxford University Press, Oxford (2004)

Hodges, W.: Model Theory. Cambridge University Press, Cambridge (1993)

Hubička, J., Nešetřil, J.: Universal structures as shadows of ultrahomogeneous structures (2008, submitted)

Immerman, N.: Languages that capture complexity classes. SIAM J. Comput. **16**, 760–778 (1987)

Kolaitis, P., Vardi, M.: Conjunctive query containment and constraint satisfaction. In: Symposium on Principles of Database Systems (PODS98), pp. 205–213 (1998)

Komárek, P.: Some new good characterizations of directed graphs. Čas. Pěst. Mat. **109**, 348–354 (1984)

Komárek, P.: Good characterizations in the class of oriented graphs. Doctoral dissertation, Praha, 1987 (in Czech)

Komjáth, P.: Some remarks on universal graphs. Discrete Math. **199**, 259–265 (1999)

Komjáth, P., Mekler, A.: J. Pach: Some universal graphs. Israel J. Math. **64**, 158–168 (1988)

Kun, G.: On the complexity of Constraint Satisfaction Problem. PhD thesis (2006) (in Hungarian)

Kun, G.: Constraints, MMSNP and expander structures. Combinatorica (2007, submitted)

Kun, G., Nešetřil, J.: NP by means of lifts and shadows. In: Proc. MFCS'07. Lecture Notes in Computer Science, vol. 4708, pp. 171–181. Springer, Berlin (2007)

Kun, G., Nešetřil, J.: Forbidden lifts (NP and CSP for combinatorists). Eur. J. Comb. **29**(4), 930–945 (2008)

Kun, G., Tardif, C.: Homomorphisms of random paths (2008, in preparation)

Ladner, R.E.: On the structure of Polynomial Time Reducibility. J. Assoc. Comput. Mach. **22**(1), 155–171 (1975)

Lovász, L.: Operations with structures. Acta Math. Hung. **18**, 321–328 (1967)

Luczak, T., Nešetřil, J.: A probabilistic approach to the dichotomy problem. SIAM J. Comput. **36**(3), 835–843 (2006)

Madelaine, F.: Constraint satisfaction problems and related logic. PhD thesis (2003)

Madelaine, F., Stewart, I.A.: Constraint satisfaction problems and related logic. Manuscript (2005)

Matoušek, J., Nešetřil, J.: Constructions of sparse graphs with given homomorphisms (2008, in preparation)

Minty, G.J.: A theorem on n-coloring the points of a linear graph. Am. Math. Monthly **69**, 623–624 (1962)

Nešetřil, J.: Teorie Grafu. SNTL, Praha (1979)

Nešetřil, J.: Aspects of structural combinatorics. Taiwan. J. Math. **3**(4), 381–424 (1999)

Nešetřil, J., Ossona de Mendez, P.: Cuts and bounds. Discrete Math. **302**(1–3), 211–224 (2005)

Nešetřil, J., Ossona de Mendez, P.: Folding. J. Comb. Theory, Ser. B **96**(5), 730–739 (2006a)

Nešetřil, J., Ossona de Mendez, P.: Tree depth, subgraph coloring and homomorphism bounds. Eur. J. Math. **27**(6), 1022–1041 (2006b)

Nešetřil, J., Ossona de Mendez, P.: Low tree-width decompositions and algorithmic consequences. In: STOC'06, Proceedings of the 38th Annual ACM Symposium on Theory of Computing, pp. 391–400. ACM Press, New York (2006c)

Nešetřil, J., Ossona de Mendez, P.: Grad and classes with bounded expansion I. Decompositions. Eur. J. Comb. **29**(3), 760–776 (2008a)

Nešetřil, J., Ossona de Mendez, P.: Grad and classes with bounded expansion II. Algorithmic aspects. Eur. J. Comb. **29**(3), 777–791 (2008b)

Nešetřil, J., Ossona de Mendez, P.: Grad and Classes with bounded expansion III—restricted dualities. Eur. J. Comb. **29**(4), 1012–1024 (2008c)

Nešetřil, J., Ossona de Mendez, P.: Structural properties of sparse graphs. In: Lovász volume, Bolyai Society and Springer (2008d)

Nešetřil, J., Ossona de Mendez, P.: First order properties on nowhere dense structures (2008e, submitted)

Nešetřil, J., Poljak, S.: Complexity of the Subgraph Problem. Comment. Math. Univ. Carol. **26**(2), 415–420 (1985)

Nešetřil, J., Pultr, A.: On classes of relations and graphs determined by subobjects and factorobjects. Discrete Math. **22**, 287–300 (1978)

Nešetřil, J., Rödl, V.: Chromatically optimal rigid graphs. J. Comb. Theory, Ser. B **46**, 133–141 (1989)

Nešetřil, J., Tardif, C.: Duality theorems for finite structures (characterising gaps and good characterizations). J. Comb. Theory, Ser. B **80**, 80–97 (2000)

Nešetřil, J., Tardif, C.: Homomorphism duality: On short answers to exponentially long questions. SIAM J. Discrete Math. **19**, 914–920 (2005)

Nešetřil, J., Tardif, C.: A dualistic approach to bounding the chromatic number of a graph. Eur. J. Comb. **29**(1), 254–260 (2008a)

Nešetřil, J., Tardif, C.: Path homomorphisms, graph colourings and boolean matrices (2008b, submitted)

Nešetřil, J., Zhu, X.: On sparse graphs with given colorings and homomorphisms. J. Comb. Theory, Ser. B **90**, 161–172 (2004)

Nešetřil, J., Pultr, A., Tardif, C.: Gaps and dualities in Heyting categories. Comment. Math. Univ. Carol. **48**, 9–23 (2007)

Pultr, A., Trnková, V.: Combinatorial, Algebraical and Topological Representations of Groups, Monoids and Categories. North-Holland, Amsterdam (1968)

Rossman, B.: Existential positive types and preservation under homomorphisms. In: 20th IEEE Symposium on Logic in Computer Science (LICS), pp. 467–476 (2005)

Roy, B.: Nombre chromatique et plus longs chemins d'un graphe. Rev. Francaise Inform. Rech. Opér. **1**, 129–132 (1967)

Simonyi, G., Tardos, G.: Local chromatic number. Ky Fan's theorem and circular colorings, Combinatorica **26**, 589–626 (2006)

Thomassen, C.: Grötzsch's 3-color theorem and its counterparts for torus and the projective plane. J. Comb. Theory, Ser. B **62**, 268–279 (1994)

Vardi, M.Y.: The complexity of relational query languages. In: Proceedings of 14th ACM Symposium on Theory of Computing, pp. 137–146 (1982)

Vitaver, L.M.: Determination of minimal coloring of vertices of a graph by means of Boolean powers of the incidence matrix. Dokl. Akad. Nauk SSSR **147**, 758–759 (1962) (in Russian)

Welzl, E.: Color families are dense. J. Theor. Comput. Sci. **17**, 29–41 (1982)

Zhu, X.: Colouring graphs with bounded generalized colouring number. Discrete Math. (2008, in press)

On the Structure of Graphs Vertex Critical
with Respect to Connected Domination

Michael D. Plummer

Summary. A dominating set of vertices S of a graph G is connected if the subgraph $G[S]$ is connected. Let $\gamma_c(G)$ denote the size of any smallest connected dominating set in G. Graph G is k-γ-connected-vertex-critical (abbreviated "$k$cvc") if $\gamma_c(G) = k$, but if any vertex v is deleted from G, then $\gamma_c(G - v) \le k - 1$.

This concept of vertex criticality stands in contrast to the concept of criticality with respect to edge addition in which a graph G is defined to be k-connected-critical if the connected domination number of G is k, but if any edge is added to G, the connected domination number falls to $k - 1$.

It is well-known that the only 1cvc graph is K_1 and the 2cvc graphs are obtained from the even complete graphs K_{2n}, with $n \ge 2$, by deleting a perfect matching. In this paper we survey some recent results for the case when $\gamma_c = 3$. In Sect. 14.2 we present some recently derived basic properties of 3cvc graphs, especially with respect to connectivity, and then present three new infinite families of 3cvc graphs. In Sect. 14.3, we present some new matching results for 3cvc graphs.

14.1 Introduction

Let G denote a finite simple undirected graph with vertex set $V(G)$ and edge set $E(G)$. A set $S \subseteq V(G)$ is a *dominating set* for G if every vertex of G either belongs to S or is adjacent to a vertex of S. A dominating set S for G is a *connected dominating set* if the subgraph induced by S is connected. The minimum cardinality of a connected dominating set in a graph G is called the *connected domination number* of G and is denoted by $\gamma_c(G)$. Note that since a graph must be connected to have a connected dominating set, henceforth in this paper, when referring to connected domination, we shall assume all graphs under consideration are connected. Moreover, note that if $G \ne K_2$ and has a vertex of degree one (usually called a *pendant* vertex), then if v is its neighbor, $G - v$ has no connected dominating set of *any* size. We will avoid this somewhat troublesome special situation in what is to follow by assuming that all graphs under consideration, except K_2, have minimum degree two.

Graph G is k-γ-*connected-vertex-critical* (abbreviated "$k$cvc") if $\gamma_c(G) = k$, but $\gamma_c(G - v) \leq k - 1$, for every vertex $v \in V(G)$. It is well-known that the only 1cvc graph is K_1 and the 2cvc graphs are obtained from the even complete graphs K_{2n}, with $n \geq 2$, by deleting the edges of a perfect matching. In this paper, we will be concerned with the first non-trivial case; that is, when $k = 3$.

We shall also need the following notions from matching theory. A graph G is *factor-critical* if $G - v$ has a perfect matching for every vertex $v \in V(G)$, *bicritical* if $G - u - v$ has a perfect matching for every pair of distinct vertices $u, v \in V(G)$ or, more generally, k-*factor-critical* if, for every set $S \subseteq V(G)$ with $|S| = k$, the graph $G - S$ contains a perfect matching. Factor-critical and bicritical graphs play important roles in a canonical decomposition theory for arbitrary graphs in terms of their matchings. The interested reader is referred to Lovász and Plummer (1986) for much more on this subject.

In Ananchuen and Plummer (2005, 2006a, 2007a) (see also Plummer 2006), it was shown that under certain assumptions regarding connectivity and minimum degree, a vertex-critical graph G with (ordinary) domination number 3 will be factor-critical (if $|V(G)|$ is odd), bicritical (if $|V(G)|$ is even) or 3-factor-critical (again if $|V(G)|$ is odd). Analogous theorems for connected domination are presented in Sect. 14.3 of the present paper. Although vertex criticality for domination and connected domination are similar in some ways, there are some interesting differences in the results obtained for each and in the proof techniques used to obtain them.

Connected domination seems to have been studied first by Sampathkumar and Waliker (1979) who attribute the terminology to Hedetniemi. (See Haynes et al. 1998a, 1998b and Hedetniemi and Laskar 1984.) The algorithmic aspects of both domination and connected domination were first discussed by Garey and Johnson in their book (Garey and Johnson 1979) where it is claimed that both domination and connected domination are NP-complete, even when the graph is planar and regular of degree 4. (In fact, approximation of connected domination seems to be hard as well in that no PTAS (polynomial-time approximation scheme) is known for the problem. Cf. Guha and Khuller 1998.) See also (Hedetniemi and Laskar 1984) in which is found the interesting observation that the problem of computing $\gamma_c(G)$ is equivalent to finding the maximum number of leaves in any spanning tree of G, since $|V(G)| - \gamma_c(G)$ is this latter number. Hence if G is $k$cvc, the maximum number of leaves in a spanning tree remains unchanged upon the deletion of any vertex.

For a nice summary concerning the algorithmic status of connected domination with respect to various special graph classes, see (Colbourn and Stewart 1990). For relations between connected domination and other graph parameters, see (Favaron and Kratsch 1991; Bo and Liu 1996; Hedetniemi and Laskar 1984).

More recently, Chen et al. (2004) began the study of graphs *edge*-critical with respect to connected domination. That is, those graphs in which the connected domination number falls upon the addition of any missing edge. They obtained some basic results most of which have previous analogs for ordinary domination.

In Ananchuen et al. (2008a), structural and matching properties of graphs *edge*-critical with respect to connected domination were explored. The present paper surveys more recent results on graphs *vertex*-critical with respect to connected domination. In addition to some new structural results, several new infinite classes of examples are presented in Sect. 14.2, while some of the matching properties of these graphs are explored in Sect. 14.3. Some of the results presented here are somewhat similar to some of those obtained in Ananchuen et al. (2008a) for edge-criticality. However, there are also interesting differences and the proofs are markedly different from those to be found in Ananchuen et al. (2008a). All proofs of results presented in Sect. 14.2 may be found in Ananchuen et al. (2008b) and the proofs for Sect. 14.3, in Ananchuen et al. (2008c).

Throughout this paper we adopt the following notation. Let G be a graph. If $S \subseteq V(G)$, then $G[S]$ denotes the subgraph of G induced by S. We denote by $\delta(G)$, the minimum degree of any vertex in G and by $N_G[v]$, the closed neighborhood of vertex v; i.e., $N_G[v] = N_G(v) \cup \{v\}$. Further, $N_H(v)$ denotes either $N_G(v) \cap V(H)$ if H is a subgraph of G or $N_G(v) \cap H$ if H is a subset of $V(G)$. Whenever v is a vertex of a $k$cvc graph G, we denote by D_v any set of $k - 1$ vertices which together dominate graph $G - v$ and such that $G[D_v]$ is connected. We write $H_1 + H_2$ for the join of graphs H_1 and H_2, that is, the graph obtained by joining every vertex of H_1 to every vertex of H_2. We shall refer to a graph G as being $K_{1,t}$-free, if it contains no $K_{1,t}$ as an induced subgraph.

14.2 General Structure

Most of the proofs of the results surveyed in this paper involve the study of the sets D_v. Hence we begin by listing some of the elementary properties of these sets.

Lemma 2.1. *Let $\gamma_c(G) = k$ and suppose G is $k$cvc. Then*

(i) *if $v \in V(G)$, $D_v \cap N_G[v] = \emptyset$,*
(ii) *if $u, v \in V(G)$ and $u \neq v$, then $D_u \neq D_v$,*
(iii) *if $v \in V(G)$, then $|D_v| = k - 1$, and*
(iv) *if $v \in V(G)$, then there is no $v_1 \neq v$ such that $N_G[v] \subseteq N_G[v_1]$.*

Theorem 2.2. *If G is a 3cvc graph, then either $G = C_5$ or G is 3-connected.*

In the case when G is *exactly* 3-connected, we can say more about its structure. We include the proof of the next result to give the reader a glimpse of the techniques used for 3cvc graphs.

Theorem 2.3. *Suppose G is a 3cvc graph and S is a vertex cutset in G with $|S| = 3$. Then*

(i) *$|E(G[S])| \leq 1$ and*
(ii) *$G - S$ consists of precisely two components.*

Proof. It is easy to see that the theorem follows if $G \cong C_5$. So we may assume that G is 3-connected. Then S is a minimum cutset.

To prove (i), suppose to the contrary that there are at least two edges in $G[S]$. Let x be a vertex in S which is adjacent to the other two members of S. Consider $G - x$. Then $D_x \cap S = \emptyset$ and so D_x cannot be connected, a contradiction.

We proceed to part (ii). Let $C_1, C_2, \ldots, C_t$ be the components of $G - S$ and let $S = \{x_1, x_2, x_3\}$. Suppose to the contrary that $t \geq 3$.

Claim. No vertex of $\bigcup_{i=1}^{t} V(C_i)$ is adjacent to every vertex of S.

Proof. Suppose to the contrary that $x \in \bigcup_{i=1}^{t} V(C_i)$ and x is adjacent to every vertex of S. Then $D_x \cap S = \emptyset$ and hence $D_x \subseteq \bigcup_{i=1}^{t} V(C_i)$. But this contradicts the connectedness of D_x and the Claim follows. □

By Theorem 2.2 and the above Claim, $|V(C_i)| \geq 2$ for $1 \leq i \leq t$. Furthermore, for $1 \leq i \leq t$ and for $1 \leq j \leq 3$, $N_{C_i}(x_j) \neq \emptyset$.

Choose $y_1 \in N_{C_1}(x_1)$ and consider $G - y_1$. Clearly, $x_1 \notin D_{y_1}$ and $D_{y_1} \cap (S - \{x_1\}) \neq \emptyset$. Without loss of generality, we may assume that $x_2 \in D_{y_1}$. Then $x_2 y_1 \notin E(G)$. Let $\{z\} = D_{y_1} - \{x_2\}$. Then $x_2 z \in E(G)$. We distinguish three cases.

Case 1: Suppose $z \in V(C_1) - \{y_1\}$. Then x_2 dominates $\bigcup_{i=2}^{t} V(C_i)$. Now choose $y_2 \in N_{C_2}(x_1)$. Then $y_2 x_1 \in E(G)$ and $y_2 x_2 \in E(G)$.

Now consider $G - y_2$. Clearly, $D_{y_2} \cap \{x_1, x_2\} = \emptyset$ and then $x_3 \in D_{y_2}$ and hence $x_3 y_2 \notin E(G)$. If x_3 dominates C_3, then $N_{C_3}(x_1) = \emptyset$ by our Claim. But this contradicts the fact that S is a *minimum* cutset. Hence x_3 does not dominate C_3.

Suppose $\{z_1\} = D_{y_2} - \{x_3\}$. Then $z_1 \in V(C_3)$ and $x_3 z_1 \in E(G)$. Hence x_3 dominates $\bigcup_{i=1}^{t} V(C_i) - (V(C_3) \cup \{y_2\})$. By our Claim, $x_1 z_1 \notin E(G)$ since x_2 dominates $V(C_3)$ and $x_3 z_1 \in E(G)$. Because $|V(C_2)| \geq 2$, there is a vertex $w \in V(C_2) - \{y_2\}$. But then w is adjacent to x_2 and x_3 since x_2 and x_3 each dominate $V(C_2) - \{y_2\}$.

Now consider $G - w$. Clearly, $D_w \cap \{x_2, x_3\} = \emptyset$. But then $x_1 \in D_w$. Let $\{z_2\} = D_w - \{x_1\}$. Then $z_2 \in \bigcup_{i=1}^{t} V(C_i) - \{w\}$. Since $x_1 z_1 \notin E(G)$, $z_2 \in V(C_3)$. Thus x_1 dominates $\bigcup_{i=1}^{t} V(C_i) - (V(C_3) \cup \{w\})$. Hence z is adjacent to every vertex of S since x_1 and x_3 each dominate $V(C_1)$ and $x_2 z \in E(G)$. But this contradicts our Claim and hence Case 1 cannot occur.

Case 2: Suppose $z \in S$. Then $z = x_3$ and thus $x_2 y_1 \notin E(G)$ and $x_3 y_1 \notin E(G)$. Furthermore, $\{x_2, x_3\}$ dominates $\bigcup_{i=1}^{t} V(C_i) - \{y_1\}$. Suppose $y_2 \in N_{C_2}(x_1)$. Then $y_2 x_2 \in E(G)$ or $y_2 x_3 \in E(G)$. Without loss of generality, assume that $y_2 x_2 \in E(G)$. By our Claim, $y_2 x_3 \notin E(G)$.

We next show that $N_G(x_1) \cap N_G(x_2) = \{y_2\}$. For suppose not. Then there exists a vertex $u \in (N_G(x_1) \cap N_G(x_2)) - \{y_2\}$. Consider $G - u$ and note that $D_u \cap \{x_1, x_2\} = \emptyset$. Clearly by part (i), $u \neq x_3$ and $x_3 \in D_u$. Suppose $\{z\} = D_u - \{x_3\}$. Then $z \in \bigcup_{i=1}^{t} V(C_i) - \{u\}$. But then z must be adjacent to both y_1 and y_2 since $x_3 y_1 \notin E(G)$ and $x_3 y_2 \notin E(G)$. But this is not possible. Hence $N_G(x_1) \cap N_G(x_2) = \{y_2\}$. Now suppose $y_3 \in N_{C_3}(x_1)$. Then $y_3 x_2 \notin E(G)$. Since $D_{y_1} = \{x_2, x_3\}$, $x_3 y_3 \in E(G)$. Now consider $G - x_1$ and note that $D_{x_1} \neq \{x_2, x_3\}$, by Lemma 2.1(ii), since $D_{y_1} = \{x_2, x_3\}$. However, $D_{x_1} \cap \{x_2, x_3\} \neq \emptyset$. So suppose $x_2 \in D_{x_1}$ and $x_3 \notin D_{x_1}$. Let $\{z_1\} = D_{x_1} - \{x_2\}$. Then $z_1 \in \bigcup_{i=1}^{t} V(C_i)$. Hence z_1 must dominate both y_1 and y_3 since $x_2 y_1 \notin E(G)$ and $x_2 y_3 \notin E(G)$. But this is not possible. Therefore, $x_2 \notin D_{x_1}$.

By a similar argument, $x_3 \notin D_{x_1}$. But this contradicts the fact that $D_{x_1} \cap \{x_2, x_3\} \neq \emptyset$. Therefore, Case 2 cannot occur.

Case 3: Suppose $z \in \bigcup_{i=2}^{t} V(C_i)$. Without loss of generality, we may assume that $z \in V(C_2)$. Then x_2 dominates $\bigcup_{i=1}^{t} V(C_i) - (V(C_2) \cup \{y_1\})$ and $x_2 z \in E(G)$. But then by our Claim, $x_1 z \notin E(G)$ or $x_3 z \notin E(G)$.

Case 3.1: Suppose $x_1 z \notin E(G)$. Then $x_1 x_2 \in E(G)$. By part (i), $x_1 x_3 \notin E(G)$ and $x_2 x_3 \notin E(G)$. Suppose $y_3 \in N_{C_3}(x_3)$. Then y_3 is adjacent to both x_2 and x_3. By our Claim, $x_1 y_3 \notin E(G)$. Consider $G - y_3$ and note that $D_{y_3} \cap \{x_2, x_3\} = \emptyset$ and $x_1 \in D_{y_3}$. Since $x_1 z \notin E(G)$ and $z \in V(C_2)$, it follows that x_1 dominates $\bigcup_{i=1}^{t} V(C_i) - (V(C_2) - \{y_3\})$. Now $(V(C_1) - \{y_1\}) \cup (V(C_3) - \{y_3\}) \subseteq N_G(x_1) \cap N_G(x_2)$. This implies that $N_{C_1}(x_3) = \{y_1\}$ and $N_{C_3}(x_3) = \{y_3\}$ by our Claim and the fact that S is a minimum cutset. Suppose $w \in V(C_1) - \{y_1\}$ and consider $G - w$. Clearly, $D_w \cap \{x_1, x_2\} = \emptyset$ and $x_3 \in D_w$. Suppose $\{z_1\} = D_w - \{x_3\}$. Then $z_1 x_3 \in E(G)$ and $z_1 \in \bigcup_{i=1}^{t} V(C_i) - \{w\}$. But then z_1 has to dominate $V(C_1) - \{y_1, w\}$ as well as $V(C_3) - \{y_3\}$.

Hence $|V(C_1)| = 2$ and $z_1 \in V(C_3)$. It then follows that $z_1 = y_3$. But then no vertex of D_w is adjacent to x_1, a contradiction. Thus Case 3.1 cannot occur.

Case 3.2: $x_3 z \notin E(G)$. By an argument similar to that used in Case 3.1, Case 3.2 cannot occur either. Therefore, Case 3 cannot occur. This completes the proof of the theorem.   □

Note that part (i) of the above theorem is best possible in the following sense. In the 3cvc graph G_0 shown in Fig. 14.1, $S_1 = \{x_1, x_2, x_3\}$ is a cutset of size three with $|E(G_0[S_1])| = 1$, whereas $S_2 = \{x_3, x_4, x_5\}$ is a cutset of size three with $|E(G_0[S_2])| = 0$.

Lemma 2.4. *Suppose G is a 3cvc graph and that S is a minimum vertex cutset in G with $|S| = 3$. Let C_1 and C_2 be the two components of $G - S$. Then either $|V(C_1)| \leq 2$ or $|V(C_2)| \leq 2$.*

Lemma 2.5. *Suppose G is a 3cvc graph and S is a minimum vertex cutset in G with $|S| = 3$. Let C_1 and C_2 be the components of $G - S$ with $|V(C_1)| \leq |V(C_2)|$. If $|V(C_1)| = 2$, then also $|V(C_2)| = 2$, and G is isomorphic to the graph G_0 in Fig. 14.1.*

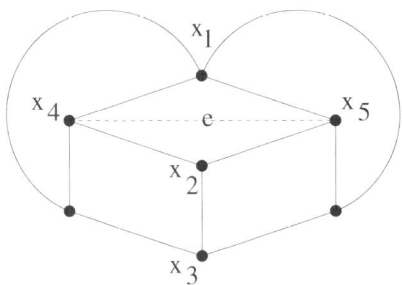

Fig. 14.1. The graphs G_0 (without edge e) and G_0' (with edge e)

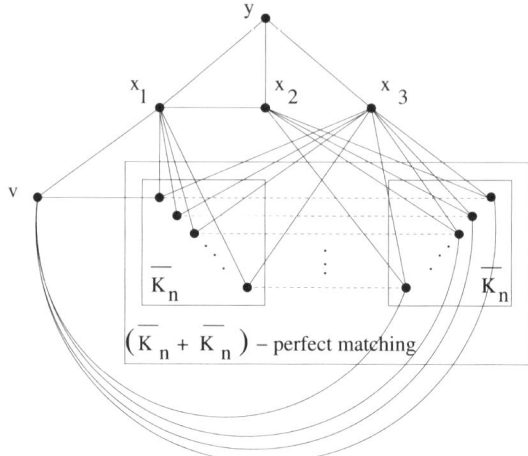

Fig. 14.2. The graph G_1

Lemma 2.6. *Suppose G is a 3cvc graph and S is a minimum vertex cutset in G with $|S| = 3$. Let C_1 and C_2 be the components of $G - S$ with $|V(C_1)| = 1$. If S is independent, then G is isomorphic to one of G_0 or G'_0, both of which are shown in Fig. 14.1.*

Theorem 2.7. *Suppose G is a 3cvc graph and S is a minimum vertex cutset in G with $|S| = 3$. Let C_1 and C_2 be the components of $G - S$ with $|V(C_1)| = 1$. If S contains an edge, then*

(i) *G has G_1 or G_2 as a spanning subgraph, where G_1 and G_2 are shown in Figs. 14.2 and 14.3, and*

(ii) *G is factor-critical.*

Corollary 2.8. *Suppose G is 3cvc and has a minimum cutset of size three. Then*

(i) *G is factor-critical and*

(ii) *G contains a vertex of degree three.*

Corollary 2.9. *If G is a 3cvc graph and $|V(G)|$ is even, then G is 4-connected.*

We now present several new infinite families of 3cvc graphs. The members of the first family contain a vertex cutset S of size 3 such that there is a component of $G - S$ with $|V(C_1)| = 1$ and in which $G[S]$ contains exactly one edge. This family will be denoted by $\{R_n\}_{n=1}^{\infty}$. Fix integer $n \geq 1$ and construct R_n as follows. Let $X = \{x_1, \ldots, x_n\}$ and $Y = \{y_1, \ldots, y_n\}$. Set $V(R_n) = X \cup Y \cup \{u_1, u_2, u_3, u_4, u_5\}$, so $|V(R_n)| = 2n+5$. Form complete graphs on X and on Y. Join u_1 to u_2, u_3 and u_4; join u_2 to u_3 and to every vertex of $X \cup \{u_5\}$; join u_3 to every vertex of $Y \cup \{u_5\}$; join u_4 and u_5 to every vertex of $X \cup Y$ and finally join each x_i to y_j, for $1 \leq i \neq j \leq n$. It is easily seen that R_n is 3cvc and has a cutset $S = \{u_2, u_3, u_4\}$ such that $G[S]$ contains exactly one edge.

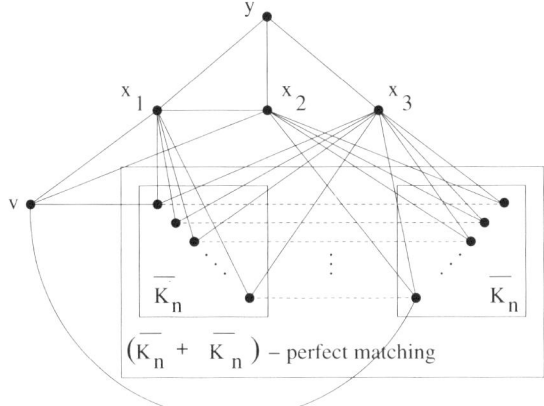

Fig. 14.3. The graph G_2

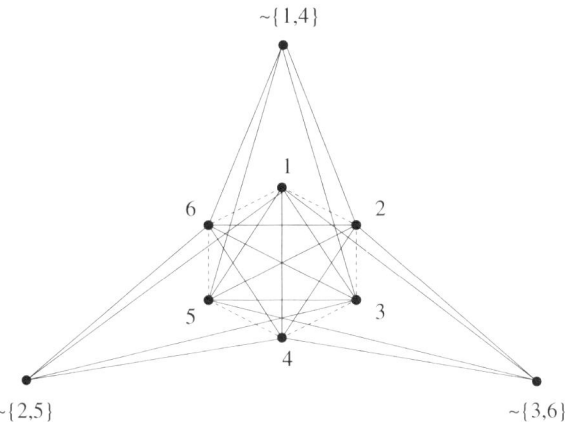

Fig. 14.4. The graph J_6

As a second example of an infinite family of 3cvc graphs consider the family $\{H_n\}_{n=2}^{\infty}$ defined as follows. Join $\overline{K}_n$ to K_n with all n^2 possible edges and then delete a perfect matching between $\overline{K}_n$ and K_n. Now add one more vertex and join it to all vertices of the $\overline{K}_n$. The resulting graph H_n has $2n + 1$ vertices and is clearly 3cvc. The smallest members of this class, H_2 and H_3 suffice to show that the conclusion of Theorem 2.2 is best possible.

We proceed to define a third infinite class of 3cvc graphs denoted by $\mathcal{J} = \{J_k\}_{k=6}^{\infty}$. Let $k \geq 6$ be a positive integer. We construct the graph J_k as follows. Let the integers $1, \ldots, k$ span a complete k-graph with the Hamilton cycle $12 \cdots k1$ removed. These k vertices will be called *central* vertices. Consider a second set of $(k(k-3)/2) - k$ *peripheral* vertices labeled with unordered pairs of distinct integers $\sim \{i, j\}, 1 \leq i < j \leq k$, except for exactly those pairs of the form $\{i, i + 1\}$ and

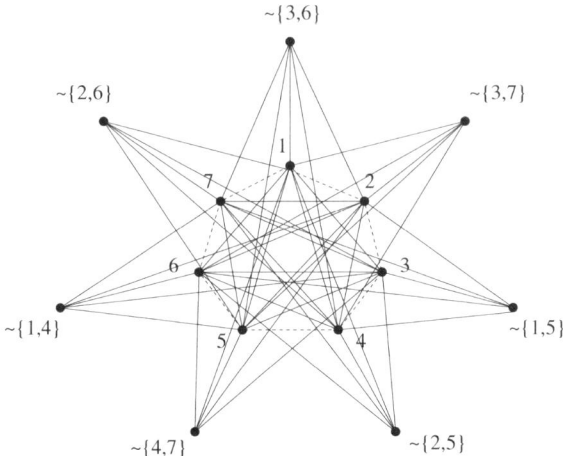

Fig. 14.5. The graph J_7

$\{i, i + 2\}$ modulo k. Now join the pair labeled $\sim \{i, j\}$ to all central vertices except i and j.

It is easy to see that $\gamma_c(J_k) = 3$. Again consider the Hamilton cycle $H = 12 \cdots k1$ removed in the construction of J_k. If vertex j is on H, then $\{j - 1, j + 1\}$ (mod k) is a connected dominating set for $J_k - j$. On the other hand, if a peripheral vertex v bears label "$\sim \{i, j\}$", then the vertices i and j form a connected dominating set for $J_k - v$.

The graphs J_6 and J_7 are shown in Figs. 14.4 and 14.5.

14.3 Matching Structure

It is known (Chen et al. 2004, Theorem 4) that every graph of even order which is *edge*-critical with respect to having $\gamma_c(G) = 3$ has a perfect matching. But this is not the case for vertex criticality. For example, it is easy to see that the 3cvc graph J_8 of order 20 constructed in Sect. 14.2 has no perfect matching. Hence it becomes of interest to investigate sufficient conditions for 3cvc graphs to have either perfect or near-perfect matchings. Previous results along this line for *edge*-criticality vis à vis both ordinary and connected domination may be found in (Ananchuen and Plummer 2003, 2004, 2006b, 2007b; Ananchuen et al. 2008a), while for vertex criticality in the case of ordinary domination see (Ananchuen and Plummer 2005, 2006a, 2007a).

In the case of *even* 3cvc graphs we are able to show the following.

Theorem 3.1. *Let G be a $K_{1,7}$-free 3cvc graph of even order. Then G contains a perfect matching.*

The preceding theorem is, in a sense, best possible in that "$K_{1,7}$-free" cannot be replaced with "$K_{1,8}$-free". To see this, we construct a graph H_1 on twenty-two vertices as follows. Begin with a complete graph on the nine vertices labelled

$1, 2, \ldots, 9$, and then delete the Hamilton cycle $12 \cdots 91$ as well as the edge 26. We now add thirteen additional vertices labelled according to the code "$\sim \{i, j\}$", where $\sim \{i, j\}$ is to be interpreted as the vertex so labeled is joined to all vertices of the set $\{1, 2, \ldots, 9\}$, except i and j. These additional vertices are $\sim \{1, 4\}$, $\sim \{1, 5\}$, $\sim \{1, 7\}$, $\sim \{2, 8\}$, $\sim \{3, 7\}$, $\sim \{3, 8\}$, $\sim \{3, 9\}$, $\sim \{4, 7\}$, $\sim \{4, 8\}$, $\sim \{4, 9\}$, $\sim \{5, 8\}$, $\sim \{5, 9\}$ and $\sim \{6, 9\}$. Finally, we add three more edges, namely, a triangle on the set $\{\sim \{2, 8\}, \sim \{3, 7\}, \sim \{3, 8\}\}$. Clearly, the 3cvc graph H_1 contains induced K_7's centered at 4 and 9 and does not contain a perfect matching, since the deletion of vertices $1, 2, \ldots, 9$ leaves eleven odd components. H_1 is, however, $K_{1,8}$-free.

Lemma 3.2. *Suppose G is a $K_{1,k}$-free 3cvc graph, where $k \geq 2$. Suppose S is a vertex cutset of G with $t = \omega(G - S) \geq k$. Then for each $x \in S$, $D_x \subseteq S$. Furthermore, $\deg_S(x) \leq |S| - 3$.*

Proof. Let $C_1, C_2, \ldots, C_t$ denote the components of $G - S$. Suppose, to the contrary, that there is a vertex $y \in S$ such that $D_y \not\subseteq S$. Since $|D_y| = 2$, and D_y is connected, it follows that $D_y \cap S \neq \emptyset$. Thus $D_y - S \subseteq \bigcup_{i=1}^{t} V(C_i)$. Let $D_y = \{y_1, y_2\}$, where $y_1 \in S$ and $y_2 \notin S$. Because D_y is connected, $y_1 y_2 \in E(G)$. Without loss of generality, we may suppose that $y_2 \in V(C_1)$. Then y_1 dominates $\bigcup_{i=2}^{t} V(C_i)$. Thus y_1 is the center of an induced $K_{1,t}$, for $t \geq k$, a contradiction.

Hence $D_x \subseteq S$ for each $x \in S$. It follows immediately that $\deg_S(x) \leq |S| - 3$, since x is not adjacent to any vertex of D_x. $\square$

An edge $e = uv$ of a 3cvc graph G will be called a *dominating edge* if $\{u, v\}$ is a connected dominating set for $G - w$, where w is some vertex of $G - \{u, v\}$. A graph G' is said to be a *virtual graph* for a 3cvc graph G if $V(G') = V(G)$ and for each pair of vertices u and v in $V(G')$, u is adjacent to v if and only if uv is a dominating edge in G. Clearly if G is a 3cvc graph, then G' is a spanning subgraph of G with $|E(G')| \geq |V(G)|$.

Lemma 3.3. *Suppose G is a $K_{1,k}$-free 3cvc graph, where $k \geq 2$. Let S be a vertex cutset in G with $t = \omega(G - S) \geq k + 1$. Then, for each $x \in V(G)$, $D_x \subseteq S$. Moreover, if G' is the virtual graph of G, then $|E(G'[S])| = |E(G')| \geq |V(G)|$.*

Proof. Let $C_1, C_2, \ldots, C_t$ be the components of $G - S$. Suppose, to the contrary, that there exists a vertex $y \in V(G)$ such that $D_y \not\subseteq S$. By Lemma 3.2, $y \notin S$. We may assume without loss of generality that $y \in V(C_1)$. Since D_y is connected, $D_y \cap S \neq \emptyset$. So $D_y - S \subseteq \bigcup_{i=1}^{t} V(C_i)$. Denote D_y by $\{y_1, y_2\}$ where $y_1 \in S$ and $y_2 \notin S$. Clearly, $y_1 y_2 \in E(G)$ and y_1 dominates at least $t - 2 \geq k - 1$ components of $G - S$, other than the components containing y and y_2. But then y_1 is the center of an induced $K_{1,t-1}$, for some $t \geq k+1$ because $y_1 y_2 \in E(G)$. But this is a contradiction.

Since G is 3cvc, for each $x \in V(G)$ there is a connected dominating set D_x. By Lemma 3.2 and the first part of the present Lemma above, $D_x \subseteq S$. Thus $|E(G'[S])| = |E(G')| \geq |V(G)|$. $\square$

In order to get an idea of the techniques needed to prove theorems on matchings in 3cvc graphs, we again include a proof.

Theorem 3.4. *Let G be a $K_{1,4}$ free 3cvc graph of even order. Then G is bicritical.*

Proof. Suppose that G is not bicritical. Then there exist vertices x and y in $V(G)$ such that $G_1 = G - \{x, y\}$ has no perfect matching. So by Tutte's 1-factor theorem and parity, it follows that there is a vertex cutset $S_1 \subseteq V(G_1)$ such that $\omega_o(G_1 - S_1) \geq |S_1| + 2$. Now let $S = S_1 \cup \{x, y\}$. Since, by Theorem 3.1, G contains a perfect matching, we have $|S| = |S_1| + 2 \leq \omega_o(G_1 - S_1) = \omega_o(G - S) \leq |S|$. This $\omega_0(G - S) = |S|$. Since G is 4-connected by Lemma 2.5, it follows that $\omega_o(G - S) = |S| \geq 4$. It then follows by Lemma 3.2 that for each $x \in S$, we have $D_x \subseteq S$ and $\deg_S(x) \leq |S| - 3$.

Now let G' be the virtual graph associated with graph G. We distinguish two cases.

Case 1: Suppose $|S| = 4$.

It is not difficult to show that $G[S] \cong 2K_2$. But then it also follows that $G'[S] \cong 2K_2$. Let $G'[S] = \{x_1y_1, x_2y_2\}$. Then $D_{x_1} = D_{y_1} = \{x_2, y_2\}$, contradicting Lemma 2.1(ii). Hence Case 1 cannot occur.

Case 2: Suppose now that $|S| \geq 5$.

Then by Lemma 3.3, for each $x \in V(G)$, $D_x \subseteq S$. Since G is $K_{1,4}$-free, $|S| = \omega_o(G - S) \leq 6$. But then by Lemma 3.3, we have $2|S| \leq |V(G)| \leq |E(G')| = |E(G'[S])| \leq (1/2)|S|(|S| - 3)$. Hence $|S| \geq 7$, a contradiction, and the Lemma is proved. □

Our hypothesis that the graph be $K_{1,4}$-free in the preceding theorem is best possible, for the graph G_3 shown in Fig. 14.6 is a 3cvc graph of order 10 containing $K_{1,4}$ as an induced subgraph. Clearly, G_3 is not bicritical since $G_3 - \{x, y\}$ has no perfect matching.

Theorem 3.5. *If G is a $K_{1,5}$-free 5-connected 3cvc graph of even order, then G is bicritical.*

Our hypotheses that the graph be $K_{1,5}$-free and that it be 5-connected in the preceding theorem are both best possible. To see this, first note that the graph G_3 shown in Fig. 14.6 is $K_{1,5}$-free and 4-connected, but not bicritical, since $G_3 - x - y$ has no perfect matching. Moreover, the graph J_7 (defined in the previous section) is an even $K_{1,6}$-free 5-connected graph which is not bicritical since $J_7 - \{1, 2\}$ contains no perfect matching.

On the other hand, for odd 3cvc graphs we have the following.

G_3: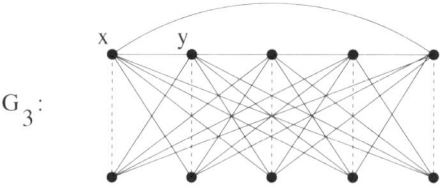

Fig. 14.6. The graph G_3

Theorem 3.6. *Let G be a $K_{1,7}$-free 3cvc graph of odd order. Then G contains a near-perfect matching.*

To see that the preceding theorem is also, in a sense, best possible, we construct a graph on twenty-seven vertices, denoted by H_2, as follows. Begin with a complete graph on ten vertices labeled $1, 2, \ldots, 9, A$ and delete the Hamilton cycle $12 \cdots 9A1$ as well as the two edges $4A$ and 59. Next, add seventeen additional vertices labeled with the code "$\sim \{i, j\}$" which is to be interpreted to say that the vertex so labeled is joined to all of the vertices $1, 2, \ldots, 9, A$, except i and j. The vertices so added are $\sim \{1, 5\}, \sim \{1, 6\}, \sim \{1, 7\}, \sim \{1, 8\}, \sim \{2, 5\}, \sim \{2, 6\}, \sim \{2, 7\}, \sim \{2, 8\},$ $\sim \{2, 9\}, \sim \{3, 6\}, \sim \{3, 7\}, \sim \{3, 8\}, \sim \{3, 9\}, \sim \{4, 7\}, \sim \{4, 8\}, \sim \{6, A\}$ and $\sim \{7, A\}$. Finally, add a triangle on vertices $\sim \{1, 5\}, \sim \{1, 6\}$ and $\sim \{1, 8\}$, as well as a triangle on $\sim \{2, 5\}, \sim \{2, 6\}$, and $\sim \{2, 8\}$. It is easily seen that the 3cvc graph H_2 contains an induced $K_{1,8}$ centered at vertex 7 and does not contain a near-perfect matching.

Theorem 3.7. *Let G be a $K_{1,6}$-free 3cvc graph of odd order. Then G is factor-critical.*

We now present a graph on twenty-seven vertices, denoted by H_3, which shows that, in a sense, Theorem 3.7 is best possible. Begin by taking a complete graph on the ten vertices labeled $1, 2, \ldots, 9, A$ and delete the Hamilton cycle $12 \cdots 9A1$, as well as edges 59 and 4A. Now add seventeen new vertices labeled via the symbol "$\sim \{i, j\}$" which is to be interpreted as saying that vertex $\sim \{i, j\}$ is adjacent to all vertices in $\{1, 2, \ldots, 9, A\}$ except i and j. The seventeen vertices to be added are $\sim \{1, 5\}, \sim \{1, 6\}, \sim \{1, 7\}, \sim \{1, 8\}, \sim \{2, 5\}, \sim \{2, 6\}, \sim \{2, 7\}, \sim \{2, 8\},$ $\sim \{2, 9\}, \sim \{3, 6\}, \sim \{3, 7\}, \sim \{3, 8\}, \sim \{3, 9\}, \sim \{4, 7\}, \sim \{4, 8\}, \sim \{6, A\}$ and $\sim \{7, A\}$. Finally, add a triangle on the vertices $\sim \{1, 5\}, \sim \{1, 6\}, \sim \{1, 8\}$, another on the vertices $\sim \{2, 5\}, \sim \{2, 6\}, \sim \{2, 8\}$ and a third on the vertices $\sim \{3, 6\},$ $\sim \{3, 8\}$ and $\sim \{3, 9\}$. It is easily checked that the 3cvc graph H_3 contains an induced $K_{1,6}$ centered at vertex 7 and that H_3 is not factor-critical.

Finally, we turn our attention to 3-factor-critical 3cvc graphs. Favaron proved the following result about k-factor-critical graphs in general.

Theorem 3.8 (Favaron 1996). *Every k-factor-critical graph of order $|V(G)| > k$ is k-connected and $(k + 1)$-edge-connected and both results are sharp.*

Hence, in particular, the hypothesis on minimum degree in the following theorem is necessary.

Theorem 3.9. *Let G be a $K_{1,3}$-free 3cvc graph of odd order. Then, if $\delta(G) \geq 4$, G is 3-factor-critical.*

To show that Theorem 3.9 is best possible we present a graph on fifteen vertices, denoted by H_4 and defined as follows. Begin with a complete graph on the eight vertices labeled $1, 2, \ldots, 8$ and delete the Hamilton cycle $12 \cdots 81$, together with the edge 15. We now add seven additional vertices again using the labeling code

"$\sim \{i, j\}$" to be interpreted as saying that the vertex so labeled will be adjacent to all of the vertices $1, 2, \ldots, 8$, except vertices i and j. We now add the vertices $\sim \{2, 6\}, \sim \{2, 7\}, \sim \{3, 6\}, \sim \{3, 7\}, \sim \{3, 8\}, \sim \{4, 7\}$ and $\sim \{4, 8\}$. The reader can easily check that the 3cvc graph H_4 contains an induced $K_{1,4}$ centered at vertex 3. Moreover, H_4 is not 3-factor-critical since $H_4 - \{1, 2, 3\}$ has no perfect matching.

14.4 Closing Remarks

In closing, we would like to remind the reader of a different type of domination, so-called *total* domination. (See, for example, Haynes et al. 1998a, 1998b.) A set of vertices $S \subseteq V(G)$ is said to be a total dominating set if every vertex in $V(G)$ is adjacent to a vertex of S. Let the size of any minimum total dominating set of G be denoted by $\gamma_t(G)$. A graph G is then said to be $k$tvc if $\gamma_t(G - v) < \gamma_t(G)$ for every vertex of G, such that v is not adjacent to any vertex of degree one. (See Goddard et al. 2004.) (Just as in the case of connected domination, the exclusion of the degree one vertices here is a natural one, since if one deletes the neighbor of such a vertex, one cannot totally dominate the resulting graph with *any* vertex set.)

It is easy to see that $\gamma_t(G) = 3$ if and only if $\gamma_c(G) = 3$. Moreover, if we assume that the minimum degree in G is at least two, it is again easy to see that if $\gamma_t(G) = \gamma_c(G) = 3$, then it follows that a vertex is critical with respect to total domination if and only if it is critical with respect to connected domination. It then follows that the results dealing with 3cvc graphs surveyed in this paper may be interpreted as results pertaining to 3tvc graphs as well. It should be pointed out that Theorem 7 in Goddard et al. (2004) is a characterization of 3tvc (and therefore 3cvc) graphs. However, this characterization is expressed in terms of another type of vertex criticality. Namely it is shown that a graph G is 3tvc if and only if the complement graph $\overline{G}$ is vertex critical with respect to having diameter two. This characterization has not proven helpful in obtaining the results surveyed in the present paper.

In the opinion of the author, any extension of the results in this paper from 3cvc to $k$cvc, with $k > 3$, will likely prove difficult. The proofs of the results surveyed in this paper depend heavily on the fact that $\gamma_c = 3$ and we predict that further advancement to $\gamma_c > 3$ will require new ideas. We also are not optimistic about extending Theorem 3.9 from 3-factor-critical to k-factor-critical for $k > 3$ even in the case when $\gamma_c = 3$.

References

Ananchuen, N., Plummer, M.D.: Some results related to the toughness of 3-domination-critical graphs. Discrete Math. **272**, 5–15 (2003). (Erratum: **306**, 292 (2006)).

Ananchuen, N., Plummer, M.D.: Matching properties in domination critical graphs. Discrete Math. **277**, 1–13 (2004)

Ananchuen, N., Plummer, M.D.: Matching in 3-vertex-critical graphs: the even case. Networks **45**, 210–213 (2005)

Ananchuen, N., Plummer, M.D.: On the connectivity and matchings in 3-vertex-critical claw-free graphs. Utilitas Math. **69**, 85–96 (2006a)

Ananchuen, N., Plummer, M.D.: Some results related to the toughness of 3-domination-critical graphs II. Utilitas Math. **70**, 11–32 (2006b)

Ananchuen, N., Plummer, M.D.: Matching in 3-vertex-critical graphs: the odd case. Discrete Math. **307**, 1651–1658 (2007a)

Ananchuen, N., Plummer, M.D.: 3-factor-criticality in domination critical graphs. Discrete Math. **307**, 3006–3015 (2007b)

Ananchuen, W., Ananchuen, N., Plummer, M.D.: Matching properties of connected domination critical graphs. Discrete Math. **308**, 1260–1267 (2008a)

Ananchuen, W., Ananchuen, N., Plummer, M.D.: Vertex criticality for connected domination, (2008b, submitted)

Ananchuen, W., Ananchuen, N., Plummer, M.D.: Connected domination: vertex criticality and matchings (2008c, submitted)

Bo, C., Liu, B.: Some inequalities about connected domination number. Discrete Math. **159**, 241–245 (1996)

Caro, Y., West, D., Yuster, R.: Connected domination and spanning trees with many leaves. SIAM J. Discrete Math. **13**, 202–211 (2000)

Chen, X.-G., Sun, L., Ma, D.-X.: Connected domination critical graphs. Appl. Math. Lett. **17**, 503–507 (2004)

Colbourn, C., Stewart, L.: Permutation graphs: connected domination and Steiner trees. Discrete Math. **86**, 179–189 (1990)

Duchet, P., Meyniel, H.: On Hadwiger's number and stability numbers. In: Graph Theory, Cambridge, 1981. North-Holland Mathematics Studies, vol. 62, pp. 71–73. North-Holland, Amsterdam (1982)

Favaron, O.: On k-factor-critical graphs. Discuss. Math., Graph Theory **16**, 41–51 (1996)

Favaron, O., Kratsch, D.: Ratios of domination parameters. In: Advances in Graph Theory, pp. 173–182. Vishwa, Gulbarga (1991)

Garey, M., Johnson, D.: Computers and Intractability—A Guide to the Theory of NP-completeness. Freeman, San Francisco (1979), p. 190

Goddard, W., Haynes, T., Henning, M., van der Merwe, L.: The diameter of total domination vertex critical graphs. Discrete Math. **286**, 255–261 (2004)

Guha, S., Khuller, S.: Approximation algorithms for connected dominating sets. Algorithmica **20**, 374–387 (1998)

Haynes, T., Hedetniemi, S., Slater, P.: Fundamentals of Domination in Graphs. Dekker, New York (1998a)

Haynes, T., Hedetniemi, S., Slater, P.: Domination in Graphs—Advanced Topics. Dekker, New York (1998b)

Hedetniemi, S., Laskar, R.: Connected domination in graphs. In: Graph Theory and Combinatorics, Cambridge, 1983, pp. 209–217. Academic Press, London (1984)

Johnson, D.: The NP-completeness column: an ongoing guide. J. Algorithms **6**, 434–451 (1985)

Lovász, L., Plummer, M.D.: Matching Theory. Annals of Discrete Math., vol. 29. North-Holland, Amsterdam (1986)

Plummer, M.: Some recent results on domination in graphs. Discuss. Math., Graph Theory **26**, 457–474 (2006)

Sampathkumar, E., Waliker, H.: The connected domination number of a graph. J. Math. Phys. Sci. **13**, 607–613 (1979)

LS-LIB: A Library of Tools for Solving Production Planning Problems

Yves Pochet, Mathieu Van Vyve, and Laurence A. Wolsey

Summary. Much progress has been made in recent years in solving certain classes of production planning problems using mixed integer programming. One of the major challenges is how to make this expertise available and relatively easy to use for the non-specialist and the practitioner. Here we describe a modeling approach and tool LS-LIB.

LS-LIB is a library of primitives to declare procedures/subroutines/global constraints in a high-level modeling language that we believe offers an interesting partial answer to this challenge. LS-LIB provides routines for problem reformulation, cut generation, and heuristic solution of instances. The user must provide an initial formulation of his problem in the chosen modeling language MOSEL. Then using knowledge of the problem the user must first classify each product or sku according to a simple three field scheme: [production type, capacity type, variant]. Then it is a simple matter to use the global constraints of LS-LIB by adding a few lines to the initial modeling language formulation to get a tightened formulation and/or call the appropriate cut generation routines. The heuristic procedures are called in a similar fashion. The result is a tool that allows researchers and end-users to improve the solution time and quality of a variety of production planning problems within minutes. The library incorporates much of the modeling knowledge concerning lot-sizing problems derived over the last twenty years, and is also easy to maintain and extend.

We illustrate the use of LS-LIB on an intractable two-level problem, and a difficult multi-level problem.

This text presents research results of the Belgian Program on Interuniversity Poles of Attraction initiated by the Belgian State, Prime Minister's Office, Science Policy Programming. The scientific responsibility is assumed by the authors.

Work of Yves Pochet carried out while at the Center of Operations Research and Econometrics (CORE) and Institut d'Administration et de Gestion (IAG), Université catholique de Louvain.

15.1 Motivation

Much progress has been made in recent years in solving certain classes of production planning problems using mixed integer programming (MIP). One of the major challenges is how to make this expertise available and easy to use for the non-specialist. In this paper we present a library called LS-LIB of primitives used to declare subroutines and global constraints in a high-level modeling language that we believe offers an interesting partial answer to this challenge.

We suppose that the user has a new production planning problem to solve, either on a one-off or regular basis, and that he believes that mixed integer programming is an appropriate approach. The scenario is typically as follows:

- the user builds an initial MIP model and, using a modeling and optimization language,
- the user runs his MIP branch-and-cut system (commercial or other) in default mode. Either the results are satisfactory (end of the story), or
- the results are unsatisfactory (i.e. finding reasonable solutions takes far too much time, and/or nothing is known about the quality of the solutions because the dual bounds are too weak).

The user then has several options:

a) try different parameter settings of the MIP system,
b) identify relaxations for which "good/improved" formulations or cutting planes are known and either use them to modify the initial model or to code separation routines so as to get stronger linear programming bounds, and hopefully faster MIP solution times or better solutions,
c) use LP-based (or other) heuristics to find good feasible solutions quickly.

The user then resolves the MIP, and iterates through a), b) and c) several times, until an appropriate solution strategy is found.

LS-LIB provides primitives for problem reformulation, cut generation, and heuristics to find good feasible solutions quickly, and thus addresses steps b) and c). The user must provide an initial formulation of the problem in the modeling language MOSEL. Then using his/her knowledge of the problem, the user first must classify each product or sku according to a simple three field scheme (production type, capacity type, variant) proposed in Wolsey (2002). Then it is a simple matter to use the new primitives offered by LS-LIB to get a tightened formulation and/or call the cut separation routines. The heuristic procedures are called in a similar fashion.

The research on reformulations and cutting planes for single item lot-sizing problems has involved numerous researchers over the last 30 years. Among others, significant contributions on reformulations have come from Krarup and Bilde (1977), Eppen and Martin (1987) for the uncapacitated lot-sizing problem, Pochet and Wolsey (1988, 1994) for the cases with backlogging, and constant production capacities respectively, van Hoesel and Kolen (1994) for the version with start-up costs and

Van Vyve (2006) for the case with both backlogging and constant capacities. In parallel, results on valid inequalities have been derived by Barany et al. (1984) for the uncapacitated case, Van Hoesel, Wagelmans and coauthors (van Hoesel et al. 1994; van Eijl and van Hoesel 1997) for the uncapacitated problem with start-ups, Pochet (1988) for backlogging, Constantino (1996, 1998) for the version with both capacities and start-ups, and Atamtürk and Munoz (2004) for the problem including bounds and fixed costs on stocks. Surveys of this area up to 2000 can be found in Pochet and Wolsey (1995) and Pochet (2001).

In addition to these polyhedral results, the recent classification scheme of Wolsey (2002) is an attempt to provide easy access to this literature, and to the useful results available, and is important for the use of LS-LIB. A second significant factor for LS-LIB is the development of approximate reformulations which lead to smaller linear programs without significant weakening of the bounds (Stadtler 2000; Van Vyve and Wolsey 2006).

Other research has dealt with the treatment and solution of multi-item and multi-level problems. An important aspect of the solution of multi-level problems is the use of echelon stocks, see Clark and Scarf (1960), and their use in Afentakis et al. (1984). The use of this research to provide an effective solution approach for a variety of problems using mixed integer programming has been demonstrated in Belvaux and Wolsey (2001).

There is also an extended literature on heuristics for multi-item lot-sizing, based on Lagrangian relaxation, column generation, etc., see for instance Tempelmeier and Derstoff (1996). Work on heuristic approaches for lot-sizing problems based on MIP and improved formulations is limited, one exception is the work of Stadtler (2003) on relax-and-fix heuristics. However new local or neighborhood search heuristic ideas for general MIP have appeared recently, see Fischetti and Lodi (2003) and Danna et al. (2005).

A development of the production planning context, including an introduction to mixed integer programming, the application of mixed integer programming to lot-sizing problems, the classification of lot-sizing problems, the theoretical results on algorithms, valid inequalities and formulations, as well as several detailed case studies can be found in Pochet and Wolsey (2006).

The outline of the paper now follows. In Sect. 15.2 we describe briefly the results on which our approach is based. We show how a multi-item problem decomposes naturally into single-item subproblems, we then present basic formulations and the classification scheme for single-item problems, and finally we give two examples of reformulations: one involving valid inequalities in the original space of variables that can be used as cutting planes, and the second an extended formulation involving additional variables.

Given the existence of many such reformulations, in Sect. 15.3 we first discuss the problem of converting this knowledge into software tools allowing the user to take advantage of these results, and specify desirable properties of a new system. In particular we describe and analyze the current tools available to support the solution of production planning problems by MIP. Based on this discussion, we list system requirements that are necessary for a new high-level modeling tool. We then present

LS-LIB: the facilities offered, how they are implemented in the MOSEL modeling language and how they satisfy the requirements listed above.

In Sect. 15.4, we present the syntax of LS-LIB in detail. In Sect. 15.5, we provide an extensive example of the use of LS-LIB on a simple multi-item bottling problem that has already appeared in the literature. We describe the approach the non-expert would follow to solve the problem with the aid of LS-LIB. He starts with the MOSEL formulation of the original problem, and classifies all the items produced. Given this, a check of the LS-LIB Tables tells him what is known and available as reformulations or cuts. We show how he modifies his MOSEL formulation to include the reformulations/cuts he wants, and the results obtained. Finally in Sect. 15.6, we present computational results using LS-LIB for an intractable two-level problem, and a hard multi-level problem.

15.2 The MIP Solution Process for Production Planning Problems

In general, MIPs are solved using branch-and-cut algorithms. The standard approach is for the user to hand an initial formulation to the MIP solver and then let the system itself add cutting planes. If this is unsuccessful, the user is required to intervene in an effort to improve the formulation. Typically this involves

1. Identification of one or more relaxations of the problem.
2. Derivation of valid inequalities for the relaxations. Ideally the inequalities describe the convex hull of the relaxation.
3. Addition of a cut generation subroutine that finds valid inequalities cutting off the present solution of the linear programming relaxation.

The effectiveness of the reformulation obtained depends, on the one hand, on how well the relaxation represents the original problem, and, on the other hand, on the strength of the valid inequalities for the relaxation itself.

Another alternative to the use of cutting planes is the use of extended formulations involving additional variables. Though such formulations are known for several classes of problems they have rarely been used in practice because the number of additional variables needed is so large that the approach is impractical. However, for many single-item lot-sizing problems, this increase in size is reasonable, so the use of extended formulations turns out to be a viable alternative.

For the most difficult problems, the standard branch-and-bound strategies will not find good feasible solutions quickly. In these cases, more specialized primal heuristics can be helpful.

To make these ideas concrete, and to motivate the development of LS-LIB, we now describe briefly

i) how a multi-item lot-sizing problem decomposes into single-item problems,
ii) provide a limited classification of single-item lot-sizing problems,

iii) indicate one example of a single-item problem with a complete description by valid inequalities and the associated cut generation algorithm, and one example of the extended reformulation approach for a different single-item problem,

iv) describe several primal heuristics that have proved to be effective for production planning problems.

15.2.1 Decomposition of Multi-Item Lot-Sizing

We consider first a typical multi-item lot-sizing problem. There are m items and n discrete time periods, typically but not necessarily of equal length. The data include the demands d_t^i for item i in period t, individual production limits C_t^i, production, storage and fixed costs p_t^i, h_t^i and f_t^i respectively, machine production rates a^i, set-up times b^i and machine capacity B_t. We define the following variables:

- x_t^i is production of item i in period t,
- s_t^i is stock of item i at the end of period t, and
- $y_t^i = 1$ if there is a set-up of item i in period t, and $y_t^i = 0$ otherwise.

Note that production can only take place if there is a corresponding set-up, so $x_t^i > 0$ implies $y_t^i = 1$.

This leads to the MIP formulation

$$\min \sum_{i,t} (p_t^i x_t^i + h_t^i s_t^i + f_t^i y_t^i)$$

$$s_{t-1}^i + x_t^i = d_t^i + s_t^i \quad \forall i, t, \tag{1}$$

$$x_t^i \leq C_t^i y_t^i \quad \forall i, t, \tag{2}$$

$$\sum_i a^i x_t^i + \sum_i b^i y_t^i \leq B_t \quad \forall t, \tag{3}$$

$$x, s \geq 0, \quad y \in \{0, 1\}. \tag{4}$$

Here (1) are the product conservation equations, (2) the variable upper bound capacity constraints imposing that a machine has to be set up for an item before it can be produced, and (3) the machine capacity constraint linking the production of different items. Two special cases often encountered are i) that in which only one item at most can be produced per period ($a^i = 0$ and $b^i = 1$ for all i, and $B_t = 1$ for all t), or ii) that in which there are no set-up times ($b^i = 0$ for all i).

We note that X can be decomposed into

$$X = \bigcap_{i=1}^{NI} X^i \cap W$$

where X^i is the single item lot-sizing region

$$s_{t-1} + x_t = d_t + s_t \quad \text{for all } t, \tag{5}$$

$$x_t \leq C_t y_t \quad \text{for all } t, \tag{6}$$

$$x, s \geq 0, \quad y \in \{0, 1\} \tag{7}$$

where the superscript i has been dropped for $i = 1, \ldots, m$. Thus the decomposition approach leads us to the problem of single-item lot-sizing treated in the following subsections.

15.2.2 Classification of Single-Item Lot-Sizing

In Wolsey (2002), a classification scheme for single item lot-sizing problems with three fields *(PROBlem, CAPacity type, VARiant)* is proposed. Below we present a significant subset of the possible variants, specifically $PROB = \{LS, WW, DLSI\}$, $CAP = \{U, CC, C\}$, $VAR = \{SC, B\}$.

The first field specifies the problem type:

- *LS* denotes the standard lot-sizing problem,
- *WW* denotes the Wagner–Whitin relaxation, explained below,
- *DLSI* denotes the discrete lot-sizing problem in which production is either zero or at full capacity, obtained from *LS* by adding the constraints $x_t = C_t y_t$ for all t.

The second field specifies the capacity type:

- *U* denotes uncapacitated production in each period,
- *CC* denotes constant production capacity in each period,
- *C* denotes production capacity varying over time.

The third field specifies the variant:

- *SC* denotes the problem with start-up variables defined below, and
- *B* denotes the problem with backlogging.

Thus with this notation, the problem of optimizing over the set (5)–(7) is classified as the problem *LS–C* and the associated set of solutions is denoted X^{LS-C}. It is often useful to view this set as the feasible region of a fixed charge network flow problem, see Fig. 15.1. The set-up variables y now indicate whether the corresponding arcs are open or not.

Note also that by substitution the objective function: $\min\{p'x + h's + qy : (x, s, y) \in X^{LS-C}\}$ can be normalized in the form: $\min\{hs + qy + Constant : (x, s, y) \in X^{LS-C}\}$. Throughout we will use the notation $d_{kt} \equiv \sum_{u=k}^{t} d_u$.

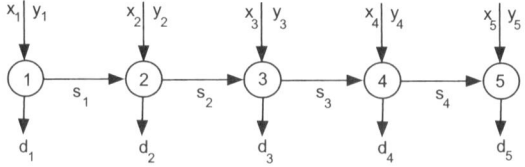

Fig. 15.1. Lot-sizing as a fixed charge network flow

Formulation of the Wagner–Whitin Relaxation

Consider again the set X^{LS-C}. Note that if we sum the equations (5) for $t = k, \dots, l$, use $s_l \geq 0$ and use (6) to replace x_u by its upper bound $C_u y_u$, we obtain the Wagner–Whitin relaxation (larger set) X^{WW-C}

$$s_{k-1} + \sum_{u=k}^{l} C_u y_u \geq d_{kt} \quad \text{for } 1 \leq k \leq t \leq n, \tag{8}$$

$$s \in \mathbb{R}_+^{n+1}, \quad y \in \{0, 1\}^n. \tag{9}$$

This relaxation is of considerable interest for two reasons. The first is stated formally below.

Theorem 2.1. *When the storage costs satisfy $h_t \geq 0$ for $t = 0, \dots, n$, the optimization problem WW–C*

$$\min\{hs + qy : (s, y) \in X^{WW-C}\}$$

has an optimal solution that solves the problem LS–C.

The second advantage of the *WW* variants is that they are easier to analyze and lead to smaller tight formulations that are typically very effective.

Formulation of the Problem with Start-up Variables LS–C–SC

Introducing binary variables $z_t = 1$ if one starts production of the item in period t (i.e., $y_t = 1$ and $y_{t-1} = 0$), and $w_t = 1$ if one stops production of the item in period t (i.e., $y_t = 1$ and $y_{t+1} = 0$), the problem *LS–C–SC* is obtained is obtained from *LS–C* by adding the following constraints

$$z_t - w_{t-1} = y_t - y_{t-1} \quad \forall t,$$
$$z_t \leq y_t \quad \forall t,$$
$$z_t \leq 1 - y_{t-1} \quad \forall t,$$
$$z_t, w_t \in \{0, 1\}^n$$

and adding a cost term $\sum_{t=1}^{n} (\alpha_t z_t + \beta_t w_t)$ to the objective function.

Formulation of the Problem with Backlogging LS–C–B

Introducing backlog variables r_t, the problem *LS–C–B* is obtained from *LS–C* by replacing the balance constraints (5) by

$$s_{t-1} - r_{t-1} + x_t = d_t + s_t - r_t \quad \forall t$$

and adding the nonnegativity constraint $r_t \geq 0$ for $t = 1, \dots, n$ and a backlogging cost term $\sum_{t=1}^{n} b_t r_t$ to the objective function.

15.2.3 Valid Inequalities and the Convex Hull for *LS–U*

In the uncapacitated case, we suppose that for all t the capacity C_t is at least as large as the sum of all the demands d_{1n}.

Observation 2.2. *Consider an interval $[k, t] = \{k, k+1, \ldots, t\}$ with $S \subset [k, t]$. The inequality*

$$s_{k-1} + \sum_{j \in S} x_j \geq \sum_{u=k}^{t} d_u \left(1 - \sum_{j \in [k,u] \setminus S} y_j \right) \quad \forall k, t, \ k \leq t$$

is valid for LS–U because, if there are no set-ups and thus there is no production in the periods $[k, u] \setminus S$, then the demand d_u must be part of the stock s_{k-1} or of the production in a period of S.

It is also easy to see that these inequality can be written in an alternative form:

$$\sum_{j \in R} x_j \leq \sum_{j \in R} d_{jt} y_j + s_t \tag{10}$$

with $R \subseteq \{1, \ldots, t\}$.

Theorem 2.3 (Barany et al. 1984). *Adding these inequalities to the constraints (5)–(6) along with the nonnegativity and bound constraints gives the convex hull of X^{LS-U}.*

Now we examine how these inequalities can be used as cutting planes. Suppose that one is given a point (x^*, s^*, y^*) satisfying the constraints (5)–(6) along with the nonnegativity and bound constraints. To check whether one of the inequalities (10) cuts it off, it suffices to check for each $t = 1, \ldots, n$ whether

$$\sum_{j=1}^{t} \max(0, x_j^* - d_{jt} y_j^*) > s_t^*.$$

Thus it is easy to add a cut generation routine for each single-item problem of the form *LS–U*.

15.2.4 An Extended Formulation for the Convex Hull of *WW–CC*

Here we consider the constant capacity problem in which $C_t = C$ for all t, and in particular the set X^{WW-CC}

$$s_{k-1} + C \sum_{u=k}^{l} y_u \geq d_{kt} \quad \forall k, t, \ k \leq t, \tag{11}$$

$$s \in \mathbb{R}_+^{n+1}, \quad y \in \{0, 1\}^n. \tag{12}$$

Let Q be the polyhedron

$$s_{k-1} = C\mu_k + C\sum_{t=k}^{n} f_{kt}\delta_{kt} \quad \forall k, t, \ k \leq t,$$

$$\mu_k + \sum_{u=k}^{t} y_u + \sum_{u\in[k,n]|f_{ku}\geq f_{kt}} \delta_{ku} \geq \left\lfloor \frac{d_{kt}}{C} \right\rfloor + 1 \quad \forall k, t, \ k \leq t,$$

$$\sum_{u\in[k,n]} \delta_{ku} \leq 1 \quad \forall k,$$

$$s \in \mathbb{R}^{n+1}, \quad \mu \in \mathbb{R}^n_+, \quad y \in [0,1]^n, \quad \delta_{kt} \in [0,1] \quad \forall k, t, \ k \leq t$$

where $f_{kt} = \frac{\delta_{kt}}{C} - \lfloor \frac{d_{kt}}{C} \rfloor$ and $[k, n]$ denotes the set/interval $\{k, k+1, \ldots, n\}$. Here $\delta_{kt} = 1$ indicates that s_{k-1} takes the value $Cf_{kt} \bmod C$, $s_{k-1} = 0 \bmod C$ when $\sum_{u\in[k,n]} \delta_{ku} = 0$ and μ_k is the integer part of s_{k-1}/C.

Theorem 2.4 (Pochet and Wolsey 1994).

i) *The linear program*

$$\min\{hs + qy : (s, y, \delta, \mu) \in Q\}$$

has an optimal solution with $y \in \{0,1\}^n$ *solving the problem WW–CC.*
ii) *$Proj_{(s,y)}(Q) = \mathrm{conv}(X^{WW-CC})$.*

Thus for any single-item problem of which WW–CC is a relaxation, it is possible to add a priori the extended formulation Q which has $O(n^2)$ new variables and $O(n^2)$ constraints.

Tables 15.1 and 15.2 in Sect. 15.4 show that similar results are known for a wide variety of variants of the single-item lot-sizing problem. Specifically Table 15.1 shows variants for which an extended formulation whose projection gives the convex hull of solutions is known. The penultimate column indicates the size of the extended formulation, and the last column gives a reference. Table 15.2 shows variants for which a cut generation algorithm is known. Here the penultimate column indicates the complexity of the cut generation routine.

15.2.5 Heuristics for Production Planning Problems

Traditionally, there are two types of heuristics, *construction heuristics* that produce a feasible solution from scratch, and *improvement heuristics* that try to improve a given feasible solution. Here, we briefly present those heuristics that appear particularly well adapted for production planning problems.

Relax-and-Fix is a construction heuristic using the idea of decomposing the problem into smaller *MIP* problems that are much easier to solve, i.e., problems involving many less integer variables. In its simplest version, the integer variables are partitioned into R disjoint sets $x^1, \ldots, x^R$ of decreasing importance. At iteration k,

$k = 1, \ldots, R$, the variables in $x^1, \ldots, x^{k-1}$ have already been fixed, the variables in x^k are considered as integer, and the variables in $x^{k+1}, \ldots, x^R$ are treated as continuous. The relatively simple MIP problem

$$\max \sum_{j=1}^{r} c^j x^j,$$

$$\max \sum_{j=1}^{r} A^j x^j \leq b,$$

$$x^j = \bar{x}^j, \quad j = 1, \ldots, k-1,$$

$$x^k \in \mathbb{Z}_+^{n_k}, \quad x^j \in \mathbb{R}_+^{n_j} \quad j = k+1, \ldots, R$$

is solved—to optimality or for a fixed amount of time—using the default *MIP* system. If it is infeasible, the heuristic algorithm has failed. Otherwise let the solution be $\tilde{x}$. One then fixes $\bar{x}^k = \tilde{x}^k$. If $k = n$, $\bar{x}$ is the heuristic solution found, and otherwise if $k < n$ one passes to iteration $k + 1$.

The LP-based *improvement heuristics* use some information from the formulation of a problem and from the best available integer solution, and try to improve the latter by searching its neighborhood. The neighborhood is typically defined in such a way that solving the *MIP* problem over it is relatively easy or fast. Then if a better (or perhaps worse) feasible solution is found, the step can be iterated.

For example, in *Local-Branching* the neighborhood consists of the solutions in which at most k integer variables take different values from those in the current best solution, see Fischetti and Lodi (2003).

In *Relaxation-Induced-Neighborhood-Search*, the neighborhood consists of the solutions in which any 0–1 variable having the same value in the optimal solution of the linear programming relaxation and in the best integer solution is fixed at this common value, see Danna et al. (2005).

Finally, we propose an *Exchange* or Fix-and-Relax heuristic that uses the same type of decomposition as relax-and-fix. Here, at each iteration, all the integer variables $(x^1, \ldots, x^{k-1}, x^{k+1}, \ldots, x^K)$ are fixed and one re-optimizes over the set of variables x^k for some $k \in \{1, \ldots, R\}$. Then, whether or not a better feasible integer solution is found, the exchange step can be iterated.

Clearly LP-based heuristics can be applied either just at the root node, or else at chosen nodes within the branch-and-cut tree. In particular, RINS and Local Branching, as well as other simple heuristics such as diving (Danna et al. 2005), are available as options in certain MIP systems.

The question we now address is how to best make these valid inequalities, separation algorithms, linear-programming reformulations, and heuristics available to the user.

15.3 LS-LIB

In this Section, we discuss software systems available to use the kind of theoretical results described in Sect. 15.2. We start by a critical description of available options prior to the implementation of LS-LIB, and identify desirable features of a system for production planning problems. We then give a high-level description of LS-LIB and discuss to what extent it satisfies the criteria identified.

15.3.1 The Need for LS-LIB

Many approaches can be taken to develop or implement a branch-and-cut system for a class of (production planning) problems, depending on

- whether the class of problems to be handled by the new system is large or narrow,
- the programming/modeling tools used by the system developer, and
- the division of the work between the end-user and the system.

A general MIP solver, such as XPRESS or CPLEX, is designed to handle the whole range of MIP problems. C/C++ library versions of these solvers are available, which makes it possible to code specific separation routines and/or heuristics. The main drawback of this approach is the very low-level, unstructured information about the model that is available to the system: just linear and integrality constraints. While this is sufficient to specify the model, this makes it very hard to (automatically) recognize the structure of the problem (multi-level? capacitated? with set-up times?, with production at full capacity?). This is redundant work, as the modeler has this higher level and structured information at the time he/she writes the model.

Recognizing this has led to the development of specialized versions of the modeling language itself. This gives access directly to the information contained in the model. Data, variables and constraints are accessible through their names and indexing sets. Thus some structural information contained in the model is available to the branch-and-cut system, making the automatic identification of high-level relaxations much easier. BC-PROD, a branch-and-cut system for production planning, makes use of this mechanism through modeling conventions, see Belvaux and Wolsey (2000). For example, $x(i, k, t)$ is the reserved name for the decision variable representing the amount of item i produced on machine k in period t.

There are several problems associated with the modeling convention approach.

- It is nearly unavoidable that soon after the development of the system, interesting new problems will arise that do not fit into the modeling conventions. For example, a multi-site problem would require a fourth index for the production variable so as to indicate the site.
- In an industrial environment, it is not easy to impose modeling conventions. In particular, rewriting an existing model is costly and error-prone, and different naming conventions might already be in place. Thus modeling conventions might themselves be problematic.
- What to do with the relaxations that are detected is left entirely to the system. For example, suppose that a single-item lot-sizing subproblem with n time periods, constant capacity and backlogging is identified. There exists an extended formulation of size $O(n^3)$ variables and constraints for this mixed-integer set. However, it might be computationally better to add the weaker, but more compact, extended formulation for the uncapacitated version of the same problem

($O(n)$ variables and $O(n^2)$ constraints). Our experience is that this kind of decision is very hard to automate, and is better seen as the outcome of the trial-and-error process a)–b)–c) outlined in the introduction.

Our general critique so far is that the modeler/developer has information about the structure of the problem that is potentially useful to the optimizer. But passing this information through modeling conventions is not always convenient or possible. On the other hand, he does not acquire (or does not want to acquire) the technical or scientific knowledge about separation algorithms, extended formulations or heuristic design, and therefore he/she is unable to exploit this structural information directly. Below we give our choice of desirable features for an efficient branch-and-cut system for production planning.

- The end-user is responsible for building the initial model and formulation, and is therefore the best person to identify sub-structures contained in the model, and their corresponding relaxations.
- The end-user should be able to control what extended formulations are added, and what separation routines are used based on the relaxations that he/she has identified.
- The decision to call heuristics should be easily integrated into the system but left to the user.
- The writing of complex extended formulations, heuristics or separation routines should be done by the system.
- The system should be computationally efficient and easy-to-use.
- The system should be easy to extend, upgrade and maintain.

15.3.2 LS-LIB: Main Features

In this section, we discuss what LS-LIB does and how it is implemented. The general idea is to provide new primitives allowing the modeler to declare global constraints. These primitives essentially define relaxations for which theoretical knowledge is available. The new primitives are implemented as procedures.

The notion of global constraint used here is very close to what is referred to as a global constraint in the Constraint Programming (CP) community. In CP, global constraints are also structures or sets of linear and non-linear constraints linking specified variables. This information is used during enumeration to reduce the range of feasible values of these variables at a node. Here, global constraints are modeled automatically as a set of linear and non-linear (integrality) constraints linking specified variables, and are used to

- strengthen the model with extended formulations, or to
- generate valid inequalities during the branch-and-bound enumeration.

Using any LS-LIB procedure involves two steps. The first step is to link the variables of the existing user's model with the abstract variables of a chosen high-level relaxation. In the simplest case, this just means specifying the name of a model variable corresponding to the abstract variable of a high-level relaxation. More generally, an abstract variable can also be defined as a linear expression in the model variables. This flexible mechanism implies that no naming or style conventions in the original model are necessary when using LS-LIB. It also gives the user the possibility to build more complex mappings between the abstract single-item relaxations and a complex, multi-level product structure of its practical problem. An example of such an echelon stock reformulation is given in Sect. 15.6.1.

The second step is to specify, for the high-level relaxation defined in the first step, which extended reformulation or family of valid inequalities will be added.

The approach that we advocate would not be possible without the facilities offered by recent high-level modeling languages that allow the integration of programming, modeling and optimization.

The features of the language that are essential for LS-LIB are the following:

- It is a typed language. Elementary types are boolean, integers, reals, strings. Compound types are arrays, sets. Optimization types are variables, linear expressions and linear constraints.
- It is a modeling-and-programming language. Basic programming constructs such as assignments, loops, conditional statements, procedures/functions and includes are available.
- It is fully interfaced with a MIP optimizer. In particular, the optimizer callbacks for branch-and-cut are directly available in the language.

LS-LIB appears to the end-user as a collection of three types of procedures, for the addition of extended formulations, cutting planes using separation routines, and primal heuristics, respectively.

The language used for the implementation of LS-LIB is the modeling language MOSEL with an interface to the MIP solver Xpress-MP (2008).

LS-LIB: Extended Formulations

Adding an extended formulation is done through the call of a MOSEL procedure. The type of relaxation (e.g., uncapacitated single-item lot-sizing with backlogging) to be declared is encoded in the name of the procedure. The arguments specify the variables (stock, backlog, production, set-up), or linear expressions thereof, that are relevant and the data defining the instance (demands, capacity). The last three arguments are always the size of the instance (e.g. the number of periods), the approximation parameter (controlling the tradeoff between the size of the reformulation and its tightness, Van Vyve and Wolsey 2006), and a boolean specifying if the constraints of the extended formulation should be added as model cuts. For example, the call

```
XFormWWUSCB(S,R,Y,Z,W,D,n,k,false)
```

declares a single-item Wagner–Whitin cost uncapacitated lot-sizing problem with backlogging and start-ups/switch-offs over n times periods with demand vector D, where S is the vector of stock variables, R is the vector of backlog variables, Y are the setup variables, and Z and W are the start-up and switch-off variables. The procedure adds an appropriate extended formulation of size $O(n)$ variables and $O(kn)$

constraints. Procedure calls to add extended formulations for other relaxations are similar, and are listed in Table 15.1.

The implementation of the extended formulations in LS-LIB is straightforward, as the scope of the objects of type MIP variables and linear constraints is always global in MOSEL. Thus, new variables and constraints declared inside the procedure are still valid after its termination.

LS-LIB: Cutting Planes

Avoiding the difficulty of the large size of extended formulations, another approach is to stay in the original variable space and add valid inequalities as cutting planes in the course of optimization.

The corresponding MOSEL procedures are very similar to those generating extended formulations. The names of the procedures begin with XCut instead of XForm, and the approximation parameter and the model cuts flag are not defined. The available procedures with their parameters and significance are listed in Table 15.2.

The implementation of the cutting planes in LS-LIB involves two phases. The procedure XCutXX only declares a global structure *XX* which holds all the information necessary to define the corresponding relaxation, i.e. its type, size, data and MIP variables, and a procedure XCut_ini controls a few parameters of cut generation (depth and frequency of the cut generation, . . .). At optimization time, within a call-back, all such global constraints are considered automatically and sequentially and an associated cut separation algorithm is invoked depending on the control parameters.

LS-LIB: Primal Heuristics

The third and last type of procedure defines and calls primal heuristics. The goal of these subroutines is to generate good feasible solutions quickly. The names of the procedures all begin with XHeur.

As an example of the calling parameters, the relax-and-fix heuristic procedure from LS-LIB is called by

```
XHeurRF(CY,SOL,NI,NT,MAXT,PAR).
```

Here *CY* is an $NI \times NT$ array of constraints, each one enforcing the integrality constraint on one binary variable. By modifying these constraints inside the procedure, it is easy to relax the integrality requirement on individual variables, or to fix these variables to specific values and re-optimize. These operations are the only ones needed to implement relax-and-fix.

The set of binary variables will be partitioned in relax-and-fix by taking columns of *CY*. Usually, these columns correspond to time periods of the production planning model. *SOL* is the name of the $NI \times NT$ array in which the heuristic solution, specified as values of the integral variables referenced by *CY*, will be found at the output of the procedure.

There are also a number of parameters to control the behavior of the procedure: *MAXT* indicates the maximum time to be spent on each sub-MIP solved during the heuristic, and *PAR* indicates into how many intervals R the time horizon $1..NT$ will be divided, which is also the number of smaller MIPs to be solved sequentially in the relax-and-fix heuristic.

The available procedures with their parameters and signification are listed in Table 15.3.

15.3.3 The LS-LIB Modeling and Optimization Process

Here we formalize somewhat the modeling and optimization prototyping process presented in the introduction.

Algorithm 1: Improved Formulations or Dual Bounds

- User builds his formulation in MOSEL. Suppose that the names chosen for the data and variables are denoted by sn^i for the stock vector, yn^i for the set-up vector, dn^i for the data vector, etc., where i refers to the item i.
- User classifies each item in his model using the classification scheme outlined in Sect. 15.2.2.
- For each item, User decides on one or more LS-LIB reformulations XFormPROD^i –CAP^i–VAR^i() and/or cut routines XCutPROD^i–CAP^i–VAR^i() to try.
- Using LS–LIB, User adds these to the model, i.e. XFormPROD^i–CAP^i–VAR^i(sn^i,yn^i,...,dn^i,...).
- User optimizes.

Algorithm 2: Improved Solutions or Primal Bounds

- User takes his original, or improved formulation.
- Given the model, the user adds a call to a construction heuristic XHeurCONSTRUCT() plus any number of calls to the improvement heuristics XHeurIMPROVE().
- The best feasible solution is kept, and possibly used in Algorithm 1. Alternatively another reformulation/cut/heuristic combination is tested.

15.4 LS-LIB: Detailed Procedures and Functions

In this section we present the procedures available in LS-LIB and their syntax.

15.4.1 Reformulations—XForm

Each reformulation concerns a single item subproblem. In Table 15.1, Columns 1 and 2 indicate the problem classification, respectively *PROB–CAP* and *VAR*.

Table 15.1. XForm

		S	R	X	Y	Z	W	D	C	NT	TK	MC	Size (cons × vars)	Ref.
LS–U		Y	–	Y	Y	–	–	Y	–	Y	Y	Y	$O(n^2) \times O(n^2)$	Krarup and Bilde (1977)
LS–U		Y	–	Y	Y	–	–	Y	–	Y	Y	Y	$O(n) \times O(n^2)$	Eppen and Martin (1987)
LS–U	B	Y	Y	Y	Y	–	–	Y	–	Y	Y	Y	$O(n^2) \times O(n^2)$	Barany et al. (1986), Pochet and Wolsey (1988)
WW–U		Y	–	–	Y	–	–	Y	–	Y	Y	Y	$O(n^2) \times O(n)$	Pochet and Wolsey (1994)
WW–U	B	Y	Y	–	Y	–	–	Y	–	Y	Y	Y	$O(n^2) \times O(n)$	Pochet and Wolsey (1994)
WW–U	SC	Y	–	–	Y	Y	–	Y	–	Y	Y	Y	$O(n^2) \times O(n)$	Pochet and Wolsey (1994)
WW–U	SC,B	Y	Y	–	Y	Y	Y	Y	–	Y	Y	Y	$O(n^2) \times O(n)$	Agra and Constantino (1999)
WW–U	LB	Y	Y	–	Y	–	–	Y	LB	Y	Y	Y	$O(n^3) \times O(n^2)$	Van Vyve (2003)
WW–U	B,LB	Y	–	–	Y	–	–	Y	LB	Y	Y	Y	$O(n^3) \times O(n^2)$	Van Vyve (2003)
WW–CC		Y	–	–	Y	–	–	Y	Y	Y	Y	Y	$O(n^2) \times O(n^2)$	Pochet and Wolsey (1994)
WW–CC	B	Y	Y	–	Y	–	–	Y	Y	Y	Y	Y	$O(n^3) \times O(n^2)$	Van Vyve (2006)
WW–C		Y	–	–	Y	–	–	Y	Y	Y	Y	Y	$O(n^2) \times O(n^2)$	Pochet and Wolsey (2008)
DLSI–CC		0	–	–	Y	–	–	Y	Y	Y	Y	Y	$O(n) \times O(n)$	Pochet and Wolsey (1994), Miller and Wolsey (2003)
DLSI–CC	B	0	Y	–	Y	–	–	Y	Y	Y	Y	Y	$O(n^2) \times O(n)$	Miller and Wolsey (2003), Van Vyve (2005)
DLS–CC	B	–	Y	–	Y	–	–	Y	Y	Y	Y	Y	$O(n) \times O(n)$	Miller and Wolsey (2003)
DLS–CC	SC	–	Y	–	Y	Y	–	Y	1	Y	Y	Y	$O(n^2) \times O(n)$	van Eijl and van Hoesel (1997)

Row 1 contains the possible procedure parameters:

S denotes the stock vector $s_0, s_1, \ldots, s_{NT}$,

R denotes the backlog vector $r_1, \ldots, r_{NT}$,

X denotes the production vector $x_1, \ldots, x_{NT}$,

Y denotes the set-up vector $y_1, \ldots, y_{NT}$,

Z denotes the start-up vector $z_1, \ldots, z_{NT}$,

W denotes the switch-off vector $w_1, \ldots, w_{NT}$,

D denotes the demand vector $d_1, \ldots, d_{NT}$,

C denotes the constant capacity C, or the capacity vector $C_1, \ldots, C_{NT}$,

NT denotes the number of time periods $n = NT$,

TK is the approximation parameter controlling the size and quality of the reformulation,

MC indicates if constraints are added as Model Cuts ($MC = 1$), or are added a priori to the formulation ($MC = 0$).

A "Y" in the table indicates that the corresponding parameter is present,
a "–" indicates that the parameter is not present,
a "0" in the "*S*" column indicates that just s_0 is present,
a "1/*LB*" in the "*C*" column indicates that the constant capacity is assumed to be $C = 1$/the constant lower bound is *LB*, and no capacity parameter is passed to the routine.
The column before last indicates the size of the formulation, and the final column gives a reference for the reformulation.

N.B. In all the examples below, it is assumed that the time horizon is represented in MOSEL as the range 1..*NT* or 0..*NT*, and the set of items/skus/products as the range 1..*NI*. If the time periods are represented as a set of strings, or as sets of integers, an appropriate translation is needed before calling the procedures.

Example 4.1. Examination of the row *WW–U–SC*, *B* in Table 15.1 indicates that we need to have declared the variables and constants marked with a "Y", and then call the reformulation for each item. We assume that the variables and data in the MOSEL problem formulation are called "sname(i,t), dname(i,t), etc."

Now, given a formulation written in the modeling language MOSEL, we show what needs to be added to call the reformulation procedure for problem *WW–U–SC*, *B* for each item.

```
! upper part, definition of the initial model:
! data, variables, constraints, objective !

!STEP1: declaration of the arguments of the procedure
begin-declaration
    NT: integer    ! Usually already declared in the model definition
    S: array(0..NT) of linctr
    R: array(1..NT) of linctr
    Y: array(1..NT) of linctr
    Z: array(1..NT) of linctr
    D: array(1..NT) of real
    TK: integer
    MC: integer
end-declaration

!REFORMULATION LOOP OVER THE ITEMS forall(i in 1..NI) do

!STEP2: computation of the arguments (for each single item i)
    S(0):= sname(i,0)
    forall (t in 1..NT) do
      S(t):= sname(i,t)
      R(t):= rname(i,t)
      Y(t):= yname(i,t)
      Z(t):= zname(i,t)
      W(t):= wname(i,t)
```

```
      D(t):= dname(i,t)
    end-do

!STEP3: call of the reformulation procedure (for each item i)
    XFormWWUSCB(S,R,Y,Z,W,D,NT,TK,MC)

!END OF DO-LOOP OVER THE ITEMS
  end-do

! ! lower part, solve instructions !
```

15.4.2 Cutting Plane Separation—XCut

To call cutting plane separation routines, the procedure arguments are shown in Table 15.2, and are essentially identical to those in Table 15.1. The penultimate column indicates the complexity of the separation routine when known.

The call to one of these cut generation routines passes the names of data and variables, and sets up the separation routines. It is implemented in the MOSEL language exactly in the same way as for extended reformulations.

In addition, there is one additional routine

$$XCut\_init(xrows,xelems,depth,freq,npassmax,eps)$$

that must be called once, before the instruction to solve the problem, to activate the cut generation routines during the optimization phase. The parameters of this procedure mean the following:

- $xrows$ denotes the number of extra rows to store cuts,
- $xelems$ denotes the number of extra matrix elements to store cuts,
- $depth$ denotes the maximum depth in the branch-and-cut tree at which cuts are generated,
- $freq$ denotes the frequency of cut generation, i.e. cut routines are called at the root node and at depths that are multiples of freq,
- $npassmax$ denotes the maximum number of re-optimization passes at each iteration of cut generation,
- eps defines the tolerance for positive cut violation.

Table 15.2. XCut

	S	R	X	Y	Z	W	D	C	NT	TK	Sep	Ref.
LS–U		Y	–	Y	Y	–	–	Y	–	Y	–	$O(n^2)$	Barany et al. (1984)
LS–C		Y	Y	Y	–	–	–	Y	Y	Y	–		Atamtürk and Munoz (2004)
WW–U		Y	–	–	Y		–	Y	–	Y	–	$O(n)$	Pochet and Wolsey (1994)
WW–U	B	Y	Y	–	Y	–	–	Y	–	Y	–	$O(n^3)$	Pochet and Wolsey (1994)
WW–CC		Y	–	–	Y	–	–	Y	Y	Y	–	$O(n^2 \log n)$	Pochet and Wolsey (1994)
DLSI–CC		0	–	–	Y	–	–	Y	Y	Y	Y	$O(n \log n)$	Miller and Wolsey (2003)
DLSI–CC	B	0	Y	–	Y	–	–	Y	Y	Y	Y	$O(n^3)$	Van Vyve (2005)

Table 15.3. XHeur

	CY	SOL	NI	NT	MAXT	PAR	Ref.
RF	Y	Y	Y	Y	Y	NS	Stadtler (1997), Wolsey (1998)
CF	Y	Y	Y	Y	Y	–	
RINS	Y	Y	Y	Y	Y	–	Danna et al. (2005)
LB	Y	Y	Y	Y	Y	–	Fischetti and Lodi (2003)
EXCH	Y	Y	Y	Y	Y	NS	Sect. 15.2.5

There are default values for all these parameters, and the non-expert may simply write

XCut_init

to activate the cut routines, which are then generated at the root node only.

15.4.3 Heuristics—XHeur

In Table 15.3 we indicate the calling parameters for the heuristics. Remember that

- *CY* denotes the constraints indexed over $1..NI$, $1..NT$ defining the y variables as binary variables,
- *SOL* indexed over $1..NI$, $1..NT$ contains as input an initial feasible solution if it is an improvement heuristic, and as output the heuristic solution found (if any).

Here RF is a relax-and-fix heuristic and CF is a cut-and-fix heuristic in which, given the LP solution, all integer variables that take integer values and at least one taking a fractional value are fixed at each iteration. The RINS, Local Branching (LB) and Exchange (EXCH) heuristics have been briefly described in Sect. 15.2.

15.5 A Detailed Example

We consider a multi-item problem with start-up and switch-off costs in which at most one item can be produced in each period.

15.5.1 The Mathematical Formulation

This has the formulation

$$\min \sum_{i,t} h_t^i s_t^i + \sum_{i,t} f_t^i y_t^i + \sum_{i,t} g_t^i z_t^i + \sum_{i,t} q_t^i w_t^i,$$

$$s_{t-1}^i + x_t^i = d_t^i + s_t^i \quad \forall i, t,$$

$$x_t^i \le C^i y_t^i \quad \forall i, t,$$

$$x_t^i \ge L^i y_t^i \quad \forall i, t,$$

$$z_t^i - w_{t-1}^i = y_t^i - y_{t-1}^i \quad \forall i, t,$$

$$z_t^i \leq y_t^i \quad \forall i, t,$$

$$\sum_i y_t^i \leq 1 \quad \forall t,$$

$$x, s \geq 0, \quad y, z \in \{0, 1\}.$$

Here x_t^i denotes the amount of item i produced per hour, and s_t^i and r_t^i are also measured in production hours. w_t^i is a switch-off variable. Note that w_t^i is precisely the slack variable in the usual constraint $z_{t+1}^i \geq y_{t+1}^i - y_t^i$ used to define z_{t+1}^i.

Each item can be classified as *WW–CC–SC*. Inspection of the XForm and XCut Tables suggests the possibility of using a reformulation of the relaxation *WW–U–SC*, and/or a reformulation or cutting planes for the relaxation *WW–CC*.

15.5.2 The Initial MOSEL Formulation

```
!-------------------------------------------------------------
model 'clbooko'
uses 'mmetc','mmxprs','mmsystem';
setparam("xprs_colorder",1)
setparam("XPRS_VERBOSE",1)

!======== ! SECTION 1: INDICES and DATA============
declarations
   NI=4                    ! # of Families
   NT=30                   ! # of Time Periods

   RMIN: array (1..NI) of real      ! Production rate [units/hour]
   H: array (1..NI) of real         ! Invent. cost of Prod in fam i
                                    !    [euro/unit,period]
   CAP: array (1..NT) of real       ! Number of production hours in period t
                                    !    [hours]
   F: array (1..NT) of real         ! Set-up cost in period t [euro]
   DEM: array (1..NI,1..NT) of real ! Demand per family per period [hours]
end-declarations

diskdata(ETC_IN,'cldemand.dat',DEM)

RMIN(1):=[807,608,1559,1622] H(1):=[0.0025,0.0030,0.0022,0.0022]
F(1):=[100,100,100,4600,100,100,100,100,100,4600,100,100,100,100,100,
       100,100,100,4600,100,100,100,100,100,4600,100,100,100,100,100]

forall(t in 1..NT) CAP(t):=16

!========== ! SECTION 2: VARIABLES !=======
 declarations
   x: array(1..NI,1..NT) of mpvar ! Production of fam i in hours
   s: array(1..NI,1..NT) of mpvar ! stock of fam i in hours at the end or
                                  !  per t
   y: array(1..NI,1..NT) of mpvar ! -1 if family i produced in per t
   z: array(1..NI,1..NT) of mpvar ! =1 if family i produced in per t
                                  !   but not in per t-1
   w: array(1..NI,1..NT) of mpvar ! =1 if family i produced in per t
                                  !   but not in per t+1
end-declarations

!============== ! SECTION 3: OBJECTIVE and CONSTRAINTS !======

COST:= SUM(i in 1..NI, t in 1..NT) H(i)*RMIN(i)*s(i,t) +
       SUM(i in 1..NI, t in 1..NT) F(t)*y(i,t) +
       SUM(i in 1..NI, t in 1..NT) 50*z(i,t) +
       SUM(i in 1..NI, t in 1..NT) 50*w(i,t)
```

```
! DEMAND SATISFACTION forall(i in 1..NI, t in 1..NT)
  AGD(i,t):= x(i,t)+ IF(t>1,s(i,t-1),0) = DEM(i,t) +s(i,t)

! VUB FOR SETUPS forall(i in 1..NI, t in 1..NT)
  SA(i,t):= CAP(t)*y(i,t) - x(i,t)>= 0

! VLB FOR SETUPS forall(i in 1..NI, t in 1..NT)
  LB(i,t):= 7*y(i,t) <= x(i,t)

! ONE PRODUCTION PER PERIOD forall(t in 1..NT)
  mode(t):= SUM(i in 1..NI) y(i,t)<= 1

! START-UPS AND SWITCH-OFFS forall(i in 1..NI, t in 2..NT)
  SYS(i,t):= y(i,t)-y(i,t-1) = z(i,t)-w(i,t-1)
forall(i in 1..NI)
  SYT(i):= y(i,1) = z(i,1)

! VARIABLE TYPES AND BOUNDS forall(i in 1..NI,t in 1..NT)
  y(i,t) is_binary
forall(i in 1..NI,t in 1..NT)
  w(i,t)<= 1
forall(i in 1..NI,t in 1..NT)
  z(i,t)<= 1


!==================== ! SECTION 4: SOLUTION !================

minimize(COST)
end-model
!----------------------------------------------------------
```

15.5.3 Adding a Reformulation

In the LS-LIB-Reformulation Table 15.1 of Sect. 15.4, we see that we can get the extended formulation for *WW–U–SC* by calling

$$XFormWWUSC(S,Y,Z,NT,TK,MC).$$

Thus we insert the following block just before the line

```
  minimize(COST)


!----------------------------------------------------------

include 'D:\\lotsizing\LSLIB-XForm.mos'

declarations
  Y: array(1..NT) of linctr
  Z: array(1..NT) of linctr
  S: array(0..NT) of linctr
  D: array(1..NT) of real
end-declarations

S(0):=0
forall(i in 1..NI) do
    forall(t in 1..NT) Y(t):=y(i,t)
    forall(t in 1..NT) Z(t):=z(i,t)
    forall(t in 1..NT) S(t):=s(i,t)
    forall(t in 1..NT) D(t):=DEM(i,t)
    XFormWWUSC(S,Y,Z,D,NT,15,0)
end-do
!----------------------------------------------------------
```

Here the first statement calls the LS-LIB library, the declarations block defines which variables and data need to be passed for each item. Then the do loop over the items, passes the names used in the MOSEL file, i.e. the set-up variable y_t^i is $y(i, t)$, the demand d_t^i is $DEM(i, t)$, etc.

15.5.4 Adding a Cut Separation Routine

In the LS-LIB-Cut Table 15.2 of Sect. 15.4, we see that we can get the cutting plane routine for *WW–CC* by calling

```
XCutWWCC(S,Y,D,C,NT).
```

Thus we insert the following block just before the line

```
minimize(COST)
```

```
!------------------------------------------------------------

include 'D:\\lotsizing\LSLIB-XCut.mos'
 declarations
  Y: array(1..NT) of linctr
  Z: array(1..NT) of linctr
  S: array(0..NT) of linctr
  D: array(1..NT) of real
end-declarations

S(0):=0 forall(i in 1..NI) do
    forall(t in 1..NT) Y(t):=y(i,t)
    forall(t in 1..NT) S(t):=s(i,t)
    forall(t in 1..NT) D(t):=DEM(i,t)
    XCutWWCC(S,Y,D,16,NT)
end-do

XCut_init
!------------------------------------------------------------
```

Here the block inserted is almost identical to that used for the reformulation. The call to XCutWWCC(S, Y, D, 16, NT) passes the names of the names of the data and variables, and sets up the separation routines. The final line XCut_init activates the routines when optimization occurs.

Note that to add both the reformulation and cutting plane routines *simultaneously*, it suffices to replace the last two lines used for the reformulation by

```
    XFormWWUSC(S,Y,Z,D,NT,15,0)
    XCutWWCC(S,Y,D,16,NT)
end-do
XCut_init
```

The results for the original formulation and the three reformulations are shown in Table 15.4, where XLP indicates the relaxation value at the root node after the addition of the Xpress and LS-LIB cuts.

Table 15.4. Results with reformulation ($TK = 15$) and/or cuts

Instance		m	n	int	LP	XLP	IP	secs	Nodes
cl-run1		510	720	120	1509.1	3972.8	4404.5	23	8155
cl-run2	XForm	1438	720	120	3775.1	4333.1	4404.5	12	196
cl-run3	XCut	510	720	120	1509.1	3972.1	4404.5	27	7659
cl-run4	XForm & XCut	1438	720	120	3775.1	4344.7	4404.5	9	45

15.5.5 Adding Some Heuristics

Even though heuristics are not necessary for this problem, we apply a relax-and-fix heuristic dividing the time horizon into $R = 2$ blocks, and allowing up to 3 seconds for the optimization of each partial problem, followed by the RINS heuristic for up to 10 seconds. We take the initial formulation of the problem and replace the *minimize(COST)* line by the following block.

```
!-----------------------------------------------------------

include 'D:\\lotsizing\LSLIB-XHeur.mos'
declarations
   CY:     array(1..NI,1..NT) of linctr
   HEURSOL:array(1..NI,1..NT) of integer
   MAXT:   integer
   hval: real
end-declarations

forall(i in 1..NI,t in 1..NT)
   CY(i,t):= y(i,t)  is_binary
MAXT   := 3
NOS    := 2

hval   :=XHeurRF(CY,HEURSOL,COST,NI,NT,MAXT,NS)

MAXT   := 10
hval   :=XHeurRINS(CY,HEURSOL,COST,NI,NT,MAXT)

!-----------------------------------------------------------
```

Here *hval* contains the value of the heuristic solution returned by the functions XHeurRF and XHeurRINS, *CY* is the name of the linear constraints indicating the 0–1 variables, *HEURSOL* contains the heuristic solution found, which is also the initial solution passed to the improvement heuristic function XHeurRINS, *COST* is the name of the linear constraint containing the objective function, the time allowed for each MIP solved during the heuristic is *MAXT* seconds, and *NS* is the number of segments into which the time horizon is divided.

Relax-and-fix finds the optimal solution after less than 1 second, and RINS cannot improve on it, terminating after 3 seconds.

15.6 Computational Results Using LS-LIB

Below we present results using LS-LIB for two intractable multi-level problems. The first is a two-level problem on which we were unable to make any significant progress for several years. The second is a multi-level assembly problem that have been discussed in several recent papers. Here there are two test instances with 78 items and two with 80 items. To our knowledge the 80 item instances have not yet been solved to optimality (with best duality gaps so far ranging from 3 to 15%).

15.6.1 Powder Production and Packing

Problem Description and Context

This problem is a simplified version of a powder production/packing problem. There are 60 types of packed powder, which are the end products. There is a known demand to be met for each of these 60 end products for 30 periods. Backlogging is allowed. The set of end products is partitioned, on the one hand, into 7 different groups sharing a common production line (resource). They are also partitioned, on the other hand, into two distinct groups sharing a common manpower resource.

The production of each end product consumes one given type among 17 available powders (bulk products). The production of the bulk product is part of the problem. There is a common resource shared by all bulks. Other complicating constraints at the bulk level are perishability (the powder must be packed at the latest one period after production), and a maximum total stock level for the powders.

None of the resources in the problem involve set-up times. Finally, there is a time-independent lower bound on production for all end and bulk products. The objective is to minimize stock and backlog, with a higher penalty or weight put on backlog.

Classification and Testing the Initial Formulation

After making minor improvements to the starting formulation provided by the company, we arrived at a starting formulation $pb1b.mos$.

We note that it has two-level product structure, so that reformulations with echelon stocks are appropriate. (The echelon stock of at item i at a given time period t is the total stock of this item either as such, or as components of other subassemblies in stock further along the production process, see Clark and Scarf 1960.) From this, we observed that the classification of the items is $\{LS-U-B, LB\}$ relative to the parameters useful for LS-LIB. This suggests the use of two relaxations $\{WW-U-B\}$ and $\{WW-U-B, LB\}$.

Using MOSEL, it is not necessary to change the original model in order to define the echelon stock reformulation. It suffices to define each echelon stock variable E_t^i via the linear expression

$$E(i, t) := s(i, t) + \sum_{j \in Suc(i)} E(j, t),$$

and use $E(i, t)$ as arguments of the LS-LIB procedures. It is worthwhile pointing out that this last equation is not an additional constraint of the model. $E(i, t)$ will be automatically replaced by its expression in terms of the original variables.

Table 15.5. pb1: Initial runs with $MAXTIME = 900$

Instance	m	n	int	LP	XLP	IP	LB	Gap
pb1b	8308	8878	2321	470.2	1062.9	2422.1	1080.4	55%
pb1c	26653	12598	2321	1547.2	1585.4	–	1585.9	–

We first carried out two runs to get some idea of the difficulty of the problem. We ran the initial problem $pb1b$, and the model $pb1c$ with the LS-LIB reformulation of $\{WW-U-B\}$ with $Tk = 5$ and with model cuts ($MC = 1$). Each run was for 900 seconds. The results are shown in Table 15.5.

For the latter run it takes 114 seconds to obtain the LP value with the model cuts from our reformulation, 193 seconds to obtain the XLP value with the automatically generated system cuts of Xpress (Covercuts $= 20$, Gomcuts $= 2$), and no feasible solution (IP value) is found within 900 seconds. We then reduced the parameter to $Tk = 3$, but still no solution is found within 900 seconds. In the Table, LB gives the lower bound obtained at the end of the computing time (i.e. after 900 seconds), and Gap measures the relative deviation from optimality and is defined as Gap $= 100 \times \frac{IP-LB}{IP}$ %.

We see that with the weak formulation $pb1b$ feasible solutions are found, but the final gap is more than 50%, so that we have no idea how good our best feasible solution really is. On the other hand with the strengthened formulation $pb1c$, we obtain a significantly stronger lower bound, but no feasible solutions are found within 900 seconds because the solutions of the linear program at each node are taking a long time due to the size of the formulation. The gap obtained by combining $pb1b$ and $pb1c$ is 33%.

Our tentative conclusion at this stage is that proving optimality for this problem is probably out of the question. So we decided to address the question of finding a feasible solution guaranteed to be within 25% of optimality or better within say 30 minutes. Specifically we consider how good a lower bound we can obtain in up to 15 minutes, and separately how good an upper bound we can obtain in the same time.

Finding Lower Bounds

Here there was no choice but to use the linear programming bounds provided by the extended formulations, as improving the lower bounds in the branch-and-cut tree was far too slow.

In Table 15.6 we show the results of runs for pb1c with different values of TK for $WW-U-B$ (see column $TK1$), and in addition the reformulation $WW-U-B$, LB for items with significant demands using $TK = 4$ (see column $TK2$).

Finding Upper Bounds

Our initial run with the tightened formulation indicated that if we wanted to use $pb1c$ for heuristics, we were obliged by the time constraint to work with a decom-

Table 15.6. pb1: Lower bounds by reformulation and LP (no branching)

Instance	TK1	TK2	m	n	LP	XLP	Secs
pb1c	3	0	20423	12598	1398.8	1452.2	110
pb1c	5	0	26653	12598	1547.2	1581.0	171
pb1c	8	0	35089	12598	1657.3	1713.2	282
pb1c	15	0	52948	12598	1692.1	1728.9	403
pb1c	8	4	134106	76094	1682.2	1717.3	3915

Table 15.7. pb1: Primal Heuristics—Upper Bounds

Instance	MAXTNs		RF val.EXCH val.EXCH val.			RF secsEXCH secsEXCH secs		
pb1b	−60	6	2254.3	2102.6	2083.3	360	360	360
pb1b	−30	7	2297.6	2140.7	2117.6	210	189	187
pb1c($TK = 8$)	60	6	2222.5	2089.1	2073.6	1569	360	360
pb1c($TK = 5$)	60	6	2158.1	2095.4	2061.2	1504	360	360

position heuristic such as relax-and-fix or exchange, which divided the problem up into significantly smaller subproblems.

In Table 15.7 we present results obtained first with the weak formulation $pb1b$, and then with the tightened formulation $pb1c$. In each case using LS-LIB, we ran relax-and-fix followed by two full runs of exchange, i.e. two runs of exchange through all $NS = R$ blocks of variables. On the tightened formulation, the time to find a feasible solution even for the smaller subproblems is long and somewhat unpredictable. Thus $MAXT = 60$ indicates that a run stops after 60 seconds if a solution has been found, and otherwise continues until a first feasible solution is discovered.

Thus we can obtain a lower bound with $TK = 8$ of 1713.2 in 282 secs, and an upper bound of 2083.3 in 360 secs. This gives a gap of 17.8%. The best bounds 1728.9 in 403 secs and 2061.2 in about 2200 secs give a gap of 15.1%.

15.6.2 Assembly of Bottling Racks

These problems are multi-level problems with assembly product structure. Many items can be produced in each time period, and the capacity constraints limiting production on each resource in each time period involve both production rates per item and set-up times for families of items. The objective is to satisfy demand without backlogging, and to minimize a combination of inventory holding costs and family set-up costs.

For these problems we adopted the initial echelon stock reformulation (44)–(50) from Wolsey (2002) for each problem instance. Then using the classification $WW–U$ for each item, we used the same parameters $TK = 5$ for the 78 item problems, and $TK = 9$ for the 80 item problems.

We adapted a similar default strategy to the one used for $pb1$. We applied relax-and-fix followed by exchange to get a good starting solution. Then we fed the heuristic value to the optimizer as a cutoff, and ran for a limited amount of time to improve

Table 15.8. cl3-Multi-Level Assembly

Instance	*heurval*	*XLP*	*LB*	Gap(*LB*)	Optimal value
cl3-78-1b	11505.4	10840.6	11164.6	3.0	11503.4
cl3-78-2b	10889.1	10515.8	10643.5	2.3	10889.1
cl3-80-1b	24913.4	21526.8	21906.0	12.1	[23238, 24544]
cl3-80-2b	26005.6	22152.8	22385.4	13.9	[22385, 25740]

the lower bound. For the 78 item instances, we set *MAXTIME* = 60 for each MIP during the heuristic (maximum 240 seconds in total), and then we ran the optimizer for the same 240 seconds. For the more difficult 80 item problems, the times were doubled. The results are shown in Table 15.8, where Gap measures the relative deviation from optimality as Gap $= 100 \times \frac{heurval - LB}{heurval}$ %, and the last column indicates the optimal or best known value for each problem instance.

For the 80 item instances we also ran the initial model without *WW–U* reformulation for the same total time of 960 seconds. In default mode Xpress-MP, version 18.00.01 gives gaps of 57% for both instances, whereas Cplex9.0 produces much stronger lower bounds and gives gaps of 32%. All these results were obtained on a 1.6 GHz, 1.00 GB laptop. With Cplex11.0 on a 2.4 GHz, 3.5 GB machine the gaps are around 26%.

15.7 Conclusions

Our goal has been to demonstrate how the arrival of high-level modeling and optimization languages, such as MOSEL, has finally enabled us to develop an easily useable platform for solving various lot-sizing problems by mixed integer programming. As researchers and consultants, we finally have a tool that can be easily maintained and extended, allows us to tackle and advise on new problems formulated in MOSEL within minutes, incorporates a considerable amount of the knowledge accumulated over the last twenty years, and does not depend on sophisticated programmers or doctoral students for its survival.

Our approach has been to enrich the modeling vocabulary of MIP (linear and integrality constraints) with higher-level lot-sizing primitives. The structured information, encoded by the modeler in these new primitives, is used by the system to automatically improve the initial MIP formulation. The end-user only works with a modeling language and has full control over extended reformulations, cutting planes and primal heuristics. However, he does not need to know, and even less to program, their mathematical or algorithmic details.

Much of LS-LIB, in particular the reformulations and cutting planes, can easily be translated into other modeling languages such as AMPL, OPL, LINGO or GAMS. At worst they can be then used in a cut-and-paste fashion, and at best, providing the appropriate modeling constructs are available, the more complete integration as procedures or subroutines should be straightforward.

Results similar to those obtained for the bottling rack assembly problem in Sect. 15.6 have been reported earlier by Belvaux and Wolsey (2000, 2001). In the first case we used the system BC-PROD based on modeling conventions that did not prove to be sufficiently flexible, and in the second the reformulations used were written explicitly into the model.

Thus what previously required the study and correct translation of an extended formulation or the programming of a separation routine is now immediately available and ready for use. It is perhaps worth adding that the idea of the conceptually trivial EXCHANGE heuristic came after implementing the relax-and-fix heuristic for LS-LIB, and, as shown on the $pb1$ problem, it quickly leads to high quality solutions that previously took twelve or more hours to find.

To date a small number of industrial end-users have confirmed that it is a useful tool, particularly for prototyping. Indeed, early and rapid prototyping is of primary importance for minimizing risks associated with projects involving MIP modeling. LS-LIB allows this crucial step to be performed faster, more completely and/or by more junior OR practitioners.

Several extensions to LS-LIB can obviously be envisaged, including treatment of various single-item variants, such as sales, piecewise concave production costs, etc. No routines for multi-item models have been incorporated into LS-LIB. This is in part because the commercial branch-and-cut MIP systems already have separation heuristics for knapsack or simple mixed integer constraints that arise as relaxations, and partly because highly effective reformulations or cutting planes have not yet been discovered.

Apart from its use to obtain improved solutions, LS-LIB has also proved useful as a research tool. Among others, it has enabled us to quickly test various hypotheses concerning tight formulations of other lot-sizing variants, leading to new results for the single-item problem with production time windows (Wolsey 2006), with non-decreasing capacities (Pochet and Wolsey 2008), with a tree of scenarios (Di Summa and Wolsey 2008) and for a multi-item problems with joint set-up costs (Anily et al. 2008).

It is an obvious question whether a similar library might be useful for other classes of MIP problems, such as network design, location, or more general supply chain problems. To date we have seen that the LS-LIB relax-and-fix routine works well on certain location problems where time intervals are replaced by geographical regions.

Finally given the sort of results that can now be obtained—good lower bounds from reformulations and cutting planes—good upper bounds from a variety of heuristics, we believe that the developers of MIP systems can now no longer avoid a crucial question: How to use a good feasible solution effectively so as to significantly speed up the branch-and-cut process?

The LS-LIB library is available at http://www.core.ucl.ac.be/LS-LIB/, or from the authors.

References

Afentakis, P., Gavish, B., Karmarkar, U.: Computationally efficient optimal solutions to the lot-sizing problem in multistage assembly systems. Manage. Sci. **30**, 222–239 (1984)

Agra, A., Constantino, M.: Lotsizing with backlogging and start-ups: The case of Wagner–Whitin costs. Oper. Res. Lett. **25**, 81–88 (1999)

Anily, S., Tzur, M., Wolsey, L.A.: Lot-sizing with family set-ups. Math. Program. (2008). doi:10.1007/s10107-007-0220-9

Atamtürk, A., Munoz, J.C.: A study of the lot-sizing polytope. Math. Program. **99**, 443–466 (2004)

Barany, I., Van Roy, T.J., Wolsey, L.A.: Uncapacitated lot sizing: The convex hull of solutions. Math. Program. **22**, 32–43 (1984)

Barany, I., Edmonds, J., Wolsey, L.A.: Packing and covering a tree by subtrees. Combinatorica **6**, 245–257 (1986)

Belvaux, G., Wolsey, L.A.: Bc-prod: a specialized branch-and-cut system for lot-sizing problems. Manage. Sci. **46**, 724–738 (2000)

Belvaux, G., Wolsey, L.A.: Modelling practical lot-sizing problems as mixed integer programs. Manage. Sci. **47**, 993–1007 (2001)

Clark, A.J., Scarf, H.: Optimal policies for multi-echelon inventory problems. Manage. Sci. **6**, 475–490 (1960)

Constantino, M.: A cutting plane approach to capacitated lot-sizing with start-up costs. Math. Program. **75**, 353–376 (1996)

Constantino, M.: Lower bounds in lot-sizing models: A polyhedral study. Math. Oper. Res. **23**, 101–118 (1998)

Danna, E., Rothberg, E., Le Pape, C.: Exploring relaxation induced neighborhoods to improve MIP solutions. Math. Program. **102**, 71–90 (2005)

Di Summa, M., Wolsey, L.A.: Stochastic lot-sizing on a tree. Oper. Res. Lett. **36**, 7–13 (2008)

Eppen, G.D., Martin, R.K.: Solving multi-item lot-sizing problems using variable definition. Oper. Res. **35**, 832–848 (1987)

Fischetti, M., Lodi, A.: Local branching. Math. Program. **98**, 23–48 (2003)

Gomory, R.E.: An algorithm for the mixed integer problem. Technical Report RM-2597, The RAND Corporation (1960)

Gu, Z., Nemhauser, G.L., Savelsbergh, M.W.P.: Lifted flow cover inequalities for mixed 0–1 integer programs. Math. Program. **85**, 439–467 (1998)

Krarup, J., Bilde, O.: Plant location, set covering and economic lot sizes: An $O(mn)$ algorithm for structured problems. In: Collatz, L., et al. (eds.) Optimierung bei Graphentheoretischen und Ganzzahligen Probleme, pp. 155–180. Birkhäuser, Basel (1977)

Miller, A., Wolsey, L.A.: Tight MIP formulations for multi-item discrete lot-sizing problems. Oper. Res. **51**, 557–565 (2003)

Nemhauser, G.L., Wolsey, L.A.: Integer and Combinatorial Optimization. Wiley, New York (1988)

Padberg, M.W., Van Roy, T.J., Wolsey, L.A.: Valid inequalities for fixed charge problems. Math. Program. **33**, 842–861 (1985)

Pochet, Y.: Valid inequalities and separation for capacitated economic lot-sizing. Oper. Res. Lett. **7**, 109–116 (1988)

Pochet, Y.: Mathematical programming models and formulations for deterministic production planning problems. In: Jünger, M., Naddef, D. (eds.) Computational Combinatorial Optimization. Lecture Notes in Computer Science, vol. 2241, pp. 57–111. Springer, Berlin (2001)

Pochet, Y., Wolsey, L.Λ.: Lot size models with backlogging: Strong formulations and cutting planes. Math. Program. **40**, 317–335 (1988)

Pochet, Y., Wolsey, L.A.: Polyhedra for lot-sizing with Wagner–Whitin costs. Math. Program. **67**, 297–324 (1994)

Pochet, Y., Wolsey, L.A.: Algorithms and reformulations for lot sizing problems. In: Cook, W., Lovasz, L., Seymour, P. (eds.) Combinatorial Optimization. DIMACS Series in Discrete Mathematics and Theoretical Computer Science, vol. 20. Am. Math. Soc., Providence (1995)

Pochet, Y., Wolsey, L.A.: Production Planning by Mixed Integer Programming. Springer, New York (2006)

Pochet, Y., Wolsey, L.A.: Single item lot-sizing with non-decreasing capacities. Math. Program. (2008). doi:10.1007/s10107-008-0228-7

Stadtler, H.: Reformulations of the shortest-route model for dynamic multi-item multi-level capacitated lot-sizing. OR-Spektrum **19**, 87–96 (1997)

Stadtler, H.: Improved rolling schedules for the dynamic single-level lot-sizing problem. Manage. Sci. **46**, 318–326 (2000)

Stadtler, H.: Multilevel lot sizing with setup times and multiple constrained resources: Internally rolling schedules with lot-sizing windows. Oper. Res. **51**, 487–502 (2003)

Tempelmeier, H., Derstoff, M.: A Lagrangean-based heuristic for dynamic multi-level multi-item constrained lot-sizing with set-up times. Manage. Sci. **42**, 738–757 (1996)

van Eijl, C.A., van Hoesel, C.P.M.: On the discrete lot-sizing and scheduling problem with Wagner–Whitin costs. Oper. Res. Lett. **20**, 7–13 (1997)

van Hoesel, C.P.M., Kolen, A.W.J.: A linear description of the discrete lot-sizing and scheduling problem. Eur. J. Oper. Res. **75**, 342–353 (1994)

van Hoesel, C.P.M., Wagelmans, A., Wolsey, L.A.: Polyhedral characterization of the economic lot-sizing problem with start-up costs. SIAM J. Discrete Math. **7**, 141–151 (1994)

Van Roy, T.J., Wolsey, L.A.: Valid inequalities for mixed 0–1 programs. Discrete Appl. Math. **14**, 199–213 (1986)

Van Vyve, M.: Lot-sizing with constant lower bounds on production. Draft 2, CORE, Université catholique de Louvain (2003)

Van Vyve, M.: The continuous mixing polyhedron. Math. Oper. Res. **30**, 441–452 (2005)

Van Vyve, M.: Linear-programming extended formulations for the single item lot-sizing problem with backlogging and constant capacity. Math. Program., Ser. A **108**, 53–77 (2006)

Van Vyve, M., Wolsey, L.A.: Approximate extended formulations. Math. Program., Ser. B **105**, 501–522 (2006)

Wolsey, L.A.: Integer Programming. Wiley, New York (1998)

Wolsey, L.A.: Solving multi-item lot-sizing problems with an MIP solver using classification and reformulation. Manage. Sci. **48**, 1587–1602 (2002)

Wolsey, L.A.: Lot-sizing with production and delivery time windows. Math. Program. **107**, 471–489 (2006)

Xpress-MP. http://www.dashoptimization.com/ (2008)

From Spheres to Spheropolyhedra: Generalized Distinct Element Methodology and Algorithm Analysis

Lionel Pournin and Thomas M. Liebling

Summary. The Distinct Element Method (DEM) is a popular tool to perform granular media simulations. The two key elements this requires are an adequate model for inter-particulate contact forces and an efficient contact detection method. Originally, this method was designed to handle spherical-shaped grains that allow for efficient contact detection and simple yet realistic contact force models. Here we show that both properties carry over to grains of a much more general shape called spheropolyhedra (Minkowski sums of spheres and polyhedra). We also present some computational experience and results with the new model.

16.1 Introduction

The Distinct Element Method (DEM) is widely used to perform granular media simulations. The two key elements this requires are an adequate model for inter-particulate contact forces and an efficient contact detection method. Originally, this method was designed to handle spherical-shaped grains that allow for efficient contact detection and simple yet realistic contact force models. We show that both properties carry over to the case of spheropolyhedra. These objects are rounded polyhedra obtained as Minkowski sums of spheres and polyhedra, thus a direct generalization of spherocylinders (Abreu et al. 2003; Pournin et al. 2005). The smoothness and convexity of these bodies are precisely the properties required by our DEM approach by which we are able to treat very general granular shapes almost as efficiently as spheres. We thus open up a whole new range of applications to DEM.

Before going into the details in the subsequent sections, we give a quick survey of existing literature on contact detection and physical contact modeling.

Most contact detection methods are found in Allen and Tildesley (1987). The most widely used among those are spacial subdivision methods. The idea underlying those methods is to cut space into cells and to check for contact inside each cell and between neighboring cells. The way the cells should be laid out and handled (cell size, adaptive subdivision) leads to a wide range of algorithms (Sigurgeirs-son et al. 2001; Samet 1984). Other methods keep track of neighbors lying inside a bounded area surrounding each particle (Schinner 1999; Muth et al. 2004, 2005). The contact detection method we use here was implemented by Pournin (2005) and is based on previous work by Müller and Liebling (1995), Müller (1996), Ferrez (2001) and Ferrez and Liebling (2002). The method relies on *constrained trian-gulations* for polygonal particles and *weighted Delaunay triangulations* (otherwise called *regular triangulations*) for discs and spheres. Those triangulations have the nice property that they provide information about inter-particle distances among a set of spheres.

Among the available methods for contact modeling, one finds *Event-Driven* meth-ods, *Molecular Dynamics* and *Contact Dynamics* as sub-categories of the *Distinct Element Method*. While the medium is treated as a sequence of instantaneous colli-sions in event driven methods, with methods using molecular dynamics, the system is driven by explicit inter-particulate forces. On the contrary, the explicit parame-ter in Contact Dynamics is displacement. In this paper, we focus exclusively on the Molecular Dynamics method. This method was originally designed to handle spher-ical particles. Here, we generalize the Distinct Element Method to a wide range of non-spherical particles. This is a subject to which researchers have been showing a growing interest over the past few years. Alternate ways to handle non-spherical par-ticles can be found in O'Connor (1996), Matuttis et al. (2000) and Muth et al. (2004, 2005).

16.2 Particle Shapes

From here on, a particle P will be the Minkowski sum of a simple convex polytope s with a ball $B(0, r) = \{x \in \mathbb{R}^2 : \|x\| \leq r\}$. Usually the polytope s is given as follows: $s = \text{conv}(s_1, \ldots, s_k)$ where $s_1, \ldots, s_k$ are elements of $\mathbb{R}^3$. In particular if s is reduced to a point, then P will be a sphere of radius r, while if s is a line segment then P is a spherocylinder. Note that these objects were already defined in Abreu et al. (2003) and Pournin et al. (2005). We say s is the skeleton of P and r its radius. Note that for a given skeleton, the smaller r, the more P will resemble the original polytope with its sharp edges and corners, whereas for growing r, P will become smoother and smoother and less and less distinguishable from a ball. Spheropoly-hedra obtained with simplicial skeletons are also called *spherosimplices*. The three non-spherical spherosimplices and a spherocube are shown in Fig. 16.1.

It is also possible to define non-convex grains in an analogous fashion, however for such particles contact detection requires more evolved machinery and will not be described here.

In computations, the surface of a spheropolyhedron is never computed explic-itly, rather any point thereon is locally considered as lying on one of its composing spheres. It is precisely this feature that allows a direct generalization of Distinct El-ement Method from spheres to spheropolyhedra.

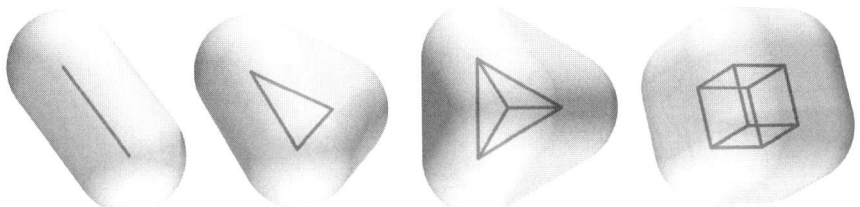

Fig. 16.1. From left to right, a spherocylinder, a spherotriangle, a spherotetrahedron, and a spherocube. The skeletons of those spheropolyhedra are represented as solid lines

16.3 Contact Models

Two particles P and Q are in contact whenever they overlap, that is if $P \cap Q \neq \emptyset$. The overlapping between P and Q models grain deformation at the contact point.

In the original Distinct Element Method, that is when P and Q are spheres, the amount of overlapping at the contact point is quantified by the *normal overlap* ξ_n as:

$$\xi_n = r_P + r_Q - \|s_Q - s_P\|,$$

where s_P and s_Q are the centers of P and Q respectively and r_P and r_Q are their radii (see Fig. 16.2). A unit vector $\boldsymbol{u}_n$ normal to the contact area can be defined as:

$$\boldsymbol{u}_n = \frac{s_Q - s_P}{\|s_Q - s_P\|}.$$

A force $\boldsymbol{f}$ is applied on the particles at contact point according to the third Newton law (that is $\boldsymbol{f} = \boldsymbol{f}_{P \to Q} = -\boldsymbol{f}_{Q \to P}$) and its normal component $f_n = \boldsymbol{f}.\boldsymbol{u}_n$ will be given as a function of ξ_n and its time derivative $\dot{\xi}_n$. The most popular model for this contact force, first introduced in Cundall and Strack (1979) reads:

$$f_n = k_n \xi_n + c_n \dot{\xi}_n. \tag{1}$$

The parameters k_n and c_n depend on the geometrical and mechanical properties of the contacting spheres (Pournin et al. 2002). Other expressions for f_n have been proposed in Walton and Braun (1986) and Pournin et al. (2002). The point x_C on either P and Q to which the total contact force is effectively applied is the intersection of segment $[s_P, s_Q]$ with the plane $\mathcal{C}$ containing the circular intersection of the boundaries of P and Q. A *tangential overlap* $\boldsymbol{\xi}_\perp$ accounting for the tangential deformation at sticky contacts can be defined as well by integrating the tangential relative speed of P and Q at the contact point x_C. The tangential component $\boldsymbol{f}_\perp = \boldsymbol{f} - f_n \boldsymbol{u}_n$ of the contact force can therefore be modeled from $\boldsymbol{\xi}_\perp$ the same way $\boldsymbol{f}_n$ has been from ξ_n. Note however that the tangential overlap is *history dependent*, as it requires an integration over time which makes it very different in nature from the normal overlap since it cannot be deduced from instantaneous particle positions alone. This tangential force can be bounded according to Coulombian friction law in order to account for slippery contacts.

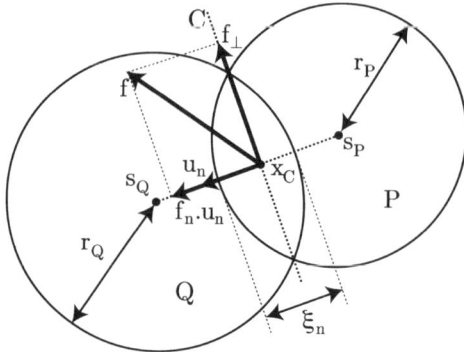

Fig. 16.2. Geometrical modeling of a contact between spherical particles P and Q in the usual DEM framework with the contact force f applied by particle P on particle Q

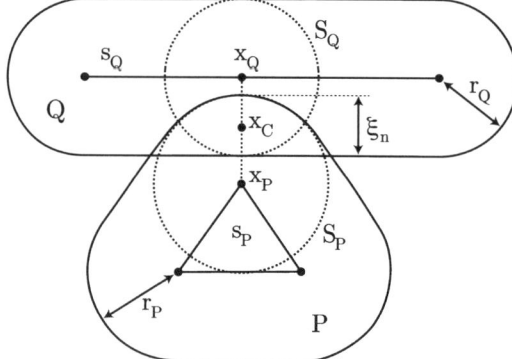

Fig. 16.3. Geometrical modeling of a contact between non-spherical particles P and Q (here, a spherotriangle and a spherocylinder). According to the model, x_P and x_Q are points in s_P and s_Q so that $\|x_Q - x_P\|$ is minimal and the contact between P and Q has same ξ_n, u_n and x_C as the imaginary contact between spheres S_P and S_Q

Now, if P and Q are not necessarily spherical, consider the set of all the pairs $(y_P, y_Q) \in s_P \times s_Q$ so that $\|y_Q - y_P\|$ is minimal:

$$M = \{(y_P, y_Q) \in s_P \times s_Q : \|y_Q - y_P\| = d(s_P, s_Q)\}.$$

In order to define a contact point x_C between P and Q, we need to choose an element $(x_P, x_Q) \in M$. As illustrated in Fig. 16.3, M may contain a single element, in which case the choice is easy. Set M, however, may be infinite (this occurs only when s_P and s_Q have parallel faces). In this case, x_P and x_Q are obtained as the barycenters of sets $m_P = \{y_P \in s_P : d(y_P, s_Q) = d(s_P, s_Q)\}$ and $m_Q = \{y_Q \in s_Q : d(y_Q, s_P) = d(s_P, s_Q)\}$ respectively. Whenever the skeletons are simplicial (polytopes with triangular facets), m_P and m_Q can be convex polygons ranging from points to hexagons. Finding m_P and m_Q amounts to solving a convex optimization problem, for which an efficient specially tailored algorithm can be given.

A normal overlap, a unit vector normal to the contact area and a contact point can be defined in this general case replacing s_P and s_Q by x_P and x_Q in the spherical case definitions of those quantities. This actually amounts to considering that the contact between P and Q actually occurs between the sphere S_P of radius r_P centered at x_P and the sphere S_Q of radius r_Q centered at x_Q.

The contact point x_C (to which the contact force is applied) will therefore be the intersection of segment $[x_P, x_Q]$ with the plane $\mathcal{C}$ containing the circular intersection of the boundaries of S_P and S_Q. The normal overlap ξ_n then is:

$$\xi_n = r_P + r_Q - \|x_Q - x_P\|,$$

and the unit vector normal to the contact surface reads:

$$\boldsymbol{u}_n = \frac{x_Q - x_P}{\|x_Q - x_P\|}.$$

The normal component of the contact force should then be computed from ξ_n according to (1) or to any other contact force model. Furthermore, the knowledge of x_C allows the computation of $\boldsymbol{\xi}_\perp$ by integration of the tangential relative speed of P and Q at point x_C and $\boldsymbol{f}_\perp$ will be found from $\boldsymbol{\xi}_\perp$ following an appropriate force model. Once the forces and their application points are known, the velocities, spins and trajectories of the particles are obtained by integration of the usual motion equations. This requires that the inertia matrix of every particle be known (Pournin 2005).

16.4 A Triangulation-Based Contact Detection Method

When simulating large populations of particles, an efficient contact detection method is required. Indeed, the quadratic complexity of testing all possible pairs of particles for contact quickly becomes prohibitive when the number of particles increases. The case is even worse when particles are no longer spheres because of the more complex individual contact computation. For spherical grains, a contact detection method using triangulations has been introduced in Müller (1996) and Ferrez and Liebling (2002). This section focuses on adapting to our particular non-spherical shapes this triangulation-based method (see also Pournin 2005).

In the following, the vertex set of a polytope p will be denoted $\mathcal{V}(p)$ and for any point $x \in \mathbb{R}^3$, we denote by $\bar{x}$ the point in $\mathbb{R}^4$ obtained by adding a 1 to x as a fourth homogenization coordinate.

Let $S_1, \ldots, S_n$ be a set of spheres in $\mathbb{R}^3$, centered at points $x_1, \ldots, x_n$ and with radii $R_1, \ldots, R_n$. The weighted Delaunay triangulation D generated by those spheres is a particular mesh of the convex hull of $\{x_1, \ldots, x_n\}$ whose basic elements are tetrahedra.

A first way to define D is to *lift* the point configuration $\mathcal{A} = \{x_1, \ldots, x_n\}$ to $\mathbb{R}^4$ using a height function $w : \mathcal{A} \to \mathbb{R}$. For each $x \in \mathcal{A}$, consider the lifted point

$x^w \in \mathbb{R}^4$ whose first three coordinates are equal to those of x and whose fourth coordinate is $w(x)$. According to the following proposition, the lower convex hull of $\mathrm{conv}(\{x^w : x \in \mathcal{A}\})$ is a polyhedral surface in $\mathbb{R}^4$ whose projection to $\mathbb{R}^3$ is a polyhedral subdivision of $\mathcal{A}$:

Proposition 4.1. *There is a unique polyhedral subdivision T of $\mathcal{A}$ so that for all $p \in T$, there exists $y \in \mathbb{R}^4$ satisfying the following two statements:*

(i) *For all $a \in \mathcal{V}(p)$, $\bar{a}.y = w(a)$,*
(ii) *For all $a \in \mathcal{A} \setminus \mathcal{V}(p)$, $\bar{a}.y < w(a)$.*

We say that w realizes T.

The triangulations obtained as polyhedral subdivisions realized by a height function are called *regular*. In particular, the weighted Delaunay triangulation D generated by spheres $(S_i)_{1 \leq i \leq n}$ is regular. It is realized by the following height function:

$$w : \begin{array}{l} \mathcal{A} \to \mathbb{R}, \\ x_i \mapsto \|x_i\|^2 - R_i^2. \end{array} \tag{2}$$

From now on, w always is height function (2).

The weighted Delaunay triangulation generated by spheres $(S_i)_{1 \leq i \leq n}$ can equivalently be defined as the geometrical dual of the power diagram generated by the same set of spheres. Consider the *power distance* π_i to sphere S_i defined on $\mathbb{R}^3$ as:

$$\pi_i : \begin{array}{l} \mathbb{R}^3 \to \mathbb{R}, \\ x \mapsto \|x - x_i\|^2 - R_i^2. \end{array}$$

Now call C_i the subset of $\mathbb{R}^3$ constituted by all points x so that for all $j \neq i$, $\pi_i(x) \leq \pi_j(x)$. The sets $(C_i)_{1 \leq i \leq n}$ are the cells of a polyhedral complex C called *power diagram* or sometimes *Laguerre complex* generated by spheres $S_1, \ldots, S_n$. For each $p \in C$, consider the polyhedron p^* obtained as the convex hull of those points x_i whose Laguerre cell C_i admit p as a face. The set $\{p^* : p \in C\}$ is then the weighted Delaunay triangulation generated by spheres $(S_i)_{1 \leq i \leq n}$.

We say that the assembly of spheres $S_1, \ldots, S_n$ is *less than orthogonal* if $\forall(i, j)$, $i \neq j$, $\|x_j - x_i\|^2 > R_i^2 + R_j^2$. A key property of triangulation D is that under the less than orthogonal condition, the line segment joining the centers of two spheres is an edge of D as soon as those spheres intersect:

Theorem 4.2 (Less than orthogonal condition). *If $S_1, \ldots, S_n$ is less than orthogonal then for all (i, j), $i \neq j$, $S_i \cap S_j \neq \emptyset \Rightarrow [x_i, x_j] \in D$.*

Proof. Suppose that S_i and S_j intersect and consider $k \in \{1, \ldots, n\} \setminus \{i, j\}$. Call x the point in $\mathrm{conv}(\{x_i, x_j\})$ so that $\pi_i(x) = \pi_j(x)$. As the pair (S_i, S_k) is less than orthogonal, we have $\|x_k - x_i\|^2 > R_i^2 + R_k^2$. Introducing x in the left hand side of this inequality and developing the scalar product, one finds $\pi_k(x) + \pi_i(x) >$

$2(x_k - x).(x_i - x)$. The same arguments hold for the pair (S_j, S_k) and we obtain the following system:

$$\begin{cases} \pi_i(x) + \pi_k(x) > 2(x_k - x).(x_i - x), \\ \pi_j(x) + \pi_k(x) > 2(x_k - x).(x_j - x). \end{cases} \tag{3}$$

Since spheres S_i and S_j intersect, x lies inside both of them, which implies that $\pi_i(x) \leq 0$ and $\pi_j(x) \leq 0$. From system (3), $\pi_k(x)$ is then strictly greater than both $2(x_k - x).(x_i - x)$ and $2(x_k - x).(x_j - x)$. Since those two quantities have opposite signs, we obtain $\pi_k(x) > 0$. As a consequence, $\pi_k(x)$ is strictly greater than both $\pi_i(x)$ and $\pi_j(x)$, and this for all $k \notin \{i, j\}$. This means that x lies in the relative interior of a common facet of C_i and C_j. According to the way the weighted Delaunay triangulation is deduced from the power diagram, $\text{conv}(\{x_i, x_j\})$ then is a face of the weighted Delaunay triangulation generated by spheres $(S_i)_{1 \leq i \leq n}$. □

Detecting contacts among spherical particles $P_1, \ldots, P_n$ then amounts to building the weighted Delaunay triangulation generated by $P_1, \ldots, P_n$ and testing for contact the pairs of particles which are linked by an edge of D. Considering that in our practical cases the number of edges of D is linear in the number n of generating spheres (see Sect. 16.7), this method allows to reduce the overall complexity of contact detection from $O(n^2)$ to $O(n \log(n))$, which is about the best one can hope for in this context. Of course, building and handling the weighted Delaunay triangulation requires special attention (see Sect. 16.5).

Now, suppose $P_1, \ldots, P_n$ are non-spherical particles with skeletons $s_1, \ldots, s_n$ and radii $r_1, \ldots, r_n$. The triangulation based contact detection method can be adapted to those non-spherical particles as follows. Cover each particle P_i with a set Σ_i of spheres having no relative motion with respect to P_i: the point-wise union of the spheres in Σ_i contains P_i for all $i \in \{1, \ldots, n\}$. Here and below, we indulge in calling spheres both the surfaces and the balls they enclose, since the interpretation is always clear from the context.

The less than orthogonal condition actually bounds the overlap of any two spheres assigned to different particles at any time during the simulation process.

Calling D the weighted Delaunay triangulation generated by the set of all covering spheres $\bigcup_{i \in \{1, \ldots, n\}} \Sigma_i$, and provided this set satisfies the less than orthogonal condition, the following holds: *if two particles P_i and P_j, $i \neq j$ intersect, then there exist two spheres $S_i \in \Sigma_i$ and $S_j \in \Sigma_j$ whose centers are linked by an edge of D.*

Therefore, detecting contacts among the non-spherical particles $P_1, \ldots, P_n$ amounts to covering each particle with a set of spheres so that the orthogonality condition is satisfied, building the weighted Delaunay triangulation generated by those spheres and testing for contact those pairs of particles having at least one covering sphere each, whose centers are linked by an edge of D. A given pair of particles P_i and P_j will be in contact as soon as $r_i + r_j - \|x_j - x_i\| > 0$, where x_i and x_j are the points in s_i and s_j respectively so that $\|x_j - x_i\|$ is minimal as defined in previous section. The actual computation of x_i and x_j is a simple, yet interesting optimization problem (see previous section).

16.5 Handling the Weighted Delaunay Triangulation

When simulating a set of moving particles with DEM, time is discretized using a time step of duration Δt and the particle positions at time $t + \Delta t$ are calculated from those at time t by integrating motion equations. The contact detection method described in previous section then requires that the weighted Delaunay triangulation generated by the covering spheres be known for the particle positions obtained at the end of each time step. Building the triangulation from scratch each time, however, is not the most efficient procedure. Indeed, from times t to $t + \Delta t$, the motion of the particles is limited and the weighted Delaunay triangulations obtained in both configurations are similar.

Here, we then choose to build the triangulation once, for the initial particle configuration, and then to modify its topology when needed by applying local transformations called *flips* or alternatively *geometric bistellar operations*.

We assume here that the point configuration $\mathcal{A} = \{x_1, \dots, x_n\}$ constituted by the covering sphere's centers is in *general position*, which means that no plane contains more than three points of $\mathcal{A}$. This assumption is not practically a limitation since the probability that a point configuration obtained from a physical setup is not in general position is 0. With this assumption, the number of possible flips reduces to 4. Those 4 flips are depicted in Fig. 16.4. Those transformations apply to any minimal affinely dependent set $Z \subset \mathcal{A}$ called *circuit*. According to Radon's theorem, such a set always admits exactly 2 triangulations. A flip then simply consists in exchanging one triangulation for the other within a triangulation of $\mathcal{A}$. Observe that flip 1 to 4 adds a vertex to a triangulation, while flip 4 to 1 removes a vertex from a triangulation. Those particular flips are unwanted in the weighted Delaunay triangulations we use for contact detection whose vertices model physical objects. This subject is discussed below, along with the statement of Theorem 5.1.

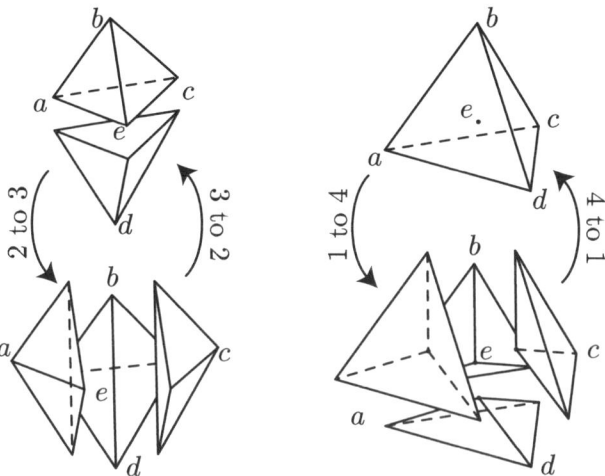

Fig. 16.4. Three-dimensional flips for point configurations in general position

Let T be a triangulation of a point configuration $\mathcal{A}$. We say a circuit Z is *flippable* in T if one of its triangulations is a subset of T.

We introduce the time-dependence of the geometrical system in the form of a functional notation. Call $x_i(t)$ the center of sphere S_i and $D(t)$ the weighted Delaunay triangulation generated by those spheres at time t. For $t' > t$, call $D(t, t')$ the set obtained from triangulation $D(t)$ if one moves its vertices from their positions at time t to their positions at time t' without changing its topology. Formally:

$$D(t, t') = \{\text{conv}(\{x_i(t') : i \in I\}) : \text{conv}(\{x_i(t) : i \in I\}) \in D(t)\}.$$

The following three cases can occur:

- $D(t, t') = D(t')$,
- $D(t, t') \neq D(t')$, but $D(t, t')$ still is a triangulation,
- $D(t, t')$ is not a triangulation any more.

In the first of above cases, no topological change in $D(t)$ is required. In the second case, topological changes are needed to obtain $D(t')$ from $D(t, t')$. This is the case we address in this section. The last of above cases where $D(t, t')$ is not a triangulation can be avoided using a tracking algorithm which will not be discussed here, but the reader can refer to Pournin (2005) for a detailed description.

In order to state the algorithm that flips $D(t, t')$ into $D(t')$, we first have to define the notion of *illegality*: let T be a triangulation of the point configuration $\mathcal{A} = \{x_1, \ldots, x_n\}$ and f a facet interior to T. Call b and c the vertices of T so that $\text{conv}(f \cup \{b\})$ and $\text{conv}(f \cup \{c\})$ are full-dimensional faces of T. Then there exists $y \in \mathbb{R}^4$ so that for all $a \in \mathcal{V}(f) \cup \{b\}$, $\bar{a}.y = w(a)$. We say that f is *w-illegal* if $\bar{c}.y > w(c)$. We may also drop the height function w and simply say that f is *illegal* when the mention of w is not needed. In the particular case where all spheres $(S_i)_{1 \leq i \leq n}$ have identical radii, facet illegality has a simple interpretation using circumspheres: f is illegal if the circumsphere of $f \cup \{b\}$ contains point c. We say that f is *flippable* in T if $Z = f \cup \{b, c\}$ is a flippable circuit in T.

Based on these definitions, Algorithm 1 attempts to regularize $D(t, t')$ into $D(t')$.

Algorithm 1 Regularization algorithm

1: $T := D(t, t')$
2: **while** T contains an illegal facet **do**
3:     **if** T admits a facet f that is simultaneously illegal and flippable **then**
4:         flip f in T
5:     **else** {no facet of T is simultaneously illegal and flippable}
6:         return failure statement
7: return T

Provided at each step an interior facet of T simultaneously illegal and flippable is found, the above algorithm will converge to $D(t')$. In order to show this, consider the polyhedral surface of $\mathbb{R}^4$ defined by $T^w = \{\text{conv}(v^w) : \text{conv}(v) \in T\}$. Every time an illegal facet of T is flipped, the volume below T^w strictly decreases. Since our regularization algorithm only flips illegal facets, it will not cycle. Still, it could fail if at step 6 a triangulation is reached that admits no simultaneously illegal and flippable facet.

There is a good case when Algorithm 1 converges to $D(t')$, obtained from the connectivity of the flip-graph of regular triangulations that share a common vertex set (see Pournin and Liebling 2007 and Gel'fand et al. 1990):

Theorem 5.1. *Let T_1 and T_2 be regular triangulations of A that have identical vertex sets and w a height function on A that realizes T_2. Then T_2 can be obtained from T_1 by only performing flips of w-illegal facets that do not add nor remove any vertex.*

Since $D(t')$ is regular, the above theorem states that Algorithm 1 converges to $D(t')$ as soon as $D(t, t')$ is regular. This is actually the usual case when regularizing triangulations in granular media simulations. Indeed, since particle motion is limited within a time step, triangulation $D(t, t')$ is similar to regular triangulations $D(t)$ and $D(t')$ and it is most often regular itself.

In some cases, however, Algorithm 1 fails to find triangulation $D(t')$. We now describe such a case: consider the seven points configuration A in $\mathbb{R}^3$ given by the following matrix whose lines correspond to the three coordinates and whose columns to points v_1 to v_7:

$$A = \begin{array}{c} x \\ y \\ z \end{array} \begin{pmatrix} 1 & \cos(\beta) & \cos(2\beta) & \cos(\alpha)\cos(\epsilon) & \cos(\alpha)\cos(\beta+\epsilon) & \cos(\alpha)\cos(2\beta+\epsilon) & 0 \\ 0 & \sin(\beta) & \sin(2\beta) & \cos(\alpha)\sin(\epsilon) & \cos(\alpha)\sin(\beta+\epsilon) & \cos(\alpha)\sin(2\beta+\epsilon) & 0 \\ 0 & 0 & 0 & \sin(\alpha) & \sin(\alpha) & \sin(\alpha) & \eta \end{pmatrix},$$

where $\beta = 2\pi/3$, $0 < \alpha < \pi/2$, $\epsilon > 0$ and $\eta < -1$. For the sake of simplicity, we denote a simplex spanned by points of A using the enumeration of its vertices' indices. If ϵ is small enough then the following set of triangles, together with their lower-dimensional faces, defines the triangulated surface S in $\mathbb{R}^3$ depicted in the left of Fig. 16.5:

$$\{125, 236, 134, 145, 256, 346, 456\}.$$

Observe that point v_7 sees the whole surface S from below. The set obtained by adding v_7 to each face of S is therefore a triangulation. But this triangulation does not completely cover the convex hull of our seven points. Indeed, all the edges of S lie in the boundary of conv(A) except 15, 26, and 34. This means that in order to cover the whole of conv(A), one has to further add tetrahedra 1245, 2356, and 1346 together with their faces. Finally the following set of tetrahedra, together with their lower-dimensional faces is the triangulation T of A sketched in Fig. 16.5:

$$\{1257, 2367, 1347, 1457, 2567, 3467, 4567, 1245, 2356, 1346\}.$$

Now imagine that points in A are the centers of seven spheres of equal radius. In this case, the illegality of interior facets of T can be checked using the circumsphere criterion. The only flippable circuits in triangulation T are $Z_1 = \{v_1, v_2, v_4, v_5, v_7\}$, $Z_2 = \{v_2, v_3, v_5, v_6, v_7\}$, and $Z_3 = \{v_1, v_3, v_4, v_6, v_7\}$. They can be flipped using

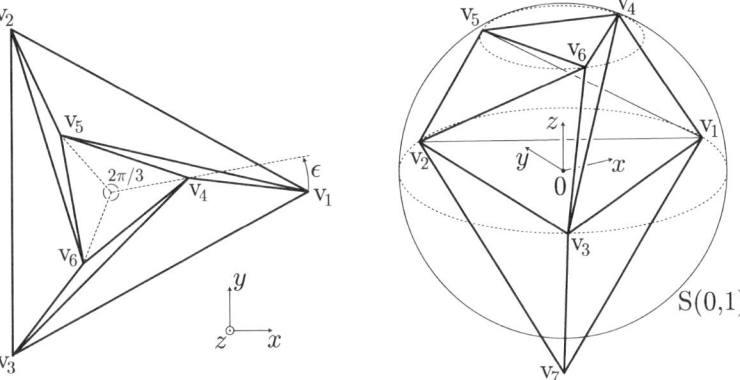

Fig. 16.5. The triangulated surface S seen from above (*left*) and a tridimensional representation of triangulation T (*right*) where edges 47, 57, 67, 16, 24 and 35 have been omitted for clarity

3 to 2 flips (see Fig. 16.4). Observe that the cases of Z_1, Z_2 and Z_3 within T are geometrically identical due to the invariance of T under a rotation of $2\pi/3$ around the z axis. Without loss of generality, we can therefore restrict to considering only Z_1 for discussing flippable facets' status regarding illegality. The flippable facets of T corresponding to circuit Z_1 are 125, 145 and 157. Tetrahedron 1245 admits the unit sphere as circumsphere. Since $\eta < -1$, point v_7 does not lie inside the unit sphere, proving by the circumsphere criterion that 125, 145 and 157 are legal in T. Triangulation T then admits no facet that is both illegal and flippable, providing a case when our regularization algorithm fails.

However, from our numerical experience over a huge sample, we can claim that such a situation is likely never to arise in practice which makes Algorithm 1 still a good choice for regularization.

16.6 Particle Covering Problem

There is some freedom left in how to actually cover particles with spheres. In this section we discuss the choice of covering spheres for the particles. If the spheres have small radii, the number of spheres needed to cover a particle will be large and the triangulation based contact test will be powerful, but also slow. This tradeoff between computational efficiency and the test power results in an optimization problem. Note also that the less than orthogonal condition bounds the radii of the covering spheres from above. Finally, it is quite easy to see that placing the centers of covering spheres within particle's skeletons does not increase the number of needed covering spheres.

Solving the covering problem exactly is hopelessly difficult in general, therefore in practice we use ad-hoc solutions to cover particles, as illustrated below.

First, we address the simple case of identical regular spherotetrahedra, which we cover each by one single sphere. Suppose P is such a spherotetrahedron with skele-

Table 16.1. Values of d for a sphere, a spherocylinder, a spherotriangle and a spherotetrahedron

Dimension of the skeleton d	
0	r_0
1	r_1
2	r_2
3	$r_3 + \frac{l_3}{3}$

ton s and radius r. Call M its mass center and l the distance between M and any vertex of s. The covering sphere S of P will be centered at point M and its radius will be $l + r$ that is, just enough for it to cover P. In such a situation, the distance between M and the boundary of P is $l/3 + r$. This means that the smallest possible distance between the centers of two covering spheres equals $2l/3 + 2r - \xi_m$, where ξ_m is an upper bound to the overlaps which will occur at any contact throughout the simulation. Therefore, if $2l/3 + 2r - \xi_m \geq \sqrt{2}(l + r)$ then the less than orthogonal condition is satisfied. Recall that the overlap accounts for grain deformation at the contact point. Hence, any realistic simulation should lead to small values for ξ_m/r. With the acceptable value of $\xi_m/r = 1/8$, we obtain that for $l \leq 0.61r$, the orthogonality condition is satisfied. These dimensions were used to produce some of the numerical experiments discussed in next section.

Now suppose the particles are all spherocylinders with equal radius r but possibly different lengths. The covering spheres will all have radius R and be centered on the particle skeletons. As in the case of spherotetrahedra, an upper bound for R is given by the less than orthogonal condition: $R \leq \sqrt{r - \xi_m/2}$ where ξ_m is again an upper bound to the overlaps which will occur at any contact throughout the simulation. Once R has been fixed, for a spherocylinder whose skeleton has length l, it is trivial to find the optimal covering which consists in equally spacing $\lceil \frac{l}{2R} \rceil$ covering spheres along the spherocylinder's skeleton.

We now address the inverse problem of dimensioning the particles in order for one only covering sphere to be needed for each of them. We consider the case of a polydisperse mixture of regular spherosimplices. For such a spherosimplex, we call r its radius, x its mass center, l the distance between x and any vertex of its skeleton and d the distance between x and its boundary. We further impose spherosimplices of a same kind to be identical, which will allow to index all those quantities when needed by the affine dimension of the spherosimplex's skeleton. We center a unique covering sphere at point x, and in order for the covering condition to be optimally satisfied, the radius of this covering sphere will be $l + r$. Observe that the triple (r, l, d) is redundant since d is a function of l and r. The values of d for each kind of spherosimplex are shown in Table 16.1.

What we need to find are the values of l and r. Observe first that if all particles are identical spheres, there is no bound on r_0. Let us address a simple bidisperse case (mixture of two different kinds of spherosimplices). If these are spheres and spherocylinders, the less than orthogonal condition yields the following inequality system:

$$\begin{cases} (r_0 + r_1 - \xi_m)^2 \geq r_0^2 + (r_1 + l_1)^2, \\ r_0 \qquad\qquad \geq \frac{\xi_m}{2-\sqrt{2}}, \\ r_1 \qquad\qquad \geq \frac{l_1}{\sqrt{2}-1} + \frac{\xi_m}{2-\sqrt{2}}. \end{cases} \qquad (4)$$

Observe that the second inequality of system (4) will be satisfied as soon as r_0 is greater than $\frac{\xi_m}{2-\sqrt{2}}$, which is not really constraining since ξ_m should be small compared to the radii of the particles. The last inequality of system (4) provides the same upper bound for l_1 as that found above in the monodisperse case. Replacing l_1 by this bound in the first inequality of system (4) gives a more constraining inequality, and admissible solutions of the following system therefore are admissible for system (4):

$$\begin{cases} r_1^2 - r_0 r_1 + \frac{\xi_m^2}{4} + r_0 \xi_m \leq 0, \\ r_0 \qquad\qquad \geq \frac{\xi_m}{2-\sqrt{2}}, \\ r_1 \qquad\qquad \geq \frac{l_1}{\sqrt{2}-1} + \frac{\xi_m}{2-\sqrt{2}}. \end{cases} \qquad (5)$$

Observe that in order for the first inequality of system (5) to have a solution in r_1, one should impose $r_0^2 - 4\xi_m r_0 - \xi_m^2 \geq 0$, that is $r_0 \geq (2 + \sqrt{5})\xi_m$. We replace the second inequality of system (5) by this more constraining inequality. Solving the first inequality of system (5) for r_1 gives $0.5(r_0 - \sqrt{r_0^2 - 4r_0\xi_m - \xi_m^2}) \leq r_1 \leq 0.5(r_0 + \sqrt{r_0^2 - 4r_0\xi_m - \xi_m^2})$, which provides the following even more constraining system:

$$\begin{cases} r_1 \geq 0.5(r_0 - \sqrt{r_0^2 - 4r_0\xi_m - \xi_m^2}), \\ r_1 \leq 0.5(r_0 + \sqrt{r_0^2 - 4r_0\xi_m - \xi_m^2}), \\ r_0 \geq (2 + \sqrt{5})\xi_m, \\ r_1 \geq \frac{l_1}{\sqrt{2}-1} + \frac{\xi_m}{2-\sqrt{2}}. \end{cases} \qquad (6)$$

Assuming that r_0 is given and satisfies the third inequality of system (6), one can find values for r_1 (given by the first two inequalities of system (6)) and l_1 (last inequality of system (6)) so that our contact detection method applies. Assuming that $\xi_m \leq r_0/8$, we deduce from system (6) that any choice within the following bounds will be admissible for our contact detection method:

$$\begin{cases} l_1/r_1 \leq \frac{15}{8\sqrt{2}} - 1 (\approx 0.325), \\ r_1/r_0 \leq \frac{1}{2} + \frac{\sqrt{31}}{16} (\approx 0.847), \\ r_1/r_0 \geq \frac{1}{2} - \frac{\sqrt{31}}{16} (\approx 0.152). \end{cases}$$

Analogous computations can be carried out for the other cases.

16.7 Results and Discussion

Among the new phenomena that can be studied with the above generalization of the traditional Distinct Elements Method let us cite spherocylinders verticalization (Pournin et al. 2005; Ramaioli et al. 2005, 2007; Ramaioli 2007). This phenomenon has been successfully replicated with our model, this providing a first validation of the approach. Still, further validations are needed and numerical experiments with more complex shapes may be conducted such as shape segregation or granular stratification (Makse et al. 1997). To close this section, we give computational results regarding the complexity of the contact detection method.

Snapshots of experiments performed with this non-spherical version of the Distinct Element Method are shown in Fig. 16.6. Those experiments consist in pouring particles into a cylindrical container of diameter 20 mm. This means that the number of particles gradually increases while the simulation performs. All particles have eight times the volume of a sphere of diameter 1 mm, independently of their shapes. Experiment 1 involves 250 spherosimplices of each kind. Experiment 2 involves 500 spheres and 500 spherotetrahedra, and experiments 3 and 4, 1000 spherocylinders each.

For the first two experiments, each particle only uses one covering sphere for contact detection. This is achieved by calibrating every shape the same way the spherotetrahedra have been calibrated at the end of Sect. 16.6. This constraint actually limits the particle sharpness (given by the ratio between the Minkowski ball radius and the radius of the smallest circumsphere of the skeleton). While such particles can be far from spherical, the complexity of contact detection is the same as for spheres. In experiment 3, particles have an elongation coefficient l/r of 3, where l is half the length

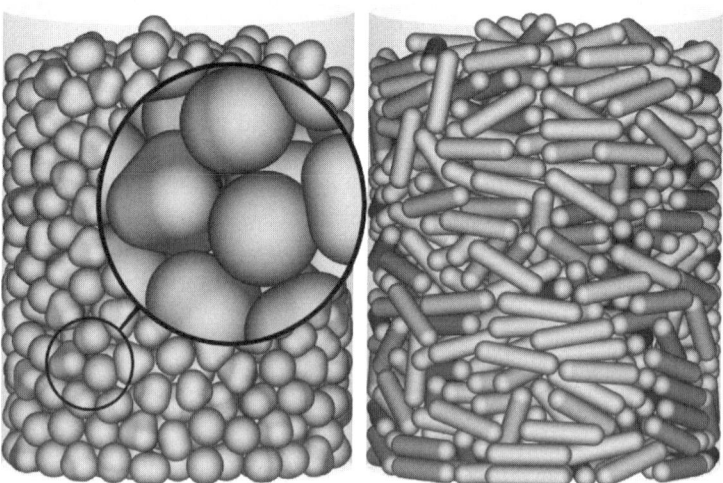

Fig. 16.6. Snapshots of final states of numerical experiments. *Left*: experiment 1, *right*: experiment 3

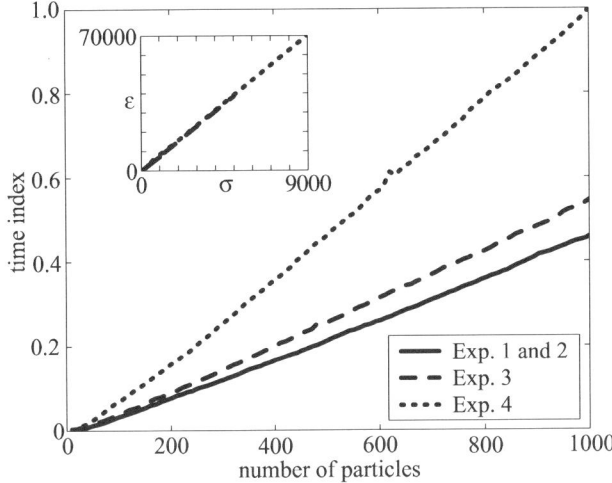

Fig. 16.7. Actual computing time of the DEM process drawn versus the number of simulated particles for all experiments. The *time index* is normalized to 1.0 for 1000 of experiment 4's spherosegments, and the corresponding computation time is approximately 4.1 s per DEM iteration on a 2.7 GHz PPC G5 PowerMac. Each point is averaged over 400 DEM iterations. Inset: number ε of edges in the triangulation versus the number σ of covering spheres for the same experiments

of the segment constituting their skeletons and r is their radius. With such elongated shapes, 5 covering spheres per particle are needed. In experiment 4, $l/r = 6$, and 9 covering spheres are needed for a particle. In every experiment, $\xi_m = 5 \times 10^{-5}$ m.

The overall computing time of the DEM process, including contact detection, is drawn on Fig. 16.7 against the number of simulated particles. In all cases it is approximately linear. Observe that the associated slope increases with particle elongation, naturally reflecting that the number of possible contacts is higher for elongated particles. The inset shows the number of edges of the triangulation, drawn versus the number of covering spheres and the four cases turn out to be identically linear. The slope of around 8 is close to the coordination number of random close-packed spheres, indicating that our method appropriately identifies the pairs of particles to be tested for contact. Observe finally that, though there exist Delaunay triangulations with a quadratic number of edges, our practical simulated setups only produce weighted Delaunay triangulations having a linear number of edges, according to Attali and Boissonnat (2004). The linear complexity of our contact detection method is a direct consequence of this crucial property.

16.8 Conclusion

This generalization of the Distinct Element Method to non-spherical particles opens numerical simulation to a wide new range of granular media. Our nearly linear triangulation-based contact detection method allows processing large populations of non-spherical particles. While many issues need to be further discussed, such as the influence of the shape of contact areas on the behaviour of a contact, the first trials to be found in Pournin et al. (2005), Pournin (2005), Ramaioli et al. (2005) and Ramaioli et al. (2007) suggest that in many cases this model successfully captures the reality of non-spherical particles.

Acknowledgements

This project was partially funded by the Swiss National Science Foundation, grant # 200020-100499/1.

References

Abreu, C.R.A., Tavares, F.W., Castier, M.: Influence of particle shape on the packing and on the segregation of spherocylinders via Monte Carlo simulations. Powder Technol. **134**(1–2), 167–180 (2003)

Allen, M.P., Tildesley, D.J.: Computer Simulation of Liquids. Clarendon, Oxford (1987)

Attali, D., Boissonnat, J.-D.: A linear bound on the complexity of the Delaunay triangulation of points on polyhedral surfaces. Discrete Comput. Geom. **31**(3), 369–384 (2004)

Cundall, P.A., Strack, O.D.L.: A discrete numerical model for granular assemblies. Géotechnique **29**(1), 47–65 (1979)

Ferrez, J.-A.: Dynamic triangulations for efficient 3D simulation of granular materials. Thèse N° 2432, EPFL (2001)

Ferrez, J.-A., Liebling, T.M.: Dynamic triangulations for efficient detection of collisions between spheres with applications in granular media simulations. Philos. Mag. B **82**(8), 905–929 (2002)

Gel'fand, I.M., Kapranov, M.M., Zelevinsky, A.V.: Discriminants of polynomials of several variables and triangulations of Newton polyhedra. Leningrad Math. J. **2**, 449–505 (1990)

Makse, H.A., Havlin, S., King, P.R., Stanley, H.E.: Spontaneous stratification in granular mixtures. Nature **386**, 379–382 (1997)

Matuttis, H.-G., Luding, S., Herrmann, H.J.: Discrete element methods for the simulation of dense packings and heaps made of spherical and non-spherical particles. Powder Technol. **109**(1–3), 278–292 (2000)

Müller, D.: Techniques informatiques efficaces pour la simulation de mileux granulaires par des méthodes d'éléments distincts. Thèse N° 1545, EPFL (1996)

Müller, D., Liebling, T.M.: Detection of collisions of polygons by using a triangulation. In: Raous, M., et al. (eds.) Contact Mechanics, pp. 369–372. Plenum, New York (1995)

Muth, B., Eberhard, P., Luding, S.: Collisions between particles of complex shapes. In: García-Rojo, R., Herrmann, H.J., McNamara, S. (eds.) Powders and Grains 2005, vol. II, pp. 1379–1383. Balkema, Rotterdam (2005)

Muth, B., Müller, M.-K., Eberhard, P., Luding, S.: Contacts between many bodies. Mach. Dyn. Prob. **28**(1), 101–114 (2004)

O'Connor, R.M.: A distributed discrete element modeling environment—algorithms, implementation and applications. Ph.D. thesis, MIT (1996)

Pournin, L.: On the behavior of spherical and non-spherical grain assemblies, its modeling and numerical simulation. Thèse N° 3378, EPFL (2005)

Pournin, L., Liebling, T.M.: Constrained paths in the flip-graph of regular triangulations. Comput. Geom. **37**(2), 134–140 (2007)

Pournin, L., Liebling, T.M., Mocellin, A.: Molecular-dynamics force models for better control of energy dissipation in numerical simulations of dense granular media. Phys. Rev. E **65**, 011302 (2002)

Pournin, L., Weber, M., Tsukahara, M., Ferrez, J.-A., Ramaioli, M., Liebling, T.M.: Three-dimensional distinct element simulation of spherocylinder crystallization. Granul. Matter **7**(2–3), 119–126 (2005)

Ramaioli, M.: Granular flow simulations and experiments for the food industry. Thèse N° 3997, EPFL (2007)

Ramaioli, M., Pournin, L., Liebling, T.M.: Numerical and experimental investigation of alignment and segregation of vibrated granular media composed of rods and spheres. In: García-Rojo, R., Herrmann, H.J., McNamara, S. (eds.) Powders and Grains 2005, vol. II, pp. 1359–1363. Balkema, Rotterdam (2005)

Ramaioli, M., Pournin, L., Liebling, T.M.: Vertical ordering of rods under vertical vibration. Phys. Rev. E **76**(2), 021304 (2007)

Samet, H.: The quadtree and related hierarchical data structures. ACM Comput. Surv. **16**, 187–260 (1984)

Schinner, A.: Fast algorithms for the simulation of polygonal particles. Granul. Matter **2**(1), 35–43 (1999)

Sigurgeirsson, H., Stuart, A., Wan, W.-L.: Algorithms for particle-field simulations with collisions. J. Comput. Phys. **172**, 766–807 (2001)

Walton, O.R., Braun, R.L.: Viscosity, granular-temperature, and stress calculations for shearing assemblies of inelastic, frictional discs. J. Rheol. **30**, 949–980 (1986)

17

Graphic Submodular Function Minimization: A Graphic Approach and Applications

Myriam Preissmann and András Sebő

Summary. In this paper we study particular submodular functions that we call "graphic". A graphic submodular function is defined on the edge set E of a graph $G = (V, E)$ and is equal to the sum of the rank-function of G and of a linear function on E. Several polynomial algorithms are known that can be used to minimize graphic submodular functions and some were adapted to an equivalent problem called "Optimal Attack" by Cunningham. We collect eight different algorithms for this problem, including a recent one (initially developed for solving a problem for physics): it consists of $|V| - 1$ steps, where the i-th step requires the solution of a network flow problem on a subgraph (with slight modifications) induced by at most i vertices of the given graph ($i = 2, \ldots, |V|$). This is a fully combinatorial algorithm for this problem: contrary to its predecessors, neither the algorithm nor its proof of validity use directly linear programming or keep any kind of dual solution. The approach is direct and conceptually simple, with the same worst case asymptotic complexity as the previous ones. Motivated by applications, we also show how this combinatorial approach to graphic submodular function minimization provides efficient solution methods for several problems of combinatorial optimization and physics.

17.1 Introduction

For basic graph theoretic notions, terminology and notation we refer to Lovász (1979). Given a finite set E, a function $f : 2^E \longrightarrow \mathbb{R}$ is called *submodular*, if for any two subsets $A, B \subseteq E$:

$$f(A) + f(B) \geq f(A \cup B) + f(A \cap B). \tag{SUB}$$

In honor of Bernhard Korte's 70-th birthday and in memory of the pioneering work and important achievements that have been reached concerning submodular functions in the Institut für Disckete Mathematik, Ökonometrie und Operations Research in the '80-s and '90-s, due to the generous visitor's program and high quality research and training of the institute led by Professor Korte, the results of which provide the background of the present work.

In this paper f will be called *modular* if the equality holds and $f(\emptyset) = 0$, that is, $f(A) = \sum_{a \in A} f(a)$. (We use here and in the following the notation a instead of $\{a\}$ when no confusion is possible.)

The submodular function f will be called *graphic* if E is the set of edges of a graph $G = (V, E)$ and moreover $f = r + w$, where $f, r, w : 2^E \longrightarrow \mathbb{R}, r = r_G$ is the rank-function of G, that is, for $X \subseteq E, r(X) := n - c_G(X)$, where $c_G(X)$ is the number of connected components of $G = (V, X)$ and w is any modular function. We will use the notation $n := |V|$ throughout.

It is well-known and easy to check that r is a submodular function, and since w is modular, f is also submodular. *Graphic submodular function minimization* is the following problem:

GRAPHIC submodular FUNCTION MINIMIZATION (GSM)
Input: graph $G = (V, E)$, $w : E \longrightarrow \mathbb{R}$
Output: $X \subseteq E$ that minimizes $r + w$.

As will be shown in the next sections, Graphic Submodular Function Minimization is useful for several applications such as computing the strength of a graph, solving the optimal reinforcement problem or approximating the partition function in the Potts model when the number of states tends to infinity. Our attention to Graphic Submodular Function Minimization was initiated by this last problem suggested by physicists. In an article that appeared in Journal of Physics A (Anglès d'Auriac et al. 2002) we already explain how it could be solved efficiently, allowing the solution of the problem for several hundred thousands of variables instead of one or two dozens provided by previous approaches. We give here the full details and mathematical background of this direct algorithm for graphic submodular function minimization. We also give a detailed analysis of the problem by comparing several (eight different) possible solution methods, their interconnections and their applications to physics or to some other problems. Notice that the previously known specific algorithms for Graphic Submodular Function Minimization were studied in the (equivalent) formulation of "Optimal Attack" (see Sect. 17.3.2).

For convenience we will most of the time consider the following equivalent problem, that we will call *Optimal Cooperation*:

$$\text{Maximize} \left\{ f_{G,w}(A) = c_G(A) + \sum_{e \in A} w(e) : A \subseteq E(G) \right\}. \tag{OC}$$

This problem could indeed arise from the following "academic" situation. In a research institute, bilateral cooperations between researchers (vertices) are encouraged by the state: the *weight of an edge represents the reward for cooperation between two researchers. Furthermore there is a unit support for each research project (set of people connected by cooperations, no cooperation is possible outside a project).* So for instance there is a loss of support when two components unite unless the benefits from cooperations between the two sum up to at least 1. The goal of the researchers is to realize the cooperations in a way which maximize the total amount of money received from the state. Cooperate optimally!

Subtracting n and replacing w by $-w$ in the expression of $f_{G,w}$ we get exactly the function of (GSM) multiplied by -1, so (OC) is equivalent to (GSM) and maximizing $f_{G,w}$ $(= n - (r + (-w)))$ is equivalent to minimizing the graphic submodular function $r + (-w)$. From this it is clear that $f = f_{G,w}$ is *supermodular*, that is,

$$f(A) + f(B) \le f(A \cup B) + f(A \cap B). \qquad \text{(SUPER)}$$

When no confusion is possible we will simply use f or f_G for $f_{G,w}$, c for c_G.

When considering (OC) we can (and we will always) suppose $0 < w(e) < 1$ for all $e \in E(G)$, because if $w(e) \le 0$, then *deleting* the edge, if $w(e) \ge 1$, then *contracting* it (identifying its two endpoints and keeping the other weights) we get an equivalent problem. We will call such w a *weight function* on the edges.

An important observation: if F is a set of edges, and X is the vertex set of a connected component of $G(F)$, then adding to F edges *induced* by X, that is with both endpoints in X, increases the value of f. Thus in an optimal solution $F^* \subseteq E$ of (OC) each connected component of F^* contains all the edges of G it induces.

Given a graph $G = (V, E)$ with edge-weighting w we will say that the *value* of a partition $\mathcal{P}$ of V is

$$g_{G,w}(\mathcal{P}) = |\mathcal{P}| + \sum \{w(xy); xy \in E(G) : x \text{ and } y \text{ are in the same class of } \mathcal{P}\}.$$

Another important observation: if $\mathcal{P}$ is a partition of V containing a set W such that the subgraph $G(W)$ induced by W is not connected, then replacing in $\mathcal{P}$ the set W by the sets of vertices of the connected components of $G(W)$ increases the value of $g_{G,w}$. Thus a partition $\mathcal{P}^*$ maximizing $g_{G,w}$ is such that each of its sets induces a connected subgraph of G.

For any subset F of edges let $w(F) = \sum_{e \in F} w(e)$ and denote by $\mathcal{P}_{G,F}$ (or simply $\mathcal{P}_F$ when no confusion is possible) the partition of $V(G)$ determined by the components of $G(V, F)$. For any set $\mathcal{S}$ of disjoint subsets of V we denote by $E(\mathcal{S})$ the subset of edges of E with both extremities in the same set of $\mathcal{S}$ and by $\delta(\mathcal{S})$ the set of edges of E joining vertices belonging to two different sets in $\mathcal{S}$. So for a partition $\mathcal{P}$ of V, $E(\mathcal{P})$ and $\delta(\mathcal{P})$ partition the edge-set E and we can write the preceding equation as:

$$g_{G,w}(\mathcal{P}) = |\mathcal{P}| + w(E(\mathcal{P})).$$

From the preceding observations it is clear that there is a one to one correspondence between subsets of edges maximum for $f_{G,w}$ and partitions of the vertices maximum for $g_{G,w}$. As will be shown in the following, it is possible to compute an optimal subset of *edges* using a greedy algorithm on the *vertices*.

By now several polynomial algorithms (Grötschel et al. 1981, 1988; Schrijver 2000; Iwata et al. 2001; Iwata 2002) are known to minimize a submodular function. However, it does not seem to be evident how these general algorithms could exploit the particularity of graphic submodular function minimization and provide an efficient algorithm for this case. Fortunately, the specific properties of our problem make possible the use of a considerably simpler and quicker algorithm.

Another possibility is to use matroids or polymatroids and "primal dual meth ods". An algorithm for finding an optimal solution of (OC) follows from Edmonds' matroid partition algorithm (Edmonds 1965), but a careless reduction to matroid partition is not sufficient for the polynomial time bound. Cunningham (1984) generalized and sharpened Edmonds' algorithm to matroid rank functions shifted by a linear function, and this provides the first algorithm not using the ellipsoid method for maximizing $f_{G,w}$ with polynomial running time, as early as 1984. Cunningham (1985) also worked out a specialized algorithm to the "Optimal Attack problem".

A crucial progress about the 'Optimal Attack' problem has been made later on by Barahona (1992) and Baïou et al. (2000), which made possible to decrease the complexity of the problem with an order of magnitude: the consideration of edges one by one is replaced by the consideration of vertices. Both Cunningham's and Barahona's approach implicitly use the extension of an intersecting submodular function to a fully submodular function called 'Dilworth truncation'. Dilworth truncation was clarified in general by Lovász (1977) and was algorithmically worked out by Frank and Tardos (1988). The present algorithm is enlightened by a knowledge of these results on the Dilworth truncation and a network flow method for computing an underlying fully submodular function based on Picard and Queyranne (1982) and Padberg and Wolsey (1983). Moreover, the here presented new ideas of combining these methods and reducing them into some simple graph-theoretic steps result in an algorithm that has the best asymptotic worst case complexity, the same as (Baïou et al. 2000). It is probably also the simplest, both conceptually and in the use of computational resources of the implementation. The sophisticated ingredients were necessary for finding the right solution, but we did not need to make explicit use of them. It consists of a greedy algorithm with n steps on the *vertices* of the graph where the steps are max-flow-min-cut computations on graphs from 3 to at most $n + 2$ vertices and very elementary manipulations on graphs. This can be seen as a greedy algorithm on a polymatroid indexed by the *vertices* of the graph where the rank oracle can be computed with network flows, and the whole procedure can be discussed in fully elementary graphic terms. The results of Barahona (1992) and Baïou et al. (2000) have already a vertex-centered framework but in a technically more complicated primal-dual context.

Let us finish this introduction by summarizing various polynomial algorithms for graphic submodular function minimization. We enumerate eight such algorithms, without mentioning possible smaller variations of each:

1–3.   The three methods handling general submodular functions: one using the ellipsoid method and two recent combinatorial algorithms (Grötschel et al. 1981, 1988; Schrijver 2000; Iwata et al. 2001; Iwata 2002).

4.   Cunningham's testing membership algorithm (Cunningham 1984) that sharpens Edmonds' matroid partition algorithm, to minimize the sum of a matroid function and of a linear function in strongly polynomial time.

5.   Grötschel, Lovász and Schrijver's variant of Khachian's linear programming algorithm (Grötschel et al. 1981) proving the equivalence of optimization and

separation, where we plug in the subroutine PQPW for separation. (PQPW is explained in Sect. 17.2.2.)

6.   Cunningham's 'Optimal Attack' algorithm (Cunningham 1985). This algorithm needs to solve $|E|$ maximum flow problems on a graph with $|V| + 2$ vertices.

7.   Barahona's 'vertex-sensitive' algorithm (Barahona 1992) needs to solve at most $|V|$ maximum flow problems on a graph with $|V| + 2$ vertices.
     Another version of this algorithm proposed by Baïou et al. (2000) consists in solving $|V|$ maximum flow problems but on graphs which have, at iteration i, exactly $i + 2$ vertices.

8.   The algorithm of Sect. 17.2.3 solving $|V|$ maximum flow problems on graphs which have, at iteration i, at most $i + 2$ vertices (but in average much less).

The last two methods implicitly use Frank and Tardos's algorithm for 'truncating' an intersecting submodular function (Frank and Tardos 1988) with a variant of PQPW (cf. Sect. 17.2.2).

In Sect. 17.2 we explain our approach and solution of the problem and restate several "classical" ones; in Sect. 17.3 we consider a few sample applications.

17.2 Solution

In this section we solve graphic submodular function minimization. In the first two subsections we provide the two main ingredients: the inductive step of extending a solution from the smaller graph that arises after the deletion of a vertex, and the known graph-theory procedure needed to execute this extension. The last two subsections make clear the complete algorithm and explain the origins of its ingredients.

17.2.1 Extension of a Solution

In this subsection we prove two theorems: the first is showing how a given graph G with weight function w and any set of edges maximizing (OC) for G minus a vertex, may be completed so that it maximizes (OC) for the graph G itself; the second shows how the optimizing set of edges changes if the weight of only one edge is increased.

The following lemma is an easy consequence of the (SUPER) property.

Lemma 2.1. *Let $G = (V, E)$ be a graph and $G' = (V, E')$ ($E' \subseteq E$) an arbitrary subgraph of E. Let $F' \subseteq E(G')$ be an optimal solution of (OC) for G'.*

There exists an optimal solution of (OC) for G which contains all edges of F'.

Proof. We first remark that the restriction of f_G to the subsets of edges of G' is equal to $f_{G'}$, so we can simply omit the subscript.

Let F be any optimal solution of (OC) for G. By the (SUPER) property one has in G:

$$f(F \cup F') + f(F \cap F') \geq f(F) + f(F').$$

Since $F \cup F' \subseteq E(G)$, $F' \subseteq E(G')$, and F, F' maximize f in G and G' respectively, we have $f(F \cup F') \leq f(F)$, $f(F \cap F') \leq f(F')$, and therefore we have equality throughout. Now we see that $F \cup F'$ satisfies the conclusion.   □

As a consequence of Lemma 2.1 we can obtain an optimal solution F^* for G by adding edges to F'. Each connected component in $G(F^*)$ will be either an element of $\mathcal{P}_{F'}$ or the union of at least two elements of $\mathcal{P}_{F'}$. The next lemma shows how the value of a partition is affected when a subset of connected components are put together.

For $U \subseteq V(G)$ we introduce the usual notation $\delta(U) := \delta(U, V \setminus U)$ for the set of edges with exactly one endpoint in U; in directed graphs $\delta(U)$ will denote the set of edges leaving U. For $\mathcal{W} \subseteq \mathcal{P}$ we will denote by $\delta(\mathcal{W})$ the set of edges of G joining vertices belonging to two different sets in $\mathcal{W}$.

Lemma 2.2. *Let $\mathcal{P}$ and $\mathcal{P}'$ be two partitions of V such that $\mathcal{P} = (\mathcal{P}' \setminus \mathcal{W}) \cup \{\bigcup X;$ $X \in \mathcal{W}\}$ for some $\mathcal{W} \subseteq \mathcal{P}'$. Then*

$$g_G(\mathcal{P}) = g_G(\mathcal{P}') - (|\mathcal{W}| - 1 - w(\delta(\mathcal{W}))).$$

In particular, if $\mathcal{P}'$ is optimal then $|\mathcal{W}| - 1 - w(\delta(\mathcal{W})) \geq 0$, and if $\mathcal{P}$ is optimal, then $|\mathcal{W}| - 1 - w(\delta(\mathcal{W})) \leq 0$.

Proof. Replacing in $\mathcal{P}$ the sets of $\mathcal{W}$ by their union decreases the cardinality of $\mathcal{P}$ by $|\mathcal{W}| - 1$. On the other hand let A and A' be the sets of edges induced by the classes of $\mathcal{P}$ and $\mathcal{P}'$ respectively, $A = A' \cup \delta(\mathcal{W})$ and so the weight of the edges increases by $w(\delta(\mathcal{W}))$. The equality is then proved and the rest follows immediately.   □

We will now use the above lemmas for proving the next two theorems.

Theorem 2.3. *Let $G = (V, E)$ be a graph with weight function w on the edges, let x be a vertex of G and let $G' = (V, E')$ be obtained from G by deleting all edges incident to x. The edges of G' keep their weight. Let $F' \subseteq E(G')$ be optimal for $f_{G'}$. For any $\mathcal{W}^* \subseteq \mathcal{P}_{F'}$ containing $\{x\}$ and minimizing $|\mathcal{W}| - 1 - w(\delta(\mathcal{W}))$ among all subsets $\mathcal{W}$ of $\mathcal{P}_{F'}$ such that $\{x\} \in \mathcal{W}$, the set $F^* = F' \cup \delta(\mathcal{W}^*)$ is optimal for f_G.*

We observe that, since x is an isolated vertex in G', the subset $\{x\}$ is an element of $\mathcal{P}_{F'}$ and F' is also optimal for $G \setminus x$.

Proof (of Theorem 2.3). By Lemma 2.1 we know that there exists F^* optimal for f_G which contains F'. Let $\mathcal{W} \subseteq \mathcal{P}_{F'}$ such that $\mathcal{W} = \bigcup_{X_i \in \mathcal{W}} X_i$ is an element of $\mathcal{P}_{F^*}$. If $\mathcal{W}$ doesn't contain $\{x\}$ then, since F' is optimal for $f_{G'}$, we get by Lemma 2.2 that $|\mathcal{W}| - 1 - w(\delta(\mathcal{W})) \geq 0$ (this value is the same in G and G'). On the other hand F^* is optimal for f_G and therefore $|\mathcal{W}| - 1 - w(\delta(\mathcal{W})) = 0$, but then $F^* \setminus \delta(\mathcal{W})$ is also optimal for f_G. Hence there exists an optimal solution F^* for f_G containing F' and such that any element of $\mathcal{P}_{F^*}$ not containing x is already in $\mathcal{P}_{F'}$.

So any $\mathcal{W}^* \subseteq \mathcal{P}_{F'}$ containing $\{x\}$ and minimizing $|\mathcal{W}| - 1 - w(\delta(\mathcal{W}))$ among all subsets $\mathcal{W}$ of $\mathcal{P}_{F'}$ such that $\{x\} \in \mathcal{W}$ will provide an optimal solution $F' \cup \delta(\mathcal{W}^*)$. (Notice that this minimum is ≤ 0 since for $\mathcal{W} = \{\{x\}\}$ we get 0.)   □

At this point we see that any way of finding an optimal $\mathcal{W}^*$ will provide a constructive algorithm for getting an optimal solution of (OC): we start with a solution for a small subgraph (for example a one vertex subgraph) and then add the vertices one by one, computing at each step an optimal solution. By Lemma 2.1 this can be done by extending the solution of the previous iteration.

Assume we are under the conditions of Lemma 2.1 and let $\{x\}, X_1, \ldots, X_k$ be the pairwise disjoint sets of $\mathcal{P}_{F'}$. We remark that the value $|\mathcal{W}| - 1 - w(\delta(\mathcal{W}))$ doesn't depend on the subgraphs induced by the subsets X_i in $\mathcal{W}$, so we may ignore these and work in a possibly smaller graph. More precisely: the result of *shrinking* $X_1, \ldots, X_k$ in G is the graph $\mathrm{shr}(G) = (V_{\mathrm{shr}}, E_{\mathrm{shr}})$ such that $V_{\mathrm{shr}} = \{x, x_1, \ldots, x_k\}$ where $x_1, \ldots, x_k$ are distinct new vertices, and the function $\mathrm{shr} : V \cup E \longrightarrow \{x, x_1, \ldots, x_k\} \cup E_{\mathrm{shr}}$ is defined by $\mathrm{shr}(v) := x_i$ if $v \in X_i$ and $\mathrm{shr}(x) := x$; the image of an edge is defined only for e with extremities a and b such that $\mathrm{shr}(a) \neq \mathrm{shr}(b)$ and $\mathrm{shr}(e) = \mathrm{shr}(a)\mathrm{shr}(b)$; edges keep their weight, that is $w_{\mathrm{shr}}(\mathrm{shr}(e)) = w(e)$; sets of vertices or of edges are replaced by the image sets. There is a one to one correspondence between the subsets $\mathcal{W}$ of $\mathcal{P}_{F'}$ containing $\{x\}$ and subsets W of vertices of $\mathrm{shr}(G)$ containing x and for any such $W = \mathrm{shr}(\mathcal{W})$ one has $|\mathcal{W}| - 1 - w(\delta(\mathcal{W})) = |W| - 1 - w_{\mathrm{shr}}(E_{\mathrm{shr}} \cap (W \times W))$. So $\mathcal{W}$ will be optimal if and only if $W = \mathrm{shr}(\mathcal{W})$ minimizes $|W| - 1 - w_{\mathrm{shr}}(E_{\mathrm{shr}} \cap (W \times W))$.

The following theorem follows also from Lemmas 2.1 and 2.2. It is useful for the *augmentation problem* that will be considered in Sect. 17.3.

Theorem 2.4. *Let $G = (V, E)$ be a graph with weight function w on the edges, let $e = xy$ be an edge of G and let w' such that $w'(e) > w(e)$ and $w'(f) = w(f)$ for any edge $f \neq e$. Let $F \subseteq E(G)$ be optimal for $f_{G,w}$. If $e \in F$ then F is also optimal for $f_{G,w'}$. If $e \notin F$ then for any $\mathcal{W}^*$ minimizing $|\mathcal{W}| - 1 - w'(\delta(\mathcal{W}))$ among all $\mathcal{W} \subseteq \mathcal{P}_F$ such that $x \in \bigcup_{X_i \in \mathcal{W}} X_i$, the set $F^* = F \cup \delta(\mathcal{W}^*)$ is optimal for $f_{G,w'}$.*

Proof. Clearly we cannot expect the optimal value to increase more than $w'(e) - w(e)$, so the case when $e \in F$ is trivial. Assume now that $e \notin F$. It is then obvious that F is optimal in $G' = G \setminus e$. So, by Lemma 2.1 we know that there exists an optimal solution F^* of $f_{G,w'}$ which contains F and we can then conclude exactly as in the proof of the previous theorem. $\square$

17.2.2 Computing the Oracle

Given a graph H with weight function w on the edges and $W \subseteq V(H)$, we can define a vertex-function $b(W) := b_{H,w}(W) := |W| - 1 - w(E(W))$, where $E(W)$ denotes the set of edges of H with both extremities in W. (To avoid confusion, some letters (like b) denote functions on the vertices, and some other letters (like w) denote functions on the edges.)

It is clear from Theorem 2.3 that a solution to the following problem solves (OC): given (H, w) and a vertex $x \in V(H)$, find a subset W^* of $V(H)$ containing x such that $b(W^*) = \min(b(W); W \subseteq V(H), x \in W)$. Luckily, precisely this variant of the problem is directly solved as a key subroutine in Picard and Queyranne (1982) or Padberg and Wolsey (1983), using a network flow model that will be described below and that will be referred to as *PQPW* in the sequel.

We first give (or remind) some definitions and well-known facts. Given a directed graph and a subset X of its vertices, $\delta(X)$ denotes the set of arcs leaving X, and is called a *cut*; if $s \in X$, $t \notin X$ it is an (s, t)-*cut*; $\delta(X, Y)$ denotes the set of arcs oriented from X to Y. A function c of nonnegative value on the arcs of a directed graph is called a *capacity* function. If F is a set of arcs then $c(F) := \sum_{e \in F} c(e)$. A *network* is a directed graph with capacity function. A first efficient algorithm for finding the minimum capacity of an (s, t)-cut is by Ford and Fulkerson (1956) (see Schrijver 2003 for the complexity)—very efficient versions are known by now, see (Ahuja et al. 1993).

We state now the algorithm.

PQPW(H, w, x),
INPUT: Arbitrary graph H, $w : E(H) \longrightarrow (0, 1)$, where $(0, 1) = \{t \in \mathbb{R} : 0 < t < 1\}$, $x \in V(H)$.
OUTPUT: $W^* \subseteq V(H)$ so that $x \in W^*$ and $b(W^*) := b_{H,w}(W^*) := |W^*| - 1 - w(E(W^*))$ is minimum under the condition $x \in W^*$.

The first three steps describe the construction of a network $(D, c) = N(H, w, x)$ that will be associated to H, w and x:

1. $V(D) := V(H) \cup \{s, t\}$ where s, t are distinct new vertices;
2. To each edge $uv \in E(H)$ we associate the arcs (u, v) and (v, u) of capacities $c((u, v)) = c((v, u)) - \frac{1}{2}w(uv)$. To each vertex $u \in V(H)$ we associate the arcs (s, u) and (u, t).
3. Define for all $u \in V(H)$:

$$p(u) := \sum_{v \in V(H),\, uv \in E(H)} c((u, v)) = \frac{1}{2} \sum_{v \in V(H),\, uv \in E(H)} w(uv),$$

and then $c((s, u)) := p(u)$, $c((u, t)) := 1$.
Note that $\sum_{u \in V(H)} p(u) = w(E(H))$.
4. Determine in the directed graph D with the capacity function c a set $S \subseteq V(D)$ so that $s, x \in S$, $t \notin S$, and $c(\delta(S))$ is minimum. Return the set $W^* := S \setminus \{s\}$.

END;

The problem that has to be solved in Step 4 is a minimum cut problem in a graph with given nonnegative capacities: $s, x \in W^*$ can be enforced by contracting sx or putting an ∞ capacity on it.

Theorem 2.5. *The output of PQPW(H, w, x) is $W^* \subseteq V(H)$ containing x and minimizing $b(W^*)$ under this condition. This optimal value can be computed by one minimum cut computation on D.*

The proof of this theorem is a direct consequence of the following lemma (using the notations of the algorithm):

Lemma 2.6. *The value $c(\delta(\{s\} \cup W)) - b_{H,w}(W)$ $(W \subseteq V(H))$ is a constant independent of W.*

Proof. Let $W \subseteq V(H)$. Then $\delta(\{s\} \cup W)$ is an (s, t)-cut of $N(H, w, x)$ and we prove:

$$c(\delta(\{s\} \cup W)) = |W| - w(E(W)) + K = b(W) + K + 1, \qquad \text{(CUT)}$$

where $K := c(\delta(s)) = w(E)$.

Indeed, let us see how the capacity of the cuts changes if we start with the set $\{s\}$ inducing an (s, t)-cut of capacity $c(\delta(s)) = K$ and then 'add' to it the vertices of W one by one:

When we add v to the side of s, the edge sv of capacity $p(v)$ disappears, and the edge vt of capacity 1 appears, whence the change corresponding to such edges is $1 - p(v)$, and

$$\sum_{v \in W} 1 - p(v) = |W| - w(E(W)) - \frac{1}{2} \sum_{xy \in E(H), x \in W, y \notin W} w(xy).$$

On the other hand the contribution of the arcs between W and $V(H) \setminus W$ is clear, at the beginning it is zero, and at the end it is:

$$c(W, V(H) \setminus W) = \frac{1}{2} \sum_{xy \in E(H), x \in W, y \notin W} w(xy).$$

The change comparing to K is provided by the sum of the two contributions, which is $|W| - w(E(W))$. □

Notice that Lemma 2.6 is verified for any network obtained from the one in the algorithm by changing the capacities of the arcs from the source s and to the sink t in such a way that $c(u, t) - c(s, u) = 1 - p(u)$ for every vertex u.

Notice also that PQPW can be easily *generalized to minimize any modular shift of b, that is, to minimize $b + m$ where m is an arbitrary modular function on the vertex set*. The special case $m(W) := |W|$ ($W \subseteq V$) worked out above is just a particular choice of a modular function. One only has to set $c(u, t) := m(u)$ instead of 1 in Step 3. The minimization of $b + m$ solved by PQPW for arbitrary m is exactly the 'oracle' that Frank and Tardos' truncation algorithm needs for 'truncating b', see (Frank and Tardos 1988), p. 526, remark. Our algorithm is a concatenation of the truncation algorithm combined with PQPW, with an elementary graphic solution of the truncation algorithm using new graphic ideas.

17.2.3 The Algorithm

In this section we state the algorithm solving the GSM problem, still in the OC form.

At each iteration a subset $U \subseteq V$ and a partition $\mathcal{P}$ of U will be at hand. In Step 0 we give trivial initial values; in Step 1, we choose an arbitrary vertex u to be added to U and through the following steps we *compute a subset $\mathcal{W}$ of $\mathcal{P} \cup \{u\}$ providing a new partition of $U \cup \{u\}$*.

OC algorithm

INPUT: A graph $G = (V, E)$, and a weight function $w: E \longrightarrow (0, 1)$.

OUTPUT: A partition $\mathcal{P}^*$ optimizing (OC): the set A^* of edges induced by the sets in $\mathcal{P}^*$ maximizes $\{f_{G,w}(A) := c_G(A) + \sum_{e \in A} w(e) : A \subseteq E\}$.

0. $U := \emptyset$, $\mathcal{P} := \emptyset$.

Do n times consecutively Steps 1 to 5, and then define $\mathcal{P}^* = \mathcal{P}$:

1. Choose a vertex $u \in V \setminus U$.
2. Define H and w_H as the result of shrinking the classes of $\mathcal{P}$ in $G(U \cup \{u\})$ with weight function w restricted to the edges of $G(U \cup \{u\})$.
3. Call PQPW(H, w_H, u). Suppose it outputs the set $W^* = \{u, x_1, \ldots, x_k\} \subseteq V(H)$ where each x_i is a vertex of H corresponding to a set $X_i \in \mathcal{P}$.
4. Define $R := \{u\} \cup X_1 \cup \cdots \cup X_k$
5. Redefine U and $\mathcal{P}$: $U := U \cup \{u\}$, $\mathcal{P} := (\mathcal{P} \setminus \{X_1, \ldots, X_k\}) \cup \{R\}$;

END.

Remarks.

– Note that W^* can be equal to $\{u\}$.
– The graphs H can also be constructed iteratively with only one shrinking in each iteration, by adding u and then shrinking the set W^*.
– The choice for u is completely free; this freedom could be used for making the computations simple.
– Since $f_{G,w}$ is a supermodular function (on the edges) we know that there exists a unique minimal optimal solution and a unique maximal optimal solution. Choosing in our $PQPW$ algorithm a minimal (respectively maximal) minimum (s, t)-cut we will obtain a minimal (respectively maximal) optimal solution. These minimum (s, t)-cuts are easy to obtain by starting from s (respectively t).

Theorem 2.7. *The output $\mathcal{P}^*$ is an optimal partition for (G, w).*

Proof. At step 0, $\mathcal{P} := \emptyset$ is an optimal solution for the subgraph of G induced by $U := \emptyset$. Assume now that $\mathcal{P}$ is an optimal solution for the subgraph of G induced by U, and show that after applying steps 1 to 5 the new partition is optimal for the subgraph of G induced by $U \cup \{u\}$. It is clear that $\mathcal{P}' = \mathcal{P} \cup \{\{u\}\}$ stays optimal when adding the vertex u but no edges. From the preceding chapter we know that $W^* = \{u, x_1, \ldots, x_k\}$ is a subset of vertices of H containing u minimizing b_{H,w_H}. But H and w_H are obtained by shrinking the classes of $\mathcal{P}'$ in $(G(U \cup \{u\}), w)$, hence $\mathcal{W} = \{u, X_1, \ldots, X_k\}$ is a subset of $\mathcal{P}'$ minimizing $|\mathcal{W}| - 1 - w(\delta(\mathcal{W}))$ and by Theorem 2.3 the partition $(\mathcal{P} \setminus \{X_1, \ldots, X_k\}) \cup \{R\}$ is then optimal. □

Notice that the algorithm consists merely of $n - 1$ network flow computations and $n - 1$ shrinking. Our approach looks lucky in the sense that the network flow model is called for a small number of vertices: *in the beginning because the number of considered vertices is small, and at the end because hopefully many vertices are identified.* Moreover we don't need computing 'primal solutions' which are present in an essential way in the previous algorithms of the same complexity and so the capacities of the networks are not affected by the change.

17.2.4 Further Explanations

As already noticed, we can rewrite (OC) as:

$$\text{Maximize} \left\{ g(\mathcal{P}) = |\mathcal{P}| + \sum_{i=1}^{|\mathcal{P}|} w(E(V_i)) : \mathcal{P} = \{V_1, \ldots, V_{|\mathcal{P}|}\} \text{ is a partition of } V \right\}.$$

Now maximizing g is equivalent to minimizing $n - g(\mathcal{P})$, moreover, noticing that $n - |\mathcal{P}| = \sum_{i=1}^{|\mathcal{P}|}(|V_i| - 1)$ we finally get the following formulation of our problem:

$$\text{Minimize} \left\{ \sum_{i=1}^{|\mathcal{P}|} (|V_i| - 1 - w(E(V_i))) : \mathcal{P} = \{V_1, \ldots, V_{|\mathcal{P}|}\} \text{ is a partition of } V \right\}.$$

(DUAL)

Recall $b(U) = b_{G,w}(U) = |U| - 1 - w(E(U))$; (DUAL) is in fact 'an integral solution to the linear programming dual' of

$$\text{Maximize} \sum_{v \in V} x_v \quad \text{under the constraints}$$

$$\sum_{v \in U} x_v \le b(U) \quad \text{for every } U \subseteq V, \ U \ne \emptyset.$$

If $x \in \mathbb{R}^V$ satisfies these inequalities we will say that it is *feasible*. Note that x is then nonpositive! (The term $|U|$ could be deleted here, this causes a modular shift which influences (DUAL) only with the additive constant n; however, it is exactly the modular shift $b(U)$ that plays a role in the formulation (FOREST) considering subsets of edges, explained below.)

The algorithm we developed determines an optimal dual solution to this linear program. Another way of looking at the problem, that provides some insight and another way to solve it (even if a less efficient one) is through Edmonds' forest polytope:

Let $G = (V, E)$ be a graph, $w \in (0, 1)^E$, and for $F \subseteq E$ define $\chi_F(e)$ be 1 if $e \in F$ and 0 otherwise. According to a basic theorem of Edmonds (1970) the forest polytope defined as:

$$\text{conv}(\chi_F : F \text{ is a forest})$$
$$= \{x \in \mathbb{R}^E : x(E(U)) \le |U| - c_G(U), \ x \ge 0, \ (U \subseteq V(G))\},$$

(FOREST)

is a particular polymatroid. Hence it has integer vertices (the forests of G), and the defining system of inequalities has an integer dual solution for any integer objective function.

Note the following:

(i) If $c_G(U) > 1$ then the corresponding inequality can be straightforwardly written as the sum of $c_G(U)$ others with $c_G(\cdot) = 1$, so *the only essential inequalities in* (FOREST) *are those with* $c_G(U) = 1$.

(ii) Edmonds (1970, 1971) proved (see also Schrijver 2003) that *a polymatroid intersected with upper bound constraints is still a polymatroid, and therefore it corresponds to a TDI system.*

(iii) The submodular function h, associated to the polymatroid defined by the constraints of (FOREST) plus the additional upper bound constraints $x \leq w$, is such that $h(\{e\}) = w(e)$ for all $e \in E$. Let us consider the objective function $\max \sum_{e \in E} x_e$ on this polymatroid. For each $e \in E$, there is a dual variable y_e corresponding to the upper bound constraint $x_e \leq w(e)$ and, for each U such that $c_G(U) = 1$ there is a dual variable y_U corresponding to the (FOREST) constraint on U. By (ii) there exists an optimal dual solution with $0 - 1$ coordinates and it is not hard to check that its value is equal to the minimum of (DUAL) + the sum of the upper bounds of all edges.

(iv) The new right hand sides are easy to express: $h(A) = \max \{\sum_{e \in A} x_e : x$ is in the forest polytope (FOREST), $x \leq w\}$, $(A \subseteq E)$. Similarly to (iii), the dual linear program provides a formula for computing h from b, and this problem is clearly the same again as GSM ("restricted to A"), so the greedy algorithm does not seem to be easily executed unless a good way of computing h recursively occurs.

Such a way exists, but first the "graphic" property has to enter the game. Since the constraints are associated to edge-sets induced by vertex-sets U, we can directly associate the new right hand sides to these vertex-sets, and this is what we did when we defined $b(U)$. The function b however is negative on the empty set, so the corresponding polymatroid would be empty. If we set $b(\emptyset) := 0$ then the new function is submodular only on intersecting pairs. Frank and Tardos (1988) compute (the rank oracle of) a new fully submodular function (which must be 0 on the $\emptyset$) that defines the same polyhedron, with an appropriate implementation of the greedy algorithm. In our case this corresponds to gradually growing U and computing "$h(E(U))$".

Algorithms for graphic submodular function minimization (Cunningham's, Barahona's and ours) can be all considered to be combinations of this greedy algorithm with a network flow model. We hope that by realizing this, and working out the details in a graphic way, the present paper provides an elementary, efficient and self-contained presentation of such an algorithm.

(v) In order to decide whether a given vector $w \in \mathbb{R}^E$ belongs to the forest polytope or not it is sufficient to know whether the minimum of $b(U) = |U| - 1 - w(E(U))$ is nonnegative for all U, $\emptyset \neq U \subseteq V$ or not. This problem is solved by PQPW. Using the solution as a separation subroutine for Grötschel, Lovász and Schrijver's equivalence of separation and optimization (Grötschel et al. 1981), and using that a dual optimum can also be computed in polynomial time see (Grötschel et al. 1988), this provides another way (mentioned in '5.' in Sect. 17.1) for solving our problem.

In short, our problem (DUAL) can be viewed as follows: one can separate from Edmonds' polytope efficiently in a combinatorial way (with PQPW) but it is not trivial to optimize on it combinatorially. However, this task can also be achieved combinatorially in a much simpler and more efficient way than general submodular function minimization, and this is exactly graphic submodular function minimization (GSM).

The algorithm has been implemented by our physicist coauthors (Anglès d'Auriac et al. 2002; Anglès d'Auriac 2004). Using the push-relabel algorithm (Goldberg and Tarjan 1988) for computing minimum cuts into PQPW, they were able to deal with grids of size 512×512 (Anglès d'Auriac et al. 2002), that is, more than 250 000 vertices.

17.3 Applications

17.3.1 Potts' Model

In this section we recall the Potts model (for a review of the huge amount of work devoted to it see Wu 1982) and show why the order of magnitude – or under a restrictive condition the approximate value – of the partition function is determined by the minimum value of the submodular function GSM if the number of 'states' tends to infinity. The partition function is an important quantity that encodes the statistical properties of a system in thermodynamic equilibrium.

A *lattice* of spins is given at each *site* of which lives a variable. Each variable $\sigma_1, \ldots, \sigma_n$ ($n \in \mathbb{N}$) can take values in a given set $\mathbb{Z}_q := \{0, \ldots, q-1\}$, $q \in \mathbb{N}$. Pairs of neighboring sites on the lattice are called *bonds*. We denote the number of sites by n, and the number of bonds by m.

Each configuration $\sigma = (\sigma_1, \ldots, \sigma_n)$ has an energy

$$E(\sigma) = \sum_{ij} K_{ij} \delta_{\sigma_i \sigma_j}, \tag{1}$$

where $\sigma_i \in \mathbb{Z}_q$, the sum runs over all bonds ij, $K_{ij} \in \mathbb{R}_+$ is a given non-negative weight of bond ij, and δ_{ab} is the Kronecker symbol (1 if $a = b$ and 0 otherwise). (A lattice is a graph, sites are vertices, bonds are edges.) The aim is to compute or approximate the partition function

$$Z(K_{ij} : ij \text{ is a bond}) := \sum_{\sigma} \exp(E(\sigma)) = \sum_{\sigma} \prod_{ij} e^{K_{ij} \delta_{\sigma_i \sigma_j}}, \tag{2}$$

where the summation runs over all assignments of values $\sigma := (\sigma_1, \ldots, \sigma_n) \in \mathbb{Z}_q^n$, and the product over all bonds ij.

We follow (Juhász et al. 2001). Note that $e^{K\delta} = 1 + (e^K - 1)\delta$ for $\delta \in \{0, 1\}$; introduce $v_{ij} = e^{K_{ij}} - 1$ and expand the product of sums:

$$\prod_{ij} e^{K_{ij} \delta_{\sigma_i \sigma_j}} - \prod_{ij} (1 + (e^{K_{ij}} - 1)\delta_{\sigma_i \sigma_j}) = \sum_{L} \prod_{ij \in L} v_{ij} \delta_{\sigma_i \sigma_j},$$

where the summation runs over all subsets of bonds L. Substituting this to (2) we get:

$$Z(K_{ij} : ij \text{ is a bond}) = \sum_\sigma \sum_L \prod_{ij \in L} v_{ij} \delta_{\sigma_i \sigma_j} = \sum_L \sum_\sigma \prod_{ij \in L} v_{ij} \delta_{\sigma_i \sigma_j}$$

$$= \sum_L q^{c(L)} \prod_{ij \in L} v_{ij}, \tag{3}$$

where the sum runs over all subsets L of bonds, and where $c(L)$ is the number of connected components of L on the set of all sites, counting also the isolated sites among the components. (By convention if L is empty then $\prod_{ij \in L} v_{ij} = 1$.) We got the last equality by counting the number of different $\sigma = (\sigma_1, \dots, \sigma_n)$ for which the product is nonzero; since it is nonzero if and only if $\sigma(i) = \sigma(j)$ for every $ij \in L$, that is, if and only if σ is constant on every connected component of L, all possible σ can be given by choosing an element of Z_q for every connected component of L independently. Therefore we have exactly $q^{c(L)}$ such σ.

Note that this sum has 2^m terms, and that in (3) q does not need to have an integer value. (This provides a way of defining the Potts model for non-integer values.) Clearly, $K_{ij} \geq 0$ is equivalent to $v_{ij} \geq 0$, and we can introduce a new set of variables α_{ij} with

$$v_{ij} = q^{\alpha_{ij}},$$

and the partition function becomes

$$Z = \sum_L q^{c(L) + \sum_{ij \in L} \alpha_{ij}}. \tag{4}$$

Finally one introduces the function

$$f(L) = c(L) + \sum_{ij \in L} \alpha_{ij} \tag{5}$$

so that

$$Z = \sum_L q^{f(L)}. \tag{6}$$

As pointed out in Juhász et al. (2001), while q tends to infinity the sum in (6) is dominated by the maximum value of $q^{f(L)}$ and the order of magnitude of the partition function depends on the maximum of the function f that we have determined. (The asymptotic value of the maximum depends also on the number of optimal solution; nevertheless we have this optimum in the important case when the optimum is unique.)

17.3.2 Optimal Cooperation, Optimal Augmentation, Strength and Reinforcement

Cunningham (1985) studied Graphic Submodular Function Minimization in the context of Optimal Attack and defense of a network. Each edge has strength $s(e)$ which is a measure of the effort required by an attacker to destroy it. On another hand there is a benefit b for each new "disconnection". The goal of the attacker is to destroy a subset of edges A minimizing $s(A) - (c_G(E \setminus A) - 1)b$. This "Optimal Attack" problem is equivalent to Optimal Cooperation but with a 'complementary' viewpoint.

In the following we go further towards other applications mentioned in Cunningham (1985), and again, we provide a fully graphic interpretation for the sake of simplicity and practical efficiency. We will follow the language of "Optimal Cooperation".

The Augmentation Problem

Let us adopt the viewpoint of an employer ready to pay to stimulate the cooperation between researchers, by increasing the benefits of some pairwise cooperations (for instance with primes or rewards), for keeping the whole working group together. (The recent policy of the authors' employer, the CNRS, encourages indeed the creation of big institutes.) A cost function $k : E \longrightarrow \mathbb{R}_+$ expresses the cost of a unit increase in sponsoring a cooperation.

We say that (G, w) (or w) is *strong* if for the weight function w, $\{V\}$ is an optimal partition (for OC), and we can now state the problem more formally.

Augmentation Problem
Input: A graph $G = (V, E)$ with weight and upper bound functions $w, u : E \longrightarrow \mathbb{R}_+$, $w \leq u \leq 1$, cost function $k : E \longrightarrow \mathbb{R}_+$.

Output: A new weight function $w' : E \longrightarrow \mathbb{R}_+$, $w \leq w' \leq u$, such that (G, w') is strong, and $\sum_{e \in E} k(e)(w'(e) - w(e))$ is minimum; or a certificate of infeasibility.

If $u := 1$ everywhere, that is, if there are no upper bounds, the augmentation problem is called *unconstrained* otherwise *constrained*.

In some sense we already know a solution for this problem: $\{V\}$ is an optimal solution for (G, w') if and only if for the all 1 objective function, (FOREST) completed with the inequalities $x(e) \leq w'(e)$ does not have a dual solution of smaller value than $|V| - 1$.

This means exactly that, $y_e := 0$ for all $e \in E$, $y_U := 0$ for all $U \subseteq V, U \neq V$, and $y_V := 1$ is an optimal dual solution. Clearly, this holds if and only if there exists a primal solution of the same value $|V| - 1$, that is, if and only if the polymatroid (FOREST) has a (noninteger) basis $w'' \leq w'$:

For (G, w') the set $\{V(G)\}$ is an optimal solution of (DUAL) *if and only if* (FOREST) *has a basis $w'' \leq w'$.*

We will say then that w' *contains a basis*. Weight functions w' which contain a basis form a "contrapolymatroid", that is, a set of vectors closed under adding nonnegative vectors, and such that the (coordinatewise) minimal elements satisfy the polymatroid basis axioms (Schrijver 2003). Equivalently, a contrapolymatroid is the set of nonnegative vectors satisfying the inequality system $x(U) \geq g(U)$ for all $U \subseteq E$, where g is a given supermodular function on 2^E.

These remarks lead to two possibilities for solving the augmentation problem. Both require to replace each edge e by two parallel edges e' and e'': e' has cost 0, weight 0 and upper bound $w(e)$, e'' has cost $k(e)$, weight 0 and upper bound $u(e) - w(e)$. We consider the polymatroid (FOREST) with the upper bound constraints associated to this new graph.

The first possibility is to start from the all 0 vector and to increase the variables one by one according to the polymatroid greedy rule (in increasing order of the costs). If at the end, the sum of the variables is equal to $|V| - 1$ then we increase the weight of each edge e by the value of the variable associated to e''; otherwise the problem is not feasible. Notice that this leads to an algorithm similar to the one of Cunningham (1985), but there is no need to introduce a new polymatroid like in Cunningham (1985).

The second possibility is to use that vectors containing a basis form a contrapolymatroid. If the vector with coordinates $x_{e'} = w(e)$, $x_{e''} = u(e) - w(e)$ for each e in E, is in this contrapolymatroid then decrease the coordinates greedily (in decreasing order of positive cost) according to the contrapolymatroid greedy algorithm (Schrijver 2003); else the problem is not feasible.

We rather work out a completely graphic version that extends naturally the approach of this paper.

Graphic Augmentation Algorithm

Input and **Output**: see above.

0. Give the initial value $w' := w$. Compute (using the OC algorithm) a maximal partition $\mathcal{P}$ such that $g_{G,w'}(\mathcal{P})$ is maximum. (That is: $\mathcal{P}$ is optimal but any partition obtained from $\mathcal{P}$ by merging elements of $\mathcal{P}$ is not. There exists a unique such partition, see the Remarks after OC algorithm)
   If $\mathcal{P} = \{V\}$, then STOP and output w', otherwise all edges induced by the classes of $\mathcal{P}$ are marked *fixed*, shrink all the classes of $\mathcal{P}$.
   H is assigned to be the arising graph, $E(H) \subseteq E(G)$ (contraction-minor).
1. Let $e = xy$ be an edge of H that is not fixed (in particular its endpoints are in different classes of $\mathcal{P}$) such that

$$k(e) = \min\{k(f) : f \in E(G), \ f \text{ is not fixed}\}.$$

   The restriction to H of the current w' will be denoted by $w'_{|H}$.
2. Set $w'(e) := u(e)$, run PQPW$(H, w'_{|H}, x)$ to get the maximal output $S_e \subseteq V(H)$. Let $b := -b_{H,w'_{|H}}(S_e)$. Fix e, and
   – In case $S_e \neq \{x\}$:
     Set $w'(e) := u(e) - b$. If $S_e = V(H)$ then STOP and output w'; else fix e and all other edges induced by S_e, shrink S_e in H, redefine H as the resulting graph and GOTO 1.
   – In case $S_e = \{x\}$:
     If furthermore every edge is fixed STOP and return the corresponding partition of G as certificate of infeasibility, otherwise keep H and w' and GOTO 1.

END

The optimality of the algorithm (stated in the theorem below) relies on the following:

Lemma 3.1. *All along the algorithm, at the beginning of Step 1, the trivial partition* $\{\{v\} : v \in V(H)\}$ *is optimal for the current* w' *in* H, *and the corresponding partition of* G *is also optimal for* (G, w'). *Furthermore all these partitions have the same value than the first one.*

Proof. Indeed, just after Step 0 this is certainly true. We must show that this is true at the end of Step 2 where w' and H are redefined when considering the edge e. By Theorem 2.4 we know that after shrinking S_e in H, the vertices of the new H correspond to an optimal partition for (G, w') for w' obtained from the previous weight function by simply increasing the weight of e to its maximum possible value $u(e)$. In the case where $S_e \neq \{x\}$ this may be true even for a smaller weight of e. The value b corresponds to the gain of increasing the weight of e. A gain of 0 is enough for us, and we decrease the weight of e to the minimum which makes it enter into the partition. So, the new partition has the same value than the previous one. In the other case there is nothing to prove since the partition stays the same. So the Lemma is proved. □

Theorem 3.2. *If the algorithm stops with infeasibility, then all edges outside the partition* $\mathcal{P}^*$ *corresponding to the output have* $w'(e) = u(e)$. *Otherwise, the algorithm stops with* $w \leq w' \leq u$ *where* w' *is strong, and* $k^\top(w' - w)$ *is smallest among any strong* w' *satisfying these bounds.*

Proof. As pointed out above, the validity of the greedy algorithm follows from the fact that (FOREST) with upper bounds w' is a polymatroid. It can also be checked directly in terms of the graph.

If, at the end of the algorithm, H has only one vertex, then according to Lemma 3.1, w' is strong. If not, then $w'(e) = u(e)$ for every edge not induced by a class of the partition $\mathcal{P}^*$ when the algorithm stops, and so this maximal partition has a bigger value than $\{V\}$ for any $w' \leq u$. □

In case of the unconstrained Augmentation Problem we get an algorithm which has the same complexity as at most n minimum cut computations on networks of decreasing sizes. Indeed, in that case, each time a new partition $\mathcal{P}'$ is computed at Step 2, its cardinality is strictly smaller than that of the previous partition $\mathcal{P}$. Note that the minimum cost spanning tree problem is a special case of the unconstrained Augmentation Problem.

Extrapolating the success of the implementation of the OC algorithm in Anglès d'Auriac et al. (2002) (solution for several hundred thousand variables) this version of the reinforcement problem should also be efficient. However the advantages that make our algorithm run quickly (the small size of the current graphs) cannot be converted into a worst-case bound. The asymptotic worst case bound for the running time of the unconstrained case is the same as that of Barahona (2006).

Unfortunately, in the case of constrained case, we may have to perform much more minimum cut computations, m in the worst case, since we increase the weights of edges one by one. For this problem, Barahona (2006) performs the computation of a sequence of at most m minimum cuts with a time-complexity of only n times the complexity of one push-relabel algorithm, by trading work for memory space: this algorithm uses a "parametric flow" technique (Gallo et al. 1989), which involves keeping in memory n copies of the original graph. In our algorithm the number of vertices of the different networks we use will decrease from $n + 2$ to 4, except if the problem is not feasible, but this last case is easy to test by running the OC algorithm separately with u as weight function. (If the maximal optimal partition is not $\{V\}$ then there exists no feasible solution.)

The reason for which we cannot use the parametric flow technique is precisely the shrinking of subsets of vertices. However in case the weights have a small common denominator q, typical in the application in physics, we may use a similar technique using "augmenting paths" instead of a "push-relabel" algorithm for computing the maximum flows and as in Barahona (2006) the complexity of the algorithm can be improved at the cost of a higher space complexity: during the Augmentation Algorithm we maintain at most n networks $N(H, w', x)$ of PQPW type obtained from the initial one by shrinking subsets of vertices and increasing the weights of some of the edges. We keep a feasible flow in each of these networks and an (s, t)-cut of weight $n = |V|$ induced by $V \setminus t$. Then the value of the maximum flow will never exceed n and so the algorithm will perform at most nq augmentations in each network and we get a complexity of $O(n^2mq)$. The way we construct the networks is as follows. At Step 2, we increase the capacities accordingly to $w'(xy) := u(xy)$. Obviously the flow we had before is still compatible. In case S_e does not contain y we just need to increase the capacities similarly in all current networks. In the other case we discard all networks $N(H, w', v)$ such that $v \in S_e \setminus y$. In $N(H, w', y)$ we simply shrink S_e into y, and keep the same flow. For all other networks $N(H, w', v)$ $v \notin S_e$ we do the following. We shrink S_e in y. We set $c(sy) := |S_e| - 1 + \sum_{ab \in E, a \in S_e, b \notin S_e} w'(ab)/2$. Since S_e was not a set of the previous maximal optimal partition we have $|S_e| - 1 > \sum_{ab \in E, a, b \in S_e} w'(ab)/2$ and so the flow obtained by setting $f(sy) := \sum_{ab \in E, a \in S_e} f(s, a)$ is a feasible flow. Notice that the new networks are not exactly as defined in PQPW but the condition $c(a, t) - c(s, a) = 1 - p(a)$ is fulfilled for every vertex a, which is sufficient. Another improvement of the m minimum cut computations algorithm of Cunningham (1985) is due to Gabow (1998) who proposed an algorithm computing a minimum-cost base of a graphic polymatroid in time $O(n^2m \log(n^2/m))$.

Strength

Cunningham (1985) defined the strength and the reinforcement of a graph—we translate these into the "Optimal Cooperation language" and show how they can be computed using graphic submodular function minimization and (constrained) augmentation.

Given an undirected graph $G = (V, E)$ where each edge e has a nonnegative strength $s(e)$, the *strength* of (G, s) is defined as

$$\sigma(G, s) := \max\{\lambda : (G, s/\lambda) \text{ is strong}\}.$$

(It is easy to see that the maximum exists.)

Cunningham showed that the strength of a graph can be computed by solving at most n Optimal Cooperation problems, that is, after executing n GSM algorithms.

Indeed, let σ be the strength of the graph $G = (V, E)$ with strength function s on the edges. By definition of σ we have $f_{G,s/\sigma}(E) \geq f_{G,s/\sigma}(\emptyset)$ and so $\sigma \leq s(E)/(n-1)$. So if E is optimal for the weights $\frac{(n-1)}{s(E)}s$ then $\sigma = s(E)/n - 1$ and we stop. Else $\sigma < s(E)/n - 1$, that is, dividing the weights by $s(E)/n - 1$ some edges are too light to enter into an optimal solution and the maximal optimal solution A is strictly included in E. By Theorem 2.4 we know that while the weights are augmented, the edges of A will stay in an optimal solution, so we can simply shrink all sets of $\mathcal{P}(A)$ in order to get a new graph G' with strength function $s_{|G'}$ of the same strength as G, s and we repeat this procedure until E becomes optimal. Notice that $|V(G')| < |V(G)|$ since else E would be optimal; and so we will stop after solving at most n Optimal Cooperation problems. This algorithm performs virtually the same actions as the one of Cunningham except that once again the sizes of the problems to be solved will be decreasing. Similarly to "parametric flow" for computing the strength (Gusfield 1991; Cheng and Cunningham 1994), we can keep at most n networks with increasing capacities and feasible flows, but in that case a small common denominator of the capacities is unlikely. However we can avoid this problem by solving the generalized Optimal Cooperation problem where the benefit of a connected component is a constant B which may be different from 1, that is

$$\text{Maximize}\left\{f_{G,w,B}(A) = Bc_G(A) + \sum_{e\in A} w(e), \ A \subseteq E(G)\right\}.$$

By dividing the weights by B we are lead to the usual Optimal Cooperation problem, but we can also solve it directly applying PQPW to a modular shift of b (as in Cheng and Cunningham 1994). In that case we obtain an algorithm of complexity $O(n^2 m B q)$. Gabow (1998) proposed an algorithm to compute the strength of a graph in time $O(n^2 m \log(n^2/m))$ and space only $O(m)$ (instead of $O(nm)$ for the other algorithms of the same complexity).

Let us finally check that our definition of the strength is the same as Cunningham's:

By definition (G, s) is of strength at least σ if and only $1 + s(E)/\sigma \geq c(A) + s(A)/\sigma$ for all $A \subseteq E$, that is, if and only if $\sigma s(E \setminus A) \geq \sigma(c(A) - 1)$. it follows that

$$\sigma(G, s) = \min\left\{\frac{s(E - A)}{c(A) - 1} : A \subseteq E, c(A) > 1\right\}.$$

Reinforcement

We consider now the minimum cost reinforcement problem:

Reinforcement
Input: A graph G with strength function $s : E \longrightarrow \mathbb{R}_+$, cost function $k : E \longrightarrow \mathbb{R}$, $k \geq 0$ and upper bound function $l \geq s$ on the edges, and $\sigma_0 > 0$.

Output: A new strength function $s' : E \longrightarrow \mathbb{N}, s < s' \leq l$, such that $\sigma(G, s') \geq \sigma_0$ and $\sum_{e \in E} k(e)(s'(e) - s(e))$ is minimum; or a certificate of infeasibility.

Clearly, according to the definition, the reinforcement problem (G, s, l, k, σ_0) can be solved by finding an optimal solution w' of the Augmentation Problem on $(G, w = \frac{s}{\sigma_0}, u = \frac{l}{\sigma_0}, k)$. Indeed, then $s' = w'\sigma_0$ is an optimal reinforcement. If there is no solution for the Constrained Augmentation Problem, then no reinforcement exists.

17.4 Conclusion

This is a thoroughly revisited version of the mathematical part of an article that appeared on this problem in Journal of Physics A. It is completed here with full details of the mathematical part of the paper, by more mathematical background, and also new results concerned by the minimization of a graphic submodular function like the problems of augmentation, strength or reinforcement.

Acknowledgements

The algorithms have been worked out for the application in physics (Anglès d'Auriac et al. 2002) with an active participation of our physicist colleagues. We thank them for helping us helping them! We also thank the referees for their thorough reading and helpful comments.

References

Ahuja, R.K., Magnanti, T.L., Orlin, J.B.: Network Flows: Theory, Algorithms and Applications. Prentice Hall, Englewood Cliffs (1993)

Anglès d'Auriac, J.-C., Iglói, F., Preissmann, M., Sebő, A.: Optimal cooperation and submodularity for computing Potts'partition functions with a large number of states. J. Phys. A, Math. Gen. **35**, 6973–6983 (2002)

Anglès d'Auriac, J.-C.: Computing the Potts free energy and submodular functions. In: Hartmann, A.K., Rieger, H. (eds.) New Optimization Algorithms in Physics, pp. 101–117. Wiley, New York (2004)

Baïou, M., Barahona, F., Mahjoub, A.R.: Separation of partition inequalities. Math. Oper. Res. **25**(2), 243–254 (2000)

Barahona, F.: Separating from the dominant of the spanning tree polytope. Oper. Res. Lett. **12**, 201–203 (1992)

Barahona, F.: Network reinforcement. Math. Program. **105**, 181–200 (2006)

Cheng, E., Cunningham, W.H.: A faster algorithm for computing the strength of a network. Inf. Process. Lett. **49**, 209–212 (1994)

Cunningham, W.H.: Testing membership in matroid polyhedra. J. Comb. Theory, Ser. B **36**, 161–188 (1984)

Cunningham, W.H.: Optimal attack and reinforcement of a network. J. Assoc. Comput. Math. **32**(3), 549–561 (1985)

Edmonds, J.: Minimum partition of a matroid into independent sets. J. Res. Natl. Bur. Stand. Sect. B **69**, 73–77 (1965)

Edmonds, J.: Submodular Functions, Matroids, and certain polyhedra. In: Guy, R., Hanani, H., Sauer, N., Schönheim, J. (eds.) Combinatorial Structures and Their Applications, pp. 69–87. Gordon and Breach, New York (1970)

Edmonds, J.: Matroids and the greedy algorithm. Math. Program. **1**, 127–136 (1971)

Ford, L.R., Fulkerson, D.R.: Maximum flow through a network. Can. J. Math. **8**, 399–404 (1956)

Frank, A., Tardos, É: Generalized polymatroids and submodular flows. Math. Program. **42**, 489–563 (1988)

Fujishige, S.: Submodular Functions and Optimization. Annals of Discrete Mathematics, vol. 47. North-Holland, Amsterdam (1991)

Gabow, H.N.: Algorithms for Graphic Polymatroids and Parametric $\bar{s}$-Sets. J. Algorithms **26**, 48–86 (1998)

Gallo, G., Grigoriadis, M.D., Tarjan, R.: A fast parametric maximum flow algorithm and applications. SIAM J. Comput. **18**(1), 30–55 (1989)

Goldberg, A.V., Tarjan, R.E.: A new approach to the maximum-flow problem. J. Assoc. Comput. Math. **35**(4), 921–940 (1988)

Grötschel, M., Lovász, L., Schrijver, A.: The ellipsoid method, and its consequences in Combinatorial Optimization. Combinatorica **1**, 169–197 (1981)

Grötschel, M., Lovász, L., Schrijver, A.: Geometric Algorithms and Combinatorial Optimization. Springer, Berlin (1988)

Gusfield, D.: Computing the strength of a graph. SIAM J. Comput. **20**(4), 639–654 (1991)

Iwata, S.: A fully combinatorial algorithm for submodular function minimization. J. Comb. Theory, Ser. B **84**, 203–212 (2002)

Iwata, S., Fleischer, L., Fujishige, S.: A combinatorial strongly polynomial algorithm for minimizing submodular functions. J. Assoc. Comput. Math. **48**, 761–777 (2001)

Juhász, R., Rieger, H., Iglói, F.: The random-bond Potts model in the large-q limit. Phys. Rev. E **64**, 56122 (2001)

Lovász, L.: Flats in matroids and geometric graphs. In: Cameron, P.J. (ed.) Combinatorial Surveys, Proceedings of the 6th British Combinatorial Conference, pp. 45–86. Academic Press, London (1977)

Lovász, L.: Combinatorial Problems and Exercises. North-Holland/Akadémiai Kiadó, Amsterdam (1979)

Padberg, M.W., Wolsey, L.A.: Trees and cuts. Ann. Discrete Math. **17**, 511–517 (1983)

Picard, J.-C., Queyranne, M.: Selected applications of minimum cuts in networks. Int. Syst. Oper. Res. **20**, 394–422 (1982)

Schrijver, A.: A combinatorial algorithm minimizing submodular functions in strongly polynomial time. J. Comb. Theory, Ser. B **80**, 346–355 (2000)

Schrijver, A.: Combinatorial Optimization. Springer, Berlin (2003)

Wu, F.Y.: The Potts model. Rev. Mod. Phys. **54**, 235–268 (1982)

18

Matroids—the Engineers' Revenge

András Recski

Summary. Matroids are applied in electric engineering for over 30 years. These applications motivated the investigation of some new, pure matroidal questions. Such results are surveyed for readers with mathematical (rather than engineering) background.

18.1 Introduction

Electrical engineering was perhaps the first area for application of graph theory. The pioneering results of Kirchhoff (1847) and Maxwell (1892) have been applied for a lot of problems of electric network theory in the first decades of the 20th century. The first "revenge" of engineers came around 1930 – for a proper understanding of the duality in electric network theory, the concept of the dual of planar graphs had to be clarified first, hence Whitney has introduced 2-isomorphism and abstract dual (Whitney 1932, 1933, 1935), and these concepts were perhaps the most important motivations for the discovery of matroids. Later matroid theory found several engineering applications, mostly between 1970–80.

We are currently witnessing the second "revenge" of the engineers: Since certain physical properties of electric networks are strongly related to the mathematical properties of the underlying graphs, and some of these latter properties can be generalized in a more or less straightforward way to matroids, network theory motivated some new, pure matroidal research in the last three decades.

In the present survey such results are collected for readers interested in matroid theory. They are arranged according to the physical concepts which inspired them, but although the subtitles suggest these physical concepts, the results are purely mathematical; no background in physics or engineering is required.

18.2 Preliminaries

Throughout the paper u stands for voltage difference and i stands for current. Suppose that an electric network consists of resistors and ideal voltage and current

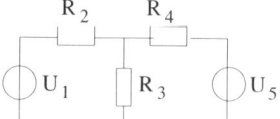

Fig. 18.1.

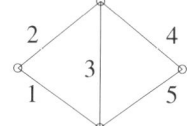

Fig. 18.2.

sources only. An ideal voltage source is described by the "equations" {u is given, i is arbitrary}, an ideal current source is similarly described by {i is given, u is arbitrary}, while a resistor is given by $u = Ri$ or $i = Yu$ where R and Y are called the resistance and the conductance of the resistor, respectively. The way how these elements are interconnected is described by the network graph where the edges correspond to the network elements. We may suppose that this graph is connected.

Recall Kirchhoff's laws that the signed sum of the voltages along any circuit of the network graph is zero and the signed sum of the currents along any cut set of the network graph is zero. The basic task of electric network analysis is that by applying Kirchhoff's laws every unknown quantity (voltage and current) should be expressed as the function of the R or Y values and of the input (voltages of the voltage sources and currents of the current sources). This is possible only if the subset of edges, corresponding to the voltage sources, is circuit free and the subset of edges, corresponding to the current sources, is cut set free. Moreover, these necessary conditions together are sufficient for the unique solvability of the network if all the resistors are passive elements, that is, if $R, Y > 0$ hold for every resistor.

As an example consider the network in Fig. 18.1. Its network graph is given in Fig. 18.2 and, for example, the current of the resistor R_3 can be expressed as

$$i_3 = \frac{R_4 u_1 + R_2 u_5}{R_2 R_3 + R_2 R_4 + R_3 R_4} = \frac{Y_2 Y_3 u_1 + Y_3 Y_4 u_5}{Y_4 + Y_3 + Y_2}.$$

The other unknown quantities would be expressed by other expressions, however, all of them are fractions with the same denominator $W_R(G) = R_2 R_3 + R_2 R_4 + R_3 R_4$ or $W_Y(G) = Y_4 + Y_3 + Y_2$ (recall that we solve a system of linear equations and, by Cramer's rule, every unknown quantity can be expressed as a ratio of two determinants where the one in the denominator does not depend on the selection of the actual unknown quantity).

Since the short circuit ($u = 0$) and the open circuit ($i = 0$) are special cases of voltage and current sources, respectively, we shall always suppose that the edges corresponding to the voltage sources are contracted and those corresponding to the

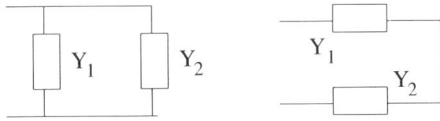

Fig. 18.3.

current sources are deleted, hence the network consists of resistors only. In what follows, this "new" network graph will be denoted by $G = (V, E)$. In case of Fig. 18.1, G consists of three parallel edges.

A fundamental observation of Kirchhoff (1847) and Maxwell (1892) was that the common denominator $W_R(G)$ (respectively $W_Y(G)$) of every unknown quantity is simply $W_R(G) = \sum_T \prod_{i \notin T} R_i$ or $W_Y(G) = \sum_T \prod_{j \in T} Y_j$, where the summations are performed over the set of spanning trees T of G. These results are often called "topological formulae" in electric network theory, to emphasize that only the structure (the "topology") of the graph plays a role in these expressions.

A strongly related result of similar character is as follows:

Theorem 2.1 (Maxwell 1892). *Suppose that resistors with conductances $Y_1, Y_2, \ldots$ are interconnected along a connected graph G. Let $a, b \in V(G)$ be two specific points of G. Add a fictitious edge $e = \{a, b\}$ to G and denote the resulting graph by G' (if the two points are already adjacent in G then e is parallel to an old edge). The effective conductance measured between a and b in the network will be*

$$Y_{a,b} = \frac{W_Y(G' \backslash e)}{W_Y(G'/e)},$$

where, as before, $W_Y(G) = \sum_T \prod_{j \in T} Y_j$ and $\backslash$ and $/$ denote deletion and contraction of an edge, respectively.

Obviously, in the above theorem $G' \backslash e = G$ and we follow the convention that empty sums are 0 and empty products are 1. For example, the effective conductance of two parallel or two series resistors respectively (Fig. 18.3) are

$$\frac{Y_1 + Y_2}{1}, \qquad \frac{Y_1 Y_2}{Y_1 + Y_2} = \frac{1}{\frac{1}{Y_1} + \frac{1}{Y_2}}$$

(observe that the denominator in the first case is one since there is one "spanning tree" (namely the empty set) in a graph consisting of loops only).

18.3 Energy

A trivial consequence of Theorem 2.1 is that if $a, b \in V(G), a \neq b$, in a connected graph G and the edges of the graph represent resistors with $Y_i > 0$ for every i then the effective conductance between a and b is also positive, hence, in particular, $W_Y(G) \neq 0$. The physical background (positive resistors are passive elements, thus their interconnection cannot produce energy) remains valid even if we use other network elements as well, like (positive) capacitors or inductors. Hence $W_Y(G) \neq 0$ remains true even if the Y_i's are complex numbers, provided that their real parts $\Re(Y_k)$ are positive for every k.

Let now $\mathcal{M}$ be an arbitrary matroid on the underlying set E and suppose that complex numbers Y_i are assigned to every element $e_i \in E$. We may define $W_Y(\mathcal{M}) = \sum_B \prod_{e_i \in B} Y_i$ where the summation is performed over the set of the bases B of the matroid. The matroid is said to have the *half-plane property* if $W_Y(\mathcal{M}) \neq 0$ holds provided that $\Re(Y_i) > 0$ for every $e_i \in E$.

Theorem 3.1 (Choe et al. 2004). *Every graphic matroid has the half-plane property.*

Although this is clear by the above physical considerations, the following purely mathematical *proof,* reproduced from Wagner (2005a), will be of importance.

Let $\mathcal{M}$ be the circuit matroid of a graph G. We may suppose that G is connected (otherwise its connected components can separately be studied), let n be the number of its vertices. Let A denote the reduced signed incidence matrix of G (obtained by deleting an arbitrary row). Since A has n 1 rows, using the theorem of Binet and Cauchy,

$$\det(AYA^*) = \sum_{S \subseteq E, |S| = n-1} \det A[S] \det A^*[S] \prod_{e_i \in S} Y_i,$$

where Y is a diagonal matrix containing the Y_i quantities and A^* stands for the adjoint (conjugate transpose) of the matrix A (which, in our case, is simply its transpose).

Let a subset S of the columns of A have cardinality $n - 1$. Then $\det A[S] = 0$ if the S-edges contain a circuit and $\det A[S] = \pm 1$ if they form a tree. Hence

$$\det(AYA^*) = \sum_T \prod_{e_i \in T} Y_i = W_Y(\mathcal{M}).$$

Therefore all what we have to prove is that AYA^* is nonsingular if $\Re(Y_i) > 0$ for every i.

Let v be an arbitrary nonzero vector from the $(n-1)$-dimensional complex space. We shall prove that $AYA^*v \neq 0$. Recall that $A^*v \neq 0$ (since the rows of A were independent), hence at least one of its coordinates, say i_0, is nonzero. Finally

$$v^*AYA^*v = \sum_i Y_i |(A^*v)_i|^2,$$

and the real part of this sum is positive (since $\Re(Y_i) > 0$ for every i, their coefficients are non-negative and at least one of them (namely the i_0th) is positive).   □

This result can be generalized in the following way:

Theorem 3.2 (Choe et al. 2004). *Every regular matroid has the half-plane property.*

Proof. Recall (Tutte 1965) that a matroid is regular (that is, representable over every field) if and only if it can be represented over the field $\mathbb{Q}$ of the rationals by a totally unimodular matrix A. A matrix is totally unimodular if all of its square submatrices have determinant 0 or ± 1. Since we may additionally suppose that the rows of the above matrix A are linearly independent, the above proof can be applied again. □

A slightly more careful analysis of the proof shows that actually we need even less. Matrix A was totally unimodular (that is, every square submatrix of it has determinant 0 or ± 1) but here we only need that if the determinant of a square submatrix is nonzero, its absolute value should be one. This property can be fulfilled by matrices over the field $\mathbb{C}$ of the complex numbers as well. In particular, a matrix is a $\sqrt[6]{1}$-matrix if all of its square submatrices are either zero or one of the sixth roots of unity. A matroid is called a *sixth-root-of-unity matroid* if it can be represented by a $\sqrt[6]{1}$-matrix over $\mathbb{C}$.

Theorem 3.3 (Choe et al. 2004). *Every sixth-root-of-unity matroid has the half-plane property.*

The importance of this class of matroids has probably not been fully recognized yet. A matroid is regular if and only if it can be represented over GF(2) and GF(3), see (Tutte 1965); similarly, a matroid is sixth-root-of-unity if and only if it can be represented over GF(3) and GF(4), see (Whittle 1997).

These relations are summarized in Fig. 18.4. Here $\boxed{2}$, $\boxed{3}$ and $\boxed{4}$ denote the sets of matroids representable over GF(2), GF(3) and GF(4), respectively, while $\boxed{A}$ is the set of regular matroids, $\boxed{A} \cup \boxed{B}$ is the set of the sixth-root-of-unity matroids and $\boxed{HPP}$ is the set of the matroids with the half-plane property.

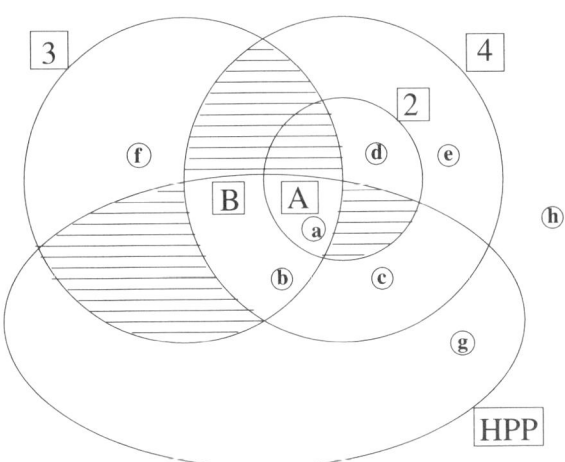

Fig. 18.4.

Theorem 3.4. *The shaded regions in Fig. 18.4 are empty subsets. The other eight subsets are non-empty, the lower case letters* **a**, **b**, . . . , **h** *in the other subsets are examples as follows:* **a**, **b**, **c** *and* **g** *are the uniform matroids* $\mathcal{U}_{3,2}$, $\mathcal{U}_{4,2}$, $\mathcal{U}_{5,2}$ *and* $\mathcal{U}_{6,2}$, *respectively.* **d** *and* **f** *are the Fano matroid* $\mathcal{F}_7$ *and the anti-Fano matroid* $\mathcal{F}_7^-$, *respectively. Finally,* **e** *and* **h** *are the direct sums* $\mathcal{F}_7 \oplus \mathcal{U}_{4,2}$ *and* $\mathcal{F}_7^- \oplus \mathcal{U}_{5,2}$, *respectively.*

Proof. Fact 1. The first statement, that the shaded regions are empty, trivially follows from the results (Choe et al. 2004) that a binary matroid has the half-plane property if and only if it is regular and a ternary matroid has the half-plane property if and only if it is a sixth-root-of-unity matroid.

Fact 2. Every uniform matroid has the half-plane property and the Fano matroid does not have it (Choe et al. 2004).

Fact 3. The uniform matroid $\mathcal{U}_{k,2}$ is known to be representable over fields with cardinality at least $k - 1$ only (folklore, see Lovász and Recski, 1973, for example).

Fact 4. The anti-Fano matroid is representable over fields of characteristic different from two and the Fano matroid over fields of characteristic two.

Fact 5. The class of matroids with the half-plane property is closed under, among others, minors and direct sum.

The positions of **a** and **b** are obvious: $\mathcal{U}_{3,2}$ is regular (in fact, graphic) and $\mathcal{U}_{4,2}$ is representable over every field except GF(2). The positions of **c**, **d** and **g** follow from Facts 2 and 3.

The anti-Fano matroid is ternary by Fact 4 and cannot have the half-plane property by Fact 1, leading to the position of **f**. Finally, the positions of **e** and **h** follow from the previous results and from Fact 5.   □

18.4 Switches

Suppose that a network contains positive resistors and a switch. The effective conductance measured between two points can only increase if we turn the switch from OFF to ON position (or, more generally, if we increase the conductance of one resistor). This observation (sometimes called *Rayleigh-monotonicity*) may lead to a trivial algebraic relation, like $Y_1 + Y_2 > Y_2$ in case of the first network of Fig. 18.5, or to a less obvious one like

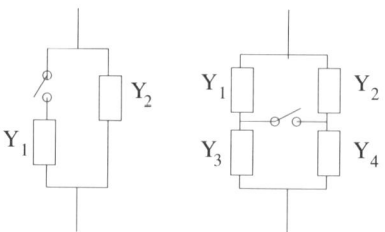

Fig. 18.5.

$$\frac{(Y_1 + Y_2)(Y_3 + Y_4)}{Y_1 + Y_2 + Y_3 + Y_4} > \frac{Y_1 Y_3}{Y_1 + Y_3} + \frac{Y_2 Y_4}{Y_2 + Y_4}$$

in case of the second network of the same figure.

Let again $\mathcal{M}$ be an arbitrary matroid on the underlying set E and suppose that positive real numbers Y_i are assigned to every element $e_i \in E$. As before, define $W_Y(\mathcal{M}) = \sum_B \prod_{e_i \in B} Y_i$ where the summation is performed over the set of the bases B of the matroid. Applying Theorem 2.1 the effective conductance between two points can be expressed as the ratio of two such sums. Since turning the switch to ON position can be expressed as the contraction of an element, one can define *Rayleigh-matroids* as those where Rayleigh-monotonicity holds.

Graphic matroids are obviously Rayleigh (see a pure combinatorial proof in Choe 2007, using the matrix tree theorem, Chaiken 1982; Chaiken and Kleitman 1978), so are the matroids having the half-plane property (hence, in particular, all the sixth-root-of-unity matroids), see (Choe and Wagner 2006).

However, not all the binary matroids are Rayleigh, in fact, a binary matroid is Rayleigh if and only if it does not contain the matroid $\mathcal{S}_8$ as a minor (Choe and Wagner 2006; Feder and Mihail 1992), where $\mathcal{S}_8$ is given by the matrix

$$\begin{pmatrix} 1 & 1 & 1 & 1 & 1 & 1 & 1 & 0 \\ 0 & 1 & 0 & 0 & 0 & 1 & 1 & 1 \\ 0 & 0 & 1 & 0 & 1 & 0 & 1 & 1 \\ 0 & 0 & 0 & 1 & 1 & 1 & 0 & 1 \end{pmatrix},$$

see also (Seymour and Welsh 1975).

This is one of the smallest counterexamples in several senses: every matroid with at most seven elements is Rayleigh, see (Choe and Wagner 2006) and every matroid of rank at most three is Rayleigh, see (Wagner 2005b).

It might be instructive to summarize the relations as

$$graphic \subseteq regular \subseteq \sqrt[6]{1} \subseteq HPP \subseteq Rayleigh.$$

All these classes are closed by taking duals and minors and all inclusions are proper:

Examples

(1) The dual of the circuit matroid of a nonplanar graph is regular but nongraphic.
(2) The uniform matroid $\mathcal{U}_{4,2}$ is sixth-root-of-unity matroid (Whittle 1997) but non-regular. (So is the matroid $\mathcal{X}$ of the next section as well.)
(3) The uniform matroid $\mathcal{U}_{5,2}$ has the half-plane property (Choe et al. 2004) but is not a sixth-root-of-unity matroid (see Theorem 3.4).
(4) The Fano matroid does not have the half-plane property (Choe et al. 2004) but it is Rayleigh (Choe and Wagner 2006).

While not related directly to the electrical engineering applications, the concept of *balanced matroids* (Feder and Mihail 1992) should also be mentioned as a proper superset of the set of Rayleigh-matroids

$$graphic \subseteq regular \subseteq \sqrt[6]{1} \subseteq HPP \subseteq Rayleigh \subseteq balanced.$$

Recent (mostly algebraic) research (Brändén 2007; Choe 2005) for properly under-standing the relation of the last three classes revealed further connections with the theory of jump systems (Bouchet and Cunningham 1995; Lovász 1997; Recski and Szabó 2006).

18.5 Control and Feedback

Networks consisting of resistors and ideal voltage and current sources can be de-scribed by graphs (see Sect. 18.2). Consider now a more complex linear network, containing controlled sources (see below) as well. The qualitative properties of such a network can be described by matroids, see (Iri and Tomizawa 1975; Narayanan 1974; Recski 1973), but these matroids will not always be graphic.

For example, both networks of Fig. 18.6 contain two current sources; in the first network both sources are "usual" ones but in the second network I_2 is not, since its describing system of equations is $\{i_2 = ki_3, u_2$ is arbitrary$\}$ rather than $\{i_2$ is given, u_2 is arbitrary$\}$. Such a device is called a *current controlled current source*.

In the first network (or in any network without controlled sources) the currents of certain devices can be prescribed independently of each other if and only if the subset of the corresponding edges contains no cut sets (that is, if the complement of this set is connected). Even the resistor R_4 could be replaced by a further independent current source I_4 since the edge set $\{1, 2, 4\}$ is cut set free. In general, the maximal subsets of edges whose currents can be prescribed independently, are just the complements of the spanning trees of the network graph G_0 (that is, they form the bases of the dual of the circuit matroid $\mathcal{G}_0$ of G_0).

However, in the second network at most two currents can be prescribed inde-pendently, for example, I_1 and I_4 (or any pair, except $\{2, 3\}$). It is still true that the maximal subsets of edges whose currents can be prescribed independently form the bases of the dual of a matroid but this matroid is not $\mathcal{G}_0$ but the union (or sum) of matroids $\mathcal{G}_0 \vee \mathcal{A}$, see (Iri and Tomizawa 1975; Narayanan 1974; Recski 1973) where a subset of edges is independent in $\mathcal{A}$ if and only if it is either the empty set or one of the single-element sets $\{2\}, \{3\}$.

This matroid $\mathcal{X} = \mathcal{G}_0 \vee \mathcal{A}$ is not graphic (it is a single element series exten-sion of the uniform matroid $\mathcal{U}_{4,2}$), in fact, this is the smallest example to illustrate that matroids, rather than graphs, are the proper tools for the study of the qualita-tive properties of active linear networks. The general question (when will the sum of two graphic matroids be graphic) seems to be difficult (see Lovasz and Recski

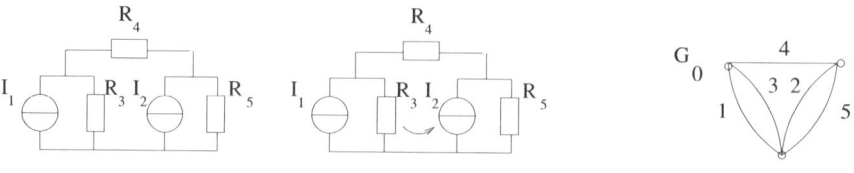

Fig. 18.6.

1973; Recski 1981 for some partial results). The following small "Kuratowski-type" observation, announced in Recski (1975) without proof, may be interesting from the physical point of view. Like in Kuratowski's theorem, series extension (or subdivision) allows iterations of single element series extensions as well.

Theorem 5.1. *The matroid $\mathcal{A}$ (describing the control) is the circuit matroid of a graph consisting of two parallel edges e_2, e_3 and loops for every other edge $e_1, e_4, e_5, \ldots$. The sum $\mathcal{M}(G) \vee \mathcal{A}$ of this matroid and the circuit matroid of a graph G is graphic if and only if G does not contain the graph G_0 of Fig. 18.6 or its series extension as a subgraph (with the two non-loop elements of $\mathcal{A}$ in the positions 2, 3).*

Proof. Suppose that G contains a subgraph isomorphic to G_0 or to its series extension. Then the sum of $\mathcal{A}$ and the circuit matroid of this subgraph is a series extension of the uniform matroid $\mathcal{U}_{4,2}$, hence it is not graphic.

On the other hand, suppose that $\mathcal{M}(G) \vee \mathcal{A}$ is not graphic for some graph G containing two specific edges e_2, e_3 corresponding to the non-loop elements in $\mathcal{A}$. We may suppose that G is connected or, at least, e_2 and e_3 are in the same connected component of G for otherwise the sum would be the circuit matroid of the graph illustrated on the right in Fig. 18.7.

Moreover, one can see using very similar arguments that these edges must not be loops or parallel to each other. This latter is illustrated by Fig. 18.8.

If these edges were series (that is, if $\{e_2, e_3\}$ were a cut set) then the sum would be the circuit matroid of the graph of $G - \{e_2, e_3\}$ extended with these edges as bridges. Hence we may conclude that there is a circuit in G containing both of these edges and also there are circuits containing exactly one of these edges. Therefore G contains the graph of Fig. 18.9 as a subgraph (where some of the paths P_1, P_2, P_4 and P_5 may be of length 0 as well).

Consider now two internal vertices from two distinct paths among $P_1, \ldots, P_5$. If all the paths connecting such pairs in G are covered by the vertices V_1, V_2 of Fig. 18.9 then, within 2-isomorphism, G looks like the first graph of Fig. 18.10 and then the sum is graphic, as illustrated by the second graph of Fig. 18.10.

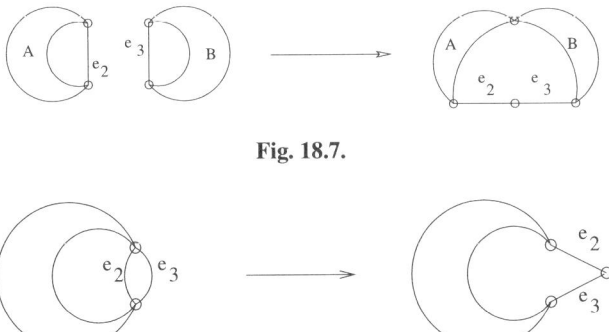

Fig. 18.7.

Fig. 18.8.

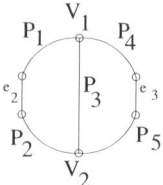

Fig. 18.9.

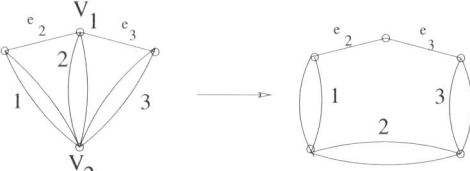

Fig. 18.10.

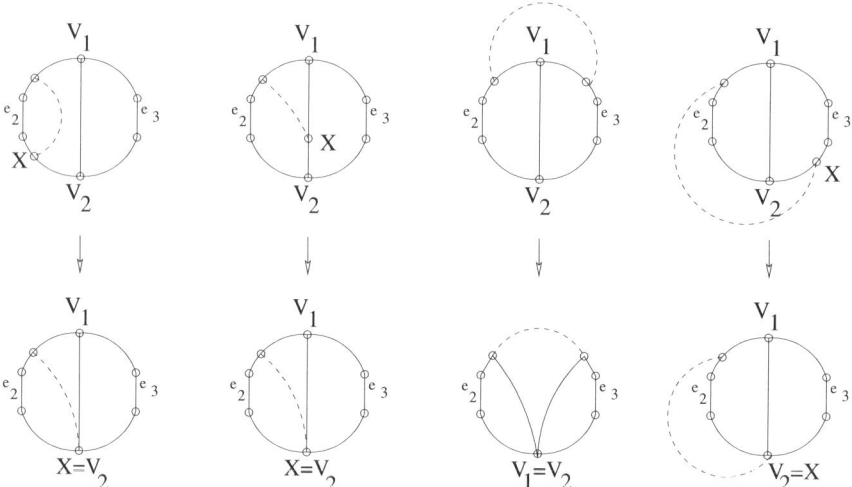

Fig. 18.11.

Hence we may suppose that there is a path connecting an internal vertex of, say, P_1 with an internal vertex of another path P_i without sharing any other vertex with the subgraph of Fig. 18.9. Then one can find the requested subgraph in every case $i = 2, 3, 4$ or 5, see Fig. 18.11. □

One may interpret this result as follows: if a network contains a single controlled source then the describing matroid will not be graphic if and only if the network has a feedback (edge 4 in the graph of Fig. 18.6).

Incidentally, the above matroid $\mathcal{X}$ is a sixth-root-of-unity matroid but it is not regular. This latter statement is obvious (it has $\mathcal{U}_{4,2}$ as a minor), for the former one observe that the following matrix coordinatizes it over GF(3) or GF(4) if x is an element in the appropriate field, differing both from 0 and 1.

$$\begin{pmatrix} 1 & x & 1 & 0 & 0 \\ 1 & 1 & 0 & 1 & 0 \\ 1 & 1 & 0 & 0 & 1 \end{pmatrix}.$$

Acknowledgements

Financial support of the Hungarian National Research Fund and the National Office for Research and Technology (Grant Number OTKA 67651) and useful comments of József Mezei, Dávid Szeszlér and Gábor Wiener are gratefully acknowledged.

References

Bouchet, A., Cunningham, W.H.: Delta-matroids, jump systems and bisubmodular polyhedra. SIAM J. Discrete Math. **8**, 17–32 (1995)

Bränden, P.: Polynomials with the half-plane property and matroid theory. Adv. Math. **216**, 302–320 (2007)

Chaiken, S.: A combinatorial proof of the all minors matrix tree theorem. SIAM J. Alg. Discrete Math. **3**, 319–329 (1982)

Chaiken, S., Kleitman, D.J.: Matrix tree theorems. J. Comb. Theory, Ser. A **24**, 377–381 (1978)

Choe, Y.B.: Polynomials with the half-plane property and the support theorems. J. Comb. Theory, Ser. B **94**, 117–145 (2005)

Choe, Y.B.: A combinatorial proof of the Rayleigh formula for graphs. Discrete Math. (2007). doi:10.1016/j.disc.2007.11.011

Choe, Y.B., Oxley, J., Sokal, A., Wagner, D.: Homogeneous multivariate polynomials with the half-plane property. Adv. Appl. Math. **32**, 88–187 (2004)

Choe, Y.B., Wagner, D.G.: Rayleigh matroids. Comb. Probab. Comput. **15**, 765–781 (2006)

Feder, T., Mihail, M.: Balanced matroids. In: Proc. of 24th STOC, pp. 26–38 (1992)

Iri, M., Tomizawa, N.: A unifying approach to fundamental problems in network theory by means of matroids. Trans. Inst. Electron. Commun. Eng. Jpn. **58-A**, 33–40 (1975)

Kirchhoff, G.: Über die Auflösung der Gleichungen, auf welche man bei der Untersuchungen der linearen Vertheilung galvanischer Ströme geführt wird. Ann. Phys. Chem. **72**, 497–508 (1847)

Lovász, L.: The membership problem in jump systems. J. Comb. Theory, Ser. B **70**, 45–66 (1997)

Lovász, L., Recski, A.: On the sum of matroids. Acta Math. Acad. Sci. Hungar. **24**, 329–333 (1973)

Maxwell, J.C.: Electricity and Magnetism. Clarendon, Oxford (1892)

Narayanan, H.: Theory of matroids and network analysis. PhD. thesis, Indian Institute of Technology, Bombay (1974)

Recski, A.: On partitional matroids with applications. Colloq. Math. Soc. János Bolyai **10**, 1169–1179 (1973)

Recski, A.: On the sum of matroids II. In: Proc. of 5th British Combinatorial Conf., Aberdeen, pp. 515–520 (1975)

Recski, A.: An algorithm to determine whether the sum of some graphic matroids is graphic. Colloq. Math. Soc. János Bolyai **25**, 647–656 (1981)

Recski, A.: Matroid Theory and Its Applications in Electric Network Theory and in Statics. Springer, Berlin (1989)

Recski, A., Szabó, J.: On the generalization of the matroid parity problem. In: Graph Theory. Trends in Mathematics, pp. 347–354. Birkhäuser, Basel (2006)

Seymour, P., Welsh, D.J.A.: Combinatorial applications of an inequality from statistical mechanics. Math. Proc. Camb. Philos. Soc. **77**, 485–495 (1975)

Tutte, W.T.: Lectures on matroids. J. Res. Natl. Bur. Stand., Ser. B **69**, 1–47 (1965)

Wagner, D.G.: Matroid inequalities from electrical network theory. Electron. J. Comb. **11** #A1 (2005a)

Wagner, D.G.: Rank-three matroids are Rayleigh. Electron. J. Comb. **12** #N81 (2005b)

Wagner, D.G., Wei, Y.: A criterion for the half-plane property. Discrete Math. (2008). doi:10.1016/j.disc.2008.02.005

Whitney, H.: Non-separable and planar graphs. Trans. Am. Math. Soc. **34**, 339–362 (1932)

Whitney, H.: 2-isomorphic graphs. Am. J. Math. **55**, 245–254 (1933)

Whitney, H.: On the abstract properties of linear dependence. Am. J. Math. **57**, 509–533 (1935)

Whittle, G.: On matroids representable over GF(3) and other fields. Trans. Am. Math. Soc. **349**, 579–603 (1997)

On the Relative Complexity of 15 Problems Related to 0/1-Integer Programming

Andreas S. Schulz

Summary. An integral part of combinatorial optimization and computational complexity consists of establishing relationships between different problems or different versions of the same problem. In this chapter, we bring together known and new, previously published and unpublished results, which establish that 15 problems related to optimizing a linear function over a 0/1-polytope are polynomial-time equivalent. This list of problems includes optimization and augmentation, testing optimality and primal separation, sensitivity analysis and inverse optimization, as well as several others.

19.1 Introduction

The equivalence of optimization and separation has been one of the most consequential results in combinatorial optimization, and beyond. For example, it gave rise to the first polynomial-time algorithms for finding maximum stable sets in perfect graphs, and for minimizing submodular functions, to mention just two of several results of this kind derived by Grötschel et al. (1981). This equivalence has also paved the way for establishing negative results. For instance, it was instrumental in showing that computing the weighted fractional chromatic number is NP-hard (Grötschel et al. 1981). For general linear optimization problems and corresponding polyhedra, the equivalence between optimization and separation holds under certain technical assumptions. For linear combinatorial optimization problems and associated 0/1-polytopes, however, these requirements are naturally satisfied. In fact, for such problems this equivalence not only means that a polynomial-time algorithm for one of the two problems implies that the other problem can be solved in polynomial time as well; if one of the two problems can be solved in strongly polynomial time, then so can the other (Frank and Tardos 1987; Grötschel et al. 1988).

The relative computational complexity of solving one problem versus another has been studied in other situations as well. For example, Papadimitriou and Steiglitz wrote in their textbook on combinatorial optimization that "An Optimization

Problem is Three Problems" (Papadimitriou and Steiglitz 1982, p. 343). The other two problems to which they referred, in addition to finding an optimal solution, are computing the cost of an optimal solution—called the evaluation problem—and, in case of minimization problems, deciding whether there exists a feasible solution of cost at most a given value. The latter problem is known as the recognition or decision problem. It is obvious that the optimization problem is at least as hard as the evaluation problem, which, in turn, is no easier than the decision problem. Papadimitriou and Steiglitz went on by asking:

> Is it the case that all these versions are roughly of the same complexity? In other words, can we solve the evaluation version by making efficient use of a hypothetical algorithm that solves the recognition version, and can we do the same with the optimization and evaluation versions, respectively? (Papadimitriou and Steiglitz 1982, p. 345)

Answers to these questions are known: Binary search over the range of possible values reduces the evaluation problem to solving a polynomial number of decision problems. The reduction from the optimization problem to the evaluation problem is oftentimes illustrated with the help of the TRAVELING SALESMAN PROBLEM: Consider the arcs of the given directed graph in order, and for each arc solve the evaluation problem for the same instance where the cost of the current arc is increased by one. If the cost of an optimal tour is the same as before, then there is an optimal tour that does not use the current arc. The cost of this arc remains incremented, and the algorithm proceeds with the next arc. If, however, the cost of an optimal tour is higher than before, then the current arc is part of an optimal tour. Its cost is changed back to the original value, and the algorithm continues.

In this chapter, we prove that 15 problems, including augmentation, decision, evaluation, inverse optimization, optimization, primal separation, sensitivity analysis, separation, and testing optimality, are equivalent in Grötschel, Lovász, and Schrijver's and Papadimitriou and Steiglitz's sense: Given a hypothetical (strongly) polynomial-time algorithm for any one of them, each of the other problems can be solved in (strongly) polynomial time as well.

The chapter is organized as follows. In Sect. 19.2, we introduce the setup, in particular the class of linear combinatorial optimization problems considered here, and necessary background information, such as oracle-polynomial time algorithms. Section 19.3 constitutes the main part of this chapter. In Sect. 19.3.1, we present the 15 problems in detail and state the main result. Section 19.3.2 is reserved for its proof, which is broken down into several separate results pertaining to solving one problem with the help of a hypothetical algorithm for another problem. This part is followed by a collection of notes and references, which attribute the individual results to their respective authors (Sect. 19.3.3). Consequences and implications of the equivalence of the 15 problems are discussed in Sect. 19.4, including a proof of the Hirsch conjecture for $0/1$-polytopes, a simplex-type algorithm for linear programming over $0/1$-polytopes that visits only polynomially many vertices, and complexity results for exact local search. Finally, Sects. 19.5 and 19.6 discuss extensions of some of the results to local search and general integer linear programming, respectively.

19.2 Preliminaries

This chapter is concerned with linear combinatorial optimization problems. A linear combinatorial optimization problem Π consists of a family of instances $(N, \mathcal{F}, c)$, where $N = \{1, 2, \ldots, n\}$ is the ground set, $\mathcal{F} \subseteq 2^N$ is the set of feasible solutions, and the vector $c \in \mathbb{Z}^n$ assigns a cost of $\sum_{i \in F} c_i$ to every feasible solution $F \in \mathcal{F}$.[1] The objective is to find a feasible solution of minimal cost.[2] The set of feasible solutions of a combinatorial optimization problem is usually not described explicitly, but is given implicitly, such as the family of all stable sets in an undirected graph, or the set of all Hamiltonian tours in a directed graph. Typically, if we have one instance of such a combinatorial optimization problem, then the same instance in which only the objective function coefficients are changed, is also an instance of the same combinatorial optimization problem. We make this our first assumption.

Assumption 2.1. *Let Π be a linear combinatorial optimization problem. If $(N, \mathcal{F}, c)$ describes an instance of Π, and $d \in \mathbb{Z}^n$, then $(N, \mathcal{F}, d)$ is an instance of Π as well.*

Our interpretation of this assumption is that certain computational properties of the problems considered here, such as polynomial-time solvability, depend only on the structure of $\mathcal{F}$, but not on that of c.[3] As a consequence of Assumption 2.1, we henceforth use Π to just denote the family of pairs $(N, \mathcal{F})$, with the understanding that each such pair together with any compatible objective function vector constitutes an instance of the associated optimization problem. This convention allows us to refer to Π, even if we consider algorithmic problems other than optimization, which are defined over the instances $(N, \mathcal{F})$ of Π.

It is well known that the optimization problem associated with Π can be stated equivalently as a 0/1-integer programming problem. Namely,

$$\min\left\{\sum_{i \in F} c_i : F \in \mathcal{F}\right\} = \min\{cx : x \in X\} = \min\{cx : x \in P\},$$

where, for a given instance $(N, \mathcal{F}, c)$, $X := \{\chi^F : F \in \mathcal{F}\}$ is the set of incidence vectors of all feasible solutions, $P := \mathrm{conv}\{X\}$ is its convex hull, and cx denotes the inner product of c and x, i.e., $cx = \sum_{i-1}^n c_i x_i$. The incidence vector χ^F of a feasible solution $F \in \mathcal{F}$ is defined as $\chi_i^F := 1$ if $i \in F$, and $\chi_i^F := 0$, otherwise. Thus, X offers an equivalent representation of feasible solutions as 0/1-points in the n-dimensional Euclidean space. Moreover, P is a 0/1-polytope, i.e., a polytope whose vertices have coordinates 0 or 1 only. Hereafter, we will use X to refer to the set of feasible solutions, as it will be convenient to work with 0/1-vectors instead of sets.

---

[1] The reason that these problems are called linear, or sometimes min-sum combinatorial optimization problems is that the cost functions are linear over the ground set.

[2] For the sake of definiteness, we consider minimization problems only. All results stated in this chapter apply equally to maximization problems.

[3] The most relevant exception to this premise are nonnegative cost coefficients, and we will comment later on which results of Sect. 19.3 remain valid if we restrict ourselves to such instances.

We assume that the reader is familiar with the concepts of polynomial-time and strongly polynomial-time algorithms, and we refer to Korte and Vygen (2008) for more information on these and other notions of relevance to combinatorial optimization, which are not introduced here. One concept that is of particular importance to this chapter is that of an oracle-polynomial time algorithm, which helps us formalize the abstraction of a "hypothetical algorithm." An oracle-polynomial time algorithm is a polynomial-time algorithm that, in addition to all standard operations, may also make a polynomial number of calls to a given oracle. In particular, the time it would take to compute the answers that the oracle provides is not counted. However, if there were a polynomial-time algorithm for finding these answers, then one could replace the oracle by this algorithm, and the oracle-polynomial time algorithm would turn into a regular polynomial-time algorithm. The definition of a strongly oracle-polynomial time algorithm is similar. For technical details, we refer to Grötschel et al. (1988, Chap. 1).

The concept of oracles and oracle-polynomial time algorithms is useful to relate the complexity of two different problems, defined over the same family Π of sets of feasible solutions, to one another. For instance, if there is an oracle-polynomial time algorithm for optimization that uses an oracle for the separation problem, then the optimization problem for Π is not harder than the separation problem for Π, at least as far as polynomial-time solvability is concerned. The concept of oracles also provides us with means to specify the way in which sets of feasible solutions are given computationally. For example, when we say that X is given by an evaluation oracle, an oracle-polynomial time algorithm for the optimization problem has no further knowledge of X than what it can acquire by calling the evaluation oracle a polynomial number of times.

We make one further assumption, which frees us from considering how to find an initial feasible solution, which often is a difficult problem in itself.

Assumption 2.2. *Let Π be a linear combinatorial optimization problem. For any instance (N, X) of Π, a feasible solution $x^0 \in X$ is explicitly known.*

Let us fix some additional notation. We use e_i to denote the i-th unit vector, which has a 1 in coordinate i and 0 elsewhere. Moreover, $\mathbb{1}$ is the all-one vector. Finally, the support of a vector $z \in \mathbb{Z}^n$ is defined by $\operatorname{supp}(z) := \{i \in N : z_i \neq 0\}$.

19.3 An Optimization Problem is 15 Problems

We start by detailing the computational problems defined over the instances (N, X) of a linear combinatorial optimization problem Π, which we will relate to one another.

19.3.1 List of Problems

Augmentation Problem (AUG). Given a feasible solution $x \in X$ and a vector $c \in \mathbb{Z}^n$, find a feasible solution $y \in X$ such that $cy < cx$, or state that x minimizes cx over X.

Component Determination Problem (CD). Given an index $i \in \{1, 2, \ldots, n\}$ and a vector $c \in \mathbb{Z}^n$, determine whether there exists an optimal solution x^* for minimizing cx over X such that $x_i^* = 1$.

Component Negation Problem (CN). Given an index $i \in \{1, 2, \ldots, n\}$, a vector $c \in \mathbb{Z}^n$, and an optimal solution x^* for minimizing cx over X, find an optimal solution for minimizing cx over $\{x \in X : x_i = 1 - x_i^*\}$, if one exists.

Decision Problem (DEC). Given a vector $c \in \mathbb{Z}^n$ and a number $K \in \mathbb{Z}$, decide whether X contains a feasible solution x with objective function value $cx \leq K$.

Evaluation Problem (EVA). Given a vector $c \in \mathbb{Z}^n$, find the cost of a solution minimizing cx over X.

Inverse Optimization Problem (INV). Given a feasible solution $x \in X$ and a vector $c \in \mathbb{Q}^n$, find a vector $d \in \mathbb{Q}^n$ such that x is optimal for minimizing dx over X and $\|c - d\|$ is minimal.[4]

Maximum Mean Augmentation Problem (MMA). Given a feasible solution $x \in X$ and a vector $c \in \mathbb{Z}^n$, find a feasible solution $y \in X$ such that $cy < cx$ and y maximizes $c(x - y)/|\text{supp}(x - y)|$ over X, or state that x minimizes cx over X.

Optimization Problem (OPT). Given a vector $c \in \mathbb{Z}^n$, find a feasible solution that minimizes cx over X.

Postoptimality Problem (POST). Given an index $i \in \{1, 2, \ldots, n\}$, a vector $c \in \mathbb{Z}^n$, and an optimal solution x^* for minimizing cx over X, compute the maximal value $\rho_i \in \mathbb{Z}_+$ such that x^* remains optimal for minimizing $(c + \rho_i e_i)x$ and $(c - \rho_i e_i)x$ over X.

Primal Separation Problem (PSEP). Given a point $z \in [0, 1]^n$ and a feasible solution $y \in X$, find a vector $a \in \mathbb{Z}^n$ and a number $\beta \in \mathbb{Z}$ such that $ax \leq \beta$ for all $x \in X$, $ay = \beta$, and $az > \beta$, or state that such a vector a and number β do not exist.

Separation Problem (SEP). Given a point $z \in [0, 1]^n$, decide whether $z \in P = \text{conv}\{X\}$, and if not, find a vector $a \in \mathbb{Z}^n$ such that $ax < az$ for all $x \in P$.

Simultaneous Postoptimality Problem (SPOST). Given an index set $I \subseteq \{1, 2, \ldots, n\}$, a vector $c \in \mathbb{Z}^n$, and an optimal solution x^* for minimizing cx over X, compute the maximal value $\rho_I \in \mathbb{Z}_+$ such that x^* remains optimal for minimizing $(c + \sum_{i \in I} \delta_i e_i)x$ over X, for all $(\delta_i)_{i \in I}$ with $|\delta_i| \leq \rho_I$.

Unit Increment Problem (INC). Given an index $i \in \{1, 2, \ldots, n\}$, a vector $c \in \mathbb{Z}^n$, and an optimal solution x^* for minimizing cx over X, compute optimal solutions for the problems of minimizing $(c - e_i)x$ and $(c + e_i)x$ over X, respectively.

Verification Problem (VER). Given a feasible solution $x \in X$ and a vector $c \in \mathbb{Z}^n$, decide whether x minimizes cx over X.

---

[4] Instead of repeating the problem statement for different norms, let us merely remark that one can use the 1-norm or the ∞-norm. They would generally lead to different answers, of course. In terms of their computational complexity, however, they behave the same.

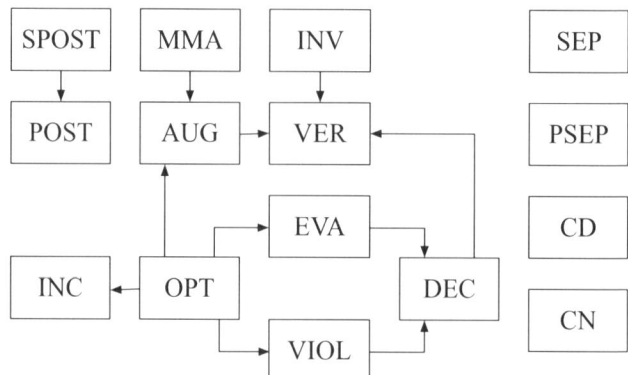

Fig. 19.1. Obvious relationships between the 15 problems

Violation Problem (VIOL). Given a vector $a \in \mathbb{Z}^n$ and a number $\beta \in \mathbb{Z}$, decide whether $ax \geq \beta$ holds for all $x \in X$, and if not, find a feasible solution $y \in P = \text{conv}\{X\}$ with $ay < \beta$.

There are some obvious relationships between these problems. For instance, if we can solve the optimization problem in polynomial time, then we can also solve the augmentation problem and the evaluation problem in polynomial time. Moreover, a polynomial-time algorithm for evaluation gives rise to polynomial-time algorithms for the decision problem and the verification problem. And the polynomial-time solvability of augmentation or inverse optimization obviously implies the same for verification. Figure 19.1 shows all trivial relationships. Here, an arrow from one problem to another indicates that the problem at the head of the arrow can be solved in oracle-polynomial time given an oracle for the problem at the tail of the arrow. The relation is transitive, but arrows implied by transitivity are omitted.

It turns out that, in fact, all these problems are equivalent, in terms of polynomial-time solvability.

Theorem 3.1. *Let Π be a linear combinatorial optimization problem. Any one of the following problems defined over the instances (N, X) of Π, augmentation, component determination, component negation, decision, evaluation, inverse optimization, maximum mean augmentation, optimization, postoptimality, primal separation, separation, simultaneous postoptimality, unit increment, verification, and violation, can be solved in oracle-polynomial time, given an oracle for any one of the other problems.*

This list is by no means exhaustive. While it contains some of the most interesting and relevant computational questions related to combinatorial optimization problems, other problems and variations of problems considered here can often proved to be equivalent by techniques similar to those illustrated in the following section.

19.3.2 Proof of Theorem 3.1

Instead of presenting a "minimal" proof of Theorem 3.1, it will be instructive to exhibit several direct reductions between different problems. In the end, there will be a "directed path," in the sense of Fig. 19.1, from any one problem to any other problem, proving Theorem 3.1.

We first introduce some additional notation. For a given set $X \subseteq \{0, 1\}^n$ of feasible solutions and an index $i \in \{1, 2, \ldots, n\}$, we define

$$X^{i,0} := \{x \in X : x_i = 0\} \quad \text{and} \quad X^{i,1} := \{x \in X : x_i = 1\}.$$

Note that either $X^{i,0}$ or $X^{i,1}$ could be empty. For a given vector $c \in \mathbb{Z}^n$, and $l \in \{0, 1\}$, let $x^{i,l}$ be an optimal solution for $\min\{cx : x \in X^{i,l}\}$, if one exists. The following observation will prove useful on several occasions.

Observation 3.2. *Let x^* be an optimal solution for minimizing cx over X.*

(a) *If $x_i^* = 1$, then $(c + \delta e_i)x^* \leq (c + \delta e_i)x$ for all $x \in X^{i,1}$ and $\delta \in \mathbb{Z}$.*
(b) *If $x_i^* = 0$, then $(c + \delta e_i)x^* \leq (c + \delta e_i)x$ for all $x \in X^{i,0}$ and $\delta \in \mathbb{Z}$.*

We start the proof of Theorem 3.1 by looking at sets of feasible solutions specified by an augmentation oracle. In addition to the trivial relationship depicted in Fig. 19.1, we have the following three lemmata.

Lemma 3.3. *Assume that $X \subseteq \{0, 1\}^n$ is given by an augmentation oracle. Then, the unit increment problem for X can be solved in oracle-polynomial time.*

Proof. Let the input to the unit increment problem be specified by x^*, c, and i. Let us first consider the case $x_i^* = 0$. Clearly, x^* stays optimal for $c + e_i$. Hence, it suffices to focus on the objective function vector $c - e_i$. If x^* is no longer optimal for this vector, then $x^{i,1}$ is. As $(c - e_i)x^* - (c - e_i)x^{i,1} = cx^* - cx^{i,1} + 1 \leq 1$, one call to the augmentation oracle with input x^* and $c - e_i$ will suffice to obtain an optimal solution for $c - e_i$. The case $x_i^* = 1$ can be handled similarly. □

Lemma 3.4. *Assume that $X \subseteq \{0, 1\}^n$ is given by an augmentation oracle. Then, there exists an oracle-polynomial time algorithm for the inverse optimization problem over X.*

Proof. Suppose we would like to solve the inverse optimization problem for x and c, and we are given access to an augmentation oracle. A first call to the oracle with input x and c will clarify whether x is optimal with respect to c. If it is, then we return c itself. Otherwise, we can formulate the inverse optimization problem for the 1-norm as the following linear program:

$$\text{minimize} \quad \sum_{i=1}^{n} \lambda_i \tag{1a}$$

$$\text{subject to} \quad \lambda_i \geq d_i - c_i \quad \text{for } i = 1, 2, \ldots, n, \tag{1b}$$

$$\lambda_i \geq c_i - d_i \quad \text{for } i = 1, 2, \ldots, n, \qquad (1c)$$

$$dx \leq dy \qquad \text{for all } y \in X. \qquad (1d)$$

Given a vector $(d, \lambda) \in \mathbb{Q}^n \times \mathbb{Q}^n$, the separation problem for the polyhedron defined by (1b)–(1d) can be solved as follows. Inequalities (1b) and (1c) can be checked directly. If one of them is violated, we obtain a separating hyperplane. If all of them are satisfied, we can call the augmentation oracle for x and d to find a violated inequality among (1d), if one exists. By the equivalence of separation and optimization, it follows that an optimal solution to the linear program (1) can be computed in polynomial time.

If we are dealing with the ∞-norm instead of the 1-norm, the argument remains essentially the same; we just replace the n variables λ_i by a single variable λ, and the objective function by λ. $\qquad\qquad \square$

The next proof presents a reduction from optimization to augmentation, which, in contrast to the previous proof, does not rely on the equivalence of optimization and separation.

Lemma 3.5. *Assume that $X \subseteq \{0, 1\}^n$ is given by an augmentation oracle. Then, the optimization problem for X can be solved in oracle-polynomial time.*

Proof. The proof is based on scaling. To simplify the exposition, we assume that the given objective function vector c has nonnegative entries. If this is not the case, we can switch x_i to $1 - x_i$ for all coordinates i with $c_i < 0$. The resulting objective function vector is nonnegative. Moreover, the augmentation oracle for the original set X of feasible solutions can be used to solve the augmentation problem for the new set of feasible solutions, after switching. So, let $c \geq 0$. We may also assume that $c \neq 0$. Otherwise, any feasible solution is optimal.

Let $C := \max\{c_i : i = 1, 2, \ldots, n\}$, and let $K := \lfloor \log C \rfloor$. For $k = 0, 1, \ldots, K$, define c^k as the objective function vector with coefficients $c_i^k := \lfloor c_i / 2^{K-k} \rfloor$, $i = 1, 2, \ldots, n$. The algorithm works in phases. In phase k, it solves the optimization problem for c^k. A detailed description is provided below. As per Assumption 2.2, $x^0 \in X$ is a given feasible solution.

1: **for** $k = 0, 1, \ldots, K$ **do**
2:     **while** x^k is not optimal for c^k **do**
3:         $x^k := \text{AUG}(x^k, c^k)$
4:     $x^{k+1} := x^k$
5: **return** x^{K+1}.

Here, $\text{AUG}(x^k, c^k)$ stands for the solution returned by the augmentation oracle when called with input x^k and c^k. The condition of whether to continue the while-loop is also checked during the same oracle call. The correctness of the algorithm follows from the fact that x^{K+1} is an optimal solution for $c^K = c$. Thus, it suffices to discuss the running time. There are only $O(\log C)$, i.e., polynomially many phases. Consequently, the analysis comes down to bounding the number of times the augmentation oracle has to be called within any given phase k. Note that, at the end of phase $k - 1$, x^k is optimal with respect to c^{k-1}, and hence for $2c^{k-1}$. Moreover, $c^k = 2c^{k-1} + c(k)$, for some 0/1-vector $c(k)$. Therefore, if x^{k+1} denotes the optimal solution for c^k at the end of phase k, we obtain

$$c^k(x^k - x^{k+1}) = 2c^{k-1}(x^k - x^{k+1}) + c(k)(x^k - x^{k+1}) \le n.$$

The inequality is a consequence of the optimality of x^k for c^{k-1}, the fact that x^k and x^{k+1} are 0/1-points, and $c(k)$ being a 0/1-vector. As a result, the algorithm determines an optimal solution with at most $O(n \log C)$ calls of the augmentation oracle.   □

The proof of the following result is somewhat reminiscent of that of the "evaluation implies optimization" construction from the introduction. It is stated here to make the importance of "component determination" explicit.

Lemma 3.6. *Assume that $X \subseteq \{0, 1\}^n$ is given by a component determination oracle. Then, the optimization problem for X can be solved in oracle-polynomial time.*

Proof. The following pseudo-code describes the reduction in detail. The condition in the if-construct of line 2 is equivalent to calling the component determination oracle.

1: **for** $i = 1$ to n **do**
2:     **if** $x_i = 1$ for some optimal solution x with respect to c **then**
3:         $c_i := c_i - 1$;
4:         $y_i := 1$
5:     **else**
6:         $y_i := 0$
7: **return** y.

The algorithm maintains the following invariant:

> After iteration i, all optimal solutions for the current objective function vector are also optimal for the original objective function vector. Moreover, all solutions $x \in X$ that are optimal for the current objective function vector satisfy $x_1 = y_1, x_2 = y_2, \ldots, x_i = y_i$.

If this holds true, then, after n iterations, y constitutes an optimal solution. Obviously, this is true initially ($i = 0$). The proof is by induction over the number of iterations. Let us assume that the induction hypothesis is true for iteration i, and consider iteration $i + 1$.

If the oracle answers "no," then the $(i + 1)$-st coordinate of all optimal solutions for the current objective function vector is 0. While the algorithm records this by setting $y_{i+1} = 0$, the objective function vector remains unchanged. It follows by induction that the invariant is true after the conclusion of iteration $i + 1$.

If the oracle's reply is "yes," the change of the $(i + 1)$-st objective function coefficient in line 3 renders all optimal solutions with $x_{i+1} = 0$ non-optimal. Optimal solutions with $x_{i+1} = 1$, of which there is at least one, remain optimal, however. The result follows.   □

We now turn to the case in which we have access to a component negation oracle.

Lemma 3.7. *Assume that* $X \subseteq \{0, 1\}^n$ *is given by a component negation oracle. Then, the unit increment problem for* X *can be solved in oracle-polynomial time.*

Proof. Let x^* and c be given, and suppose that $cx^* \leq cx$ for all $x \in X$. There are two cases. If $x_i^* = 1$, then x^* continues to be optimal for $c - e_i$. In order to determine an optimal solution for $c + e_i$, we call the component negation oracle with input x^*, c, and i. If there is no feasible solution $x \in X$ with $x_i = 0$, then x^* remains optimal. Otherwise, let $x^{i,0}$ be the solution returned by the oracle. Comparing the objective function values of x^* and $x^{i,0}$ with respect to $c + e_i$ yields an optimal solution. The case $x_i^* = 0$ is similar.   □

Since we will later show that an algorithm for the optimization problem can be designed with the help of an oracle for the unit increment problem (Lemma 3.13), the following lemma is not needed, strictly speaking. However, the techniques involved are quite different and worth recording.

Lemma 3.8. *Let* $X \subseteq \{0, 1\}^n$ *be given by a component negation oracle. Then, the optimization problem for* X *can be solved in oracle-polynomial time.*

Proof. Let x^0 be the given feasible solution. As usual, c is the objective function vector. The following algorithm determines an optimal solution x^*:

1: $x^* := x^0$;
2: $L := \emptyset$;
3: $d := c$;
4: **for** $i = 1, 2, \ldots, n$ **do**
5:     **if** $(x_i^* = 0$ and $c_i < 0)$ or $(x_i^* = 1$ and $c_i > 0)$ **then**
6:         $L := L \cup \{i\}$;
7:         $d_i := 0$
8: **for all** $i \in L$ **do**
9:     call CN with input x^*, d, and i;
10:     $d_i := c_i$;
11:     **if** CN returns a solution y^* **then**
12:         **if** $dy^* < dx^*$ **then**
13:             $x^* := y^*$
14: **return** x^*.

The idea is the following. We first modify c such that x^0 becomes optimal (lines 4–7). Let d be the resulting vector. The algorithm maintains an optimal solution x^* for the modified objective function vector d, while incrementally changing it back to c. Initially, $x^* = x^0$. We then call the component negation oracle for each objective function coefficient d_i that differs from c_i. In each such iteration, the solution returned by the oracle, y^*, is optimal for d among all feasible solutions x with $x_i = y_i^* = 1 - x_i^*$. Moreover, x^* is optimal for d; in particular, x^* is optimal for d among all feasible solutions x with $x_i = x_i^*$. So when we change d_i to c_i (line 10), it follows that x^* or y^* is an optimal solution for minimizing dx over X. The algorithm makes the appropriate choice, and repeats the same argument with the next coefficient.   □

The following proposition is not only quite useful in establishing relationships between the postoptimality problem and other problems; it also provides an interesting, if simple structural insight. Recall that ρ_i is the maximal value by which a coefficient c_i can be changed in either direction without causing a given optimal solution x^* to become non-optimal.

Proposition 3.9. *Let* x^* *be an optimal solution for minimizing* cx *over* X. *Let* $i \in \{1, 2, \ldots, n\}$. *If* $x_i^* = 1$ *and* $X^{i,0} \neq \emptyset$, *then* $\rho_i = cx^{i,0} - cx^*$. *If* $x_i^* = 0$ *and* $X^{i,1} \neq \emptyset$, *then* $\rho_i = cx^{i,1} - cx^*$. *Otherwise,* $\rho_i = \infty$.

Proof. Let us consider the case $x_i^* = 1$ first. Note that x^* remains optimal if we decrement c_i. Hence, ρ_i is only constrained by values that are greater than c_i. Clearly, x^* stays optimal for $c + \delta e_i$ if and only if $(c + \delta e_i)x^* \leq (c + \delta e_i)x$ for all $x \in X^{i,0}$. This is the case if and only if $\delta \leq cx - cx^*$ for all $x \in X^{i,0}$, which is equivalent to $\delta \leq cx^{i,0} - cx^*$, if $X^{i,0} \neq \emptyset$.

A similar argument yields $\rho_i = cx^{i,1} - cx^*$ when $x_i^* = 0$ and $X^{i,1} \neq \emptyset$.   □

Proposition 3.9 renders the proof of our next result rather easy.

Lemma 3.10. *Let* $X \subseteq \{0, 1\}^n$ *be given by a component negation oracle. Then, the postoptimality problem for* X *can be solved in oracle-polynomial time.*

Proof. Let x^*, c, and i be as specified in the description of the postoptimality problem. Use the component negation oracle to compute $x^{i,1-x_i^*}$, if it exists. If it does not exist, then $\rho_i = \infty$. Otherwise, $\rho_i = cx^{i,1-x_i^*} - cx^*$.   □

The following result completes the formal proof of the "evaluation implies optimization" claim from the introduction; the second part, i.e., the step from component determination to optimization, is provided by Lemma 3.6.[5]

Lemma 3.11. *Assume that* $X \subseteq \{0, 1\}^n$ *is given by an evaluation oracle. Then, the component determination problem for* X *can be solved in oracle-polynomial time.*

Proof. Solving the component determination problem with input $c \in \mathbb{Z}^n$ and $i \in \{1, 2, \ldots, n\}$, requires just two calls of the evaluation oracle, one with the original vector c as input, and one with the modified vector $c - e_i$ as input. If the returned values are the same, the answer to the component determination problem is "no." Otherwise, it is "yes."   □

We continue with another problem that can be solved with the help of an evaluation oracle.

Lemma 3.12. *The postoptimality problem for* $X \subseteq \{0, 1\}^n$ *given by an evaluation oracle, can be solved in oracle-polynomial time.*

---

[5] The procedure sketched in the introduction implicitly uses a version of the component determination problem in which we query the existence of an optimal solution whose i-th component is 0, which can easily shown to be equivalent to the variant considered here.

Proof. Let x^*, c, and i be the specified input of the postoptimality problem. Define $M := \sum_{k=1}^{n} |c_k| + 1$. We distinguish two cases. If $x_i^* = 1$, then we call the evaluation oracle with the objective function vector $c + Me_i$. Let $V(0)$ be the corresponding optimal value. Note that the i-th coordinate of any optimal solution with respect to $c + Me_i$ has to be 0, as long as there exists a feasible solution $x \in X$ with $x_i = 0$. Therefore, if $V(0) = cx^* + M$, then we return $\rho_i = \infty$. Otherwise, it follows from Proposition 3.9 that $\rho_i = V(0) - cx^*$.

If $x_i^* = 0$, then we feed the vector $c - Me_i$ into the evaluation oracle. Let $V(1)$ be the value returned by the oracle. We return $\rho_i = \infty$ if $V(1) = cx^*$, and $\rho_i = V(1) + M - cx^*$, otherwise.   □

The algorithm establishing the next result differs only slightly from that presented in the proof of Lemma 3.5. Notwithstanding, we include its description here because it remains valid for general integer programs, as we will discuss in Sect. 19.6.

Lemma 3.13. *Assume that $X \subseteq \{0, 1\}^n$ is given by a unit increment oracle. Then, there exists an oracle-polynomial time algorithm that solves the optimization problem for X.*

Proof. As we did in the proof of Lemma 3.5, we may assume that the given objective function vector is nonnegative, applying the switching operation to coordinates with negative coefficients, if necessary. We also use the same notation as in the proof of Lemma 3.5. The algorithm proceeds in phases. Consider phase k, and let x^k be the optimal solution from the previous phase; i.e., $c^{k-1}x^k \leq c^{k-1}x$ for all $x \in X$. Recall that c^k arises from c^{k-1} as $c^k = 2c^{k-1} + c(k)$. In phase k, we consider all indices i with $c_i(k) = 1$ in order. Suppose $\{i : c_i(k) = 1\} = \{i_1, i_2, \ldots, i_l\}$. Then, $c^k = 2c^{k-1} + \sum_{h=1}^{l} e_{i_h}$. We therefore call the unit increment oracle in turn for the current optimal solution and $2c^{k-1} + e_{i_1}$, for the new optimal solution and $2c^{k-1} + e_{i_1} + e_{i_2}$, and so on. The optimal solution at the end of the phase is optimal for c^k, as required.   □

We now make the connection to maximum mean augmentation.

Lemma 3.14. *Assume that $X \subseteq \{0, 1\}^n$ is given by an optimization oracle. Then, there exists an oracle-polynomial time algorithm that solves the maximum mean augmentation problem for X.*

Proof. Let the input of the maximum mean augmentation problem be composed of x and c. A first call to the optimization oracle clarifies whether x minimizes cx over X. From now on, we assume it does not. We define $S := \text{supp}(x)$, and we denote the value of the maximum ratio by μ^*. We are looking for a feasible point $y \in X$ such that $cy < cx$ and $c(x - y)/|\{i : x_i \neq y_i\}| = \mu^*$. Since x is not optimal, $\mu^* > 0$, and $cy < cx$ will be satisfied automatically if y is a feasible solution maximizing this ratio. Note that $\mu^* \leq C := \max\{|c_i| : i = 1, 2, \ldots, n\}$. We perform binary search over the interval $(0, C]$. For some value $0 < \mu \leq C$, we define an objective function vector c^μ as follows:

$$c_i^\mu := \begin{cases} c_i - \mu & \text{if } i \in S, \\ c_i + \mu & \text{if } i \notin S. \end{cases}$$

Suppose we call the optimization oracle with input c^μ; let x^μ be the output. There are three possible outcomes.

Case 1: $x^\mu = x$. Therefore, $c^\mu x \le c^\mu z$ for all $z \in X$. Spelled out in detail, this means that $c(x - z) \le \mu(|S| - \sum_{i \in S} z_i + \sum_{i \notin S} z_i)$, or, equivalently,

$$\frac{c(x - z)}{|\{i : x_i \ne z_i\}|} \le \mu \quad \text{for all } z \in X \setminus \{x\}.$$

Accordingly, μ is an upper bound on μ^*.

Case 2: $x^\mu \ne x$ and $c^\mu x^\mu = c^\mu x$. This implies again that

$$\frac{c(x - z)}{|\{i : x_i \ne z_i\}|} \le \mu \quad \text{for all } z \in X \setminus \{x\}.$$

However, x^μ satisfies this inequality with equality, and thus: $\mu = \mu^*$, and x^μ is the desired solution.

Case 3: $x^\mu \ne x$ and $c^\mu x^\mu < c^\mu x$. In this case,

$$\mu < \frac{c(x - x^\mu)}{|\{i : x_i \ne x_i^\mu\}|} \le \mu^*.$$

Consequently, μ is a strict lower bound of μ^*.

Note that the absolute value of the difference of any two distinct ratios is at least $1/n^2$. Hence, after $O(\log(nC))$ steps and calls of the optimization oracle, binary search yields the optimal value μ^*. (There is a subtle technical detail: If the binary search procedure terminates with case 2, we get a corresponding solution automatically. Otherwise, the last lower bound resulted from case 1, and we do not explicitly have an optimal solution for c^μ that is different from the original point x. In this case, we perturb the objective function vector, so as to make sure that x is no longer optimal. An optimal solution x^μ different from x can then be attained by at most n additional oracle calls.) $\square$

We follow up with another reduction relying on an optimization oracle.

Lemma 3.15. *Let $X \subseteq \{0, 1\}^n$ be given by an optimization oracle. Then, the simultaneous postoptimality problem for X can be solved in oracle-polynomial time.*

Proof. In order to determine ρ_I, for $I \subseteq \{1, 2, \ldots, n\}$, we need to find the largest value of ρ such that $\sum_{i=1}^n (c_i + \delta_i) x_i^* \le \sum_{i=1}^n (c_i + \delta_i) x_i$ for all $x \in X$ and $\delta_i = 0$ for $i \notin I$, $|\delta_i| \le \rho$ for $i \in I$. Equivalently, $\sum_{i \in I} \delta_i (x_i^* - x_i) \le \sum_{i=1}^n c_i (x_i - x_i^*)$. Let us fix $x \in X$ and ρ for a moment. Then, the right-hand side of the last inequality is constant, and the inequality holds if and only if it holds for a vector δ with $|\delta_i| \le \rho$ that maximizes the left-hand side. Consider some $i \in I$. If $x_i^* = 0$, then $\delta_i (x_i^* - x_i) = -\delta_i x_i$, which is maximized for $\delta_i = -\rho$, regardless of the value of x_i. We define $d_i := 1$. If $x_i^* = 1$, then $\delta_i (x_i^* - x_i) = \delta_i (1 - x_i)$, which is maximized for $\delta_i = \rho$, regardless of the value of x_i. We set $d_i := -1$. We also let $d_i := 0$ for $i \notin I$. Combining the pieces, we obtain that

$$\sum_{i\in I}\delta_i(x_i^* - x_i) \le \sum_{i=1}^{n}c_i(x_i - x_i^*) \quad \text{iff} \quad \rho\sum_{i\in I}x_i^* + \sum_{i=1}^{n}\rho d_i x_i \le \sum_{i=1}^{n}c_i(x_i - x_i^*).$$

Thus, we need to find the largest ρ for which

$$\min_{x\in X}\sum_{i=1}^{n}(c_i - \rho d_i)x_i \ge \sum_{i=1}^{n}c_i x_i^* + \rho\sum_{i\in I}x_i^*. \tag{2}$$

Let us denote the left-hand side of (2) by $v(\rho)$. This function is piecewise linear and concave. The slope of $v(\rho)$ for any given ρ is equal to $-\sum_{i=1}^{n}d_i x_i$, for some $x \in X$. Since $d_i \in \{0, \pm 1\}$ and $x_i \in \{0, 1\}$, there can be at most $2n + 1$ different slopes. In other words, $v(\rho)$ has a small number of linear pieces. Before we discuss how to construct the relevant part of $v(\rho)$, let us quickly address over which interval we should do so.

Apparently, $\rho_I \ge 0$. If $x_i = x_i^*$ for all $i \in I$ and $x \in X$, then $\rho_I = \infty$. Note that one can easily check this condition by calling the optimization oracle with objective function vector $-\sum_{i\in I}(-1)^{x_i^*}c_i$. If x^* is optimal, then all $x \in X$ satisfy $x_i - x_i^*$, for all $i \in I$. Otherwise, ρ_I is finite, and $\rho_u := \sum_{i=1}^{n}|c_i| + 1$ is an upper bound on its value. From now on, we assume that ρ_I is finite.

We start by computing $v(\rho_u)$, which amounts to calling the optimization oracle once. If (2) is satisfied for $\rho = \rho_u$, we are done. Otherwise, the optimal solution associated with $v(\rho_u)$ defines a linear function on $[0, \rho_u]$. We compute the largest value of ρ for which the value of this linear function is greater than or equal to the right-hand side of (2). We denote this value by ρ_1 and compute $v(\rho_1)$. If $v(\rho_1)$ coincides with the value of the linear function at the point ρ_1, then $\rho_I = \rho_1$. Otherwise, the optimal solution corresponding to $v(\rho_1)$ defines a second linear function. We then compute the largest value of ρ for which this linear function is greater than or equal to the right-hand side of (2); let this value be ρ_2. Note that $\rho_2 < \rho_1$. We repeat the same procedure for ρ_2, and so on. Because $v(\rho)$ has $O(n)$ linear pieces, ρ_I is found after as many steps. □

The following two results are of a more polyhedral nature. Let us establish first that primal separation is no harder than ordinary separation.

Lemma 3.16. *Assume that $X \subseteq \{0, 1\}^n$ is given by a separation oracle. Then, the primal separation problem for X can be solved in oracle-polynomial time.*

Proof. Let $z \in [0, 1]^n$ be the given point, and let $y \in X$ be the given feasible solution, which together form the input of the primal separation problem. Without loss of generality, we may assume that $y = 0$. Let C be the polyhedral cone that is defined by the inequalities of $P := \text{conv}\{X\}$ that are satisfied with equality by y. The key observation is that the primal separation problem for P and y is essentially equivalent to the separation problem for C. The only difference is that a hyperplane $ax = \beta$ that separates z from C may not contain y. However, this can be fixed by pushing the hyperplane towards C until they touch. Put differently, $ax = 0$ is a separating hyperplane, too.

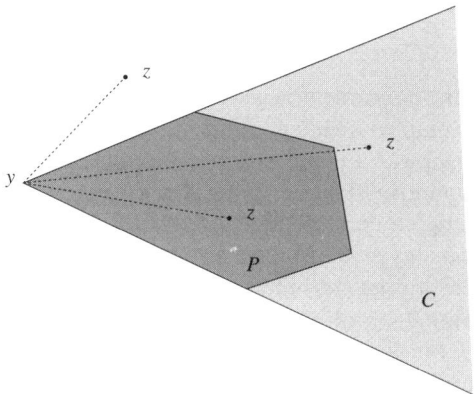

Fig. 19.2. Schematic drawing of P, C, and potential locations of z relative to y

As we are dealt a separation oracle for P, and not for C, it remains to show how we can use the given oracle to emulate one for C. Figure 19.2 provides a schematic picture of the situation. If we called the separation oracle for the original point z, it could happen that $z \in C \setminus P$, and the oracle would return a separating hyperplane when, in fact, $z \in C$. By choosing a point $\bar{z}$ on the line between y and z, close enough to y, this can be prevented. We now prove that $\bar{z} := (1 - \frac{1}{n})y + \frac{1}{n}z \in P$, if $z \in C$. Let $y^1, y^2, \ldots, y^k$ be the vertices of P that are adjacent to y. The distance of y, measured in terms of the 1-norm, to any of its adjacent vertices is at least 1. This implies that the distance of y to any point in C that is not a convex combination of $y, y^1, y^2, \ldots, y^k$, is strictly greater than one. However, $z \leq \mathbb{1}$ implies $\|\bar{z}\|_1 \leq 1$. Therefore, $\bar{z}$ is contained in P, as we wanted to show.

At last, observe that when z does not belong to C, $\bar{z}$ does not belong to C, either. This completes the proof.   □

The next proof relies on the equivalence of optimization and separation.

Lemma 3.17. *Let $X \subseteq \{0, 1\}^n$ be given by a primal separation oracle. Then, the verification problem for X can be solved in oracle-polynomial time.*

Proof. Recall that verification means deciding whether a given feasible solution $y \in X$ is optimal with respect to a given objective function vector $c \in \mathbb{Z}^n$. Let $P := \text{conv}\{X\}$, and let C be the polyhedron defined by the linear inequalities defining P that are active at y. It follows from linear programming duality that y minimizes cy over P if and only if it minimizes cy over C. By the equivalence of optimization and separation, minimizing cy over C is equivalent to solving the separation problem for C. As explained in the proof of Lemma 3.16, the separation problem for C can be reduced to the primal separation problem for P and y. Hence, we can use the primal separation oracle to optimize over C. If the resulting value is finite, then y is an optimal solution. Otherwise, y is not optimal.   □

We proceed with the postoptimality oracle, and discuss how it can be used to solve the component negation problem.

Lemma 3.18. *The component negation problem for $X \subseteq \{0, 1\}^n$ given by a postoptimality oracle, can be solved in oracle-polynomial time.*

Proof. The input of the component negation problem is specified by an objective function vector c, an optimal solution x^*, and an index i. Proposition 3.9 implies that a single call to the postoptimality oracle suffices to compute the optimal value of the problem of minimizing cx over $\{x \in X : x_i = 1 - x_i^*\}$. However, we are interested in determining a corresponding optimal solution, not just its value. Of course, if the postoptimality oracle outputs $+\infty$, then there exists no solution $x \in X$ with $x_i = 1 - x_i^*$, and we can stop. Otherwise, the algorithm below describes how this can be done. We use CN-VAL(x^*, c, i) to denote the optimal value of $\min\{cx : x \in X, x_i = 1 - x_i^*\}$. For the sake of simplifying the exposition, we assume that $x^* = 0$. Similarly to the proof of Lemma 3.5, this can be achieved by switching coordinates of value 1 to 0, and by observing that the postoptimality oracle for the original problem can be used to solve the postoptimality problem for the transformed problem.

```
 1: V := CN-VAL(x*, c, i);
 2: y_i* := 1;
 3: for all k ∈ {1, 2, ..., n} \ {i} do
 4:     c_k := c_k + 1;
 5:     if CN-VAL(x*, c, i) > V then
 6:         y_k* := 1;
 7:         c_k := c_k - 1
 8:     else
 9:         y_k* := 0
10: return y*.
```

Basically, this algorithm imitates the evaluation-to-optimization procedure from the introduction, except that it has to use the postoptimality oracle instead of the evaluation oracle. In particular, we need to ensure that x^* remains an optimal solution for the current objective function vector c at all times. Otherwise, we would not be able to make calls to the postoptimality oracle in line 5. However, this is obvious because we are only raising c_k if $x_k^* = 0$. The rest of the proof is similar to that of showing that a polynomial-time algorithm for evaluation can be used to solve the optimization problem in polynomial time as well. □

Finally, we turn to two problems that can be directly solved by a verification oracle.

Lemma 3.19. *The augmentation problem for $X \subseteq \{0, 1\}^n$ given by a verification oracle, can be solved in oracle-polynomial time.*

Proof. Let x and c specify the input to the augmentation problem. We may assume that $x = \mathbb{1}$. Indeed, if $x \neq \mathbb{1}$, we can replace all coordinates of the form $x_i = 0$ by $\bar{x}_i := 1 - x_i$. Formally, applying this switching operation transforms the given set of feasible solutions, X, into another set of feasible solutions, $\bar{X}$. By switching the signs of the corresponding entries c_i in the objective function, it is easy to see that x is optimal with respect to c if and only if $\mathbb{1}$ is optimal for the new objective function vector. The same argument shows that a verification oracle for X yields one for $\bar{X}$.

A first call of the verification oracle with $x = \mathbb{1}$ and c as input will determine whether x is optimal with respect to c. If it is not, we identify a better feasible solution with the help of the following algorithm.

```
1:  M := Σⁿ_{i=1} |cᵢ| + 1;
2:  for i = 1 to n do
3:      cᵢ := cᵢ − M;
4:      call the verification oracle with input x and c;
5:      if x is optimal then
6:          yᵢ := 0;
7:          cᵢ := cᵢ + M
8:      else
9:          yᵢ := 1
10: return y.
```

Because c may change during the course of the algorithm, we introduce the following terminology to simplify the subsequent exposition. We say that a feasible solution is "better" than x if the objective function value of that solution with respect to the original vector c is smaller than that of x. Moreover, we will use c^i to denote the objective function defined by line 3 in iteration i of the algorithm.

It is instructive to go carefully over the first iteration, not only because it will serve as the base case for the inductive proof, but also because it lays out the kind of arguments that we will use in the inductive step. For $i = 1$, the verification oracle is called with an objective function vector whose first coordinate has been changed to $c_1 - M$. If x is optimal for c^1, we claim that the first component of *all* feasible solutions that are better than x has to be 0. Suppose not. Then there exists a solution $z \in X$ with $cz < cx$ and $z_1 = 1$. But then, $c^1z = cz - M < cx - M = c^1x$, which is a contradiction. We record this by setting $y_1 = 0$ in line 6 of the algorithm. On the other hand, we claim that, if x is not optimal for c^1, then there must exist *some* better feasible solution whose first component is 1. Again, suppose this was not the case. Note first that, if $z \in X$ is any feasible solution with $cx \leq cz$, then $c^1x \leq c^1z$. Now, let $z \in X$ be an arbitrary solution that is better than x; i.e., $cz < cx$. According to our assumption, $z_1 = 0$. Note that $cx - cz \leq \sum_{i=1}^{n} |c_i| < M$. Therefore, $c^1x = cx - M < cz = c^1z$, a contradiction. Hence, there has to be a better solution whose first coordinate is 1. We keep this in mind by setting $y_1 = 1$ in line 9. Moreover, in this case we do *not* revert c_1 to its original value. Keeping $c_1 - M$ ensures that, from now on, we are only looking for solutions better than x whose first component is 1.

We now continue by induction over the number of iterations. Our induction hypothesis is the following:

> After iteration i, there exists a better feasible solution whose first i components are equal to $y_1, y_2, \ldots, y_i$, respectively.

It follows from the discussion above that the induction hypothesis is satisfied after the first iteration. So let us assume that it is true after i iterations, and consider iteration

$i + 1$. Note that

$$c^{i+1} = c - \sum_{k=1}^{i} y_k M e_k - M e_{i+1}.$$

Let us first consider the case that x is optimal for c^{i+1}. Suppose there exists a better solution $z \in X$ with $z_k = y_k$ for $k = 1, 2, \ldots, i$, and $z_{i+1} = 1$. Then,

$$c^{i+1} z = cz - M \left(\sum_{k=1}^{i} y_k + 1 \right) < cx - M \left(\sum_{k=1}^{i} y_k + 1 \right) = c^{i+1} x.$$

Hence, all solutions z with $z_k = y_k$ for $k = 1, 2, \ldots, i$, which are better than x, have $z_{i+1} = 0$. If x is not optimal with respect to c^{i+1}, let z be a better solution such that $z_k = y_k$ for $k = 1, 2, \ldots, i$, and $z_{i+1} = 0$. Since $cx < cz + M$, we obtain

$$c^{i+1} x = cx - M \left(\sum_{k=1}^{i} y_k + 1 \right) < cz - M \left(\sum_{k=1}^{i} y_k + 1 \right) + M = c^{i+1} z.$$

Consequently, there exists a better solution z such that $z_k = y_k$ for $k = 1, 2, \ldots, i$, and $z_{i+1} = 1$.

It follows that, after n iterations, y is a better feasible solution than x.   □

The next result is much simpler to obtain.

Lemma 3.20. *Assume that $X \subseteq \{0, 1\}^n$ is given by a verification oracle. Then, the postoptimality problem for X can be solved in oracle-polynomial time.*

Proof. An input of the postoptimality problem is specified by a vector c, an optimal solution x^*, and an index i. We discuss here the case that $x_1^* = 1$. The other case, $x_i^* = 0$, is similar. Let $M := \sum_{k=1}^{n} |c_k| + 1$, and $C := \max\{|c_k| : k = 1, 2, \ldots, n\}$. We perform binary search over the interval $[0, M]$ to identify the largest value of δ such that x^* remains optimal for $c + \delta e_i$. As all data are integral, this requires $O(\log(nC))$ calls of the verification oracle. Moreover, if x^* is optimal for $c + M e_i$, then x^* continues to be optimal for all values of δ; i.e., $\rho_i = \infty$. This completes the proof.   □

Actually, this completes the proof not only of Lemma 3.20, but also of Theorem 3.1. We refer to Grötschel et al. (1988, Theorem 6.4.9) for a proof of the missing equivalence between optimization, separation, and violation.

19.3.3 Notes and References

All primal algorithms[6] for solving specific combinatorial optimization problems, i.e., algorithms that move from one feasible solution to the next while improving

---

[6] Primal algorithms are sometimes also referred to as exact local search algorithms, see Sect. 19.4.2 below.

the objective function value, solve, by definition, the augmentation problem in each iteration. This includes well-known algorithms such as cycle-canceling algorithms for min-cost flow problems and augmenting path algorithms for matching problems. However, the design of these algorithms often seems to require finding particular augmenting structures, such as cycles of minimum mean cost, in order to guarantee that the total number of iterations is polynomial in the input size. The universal applicability of the bit-scaling framework in the proof of Lemma 3.5 was not realized before 1995, when Grötschel and Lovász (1995) and Schulz et al. (1995) first published this result. Grötschel and Lovász underlined its usefulness by devising a simple polynomial-time algorithm for finding a maximum-weight Eulerian subdigraph in an arc-weighted directed graph. The paper by Schulz, Weismantel, and Ziegler contained some other equivalences, including that between optimization and "irreducible" augmentation, a concept relevant to test sets in integer programming (Weismantel 1998), and that of optimization and maximum mean augmentation (Lemma 3.14), parts of which were later extended to general integer programming; see Sect. 19.6 below.

The verification and the component determination problem both go back to a paper by Papadimitriou and Steiglitz (1977), in which they showed that the following two problems are NP-hard for the TRAVELING SALESMAN PROBLEM:

Given an instance and a tour x, is x suboptimal?

Given an instance and an edge e, does e not appear in any optimal tour?

Lemmata 3.6 and 3.19, first brought up by Schulz et al. (1997), imply that these two problems are in fact NP-hard for every NP-hard linear combinatorial optimization problem.

As mentioned in the introduction, the equivalence of decision, evaluation, and optimization is widely known, but it appears difficult to pin down its original source. The direction from evaluation to optimization has sometimes been proved using the self-reducibility that certain problems, such as MAXIMUM SATISFIABILITY, exhibit (Vazirani, 2001, Chap. A.5). Exploiting this property effectively amounts to the explicit fixing of variables, which can also be achieved by modifying objective function coefficients accordingly, without necessitating a change of the set X of feasible solutions. If we look beyond the realm of linear combinatorial optimization problems, the question whether evaluation is as hard as optimization has been settled in the affirmative for all optimization problems whose associated decision problems are NP-complete (Ausiello et al., 1999, Chap. 1.4.4). In related work, Crescenzi and Silvestri (1990) provided sufficient and necessary conditions for the existence of optimization problems for which obtaining an optimal solution is harder than computing the optimal cost.

The unit increment problem, which together with Lemmata 3.3 and 3.13 is put forward in Orlin et al. (2008), is included here because Lemma 3.13 remains valid for general integer linear programming problems. Section 19.6 has more on this topic as well as on the equivalence of evaluation and optimization in this broader context.

Ahuja and Orlin (2001) showed that if a linear combinatorial optimization problem can be solved in polynomial time, then so can its inverse problem. Lemma 3.4

sharpens their result a bit, at the same time establishing a direct connection between augmentation and separation, which was speculated on earlier by Schulz et al. (1997).

The origins of primal separation trace back to primal cutting plane algorithms for integer programming; see, e.g., Young (1968) and Padberg and Hong (1980) for earlier studies, and Letchford and Lodi (2002) for a more recent account. Padberg and Grötschel (1985, Exercise 7) noted that the primal separation problem can be transformed into the standard separation problem; the proof of Lemma 3.16 uses the particularities of $0/1$-polytopes to simplify the argument. The reverse direction, especially that primal separation implies augmentation, was established by Eisenbrand et al. (2003). Their original proof is based on solving a constrained version of the verification problem. In this variant, an index set $I \subseteq \{1, 2, \ldots, n\}$ is added to the input, and the question is whether x minimizes cy over $\{y \in X : y_i = x_i$ for all $i \in I\}$. However, it is not difficult to see that this version is polynomial-time equivalent to the ordinary verification problem. In fact, this is true for similar variants of other problems, such as component determination.

The component negation problem was introduced by Ramaswamy and Chakravarti (1995) in their proof of the equivalence of optimization and postoptimality. The proof presented here is a streamlined version of the original proof; in particular, Lemma 3.8, Proposition 3.9, Lemma 3.10, Lemma 3.12, and Lemma 3.18 are all contained, either explicitly or implicitly, in their paper. Chakravarti and Wagelmans (1998) generalized this result to allow for simultaneous changes of more than one objective function coefficient. In particular, they proved Lemma 3.15 and also showed that a similar approach works if the deviation from the original objective function coefficients is measured in a relative sense instead of in terms of absolute differences. They also addressed situations in which the objective function coefficients are restricted to be nonnegative, as was previously done for the single variable postoptimality problem by Ramaswamy and Chakravarti (1995) and van Hoesel and Wagelmans (1999).

More generally, a second look at the proofs in Sect. 19.3.2 reveals that most results remain valid if the objective function coefficients are restricted to be nonnegative for both the given oracle and the problem to be solved. Yet, some problem statements have to be adjusted accordingly. For instance, if we do not restrict the objective function coefficients in any way, it is straightforward to show that the version of the component determination problem considered here is polynomial-time equivalent to the one in which we inquire about the existence of an optimal solution whose i-th coordinate is 0. For nonnegative cost coefficients, however, this is not true. While the latter version remains equivalent with optimization and evaluation, it is NP-complete to decide, for example, whether a given directed graph with nonnegative arc costs has a shortest s-t-path that contains a specific arc (Fortune et al. 1980).

19.3.4 Weakly vs. Strongly Polynomial Time

Some of the reductions in the proof of Theorem 3.1 lead to weakly polynomial time algorithms, and one may wonder whether the 15 problems are strongly polynomial time equivalent. As far as linear combinatorial optimization problems are concerned, Frank and Tardos (1987) provided a rounding scheme that resolves this matter for good:

Lemma 3.21. *There exists an algorithm that, for a given positive integer $Q \in \mathbb{Z}_+$ and a rational vector $c \in \mathbb{Q}^n$, finds an integer vector $d \in \mathbb{Z}^n$ such that $\|d\|_\infty \leq 2^{4n^3} Q^{n(n+2)}$ and $\mathrm{sign}(cz) = \mathrm{sign}(dz)$ whenever z is an integer vector with $\|z\|_1 \leq Q - 1$. Moreover, the running time of this algorithm is polynomial in n and $\log Q$.*

For example, with $Q := n + 1$, given an objective function vector $c \in \mathbb{Q}^n$, we obtain another objective function vector $d \in \mathbb{Z}^n$ such that $cx \leq cy$ if and only if $dx \leq dy$, for all $x, y \in \{0, 1\}^n$. Moreover, the size of d is polynomial in n: $\log \|d\|_\infty = O(n^3)$. For instance, if we apply this rounding procedure prior to the bit-scaling algorithm described in the proof of Lemma 3.5, then it follows that the algorithm described therein runs in strongly oracle-polynomial time. Similar preprocessing is possible in other cases, yielding:

Corollary 3.22. *Let Π be a linear combinatorial optimization problem. If any one of the 15 problems listed in Theorem 3.1 can be solved in strongly polynomial time, then so can any other problem from this list.*

19.4 Further Consequences

There are two ways of looking at Theorem 3.1. On the one hand, it is a tool to prove negative results. If a linear combinatorial optimization problem is NP-hard, then so are several other computational aspects related to that problem. Sensitivity analysis is hard, finding another optimal solution for a nearby objective function is hard, and so is recognizing an optimal solution. On the other hand, if the problem can be solved in polynomial time, it is convenient to know, for instance, that the simultaneous postoptimality problem can be solved in polynomial time as well. Moreover, Theorem 3.1 provides a variety of means to show that a linear combinatorial optimization problem can be solved in polynomial time, giving the algorithm designer some choice. Depending on the circumstances, one way of attack may prove to be simpler than another. For example, solving the primal separation problem of the PERFECT MATCHING PROBLEM in some graph $G = (V, E)$ requires only $|V|/2$ elementary maximum flow computations (Eisenbrand et al. 2003), compared to more complicated primal augmentation algorithms or minimum odd cut algorithms for solving the ordinary separation problem.

We continue by exploring some other implications of Theorem 3.1.

19.4.1 0/1-Polytopes and Linear Programming

As discussed earlier, optimizing cx over $X \subseteq \{0, 1\}^n$ is equivalent to optimizing cx over $P := \mathrm{conv}\{X\}$. By definition, P is a 0/1-polytope; i.e., all its vertices have coordinates 0 or 1. Two vertices of P are adjacent if they belong to the same edge of P. Thanks to the following well-known fact from linear programming, Lemma 3.5 has implications for the combinatorial structure of $0/1$-polytopes, and for linear programming over $0/1$-polytopes, which are thus far elusive for general polytopes.

Fact 4.1. *Let P be a polytope, and consider $\min\{cx : x \in P\}$, for some $c \in \mathbb{Z}^n$. If z is a non-optimal vertex of P, then there exists an adjacent vertex $y \in P$ such that $cy < cz$.*

The vertices and edges of a polytope P form an undirected graph. The diameter of P is the diameter of that graph; that is, it is the smallest number δ such that any two vertices of P are connected by a path consisting of at most δ edges. Lemma 3.5 yields an alternate proof of the following result, due to Naddef (1989).

Lemma 4.2. *The diameter of a $0/1$-polytope is at most its dimension.*

Proof. Let $P \subseteq [0, 1]^n$ be a $0/1$-polytope. Without loss of generality, we may assume that P is full-dimensional (see, e.g., Ziegler 1995, Chap. 3.3). Let y and z be two distinct vertices of P. If we define the vector c by setting

$$c_i := \begin{cases} 1 & \text{if } y_i = 0, \\ -1 & \text{if } y_i = 1, \end{cases} \quad \text{for } i = 1, 2, \ldots, n,$$

then y is the unique minimum of cx over P. According to the proof of Lemma 3.5, starting from z, we obtain y after at most n calls of the augmentation oracle. By Fact 4.1, we may assume that the oracle, when fed with a non-optimal vertex x, outputs a vertex that is adjacent to x. The vertices returned by the oracle form the required path of length at most n. $\qquad\square$

Naddef (1989) went on to show that Lemma 4.2 implies that the Hirsch conjecture is true for $0/1$-polytopes:

Corollary 4.3. *The diameter of a $0/1$-polytope is bounded by the number of its facets minus its dimension.*

Proof. We include a proof for the sake of completeness. Let P be a $0/1$-polytope, and let f be its number of facets, and n its dimension. If $f \geq 2n$, we are done, by Lemma 4.2. If $f < 2n$, any two vertices of P lie on a common facet, and the result follows by induction over the dimension. $\qquad\square$

It is well known that the diameter of a polytope is a lower bound for the number of iterations needed by the simplex algorithm, regardless of the pivot rule used. The following result provides a corresponding upper bound, for a particular variant of the simplex algorithm.

Theorem 4.4. *Let $A \in \mathbb{Z}^{m \times n}$, $b \in \mathbb{Z}^m$, $c \in \mathbb{Z}^n$, and $x^0 \in \{0, 1\}^n$ with $Ax^0 \geq b$ be given. Moreover, assume that $P := \{x \in \mathbb{R}^n : Ax \geq b\}$ is a $0/1$-polytope. Then, there exists a variant of the simplex algorithm that solves $\min\{cx : x \in P\}$ by visiting at most $O(n^4)$ vertices of P.*

Proof. In a first step, we use the polynomial-time algorithm of Lemma 3.21 with input c and $Q := n + 1$ to replace the given objective function vector c by another objective function vector $d \in \mathbb{Z}^n$. As a result, $\log \|d\|_\infty = O(n^3)$, and, for any two points $x, y \in \{0, 1\}^n$, $cx \le cy$ if and only if $dx \le dy$. According to the algorithm given in the proof of Lemma 3.5, we then proceed in phases. In phase k, we minimize the objective function vector obtained by only considering the k most significant bits of every coefficient of d. Each phase starts with the optimal solution from the previous phase, with the exception of the first phase, which starts with x^0. In each phase, we use the simplex algorithm with any monotone pivot rule to find an optimum. As all vertices of P are 0/1-points, Lemma 3.5 implies the result. $\qquad\square$

We conclude this section by pointing out that the path of vertices traced by the algorithm described in the proof of Theorem 4.4 is not necessarily monotone in d, nor in c.

19.4.2 Exact Local Search

The simplex algorithm has often been rightly characterized as a local search algorithm, where the neighborhood of a vertex contains its adjacent vertices (e.g., Papadimitriou and Steiglitz 1982). By Fact 4.1, this neighborhood is exact.[7] Conversely, the unique minimal exact neighborhood of a linear combinatorial optimization problem Π is precisely the one that would be searched by the simplex algorithm on $P = \text{conv}\{X\}$ (Savage 1973). It turns out that Theorem 3.1 subsumes some generalizations of problem-specific results pertaining to exact neighborhoods.

After Savage et al. (1976) had shown that any exact neighborhood for the TRAVELING SALESMAN PROBLEM has to have exponential size, Papadimitriou and Steiglitz (1977) formulated the following result for the TSP.

Corollary 4.5. *If $P \ne NP$, no NP-hard linear combinatorial optimization problem can have an exact neighborhood that can be searched in polynomial time for an improving solution.*

Proof. Let Π be an NP-hard linear combinatorial optimization problem, and let N be an exact neighborhood. If one can check in polynomial time whether the current feasible solution is locally (= globally) optimal, and, if not, produce a better feasible solution, then there is a polynomial-time algorithm for solving the augmentation problem. Hence, by Lemma 3.5, Π can be solved in polynomial time, which would imply P = NP. $\qquad\square$

For linear combinatorial optimization problems, Corollary 4.5 refines earlier results of Tovey (1985) and Yannakakis (1997, Theorem 18), which needed the stronger assumption of either NP $\ne$ co-NP, or that the problem be strongly NP-hard.

---

[7] In general, a neighborhood function N for a linear combinatorial optimization problem Π is called exact, if, for any instance X, any locally optimal solution is already globally optimal; i.e., $cx \le cy$ for all $y \in N(x)$ implies $cx \le cy$ for all $y \in X$, for all $c \in \mathbb{Z}^n$.

19.5 Local vs. Global Optimality

In light of the results for exact neighborhoods, it is natural to ask whether some of them remain valid in situations where local optima are not necessarily globally optimal. In particular, it often is the case that a neighborhood can be searched fast, either because it is of polynomial size (examples include the k-exchange neighborhood for the TRAVELING SALESMAN PROBLEM for some constant k, the flip neighborhood for the MAXIMUM CUT PROBLEM, and the swap neighborhood for the GRAPH PARTITIONING PROBLEM), or because there is an efficient algorithm to identify an improving solution in the neighborhood, if one exists (examples of neighborhoods of exponential size that can be searched in polynomial time include the twisted sequences neighborhood, the pyramidal tours neighborhood, and the permutation tree neighborhood for the TRAVELING SALESMAN PROBLEM). Put differently, given a linear combinatorial optimization problem Π together with a neighborhood N, the "local" version of the augmentation problem can typically be solved in polynomial time.[8]

Local Augmentation Problem. Given a feasible solution $x \in X$ and a vector $c \subset \mathbb{Z}^n$, find a feasible solution $y \in N(x)$ such that $cy < cx$, or state that x is locally optimal with respect to N and c.

However, an analogous result to Lemma 3.5 is not known. That is, given a local augmentation oracle, we do not have an oracle-polynomial time algorithm for computing a local optimum. In fact, no polynomial-time algorithm for finding a local optimum is known for any of the above-mentioned problems with neighborhoods of polynomial size,[9] nor for many others. It is instructive to go back to the proof of Lemma 3.5 and see where it fails when one replaces "augmentation oracle" with "local augmentation oracle" and "optimum" with "local optimum." We will now reproduce an argument by Orlin et al. (2004), which actually shows that a similar result cannot hold, unless one uses additional information.

More specifically, let A be any algorithm for finding a local optimum, which only uses the following information: the objective function vector c, an initial feasible solution $x^0 \in X$, a local augmentation oracle, and a feasibility oracle. The feasibility oracle accepts as input any point $x \in \{0, 1\}^n$, and it answers "yes" or "no," depending on whether x belongs to X or not.

Suppose the coefficients of the objective function vector c are $c_i = 2^{i-1}$, for $i = 1, 2, \ldots, n$, and the initial feasible solution is $x^0 = \mathbb{1}$. Neither the neighborhood nor the set of feasible solutions is fully specified in advance; an adversary will adjust them dynamically, according to the oracle calls made by the algorithm. Let us denote the other potential feasible solutions by $x^1, x^2, \ldots, x^{2^n-1} \in \{0, 1\}^n$, with the understanding that $cx^1 > cx^2 > \cdots > cx^{2^n-1}$. The neighborhood of a feasible solution

---

[8] Linear combinatorial optimization problems and their neighborhoods for which this is the case belong to the complexity class PLS, introduced by Johnson et al. (1988).

[9] Indeed, these problems are PLS-complete; i.e., the existence of a polynomial-time algorithm that finds local optima for any one of them would imply the existence of such an algorithm for all problems in PLS (Johnson et al. 1988; Yannakakis 1997).

x^i consists of x^j, where j is the smallest index larger than i such that x^j is feasible. If there is no feasible solution x^j with $i < j$, then the neighborhood of x^i is empty.

This neighborhood is exact for the objective function vector c specified above, but it is not necessarily exact for other objective function vectors. The adversary acts as follows. If algorithm A calls the feasibility oracle with some point $x^i \in \{0, 1\}^n$ that has not yet been proclaimed feasible, then x^i is declared infeasible. If the algorithm calls the local augmentation oracle with input $x^i \in X$ and $d \in \mathbb{Z}^n$, we distinguish two cases. Let $j > i$ be the smallest index such that x^j has not been labeled infeasible. If such an index exists, then x^j is marked as feasible (but A is not necessarily told). If such an index does not exist or if $dx^i \le dx^j$, then the oracle confirms x^i as locally optimal with respect to d. Otherwise, it returns x^j.

We claim that A needs to touch every single point $x^0, x^1, \ldots, x^{2^n-1}$ before it can announce the correct solution. Indeed, if A never uses the feasibility oracle, then x^{2^n-1} is the only local optimum, and the local augmentation oracle with input x^i and d will either return x^{i+1}, or assert that x^i is a local optimum for d. If A does make use of the feasibility oracle, the local augmentation oracle may return feasible solutions x^j with $j - i > 1$, but only if A previously checked all solutions $x^{i+1}, x^{i+2}, \ldots, x^{j-1}$ for feasibility. Moreover, the unique local optimum may be attained by some x^i with $i < 2^n - 1$, but only if A previously called the feasibility oracle for all points $x^{i+1}, x^{i+2}, \ldots, x^{2^n-1}$. In either case, A requires exponential time.

While the prospects of computing a local optimum in oracle-polynomial time are bleak, it is possible to find a solution that is "nearly" locally optimal in polynomial time (Orlin et al. 2004):

Theorem 5.1. *Let $x^0 \in X$, $c \in \mathbb{Z}_+^n$, and $\epsilon > 0$ be given. Assume that the set $X \subseteq \{0, 1\}^n$ of feasible solutions and the neighborhood function $N : X \to 2^N$ are specified via a local augmentation oracle. Then, there exists an oracle-polynomial time algorithm that computes a solution $x^\epsilon \in X$ such that*

$$cx^\epsilon \le (1 + \epsilon)cx \quad \text{for all } x \in N(x^\epsilon). \tag{3}$$

A solution x^ϵ that satisfies (3) is called an ϵ-local optimum.

Proof. The algorithm works in phases. Let x^{k-1} be the current feasible solution at the end of the previous phase. At the beginning of phase k, each objective function coefficient is rounded up to the nearest multiple of q, where $q := (cx^{k-1}\epsilon)/(2n(1 + \epsilon))$. Let c^k be the resulting objective function vector. In phase k, we call the local augmentation oracle repeatedly, feeding it with c^k and its own output. The k-th phase ends either when the oracle declares the current feasible solution x locally optimal for c^k, or if the current feasible solution x satisfies $cx \le cx^{k-1}/2$, whichever happens first. In the former case, the algorithm terminates and returns $x^\epsilon := x$; in the latter case, $x^k := x$, and we start phase $k + 1$.

Assume that the algorithm stops during phase k. Let x^ϵ be the solution returned by the algorithm. We will show that x^ϵ is an ϵ-local optimum. As $c \le c^k$, we have $cx^\epsilon \le c^k x^\epsilon$. Since x^ϵ is locally optimal with respect to c^k, we get $c^k x^\epsilon \le c^k x$, for all $x \in N(x^\epsilon)$. By the definition of c^k, we have $c^k x = \sum_{i=1}^{n} \lceil \frac{c_i}{q} \rceil q x_i \leq \sum_{i=1}^{n} q(\frac{c_i}{q} + 1) x_i \leq cx + nq$. The definition of q and the fact that $cx^\epsilon \geq cx^{k-1}/2$ help to make this chain of inequalities complete, yielding $cx^\epsilon \leq cx + \frac{\epsilon}{1+\epsilon} cx^\epsilon$, for all $x \in N(x^\epsilon)$. It follows that x^ϵ is an ϵ-local optimum.

As for the running time, note that the objective function values of any two consecutive solutions in any given phase differ by at least q. Hence, there are at most $O(n/\epsilon)$ calls of the oracle within each phase. The number of phases is itself bounded by $O(\log(cx^0))$. Thus, the algorithm runs in oracle-polynomial time.   □

One can actually show that the number of oracle calls made by the algorithm is bounded by $O(\epsilon^{-1} n^2 \log n)$. For details on how to prove this bound as well as other aspects of ϵ-local optima, we refer to Orlin et al. (2004).

19.6 General Integer Programming

So far, our discussion has revolved around combinatorial optimization problems that can be naturally formulated as 0/1-integer programs. It is natural to wonder which of the results carry forward to general integer programs, in which variables are still integer, but not necessarily binary. Although this topic has received less attention in the literature, there are a few results worth mentioning.

Throughout this section, we assume that upper bounds $u_i \in \mathbb{Z}_+$ on the variables x_i are explicitly known. Thus, a typical feasible region X is of the form $X \subseteq \{0, 1, \ldots, u_1\} \times \{0, 1, \ldots, u_2\} \times \cdots \times \{0, 1, \ldots, u_n\}$. As before, for a given objective function vector $c \in \mathbb{Z}^n$, let $C := \max\{|c_i| : i = 1, 2, \ldots, n\}$; in addition, $U := \max\{u_i : i = 1, 2, \ldots, n\} + 1$. We also assume that a feasible point $x^0 \in X$ is given.

MMA → OPT

The first result concerns sets of feasible solutions X that are given by a maximum mean augmentation oracle. Let x^* be an optimal solution of $\min\{cx : x \in X\}$. Starting from x^0, we can iteratively call the oracle with input x^i and c. If x^i is not yet optimal, let $x^{i+1} \in X$ be the feasible solution of lower cost returned by the oracle. It is not difficult to show that, in every iteration i, we have $c(x^i - x^{i+1}) \geq c(x^i - x^*)/n$. Hence, we get the following theorem of Schulz and Weismantel (2002).

Theorem 6.1. *Let X be given by a maximum mean augmentation oracle. Then, there exists an oracle-polynomial time algorithm that solves the optimization problem for X. The number of calls to the oracle is $O(n \log(nCU))$.*

Schulz and Weismantel (2002) consider other maximum ratio oracles as well, including "Wallacher's ratio," which was inspired by barrier functions used in interior-point algorithms for linear programming (Wallacher 1992). This ratio, specified in equation (4) below, is instrumental in proving the next result.

$AUG_\pm \to OPT$

Here, the subindex "$\pm$" alludes to a distinction between coordinates that are increased and those that are decreased during augmentation, which is immaterial for 0/1-integer programs. For a vector $z \in \mathbb{Z}^n$, we use z^+ to denote its positive part, and z^- for its negative part. That is, $z_i^+ := \max\{z_i, 0\}$, $z_i^- := \max\{-z_i, 0\}$, for $i = 1, 2, \ldots, n$, and, therefore, $z = z^+ - z^-$. We can now define the following refinement of the augmentation problem, which was introduced by Schulz and Weismantel (2002).

Directed Augmentation Problem ($AUG_\pm$). Given a feasible solution $x \in X$, and two vectors $c, d \in \mathbb{Z}^n$, find a direction $z \in \mathbb{Z}^n$ such that $x + z \in X$ and $cz^+ + dz^- < 0$, or assert that such a vector z does not exist.

For example, in min-cost flow problems, the directed augmentation problem corresponds to the ordinary augmentation problem for the residual graph.

For any feasible solution $x \in X$ and any coordinate $i \in \{1, 2, \ldots, n\}$, we define $p_i(x) := 1/(u_i - x_i)$ and $n_i(x) := 1/x_i$, with the understanding that $1/0 = \infty$. Binary search can be used to solve the following maximum ratio augmentation problem in oracle-polynomial time, given a directed augmentation oracle.

Maximum Ratio Augmentation Problem (MRA). Given a feasible solution $x \in X$, and a vector $c \in \mathbb{Z}^n$, find a direction $z = z^+ - z^-$ such that $x + z \in X$, $cz < 0$, and z maximizes the ratio

$$\frac{|cz|}{p(x)z^+ + n(x)z^-}, \tag{4}$$

or assert that x is optimal.

For the following result, originally proved by Schulz and Weismantel (2002), we need the additional technical assumption that X is of the form $X = \{x \in \mathbb{Z}^n : Ax = b, 0 \le x \le u\}$, even though neither the matrix A nor the vector b need to be known explicitly.

Theorem 6.2. *Let X be given by a directed augmentation oracle. Then, there exists an algorithm that solves the optimization problem for X in oracle-polynomial time.*

Proof. We have already sketched the reduction from the maximum ratio augmentation problem to the directed augmentation problem. It remains to show that the optimization problem can be solved in oracle-polynomial time with the help of a maximum ratio augmentation oracle. The algorithm is quite similar to the one used in the proof of Theorem 6.1. That is, we simply call the maximum ratio oracle repeatedly, feeding it with its own output from the previous iteration. The only difference is that we "stretch" the feasible direction z returned by the oracle, if necessary. Namely, we make sure that $x + z$ is feasible, but $x + 2z$ is not. Note that this implies that there exists a coordinate i such that either $z_i^+ > (u_i - x_i)/2$, or $z_i^- > x_i/2$. Moreover, if z maximizes the ratio (4), then so does any positive multiple of z.

To analyze the running time of this algorithm, let x be the feasible point at the start of the current iteration, and let x^* be an optimal solution. Moreover, let $z^* := x^* - x$. Using the previous observation and $p(x)(z^*)^+ + n(x)(z^*)^- \leq n$, we obtain that $|cz| \geq |cz^*|/(2n)$. Since the difference in objective function values between the initial feasible solution x^0 and the optimal solution x^* is $O(nCU)$, it follows that the algorithm terminates with an optimal solution after $O(n \log(nCU))$ calls of the maximum ratio oracle.   □

In total, one needs $O(n^2 \log^2(nCU))$ calls of the directed augmentation oracle to solve the optimization problem. A more efficient algorithm, which gets by with $O(n \log(nCU))$ oracle calls, is described in Schulz and Weismantel (2002).

INC → OPT and EVA → OPT

Interestingly, Lemma 3.13 and its proof remain valid for general integer programs.[10] Consequently, given a unit increment oracle, one can solve the optimization problem in oracle polynomial time. In fact, one can show something stronger. To facilitate the discussion, we restrict ourselves to nonnegative objective function vectors and introduce the following evaluation version of the unit increment problem.

Unit Increment Evaluation Problem (INC-EVA). Given a vector $c \in \mathbb{Z}_+^n$, and an index $i \in \{1, 2, \ldots, n\}$, compute the difference of the optimal objective function values for $\min\{(c + e_i)x : x \in X\}$ and $\min\{cx : x \in X\}$.

Accordingly, the output is a single number. Obviously, given an oracle for the evaluation problem, one can solve the unit increment evaluation problem in oracle-polynomial time. In fact, the reverse is also true, as can easily be derived from the proof of Lemma 3.13. The following theorem, taken from Orlin et al. (2008), shows that either oracle suffices to solve the optimization version of general integer programming problems.

Theorem 6.3. *Assume that X is given by a (unit increment) evaluation oracle. Then one can solve the optimization problem for X in oracle-polynomial time.*

Proof. Here is an outline of the proof. Given $c \in \mathbb{Z}_+^n$, let y be the lexicographically smallest optimal solution of $\min\{cx : x \in X\}$. We define a new objective function vector d that has the following properties:

(a) The solution y is optimal with respect to d.
(b) The solution y is optimal with respect to $d + e_i$, for all $i = 1, 2, \ldots, n$.

Obviously, if this is the case, one can use the oracle to compute $(d + e_i)y - dy$, and recover y via $y_i = (d + e_i)y - dy$, for $i = 1, 2, \ldots, n$.

It remains to define the objective function vector d:

$$d_i := nU^{2n}c_i + (n - i + 1)U^{2(n-i)} \quad \text{for } i = 1, 2, \ldots, n.$$

The verification of (a) and (b) is left to the reader.   □

---

[10] The necessary modification of the switch operation for coordinates with negative objective function coefficients is straightforward: $x_i \mapsto u_i - x_i$.

Acknowledgements

The author is grateful to Juliane Dunkel, Berit Johannes, and an anonymous referee for several comments on earlier versions of this chapter, which helped to improve the presentation. He also likes to thank Jens Vygen for all his help and support.

References

Ahuja, R.K., Orlin, J.B.: Inverse optimization. Oper. Res. **49**, 771–783 (2001)

Ausiello, G., Crescenzi, P., Gambosi, G., Kann, V., Marchetti-Spaccamela, A., Protasi, M.: Complexity and Approximation. Springer, Berlin (1999)

Chakravarti, N., Wagelmans, A.P.M.: Calculation of stability radii for combinatorial optimization problems. Oper. Res. Lett. **23**, 1–7 (1998)

Crescenzi, P., Silvestri, R.: Relative complexity of evaluating the optimum cost and constructing the optimum for maximization problems. Inf. Process. Lett. **33**, 221–226 (1990)

Eisenbrand, F., Rinaldi, G., Ventura, P.: Primal separation for 0/1 polytopes. Math. Program. **95**, 475–491 (2003)

Fortune, S., Hopcroft, J.E., Wyllie, J.: The directed subgraph homeomorphism problem. Theor. Comput. Sci. **10**, 111–121 (1980)

Frank, A., Tardos, É.: An application of simultaneous Diophantine approximation in combinatorial optimization. Combinatorica **7**, 49–65 (1987)

Grötschel, M., Lovász, L.: Combinatorial optimization. In: Graham, R.L., Grötschel, M., Lovász, L. (eds.) Handbook of Combinatorics, vol. 2, chapter 28, pp. 1541–1597. Elsevier, Amsterdam (1995)

Grötschel, M., Lovász, L., Schrijver, A.: The ellipsoid method and its consequences in combinatorial optimization. Combinatorica **1**, 169–197 (1981)

Grötschel, M., Lovász, L., Schrijver, A.: Geometric Algorithms and Combinatorial Optimization. Springer, Berlin (1988)

Johnson, D.S., Papadimitriou, C.H., Yannakakis, M.: How easy is local search? J. Comput. Syst. Sci. **37**, 79–100 (1988)

Korte, B., Vygen, J.: Combinatorial Optimization: Theory and Algorithms, 4th edn. Springer, Berlin (2008)

Letchford, A.N., Lodi, A.: Primal cutting plane algorithms revisited. Math. Methods Oper. Res. **56**, 67–81 (2002)

Naddef, D.: The Hirsch conjecture is true for (0, 1)-polytopes. Math. Program. **45**, 109–110 (1989)

Orlin, J.B., Punnen, A.P., Schulz, A.S.: Approximate local search in combinatorial optimization. SIAM J. Comput. **33**, 1201–1214 (2004)

Orlin, J.B., Punnen, A.P., Schulz, A.S.: In preparation, 2008

Padberg, M.W., Grötschel, M.: Polyhedral computations. In: Lawler, E.L., Lenstra, J.K., Rinnooy Kan, A.H.G., Shmoys, D.B. (eds.) The Traveling Salesman Problem: A Guided Tour of Combinatorial Optimization, pp. 307–360. Wiley, New York (1985)

Padberg, M.W., Hong, S.: On the symmetric travelling salesman problem: A computational study. Math. Program. Study **12**, 78–107 (1980)

Papadimitriou, C.H., Steiglitz, K.: On the complexity of local search for the traveling salesman problem. SIAM J. Comput. **6**, 76–83 (1977)

Papadimitriou, C.H., Steiglitz, K.: Combinatorial Optimization: Algorithms and Complexity. Prentice-Hall, Englewood Cliffs (1982)

Ramaswamy, R., Chakravarti, N.: Complexity of determining exact tolerances for min-sum and min-max combinatorial optimization problems. Working Paper WPS-247/95, Indian Institute of Management, Calcutta, India (1995)

Savage, S.L.: The solution of discrete linear optimization problems by neighborhood search techniques. Doctoral dissertation, Yale University, New Haven, CT (1973)

Savage, S., Weiner, P., Bagchi, A.: Neighborhood search algorithms for guaranteeing optimal traveling salesman tours must be inefficient. J. Comput. Syst. Sci. **12**, 25–35 (1976)

Schulz, A.S., Weismantel, R.: The complexity of generic primal algorithms for solving general integer programs. Math. Oper. Res. **27**, 681–692 (2002)

Schulz, A.S., Weismantel, R., Ziegler, G.M.: 0/1-integer programming: Optimization and augmentation are equivalent. In: Lecture Notes in Computer Science, vol. 979, pp. 473–483. Springer, Berlin (1995)

Schulz, A.S., Weismantel, R., Ziegler, G.M.: An optimization problem is nine problems. Talk presented by Andreas S. Schulz at the 16th International Symposium on Mathematical Programming, Lausanne, Switzerland (1997)

Tovey, C.A.: Hill climbing with multiple local optima. SIAM J. Alg. Discrete Methods **6**, 384–393 (1985)

van Hoesel, S., Wagelmans, A.: On the complexity of postoptimality analysis of 0/1 programs. Discrete Appl. Math. **91**, 251–263 (1999)

Vazirani, V.V.: Approximation Algorithms. Springer, Berlin (2001)

Wallacher, C.: Kombinatorische Algorithmen für Flußprobleme und submodulare Flußprobleme. Doctoral dissertation, Technische Universität Carolo-Wilhelmina zu Braunschweig, Germany (1992)

Weismantel, R.: Test sets of integer programs. Math. Methods Oper. Res. **47**, 1–37 (1998)

Yannakakis, M.: Computational complexity. In: Aarts, E., Lenstra, J.K. (eds.) Local Search in Combinatorial Optimization, pp. 19–55. Wiley, New York (1997)

Young, R.D.: A simplified primal (all-integer) integer programming algorithm. Oper. Res. **16**, 750–782 (1968)

Ziegler, G.M.: Lectures on Polytopes. Springer, Berlin (1995)

Single-Sink Multicommodity Flow with Side Constraints

F. Bruce Shepherd

Summary. In recent years, several new models for network flows have been analyzed, inspired by emerging telecommunication technologies. These include models of *resilient flow*, motivated by the introduction of high capacity optical links, *coloured flow*, motivated by Wavelength-Division-Multiplexed optical networks, *unsplittable flow* motivated by SONET networks, and *confluent flow* motivated by next-hop routing in internet protocol (IP) networks. In each model, the introduction of new side-constraints means that a max-flow min-cut theorem does not necessarily hold, even in the setting where all demands are destined to a common node (sink) in the network. In such cases, one may seek bounds on the "flow-cut gap" for the model. Such approximate max-flow min-cut theorems are a useful measure for bounding the impact of new technology on congestion in networks whose traffic flows obey these side constraints.

20.1 Introduction

We study several network flow models inspired by recent telecommunication technologies. Rather than presenting the models as questions of independent mathematical interest, as they usually are, we also attempt to explain enough about each technology to see how the model was motivated. We are by no means describing engineering solutions to practical problems, but rather we attempt to illustrate how mathematical theory (in most cases still only partially complete) can shed light on the principles underlying network engineering decisions. We believe the theory ultimately gives important intuition to the practitioner as well. As for our choice of topics, we adopt a view expressed by Ford and Fulkerson in their 1962 book on Network Flows (Ford and Fulkerson 1962), "...we have simply written about mathematics which has interested us, pure or applied."

We start by stating the Max-Flow Min-Cut Theorem; we do this formally in order to fix some of our notation. We are given a directed graph $D = (V, A)$ where V is the set of n *nodes*, A is the set of m *arcs* (sometimes called *links*); we may also be supplied with (usually integer) arc capacities $u(a) \geq 0$. Given a specified *origin* node s, and *destination* node t, one asks how much "flow" can be *routed* from s to t? For our purposes, an *st flow* is usually defined as an assignment of nonnegative values $f(P)$ to simple directed paths P from s to t. The *value* of the flow is then $\sum_P f(P)$ and the flow is *feasible* if for each arc a, we have: $\sum_{P:a \in P} f(P) \leq u(a)$. A set $S \subseteq V$ is an *st set* if $s \in S, t \notin S$; any such set determines an *st cut* $\delta^+(S)$ consisting of all arcs with their tail in S, and head in $V - S$. The *capacity* of such a cut, denoted by $u(S)$ is then $\sum_{a \in \delta^+(S)} u(a)$. The *Max-Flow Min-Cut Theorem* states that there is a flow of value k if and only if the capacity of each *st* cut is at least k. In Menger (1927) the essence of a proof for this theorem is given, although the theorem is often ascribed to Ford and Fulkerson (1956), as they incorporated capacities and costs to network flows and studied the topic from an algorithmic perspective. Ford and Fulkerson also studied the *multicommodity flow* case, where many pairs of nodes (s_i, t_i) wish to simultaneously use the network capacity to route their individual flow demands. If this is not a unit demand, we often use $d_{s_i t_i}$ to denote the amount requested.[1] We also consider flows in undirected networks, in which case analogous definitions will apply; in that case we refer to a graph $G = (V, E)$ with edge set E, and use $\delta(S)$ to denote the undirected edges in a cut induced by S.

Network flows are the focus of a vast amount of research in combinatorics, optimization, operations research and algorithms—see (Ahuja et al. 1993) for a comprehensive introduction. Apart from being mathematically rich however, they also form the basis for practical contributions to a wide range of applications. Many of the applications are quite natural, such as transportation problems. In that setting, *flow paths*, i.e., paths P such that $f(P) > 0$, have a natural interpretation as shipments through a network. However, flow problems can also arise in unexpected contexts as critical subproblems whose solutions are part of a larger (practical or theoretical) agenda. This ubiquity establishes flow in networks as one of the fundamental problems in optimization.

Switch-Sensitive Flows

Network flow theory makes a key assumption on how the computed flow paths are implemented in practice. Consider the instance displayed in Fig. 20.1(a) with three unit-flow paths indicated by bold, dotted and dashed lines (ignore the labels a_i, b_i for now). The flow depicted is feasible as long as each arc has capacity 1, but in practice there may also be constraints imposed by the nodes. We study several flow models where new constraints are imposed by telecom applications or equipment. In this setting, nodes are typically much more complicated objects. For instance, in a network design process, a node may represent a regional subnetwork or an office. One initially designs the "higher layer", under the assumption that computed traffic flows can ultimately be implemented in the subnetworks when we "open up" nodes

---

[1] We need not consider the case of parallel demands between the same nodes.

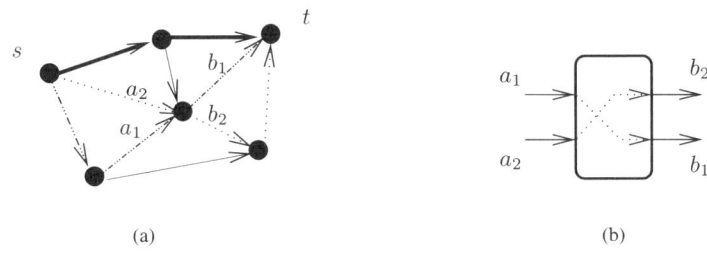

(a)   (b)

Fig. 20.1.

and look inside. This *hierarchical approach* thus defers or *outsources* the work on subnetworks (i.e., inside the nodes) to a second phase.[2]

A "base case" for such a hierarchical design is when each node simply represents a network switch. This approach assumes to some (approximate) extent, that each node is able to *switch*, any two paths that cross it. For instance, in the Fig. 20.1(b), we assume that flow from arc a_1 can be connected to b_1, and flow from a_2 is connected to b_2. A *crossbar* is the term used in telecommunication engineering to refer to a switch that is capable of handling any possible demand from its inputs to its outputs, that is, for any matching M from some inputs s_i to some outputs t_i, the node is capable of simultaneously connecting each s_i to its t_i. Crossbars were historically the prevalent architecture within the telephone network and their internal switching networks were quite sophisticated—see Fig. 20.2. Due to the central role played by crossbars, there is a significant literature (Hwang 1998; Pippenger 1978, 1991; cf. also with computer interconnection networks, Leighton 1992) on the design of switching networks with the crossbar property, and stronger *non-blocking properties* that allow input-output connections to be satisfied dynamically regardless of the order in which requests arrive and leave. The crossbar notion has also figured prominently in theory. One such example dates back to Valiant's work on circuit complexity lower bounds which led him to so-called super-concentrators possessing a crossbar property. In Sect. 20.4 we discuss further applications to the theory of disjoint paths.

The dominance of crossbars probably explains why network flow theory has not focused on more complicated *switch-sensitive* flows. This is in contrast to other well-studied side constraints such as arc gains and losses (*generalized flow*) or delay (*flows over time*). However, modern optical and data network switches are imposing new and sometimes unusual constraints on how traffic can traverse network nodes. One simple example suffices to explain. Suppose that all traffic in a network destined to a node t, must obey the following *confluence* property: any flow that enters a node v must exit on a single arc. This is a constraint imposed by early standards for IP (internet protocol) routing. This constraint has a pronounced affect on the basic structure of flow, since any such flow must have a tree-structure—see Fig. 20.5(a). In

---

[2] This approach is often reasonable for network design, but it may not viable for network management where operational end-to-end signals must be established in real time.

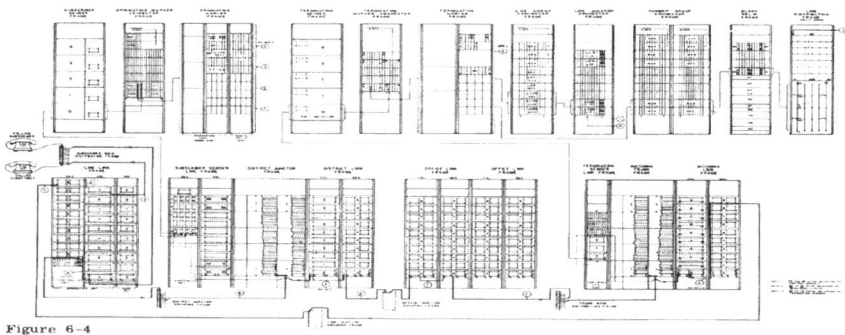

Figure 6-4

Fig. 20.2. "A Crossbar Switch and the call path of a single simple intra-office call (e.g., to a next door neighbor) within a Crossbar No. 1 office is diagrammed above. Once your dialed digits were collected, your voice path might be built in a third of a second, and involve the split-second operation of more than ten thousand relays. In practice this was quite a complex process. To create a voice path across a state or across the country required that many additional offices all work in concert to build a specific call path from among the incalculable possibilities; an almost unbelievable achievement of electromechanical computing, executed in fractions of seconds." *From the Vintage Telephone Equipment Museum:* http://simplethinking.com/photo/phone/vintage/

the next sections we discuss the confluent flow problem and other examples, with an eye to understanding the impact of these constraints from a theoretical vantage point.

Flow-Cut Gaps and Single-Sink Multicommodity Flow

We study several examples where network flows must obey additional constraints imposed by a communication technology. In each case, the new technology has certain advantages independent of its affect on network flows. For instance, optical technologies offer huge amounts of cost-effective bandwidth, while IP routers are operationally simple and inexpensive to deploy. Our interest, however, is in the question: *How should we evaluate the impact of new flow constraints on required resources in the (idealized) network?* We adopt two simplifications that allow fairly clean theoretical answers. First, we focus on the commonly used measure of *congestion*. Hence, for us, the "cost" of a new technology will be measured as how much we must uniformly augment the network capacity in order to guarantee a side-constrained flow.

Our second simplification is to focus on the classical single-commodity flow problem. Actually, we consider the extension of single-commodity network flows to so-called *single-sink multicommodity flows* where each node $v \neq t$ in the network has a rational demand $d_{vt} \geq 0$ it wishes to route to the destination (or sink) t. Using such single-sink multicommodity flows as a benchmark for flow problems with side-constraints is inspired by the work of Kleinberg who introduced the Single-Sink Unsplittable Flow Problem.[3] For standard flows, a reduction to the single-commodity case is trivial. Add a new node $s \notin V$, and for each $v \in V - t$, add an arc $a_v = (s, v)$ with a capacity of d_{vt}. Evidently, a single-sink multicommodity flow exists if and only if the augmented graph has an st flow of value $\sum_{v \in V - t} d_{vt}$. In many cases, however, constrained network flows are not amenable to this simple reduction to a single commodity flow. For instance, in the *unsplittable network flow* problem, each demand from a node v is required to route its entire flow of d_{vt} on a single path P. While we could add the new node s, we would still need to identify multiple demands between s and t in order to represent the unsplittable flow problem. Hence there is no advantage gained by the reduction; in this sense, the problem, at its heart, is a multicommodity flow problem.

Consider a (unit capacity) single-sink multicommodity flow instance consisting of a digraph $D = (V, A)$ (or undirected graph $G = (V, E)$) and demand pairs s_i, t; there is a demand of $d_{s_i t}$ from s_i to t. The *density* of a cut $\delta^+(S)$ is defined as $\frac{|\delta^+(S)|}{d(sep(S))}$; here $sep(S) = \sum_{s_i \in S} d_{s_i t}$, and it is understood $t \notin S$. Clearly if there is a feasible fractional multicommodity flow, then the density of each cut is at least one, and the *cut condition* is said to hold for the instance. Let Δ denote the density of a minimum density cut. As long as $\Delta \geq 1$, a standard network flow exists. For the problems we discuss, this will not always be the case, and hence we seek $\phi > 0$, the largest value such that there is a (constrained) feasible flow for all *scaled* demand values $\phi d_{s_i t}$. The *flow-cut gap* for an instance is defined as the value $\frac{\Delta}{\phi}$ (there are natural extensions of these definitions to the undirected case and to the case of general multicommodity demands). Two classical results state that (i) for any single-sink (directed or undirected) multicommodity flow instance the flow-cut gap is 1 (Ford and Fulkerson 1956), and (ii) for any undirected multicommodity flow instance the flow-cut gap is $O(\log n)$ and this is tight (Leighton and Rao, 1999; Linial et al., 1995).

---

[3] Actually he looked at single *source* flows. Unsplittable flows had been studied earlier by Cosares and Saniee, however Kleinberg coined the actual term "unsplittable."

For our purposes, the correct interpretation of a flow-cut gap of α is that if you only build your network to satisfy the cut condition, then α is the amount of congestion you can expect on some link in your network. In our presentation we also consider "additive" flow-cut gaps (Theorem 2.4), as well as gaps based on node as opposed to link congestion (Theorem 3.1). In most cases, we use the flow-cut gap to measure the impact or "cost" of forcing our flows to satisfy a new constraint.

20.2 Optical Networks

20.2.1 Resilient Flows

The first deployments of optics in telecommunications involved simple point-to-point optical links, such as transatlantic cables. One important side effect of the enormous bandwidth of fibre was that the loss of a single cable could now mean the loss of an unacceptable amount of traffic (think in terms of a city sector versus a few hundred phone calls). The design of *resilient* networks that could withstand failures of its elements, was elevated as an important design objective.

A woeful fact of practical network design is that many initial cost estimates are produced as follows. Suppose we want to carry d_{ij} units of traffic between each node pair i, j. For each i, j find a shortest available i–j path P in the network. Increase the capacity on the links of P by d_{ij}. Repeat. Apart from ignoring existing capacity and upper bounds on link capacity,[4] this approach has a more serious drawback even as a rough cost estimator. Traffic between nodes i, j is completely vulnerable to a single link failure. In this section we study a resilient version of the shortest st path problem that, for instance, could be used in the greedy approach above to give a an cost estimate for a resilient network design. We motivate the ideas first with a simple example.

My first engagement with a network operator was with British Telecom (BT) in the early 1990's. In one consultation, Graham Brightwell and I (Brightwell and Shepherd 1996) were asked to examine the following simple resilience problem. One is given n links, between nodes s and t say, where the per-unit cost of reserving bandwidth on link l is c_l. We are also given a threshold T. Find the optimal allocation of capacities to the links such that after deleting any k links, there is still at least T units of capacity between s and t. Linear programming gives a simple answer which we briefly describe.

Consider the case $k = 1$ (the general case is analogous). Let $A = J - I$ be the $n \times n$ matrix (J is the matrix of all 1's) with all 1's except 0's on the diagonal. We require a solution to $\min\{\sum_l c_l x_l : Ax \geq T\mathbf{j}, x \geq 0\}$; $\mathbf{j}$ denotes the column vector of all 1's. Assume $0 \leq c_1 \leq c_2 \leq \cdots \leq c_n$. An easy argument shows that any basic solution x^* (of positive and minimum total cost) has the following form: for some integer $p \geq 2$, $x_l^* = \frac{T}{p-1}$ for $l = 1, 2, \ldots, p$, and each other component is 0. Furthermore, one can also

---

[4] One is often just "lighting" up equipment that already exists in the network

find the optimal integer solution by a procedure called <u>up-down rounding</u>. *Take the optimal fractional solution x^*; without loss of generality $\frac{T}{p-1}$ is not an integer or we are done. We create two integer solutions z^r: one for each choice of $r \in \{\lceil \frac{T}{p-1} \rceil, \lfloor \frac{T}{p-1} \rfloor\}$, as long as $r > 0$. We define $l = \lceil \frac{T}{r} \rceil + k + 1$ and set $z_1^r = z_2^r = \cdots = z_{l-1}^r = r$, and $z_l^r = T - (l-2)r$. All other components of z^r are 0. One may use unimodality of the cost function to establish that one of these two solutions is optimal* (Bartholdi et al. 1980; Brightwell et al. 2001).

This naturally motivates the question of how to plan for resilience in general networks. We now discuss the models proposed in Brightwell et al. (2001); the above procedure plays an important role in the answer.

In the general setting, how does one plan for resilience. Specifically, how could one define a *shortest "resilient" st path*? The answer depends crucially on how the network responds to a link failure. Two broad restoration categories are: *local restoration* where the endpoints of a link recognize its failure and react to re-route lost traffic, and *endpoint restoration* where traffic between s, t is always restored by the endnodes s, t regardless of where the failed link lies in the network. Both types of restoration exist in the first generation optical networks.

It is yet uncertain what strategies will be adopted in future all-optical networks. A simple and attractive approach is to force traffic between s, t to be routed along disjoint *st* paths. This scheme is called *diverse path routing* and generalizes the commonly studied $1 + 1$ scheme where capacity is simply reserved on two (edge or node) disjoint paths; hence if any link fails, the other path is still available. This approach is attractive since it allows easy restoration at the endpoints s, t; some link fails, s and t simply shunt traffic on to one of the nonfailed paths.

If we use such diverse path routing, the resilient network design (or resilient shortest path) problem becomes: find disjoint *st* paths $P_1, P_2, P_3, \ldots$ and capacities u_i for each path P_i, such that if any path fails, the capacity assigned to the remaining paths is at least T (again, the case of allowing $k > 1$ link failures is analogous). Consider such a collection of paths, and let c_i denote the per-unit cost of reserving capacity along the whole path P_i. Suppose the optimal solution assigns positive capacities to a collection of paths $P_1, P_2, \ldots, P_k$, then these paths simply look like direct links connecting s and t. Thus our previous analysis shows that there is an optimal solution that assigns $\frac{T}{k-1}$ capacity to each P_i, and hence the cost of the solution is $\frac{T}{k-1} \sum_i c_i$. It now follows that the best possible choice for the paths P_i is to find a minimum cost *st* flow of size k. This gives a complete solution for finding the optimal T-resilient diverse path reservation as we now describe.

Find the cheapest pair P_1, P_2 of edge disjoint *st* paths. Let $C_2 = c(P_1) + c(P_2)$ and x^2 be the solution that assigns T units of capacity to each P_i. For each $k = 3, \ldots$, find a minimum cost k-flow $P_1^k, P_2^k, \ldots, P_k^k$ between s and t. Set $C_k = \sum_i c(P_i^k)/(k-1)$. If $C_k > C_{k-1}$, then output x^{k-1}, it is the optimal solution. Otherwise, define x^k by assigning $\frac{T}{k-1}$ capacity to each P_i^k.

This simple procedure finds the optimal fractional resilient diverse-path reservation by solving a sequence of minimum cost flow problems. Hence it can be reduced to solving a sequence of shortest path problems, using the successive shortest path algorithm (Iri 1960). But what if we want an optimal *integral* solution? One natural heuristic is suggested by the algorithm above. For each computed x^k, apply the up-down rounding procedure. This gives potentially two integral diverse-path reservations $z^{k,+}$ and $z^{k,-}$. Amongst all of these solutions simply choose the cheapest. It turns out that this provides a solution with integral capacities, that is within a factor of $\frac{15}{14}$ of the optimal integral solution. Somewhat bizarrely, this is also best possible!

Theorem 2.1 (Brightwell et al. 2001). *For any $\epsilon > 0$, it is NP-hard to find an integral diverse-path reservation that is within a multiplicative factor of $\frac{15}{14} - \epsilon$ from the optimal. Moreover, there is a polytime $\frac{15}{14}$-approximation algorithm for this problem.*

We now discuss an alternative, much weaker, resilient shortest path model that completely ignores any restoration constraints imposed by the network. Indeed, perhaps the most mathematically natural extension of shortest paths that incorporates resilience, is simply to seek arc capacities such that each st cut has capacity at least T, even after a single arc is deleted. This amounts to studying the polyhedron:

$$\mathcal{R}(D, s, t, T) = \{x \in \mathbb{R}^A : x(\delta^+(S) - a) \geq T \text{ for each } st \text{ set } S, \text{ and } a \in \delta^+(S)\}$$

While one may optimize over $\mathcal{R}$ in polytime (indeed it has a simple but large compact formulation), there is no combinatorial method known for this problem. Not surprisingly, there is then not much known about the structure of its extreme points. Several examples are given in Brightwell et al. (2001)—two examples are depicted in Fig. 20.3.

As mentioned, there are many possible, practical models for providing resilience; we discussed one based on diverse paths. The polyhedron $\mathcal{R}$ represents the weakest possible condition one could impose for resilience. Hence it is a useful benchmark

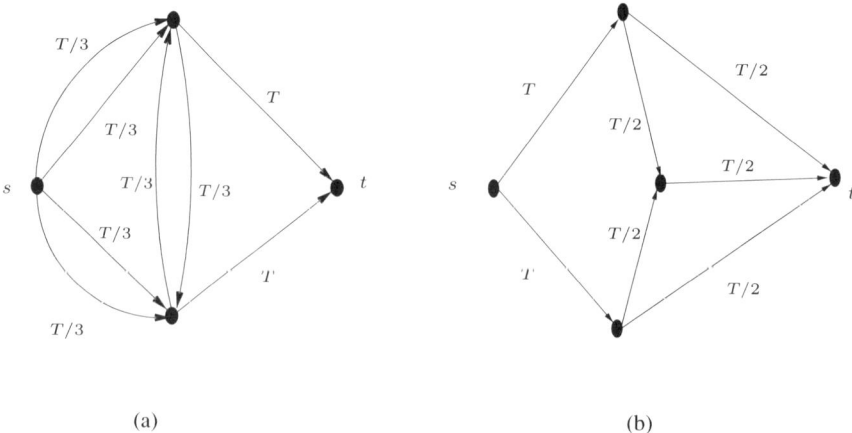

(a)   (b)

Fig. 20.3. (a) Vertex of $\mathcal{R}(D)$ with a cycle in the support; **(b)** basic solution for both arc-failure and node-failure resilience

for lower bounds. One can show that its integrality gap is "small". More significantly for us, it does not yield better than a factor 2 improvement on the cheapest diverse path solution; the proof of this is a fun exercise.

Theorem 2.2 (The cost of diverse path reservations; Brightwell et al. 2001). *For any nonnegative arc costs c, the cost of reserving T units of capacity on a cheapest pair of disjoint paths, is at most twice the minimum cost solution in $\mathcal{R}(D, s, t, T)$.*

We close by mentioning that in the single-sink flow case with multiple sources, or when the network has arc capacities, the picture is far from complete. The reader is referred to Brightwell et al. (2003) for open problems.

20.2.2 Unsplittable Flows

The second wave of optics in the telecommunication network began being deployed in the early 1990's. Instead of just single point to point optical links, the new networks consisted of rings of nodes (switches) with optical links between each pair of neighbhouring nodes on the ring. Most importantly, nodes on a ring worked together in terms of detecting network failures and restoring lost traffic across the ring. In America, the standard adopted was called SONET (synchronous optical networks) whereas the standard in most of the rest of the world was SDH. In both cases, nodes were able to terminate flow paths (communication paths) whose capacity was from a set of bandwidths prescribed by the standard (e.g., 155, 622 or 2488 Mb/s).

Building the SONET network took place "from scratch" and this was a monumental task. It was also one of the most capital intensive network projects and hence arguably one of the most important network optimizations to have been undertaken. The optimization problem was also enormously complex: where are rings installed? how large is each ring? how is demand routed across multiple rings, before and after network failures? Most American network operators used software developed by Bell Communications Research (Bellcore), called the SONET Toolkit.

In this effort, Cosares and Saniee (1994) introduced the following unsplittable flow problem which they called the *Ring Loading Problem*. They are given an undirected ring (cycle) G on nodes $0, 1, 2, \ldots, n-1$ (with edges $i(i+1)$ for each i) and integer demand matrix $[d_{ij}]$. Routing the demand matrix *unsplittably* means that for each $i < j$, with $d_{ij} > 0$, we choose to route this entire demand either clockwise on the path $i, i+1, \ldots, j$ or counterclockwise on the path $j, j+1, \ldots, i$. We seek the minimum integer L such that there is an unsplittable routing where the maximum load on an edge of G is at most L. They consider the fractional relaxation and define L^* to be the optimal ring load if demands can be fractionally routed, i.e., demands may split their flow across the two directions around the ring. They prove that one may always obtain an unsplittable routing with a load of at most $2L^*$.

Subsequently, Schrijver, Seymour and Winkler (SSW)[5] strengthened the result considerably by showing that an unsplittable routing can be obtained with maximum edge load of $L^* + \frac{3}{2}d_{max}$ ($d_{max} = \max_{i,j} d_{ij}$)—see (Schrijver et al. 1998). They

---

[5] The last two authors were also at Bellcore at the time.

prove this bound via the following result. Consider $2m$ nonnegative reals u_i, v_i for $i = 1, 2 \ldots, m$ such that $u_i + v_i \leq 1$. A *signing* of these numbers is a sequence $z_1, \ldots, z_m$ such that for each i, $z_i \in \{v_i, -u_i\}$. Find the minimum value β such that for any given collection of u_i, v_i's there is a signing such that the following holds for each k:

$$\left| \sum_{i=1}^{k} z_i - \sum_{i=k+1}^{m} z_i \right| \leq \beta.$$

They show that one may always obtain an unsplittable ring routing with a maximum edge load of $L^* + \beta d_{\max}$. They also prove that $\beta \leq \frac{3}{2}$, thus establishing their bound for the ring loading problem. They actually exhibit an instance of u_i, v_i's where $\beta > 1$. However, this does not refute the following conjecture which remains open. It also seems temptingly similar to the results we discuss next for single-sink flows.

Conjecture 2.3 (Schrijver et al. 1998) For any instance G, $[d_{i,j}]$ of the ring loading problem, there is an unsplittable routing with maximum edge load at most $L^* + d_{\max}$.

We mention that the algorithm of SSW is also simple to implement and was incorporated in the SONET Toolkit; for a given instance, it was an important subroutine that was solved tens of thousands of times to get quick estimates on the exponentially many possible ring placements.

In 1994, Kleinberg coined the term "unsplittable" flow and initiated the study of unsplittable flows for the single-sink multicommodity case (Kleinberg 1996) in directed graphs $D = (V, A)$ with arc capacities $u(a)$. By considering a network with two nodes and parallel links, this includes the special case where items can be fractionally packed into bins (of varying capacities) and we wish to integrally pack them without increasing the bin capacities by too much. Kleinberg established that this version of unsplittable flow has a constant flow-cut gap and this was strengthened by Kolliopoulos and Stein (1997). Dinitz, Garg, and Goemans ultimately proved the following tight result.

Theorem 2.4 (The cost of unsplittability; Dinitz et al. 1999). *Suppose we have an instance of the single-sink unsplittable flow problem for which there is a feasible fractional flow, then there is an unsplittable flow where the maximum load on each edge e is at most $u(a) + d_{\max}$.*

The situation for the minimum cost unsplittable flow is less developed. Goemans (see Costain et al., 2007; cf. Skutella, 2002) has conjectured that there is an unsplittable flow using the capacities $(u(a) + d_{\max})$ that costs no more than OPT, the cost of an optimal fractional flow using capacities $u(a)$. To date, the strongest result known in this direction is due to Skutella (2002) who shows the existence of an unsplittable flow of cost OPT using capacities $2u(a) + d_{\max}$.

20.2.3 Coloured Flows

In existing SONET networks, optical connections only exist between neighbhouring nodes on rings. Incoming information is optically decoded and converted to an electronic signal. Any traffic that is destined for this node is stripped, and new traffic is added before the traffic is rebundled and lasers are then used to transmit a new optical signal. Such an OEO conversion is expensive in terms of delays as well as the equipment need to achieve the individual tasks. Existing optical networks have also incorporated point-to-point *Wavelength Division Multiplexed* (WDM) fibres or line systems. A WDM fibre is simply one where many different wavelengths (channels) can be transmitted simultaneously. This capability is enabled by installing (moderately expensive) optical terminals within nodes at either end of a link. Typical current WDM fibres may carry anywhere between roughly 4 and 64 channels (although, 1000's of wavelengths have been achieved in laboratory experiments). As the demand for capacity increases, the future optical networks will increasingly take advantage of nodes that can switch paths optically without the need for conversion to electronics, thus producing a genuine all-optical network.

All-optical networking is in its early stages, and it is not entirely clear how future networks will look. These networks will no longer be ring-based as in SONET, but will rather look like general graphs. The most important question is the switching behaviour of network nodes. If nodes are to act as crossbars, as in traditional telecom, then a node should have the ability to make *wavelength translation*; this is a technology that allows a lightpath to change wavelength without conversion to electronics. This point is illustrated in the simple example Fig. 20.4(a) where each link supports two wavelengths, red and blue. There are three paths and no edge is used by more than two paths. If we do not have wavelength translation, then each path must be assigned an end-to-end wavelength (the same wavelength for each of its edges). Obviously this is not possible since each pair of paths intersect in some edge, but we only have two colours. This simple example shows that either we would need at least one node to perform wavelength translation, or more colours per fiber, or we would have to "leave" the optical layer (i.e., convert the signal along some path into electronics, as was done in SONET). Since wavelength translators are extremely expensive, this solution seems unlikely to be adopted as a viable all-optical strategy.

The leading switching alternative are nodes that only do *spatial switching*. This means that a node may connect signals between distinct fibers, as long as the signals

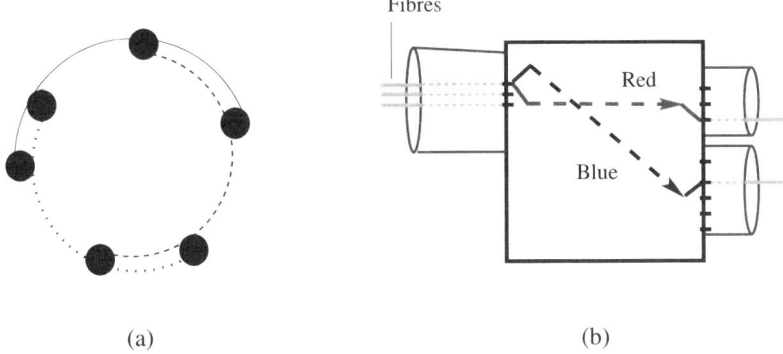

| (a) | (b) |

Fig. 20.4. (**a**) One needs 3 colours for paths. (**b**) A node with spatial switching

are using the same wavelength—see Fig. 20.4(b).[6] Consider the *minimum fibre* MIN FIB network design problem that arises in this setting; Winkler and Zhang (2003) studied the special case for path graphs, introduced due to its relevance in the design of so-called optical line systems.

We have a directed graph $D = (V, A)$ and for each arc a we may install WDM links each at a cost of $c(a)$, and each supporting λ wavelengths. There is also have a collection of unit demands, given as a set F of ordered node pairs (u, v). Our goal is to find an optimal set of WDM fibers to activate. That is, we want to install $u(a)$ fibres on each (directed) link, such that the network has enough capacity to route every demand $f \in F$ on a path P_f, and assign this path a wavelength $\lambda_f \in \{1, 2, \ldots, \lambda\}$. Since nodes can switch paths between different fibres this problem is potentially much easier than a standard path colouring problem in graphs (which is at least as hard as graph colouring). Indeed we only need to guarante that the path assignments (colouring) satisfy: $\sum_{f:a \in P_f, \lambda_f = k} 1 \leq u(a)$ for each colour k. Aggregating these constraints, we obtain an obvious necessary condition for the routing P_f and installed capacities $u(a)$:

$$\sum_{f:a \in P_f} 1 \leq \lambda u(a). \tag{1}$$

In general, one would not expect this condition to be sufficient. Surprisingly, one can show that in general networks, a "fractional implies integral" type result holds if we restrict to single-sink multicommodity flow instances. This result can be deduced by seeing that a certain polyhedron has the *integer decomposition property* (IDP) (Baum and Trotter 1978). A polyhedron P has the IDP if for any $x \in P$, and any integer k such that kx is integral, we have $kx = \sum_{i=1}^{k} x^i$ where each x^i is an integral vector in P. Trivially, any polyhedron with the IDP is integral, i.e., each of its vertices is integral. One can make use of the following characterization of total unimodularity[7] due to Baum and Trotter (1978).

Theorem 2.5 (Baum and Trotter 1978). *A matrix A is totally unimodular if and only if $\{x : Ax \leq b, x \geq 0\}$ has the integer decomposition property for every integer vector b.*

Consider an arbitrary directed supply digraph $D = (V, A)$ and a single-sink set F of unit demands of the form (s_i, t) for some destination node t and nodes $s_1, s_2, \ldots, s_k$. We wish to determine the number $x(a)$ of fibers to light in each arc a. (The remarks equally apply to the undirected version.) The following flow-gap 1 type result can be established.

Theorem 2.6 (The cost of colouring; McGregor and Shepherd 2007). *If there is a single-sink network integral flow such that the load on any arc a is at most $u(a)\lambda$,*

---

[6] Currently there are quite restrictive bounds on the number of fibers that a node is able to terminate. This leads to even more complicated network design problems (Winkler and Zhang 2003; Anshelevich and Zhang 2004; Fortune et al. 2004; McGregor and Shepherd 2007).

[7] A $0, +1, -1$ matrix is totally unimodular if each square submatrix has determinant $0, 1$ or -1.

then there is a wavelength-assigned such flow that uses at most $u(a)$ fibers on each arc a.

Proof. Consider the node-arc incidence matrix B for D: for each node v and arc a, $B_{v,a}$ is 1 if a has tail v, -1 if it has head v, and otherwise it is 0. Since B is totally unimodular, we have that $P = \{f : Af \leq u, f \geq 0\}$ has the IDP. Thus if f' is a flow vector whose total load on any arc a is at most $\lambda u(a)$, then $\frac{f'}{\lambda} \in P$, and so by the result of Baum and Trotter, we may decompose f' as $\sum_{i=1}^{\lambda} f^i$ where each f^i is an integral vector in P. As the flow paths determined by any f^i place a load of at most $u(a)$ on any link, there are enough fibers to route these demands on the same wavelength, and the proof is complete.   □

We now describe an alternative proof based on a simple combinatorial algorithm (see McGregor and Shepherd 2007). A flow f is called *k-ready* (for arc capacities $u(a)$) if the total load $f(a)$ on any arc a is at most $ku(a)$. For each arc a in a k-ready flow f, we define its *requirement $r(a)$* as: $\max\{0, f(a)-(k-1)u(a)\}$. The main idea is to find a flow f' satisfying some of our demands, so that each arc has $f'(a) \geq r(a)$ and hence we can "reduce" f, roughly stated as $f - f'$, to obtain a new $(k-1)$-ready flow. If we can choose f' so that $f'(a) \leq u(a)$, then all of the demands routed by f' can use the same wavelength, and hence we may greedily repeat this process to route all demands using k colours on each fibre in the network (i.e., lighting $u(a)$ fibers for each arc a suffices if each fiber has k colours). We now describe this procedure in more detail.

Call an arc a *critical* if $r(a) > 0$. We create an auxiliary digraph as follows. For each arc a, we include a forward auxiliary arc with capacity (upper bound) of $u(a)$ and a lower bound of $r(a)$. For each arc with a load of $(k-1)u(a) - \sigma$ for some $\sigma > 0$ (i.e., its load is less than $(k-1)u(a)$) we include a reverse arc with capacity σ and lower bound of 0. In addition, from each terminal s_i, that sends $d(s_i)$ units of flow to t, we include an arc from t back to s_i with capacity $d(s_i)$. Checking several cases, one deduces that if f is a k-ready flow for u, then in the auxiliary digraph H, we have that for each proper node subset S, the sum of lower bounds on arcs in $\delta_H^+(S)$ is at most the sum of upper bounds on arcs in $\delta_H^-(S)$. Hence Hoffman's Circulation Conditions (Hoffman 1960) imply that there is an integral circulation obeying the lower and upper bounds. Such a circulation corresponds to a routing f' of some of our demands that obeys the capacities $u(a)$. In addition, it is easily checked a $(k-1)$-ready flow f'' is produced by performing the standard symmetric difference on the flows f and f'.

The circulation f' determines a set of flow paths that can be simultaneously routed on a single wavelength (e.g., k) since we had upper bounds of $u(a)$. We can thus remove these demands, and repeat the process with the $(k-1)$-ready flow f''. Note that each circulation problem can be solved simply (and efficiently) by a sequence of shortest path problems, using the successive shortest path algorithm (Iri 1960).

The previous theorem (and algorithm) also generalizes the so-called *p-multi-route flow problem* (Aggarwal and Orlin 2002). Indeed multi-route flows correspond to the special case where $u(a) = 1$ for all a. To find a multi-route flow of size say kp, one may compute a k-ready flow (for unit arc capacities) of size kp and then apply the procedure to decompose into the desired k "protected" flows, each of size p. The interested reader is referred to Aggarwal and Orlin (2002) for more background on this problem.

We close by mentioning that the approaches above also give a slight extension of the results for path graphs in Winkler and Zhang (2003) to certain directed trees; the key is the total unimodularity of the underlying path-edge incidence matrix. The related problem where one is not given λ, the number of wavelengths per fiber, but wishes to minimize λ is studied for single-sink (multicast) problems in Beauquier et al. (1998); the above approaches also apply in that setting.

20.3 Data Networks

The opening of the Internet to commercial interests in the early 1990's and the advent of browsers led to an explosive growth in terms of data traffic (which now far exceeds the world's voice traffic). The workhorse for moving this traffic is the IP router. Part of the success of the Internet is due to the fact that IP routers are relatively inexpensive and operationally simple, compared for instance, to traditional large-scale telecom switches such as the 5ESS whose "five 9's" reliability is legendary (it was operational 99.999% of the time). Examining the behaviour of such routers inspires the next two flow models we consider.

20.3.1 Confluent Flows

The strength of IP routing is in the simplicity of its operation which, to a first order of approximation, can be described as follows. A router first builds a *next hop table* that contains an entry for each possible IP address t; this entry contains the *next hop* for t denoted by *next(t)*. This table is built by means of a protocol (distributed algorithm) which we do not discuss in detail. One a next hop table is established, upon receiving a packet destined for t, a router simply looks in the table, and forwards the packet to the IP router *next(t)*. It follows that all flow destined to t that passes through v must exit v on a single link—see Fig. 20.5(a). A flow in a digraph with this property is called *confluent* and was introduced by Chen et al. (2003). They proposed that an important indicator of the impact of IP routing on network performance was to measure the degree to which confluent flows force node congestion to increase. It turns out that this congestion could be unbounded. Indeed the example in Fig. 20.5(b) is an instance from Chen et al. (2004) which shows that congestion may grow as large as $\Omega(\log n)$ at a node. On the positive side, the congestion is no worse than this.

Theorem 3.1 (The cost of confluence; Chen et al. 2004). *If a single-sink multicommodity flow instance has a fractional flow with a maximum node load of U, then there is a confluent flow whose maximum node load is $O((\log n)U)$. Moreover, it is NP-hard to determine the optimal node congestion for a confluent flow to within a factor less than $\frac{1}{2}\log n$.*

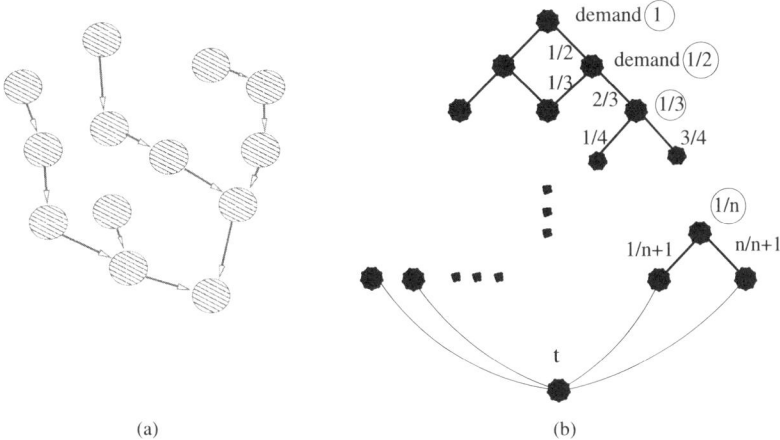

Fig. 20.5. (a) A confluent flow. **(b)** An example with $\log n$ congestion. The fractional network flow displayed has a maximum node load of 1; however any confluent flow must route the flow from the top node along a path whose nodes have demands $1, \frac{1}{2}, \frac{1}{3}, \ldots$

Several interesting directions remain. For instance, what can one say about instances where the nodes have capacities?[8] Note that the above result applies only to the uniform capacity case since the bound is in terms of the maximum load U. Another direction is to analyze the behaviour of confluent flows for general multicommodity instances. The algorithm which proves the above theorem is centralized. It would be interesting to understand the ramifications of making a distributed implementation.

20.3.2 Bifurcated and d-Furcated Flows

In our discussion of IP routers, we considered the case where traffic destined to node t must use a single next hop. What happens if we allow a node to send its outgoing traffic on two next hops? Or d next hops? Allowing multiple next hops is a strategy that many network operators allow. We call a single-sink multicommodity flow *bifurcated* (or *d-furcated*) if the support of the flow has the property that the out-degree of any node $v \neq t$ is at most 2 (resp. d). Interestingly, allowing an extra next hop eliminates the unbounded node congestion that occurs with confluent flows. In the following we discuss this result and some of its consequences.

Theorem 3.2 (The cost of bifurcation; Donovan et al. 2007). *If a single-sink multicommodity flow has a fractional flow with a maximum node load of U, then there is a bifurcated flow whose maximum node load is $U + d_{\max} \leq 2U$. Moreover, the bound of 2 is best possible.*

---

[8] A motivation based on interdomain routing is suggested in Shepherd and Wilfong (2005).

In practice, operators may deem it undesirable to split traffic from a common IP address since this leads to the need to perform resequencing of packets at the destination. One common approach is to split traffic using source IP address hashing. In this setting, we then have an additional restriction that requires that traffic from a single source s_i, not be split when it leaves a node. One can handle this refinement as follows. First, apply the methods above to obtain a bifurcated flow. Second, apply Theorem 2.4 to obtain an unsplittable flow.

Theorem 3.3 (Cost of bifurcation with unsplit routes from source; Donovan et al. 2007). *If a single-sink multicommodity flow has a fractional flow with a maximum node load of U, then there is a bifurcated flow whose maximum node load is $2U + d_{max}$ (at most $3U$ under the no bottleneck assumption) with the following properties. If a source node s_i routes any demand through a node v, then all of the demand d_i arrives at v, and exits v on a unique arc.*

As one might expect, if the outdegree (number of next hops) d is allowed to grow, the worst case node congestion improves. The same methods used to prove Theorem 3.2 also show that if there is a feasible fractional flow with node congestion one, then there is a d-furcated flow with node congestion at most $1 + \frac{1}{d-1}$, for $d \geq 2$.

Given that an extra next hop ($d = 2$) eliminates unbounded congestion, it is natural to ask what is happening between $d = 1$ (confluence) and $d = 2$ (bifurcation). Again, the proof techniques used for Theorem 3.2 help analyze this question. For $\beta \in [0, 1]$, call a flow β-*confluent* if at each node v, at least a β fraction of its total flow exits on a single arc, and any remaining flow uses at most one more arc. Thus 1-confluent flows are standard confluent flows, and for any $\beta \leq .5$, a β-confluent flow is just a bifurcated flow. For $\beta \in (.5, 1]$ there is a spectrum of problems. For $\beta = 1$, the confluent flow problem has unbounded node congestion but this is not the case for any *fixed* $\beta < 1$.

Theorem 3.4 (The cost of β-confluence; Donovan et al. 2007). *Let $\beta \in [\frac{1}{2}, 1)$. Suppose there is a fractional flow for a single-sink multicommodity instance such that for each node v, its load $f(v)$ is at most U. Then there is a β-confluent flow such that for each node v, its load is at most $f(v) + \frac{\beta}{1-\beta}U$.*

Several interesting directions should be explored. As with confluent flows, the situation for general multicommodity flows, non-uniform capacities, costs, or extensions to distributed algorithms have not yet been addressed. It is also natural to ask about a special class of bifurcated flows where each node must split its flow equally along its outgoing arcs. For instance, in practice, this would be the case if an IP router chose its next hop for each packet based on a simple round-robin scheme. We call such flows *halfluent*.

20.3.3 Network Coding

Consider a single source (we now have a single source s not sink t) multicommodity flow problem, where there are some k destination nodes $t_1, t_2, \ldots, t_k$. The source itself is the location of several bit streams $b_1, b_2, \ldots, b_q$; the bits in each stream must be transmitted to each destination t_i. Viewed as a flow problem, this gives rise to a commodity for each pair b_j, t_i. Creating a commodity for each pair, however, is wasteful since it does not take advantage of capacity sharing for a fixed stream b_j. Instead one should think of reserving a tree (or rather an arborescence) on which the bits in b_j are routed. This is referred to as a *multicast* routing problem. One thus naturally seeks q disjoint arborescences rooted at s, each of which includes all of the terminals t_i. It turns out, however, that sometimes we may not even need q disjoint trees! We show that this is sometimes the case when our network's nodes are allowed extra processing power (this section is less about adding node constraints to flow, but rather *removing* them!). We obviously require the following *cut condition* be satisfied: if S contains s but not all of the t_i's, then $|\delta^+(S)| \geq q$.[9]

In traditional *store-and-forward* networks, packets arriving at an internal node are stored, and a copy is forwarded to the next node. Naturally, the number of packets placed on a link between two nodes cannot exceed its channel capacity, and hence this gives rise to a standard integer multicommodity flow problem. Prevailing wisdom had suggested (cf. Yeung et al. 2006) that there was no need for data processing within network nodes beyond the capability to replicate packets. In Ahlswede et al. (2000) a surprisingly simple example showed that this was not valid. Consider the digraph of Fig. 20.6 where there are two bit streams, denoted by a, b, at node s, to be transmitted to nodes t_1, t_2.

In the first network, we simply reserve two disjoint arborescences rooted at s, one for each bitstream. In the second network, one easily checks that there does not exist two disjoint arborescences. We are still able, however, to transmit all bits to their destinations t_i, by means of coding (in this case, performing an XOR) at an internal node, and decoding at nodes t_1, t_2. This example has been proved to extend to any instance where a collection of bitstreams are co-located at a single origin node. Namely, if the cut condition holds then, possibly with coding/decoding at nodes, all streams can be transmitted to the destinations (Ahlswede et al. 2000). This surprising result has sparked a huge amount of work in the new area of network coding. There are several interesting open questions in this area. One that seems most relevant to flow theory is whether "coding helps" (as it does in the above example) in the case of undirected multicommodity instances. We have only roughly touched on the topic of network coding; the interested reader is referred to (Yeung et al. 2006; Fragouli and Soljanin 2007) for a more thorough background.

20.4 Multicommodity Flows via Crossbars and Single-Sink Flows

We close with some comments on how the combination of single-sink flows and crossbars have played a role for general multicommodity flow and disjoint path problems.

---

[9] The cut condition is not normally sufficient to yield a packing of arborescences, a problem which is NP-hard. There is a sharp characterization, however, when the set of terminals is the set of all nodes V (Edmonds 1973). Namely, there exists q arc-disjoint arborescences rooted at s if and only if the cut condition holds.

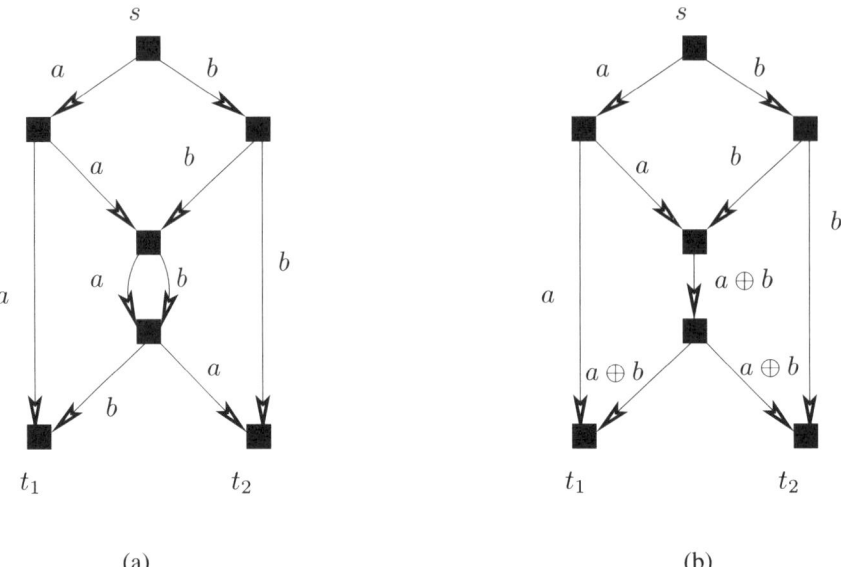

Fig. 20.6. (**a**) Without network coding; (**b**) with network coding

Our introductory remarks described traditional network flows as assuming that nodes can behave as crossbars, i.e., that any collection of paths crossing through a node is feasible, as long as arc capacities are satisfied. The problem of how the interconnection of paths is actually achieved within a node is ignored, or delayed to a later subproblem. This concept of "outsourcing" work to crossbars has been used in a number of theoretical routing results. Probably the most celebrated example is the work of Robertson and Seymour (1990, 1995), and specifically their algorithm for the undirected disjoint path problem with a fixed number of pairs s_i, t_i. An important scheme in their work is to look for a well-connected subgraph that behaves like a crossbar. The terminals s_i, t_i are then routed to the crossbar along disjoint paths (this amounts to a single-sink flow problem). If the pairs are able to reach the crossbar, then they can be matched up appropriately using the crossbar. The most basic crossbar is simply a single node. Another classical example is a grid since an $n \times n$ grid can act as a crossbar for the nodes (terminals) in any fixed grid row. No matter how we place $n/2$ pairs s_i, t_i on distinct nodes in a row, one may connect them using edge disjoint paths. A clique is another simple example of a crossbar. Routing(outsourcing) to a clique minor is used in the Robertson–Seymour scheme. A similar approach is used by Kleinberg (1998) to route a super-constant number of pairs in a half-disjoint fashion. Routing to a single node crossbar is the basis for an $O(\sqrt{n}\log n)$-approximation for the maximum arc-disjoint path problem in acyclic directed graphs (Chekuri and Khanna 2003). This is further developed to show that the natural LP relaxation for maximum edge-disjoint paths in undirected graphs is $O(\sqrt{n})$ (Chekuri et al. 2006b) (and hence $\Theta(\sqrt{n})$). Routing to a grid is used by

Chekuri et al. (2004b) to give a polylogarithmic approximation for the maximum edge-disjoint path problem in planar graphs; an extension for even graphs is given in Kleinberg (2005). Frank (2005) indicated that a similar idea of outsourcing to a crossbar appears in early work of Jung on disjoint paths in directed graphs (Jung 1970).

This same general approach has also been used for (fractional) multicommodity flows. In that setting, a fundamental crossbar is given by the results of Leighton and Rao (1999) on uniform (or product) multicommodity flow. They show that for any undirected graph $G = (V, E)$ and any set of *terminals* W, if $|\delta(S)| \geq \min\{|W \cap S|, |(V - S) \cap W|\}$, then there is a *uniform multicommodity flow* with $O(\log n)$ edge congestion. In other words, one can route $\frac{1}{|W|}$ between each pair of nodes in W, and send at most $O(\log n)$ flow on any edge. One deduces that given any matching s_i, t_i on the nodes of W, one may fractionally satisfy these demands using the uniform multicommodity flow above (this is also the essential idea behind Valiant's randomized load balancing). Hence G behaves like a *fractional crossbar* for the terminals W. Routing to such fractional crossbars has been an important ingredient in various places. These include Räcke (2002) breakthrough on low-congestion oblivious routings, and the work (Chekuri et al. 2004a) on finding a polylogarithmic approximation for the *all-or-nothing multicommodity flow problem*. Both of these also rely on single-sink flows when routing to a fractional crossbar.

Acknowledgements

I am grateful for stimulating discussions and input from colleagues Steve Fortune, David Johnson, Iraj Saniee, Peter Winkler, Gord Wilfong, and Francis Zane. This paper has benefitted from many suggestions and enjoyable collaborations with Chandra Chekuri. I also thank the two anonymous reviewers whose detailed input pointed out numerous ways for improving the article. Finally, I wish to give thanks to Bernhard Korte for supervising my postdoctoral work in Bonn in 1991. His view that there were plenty of practical problems that needed mathematical development apparently rubbed off on me.

References

Aggarwal, C., Orlin, J.: On multi-route maximum flows in networks. Networks **39**, 43–52 (2002)

Ahlswede, R., Cai, N., Li, S.-Y.R., Yeung, R.W.: Network information flow. IEEE Trans. Inf. Theory **46**, 1204–1216 (2000)

Ahuja, R., Magnanti, T., Orlin, J.: Network Flows: Theory, Algorithms, and Applications. Prentice Hall, Englewood Cliffs (1993)

Anshelevich, E., Zhang, L.: Path decomposition under a new cost measure with applications to optical network design. In: European Symposium on Algorithms (ESA), pp. 28–39 (2004)

Bartholdi, J.J., Orlin, J.B., Ratliff, H.D.: Cyclic scheduling via integer programs with circular ones. Oper. Res. **28**, 1074–1085 (1980)

Baum, S., Trotter, L.E. Jr.: Integer rounding and polyhedral decomposition for totally uni-modular systems. In: Henn, R., Korte, B., Oettli, W. (eds.) Optimization and Operations Research (Proceedings of Workshop Bad Honnef, 1977). Lecture Notes in Economics and Mathematical Systems, vol. 157, pp. 15–23. Springer, Berlin (1978)

Beauquier, B., Hell, P., Perennes, S.: Optimal wavelength-routed multicasting. Discrete Appl. Math. **84**, 15–20 (1998)

Beauquier, B., Bermond, J.C., Gargano, L., Hell, P., Perennes, S., Vaccaro, U.: Graph problems arising from wavelength-routing in all-optical networks. Theor. Comput. Sci. **233**(1–2), 165–189 (2000)

Bienstock, D., Muratore, G.: Strong inequalities for capacitated survivable network design problems. Math. Program., Ser. A **89**, 127–147 (2000)

Brightwell, G., Shepherd, F.B.: Consultancy report: Resilience strategy for a single source-destination pair. LSE CDAM Report 96-22 (August 1996)

Brightwell, G., Oriolo, G., Shepherd, B.: Reserving resilient capacity in a network. SIAM J. Discrete Math. **14**, 524–539 (2001)

Brightwell, G., Oriolo, G., Shepherd, B.: Reserving resilient capacity for a single commodity with upper bound constraints. Networks **41**(2), 87–96 (2003)

Chekuri, C., Khanna, S.: Edge disjoint paths revisited. In: Proc. of the ACM–SIAM Symposium on Discrete Algorithms (SODA) (2003)

Chekuri, C., Khanna, S., Shepherd, F.B.: The all-or-nothing multicommodity flow problem. In: Proc. of the ACM Symposium on Theory of Computing (STOC) (2004a)

Chekuri, C., Khanna, S., Shepherd, F.B.: Edge-disjoint paths in planar graphs. In: Proc. of IEEE Foundations of Computer Science (FOCS) (2004b)

Chekuri, C., Claisse, P., Essiambre, R., Fortune, S., Kilper, D., Nithi, K., Lee, W., Saniee, I., Shepherd, B., Wilfong, G., White, C., Zhang, L.: Design tools for transparent optical networks. Bell Labs Techn. J. **11**(2), 129–143 (2006a)

Chekuri, C., Khanna, S., Shepherd, F.B.: An $O(\sqrt{n})$ approximation and integrality gap for disjoint paths and unsplittable flow. Theory Comput. **2–7**, 137–146 (2006b)

Chen, J., Rajaraman, R., Sundaram, R.: Meet and merge: approximation algorithms for confluent flow. In: Proceedings of the 35th ACM Symposium on Theory of Computing (STOC), pp. 373–382 (2003)

Chen, J., Kleinberg, R., Lovasz, L., Rajaraman, R., Sundaram, R., Vetta, A.: (Almost) tight bounds and existence theorems for confluent flows. In: Proceedings of the 36th ACM Symposium on Theory of Computing (STOC), pp. 529–538 (2004)

Cosares, S., Saniee, I.: An optimization problem related to balancing loads on SONET rings. Telecommun. Syst. **3**, 165–181 (1994)

Costain, G., Kennedy, S., Meagher, C. (eds.): Bellairs Combinatorial Optimization Open Problems. http://www.math.mcgill.ca/~bshepherd/Bellairs/bellairs2007.pdf

Dinitz, Y., Garg, N., Goemans, M.: On the single-source unsplittable flow problem. Combinatorica **19**, 17–41 (1999)

Donovan, P., Shepherd, F.B., Vetta, A., Wilfong, G.T.: Degree constrained network flows. In: Proc. of the ACM Symposium on Theory of Computing (STOC) (2007)

Edmonds, J.: Edge-disjoint branchings. In: Rustin, R. (ed.) Combinatorial Algorithms, Courant Computer Science Symposium 9, Monterey, CA, 1972, pp. 91–96. Academic Press, New York (1973)

Fong, J., Gilbert, A.C., Kannan, S., Strauss, M.: Better alternatives to OSPF routing. Algorithmica **43**(1–2), 113–131 (2005)

Ford, L.R. Jr., Fulkerson, D.R.: Maximal flow through a network. Can. J. Math. **8**, 399–404 (1956)

Ford, L.R. Jr., Fulkerson, D.R.: Flows in Networks. Princeton University Press, Princeton (1962)

Fortune, S., Sweldens, W., Zhang, L.: Line system design for DWDM networks. In: Proceedings of the 11th International Telecommunications Network Strategy and Planning Symposium (Networks), pp. 315–320 (2004)

Fortz, B., Thorup, M.: Optimizing OSPF/IS-IS weights in a changing world. IEEE J. Sel. Areas Commun. **20**(4), 756–767 (2002)

Fragouli, C., Soljanin, E.: Network coding fundamentals. Found. Trends Netw. **2**(1), 1–133 (2007)

Frank, A.: Personal communication (2005)

Hoffman, A.J.: Some recent applications of the theory of linear inequalities to extremal combinatorial analysis. In: Bellman, R., Hall, M. (eds.) Combinatorial Analysis, pp. 113–128. Am. Math. Soc., Providence (1960)

Hwang, F.K.: The Mathematical Theory of Nonblocking Switching Networks. Series on Applied Mathematics, vol. 11. World Scientific, River Edge (1998)

Iri, M.: A new method of solving transportation-network problems. J. Oper. Res. Soc. Jpn. **3**, 27–87 (1960)

Jung, H.A.: Einer Verallgemeinerung des n-fachen Zusammenhangs für Graphen. Math. Ann. **187**, 95–103 (1970)

Kleinberg, J.: Single-source unsplittable flow. In: Proceedings of the 37th Symposium on Foundations of Computer Science (FOCS), pp. 68–77 (1996)

Kleinberg, J.M.: Decision algorithms for unsplittable flow and the half-disjoint paths problem. In: Proc. of the ACM Symposium on Theory of Computing (STOC), pp. 530–539 (1998)

Kleinberg, J.M.: An approximation algorithm for the disjoint paths problem in even-degree planar graphs. In: Proc. of the Foundations of Computer Science (FOCS) (2005)

Kolliopoulos, S., Stein, C.: Improved approximation algorithms for unsplittable flow problems. In: Proceedings of the 38th Symposium on Foundations of Computer Science (FOCS), pp. 426–435 (1997)

Leighton, F.T.: Introduction to Parallel Algorithms and Architectures: Arrays, Trees, Hypercubes. Kauffman, Los Altos (1992)

Leighton, T., Rao, S.: Multicommodity max-flow min-cut theorems and their use in designing approximation algorithms. J. Assoc. Comput. Mach. **46**(6), 787–832 (1999); Preliminary version in Proc. of the Foundations of Computer Science (FOCS) (1988)

Linial, N., London, E., Rabinovich, Y.: The geometry of graphs and some of its algorithmic applications. Combinatorica **15**(2), 215–245 (1995); Preliminary version in Proc. of IEEE Foundations of Computer Science (FOCS) (1994)

McGregor, A., Shepherd, F.B.: Island hopping and path colouring, with applications to WDM network design. In: Proceedings of the 18th Annual ACM–SIAM Symposium on Discrete Algorithms (SODA) (2007)

Menger, K.: Zur allgemeinen Kurventheorie. Fund. Math. **10**, 96–115 (1927)

Pippenger, N.: Telephone switching networks. AMS Proc. Symp. Appl. Math. **26**, 101–133 (1978)

Pippenger, N.: Communication Networks, Handbook of Theoretical Computer Science (vol. A): Algorithms and Complexity. MIT Press, Cambridge (1991)

Räcke, H.: Minimizing congestion in general networks. In: Proc. of the Foundations of Computer Science (FOCS) (2002)

Robertson, N., Seymour, P.D.: Outline of a disjoint paths algorithm. In: Korte, B., Lovász, L., Prömel, H.J., Schrijver, A. (eds.) Paths, Flows and VLSI-Layout. Springer, New York (1990)

Robertson, N., Seymour, P.D.: Graph minors XIII: the disjoint paths problem. J. Comb. Theory, Ser. B **63**(1), 65–110 (1995)

Schrijver, A.: Theory of Linear and Integer Programming. Wiley, New York (1986)

Schrijver, A., Seymour, P., Winkler, P.: The ring loading problem. SIAM J. Discrete Math. **11**(1), 1–14 (1998); Reprinted in SIAM Rev. **41**(1) 777–791 (1999)

Shepherd, F.B., Vetta, A.: Visualizing, finding and packing dijoins. In: Avis, D., Hertz, A., Marcotte, O. (eds.) Graph Theory and Combinatorial Optimization, pp. 219–254. Kluwer, New York (2005)

Shepherd, F.B., Wilfong, G.T.: Multilateral transport games. In: INOC (2005)

Skutella, M.: Approximating the single source unsplittable min-cost flow problem. Math. Program., Ser. B **91**, 493–514 (2002)

Winkler, P., Zhang, L.: Wavelength assignment and generalized interval graph coloring. In: Proc. of the ACM–SIAM Symposium on Discrete Algorithms (SODA), pp. 830–831 (2003)

Yeung, R.R., Li, S.-Y.R., Cai, N., Zhang, Z.: Network Coding Theory. Now Publishers, Boston (2006)

An Introduction to Network Flows over Time

Martin Skutella

Summary. We give an introduction into the fascinating area of flows over time—also called "dynamic flows" in the literature. Starting from the early work of Ford and Fulkerson on maximum flows over time, we cover many exciting results that have been obtained over the last fifty years. One purpose of this chapter is to serve as a possible basis for teaching network flows over time in an advanced course on combinatorial optimization.

Flow variation over time is an important feature in network flow problems arising in various applications such as road or air traffic control, production systems, communication networks (e. g., the Internet), and financial flows. In such applications, flow values on arcs are not constant but may change over time. Moreover, there is a second temporal dimension in these applications. Usually, flow does not travel instantaneously through a network but requires a certain amount of time to travel through each arc. In particular, when routing decisions are being made in one part of a network, the effects can be seen in other parts of the network only after a certain time delay. Not only the amount of flow to be transmitted but also the time needed for the transmission plays an essential role.

The above mentioned aspects of network flows are not captured by the classic *static* network flow models. This is where *network flows over time* come into play. They include a temporal dimension and therefore provide a more realistic modeling tool for numerous real-world applications. Only few textbooks on combinatorial optimization and network flows, however, mention this topic at all; see, e.g., Ford and Fulkerson (1962, Chapter III.9) Ahuja et al. (1993, Chapter 19.6) Korte and Vygen (2008, Chapter 9.7) and Schrijver (2003, Chapter 12.5c).

The following treatment of the topic has been developed for the purpose of teaching in an advanced course on combinatorial optimization. We concentrate on flows over time (also called "dynamic flows" in the literature) with finite *time horizon* and constant capacities and constant *transit times* in a *continuous time model*. For

This work was supported by DFG Research Center MATHEON in Berlin.

a broader overview of flows over time we refer to the survey papers by Aronson (1989), Powell et al. (1995), Kotnyek (2003), and Lovetskii and Melamed (1987). We also refer to the PhD thesis of Hoppe (1995) for an easily accessible and detailed treatment of the topic based on the *discrete time model*.

The paper is organized as follows. In Sect. 21.1 we introduce notation and shortly repeat some basic definitions and results from static network flow theory that are of particular importance for our purposes. For a more detailed treatment of those issues we refer to standard textbooks such as, e.g., (Ahuja et al. 1993) and (Korte and Vygen 2008). In Sect. 21.2 we present a classical result of Ford and Fulkerson on the Maximum Flow over Time Problem. This problem can be reduced to a static min-cost flow computation. Section 21.3 is devoted to a special class of flows over time called *earliest arrival flows* which can be used to model evacuation scenarios. We present a classical result that shows how to compute an earliest arrival flow with the Successive Shortest Path Algorithm. In Sect. 21.4 we consider flows over time with costs on the arcs, discuss their complexity, and introduce *time-expanded networks*. With the help of these networks, many flow over time problems can be reduced to static flow problems at the cost of a considerable increase in the size of the network. Finally, in Sect. 21.5 we discuss multi-commodity flows over time and present a recent approximation result. Pointers to the literature are discussed at the end of each section. We do not claim, however, to give a complete review of all related literature. Several exercises are given in the very end.

21.1 Basic Notions and Results on Static Network Flows

In this section we compile basic notions and results from the area of static network flows that are needed in the remainder of the chapter.

Network. Let $G = (V, E)$ be a network (directed graph) with a *source node* $s \in V$ and a *sink node* $t \in V$. Each arc $e \in E$ has an associated *capacity* u_e and a *transit time* (or *length*) $\tau_e \geq 0$. In the setting with costs, each arc e also has a *cost coefficient* c_e, which determines the cost for sending one unit of flow through the arc. An arc e from node v to node w is sometimes also denoted (v, w); in this case, we write head$(e) = w$ and tail$(e) = v$. To avoid confusion, we assume without loss of generality that there is at most one arc between any pair of nodes in G and that there are no loops.

Let $\delta^+(v)$ and $\delta^-(v)$ denote the set of arcs $e \in E$ leaving node v (tail$(e) = v$) and entering node v (head$(e) = v$), respectively. We sometimes also write $\delta_E^+(v)$ and $\delta_E^-(v)$ to emphasize the underlying set of arcs E. For a subset of nodes $X \subseteq V$, let

$$\delta^+(X) := \{(v, w) \in E \mid v \in X \wedge w \in V \setminus X\}.$$

For technical reasons we assume that there is an s-v-path and a v-t-path in G for every node $v \in V$. Notice that nodes violating this condition (and their incident arcs) are useless when it comes to sending flow from s to t and can therefore be deleted.

Network flow. A (static) *network flow* x assigns a non-negative flow value x_e to each arc $e \in E$. The flow x is *feasible* if it obeys the *capacity constraints*: $x_e \leq u_e$ for each $e \in E$. An s-t-flow x satisfies *flow conservation* at each node $v \in V \setminus \{s, t\}$:

$$\sum_{e \in \delta^-(v)} x_e - \sum_{e \in \delta^+(v)} x_e = 0. \tag{1}$$

The *value* $|x|$ of an s-t-flow x is

$$|x| := \sum_{e \in \delta^-(t)} x_e - \sum_{e \in \delta^+(t)} x_e = \sum_{e \in \delta^+(s)} x_e - \sum_{e \in \delta^-(s)} x_e.$$

A *circulation* is a flow obeying flow conservation (1) at each node $v \in V$. The *cost* of flow x is

$$c(x) := \sum_{e \in E} c_e \cdot x_e.$$

Residual network. For an arc $e = (v, w) \in E$ we denote the corresponding *backward arc* (w, v) by $\overleftarrow{e} := (w, v)$. Notice that $e \in E$ implies $\overleftarrow{e} \notin E$ due to our assumption that there is at most one arc between any pair of nodes in G. The *bidirected network* $\overleftrightarrow{G} = (V, \overleftrightarrow{E})$ corresponding to $G = (V, E)$ is defined by $\overleftrightarrow{E} := E \dot\cup \{\overleftarrow{e} \mid e \in E\}$. The transit time of a backward arc $\overleftarrow{e}$ with $e \in E$ is $\tau_{\overleftarrow{e}} := -\tau_e$. In particular, the transit time of a backward arc can be negative.

Given a feasible flow x in G, the *residual capacity of arc* $e \in E$ is $u_e - x_e$ and the *residual capacity of the corresponding backward arc* $\overleftarrow{e}$ is x_e. The *residual network* $G_x = (V, E_x)$ consists of all arcs in $\overleftrightarrow{E}$ with positive residual capacity. For $v, w \in V$ we denote the transit time of a shortest v-w-path in G_x by $\text{dist}_x(v, w)$.

Flow decomposition. Let $\mathcal{P}$ and $\mathcal{C}$ denote the sets of all simple s-t-paths and all simple cycles in G, respectively. For $P \in \mathcal{P} \cup \mathcal{C}$ we write $e \in P$ and $v \in P$ to indicate that arc $e \in E$ and node $v \in V$, respectively, lie on P. The well-known Flow Decomposition Theorem states that any static s-t-flow x has a *flow decomposition* $(x_P)_{P \in \mathcal{P} \cup \mathcal{C}}$ where $x_P \geq 0$ for each $P \in \mathcal{P} \cup \mathcal{C}$ and

$$x_e = \sum_{P \in \mathcal{P} \cup \mathcal{C} : e \in P} x_P \quad \text{for each } e \in E.$$

Moreover, there always exists a flow decomposition for which the number of *flow-carrying* paths and cycles P with $x_P > 0$ is bounded by the number of arcs $|E|$. A flow decomposition with $x_P = 0$ for all cycles $P \in \mathcal{C}$ is also called *path decomposition*.

The set of s-t-paths in $\overleftrightarrow{G}$ is denoted by $\overleftrightarrow{\mathcal{P}}$. For a path or cycle $P \in \overleftrightarrow{\mathcal{P}} \cup \mathcal{C}$ we set $\tau(P) := \sum_{e \in P} \tau_e$. If node v is contained in an s-t-path $P \in \overleftrightarrow{\mathcal{P}}$, we denote the subpath from s to v by $P_{s,v}$ and the subpath from v to t by $P_{v,t}$.

The collection $(x_P)_{P \in \overleftrightarrow{\mathcal{P}}}$ is a *generalized path decomposition* of a given s-t-flow x if $x_P \geq 0$ for each $P \in \overleftrightarrow{\mathcal{P}}$ and

$$x_e = \sum_{P \in \overleftrightarrow{\mathcal{P}}: e \in P} x_P - \sum_{P \in \overleftrightarrow{\mathcal{P}}: \overleftarrow{e} \in P} x_P \quad \text{for each } e \in E.$$

We mention that there are s-t-flows that do not have a generalized path decomposition since, in general, cycles are also needed in a decomposition.

Multi-commodity flow. Every commodity $i = 1, \ldots, k$ has a source node $s_i \in V$, a sink node $t_i \in V$, and a demand $d_i \geq 0$. A (static) *multi-commodity flow* x consists of k single-commodity flows x^i, $i = 1, \ldots, k$. We call x *feasible* if it satisfies the capacity constraints $\sum_{i=1}^{k} x_e^i \leq u_e$ for each $e \in E$ and if x^i is an s_i-t_i-flow for each commodity $i = 1, \ldots, k$. A feasible multi-commodity flow x *satisfies demands* $d_1, \ldots, d_k$ if $|x^i| \geq d_i$ for $i = 1, \ldots, k$.

21.2 Maximum Flows over Time

We consider flows over time with a fixed time horizon $T \geq 0$.

Definition 2.1 (Flow over time). *A flow over time f with time horizon T consists of a Lebesgue-integrable function $f_e : [0, T) \to \mathbb{R}_{\geq 0}$ for each arc $e \in E$; moreover $f_e(\theta) = 0$ must hold for $\theta \geq T - \tau_e$. To simplify notation, we sometimes consider f_e as a function with domain $\mathbb{R}$; in this case we set $f_e(\theta) := 0$ for all $\theta \notin [0, T)$.*

We say that $f_e(\theta)$ is the *rate of flow* (i.e., amount of flow per time unit) entering arc e at time θ. The flow particles entering arc e at its tail at time θ arrive at the head of e exactly τ_e time units later at time $\theta + \tau_e$. In particular, the *outflow rate* at the head of arc e at time θ equals $f_e(\theta - \tau_e)$. Definition 2.1 ensures that all flow has left arc e at time T as $f_e(\theta) = 0$ for $\theta \geq T - \tau_e$.

In order to gain an intuitive understanding of flows over time, one can associate arcs of the network with pipes in a pipeline system for transporting some kind of fluid. The length of each pipeline determines the transit time of the corresponding arc while the width determines its capacity.

Example 2.2. To illustrate the described model of flows over time, we consider the following simple example with two nodes s and t that are connected by an arc $e = (s, t)$ with transit time $\tau_e = 3.5$. We can, for example, send two units of flow from s to t as follows: At time 0 we start to pump flow at rate 1 into arc e and continue to do so until time 2. More precisely, we define a flow over time f with time horizon $T := 5.5$ by

$$f_e(\theta) := \begin{cases} 1 & \text{for } \theta \in [0, 2), \\ 0 & \text{otherwise.} \end{cases}$$

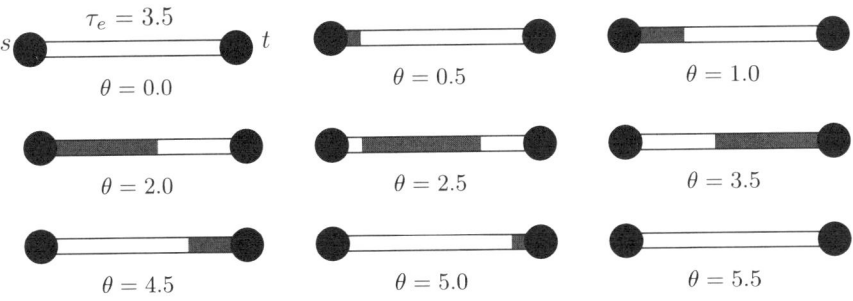

Fig. 21.1. Snapshots of the simple flow over time discussed in Example 2.2 for several points in time. At time 0, the arc $e = (s, t)$ (directed from left to right) is empty and we start to pump flow at rate 1 into the arc. At time 0.5 we have pumped half a unit of flow into the arc, at time 2 there are two flow units and we stop pumping. The first flow particles of the first flow unit reach the head node t at time $\tau_e = 3.5$. At time 4.5 the first flow unit has arrived at node t while the second flow unit is still on the arc. Finally, at time 5.5 the last flow particles have arrived at node t and the arc is empty again

The amount of flow that we have sent into arc e is obtained by integrating the flow rate f_e over time:

$$\int_0^T f_e(\theta)\, d\theta = 2.$$

Since the last flow particle enters arc e shortly before time 2, it arrives at node t exactly 3.5 time units later, that is, shortly before time 5.5. More precisely, the outflow rate $f_e(\theta - \tau_e)$ of arc e at its head node t is

$$f_e(\theta - \tau_e) = \begin{cases} 1 & \text{for } \theta \in [3.5, 5.5), \\ 0 & \text{otherwise.} \end{cases} \tag{2}$$

An illustration of this example is given in Fig. 21.1.

As already mentioned above, we work with the so-called *continuous time model* where a flow over time assigns a *flow rate* $f_e(\theta)$ to every point in time θ in the interval $[0, T)$. The original definition of flows over time given by Ford and Fulkerson is based on a so-called *discrete time model*. For integral transit times $(\tau_e)_{e \in E}$ and an integral time horizon T, the flow along an arc e is described by a function $g_e : \{0, 1, 2, \ldots, T\} \to \mathbb{R}_{\geq 0}$ where $g_e(\theta)$ denotes the *amount of flow* sent at time θ into arc e and arriving at the head node of e at time $\theta + \tau_e$. Both models have their advantages and disadvantages. More details can be found in the references given at the end of this section.

Definition 2.3 (Capacity, excess, flow conservation, s-t-flow over time). *Let f be a flow over time with time horizon T.*

(a) *The flow over time f fulfills the* capacity constraints *(and is called* feasible*) if $f_e(\theta) \leq u_e$ for each $e \in E$ and all $\theta \in [0, T)$.*

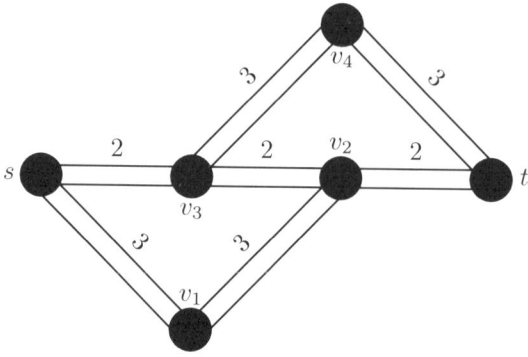

Fig. 21.2. A network consisting of arcs (pipelines) with unit capacities. The arcs are all directed from left to right towards the sink t. The numbers at arcs indicate transit times

(b) *For $v \in V$, the* excess *for node v at time θ is the net amount of flow that enters node v up to time θ, that is,*

$$\mathrm{ex}_f(v, \theta) := \sum_{e \in \delta^-(v)} \int_0^{\theta - \tau_e} f_e(\xi) \, d\xi - \sum_{e \in \delta^+(v)} \int_0^{\theta} f_e(\xi) \, d\xi.$$

(c) *The flow over time f fulfills the* weak flow conservation constraints *if $\mathrm{ex}_f(v, \theta) \geq 0$ for each $v \in V \setminus \{s\}$ and all $\theta \in [0, T)$. Moreover, $\mathrm{ex}_f(v, T) = 0$ must hold for each $v \in V \setminus \{s, t\}$.*

(d) *A flow over time obeying the weak flow conservation constraints is an s-t-flow over time. The* value *of an s-t-flow over time with time horizon T is $|f| := \mathrm{ex}_f(t, T)$.*

(e) *An s-t-flow over time f fulfills the* strict flow conservation constraints *if $\mathrm{ex}_f(v, \theta) = 0$ for all $v \in V \setminus \{s, t\}$ and $\theta \in [0, T]$. The strict flow conservation constraints say that flow must not be stored at intermediate nodes.*

Notice that the *weak* flow conservation constraints allow to store flow at intermediate nodes for some time as long as the flow has left the node again before the time horizon is over. In contrast to this, the *strict* flow conservation constraints ensure that flow entering an intermediate node must leave the node again *immediately*.

In Fig. 21.2 we give an example of a network. An illustration of a feasible s-t-flow over time in this network fulfilling the strict flow conservation constraints is given in Fig. 21.3.

We study the following basic optimization problem for flows over time.

MAXIMUM FLOW OVER TIME PROBLEM

Given: A network $G = (V, E)$ with capacities and transit times on the arcs, a source node $s \in V$, a sink node $t \in V$, and a time horizon $T \geq 0$.
Task: Find a feasible s-t-flow over time f with time horizon T and maximum value $|f|$.

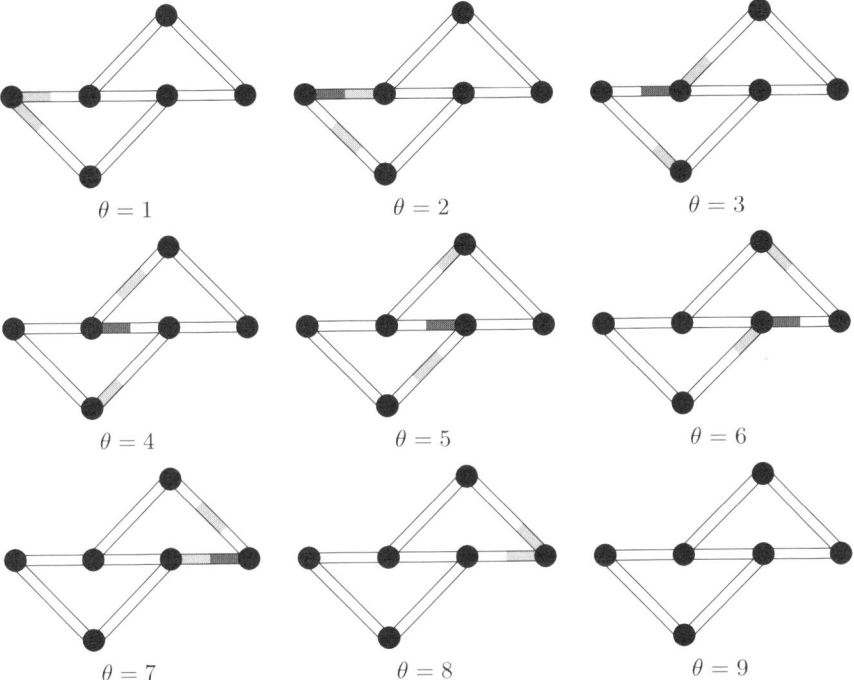

$\theta = 1$ $\theta = 2$ $\theta = 3$

$\theta = 4$ $\theta = 5$ $\theta = 6$

$\theta = 7$ $\theta = 8$ $\theta = 9$

Fig. 21.3. Snapshots of a feasible s-t-flow over time with time horizon $T = 9$ and value 3 in the network depicted in Fig. 21.2. In order to distinguish flow units traveling one after another along an arc, different shadings are used for the flow units

We consider a special class of feasible s-t-flows over time that are induced by static s-t-flows and feature a very simple structure. The intuition behind the following definition is to take a path decomposition (x_P) of a static s-t-flow x. At time 0 the resulting s-t-flow over time starts to send flow on each s-t-path P with flow rate x_P. It keeps sending flow along each path as long as there is enough time left for the flow along the path to arrive at the sink by time T. We first give a formal definition and afterwards, in Observation 2.5, a precise interpretation and intuition.

Definition 2.4 (Temporally repeated flow). *Let x be a static s-t-flow with some flow decomposition $(x_P)_{P \in \mathcal{P} \cup \mathcal{C}}$. The corresponding temporally repeated flow f with time horizon T is defined by*

$$f_e(\theta) := \sum_{P \in \mathcal{P}_e(\theta)} x_P \quad for\ e = (v, w) \in E,\ \theta \in [0, T), \tag{3}$$

where

$$\mathcal{P}_e(\theta) := \{P \in \mathcal{P} : e \in P \ \wedge\ \tau(P_{s,v}) \le \theta \ \wedge\ \tau(P_{v,t}) < T - \theta\}.$$

Notice that the flow x_P on path $P \in \mathcal{P}$ contributes to the flow rate f_e on arc $e = (v, w) \in P$ within the time interval starting at time $\tau(P_{s,v})$ and ending before time $T - \tau(P_{v,t}) = T - \tau(P) + \tau(P_{s,v})$. Thus, an alternative "path-based" description of the temporally repeated flow f is as follows.

Observation 2.5. *The temporally repeated flow f in Definition 2.4 can be obtained as follows: For each path $P \in \mathcal{P}$, send flow at rate x_P into P from the source s during the time interval $[0, T - \tau(P))$ and let the flow progress towards the sink without any delay at intermediate nodes. In particular, all flow reaches the sink by time T.*

Notice that the s-t-flow over time depicted in Fig. 21.3 is not temporally repeated since it does not have the structure described in Observation 2.5. This can, for example, be seen as follows. The flow over time in Fig. 21.3 sends flow along the shortest s-t path of length 6. However, it only starts to send flow into this path at time 1 and stops again at time 2 in order to avoid collisions with the remaining two flow units traveling along paths of length 8. In contrast, a temporally repeated flow of value 3 and time horizon 9 on this network sends all 3 units of flow along the shortest path. That is, it starts to send flow at rate 1 into this path at time 0 and stops before time 3 such that the last flow particles arrive at the sink before time 9.

Lemma 2.6. *Let x be a feasible static s-t-flow with flow decomposition $(x_P)_{P \in \mathcal{P} \cup \mathcal{C}}$. Then the corresponding temporally repeated flow f is a feasible s-t-flow over time with time horizon T that satisfies the strict flow conservation constraints.*

Proof. By construction (see Observation 2.5), flow entering an intermediate node leaves this node again immediately. Thus f is an s-t-flow over time with time horizon T that satisfies the strict flow conservation constraints. By (3) the feasibility of x implies

$$f_e(\theta) \leq \sum_{P \in \mathcal{P}: e \in P} x_P \leq x_e \leq u_e \quad \text{for } e \in E, \theta \in [0, T).$$

Thus f is feasible. □

The next lemma gives an indication how the static s-t-flow x should be chosen to get a temporally repeated flow with large value.

Lemma 2.7. *Let x be a feasible static s-t-flow with flow decomposition $(x_P)_{P \in \mathcal{P} \cup \mathcal{C}}$ such that $x_P = 0$ for all $P \in \mathcal{P}$ with $\tau(P) > T$ and for all $P \in \mathcal{C}$. Then the value of the corresponding temporally repeated flow f is equal to*

$$|f| = T \cdot |x| - \sum_{e \in E} \tau_e \cdot x_e.$$

In particular, the value of f does not depend on the chosen path decomposition of x.

Proof. By Observation 2.5 we get

$$|f| = \sum_{P \in \mathcal{P}} (T - \tau(P)) \cdot x_P$$

$$= T \cdot \sum_{P \in \mathcal{P}} x_P - \sum_{P \in \mathcal{P}} \sum_{e \in P} \tau_e \cdot x_P$$

$$= T \cdot |x| - \sum_{e \in E} \tau_e \cdot \sum_{P \in \mathcal{P}: e \in P} x_P$$

$$= T \cdot |x| - \sum_{e \in E} \tau_e \cdot x_e.$$

This concludes the proof. □

We mention the following corollary only for later use in Sect. 21.5.

Corollary 2.8. *Let x be a feasible static s-t-flow with flow decomposition $(x_P)_{P \in \mathcal{P} \cup \mathcal{C}}$. Then the value of the corresponding temporally repeated flow f is at least*

$$|f| \geq T \cdot |x| - \sum_{e \in E} \tau_e \cdot x_e.$$

Proof. We modify x by deleting flow on all paths $P \in \mathcal{P}$ with $\tau(P) > T$ and on all cycles $P \in \mathcal{C}$. More precisely, we set

$$\tilde{x}_P := \begin{cases} x_P & \text{if } P \in \mathcal{P} \text{ and } \tau(P) \leq T, \\ 0 & \text{otherwise,} \end{cases} \quad \text{for } P \in \mathcal{P} \cup \mathcal{C}.$$

By Observation 2.5, the temporally repeated flows f and $\tilde{f}$ corresponding to $(x_P)_{P \in \mathcal{P} \cup \mathcal{C}}$ and $(\tilde{x}_P)_{P \in \mathcal{P} \cup \mathcal{C}}$, respectively, are identical. Moreover, since flow along cycles and flow along s-t-paths of length at least T make a non-positive contribution to

$$T \cdot |x| - \sum_{e \in E} \tau_e \cdot x_e,$$

we get by Lemma 2.7

$$|f| = |\tilde{f}| = T \cdot |\tilde{x}| - \sum_{e \in E} \tau_e \cdot \tilde{x}_e \geq T \cdot |x| - \sum_{e \in E} \tau_e \cdot x_e.$$

This concludes the proof. □

As a consequence of Lemma 2.7, a maximum temporally repeated flow can be obtained as follows.

FORD–FULKERSON ALGORITHM

Input: A network $G = (V, E)$ with capacities and transit times on the arcs, a source node $s \in V$, a sink node $t \in V$, and a time horizon $T \geq 0$.
Output: A temporally repeated flow with time horizon T.

1. Compute a feasible static s-t-flow x maximizing

$$T \cdot |x| - \sum_{e \in E} \tau_e \cdot x_e.$$

2. Compute a flow decomposition $(x_P)_{P \in \mathcal{P} \cup \mathcal{C}}$ of x.
3. Output the corresponding temporally repeated flow f.

We next discuss how to obtain the static s-t-flow in Step 1 of the Ford–Fulkerson Algorithm. We consider an *extended network* $G' = (V, E')$ where the set of arcs E' is obtained by adding the *artificial arc* (t, s) to E with $u_{(t,s)} := \infty$ and $\tau_{(t,s)} := -T$. Any static s-t-flow x in G naturally induces a static circulation in G' by assigning flow value $|x|$ to the artificial arc (t, s). Conversely, any static circulation in G' naturally induces a static s-t-flow x in G whose value $|x|$ is equal to the flow value on the artificial arc (t, s). If we interpret transit times as cost coefficients ($c_e := \tau_e$ for $e \in E'$), the problem of finding a static s-t-flow x maximizing

$$T \cdot |x| - \sum_{e \in E} \tau_e \cdot x_e$$

is equivalent to finding such a flow minimizing the negated objective

$$-T \cdot |x| + \sum_{e \in E} \tau_e \cdot x_e = c_{(t,s)} \cdot |x| + \sum_{e \in E} c_e \cdot x_e.$$

The latter problem is apparently a static minimum cost circulation problem on the extended network G'.

Observation 2.9. *The static s-t-flow in Step 1 of the Ford–Fulkerson Algorithm can be obtained from a static minimum cost circulation computation in the extended network $G' = (V, E')$.*

As a consequence of this interpretation, we obtain the following insights on the path decomposition computed in Step 2.

Lemma 2.10. *Take any flow decomposition $(x_P)_{P \in \mathcal{P} \cup \mathcal{C}}$ computed in Step 2 of the Ford–Fulkerson Algorithm. Then the following holds for all $P \in \mathcal{P} \cup \mathcal{C}$:*

$$\text{If } x_P > 0, \quad \text{then } \tau(P) \begin{cases} = 0 & \text{if } P \in \mathcal{C}, \\ \leq T & \text{if } P \in \mathcal{P}. \end{cases} \tag{4}$$

In particular, all flow along cycles in x can be canceled without changing the objective function value of x. The resulting s-t-flow x is still feasible and optimal and satisfies the requirements of Lemma 2.7.

Proof. Notice that a path or cycle $P \in \mathcal{P} \cup \mathcal{C}$ violating (4) yields a negative cost cycle in the residual network G'_x. This contradicts the optimality of x.   □

Since the static s-t-flow x computed in Step 1 satisfies the requirements of Lemma 2.7, the flow value of the temporally repeated flow equals the optimum objective function value of x.

Corollary 2.11. *Let x be the static s-t-flow computed in Step 1 of the Ford–Fulkerson Algorithm and f the temporally repeated flow obtained in Step 3. Then*

$$|f| = T \cdot |x| - \sum_{e \in E} \tau_e \cdot x_e.$$

We conclude from Corollary 2.11 that the s-t-flow over time f computed by the Ford–Fulkerson Algorithm has maximum value among all temporally repeated flows. The following theorem states that it is even a maximum s-t-flow over time.

Theorem 2.12. *The temporally repeated flow computed by the Ford–Fulkerson Algorithm is a maximum s-t-flow over time with time horizon T. The running time of the algorithm is dominated by the static min-cost flow computation in Step 1.*

A proof can be found below. As a result of this theorem we know that computing a maximum s-t-flow over time is at most as difficult as computing a (static) min-cost flow.

The proof of the theorem goes along the same lines as the proof of the corresponding theorem for maximum static s-t-flows. The idea is to come up with an s-t-cut whose capacity is an upper bound on the maximum flow value and matches the value of the computed flow. In the static setting, an s-t-cut is defined by a subset of nodes $X \subseteq V$ with $s \in X$ and $t \notin X$. It consists of the arcs in $\delta^+(X)$ and its capacity is $\sum_{e \in \delta^+(X)} u_e$. For the case of flows over time, however, a more elaborate definition of an s-t-cut and its capacity is required in order to find a tight upper bound on the maximum flow value (see also Exercise 6.1).

Definition 2.13 (s-t-cut over time). *An s-t-cut over time with time horizon T is specified by threshold values $\alpha_v \in \mathbb{R}$ for each $v \in V$ with $\alpha_s = 0$ and $\alpha_t \geq T$. The capacity of an s-t-cut over time is defined as*

$$\sum_{e=(v,w) \in E} \max\{0, \alpha_w - \tau_e - \alpha_v\} \cdot u_e.$$

We say that node $v \in V$ belongs to the *t-side of the s-t-cut over time* up to time α_v and to the *s-side of the s-t-cut over time* from time α_v on. Thus, we can think of the s-t-cut over time as a sequence of s-t-cuts moving through the network over time towards the sink. An illustration is given in Fig. 21.4.

The definition of the capacity of an s-t-cut over time is motivated as follows: Every flow particle moving from s to t within the time interval $[0, T)$ will eventually cross the considered s-t-cut, i.e., move from the s-side to the t-side of the cut.

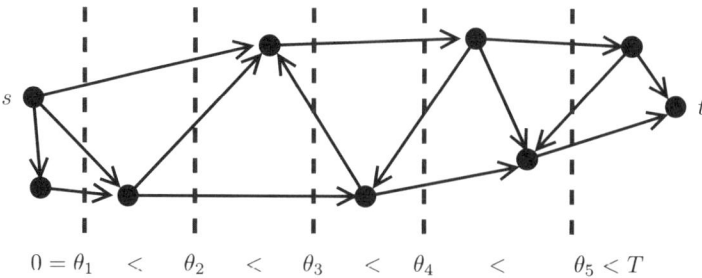

Fig. 21.4. An *s-t*-cut over time moves through the network over time from the source towards the sink

A particle moving along some arc $e = (v, w)$ crosses the cut if it enters the arc at node v while v is on the s-side of the cut and leaves the arc at node w while w is on the t-side of the cut. Thus, any particle entering arc e within the *critical time interval* $[\alpha_v, \alpha_w - \tau_e)$ crosses the cut while traveling along this arc. The total amount of flow that can cross the cut on arc e is thus bounded by u_e times the size of the critical time interval. The capacity of the s-t-cut over time is the sum of those values.

This motivates the following result.

Lemma 2.14. *The capacity of an s-t-cut over time with time horizon T is an upper bound on the value of any feasible s-t-flow over time with time horizon T.*

Proof. Let f be a feasible s-t-flow over time with time horizon T and let $(\alpha_v)_{v \in V}$ define an s-t-cut over time with time horizon T. By Definition 2.3 we get

$$|f| = \operatorname{ex}_f(t, \alpha_t) \leq \sum_{v \in V} \operatorname{ex}_f(v, \alpha_v)$$

$$= \sum_{v \in V} \left(\sum_{e \in \delta^-(v)} \int_0^{\alpha_v - \tau_e} f_e(\theta)\, d\theta - \sum_{e \in \delta^+(v)} \int_0^{\alpha_v} f_e(\theta)\, d\theta \right).$$

For the inequality we also use the fact that $\operatorname{ex}_f(s, 0) = 0$. Exchanging the order of summation in the right hand side term yields

$$|f| \leq \sum_{e=(v,w)\in E} \left(\int_0^{\alpha_w - \tau_e} f_e(\theta)\, d\theta - \int_0^{\alpha_v} f_e(\theta)\, d\theta \right)$$

$$= \sum_{e=(v,w)\in E} \int_{\alpha_v}^{\alpha_w - \tau_e} f_e(\theta)\, d\theta$$

$$\leq \sum_{e=(v,w)\in E} \max\{0, \alpha_w - \tau_e - \alpha_v\} \cdot u_e.$$

This concludes the proof.   □

We are now ready to prove Theorem 2.12.

Proof (of Theorem 2.12). It follows from Observation 2.9 and Corollary 2.11 that the value of the temporally repeated flow computed by the Ford–Fulkerson Algorithm is equal to the optimum solution value of the following linear programming formulation of a min-cost circulation problem on the extended network $G' = (V, E')$.

$$\max \quad T \cdot x_{(t,s)} - \sum_{e \in E} \tau_e \cdot x_e$$

$$\text{s.t.} \quad \sum_{e \in \delta^+_{E'}(v)} x_e - \sum_{e \in \delta^-_{E'}(v)} x_e = 0 \quad \text{for all } v \in V,$$

$$x_e \leq u_e \qquad\qquad\qquad \text{for all } e \in E,$$

$$x_e \geq 0 \qquad\qquad\qquad \text{for all } e \in E'.$$

The dual linear program looks as follows.

$$\min \quad \sum_{e \in E} u_e y_e$$

$$\text{s.t.} \quad y_e + \alpha_v - \alpha_w \geq -\tau_e \quad \text{for all } e = (v, w) \in E,$$

$$\alpha_t - \alpha_s \geq T,$$

$$y_e \geq 0 \qquad\qquad\qquad \text{for all } e \in E.$$

Let $(\alpha_v)_{v \in V}$, $(y_e)_{e \in E}$ be an optimum dual solution. We can assume without loss of generality that $\alpha_s = 0$ (otherwise, replace α_v with $\alpha_v - \alpha_s$ for each $v \in V$). Thus, the values $(\alpha_v)_{v \in V}$ define an s-t-cut over time. Moreover, for an optimum solution it holds that $y_e := \max\{0, \alpha_w - \tau_e - \alpha_v\}$. Therefore the optimum dual solution value equals the capacity of the cut defined by the α_v's. On the other hand, the value of the temporally repeated flow computed by the Ford–Fulkerson Algorithm equals the optimum primal solution value. The result follows from strong linear programming duality. □

Corollary 2.15. *An s-t-cut over time with time horizon T and minimum capacity can be obtained by a static min-cost flow computation.*

Proof. We consider again the pair of linear programs in the proof of Theorem 2.12. It is well-known (and easy to observe using complementary slackness) that an optimum dual solution can be obtained from an optimum primal solution (min-cost circulation) x as follows. For each $v \in V$, let α_v be the length of a shortest s-v-path in the residual network G'_x. □

As another corollary we get the following Max-Flow-Min-Cut Theorem for flows and cuts over time.

Theorem 2.16 (Max-Flow-Min-Cut Theorem). *The maximum value of an s-t-flow over time with time horizon T equals the minimum capacity of an s-t-cut over time with time horizon T.*

Pointers to the Literature

The results on the Maximum Flow over Time Problem presented above are due to Ford and Fulkerson (1958, 1962). More precisely, these results were originally developed for the discrete time model that we already shortly mentioned above. That is, time is discretized into steps of unit length. In each time step, flow can be sent from a node v through an arc (v, w) to the adjacent node w, where it arrives $\tau_{(v,w)}$ time steps later. In particular, the time-dependent flow on an arc is represented by a time-indexed vector in this model. Fleischer and Tardos (1998) point out a strong connection between the two models. They show that many results and algorithms which have been developed for the discrete time model can be carried over to the continuous time model. The continuous time version of Theorem 2.12 has first been observed by Anderson and Philpott (1994).

The concept of s-t-cuts over time was introduced by Anderson et al. (1982) (see also Anderson and Nash 1987) for the case of zero transit times and later extended to arbitrary transit times by Philpott (1990). In particular, Lemma 2.14 is due to Anderson et al. (1982), Philpott (1990).

A problem closely related to the Maximum Flow over Time Problem is the Quickest s-t-Flow Problem: Send a given amount of flow from the source to the sink in the shortest possible time. This problem can be solved in polynomial time by incorporating the Ford–Fulkerson Algorithm in a binary search framework; see also (Fleischer and Tardos 1998). Using Megiddo's method of parametric search (Megiddo 1979), Burkard et al. (1993) present a faster algorithm which solves the Quickest s-t-Flow Problem in strongly polynomial time.

An interesting generalization of the Quickest s-t-Flow Problem is the Quickest Transshipment Problem: Given a vector of supplies and demands at the nodes, the task is to find a flow over time that satisfies all supplies and demands within minimal time. Unlike the situation for standard (static) network flow problems, this multiple source, multiple sink, single commodity flow over time problem is not equivalent to a maximum s-t-flow over time problem (see Exercise 6.3). Hoppe and Tardos describe the first polynomial-time algorithm to solve this problem (Hoppe and Tardos 2000; Hoppe 1995). They introduce a generalized class of temporally repeated flows which can also be compactly encoded as a collection of paths. However, in contrast to temporally repeated flows, these paths may also contain backward arcs. Therefore, a careful analysis is necessary to show feasibility of the resulting flows over time (see also Sect. 21.3). Moreover, the algorithm of Hoppe and Tardos is not practical as it requires a submodular function minimization oracle for a subroutine.

21.3 Earliest Arrival Flows

In the last section we have considered a given, fixed time horizon T and the problem to send as much flow as possible from source s to sink t by time T. A possible application scenario for this problem is evacuation planning where one wants to get as many people as possible out of an endangered building or area. Since it is usually not clear a priori how long a building can withstand a fire before it collapses or how long a dam can resist a flood before it breaks, the exact time horizon T is not known in such a setting. It is therefore advisable to organize an evacuation such that as much as possible is saved no matter when the catastrophe will actually happen.

Coined in terms of s-t-flows over time, the goal is to find a single s-t-flow over time that simultaneously maximizes the amount of flow reaching the sink t up to any time $\theta \geq 0$.

Definition 3.1 (Earliest arrival flow). *A feasible s-t-flow over time f with time horizon T has the earliest arrival property and is called earliest arrival flow if it maximizes $\mathrm{ex}_f(t, \theta)$ simultaneously for all $\theta \in [0, T]$.*

At first sight, the existence of earliest arrival flows is not evident. One can indeed show that the important class of temporally repeated flows does, in general, not contain an earliest arrival flow (see Exercise 6.4). The situation changes if we consider a slightly more general class of s-t-flows over time.

Definition 3.2 (Generalized temporally repeated flow). *Let x be a static s-t-flow with generalized path decomposition $(x_P)_{P \in \overleftrightarrow{\mathcal{P}}}$. The corresponding generalized temporally repeated flow f with time horizon T is defined by*

$$f_e(\theta) := \sum_{P \in \overrightarrow{\mathcal{P}}_e(\theta)} x_P - \sum_{P \in \overleftrightarrow{\mathcal{P}}_{\overleftarrow{e}}(\theta)} x_P \quad \text{for each } e = (v, w) \in E, \theta \in [0, T), \quad (5)$$

where

$$\overrightarrow{\mathcal{P}}_e(\theta) := \left\{ P \in \overleftrightarrow{\mathcal{P}} \mid e \in P \wedge \tau(P_{s,v}) \leq \theta \wedge \tau(P_{v,t}) < T - \theta \right\}$$

and analogously

$$\overleftrightarrow{\mathcal{P}}_{\overleftarrow{e}}(\theta) := \left\{ P \in \overleftrightarrow{\mathcal{P}} \mid \overleftarrow{e} \in P \wedge \tau(P_{s,v}) \leq \theta \wedge \tau(P_{v,t}) < T - \theta \right\}.$$

Notice that a generalized temporally repeated flow f is not necessarily a proper flow over time since $f_e(\theta)$ can, in general, be negative. More precisely, and in analogy to Observation 2.5, f can be interpreted as follows.

Observation 3.3. *The generalized temporally repeated flow f in Definition 3.2 can be obtained as follows: For each path $P \in \overleftrightarrow{\mathcal{P}}$, send flow into P at rate x_P during the time interval $[0, T - \tau(P))$ and let the flow progress towards the sink without any delay at intermediate nodes. In particular, if some path P contains a backward arc $\overleftarrow{e} = (w, v)$ with negative transit time $-\tau_e$, flow traveling along arc $\overleftarrow{e}$ goes back in time. A flow particle entering $\overleftarrow{e}$ in node w at time θ arrives in node v at time $\theta - \tau_e$. In (5) this anomaly is captured by a negative flow rate $f_e(\theta - \tau_e)$ on the corresponding forward arc e.*

Under certain circumstances, however, going back in time can be justified. If there is another flow particle at node v that is about to enter arc e at time $\theta - \tau_e$, the two particles traveling along e in opposite (also with respect to time) directions cancel out, that is, they exchange their identity. In (5) this is reflected by the fact that the positive terms in the first sum compensate for a negative contribution of the second sum.

The following example illustrates the intuition behind Observation 3.3.

Example 3.4. Consider again the network depicted in Fig. 21.2. Let x be the feasible s-t-flow that sends one unit of flow across each arc except (v_3, v_2) where the flow value is zero. A generalized path decomposition of x is obtained by sending one unit of flow on the shortest s-t-path $P^1 = s, v_3, v_2, t$ and one unit along the path $P^2 = s, v_1, v_2, v_3, v_4, t$. The corresponding generalized temporally repeated flow f with time horizon $T = 11$ is illustrated in Fig. 21.5. Since $\tau(P^1) = 6$, the flow over time f sends $5 = 11 - 6$ units of flow through path P^1 (dark flow in Fig. 21.5). Moreover, since $\tau(P^2) = 10$, it sends $1 = 11 - 10$ flow units through path P^2 (light flow in Fig. 21.5). As can be seen in Fig. 21.5, f is a feasible s-t-flow over time with time horizon $T = 11$ although the light flow unit goes back in time when traveling through the backward arc (v_2, v_3) with transit time -2. As a result, we see two copies of this flow unit in the network at times $\theta = 5$ and $\theta = 6$. Since the light flow unit and the third dark flow unit cancel out on arc (v_3, v_2), this dark flow unit has disappeared at times $\theta = 5$ and $\theta = 6$. It reappears at time 7 on arc (v_2, t) when the light flow unit has arrived at node v_2.

In the following we interpret transit times also as cost coefficients, that is, we set $c_e := \tau_e$ for each $e \in E$. In particular, $\text{dist}_x(v, w)$ refers to the minimum transit time of a v-w-path in the residual network G_x. Notice that the two paths P^1 and P^2 considered in Example 3.4 are exactly the augmenting paths chosen by the Successive Shortest Path Algorithm. Moreover, it can be easily checked that the resulting s-t-flow over time f is an earliest arrival flow. We show that the augmenting paths chosen by the Successive Shortest Path Algorithm always yield a generalized temporally repeated flow that is feasible and has the earliest arrival property.

EARLIEST ARRIVAL ALGORITHM

Input: A network $G = (V, E)$ with capacities and transit times on the arcs, a source node $s \in V$, a sink node $t \in V$, and a time horizon $T \geq 0$.

Output: A generalized temporally repeated flow with time horizon T.

1. Let $x_P := 0$ for all $P \in \overleftrightarrow{\mathcal{P}}$ and let x denote the static s-t-flow with generalized path decomposition $(x_P)_{P \in \overleftrightarrow{\mathcal{P}}}$;
2. While $\text{dist}_x(s, t) < T$, find a shortest s-t-path P in G_x and increase x_P by the residual capacity of P.
3. Output the generalized temporally repeated flow f with time horizon T corresponding to $(x_P)_{P \in \overleftrightarrow{\mathcal{P}}}$.

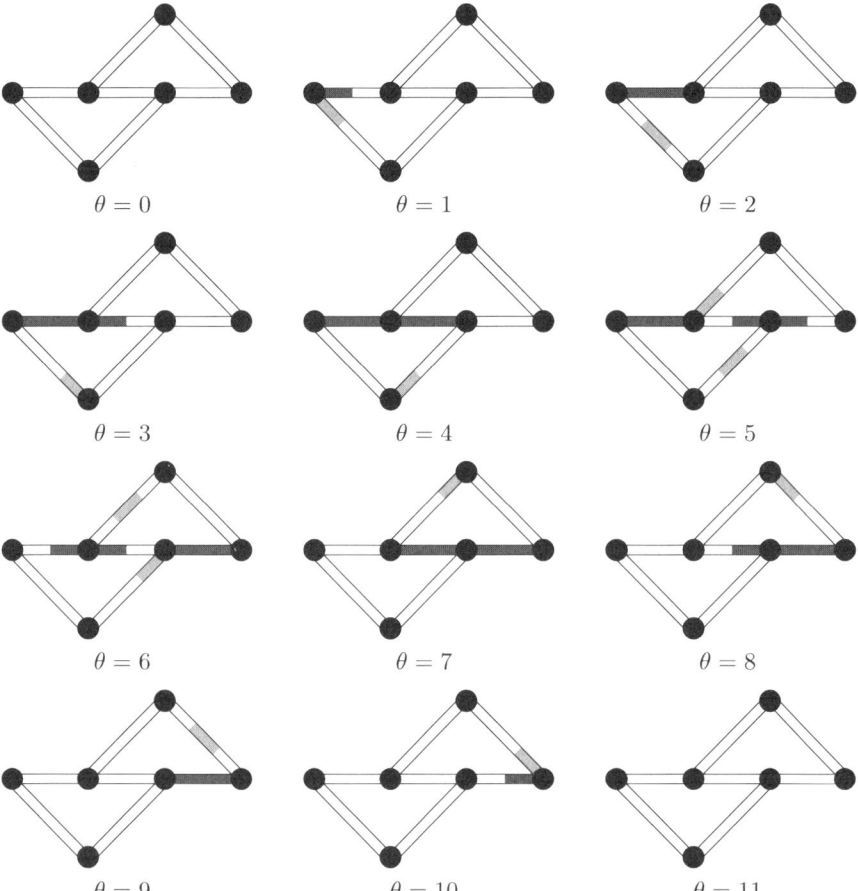

$\theta = 0$ $\theta = 1$ $\theta = 2$

$\theta = 3$ $\theta = 4$ $\theta = 5$

$\theta = 6$ $\theta = 7$ $\theta = 8$

$\theta = 9$ $\theta = 10$ $\theta = 11$

Fig. 21.5. Snapshots of the generalized temporally repeated flow f with time horizon $T = 11$ described in Example 3.4

We assume that Step 2 of the Earliest Arrival Algorithm terminates after q iterations. For $i = 1, \ldots, q$ let x^i denote the feasible s-t-flow x before iteration i and let P^i be the shortest path found in iteration i. In particular, $(x_{P^i})_{i=1,\ldots,k}$ is a generalized path decomposition of x^{k+1}. Since P^i is a shortest s-t-path in the residual network G_{x^i}, we get for each node $v \in P^i$

$$\tau(P^i_{s,v}) = \text{dist}_{x^i}(s, v) \quad \text{and} \quad \tau(P^i_{v,t}) = \text{dist}_{x^i}(v, t).$$

In order to show that f is a feasible flow over time, we need the following result from the theory of static network flows.

Lemma 3.5. *The values* $\text{dist}_{x^i}(s, v)$ *and* $\text{dist}_{x^i}(v, t)$ *are monotonically increasing in* i.

Proof (sketch). It is well-known (see, e.g., (Ahuja et al. 1993, Chapter 9.7) or (Korte and Vygen 2008, Chapter 9.4)) that the flows x^i are min-cost flows in G. A certificate of optimality for x^i is the feasible node potential $\pi_i(v) := \mathrm{dist}_{x^i}(s, v)$ (remember that distances are measured with respect to transit times of arcs). After augmenting flow along the shortest s-t-path P^i in G_{x^i}, the node potential $\pi_i(v)$ is still feasible. This implies that the shortest path distances have not decreased.   □

We show that the flow over time computed by the Earliest Arrival Algorithm is feasible.

Lemma 3.6. *The generalized temporally repeated flow f computed by the Earliest Arrival Algorithm is a feasible s-t-flow over time with time horizon T.*

Proof. We first show that f is indeed a feasible flow over time, i.e., $0 \le f_e(\theta) \le u_e$ for $e = (v, w) \in E, \theta \in [0, T)$. For fixed e and θ, let

$$k := \max\{i \mid \mathrm{dist}_{x^i}(s, v) \le \theta \wedge \mathrm{dist}_{x^i}(v, t) < T - \theta\}.$$

By the definition of f in (5) and by Lemma 3.5 we get

$$f_e(\theta) = \sum_{P \in \overleftrightarrow{\mathcal{P}_e}(\theta)} x_P - \sum_{P \in \overleftrightarrow{\mathcal{P}_{\overleftarrow{e}}}(\theta)} x_P$$

$$= \sum_{i \in \{1,\dots,k\}: e \in P^i} x_{P^i} - \sum_{i \in \{1,\dots,k\}: \overleftarrow{e} \in P^i} x_{P^i}$$

$$= x^{k+1}(e) \in [0, u_e].$$

Finally, it follows from Observation 3.3 that f fulfills strict flow conservation and has time horizon T.   □

Theorem 3.7. *The generalized temporally repeated flow f computed by the Earliest Arrival Algorithm is an earliest arrival flow with time horizon T.*

Proof. We prove that $\mathrm{ex}_f(t, \theta)$ is maximal for all $\theta \in [0, T]$. For a fixed θ let

$$k := \max\{i \mid \mathrm{dist}_{x^i}(s, t) \le \theta\}.$$

Since $\tau(P^i) = \mathrm{dist}_{x^i}(s, t)$ is monotonically increasing in i, Observation 3.3 yields

$$\mathrm{ex}_f(t, \theta) = \sum_{i=1}^{k} (\theta - \tau(P^i)) \cdot x_{P^i} = \theta \cdot |x^{k+1}| - \sum_{e \in E} \tau_e \cdot x_e^{k+1}.$$

By our choice of k, the s-t-flow x^{k+1} is not only a min-cost s-t-flow in G (with respect to cost coefficients $c_e := \tau_e$ for each $e \in E$) but also induces a min-cost circulation in the extended network that is obtained by adding an artificial arc (t, s) with transit time $\tau_{(t,s)} := -\theta$ (see also Sect. 21.2). The artificial arc (t, s) and its backward arc do not induce a negative cycle in the residual graph. The reason is that by our choice of k the length of each flow-carrying s-t-path in x^{k+1} is bounded by θ; moreover, the length of each s-t-path in $G_{x^{k+1}}$ is greater than θ. Therefore x^{k+1} maximizes the objective function

$$\theta \cdot |x| - \sum_{e \in E} \tau_e \cdot x_e$$

and $\mathrm{ex}_f(t, \theta)$ is therefore maximal (see Sect. 21.2). □

Notice that the running time of the Earliest Arrival Algorithm is not polynomially bounded in the input size as the Successive Shortest Path Algorithm in Step 2 requires an exponential number of iterations in the worst case (also see references below). As a consequence, the function $\theta \mapsto \mathrm{ex}_f(t, \theta)$ can have exponentially many breakpoints for an earliest arrival flow f. It is therefore unlikely that an algorithm exists that computes f and whose running time is polynomially bounded in the input size.

Pointers to the Literature

Shortly after Ford and Fulkerson introduce flows over time, Gale (1959) shows that earliest arrival s-t-flows always exist (they are also called "universally maximum dynamic flows" or "universally quickest flow" in the literature). The Earliest Arrival Algorithm is due to Minieka (1973) and Wilkinson (1971). While Gale, Wilkinson, and Minieka all work in the discrete time model, the existence of earliest arrival flows in the continuous time model is first observed by Philpott (1990). Fleischer and Tardos (1998) finally discuss the Earliest Arrival Algorithm in the continuous time setting.

Zadeh (1973) presents a class of instances for which the Successive Shortest Path Algorithm and thus also the Earliest Arrival Algorithm requires an exponential number of iterations. Those instances also show that the piece-wise linear and convex function $\theta \mapsto \mathrm{ex}_f(t, \theta)$ has exponentially many breakpoints in the worst case.

Hoppe and Tardos (1994) present a fully polynomial-time approximation scheme for the earliest arrival s-t-flow problem that is based on a clever scaling trick. For a fixed $\epsilon > 0$, the computed s-t-flow over time f has the following property. For all θ, the excess $\mathrm{ex}_f(t, \theta)$ is at least $(1 - \epsilon)$ times the value of a maximum s-t-flow over time with time horizon θ.

In a network with several sources and sinks with given supplies and demands, flows over time having the earliest arrival property do not necessarily exist (Fleischer 2001; see Exercise 6.6). For the case of several sources with given supplies and a single sink, however, earliest arrival transshipments do always exist. This follows, for example, from the existence of lexicographically maximal flows in time-expanded networks; see, e.g., (Minieka 1973) and (Megiddo 1974). Hajek and Ogier (1984) give the first polynomial time algorithm for computing earliest arrival flows in networks with several sources and zero transit times. Fleischer (2001) gives an algorithm with improved running time.

Baumann and Skutella (2006) give an algorithm that computes earliest arrival flows for the case of several sources and arbitrary transit times and whose running time is polynomially bounded in the input plus output size. Fleischer and Skutella (2007) use condensed time-expanded networks to approximate such earliest arrival flows. They give a fully polynomial-time approximation scheme that approximates the time delay as follows: For every time $\theta \geq 0$, the amount of flow that should have reached the sink in an earliest arrival flow by time θ, reaches the sink at latest at time $(1 + \epsilon)\theta$. Tjandra (2003) shows how to compute earliest arrival transshipments in networks with time dependent supplies and capacities in time polynomial in the time horizon and the total supply at sources.

Earliest arrival flows and transshipments are motivated by applications related to evacuation. In the context of emergency evacuation from buildings, Berlin (1979) and Chalmet et al. (1982) study the quickest transshipment problem in networks with multiple sources and a single sink. Jarvis and Ratliff (1982) show that three different objectives of this optimization problem can be achieved simultaneously: (i) Minimizing the total time needed to send the supplies of all sources to the sink, (ii) fulfilling the earliest arrival property, and (iii) minimizing the average time for all flow needed to reach the sink. Hamacher and Tufecki (1987) study an evacuation problem and propose solutions which further prevent unnecessary movement within a building.

21.4 Minimum Cost Flows over Time

In this section we consider flows over time in networks with additional cost coefficients $c_e \geq 0$ on the arcs $e \in E$. As in the case of static flows, c_e is the cost for sending one flow unit along arc e. Thus, the cost of a given flow over time f with time horizon T is

$$c(f) := \sum_{e \in E} c_e \cdot \int_0^T f_e(\theta) \, d\theta. \tag{6}$$

We consider the following minimum cost flow over time problem.

MINIMUM COST s-t-FLOW OVER TIME PROBLEM

Given: A network $G = (V, E)$ with capacities, transit times, and costs on the arcs, a source node $s \in V$, a sink node $t \in V$, a time horizon $T \geq 0$, and a demand $d \geq 0$.
Task: Find a feasible s-t-flow over time f with time horizon T, value d, and minimum cost.

Surprisingly, and in contrast to static min-cost flow problems, this problem is already NP-hard.

Theorem 4.1. *The Minimum Cost s-t-Flow over Time Problem is weakly NP-hard.*

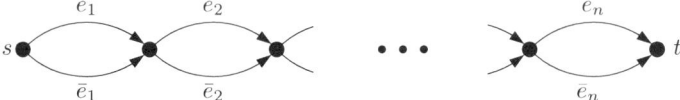

Fig. 21.6. The network obtained from an instance of the Partition Problem in the proof of Theorem 4.1

Proof. We consider the corresponding decision problem that asks for a feasible s-t-flow over time f with time horizon T, value d, and cost at most b for some given cost bound b. We reduce the well-known weakly NP-complete Partition Problem to this decision problem.

PARTITION PROBLEM

Given: $a_1, \ldots, a_n \in \mathbb{Z}_{>0}$ with $A := \sum_{i=1}^{n} a_i$ even.
Question: Is there a subset $I \subseteq \{1, \ldots, n\}$ with $\sum_{i \in I} a_i = \sum_{i \notin I} a_i$.

We denote the complement of $I \subseteq \{1, \ldots, n\}$ by $\bar{I} := \{1, \ldots, n\} \setminus I$.

Given an instance of the Partition Problem, we build up the network consisting of unit capacity arcs[1] depicted in Fig. 21.6.

For $i = 1, \ldots, n$, arc e_i has transit time $\tau_{e_i} := a_i$ and cost $c_{e_i} := 0$, while arc $\bar{e}_i$ has transit time $\tau_{\bar{e}_i} := 0$ and cost $c_{\bar{e}_i} := a_i$. The demand is set to $d := 2$, the time horizon is $T := 1 + A/2$ and the cost bound is $b := A$. We get a natural bijection between subsets $I \subseteq \{1, \ldots, n\}$ and s-t-paths by defining P_I to be the unique s-t-paths with $e_i \in P_I$ if and only if $i \in I$. Notice that P_I and $P_{\bar{I}}$ are arc-disjoint s-t-paths with

$$\tau(P_I) = \sum_{i \in I} a_i, \qquad c(P_I) = \sum_{i \notin I} a_i,$$

$$\tau(P_{\bar{I}}) = \sum_{i \notin I} a_i, \qquad c(P_{\bar{I}}) = \sum_{i \in I} a_i.$$

We argue that the Partition instance is a yes-instance if and only if there exists a feasible s-t-flow over time with time horizon T, value d, and cost at most b.

If the given Partition instance is a yes-instance with solution I, the two arc-disjoint paths P_I and $P_{\bar{I}}$ have transit time and cost $A/2$ and can thus be used to send two units of flow (one on each path) to the sink within time $T = 1 + A/2$ and cost $b = 2 \cdot A/2$.

We now assume by contradiction that the given instance of the Partition Problem is a no-instance but f is a feasible s-t-flow over time with time horizon T, value d,

---

[1] The network violates our general assumption that there is at most one arc between any pair of nodes. Notice, however, that this can be avoided if we split arcs by introducing intermediate nodes.

and cost at most b. Since all flow arrives strictly before time T in the sink, flow can only be sent along s-t-paths with transit time at most $T - 1 = A/2$. Moreover, by our assumption, no path with transit time $A/2$ exists such that all flow travels along paths with transit time at most $A/2 - 1$. Those paths, however, have cost at least $A/2+1$ such that the total cost of f is at least $2(A/2+1) > b$. This contradicts our assumption and thus proves the result.   □

On the positive side, there is an algorithm that solves the Minimum Cost s-t-Flow over Time Problem in pseudo-polynomial time.

Theorem 4.2. *The Minimum Cost s-t-Flow over Time Problem can be solved in pseudo-polynomial time.*

The proof of this theorem is given below. It relies on the very general concept of *time-expanded networks* which we introduce next.

Definition 4.3 (Time-expanded network). *Let $G = (V, E)$ be a network with capacities u, non-negative integral transit times τ, and costs c on the arcs. For a given time horizon $T \in \mathbb{Z}_{>0}$, the corresponding* time-expanded network $G^T = (V^T, E^T)$ *with capacities and costs on the arcs is defined as follows. For each node $v \in V$ we create T copies $v_0, v_1, \ldots, v_{T-1}$, that is,*

$$V^T := \{v_\theta \mid v \in V,\ \theta = 0, 1, \ldots, T - 1\}.$$

For each arc $e = (v, w) \in E$, there are $T - \tau_e$ copies $e_0, e_1, \ldots, e_{T-1-\tau_e}$ where arc e_θ connects node v_θ to node $w_{\theta+\tau_e}$. Arc e_θ has capacity $u_{e_\theta} := u_e$ and cost $c_{e_\theta} := c_e$. Moreover, E^T contains holdover arcs $(v_\theta, v_{\theta+1})$ *for $v \in V$ and $\theta = 0, \ldots, T-2$. The capacity of holdover arcs is infinite and they have zero cost. Summarizing, the set of arcs E^T is given by*

$$E^T := \{e_\theta = (v_\theta, w_{\theta+\tau_e}) \mid e = (v, w) \in E,\ \theta = 0, 1, \ldots, T - 1 - \tau_e\}$$
$$\cup \{(v_\theta, v_{\theta+1}) \mid v \in V,\ \theta = 0, 1, \ldots, T - 2\}.$$

An example of a time-expanded network is given in Fig. 21.7. Notice that the size of the time-expanded network G^T is linear in T and therefore only pseudo-polynomial in the input size.

Lemma 4.4. *Let $G = (V, E)$ be a network with capacities u, non-negative integral transit times τ, and costs c on the arcs. For a given time horizon $T \in \mathbb{Z}_{>0}$, a feasible static s_0-t_{T-1}-flow x in G^T yields a feasible s-t-flow over time f in G with time horizon T, cost $c(x) = c(f)$, and value $|f| = |x|$. The reverse direction is also true.*

Proof. Given x, we define f by

$$f_e(\xi) := x_{e_\theta} \quad \text{for } e \in E, \xi \in [\theta, \theta + 1), \theta = 0, 1, \ldots, T - 1 - \tau_e.$$

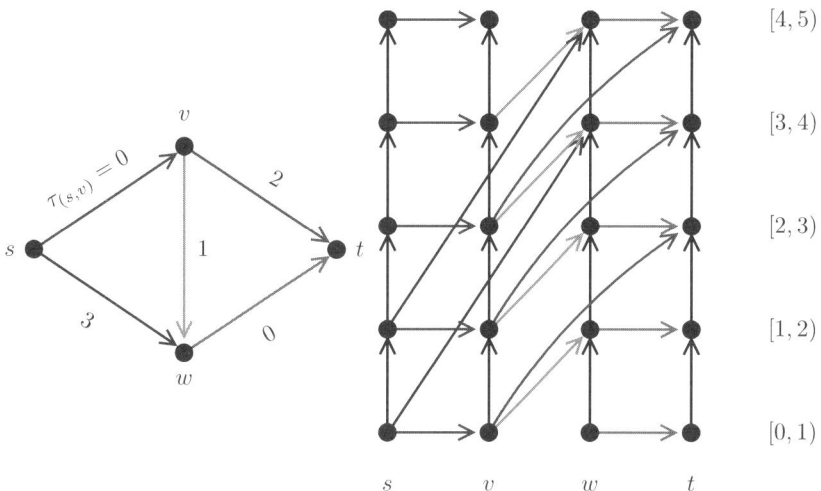

Fig. 21.7. On the left hand side a network G with transit times on the arcs is given. On the right hand side the time-expanded network G^T with time horizon $T = 5$ is depicted. There is one copy of each node for each time interval $[\theta, \theta + 1)$, for $\theta = 0, 1, \ldots, T - 1$

It is straightforward to verify that f obeys capacity and weak flow conservation constraints, has cost $c(f) = c(x)$, and value $|f| = |x|$. Conversely, given f, we define x by

$$x_{e_\theta} := \int_\theta^{\theta+1} f_e(\xi)\, d\xi \quad \text{for } e \in E, \theta = 0, 1, \ldots, T - 1 - \tau_e.$$

The flow values on holdover arcs are defined by

$$x_{(v_\theta, v_{\theta+1})} := \sum_{e \in \delta^-(v)} \int_0^{\theta+1-\tau_e} f_e(\xi)\, d\xi - \sum_{e \in \delta^+(v)} \int_0^{\theta+1} f_e(\xi)\, d\xi,$$

for $v \in V \setminus \{s\}, \theta = 0, 1, \ldots, T - 2$. For the special case of the source node s, we set

$$x_{(s_\theta, s_{\theta+1})} := \sum_{e \in \delta^+(s)} \int_{\theta+1}^T f_e(\xi)\, d\xi - \sum_{e \in \delta^-(s)} \int_{\theta+1-\tau_e}^T f_e(\xi)\, d\xi,$$

for $\theta = 0, 1, \ldots, T - 2$. It is again straightforward to verify that x obeys capacity and flow conservation constraints, has cost $c(x) = c(f)$, and value $|x| = |f|$. $\square$

We can finally prove Theorem 4.2.

Proof (of Theorem 4.2). As a result of Lemma 4.4, a min-cost s-t-flow over time in G can be obtained in pseudo-polynomial time by computing a min-cost s_0-t_{T-1}-flow

x in the time expanded network G^T. This result relies on the assumption that all transit times and the time horizon are integral. In the more general setting with arbitrary rational transit times and time horizon, integrality can be achieved if we scale time by the least common multiple of the denominators of those rational numbers. This concludes the proof.   □

Pointers to the Literature

Time-expanded networks have already been introduced by Ford and Fulkerson (1958, 1962). Unfortunately, due to the time expansion, the size of the network grows linearly in T which is, in general, exponential in the input size of the problem. This difficulty has already been pointed out by Ford and Fulkerson. On the other hand, the advantage of a time-expanded network is that it turns the problem of determining an optimal flow over time into a static network flow problem. Moreover, time-expanded networks are very flexible since they can also model time-dependent capacities, costs, transit times etc. This approach is also used in practice to solve flow over time problems. In many cases, however, the size of these networks makes the problem solution prohibitively expensive.

The NP-hardness result in Theorem 4.1 and the presented reduction are due to Klinz and Woeginger (2004). They also show that computing a temporally repeated flow with minimum cost is strongly NP-hard (via a reduction of the strongly NP-complete 3-Partition Problem). Fleischer and Skutella (2007) show that there always exists a minimum cost flow over time obeying the strict flow conservation constraints, even for the case of multiple sources and sinks with given supplies and demands, respectively. Moreover, Fleischer and Skutella (2007) introduce condensed time expanded networks of polynomial size that lead to fully polynomial-time approximation schemes for several NP-hard flow over time problems including min-cost flows over time. The idea is to partition time into intervals of size $\epsilon^2 \cdot T/n$ and to round transit times accordingly. For the case that arc costs are proportional to transit times, Fleischer and Skutella (2003) describe a very simple fully polynomial-time approximation scheme based on capacity scaling for the Minimum Cost s-t-Flow over Time Problem. They observe that optimal solutions to this problem are flows over time satisfying the earliest arrival and latest departure property (see Exercise 6.7 for the definition of latest departure flows). Their algorithm runs directly on the original network (i.e., no time expansion).

21.5 Multi-Commodity Flows over Time

In the preceding sections we have considered single-commodity flows in the network G with one source node s and one sink node t. In this section we consider the situation where multiple commodities have to be shipped through a common network G and thus have to share the capacities of arcs. Each commodity i has a source node $s_i \in V$ and a sink node $t_i \in V$.

Definition 5.1 (Multi-commodity flow over time). *A multi-commodity flow over time f with k commodities and time horizon T is a collection of k single-commodity flows over time f^i, $i = 1, \ldots, k$, with time horizon T. We call f feasible if it satisfies the capacity constraints*

$$\sum_{i=1}^{k} f_e^i(\theta) \leq u_e \quad \text{for } e \in E, \theta \in [0, T),$$

and f^i is an s_i-t_i-flow over time for each commodity $i = 1, \ldots, k$. A feasible multi-commodity flow over time f satisfies given demands $d_1, \ldots, d_k$ if $|f^i| \geq d_i$, for $i = 1, \ldots, k$.

MULTI-COMMODITY FLOW OVER TIME PROBLEM

Given: A network $G = (V, E)$ with capacities and transit times on the arcs; k commodities $i = 1, \ldots, k$, each with a source node $s_i \in V$, a sink node $t_i \in V$, and a demand $d_i \geq 0$; a time horizon $T \geq 0$.
Task: Find a feasible multi-commodity flow over time with time horizon T satisfying the given demands.

We mention the following theorem on the complexity of the Multi-Commodity Flow over Time Problem without proof (see references at the end of this section).

Theorem 5.2. *For $k \geq 2$ commodities, the Multi-Commodity Flow over Time Problem is weakly NP-hard.*

It is not clear whether the Multi-Commodity Flow over Time Problem is contained in the class NP since the encoding size of a feasible solution might be exponential in the input size. This is an interesting open problem for future research.

On the other hand, the concept of time-expanded networks yields the following positive result on the complexity of the problem.

Theorem 5.3. *The Multi-Commodity Flow over Time Problem can be solved in pseudo-polynomial time.*

Proof. The Multi-Commodity Flow over Time Problem on network G can be reduced to a static multi-commodity flow problem on the time-expanded network G^T (see Definition 4.3). Static multi-commodity flows can be computed by general linear programming techniques in time polynomial in the size of the network.   □

As for the case of single-commodity flows over time, we consider multi-commodity flows over time with a simple structure and generalize the concept of temporally repeated flows.

Definition 5.4 (Temporally repeated multi-commodity flow). *A temporally repeated multi-commodity flow f with time horizon T is a collection of temporally repeated flows f^i with time horizon T, for each commodity $i = 1, \ldots, k$, such that the underlying s_i-t_i-flows x^i form a feasible multi-commodity flow x, that is, $\sum_{i=1}^{k} x_e^i \leq u_e$ for each $e \in E$.*

Lemma 5.5. *A temporally repeated multi-commodity flow f with time horizon T is a feasible multi-commodity flow over time with time horizon T.*

Proof. By definition, f^i is an s_i-t_i-flow over time with time horizon T, for $i = 1, \ldots, k$. Moreover, capacity constraints are satisfied since $f_e^i(\theta) \leq x_e^i$, for $e \in E$, $\theta \in [0, T)$, and thus

$$\sum_{i=1}^{k} f_e^i(\theta) \leq \sum_{i=1}^{k} x_e^i = x_e \leq u_e \quad \text{for } e \in E, \theta \in [0, T).$$

This concludes the proof.   □

We obtain the following approximation result.

Theorem 5.6. *If there is a feasible multi-commodity flow over time f with time horizon T satisfying demands $d_1, \ldots, d_k$, then there exists a temporally repeated multi-commodity flow with time horizon $2T$ satisfying demands $d_1, \ldots, d_k$. Moreover, such a temporally repeated multi-commodity flow can be computed in polynomial time.*

Proof. Consider a feasible multi-commodity flow over time f with time horizon T satisfying demands $d_1, \ldots, d_k$. We define a static multi-commodity flow $\bar{x}$ by averaging f over the entire time horizon, that is,

$$\bar{x}_e^i := \frac{1}{T} \cdot \int_0^T f_e^i(\theta) \, d\theta \quad \text{for } i = 1, \ldots, k, e \in E. \tag{7}$$

The feasibility of f implies that the capacity constraint $\sum_{i=1}^{k} f_e^i(\theta) \leq u_e$ holds for $e \in E$, $\theta \in [0, T)$. As a consequence, $\bar{x}$ also obeys capacity constraints:

$$\sum_{i=1}^{k} \bar{x}_e^i = \frac{1}{T} \cdot \int_0^T \sum_{i=1}^{k} f_e^i(\theta) \, d\theta \leq \frac{1}{T} \cdot \int_0^T u_e \, d\theta = u_e.$$

Similarly, since $\operatorname{ex}_{f^i}(v, T) = 0$ for $v \in V \setminus \{s_i, t_i\}$, $i = 1, \ldots, k$ (see Definition 2.3(c)) it follows that $\bar{x}^i$ obeys flow conservation:

$$\sum_{e \in \delta^-(v)} \bar{x}_e^i - \sum_{e \in \delta^+(v)} \bar{x}_e^i = \frac{1}{T} \cdot \left(\sum_{e \in \delta^-(v)} \int_0^T f_e^i(\theta) \, d\theta - \sum_{e \in \delta^+(v)} \int_0^T f_e^i(\theta) \, d\theta \right)$$

$$= \frac{\operatorname{ex}_{f^i}(v, T)}{T} = 0$$

for $v \in V \setminus \{s_i, t_i\}$, $i = 1, \ldots, k$. Applying the same argument for node t implies that $|\bar{x}^i| = |f^i|/T$ for $i = 1, \ldots, k$.

We have thus shown that $\bar{x}$ is a feasible multi-commodity flow with $|\bar{x}^i| = |f^i|/T$, for $i = 1, \ldots, k$. Moreover, for $i = 1, \ldots, k$,

$$\sum_{e \in E} \tau_e \cdot \bar{x}_e^i = \frac{1}{T} \cdot \sum_{e \in E} \tau_e \cdot \int_0^T f_e^i(\theta) \, d\theta. \tag{8}$$

If we set $c_e := \tau_e$ for each $e \in E$, the term on the right hand side of (8) is equal to $1/T$ times the cost $c(f^i)$ of the flow over time f^i; see (6). Since f^i has time horizon T, flow can only travel along paths of cost at most T. This means that each flow unit causes cost at most T and the total cost $c(f^i)$ is thus upper bounded by $T \cdot |f^i|$.

Putting things together we show that the temporally repeated multi-commodity flow with time horizon $2T$ corresponding to the feasible multi-commodity flow $\bar{x}$ satisfies demands $d_1, \ldots, d_k$. By Lemma 2.7 the flow value of the ith commodity in the temporally repeated multi-commodity flow is equal to

$$2T \cdot |\bar{x}^i| - \sum_{e \in E} \tau_e \cdot \bar{x}_e^i = 2 \cdot |f^i| - \frac{1}{T} \cdot c(f^i) \geq |f^i| \geq d_i,$$

for $i = 1, \ldots, k$. Thus, the temporally repeated multi-commodity flow with time horizon $2T$ corresponding to the feasible multi-commodity flow $\bar{x}$ has the desired properties. This concludes the proof of existence.

It remains to show that such a temporally repeated multi-commodity flow can be computed in polynomial time. The given proof of existence requires knowledge of the multi-commodity flow over time f which is NP-hard to compute. We can, however, compute a static multi-commodity flow x which mimics $\bar{x}$ and still yields the desired temporally repeated multi-commodity flow with time horizon $2T$. The task is to find a feasible multi-commodity flow x such that x^i is an s_i-t_i-flow with

$$2T \cdot |x^i| - \sum_{e \in E} \tau_e \cdot x_e^i \geq d_i, \tag{9}$$

for all $i = 1, \ldots, k$. This is a multi-commodity flow problem with linear cost constraints and can easily be formulated as a linear program of polynomial size. Moreover, as argued above, the temporally repeated multi-commodity flow with time horizon $2T$ corresponding to the feasible multi-commodity flow x has the desired properties and can be obtained from x in polynomial time. $\square$

The result in Theorem 5.6 implies an approximation algorithm with performance guarantee 2 for the Quickest Multi-Commodity Flow Problem.

QUICKEST MULTI-COMMODITY FLOW PROBLEM

Given: A network $G = (V, E)$ with capacities and transit times on the arcs; k commodities $i = 1, \ldots, k$, each with a source node $s_i \in V$, a sink node $t_i \in V$, and a demand $d_i \geq 0$.

Task: Find a feasible multi-commodity flow over time with minimum time horizon T satisfying the given demands.

Corollary 5.7. *There is an approximation algorithm with performance guarantee 2 for the Quickest Multi-Commodity Flow Problem.*

Proof (sketch). If one knew the minimum time horizon T^*, the algorithm described in the proof of Theorem 5.6 immediately yields a 2-approximate solution. The remaining problem is thus to find an appropriate estimate of T^*. The algorithm described in the proof of Theorem 5.6 can, for example, be embedded into a binary search for the minimum time horizon T such that a feasible static multi-commodity flow x satisfying (9) exists. It follows from the proof of Theorem 5.6 that this minimum value T is a lower bound on T^*. Standard binary search can, however, only determine the minimum value T up to a certain finite precision ϵ (notice that neither T nor T^* are in general integral). This yields an approximation algorithm with performance guarantee $2 + \epsilon$ for any fixed $\epsilon > 0$. In order to get rid of the additional ϵ in the performance guarantee, one can replace the binary search by a parametric search. We omit further details.   □

Pointers to the Literature

The NP-hardness result in Theorem 5.2 is due to Hall et al. (2007). They also show that the Multi-Commodity Flow over Time Problem with simple flow paths and strict flow conservation is strongly NP-hard. The presented approximation result for the Quickest Multi-Commodity Flow Problem is due to Fleischer and Skutella (2007). The described approach also works in a more general setting with costs by using length-bounded static flow computations. Martens and Skutella (2006) make use of this approach to approximate s-t-flows over time with a bound on the number of flow-carrying paths (k-*splittable flows over time*). Moreover, fully polynomial-time approximation schemes for various quickest flow problems can be obtained by using condensed time-expanded networks (Fleischer and Skutella 2007). More results on single-commodity quickest flows are described at the end of Sect. 21.2.

21.6 Exercises

Exercise 6.1. A simple upper bound on the value of a maximum s-t-flow over time with time horizon T can be obtained by multiplying the minimum capacity of a (static) s-t-cut by T. Prove that this upper bound can diverge from the maximum value of a feasible s-t-flow over time by an arbitrarily large factor.

Exercise 6.2. Consider the network G depicted in Fig. 21.8.

(a) Use the Ford–Fulkerson Algorithm in order to determine a maximum s-t-flow over time with time horizon $T = 18$.
(b) Determine a minimum capacity s-t-cut over time with time horizon $T = 18$.

Exercise 6.3. Consider a transshipment problem with a given vector of supplies and demands at the nodes of the network G. It is well-known that the static flow problem to satisfy all supplies and demands can be reduced to a maximum s-t-flow problem by introducing a super source s and a super sink t. Try to find a similar reduction for the case of flows over time and discuss the principal difficulties that occur.

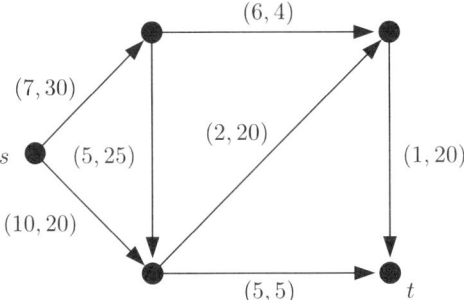

Fig. 21.8. An instance of the Maximum Flow over Time Problem. The arc labels indicate transit times and capacities, i.e., the label at arc e is (τ_e, u_e). The time horizon is $T = 18$

Exercise 6.4. Find a network such that no temporally repeated flow has the earliest arrival property.

Exercise 6.5. Use the Earliest Arrival Algorithm in order to determine an earliest arrival flow f with time horizon $T = 18$ in the network depicted in Fig. 21.8. Draw the graph of the function $\theta \mapsto \mathrm{ex}_f(t, \theta)$.

Exercise 6.6. Definition 3.1 of an earliest arrival flow can be generalized straightforwardly to a setting with multiple sources, each with a given supply. It turns out that an earliest arrival flow still exists in this case. On the other hand, we run into problems when there are multiple sinks $t_1, \ldots, t_k$ with given demands $d_1, \ldots, d_k$ that bound the excess of the sinks. That is, $\mathrm{ex}_f(t_i, \theta) \le d_i$ must hold for all i and θ. Construct an example with two sinks and one source for which no flow over time f maximizes $\mathrm{ex}_f(t_1, \theta) + \mathrm{ex}_f(t_2, \theta)$ for all θ simultaneously.

Exercise 6.7. An s-t-flow over time f with time horizon T is a *latest departure flow*, if it maximizes the amount of flow leaving the source after time θ for all $\theta \in [0, T]$ (subject to the constraint that all flow must arrive at t before time T). Prove that the flow over time computed by the Earliest Arrival Algorithm is a latest departure flow.

Exercise 6.8. We consider temporally repeated flows in the context of min-cost flows over time.

(a) Construct a network with two temporally repeated flows f and f' corresponding to two different path decompositions of the same static s-t-flow x such that $c(f) \ne c(f')$.
(b) Construct an example in which the cost of any temporally repeated flow with time horizon T and value d is larger than the minimum cost of an s-t-flow over time with time horizon T and value d.

Exercise 6.9. Show that an arbitrary s_0-$t_{(T-1)}$-cut with capacity $\delta < \infty$ in the time-expanded network G^T naturally induces an s-t-cut over time with time horizon T and capacity δ in G.

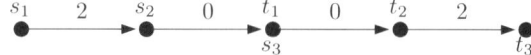

Fig. 21.9. An example with $k = 3$ commodities. Commodities 1 and 3 each have demand 1, commodity 2 has demand 2. The numbers at the arcs indicate transit times; all arcs have unit capacity

Exercise 6.10. It follows from Theorem 2.12 that there always exists a maximum s-t-flow over time obeying the strict flow conservation constraints. The purpose of this exercise is to show that this result does not hold in the more general setting with multiple commodities.

(a) Find an instance of the Multi-Commodity Flow over Time Problem such that any multi-commodity flow over time with time horizon T satisfying all demands uses storage of flow at intermediate nodes for at least one commodity.
   Hint: Consider the instance given in Fig. 21.9.
(b) Try to come up with an instance of the Quickest Multi-Commodity Flow Problem such that the ratio of the minimal possible time horizon with and without storage of flow at intermediate nodes, respectively, is as large as possible.
   Remark: This is an open research question. There is no instance known which achieves a larger ratio than the one depicted in Fig. 21.9. On the positive side, Theorem 5.6 implies that the ratio is always upper bounded by 2.

Acknowledgement

We thank an anonymous referee for many useful comments that helped to improve the presentation of the paper.

References

Ahuja, R.K., Magnanti, T.L., Orlin, J.B.: Network Flows. Theory, Algorithms, and Applications. Prentice Hall, Englewood Cliffs (1993)

Anderson, E.J., Nash, P.: Linear Programming in Infinite-Dimensional Spaces. Wiley, New York (1987)

Anderson, E.J., Nash, P., Philpott, A.B.: A class of continuous network flow problems. Math. Oper. Res. **7**, 501–514 (1982)

Anderson, E.J., Philpott, A.B.: Optimisation of flows in networks over time. In: Kelly, F.P. (ed.) Probability, Statistics and Optimisation, pp. 369–382. Wiley, New York (1994)

Aronson, J.E.: A survey of dynamic network flows. Ann. Oper. Res. **20**, 1–66 (1989)

Baumann, N., Skutella, M.: Solving evacuation problems efficiently: Earliest arrival flows with multiple sources. In: Proceedings of the 47th Annual IEEE Symposium on Foundations of Computer Science, Berkeley, CA, pp. 399–408 (2006)

Berlin, G.N.: The use of directed routes for assessing escape potential. National Fire Protection Association, Boston, MA (1979)

Burkard, R.E., Dlaska, K., Klinz, B.: The quickest flow problem. ZOR, Z. Oper.-Res. **37**, 31–58 (1993).

Chalmet, L.G., Francis, R.L., Saunders, P.B.: Network models for building evacuation. Manage. Sci. **28**, 86–105 (1982)

Fleischer, L., Skutella, M.: Minimum cost flows over time without intermediate storage. In: Proceedings of the 14th Annual ACM–SIAM Symposium on Discrete Algorithms, Baltimore, MD, pp. 66–75 (2003)

Fleischer, L., Skutella, M.: Quickest flows over time. SIAM J. Comput. **36**, 1600–1630 (2007)

Fleischer, L.K.: Faster algorithms for the quickest transshipment problem. SIAM J. Optim. **12**, 18–35 (2001)

Fleischer, L.K., Tardos, E.: Efficient continuous-time dynamic network flow algorithms. Oper. Res. Lett. **23**, 71–80 (1998)

Ford, L.R., Fulkerson, D.R.: Constructing maximal dynamic flows from static flows. Oper. Res. **6**, 419–433 (1958)

Ford, L.R., Fulkerson, D.R.: Flows in Networks. Princeton University Press, Princeton (1962)

Gale, D.: Transient flows in networks. Mich. Math. J. **6**, 59–63 (1959)

Hajek, B., Ogier, R.G.: Optimal dynamic routing in communication networks with continuous traffic. Networks **14**, 457–487 (1984)

Hall, A., Hippler, S., Skutella, M.: Multicommodity flows over time: Efficient algorithms and complexity. Theor. Comput. Sci. **379**, 387–404 (2007)

Hamacher, H.W., Tufecki, S.: On the use of lexicographic min cost flows in evacuation modeling. Nav. Res. Logist. **34**, 487–503 (1987)

Hoppe, B.: Efficient dynamic network flow algorithms. PhD thesis, Cornell University (1995)

Hoppe, B., Tardos, E.: Polynomial time algorithms for some evacuation problems. In: Proceedings of the 5th Annual ACM–SIAM Symposium on Discrete Algorithms, Arlington, VA, pp. 433–441 (1994)

Hoppe, B., Tardos, E.: The quickest transshipment problem. Math. Oper. Res. **25**, 36–62 (2000)

Jarvis, J., Ratliff, H.: Some equivalent objectives for dynamic network flow problems. Manage. Sci. **28**, 106–108 (1982)

Klinz, B., Woeginger, G.J.: Minimum-cost dynamic flows: The series-parallel case. Networks **43**, 153–162 (2004)

Korte, B., Vygen, J.: Combinatorial Optimization: Theory and Algorithms, 4th edn. Springer, Berlin (2008)

Kotnyek, B.: An annotated overview of dynamic network flows. Rapport de recherche 4936, INRIA Sophia Antipolis (2003)

Lovetskii, S.E., Melamed, I.I.: Dynamic network flows. Autom. Remote Control **48**, 1417–1434 (1987). Translated from Avtomatika i Telemekhanika **11**, 7–29 (1987)

Martens, M., Skutella, M.: Length-bounded and dynamic k-splittable flows. In: Haasis, H.-D., Kopfer, H., Schönberger, J. (eds.) Operations Research Proceedings 2005, pp. 297–302. Springer, Berlin (2006)

Megiddo, N.: Optimal flows in networks with multiple sources and sinks. Math. Program. **7**, 97–107 (1974)

Megiddo, N.: Combinatorial optimization with rational objective functions. Math. Oper. Res. **4**, 414–424 (1979)

Minieka, E.: Maximal, lexicographic, and dynamic network flows. Oper. Res. **21**, 517–527 (1973)

Philpott, A.B.: Continuous-time flows in networks. Math. Oper. Res. **15**, 640–661 (1990)

Powell, W.B., Jaillet, P., Odoni, A.: Stochastic and dynamic networks and routing. In: Ball, M.O., Magnanti, T.L., Monma, C.L., Nemhauser, G.L. (eds.) Network Routing. Handbooks in Operations Research and Management Science, vol. 8, Chap. 3, pp. 141–295. North-Holland, Amsterdam (1995)

Schrijver, A.: Combinatorial Optimization: Polyhedra and Efficiency. Springer, Berlin (2003)

Tjandra, S.: Dynamic network optimization with application to the evacuation problem. PhD thesis, Universität Kaiserslautern, Shaker Verlag, Aachen (2003)

Wilkinson, W.L.: An algorithm for universal maximal dynamic flows in a network. Oper. Res. **19**, 1602–1612 (1971)

Zadeh, N.: A bad network problem for the simplex method and other minimum cost flow algorithms. Math. Program. **5**, 255–266 (1973)

Edge-Connectivity Augmentations
of Graphs and Hypergraphs

Zoltán Szigeti

Summary. A. Frank (Augmenting graphs to meet edge-connectivity requirements, SIAM J. Discrete Math. **5**(1), 22–53, 1992) developed a method to solve edge-connectivity augmentation problems. His paper has stimulated further research in a number of directions, including many interesting generalizations.

This paper surveys the current State of the Art on the edge-connectivity augmentation problem. Recent extensions of the problem are presented for undirected graphs, hypergraphs and more generally for set functions. Shortened proofs are provided for some of the results. A list of open problems is also presented.

22.1 Introduction

In this paper all graphs and hypergraphs are undirected, directed versions of the problems will not be treated here. By graphs and hypergraphs we mean multi-graphs and multi-hypergraphs, that is, parallel edges and parallel hyperedges are allowed, however loops are forbidden. All the problems here will concern edge-connectivity, that is, vertex-connectivity will not be considered. We have to emphasize immediately that in our problems we may add edges between adjacent vertices, we may even add parallel edges. We remark that without these assumptions, that is, when the starting and also the resulting graph must be simple, the problem is NP-complete (Jordán 1997). The optimization problems will always be unweighted, the weighted version of the simplest problem already being NP-complete (Eswaran and Tarjan 1976).

The basic problem: The starting point for introducing edge-connectivity problems is the problem of increasing the reliability of a telephone network. We may associate a graph to the network: the telephone centers and the connections between them

Some part of this work was done while the author was visiting the Research Institute for Discrete Mathematics, University of Bonn, Lennéstrasse 2, 53113. Bonn, Germany by an Alexander von Humboldt fellowship.

are the vertices and the edges of the graph and the reliability of the network corresponds to the edge-connectivity of this graph. As a natural requirement, we may wish to increase the reliability of the network by constructing new connections between the centers. The optimization problem in the language of graphs is the global edge-connectivity augmentation problem in graphs, namely: Given a graph $G = (V, E)$ and a positive integer k, what is the minimum number γ of new edges whose addition results in a k-edge-connected graph? We show how to solve this problem, in doing so we introduce the key ideas to be applied throughout this paper.

The lower bound: First we provide a lower bound on γ. Suppose that G is not k-edge-connected. This is because there is a set X of degree $d(X)$ less than k. Then the deficiency of X is $k - d(X)$, that is, we must add at least $k - d(X)$ edges between X and $V - X$. Let $\{X_1, \ldots, X_l\}$ be a subpartition of V. The deficiency of this subpartition is the sum of the deficiencies of the X_i's. By adding a new edge we may decrease the deficiency of at most two X_i's so we may decrease the deficiency of the subpartition by at most two, hence we obtain the following lower bound:

$$\gamma \geq \alpha := \lceil \text{half of the maximum deficiency of a subpartition of } V \rceil. \qquad (1)$$

The minimax theorem (see Theorem 4.2), due to Watanabe and Nakamura (1987), saying that this lower bound can always be achieved, can be proved as follows:

Frank's algorithm: (1) *Minimal extension:* First, add a new vertex s to G and connect it to each vertex of G by k edges. The resulting graph is k-edge-connected in V. Secondly, delete as many new edges as possible preserving k-edge-connectivity in V. The graph obtained is denoted by $G' = (V, F')$. If the degree of s in G' is odd, then add an arbitrary new edge incident to s. Then we have a graph $G'' = (V + s, E \cup F'')$ that is k-edge-connected in V so that the degree of s is even.

(2) *Splitting off:* Now we will use the main operation of this paper, called splitting off. Splitting off a pair of edges sr, st incident to s means that we delete these two edges and we add a new edge rt. Applying Lovász's theorem 4.1(a), split off edge pairs incident to s, preserving k-edge-connectivity in V, as far as the degree of s becomes zero. This way we obtain a k-edge-connected graph $G^* = (V, E \cup F)$ with $|F| = \frac{|F''|}{2} = \lceil \frac{|F'|}{2} \rceil$.

Optimality: The optimality will be proved by the existence of a subpartition of V whose deficiency provides equality in (1). In G', no edge incident to s can be deleted without violating k-edge-connectivity in V, so each edge $e \in F'$ enters a maximal proper subset X_e in V of degree k, that is, $d_G(X_e) + d_{F'}(X_e) = k$. By the submodularity of the degree function d, these sets provide a subpartition $\{X_1, \ldots, X_l\}$ of V, for which we do have equality in (1):

$$\gamma \leq |F| = \left\lceil \frac{|F'|}{2} \right\rceil = \left\lceil \frac{1}{2} \sum_{1}^{l} d_{F'}(X_i) \right\rceil = \left\lceil \frac{1}{2} \sum_{1}^{l} (k - d_G(X_i)) \right\rceil \leq \alpha \leq \gamma.$$

Minimal extension: The above method of Frank has two main phases: the first one—minimal extension—consists of the first two steps, while the second one is the splitting off.

The first phase is in fact the construction of a graph $H = (V + s, F)$ with a minimum number of edges such that each edge of H is incident to s and H covers the deficiency function of G, that is, for every vertex set $X \subset V$, the number of edges of H leaving X is at least the deficiency of X in G with respect to k.

Frank (1992a) proved that this can be done not only for the deficiency function of a graph: such an optimal graph H can be constructed that covers a symmetric skew-supermodular function. This result (Theorem 3.2 in this paper) is not explicitly presented in Frank (1992a), it was published in Bang-Jensen et al. (1995).

This general result on extension implies that for an edge-connectivity augmentation problem, if the corresponding splitting off exists, then the optimization problem can be solved. Thus we will concentrate on splitting off results in this paper.

Generalizations: The above mentioned basic problem (that is the problem of augmenting global edge-connectivity of a graph by adding graph edges) and its solution capture already the most important ingredients of the theory and they provide a point of departure for studying more complex edge-connectivity augmentation problems. Lots of generalizations of Watanabe and Nakamura's result will be presented here. This paper is divided in three parts: results on graphs, on hypergraphs and on set functions. The results in the different parts are intimately related, we will see that a great number of results on graphs can be generalized for hypergraphs, which in turn can sometimes be further extended to "connectivity" functions.

The first part contains the following generalizations in *graphs*:

- local edge-connectivity augmentation (Frank 1992a),
- global edge-connectivity augmentation over symmetric parity families (Szigeti 2008b),
- node to area global edge-connectivity augmentation (Ishii and Hagiwara 2006),
- global edge-connectivity augmentation by attaching stars (B. Fleiner 2005),
- local edge-connectivity augmentation by attaching stars (Jordán and Szigeti 2003),
- global edge-connectivity augmentation with partition constraint (Bang-Jensen et al. 1999).

We present in the second part generalizations in *hypergraphs*:

- global edge-connectivity augmentation in hypergraphs by adding graph edges (Bang-Jensen and Jackson 1999),
- global edge-connectivity augmentation in hypergraphs by adding uniform hyperedges (T. Király 2004b),
- local edge-connectivity augmentation in hypergraphs by adding graph edges (we mention at once that this problem is NP-complete, Cosh et al. 2008),
- local edge-connectivity augmentation in hypergraphs by adding a hypergraph of minimum total size (Szigeti 1999).

The deficiency function with respect to global (resp. local) edge-connectivity in graphs and in hypergraphs is symmetric and crossing supermodular (resp. skew-supermodular). The third part is devoted to generalizations on such *set functions*:

- covering a symmetric crossing supermodular set function by a graph (Benczúr and Frank 1999),
- covering a symmetric crossing supermodular set function by a uniform hypergraph (T. Király 2004b),
- covering a symmetric skew-supermodular set function by a graph (this problem is NP-complete, Z. Király 2001),
- covering a symmetric semi-monotone set function by a graph (Ishii 2007),
- covering a symmetric skew-supermodular set function by a hypergraph of minimum total size (Szigeti 1999).

The main contribution of this paper is to call the reader's attention to Theorem 3.2 of Frank, to survey the results on the edge-connectivity augmentation problem, to provide short proofs for Theorem 4.11 on detachments satisfying local edge-connectivity requirements, and for Theorem 4.13 on partition constrained splitting off preserving global edge-connectivity, and finally, to present some open problems of the theory.

We finish this introduction by emphasizing that we do not attempt to cover all topics of the field, e.g. we do not focus on the design of efficient algorithms, but instead, we concentrate on minimax results of the area.

For further topics, such as Local edge-connectivity augmentation of mixed graphs, Global edge-connectivity augmentation preserving simplicity, Global edge-connectivity augmentation in a graph by adding edges within the members of a partition, Successive edge-connectivity augmentation, Simultaneous global edge-connectivity augmentation, we refer to Bang-Jensen et al. (1995), Bang-Jensen and Jordán (1998, 2000), Cheng and Jordán (1999) and Jordán (2003).

22.2 Definitions

This section is divided in three parts: definitions on graphs, on hypergraphs and finally on set functions.

Graph: Let $G = (V, E)$ be a graph. Recall that parallel edges are allowed. The set of all subpartitions of V will be denoted by $\mathcal{S}(V)$. For a vertex $v \in V$, $\Gamma(v)$ denotes the neighbours of v. For $X, Y \subset V$, G/X denotes the graph obtained from G by contracting X into one vertex, while the resulting parallel edges are kept, $d(X, Y)$ (resp. $\overline{d}(X, Y)$) denotes the number of edges between $X - Y$ and $Y - X$ (resp. $X \cap Y$ and $V - (X \cup Y)$), $d(X) = d(X, V - X)$. The set of edges leaving X is called a *cut* and is denoted by $\delta(X)$, that is, $d(X) = |\delta(X)|$. It is well-known and easy to check that, for all $X, Y \subseteq V$, (2) and (3) are satisfied.

$$d(X) + d(Y) = d(X \cap Y) + d(X \cup Y) + 2d(X, Y), \tag{2}$$

$$d(X) + d(Y) = d(X - Y) + d(Y - X) + 2\overline{d}(X, Y). \tag{3}$$

The *local edge-connectivity* between two different vertices x and y of G is defined by $\lambda_G(x, y) = \min\{d_G(X) : x \in X, y \notin X\}$, while $\lambda_G(x, x) = +\infty$. By

Menger's theorem, $\lambda_G(x, y)$ is the maximum number of edge disjoint paths in G between x and y. Let $G = (U, E)$ be a graph. For $X \subset U$, $x, y \in U - X$, $s \in U$, $u, v \in U - s$ we have

$$\lambda_{G/X}(x, y) \geq \lambda_G(x, y), \tag{4}$$

$$\lambda_{G-s}(u, v) \geq \lambda_G(u, v) - \left\lfloor \frac{d_G(s)}{2} \right\rfloor. \tag{5}$$

Indeed, if a cut Q separates x and y in G/X then Q also separates them in G and (4) follows. On the other hand, if $\mathcal{P}$ is a set of $\lambda_G(u, v)$ edge-disjoint paths in G then at most $\lfloor \frac{d_G(s)}{2} \rfloor$ of them may contain the vertex s and hence the other paths of $\mathcal{P}$ belong to $G - s$ and (5) follows.

A graph $G = (V, E)$ is called *k-edge-connected in U* (for some $k \in \mathbb{Z}_+$ and $U \subseteq V$) if $\lambda_G(x, y) \geq k \ \forall x, y \in U$. This definition will be usually used for $U = V$ or $V - s$ with a specified vertex s of V, in which case it is equivalent to (6). $\lambda(G)$ denotes the global edge-connectivity of G that is the maximum integer k so that G is k-edge-connected. Given a symmetric function $r : V \times V \to \mathbb{Z}_+$, we say that G is *r-edge-connected* if (7) is satisfied.

$$d_G(X) \geq k \qquad \forall \emptyset \neq X \subset U, \tag{6}$$

$$\lambda_G(u, v) \geq r(u, v) \quad \forall u, v \in V. \tag{7}$$

Let $G = (V + s, E)$ be a graph. By *splitting off* two edges sr, st we mean the operation that replaces sr, st by a new edge rt, the new graph will be denoted by G_{rt}. Note that we do not allow loops so if $r = t$, then the resulting loop must be deleted. A pair sr, st is called *k-admissible* if G_{rt} is k-edge-connected in V and it is called *λ-admissible* if (8) is satisfied. On several occasions, the splitting off will be called k-admissible instead of the edge pair. A sequence of splittings off is said to be *complete* if the degree of s becomes 0. A complete splitting off is called *k-admissible* (*λ-admissible*) if each splitting off in its sequence is k-admissible (λ-admissible, respectively).

$$\lambda_{G_{rt}}(x, y) \geq \lambda_G(x, y) \quad \forall x, y \in V. \tag{8}$$

Hypergraph: Let $\mathcal{G} := (V, E)$ be a *hypergraph*, that is E is a multiset of subsets of V. The element of E are called *hyperedges*. Note that parallel hyperedges are allowed. The above definitions of the degree function, local edge-connectivity and k-edge-connectivity can be naturally generalized for hypergraphs. Indeed, let $\mathcal{G} := (V, E)$ be a hypergraph. The degree $d_{\mathcal{G}}(X)$ of a vertex set X is defined as the number of hyperedges intersecting X and $V - X$. The *local edge-connectivity* between two different vertices x and y of $\mathcal{G}$ is defined by $\lambda_{\mathcal{G}}(x, y) = \min\{d_{\mathcal{G}}(X) : x \in X, y \notin X\}$, while $\lambda_{\mathcal{G}}(x, x) = +\infty$. $\mathcal{G}$ is called *k-edge-connected in U* (for some $k \in \mathbb{Z}_+$ and $U \subseteq V$) if $\lambda_{\mathcal{G}}(x, y) \geq k \ \forall x, y \in U$. We say that a hypergraph $\mathcal{H}$ *covers* a function p if

$$d_{\mathcal{H}}(X) \geq p(X) \quad \forall \emptyset \neq X \subset V. \tag{9}$$

Call a set X of V *tight* if $d_{\mathcal{H}}(X) = p(X)$. $c(\mathcal{G})$ denotes the number of connected components of $\mathcal{G}$. A hypergraph is *r-uniform* if each hyperedge is of size r. A graph is a 2-uniform hypergraph. $\mathcal{G}$ is called a *2–3 hypergraph* if each hyperedge is of size two or three. The operation $\Delta - Y$ replaces a given 3-hyperedge abc by the star qa, qb, qc of a new vertex q of degree three. The operation $Y - \Delta$ replaces the star qa, qb, qc of a given vertex q of degree three by a new 3-hyperedge abc. It is easy to check that the local edge-connectivities between the original vertices do not change after a $\Delta - Y$ or a $Y - \Delta$ operation.

Set function: Let $p : 2^V \to \mathbb{Z} \cup \{-\infty\}$ be a set function. The function p is called *supermodular* (*crossing supermodular*) if (10) holds for all $X, Y \subseteq V$ (for all $X, Y \subseteq V$ that are *crossing* that is $X \cap Y, X - Y, Y - X, V - (X \cup Y) \neq \emptyset$) and p is called *skew-supermodular* if at least one of (10) and (11) hold for all $X, Y \subseteq V$. We say that p is *symmetric* if (12) is satisfied for each $X \subseteq V$. A function $b : 2^V \to \mathbb{Z}$ is called *submodular* if $-b$ is supermodular. Note that the degree function $d_G(X)$ of a graph G is symmetric and, by (2), it is submodular.

$$p(X) + p(Y) \leq p(X \cap Y) + p(X \cup Y), \tag{10}$$

$$p(X) + p(Y) \leq p(X - Y) + p(Y - X), \tag{11}$$

$$p(X) = p(V - X). \tag{12}$$

Given a symmetric function $r : V \times V \to \mathbb{Z}_+$, let us define $R(X) := \max\{r(x, y): x \in X, y \in V - X\}$. It is known that R is skew-supermodular (see e.g. Frank 1992b).

Symmetric crossing supermodular functions generalize the deficiency function of a graph G concerning global edge-connectivity (that is the function $k - d_G(X)$ is symmetric crossing supermodular) and symmetric skew-supermodular functions generalize the deficiency function of a graph G concerning local edge-connectivity (that is the function $R(X) - d_G(X)$ is symmetric skew-supermodular).

22.3 Minimal Extension

We start this section with a typical lemma that shows how to use the skew-supermodularity of a function. Recall that if a graph H covers a set function p, then a set X of V is called tight if $d_H(X) = p(X)$.

Lemma 3.1. *Let $p : 2^V \to \mathbb{Z} \cup \{-\infty\}$ be a symmetric skew-supermodular set function. Let $H = (V + s, E)$ be a graph that covers p. If X and Y are tight sets and $X \cap Y \neq \emptyset$, then either (a) $X \cap Y$ and $X \cup Y$ are tight, or (b) $X - Y$ and $Y - X$ are tight and $\bar{d}(X, Y) = 0$.*

Proof. We may suppose that $X - Y \neq \emptyset \neq Y - X$ because otherwise (a) is trivially satisfied.

If p satisfies (11) for X and Y then, by (3) and (9),

$$p(X) + p(Y) = d_H(X) + d_H(Y)$$
$$= d_H(X - Y) + d_H(Y - X) + 2\overline{d}(X, Y)$$
$$\geq p(X - Y) + p(Y - X) + 0$$
$$\geq p(X) + p(Y),$$

so we have equality everywhere, implying (b).

Otherwise, $X \cup Y \neq V$ and p satisfies (10) for X and Y. Then, by (2) and (9),

$$p(X) + p(Y) = d_H(X) + d_H(Y)$$
$$\geq d_H(X \cap Y) + d_H(X \cup Y)$$
$$\geq p(X \cap Y) + p(X \cup Y)$$
$$\geq p(X) + p(Y),$$

so we have equality everywhere, implying (a). □

All the augmentation results of this paper will be obtained by applying the following general result of Frank and some suitable splitting off theorem. For the sake of completeness we provide the proof of this theorem.

Theorem 3.2 (Frank 1992a; Bang-Jensen et al. 1995). *Let $p : 2^V \to \mathbb{Z} \cup \{-\infty\}$ be a symmetric skew-supermodular function. Then the edgeless graph on V can be extended to a graph H by adding a new vertex s and γ edges incident to s so that H covers p if and only if*

$$\sum_{X \in \mathcal{X}} p(X) \leq \gamma \quad \forall \mathcal{X} \in \mathcal{S}(V). \tag{13}$$

Proof. Suppose that $H = (V + s, E)$ covers p, each edge of H is incident to s and $d_H(s) \leq \gamma$. Then, for any subpartition $\mathcal{X}$ of V, (13) is satisfied by

$$\sum_{X \in \mathcal{X}} p(X) \leq \sum_{X \in \mathcal{X}} d_H(X) \leq d_H(s) \leq \gamma.$$

Now suppose that (13) is satisfied. The desired graph is constructed as follows. First add a new vertex s to V and connect it to each vertex of V by $\max\{p(X) : X \subset V\}$ edges. Then, of course, this graph covers p. Secondly, delete as many edges as possible preserving that p is covered. Let H be the graph obtained. It remains to show that $d_H(s) \leq \gamma$.

No edge of H can be deleted, that is, each edge of H enters a tight set. Thus there exists a set $\mathcal{X} := \{X_1, \ldots, X_l\}$ of tight sets so that each edge enters some set X_i and $\sum_1^l |X_i|$ is minimal.

We claim that $\mathcal{X}$ is a subpartition of V. Suppose that $X_i \cap X_j \neq \emptyset$ for some $X_i, X_j \in \mathcal{X}$. By Lemma 3.1, either $X_i \cup X_j$ is tight, hence X_i and X_j can be replaced by $X_i \cup X_j$ or $X_i - X_j$ and $X_j - X_i$ are tight and $\overline{d}(X_i, X_j) = 0$ (that is no edge enters $X_i \cap X_j$), so X_i and X_j can be replaced by $X_i - X_j$ and $X_j - X_i$. In both cases we obtained a contradiction to the fact that $\sum_1^l |X_i|$ is minimal.

Thus, $\mathcal{X} \in \mathcal{S}(V)$, so we are done because, by (13),

$$d_H(s) = \sum_{X \in \mathcal{X}} d_H(X) = \sum_{X \in \mathcal{X}} p(X) \leq \gamma. \quad \square$$

22.4 Graphs

In this section we will present the problems on edge-connectivity augmentation in graphs and their solutions. As we mentioned in the introduction, the basic tool is splitting off. Each subsection is devoted to one problem, and it is divided in two parts: results on splitting off and then, applying Theorem 3.2, we read out the minimax result on augmentation.

22.4.1 Global Edge-Connectivity I

This section is about k-edge-connectivity for some $k \geq 2$. First we consider the operation splitting off: here the graph $G = (V + s, E)$ is k-edge-connected in V and we wish to reduce the graph (by splitting off an edge pair incident to s) in such a way that the graph remains k-edge-connected in V. Then we consider the augmentation problem where the aim is to make a graph k-edge-connected by adding new edges.

Splitting off Preserving Global Edge-Connectivity

The first result on splitting off is due to Lovász (1979). It concerns global edge-connectivity, namely it provides a sufficient condition for the existence of a k-admissible splitting off. More precisely, Lovász showed that if $k \geq 2$ and the degree of the special vertex s is even, then each edge belongs to a k-admissible pair at s. In Bang-Jensen et al. (1999) it was shown that each edge belongs to many k-admissible pairs.

Theorem 4.1. *Let $G = (V + s, E)$ be a k-edge-connected graph in V with $k \geq 2$ and $d(s)$ is even. Then:*

(a) (Lovász 1979) *each edge st belongs to a k-admissible pair at s.*
(b) (Bang-Jensen et al. 1999) *each edge st belongs to at least $\frac{d(s)}{2}$ (resp. $\frac{d(s)}{2} - 1$) k-admissible pairs at s if k is even (resp. odd).*

B. Fleiner (2005), and independently Bang-Jensen and Jackson (1999) proved that Theorem 4.1(a) is true for 2–3 hypergraphs containing no 3-hyperedges incident to s. The special case of Theorem 4.1(b), when G is Eulerian, was proved earlier by Jackson (1988).

We must emphasize that the above theorem is true only if $k \geq 2$. Let us consider the following example for $k = 1$: let the graph $G = (V + s, E)$ be the star of the vertex s of degree four. Then G is connected in V and there is no complete 1-admissible splitting off. To avoid this problem we will usually suppose that the connectivity requirement is at least two.

Augmentation of Global Edge-Connectivity by Adding Edges

The problem of *global edge-connectivity augmentation in graphs,* already introduced and also solved in the introduction, is the following: *Given a graph $G = (V, E)$ and an integer k, what is the minimum number γ of new edges whose addition results in a k-edge-connected graph?* In other words, we are looking for

$$\gamma := \min\{|E'| : d_{G+E'}(X) \geq k \ \forall \emptyset \neq X \subset V\}$$
$$= \min\{|E'| : d_{(V,E')}(X) \geq k - d_G(X) \ \forall \emptyset \neq X \subset V\}.$$

As the function $p(X) = k - d_G(X)$ is symmetric and, by (2), skew-supermodular, Theorems 3.2 and 4.1(a) imply the following theorem. (Cai and Sun 1989 also proved this result later using some splitting off technique.)

Theorem 4.2 (Watanabe and Nakamura 1987). *Let $G = (V, E)$ be a graph and $k \geq 2$. Then G can be made k-edge-connected by adding at most γ new edges if and only if*

$$\sum_{X \in \mathcal{X}} (k - d_G(X)) \leq 2\gamma \quad \forall \mathcal{X} \in \mathcal{S}(V). \tag{14}$$

We point out again that the case when $k = 1$ does not fit into this framework. It is obvious that in this case we have to add $l - 1$ new edges, where l is the number of connected components of the graph.

22.4.2 Local Edge-Connectivity I

In the above section we were interested in k-edge-connectivity, that is, in the minimum value of the local edge-connectivities over all pairs of vertices. This section concerns the problem where each value counts not just the minimum, that is, we wish to add new edges to a graph so that the local edge-connectivity be greater than or equal to a given requirement for each pair of vertices.

Splitting off Preserving Local Edge-Connectivity

In this section we summarize results on splitting off preserving local edge-connectivity. Mader (1978) generalized Lovász's result Theorem 4.1(a) for local edge-connectivity by showing that a λ-admissible pair always exists if the degree of the vertex s is different from 3 and roughly speaking G is 2-edge-connected. This result implies that at most three edges incident to s belong to no λ-admissible pair. Frank (1992b) improved this by showing that at most one edge incident to s belongs to no λ-admissible pair. In Szigeti (2008a) we characterized this edge. Since not every edge belongs to a λ-admissible pair, the best we may hope for is that there exists at least one edge that belongs to many λ-admissible pairs. In Szigeti (2008a) it is shown that the correct number is $\lfloor \frac{d(s)}{3} \rfloor$ and examples show that this result is best possible.

Theorem 4.3. *Let $G = (V + s, E)$ be a connected graph so that $d(s) \neq 3$ and no cut edge is incident to s.*

(a) (Mader 1978) *There exists a λ-admissible pair at s.*

(b) (Frank 1992b) *There exist $\lfloor \frac{d(s)}{2} \rfloor$ disjoint λ-admissible pairs at s. (Hence at most one edge incident to s belongs to no λ-admissible pair.)*

(c) (Szigeti 2008a) *An edge st belongs to no λ-admissible pair if and only if $d(s)$ is odd and there exist two disjoint sets $C_1, C_2 \subset V - t$ such that $d(C_i) = R(C_i)$ and $d(s, C_i) = \frac{d(s)-1}{2}$ for $i = 1, 2$. Moreover, for every $c_1 \in C_1 \cap \Gamma(s), c_2 \in C_2 \cap \Gamma(s)$, $\{sc_1, sc_2\}$ is a λ-admissible pair.*

(d) (Szigeti 2008a) *There exists an edge belonging to at least $\lfloor \frac{d(s)}{3} \rfloor$ λ-admissible pairs at s.*

Augmentation of Local Edge-Connectivity by Adding Edges

The problem of *local edge-connectivity augmentation in graphs* is defined as follows: *Given a graph G and a symmetric requirement function $r : V \times V \to \mathbb{Z}_+$, what is the minimum number γ of new edges whose addition results in an r-edge-connected graph?*

Note that, by taking r to be equal to k for each pair of vertices, this problem contains, as a special case, the global edge-connectivity augmentation problem in graphs.

Recall that $R(X) = \max\{r(x, y) : x \in X, y \in V - X\}$ is symmetric and skew-supermodular. We can reformulate the problem as follows: we look for

$$\gamma := \min\{|E'| : \lambda_{G+E'}(u, v) \geq r(u, v) \; \forall u, v \in V\}$$
$$= \min\{|E'| : d_{G+E'}(X) \geq R(X) \; \forall \emptyset \neq X \subset V\}$$
$$= \min\{|E'| : d_{(V,E')}(X) \geq R(X) - d_G(X) \; \forall \emptyset \neq X \subset V\}.$$

As $p(X) = R(X) - d_G(X)$ is a symmetric, skew-supermodular function, Theorems 3.2 and 4.3 imply the following theorem.

Theorem 4.4 (Frank 1992a). *Let $G = (V, E)$ be a graph and $2 \leq r(u, v) \in \mathbb{Z} \; \forall u, v \in V$. Then G can be made r-edge-connected by adding at most γ new edges if and only if*

$$\sum_{X \in \mathcal{X}} (R(X) - d_G(X)) \leq 2\gamma \quad \forall \mathcal{X} \in \mathcal{S}(V). \tag{15}$$

Note that the special case of Theorem 4.4 when $r(u, v) = k \; \forall u, v \in V$ is exactly Theorem 4.2.

22.4.3 Symmetric Parity Families

A family $\mathcal{F}$ of subsets of V is called a *symmetric parity family* if it satisfies the following three properties. (i) $\emptyset, V \notin \mathcal{F}$, (ii) if $A \in \mathcal{F}$, then $V - A \in \mathcal{F}$, (iii) if

$A, B \notin \mathcal{F}$ and $A \cap B = \emptyset$, then $A \cup B \notin \mathcal{F}$. Let $T \subseteq V$ be a set of even cardinality. A set X is called *T-odd* if $|T \cap X|$ is odd. If $G = (V, E)$ is a connected graph then (G, T) is called a *graft*. A cut $\delta_G(X)$ is called a *T-cut* if X is T-odd, more generally $\delta_G(X)$ is called an *$\mathcal{F}$-cut* if $X \in \mathcal{F}$. The most important examples of parity families are $\mathcal{F} := 2^V - \{\emptyset, V\}$ and $\mathcal{F} := \{X \subseteq V : X \text{ is T-odd}\}$.

In this section we will deal with $\mathcal{F}$-cuts.

Splitting off Preserving Global Edge-Connectivity over Symmetric Parity Families

Theorem 4.3(a) implies easily the following.

Theorem 4.5 (Szigeti 2008b). *Let $G = (V + s, E)$ be a graph so that $d(s) > 0$ is even and let $\mathcal{F}$ be a symmetric parity family on V. Suppose that for some $k \geq 2$,*

$$d(X) \geq k \quad \forall X \in \mathcal{F}. \tag{16}$$

Then there exists a pair of edges incident to s that can be split off without violating (16).

Augmentation of Global Edge-Connectivity over Symmetric Parity Families

In this section we solve the following *global edge-connectivity augmentation problem over a symmetric parity family: Given a graph $G = (V, E)$, a symmetric parity family $\mathcal{F}$ on V and an integer k, what is the minimum number γ of edges whose addition results in a graph in which each $\mathcal{F}$-cut contains at least k edges?*

A special case is the minimum T-cut augmentation problem: how many new edges must be added to a graph so that the minimum T-cut contains at least k edges? It also contains, as a special case, the global edge-connectivity augmentation problem in graphs.

For a symmetric parity family $\mathcal{F}$, a graph $G = (V, E)$ and $k \in \mathbb{Z}^+$, let $p(X) := k - d_G(X)$ if $X \in \mathcal{F}$, and $-\infty$ otherwise. The problem of edge-connectivity augmentation over a symmetric parity family can be reformulated as follows: what is the value

$$\gamma := \min\{|E'| : d_{G+E'}(X) \geq k \; \forall X \in \mathcal{F}\}$$
$$= \min\{|E'| : d_{(V,E')}(X) \geq p(X) \; \forall X \subseteq V\}?$$

It is easy to see that p is a symmetric skew-supermodular function, so Theorems 3.2 and 4.5 provide at once the following theorem.

Theorem 4.6 (Szigeti 2008b). *For a graph $G = (V, E)$, a symmetric parity family $\mathcal{F}$ on V and an integer $k \geq 2$, the minimum cardinality of an $\mathcal{F}$-cut can be augmented to k by adding at most γ edges if and only if*

$$\sum_{i=1}^{l} (k - d(X_i)) \leq 2\gamma \quad \forall \{X_1, \ldots, X_l\} \in \mathcal{S}(V) \text{ with } X_i \in \mathcal{F}. \tag{17}$$

By applying Theorem 4.6, for $\mathcal{F} = 2^V - \{\emptyset, V\}$ we get Theorem 4.2, and, for $\mathcal{F}$ being the set of T-odd subsets of V, we get the following theorem on T-cuts.

Theorem 4.7 (Szigeti 2008b). *For any graft (G, T), the minimum cardinality of a T-cut can be augmented to $k \geq 2$ by adding at most γ edges if and only if $\sum_{X \in \mathcal{X}} (k - d(X)) \leq 2\gamma$ for each subpartition $\mathcal{X}$ of V into T-odd sets.*

22.4.4 Node to Area Edge-Connectivity

The *node to area global edge-connectivity augmentation* problem can be defined as follows: *Given a graph $G = (V, E)$, a family $\mathcal{A}$ of sets $A \subseteq V$ (called areas), and a requirement function $r : \mathcal{A} \to \mathbb{Z}_+$, add a minimum number $Opt(r, G)$ of new edges to G so that the resulting graph contains $r(A)$ edge-disjoint paths from any area A to any vertex $v \notin A$.*

Note that, by taking just one vertex as the family of areas and k as the requirement for this vertex, then we get as a special case the global edge-connectivity augmentation problem in graphs.

Let us define $P_{\mathcal{A}}(X) = \max\{r(A) : A \in \mathcal{A}, A \cap X = \emptyset$ or $A \subseteq X\}$ if $V \neq X \neq \emptyset$ and $P_{\mathcal{A}}(V) = P_{\mathcal{A}}(\emptyset) = 0$ and $Q_{\mathcal{A}}(G) := \max\{\sum_{X \in \mathcal{X}} q_{\mathcal{A}}(X) : \mathcal{X} \in \mathcal{S}(V)\}$, where $q_{\mathcal{A}}(X) = P_{\mathcal{A}}(X) - d_G(X)$.

Now we can provide a lower bound for the optimal value, namely $Opt(r, G) \geq \lceil \frac{Q_{\mathcal{A}}(G)}{2} \rceil$. The question is whether we have always equality here or not. Usually equality will hold, unless the graph contains a special configuration.

The node to area global edge-connectivity augmentation problem is a special case of the symmetric semi-monotone function covering problem treated in Sect. 22.6.2, so Theorem 6.6 implies the following. We mention that the above defined $P_{\mathcal{A}}(X)$ is the symmetric semi-monotone function to be covered.

Theorem 4.8 (Ishii and Hagiwara 2006). *Let $G = (V, E)$ be a graph, $\mathcal{A}$ a family of sets $A \subseteq V$, and $r : \mathcal{A} \to \mathbb{Z}_+$ a requirement function so that $r(A) \neq 1 \, \forall A \in \mathcal{A}$. If G contains no $\mathcal{A}$-configuration, then $Opt(r, G) = \lceil \frac{Q_{\mathcal{A}}(G)}{2} \rceil$, otherwise $Opt(r, G) = \lceil \frac{Q_{\mathcal{A}}(G)}{2} \rceil + 1$.*

The definition of a $\mathcal{A}$-configuration can be found in Ishii and Hagiwara (2006), where it is called P-property. We mention that if $\mathcal{A} = \{v\}$ and $r(v) = k$, then no $\mathcal{A}$-configuration can exist, so Theorem 4.8 implies Theorem 4.2. We remark that without the condition $r(A) \neq 1 \, \forall A \in \mathcal{A}$, the problem is NP-complete, see Sect. 22.6.2.

22.4.5 Global Edge-Connectivity II

Let us return to global edge-connectivity. First we generalize the operation splitting off, by introducing detachment, and then we consider the problem where we wish again to make a graph k-edge-connected, but this time by attaching stars. The essential tool to solve this problem is exactly the operation detachment.

Detachments Preserving Global Edge-Connectivity

In this section we generalize the operation splitting off.

Let $G = (V + s, E)$ be a graph. A *degree specification* for s is a sequence $f(s) = (d_1, \ldots, d_p)$ of positive integers with $\sum_{j=1}^{p} d_j = d_G(s)$. An $f(s)$-*detachment* of G at s is the graph G' obtained from G by replacing s by a set $s_1, \ldots, s_p$ of independent vertices and distributing the edges incident to s among them in such a way that $d_{G'}(s_i) = d_i$ ($1 \leq i \leq p$). Note that all the other ends of the edges in G remain the same.

Let us mention that a splitting off can really be considered as a special case of detachment, namely if after a splitting off we subdivide the new edge then the new graph is a $(2, d(s) - 2)$-detachment and this operation does not change the local edge-connectivities between the original vertices.

The following beautiful theorem of B. Fleiner characterizes graphs that have a k-edge-connected $f(s)$-detachment.

Theorem 4.9 (B. Fleiner 2005). *Let $G = (V + s, E)$ be a graph, $2 \leq k \in \mathbb{Z}$ and $f(s) = (d_1, \ldots, d_p)$ a degree specification for s with $d_i \geq 2 \ \forall i$. Then there exists an $f(s)$-detachment of G that is k-edge-connected in V if and only if*

$$G \ is \ k\text{-}edge\text{-}connected \ in \ V, \tag{18}$$

$$G - s \ is \ (k - \sum_{1}^{p} \lfloor \tfrac{d_i}{2} \rfloor)\text{-}edge\text{-}connected. \tag{19}$$

We note that the special case of Theorem 4.9, when each d_i is even, is equivalent to Theorem 4.1(a). Indeed, in this case the condition (19) is automatically satisfied by (5) and hence Theorem 4.9 implies Theorem 4.1(a). On the other hand, by Theorem 4.1(a), there exists a complete k-admissible splitting off, subdividing each new edge by a vertex and combining the suitable number of new vertices, an $f(s)$-detachment is obtained which is k-edge-connected in V by (4), and hence Theorem 4.1(a) implies this special case of Theorem 4.9.

The proof technique of B. Fleiner (2005) needed the more general framework of 2–3 hypergraphs. He proved that Theorem 4.9 is true for 2–3 hypergraphs containing no 3-hyperedges incident to s.

We give a generalization of this result in Sect. 22.4.6, for which we will provide a short proof in the appendix, much shorter than the original proof of B. Fleiner of Theorem 4.9.

Augmentation of Global Edge-Connectivity by Attaching Stars

By *attaching a star* of degree d to a graph G, we mean adding a new vertex and connecting it to some vertices of G so that the degree of the new vertex becomes d. The problem of this section is the *global edge-connectivity augmentation problem of a graph by attaching stars* that can be defined as follows: *Given a graph $G = (V, E)$ and integers $k, d_1, \ldots, d_p$, decide whether it is possible to attach p stars to G with degrees $d_1, \ldots, d_p$ to have a k-edge-connected graph in V.*

Note that, by taking each d_i to be equal to 2, we get as a special case the global edge-connectivity augmentation problem in a graph.

The solution of the problem of this section is given in the following theorem of B. Fleiner which is implied by Theorems 3.2 and 4.9. It might be advantageous to notice that condition (20) and the fact that $k - d(X)$ is symmetric skew-supermodular guarantees that the minimal extension can be made by $\sum_{j=1}^{p} d_j$ edges, while condition (21) allows us to find the suitable detachment.

Theorem 4.10 (B. Fleiner 2005). *A graph $G = (V, E)$ can be made k-edge-connected ($k \geq 2$) by attaching p stars with degrees $d_1, \ldots, d_p$ ($d_i \geq 2 \ \forall i$) if and only if*

$$\sum_{X \in \mathcal{X}} (k - d(X)) \leq \sum_{j=1}^{p} d_j \quad \forall \mathcal{X} \in \mathcal{S}(V), \tag{20}$$

$$k - \sum_{j=1}^{p} \left\lfloor \frac{d_j}{2} \right\rfloor \leq \lambda(G). \tag{21}$$

Note that, for $d_i = 2 \ \forall i$, (21) is satisfied by (20). Indeed, for $X \subset V$ with $d(X) = \lambda(X)$, by (20) for $\{X, V - X\}$, $(k - d(X)) + (k - d(V - X)) \leq 2p$, so by the symmetry of $d(X)$, $k - p \leq d(X) = \lambda(G)$, that is, (21) is satisfied. Hence Theorem 4.10 implies Theorem 4.2.

22.4.6 Local Edge-Connectivity II

In this section we generalize the results of the previous section. First we extend the Theorem of B. Fleiner to the case when local edge-connectivity is involved, and then we solve the problem where we wish to attach stars of given degree to a graph so that the local edge-connectivity be greater than or equal to a given requirement for each pair of vertices.

Detachments Preserving Local Edge-Connectivity

Recall that $G = (V + s, E)$ is r-edge-connected if $\lambda_G(u, v) \geq r(u, v) \ \forall u, v \in V$, where $r : V \times V \to \mathbb{Z}_+$ is a symmetric requirement function. In the next result we characterize graphs that have an r-edge-connected $f(s)$-detachment.

Note that, by taking r to be equal to k for each pair of vertices, we get a characterization of graphs having a k-edge-connected $f(s)$-detachment.

Theorem 4.11 (Jordán and Szigeti 2003). *Let $G = (V + s, E)$ be a graph, r a symmetric requirement function on V with $r(u, v) \geq 2 \ \forall u, v \in V$ and $f(s) = (d_1, \ldots, d_p)$ a degree specification for s with $d_i \geq 2 \ \forall i$. Let $\varphi = \sum_{1}^{p} \lfloor \frac{d_i}{2} \rfloor$. Then there exists an r-edge-connected $f(s)$-detachment of G if and only if*

$$G \text{ is } r\text{-edge-connected}, \tag{22}$$

$$G - s \text{ is } (r - \varphi)\text{-edge-connected}. \tag{23}$$

The special case of Theorem 4.11, when $r(u, v) = k \; \forall u, v \in V$, is Theorem 4.9, and when $r(u, v) = \lambda_G(u, v) \; \forall u, v \in V$, provides a characterization of the existence of an $f(s)$-detachment that preserves local edge-connectivities, while when at most one d_i is odd, is equivalent to Theorem 4.3(a).

We will provide a short proof of the above theorem in Sect. A.1. It will use Theorem 4.3(a) so it does not provide a new proof for Theorem 4.3(a). Note that it provides a short proof for Theorem 4.9.

Augmentation of Local Edge-Connectivity by Attaching Stars

In this section we solve the following *global edge-connectivity augmentation problem of a graph by attaching stars*: *Given a graph $G = (V, E)$, a symmetric requirement function $r : V \times V \to \mathbb{Z}_+$ and integers $d_1, \ldots, d_p$, decide whether it is possible to attach p stars to G with degrees $d_1, \ldots, d_p$, so as to obtain an r-edge-connected graph in V.*

Note that, by taking r to be equal to k for each pair of vertices, we get as a special case the global edge-connectivity augmentation problem of a graph by attaching stars, and by taking each d_i to be equal to 2 but r being arbitrary, we get the local edge-connectivity augmentation problem.

Recall that $R(X) = \max\{r(x, y) : x \in X, y \in V - X\}$ and that the function $R(X) - d(X)$ is symmetric skew-supermodular. Then Theorems 3.2 and 4.11 imply at once the following theorem. As for the global case, we may notice that condition (24) guarantees that the minimal extension can be made by $\sum_{j=1}^{p} d_j$ edges, while condition (25) allows us to find the suitable detachment.

Theorem 4.12 (Jordán and Szigeti 2003). *Let $G = (V, E)$ be a graph, $r : V \times V \to \mathbb{Z}_+$ a symmetric requirement function with $r(u, v) \geq 2 \; \forall u, v \in V$. Then G can be made r-edge-connected by attaching p stars with degrees $d_1, \ldots, d_p$ ($d_i \geq 2 \; \forall i$) if and only if*

$$\sum_{X \in \mathcal{X}} (R(X) - d(X)) \leq \sum_{j=1}^{p} d_j \qquad \forall \mathcal{X} \in \mathcal{S}(V), \tag{24}$$

$$r(u, v) - \sum_{j=1}^{p} \left\lfloor \frac{d_j}{2} \right\rfloor \leq \lambda_G(u, v) \quad \forall u, v \in V. \tag{25}$$

Note that, if $r(u, v) = k \; \forall u, v \in V$, then (24) and (25) are equivalent to (20) and (21), so Theorem 4.12 implies Theorem 4.10, and if $d_i = 2 \; \forall i$, then (24) implies (25) (that can be shown similarly, as it was shown for the global case) and hence Theorem 4.12 implies Theorem 4.4. Hence, our result is a common generalization of B. Fleiner's theorem on global edge-connectivity augmentation by attaching stars and Frank's theorem on local edge-connectivity augmentation by adding edges.

22.4.7 Global Edge-Connectivity with Partition Constraints

The aim of this section is to present the solution of the problem of global edge-connectivity augmentation in bipartite graphs. In fact, we consider a more general setting, namely the *global edge-connectivity augmentation problem of a graph with partition constraints: we want to make an arbitrary graph k-edge-connected by adding a minimum number of new edges between different members of a given partition of the vertex set.*

Splitting off Preserving Global Edge-Connectivity with Partition Constraints

Let $G = (V + s, E)$ be a graph with a specified vertex s of even degree, and $\mathcal{P} = \{P_1, \ldots, P_r\}$ a partition of V (resp. $\delta(s)$). If we have a partition of V, then obviously we may define a partition of $\delta(s)$, hence the second form is more general than the first one. A splitting off $\{su, sv\}$ is called $\mathcal{P}$-*allowed* if u and v (resp. su and sv) belong to different members of $\mathcal{P}$. A k-admissible $\mathcal{P}$-allowed splitting off will be called *allowed*. We wish to characterize graphs and partitions for which there exists a complete allowed splitting off, that is, we are interested in a complete splitting off, that is, at the same time $\mathcal{P}$-allowed and k-admissible. Notice that a complete $\mathcal{P}$-allowed splitting off exists if and only if $d(s, P_i) \leq \frac{d(s)}{2}$ $\forall 1 \leq i \leq r$, while, as it is already known for us, a complete k-admissible splitting off exists if and only if G is k-edge-connected in V. Are these conditions together sufficient to have an allowed complete splitting off? We will answer this question in this section.

To show the difficulties of our problem, suppose we wish to make 3-edge-connected a 4-cycle C_4. Clearly, this can be done by adding two edges. Note that the optimal solution is unique, we have to transform the graph into a K_4. Now suppose that we have an additional condition, we must maintain the bipartiteness of C_4. Then this solution is not feasible any more. In this case we have to add 3 edges. Let us reformulate this difficulty in terms of splitting off. Let G be the graph obtained from C_4 by adding a new vertex s and connecting s to all the four vertices. Let $\mathcal{P}$ be the bipartition of C_4. Then G admits no complete 3-admissible $\mathcal{P}$-allowed splitting off. The essential properties of this example are kept in a more general structure called C_4-obstacle, defined as follows.
A partition $\{A_1, A_2, A_3, A_4\}$ of V is called a C_4-*obstacle* of G if k is odd and

$$d(A_i) = k \qquad \forall 1 \leq i \leq 4, \tag{26}$$

$$d(A_i, A_{i+2}) = 0 \qquad \forall 1 \leq i \leq 2, \tag{27}$$

$$|P_l| = \frac{d(s)}{2} \qquad \exists 1 \leq l \leq r, \tag{28}$$

$$\delta(A_j \cup A_{j+2}) \cap \delta(s) = P_l \qquad \exists 1 \leq j \leq 2. \tag{29}$$

Another difficulty may turn up if the partition contains more than two members. Suppose we wish to make 3-edge-connected a 6-cycle C_6. Clearly, this can be done by adding three edges. Note that, though the optimal solution is not unique, we have to add at least one diagonal edge. Now suppose that we have the additional condition that we must maintain the non-adjacency of the opposite vertices of C_6. Then this solution is not feasible any more. In this case we have to add 4 edges. Let us reformulate this difficulty in terms of splitting off. Let G be the graph obtained from C_6 by adding a new vertex s and connecting s to all the six vertices. Let $\mathcal{P}$ the 3-partition of C_6 where the sets consist of the opposite vertices of C_6. Then G admits no complete 3-admissible $\mathcal{P}$-allowed splitting off. As before, the essential properties of this example are kept in a more general structure called C_6-obstacle, defined as follows.

A partition $\{A_1, \ldots, A_6\}$ of V is called a C_6-obstacle of G if k is odd and

$$d(A_i) = k \qquad \forall 1 \le i \le 6, \tag{30}$$

$$d(A_i, A_{i+1}) = \frac{k-1}{2} \qquad \forall 1 \le i \le 6 \ (A_7 = A_1), \tag{31}$$

$$d(s, A_i) = 1 \qquad \forall 1 \le i \le 6, \tag{32}$$

$$\delta(A_j \cup A_{j+3}) \cap \delta(s) = P_{l_j} \qquad \forall 1 \le j \le 3, \exists 1 \le l_j \le r. \tag{33}$$

We must emphasize that these difficulties may exist only if the target edge-connectivity is odd, and if one of them exists, then no complete allowed splitting off may exist. Now we are in a position to provide a characterization of the existence of a complete allowed splitting off.

Theorem 4.13 (Bang-Jensen et al. 1999). *Let $G = (V + s, E)$ be a graph with $d(s)$ even, $2 \le k \in \mathbb{Z}$, and $\mathcal{P} = \{P_1, \ldots, P_r\}$ a partition of V. Then there exists a complete k-admissible $\mathcal{P}$-allowed splitting off at s if and only if*

$$G \text{ is } k\text{-edge-connected in } V, \tag{34}$$

$$d(s, P_i) \le \frac{d(s)}{2} \qquad \forall 1 \le i \le r, \tag{35}$$

$$G \text{ contains no } C_4\text{- or } C_6\text{-obstacle.} \tag{36}$$

In the special case, when each element of $\mathcal{P}$ is a singleton, no C_4 or C_6-obstacle can exist, thus Theorem 4.13 implies Theorem 4.1(a), while when $|\mathcal{P}| = 2$, it provides (with Theorem 3.2) a solution for the problem of global edge-connectivity augmentation in bipartite graphs. We may observe that in this case no C_6-obstacle can exist.

The following result from (Szigeti 2004b) is a slight generalization of Theorem 4.13. The motivation of this form is that it allows us to contract tight sets and hence it enables us to simplify the proof that will be presented in Sect. A.2.

Theorem 4.14 (Szigeti 2004b). *Let $G = (V + s, E)$ be a graph with $d(s)$ even, $2 \le k \in \mathbb{Z}$, and $\mathcal{P} = \{P_1, \ldots, P_r\}$ a partition of $\delta(s)$. Then there exists a complete k-admissible $\mathcal{P}$-allowed splitting off at s if and only if (34), (36) and the following condition are satisfied.*

$$|P_i| \le \frac{d(s)}{2} \qquad \forall 1 \le i \le r. \tag{37}$$

Augmentation of Global Edge-Connectivity with Partition Constraints

We present a more precise reformulation of the problem of this section: *Given a graph $G = (V, E)$, an integer k and a partition $\mathcal{P} = \{P_1, \ldots, P_r\}$ of V, what is the minimum number $OPT^k_{\mathcal{P}}$ of $\mathcal{P}$-allowed edges whose addition results in a k-edge-connected graph?*

Note that, by taking the partition of all singletons in this problem, we get as a special case the global edge-connectivity augmentation problem in graphs.

The following theorem answers this problem. Let $\Phi := \max\{\alpha, \beta_1, \ldots, \beta_r\}$ where

$$\alpha := \max \left\{ \left\lceil \sum_{X \in \mathcal{X}} \frac{k - d(X)}{2} \right\rceil : \mathcal{X} \in \mathcal{S}(V) \right\},$$

$$\beta_j := \max \left\{ \sum_{Y \in \mathcal{Y}} (k - d(Y)) : \mathcal{Y} \in \mathcal{S}(P_j) \right\} \quad \forall 1 \leq j \leq r.$$

It is crucial to point out that Φ is a lower bound for the optimal value, that is,

$$OPT^k_{\mathcal{P}} \geq \Phi. \tag{38}$$

Indeed, by adding an edge to G we may decrease the deficiency $k - d(Z)$ of at most two sets in $\mathcal{X}$ and of at most one set in each $\mathcal{Y}$, and hence we can decrease Φ by at most one.

We will have equality in (38), unless G contains one of the following two configurations. These are the structures that force us to have a C_4- or C_6-obstacle in any optimal extension.

A partition $\{A_1, A_2, A_3, A_4\}$ of V is called a C_4-*configuration* of G if k is odd and

$$d(A_i) < k \qquad \forall 1 \leq i \leq 4, \tag{39}$$

$$d(A_i, A_{i+2}) = 0 \qquad \forall 1 \leq i \leq 2, \tag{40}$$

$$\sum_{X \in \mathcal{X}_i} (k - d(X)) = k - d(A_i) \quad \exists \mathcal{X}_i \in \mathcal{S}(A_i) \, \forall 1 \leq i \leq 4, \tag{41}$$

$$\mathcal{X}_j \cup \mathcal{X}_{j+2} \in \mathcal{S}(P_l) \qquad \exists 1 \leq l \leq r \, \exists 1 \leq j \leq 2, \tag{42}$$

$$k - d(A_i) + k - d(A_{i+2}) = \Phi \qquad \forall 1 \leq i \leq 2. \tag{43}$$

A partition $\{A_1, A_2, \ldots, A_6\}$ of V is called a C_6-*configuration* of G if k is odd and

$$d(A_i) = k - 1 \quad \forall 1 \leq i \leq 6, \tag{44}$$

$$d(A_i, A_{i+1}) = \frac{k - 1}{2} \quad \forall 1 \leq i \leq 6 \, (A_7 = A_1), \tag{45}$$

$$\Phi = 3, \tag{46}$$

$$d(A'_i) = k - 1 \quad \exists 1 \leq j_1, j_2, j_3 \leq r \, \forall 1 \leq i \leq 6,$$
$$\exists A'_i \subseteq A_i \cap P_{j_{i-3\lfloor \frac{(i-1)}{3} \rfloor}}, \tag{47}$$

where the j_i's must be different in (47).

Since these configurations force us to have an obstacle in any optimal extension, in which case there exists no complete allowed splitting off in the extended graph, the existence of a configuration implies that the optimal solution must contain at least $\Phi + 1$ edges. We show that this can be achieved. Using Theorems 3.2 and 4.13 we can prove with some effort the following result.

Theorem 4.15 (Bang-Jensen et al. 1999). *Let $G = (V, E)$ be a connected graph, $\mathcal{P} = \{P_1, \ldots, P_r\}$ a partition of V and $k \geq 2$. Then G can be made k-edge-connected by adding Φ $\mathcal{P}$-allowed edges (that is $OPT_{\mathcal{P}}^k = \Phi$) unless G contains a C_4- or C_6-configuration when we need one more edge (that is $OPT_{\mathcal{P}}^k = \Phi + 1$).*

As the global edge-connectivity augmentation problem *without partition constraints* can be considered as one with partition constraints where each element of $\mathcal{P}$ is a singleton, in which case no C_4- or C_6-configuration can exist and $\Phi = \alpha$, Theorem 4.15 implies Theorem 4.2. We note that if we want to augment the global edge-connectivity of a *bipartite graph* then $|\mathcal{P}| = 2$ and hence a C_6-configuration can not exist.

22.5 Hypergraphs

In this section we wish to present problems on edge-connectivity augmentation in hypergraphs and their solutions. We continue with the same structure as before: each subsection is devoted to one problem, and it is divided in two parts; results on splitting off and then the minimax result on augmentation. We mention that some of the problems that were easy in graphs turn out to be already too difficult in hypergraphs.

22.5.1 Augmentation of Global Edge-Connectivity in a Hypergraph by Adding Graph Edges

In this section we consider the problem of *global edge-connectivity augmentation in hypergraphs by adding graph edges*, that can be formulated as follows: *Given a hypergraph $\mathcal{G}$ and $k \in \mathbb{Z}^+$, what is the minimum number of graph edges (hyperedges of size two) whose addition results in a k-edge-connected hypergraph?*

Of course, if the hypergraph is in fact a graph, then the problem reduces to the global edge-connectivity augmentation problem in graphs.

One of the main difficulties with hypergraphs is that we have to handle the case when $k = 1$. Why is it so? Because by deleting a hyperedge the number of connected components may increase by a large value. This discussion shows the necessity of condition (49).

The corresponding splitting off result is the following.

Theorem 5.1 (Bang-Jensen and Jackson 1999). *Let $\mathcal{G} = (V + s, \mathcal{E})$ be a hypergraph and $\gamma, k \in \mathbb{Z}^+$ so that $d(s) = 2\gamma$ and each edge incident to s is of size two. Then there is a complete k-admissible splitting off at s if and only if*

$$\mathcal{G} \text{ is } k\text{-edge-connected in } V, \tag{48}$$

$$c(\mathcal{G} - s - \mathcal{H}) - 1 \leq \gamma \quad \forall \mathcal{H} \subseteq \mathcal{E}, |\mathcal{H}| \leq k - 1. \tag{49}$$

If $\mathcal{G}$ is a graph, then the second condition of the above theorem is satisfied (after deleting $k - 1$ edges in a k-edge-connected graph, the graph remains connected) and hence Theorem 5.1 implies Theorem 4.1(a).

Theorems 3.2 and 5.1 imply the following theorem that solves the problem of this section. We notice that (50) and the fact that the function $k - d_{\mathcal{H}}(X)$ is symmetric skew-supermodular guarantee that the optimal extension (with respect to k) can be made by 2γ edges and (51) allows us to find the complete k-admissible splitting off.

Theorem 5.2 (Bang-Jensen and Jackson 1999). *Let $\mathcal{G}$ be a hypergraph and $k \in \mathbb{Z}^+$. Then $\mathcal{G}$ can be made k-edge-connected by adding at most γ new edges (hyperedges of size two) if and only if*

$$\sum_{X \in \mathcal{X}} (k - d_{\mathcal{G}}(X)) \leq 2\gamma \quad \forall \mathcal{X} \in \mathcal{S}(V), \tag{50}$$

$$c(\mathcal{G} - \mathcal{H}) - 1 \leq \gamma \quad \forall \mathcal{H} \subseteq E(\mathcal{G}), |\mathcal{H}| \leq k - 1. \tag{51}$$

As above, if $\mathcal{G}$ is a graph, then (51) is implied by (50) and hence Theorem 5.2 implies Theorem 4.2.

22.5.2 Augmentation of Global Edge-Connectivity of a Hypergraph by Adding a Uniform Hypergraph

As a natural generalization of the problem of the previous section, one can consider the *global edge-connectivity augmentation problem of a hypergraph by adding hyperedges of the same size*, namely: *Given a hypergraph $\mathcal{G}$ and $k, r \in \mathbb{Z}^+$, what is the minimum number of hyperedges of size r whose addition results in a k-edge-connected hypergraph?*

Since the function $k - d_{\mathcal{H}}(X)$ is symmetric crossing supermodular, the problem of covering a symmetric crossing supermodular function by a uniform hypergraph contains this problem as a special case, thus Theorem 6.3 implies the following theorem. We mention that it is not easy to see that (54) implies (65) when $p(X) = k - d_{\mathcal{G}}(X)$.

Theorem 5.3 (T. Király 2004b). *Let $\mathcal{G} = (V, \mathcal{E})$ be a hypergraph and $k, r \in \mathbb{Z}_+$, $r \leq |V|$. Then $\mathcal{G}$ can be made k-edge-connected by adding at most γ new hyperedges of size r if and only if*

$$\sum_{X \in \mathcal{X}} (k - d_{\mathcal{G}}(X)) \leq r\gamma \qquad \forall \mathcal{X} \in \mathcal{S}(V), \tag{52}$$

$$k - d_{\mathcal{G}}(X) \leq \gamma \qquad \forall X \subseteq V, \tag{53}$$

$$c(\mathcal{G} - \mathcal{H}) - 1 \leq (r - 1)\gamma \quad \forall \mathcal{H} \subseteq \mathcal{E}, |\mathcal{H}| = k - 1. \tag{54}$$

The special case when we want to augment the edge-connectivity from k to $k + 1$ was solved earlier in Fleiner and Jordán (1999). Note that, when $r = 2$, then (52) and (54) reduce to (50) and (51), and (53) is implied by (52), so Theorem 5.3 implies Theorem 5.2.

22.5.3 Augmentation of Local Edge-Connectivity of a Hypergraph by Adding Graph Edges

This section is the devoted to the following problem.

HYPERGRAPH LOCAL EDGE-CONNECTIVITY AUGMENTATION BY A GRAPH
Instance: A hypergraph $\mathcal{H}$ on V, a symmetric requirement function $r(u, v) \in \mathbb{Z}$ $\forall u, v \in V$, and $\gamma \in \mathbb{Z}_+$.
Question: Does there exist a graph $G = (V, E)$ with at most γ edges so that $\lambda_{\mathcal{H}+G}(u, v) \geq r(u, v) \ \forall u, v \in V$?

The following theorem shows that this problem is already too complicated.

Theorem 5.4 (Cosh et al. 2008). *The problem* HYPERGRAPH LOCAL EDGE-CONNECTIVITY AUGMENTATION BY A GRAPH *is NP-complete.*

We mention that the above problem remains NP-complete if the hypergraph contains only just one hyperedge of size greater than 2.

We remark that the special case of the problem HYPERGRAPH LOCAL EDGE-CONNECTIVITY AUGMENTATION BY A GRAPH for 2–3 hypergraphs is tractable. This is because of the fact that a 2–3 hypergraph can be transformed into a graph with the same local edge-connectivities by $\Delta - Y$ operations and vice versa. Thus Theorem 4.11 implies the following.

Theorem 5.5 (Jordán and Szigeti 2003). *Let r be a symmatric requirement function with $2 \leq r(u, v) \in \mathbb{Z} \ \forall u, v \in V$. A 2–3 hypergraph $G = (V, E)$ can be made r-edge-connected by adding γ edges and γ' 3-hyperedges if and only if*

$$\sum_{X \in \mathcal{X}} (R(X) - d(X)) \leq 2\gamma + 3\gamma' \quad \forall \mathcal{X} \in \mathcal{S}(V), \tag{55}$$

$$G \text{ is } r - (\gamma + \gamma')\text{-edge-connected.} \tag{56}$$

22.5.4 Bipartite Constrained Augmentation of Local Edge-Connectivity of a Hypergraph by Adding Graph Edges

In this section we consider a restricted version of the main problem of the preceding section.

BIPARTITION CONSTRAINED HYPERGRAPH LOCAL EDGE-CONNECTIVITY AUGMENTATION BY A GRAPH
Instance: A hypergraph $\mathcal{H}$ on V, a bipartition $\{A, B\}$ of V, a symmetric requirement function $r(u, v) \in \mathbb{Z} \ \forall u, v \in V$, and $\gamma \in \mathbb{Z}_+$.

Question: Does there exist a bipartite graph $G = (A, B; E)$ with colour classes A and B with at most γ edges so that $\lambda_{\mathcal{H}+G}(u, v) \geq r(u, v) \ \forall u, v \in V$?

By checking the proof of Theorem 5.4 given in Cosh et al. (2008) we may observe that it also provides the NP-completeness of the above problem.

Theorem 5.6. *The problem* BIPARTITION CONSTRAINED HYPERGRAPH LOCAL EDGE-CONNECTIVITY AUGMENTATION BY A GRAPH *is NP-complete.*

22.5.5 Augmentation of Local Edge-Connectivity of a Hypergraph by Adding Hyperedges

In this section we want to augment a hypergraph by adding hyperedges to satisfy local edge-connectivity requirements. The problem of minimizing the number of hyperedges is trivial since we can add the whole vertex set as many times as needed. What we want to minimize is the total size of the hypergraph to be added, that is, the sum of the sizes of the hyperedges. More precisely, the *local edge connectivity augmentation problem of a hypergraph by adding hyperedges* is the following: *Given a hypergraph $\mathcal{G}$ and a symmetric requirement function $r : V \times V \to \mathbb{Z}_+$, what is the minimum total size $\sum_{H \in \mathcal{H}} |H|$ of new hyperedges $H \in \mathcal{H}$ whose addition results in an r-edge-connected hypergraph?*

Recall that $R(X) = \max\{r(x, y) : x \in X, y \in V - X\}$ and that the function $R(X) - d_{\mathcal{G}}(X)$ is symmetric skew-supermodular. Then the problem of covering a symmetric skew-supermodular function by a hypergraph contains this problem as a special case thus Theorem 6.8 implies the following theorem.

Theorem 5.7 (Szigeti 1999). *Let $\mathcal{G}$ be a hypergraph on V and $r(u, v) \in \mathbb{Z}_+ \ \forall u, v \in V$ a symmetric requirement function. Then there exists a hypergraph $\mathcal{H}$ on V with $\sum_{H \in \mathcal{H}} |H| \leq \gamma$ so that $\lambda_{\mathcal{G}+\mathcal{H}}(u, v) \geq r(u, v) \ \forall u, v \in V$ if and only if*

$$\sum_{X \in \mathcal{X}} (R(X) - d_{\mathcal{G}}(X)) \leq \gamma \quad \forall \mathcal{X} \in \mathcal{S}(V). \tag{57}$$

22.6 Abstract Forms

In this section we present results on "connectivity" set functions, that generalize the results presented in the previous sections. We start with a generalization of splitting off. Let $p : 2^V \to \mathbb{Z}$ be a function and $H = (V + s, E)$ a graph so that each edge is incident to s. By a *complete hypergraph splitting off* we mean a hypergraph $\mathcal{H}$ on V such that $\mathcal{H}$ covers p and the following is satisfied:

$$d_{\mathcal{H}}(v) = d_H(v) \quad \forall v \in V. \tag{58}$$

In other words, it is a degree constrained hypergraph that covers p. Note that there is no restriction on the size of the hyperedges. If we pose the restriction that each hyperedge must be of size at most two, then we are back to the definition of the usual complete splitting off.

22.6.1 Symmetric Crossing Supermodular Functions

The problem of covering a symmetric crossing supermodular function can be considered as a generalization of the global edge-connectivity augmentation problem. In this section we present the result of Benczúr and Frank on covering a symmetric crossing supermodular function by a graph (that generalizes Theorem 5.1) and then its generalization due to T. Király on covering such a function by an r-uniform hypergraph (that also generalizes Theorem 5.3).

Complete Uniform Hypergraph Splitting off

The splitting off results that will solve (together with Theorem 3.2) the above mentioned problems are the following. We call the attention to the fact that in Theorem 6.1 a hyperedge may contain the same vertex many times and the size of the hyperedge is meant by multiplicities.

A partition $\{V_1, \ldots, V_l\}$ of V is called an *l-partition*. Let $r \in \mathbb{Z}_+$, $H = (V + s, E)$ a graph so that each edge is incident to s and r divides $d_H(s)$ and $p : 2^V \to \mathbb{Z}_+$ a set function. An *l*-partition is called *p-full* if $l > r$ and $p(\bigcup_{i \in I} V_i) > 0 \ \forall \emptyset \neq I \subset \{1, \ldots, l\}$. A *p*-full partition is called *deficient* if $\frac{l-1}{r-1} > \frac{d_H(s)}{r}$. We say that the set function p is *positively crossing supermodular* if (10) is satisfied if $X, Y \subseteq V$ are crossing with $p(X), p(Y) > 0$.

Theorem 6.1 (T. Király 2004b). *Let $p : 2^V \to \mathbb{Z}_+$ be a symmetric, positively crossing supermodular set function, $r \geq 2$ an integer, and $H = (V + s, E)$ a graph so that each edge is incident to s with r divides $d_H(s)$. Then there exists a complete r-uniform hypergraph splitting off if and only if*

$$\min\{d_H(X), d_H(s)/r\} \geq p(X) \quad \forall X \subseteq V, \tag{59}$$

$$\textit{there are no deficient partitions.} \tag{60}$$

The special case when $r = 2$ was proved earlier by Benczúr and Frank (1999).

Covering a Symmetric Crossing Supermodular Function by a Graph

As a generalization of the global edge-connectivity augmentation in hypergraphs by adding graph edges, Benczúr and Frank considered the problem of *covering a symmetric crossing supermodular function by a graph*, namely: *Given a symmetric, positively crossing supermodular set function p, what is the minimum number of edges that cover p?*

As a generalization of Theorem 5.2 they proved in Benczúr and Frank (1999) the following result by applying Theorem 3.2 and their splitting off result which is the $r = 2$ special case of Theorem 6.1.

Theorem 6.2 (Benczúr and Frank 1999). *Let $p : 2^V \to \mathbb{Z}_+$ be a symmetric, positively crossing supermodular set function. Then there exists a graph on V with γ edges that covers p if and only if*

$$\sum_{X \in \mathcal{X}} p(X) \leq 2\gamma \quad \forall \mathcal{X} \in \mathcal{S}(V), \tag{61}$$

$$l - 1 \leq \gamma \quad \text{if a } p\text{-full } l\text{-partition exists.} \tag{62}$$

For a hypergraph $\mathcal{G}$, the function $k - d_{\mathcal{G}}(X)$ is symmetric, positively crossing supermodular, thus Theorem 6.2 implies Theorem 5.2.

Covering a Symmetric Crossing Supermodular Function by a Uniform Hypergraph

As a generalization of the problems of covering a symmetric crossing supermodular function by graph edges and of the global edge-connectivity augmentation in hypergraphs by r-uniform hyperedges, T. Király considered the problem of *covering a symmetric crossing supermodular function by an r-uniform hypergraph*, namely: *Given a symmetric, positively crossing supermodular set function p and $r \in \mathbb{Z}^+$, what is the minimum number of hyperedges of size r that cover p?*

As a generalization of Theorems 6.2 and 5.3 he obtained in T. Király (2004b) the following result by applying Theorems 3.2 and 6.1.

Theorem 6.3 (T. Király 2004b). *Let $p : 2^V \to \mathbb{Z}_+$ be a symmetric, positively crossing supermodular set function, $2 \leq r \leq |V|$ an integer. Then there exists an r-uniform hypergraph on V with γ hyperedges that covers p if and only if*

$$\sum_{X \in \mathcal{X}} p(X) \leq r\gamma \qquad \forall \mathcal{X} \in \mathcal{S}(V), \tag{63}$$

$$p(X) \leq \gamma \qquad \forall X \subseteq V, \tag{64}$$

$$l - 1 \leq (r - 1)\gamma \quad \text{if a } p\text{-full } l\text{-partition exists.} \tag{65}$$

Note that Theorem 6.3 implies Theorem 6.2 (when $r = 2$) and Theorem 5.3 (when the function is the deficiency function of a hypergraph).

22.6.2 Symmetric Skew-Supermodular Functions

The problem of covering a symmetric skew-supermodular function can be considered as a common generalization of many of the preceding problems. In this section we provide a nice proof of the NP-completeness of the problem of covering a symmetric skew-supermodular function by a graph (which we know already since it generalizes the problem of local edge-connectivity augmentation of a hypergraph by graph edges) while the problem of covering such a function by a hypergraph is solvable in a certain sense.

Covering a Symmetric Skew-Supermodular Function by a Graph

The problem of this section can be formulated as follows.

MINIMUM COVER OF A SYMMETRIC, SKEW-SUPERMODULAR FUNCTION BY A GRAPH

Instance: A symmetric skew–supermodular function p on V and $\gamma \in \mathbb{Z}^+$.
Question: Does there exist a graph on V with at most γ edges that covers p?

Note that, as we already mentioned, the NP-complete problem HYPERGRAPH LOCAL EDGE-CONNECTIVITY AUGMENTATION BY A GRAPH is a special case of the problem MINIMUM COVER OF A SYMMETRIC, SKEW-SUPERMODULAR FUNCTION BY A GRAPH and hence this last one is also NP-complete, which we prove here in an elegant way. The proof is due to Z. Király (2001) and independently to Nutov (2005).

Theorem 6.4 (Z. Király 2001). *The problem* MINIMUM COVER OF A SYMMETRIC, SKEW-SUPERMODULAR FUNCTION BY A GRAPH *is NP-complete.*

Proof (Z. Király). We reduce 3DM to MINIMUM COVER OF A SYMMETRIC, SKEW-SUPERMODULAR FUNCTION BY A GRAPH. Let $\mathcal{H}$ be a 3-uniform hypergraph on V. Let $n := |V|$. Let $p(X) := 1$ if $|X| \in \{1, 2, n-1, n-2\}$, or $X \in \overline{\mathcal{H}}$ or $V - X \in \overline{\mathcal{H}}$ and 0 otherwise. It is easy to verify that $p(X)$ is a symmetric skew-supermodular set function. The following completes the proof. *There exists a graph on V with at most $2n/3$ edges that covers p if and only if $\mathcal{H}$ contains a 3-dimensional matching.* Indeed, first suppose that $H_1, \ldots, H_{n/3}$ is a 3-dimensional matching. For each H_i, let us choose two edges on $V(H_i)$, and let F be the union of these edges. Then $|F| = 2n/3$ and, clearly, F covers p. Now suppose that the graph F covers p and $|E(F)| \le 2n/3$. Let $F_1, \ldots, F_l$ be the connected components of F. Since F covers p, $|V(F_i)| \ge 3$ for $1 \le i \le l$ thus $l \le n/3$. Then $2n/3 \ge |E(F)| \ge n - l \ge 2n/3$ so $l = n/3$ and $|V(F_i)| = 3$ for $1 \le i \le l$. Since F covers p, each F_i belongs to $\mathcal{H}$, that is, $F_1, \ldots, F_l$ is a 3-dimensional matching. $\quad\square$

By the following theorem, which is an easy corollary of Theorem 6.8, MINIMUM COVER OF A SYMMETRIC, SKEW-SUPERMODULAR FUNCTION BY A GRAPH can be solved if $p(X)$ is even for every $X \subseteq V$.

Theorem 6.5 (Szigeti 1999). *Let p be a symmetric, skew–supermodular, "even integer" valued function on V. Then there exists a graph on V with at most γ edges that covers p if and only if (61) is satisfied.*

Proof. To prove the difficult part of Theorem 6.5, suppose that (61) is satisfied and let $p'(X) = p(X)/2$. Then, by the assumptions on p, p' is a symmetric, skew–supermodular, integer valued function on V. Note that, since p satisfies (61), p' satisfies (13). Then, by Theorem 6.8, there exists a hypergraph $\mathcal{H}$ on V so that $\sum_{H \in \mathcal{H}} |H| \le \gamma$ and $\mathcal{H}$ covers p', that is, $d_{\mathcal{H}}(X) \ge p'(X)$. Let $G := (V, \bigcup_{H \in \mathcal{H}} E_H)$ where E_H is an arbitrary cycle on $V(H)$ for each $H \in \mathcal{H}$. Then $|E(G)| = \sum_{H \in \mathcal{H}} |E_H| = \sum_{H \in \mathcal{H}} |H| \le \gamma$ and $d_G(X) \ge 2d_{\mathcal{H}}(X) \ge 2p'(X) = p(X)$, that completes the proof. $\quad\square$

Note that the problem MINIMUM COVER OF A SYMMETRIC, SKEW-SUPER-MODULAR FUNCTION BY A GRAPH contains as a special case: the global and local edge-connectivity augmentation in graphs and in hypergraphs, the global edge-connectivity augmentation over symmetric parity families, the node to area edge-connectivity augmentation, and the problem of covering a symmetric crossing supermodular function by a graph.

Covering a Symmetric Semi-Monotone Function by a Graph

In this section, we mention another special case of the NP-complete problem MINIMUM COVER OF A SYMMETRIC, SKEW-SUPERMODULAR FUNCTION BY A GRAPH, that can be solved in polynomial time.

We call a function $P : 2^V \to \mathbb{Z}$ *semi-monotone* if $P(\emptyset) = P(V) = 0$ and for each set $\emptyset \neq X \neq V$, $0 \leq P(X) \leq P(X')$ either for all $\emptyset \neq X' \subseteq X$ or for all $\emptyset \neq X' \subseteq V - X$. We note that a symmetric semi-monotone function is skew-supermodular.

We consider the problem of *Covering a symmetric semi-monotone function by a graph*: *Given a graph $G = (V, E)$ and a symmetric semi-monotone function P on V, add a minimum number $Opt(P, G)$ of new edges to G to get a covering of P.*

We have already seen in Sect. 22.4.4 that the node to area global edge-connectivity augmentation problem is a special case of this problem. On the other hand, any instance of the symmetric semi-monotone function covering problem can easily be formulated as a node to area global edge-connectivity augmentation problem. This discussion establishes the surprising equivalence between two seemingly not closely related problems, showing that these are two alternative models of the same problem. (see Ishii 2007; Grappe and Szigeti 2008)

We have to emphasize that this problem in general is NP-complete because the function defined in the proof of Theorem 6.4 is semi-monotone. However, if the function does not take the value 1, then it can be solved.

Let us define $Q(G) := \max\{\sum_{X \in \mathcal{X}} q(X) : \mathcal{X} \in \mathcal{S}(V)\}$, where $q(X) = P(X) - d_G(X)$. Now we have a lower bound for the optimal value, namely $Opt(P, G) \geq \lceil \frac{Q(G)}{2} \rceil$. Usually equality will hold, unless the graph contains a special configuration.

Since the above defined function $q(X)$ is symmetric skew-supermodular, Theorem 3.2 and a suitable splitting result (Grappe and Szigeti 2008) provide the solution of the covering problem.

Theorem 6.6 (Ishii 2007; Grappe and Szigeti 2008). *Let $G = (V, E)$ be a graph and P a symmetric semi-monotone function on V so that $P(X) \neq 1 \; \forall X \subseteq V$. If G contains no configuration, then $Opt(P, G) = \lceil \frac{Q(G)}{2} \rceil$, otherwise $Opt(P, G) = \lceil \frac{Q(G)}{2} \rceil + 1$.*

The definition of the configuration (which is fairly complicated) and a short proof of Theorem 6.6 can be found in Grappe and Szigeti (2008). Notice that Theorem 6.6 implies Theorem 4.8.

Complete Hypergraph Splitting off

For symmetric skew-supermodular functions we have the following splitting off result.

Theorem 6.7 (Szigeti 1999). *Let $p : 2^V \to \mathbb{Z}$ be a symmetric skew–supermodular function. Let $H = (V + s, E)$ be a graph so that each edge is incident to s. Then there exists a complete hypergraph splitting off if and only if H covers p.*

T. Király (2004a) has recently found a very short proof for a slight extension of Theorem 6.7.

Covering a Symmetric Skew-Supermodular Function by a Hypergraph

In this section we provide the solution for the problem of *covering a symmetric skew-supermodular function by a hypergraph*, that is, a generalization of the problem of hypergraph local edge-connectivity augmentation by hyperedges, and that can be formulated as follows: *Given a symmetric skew–supermodular function p, what is the minimum total size $\sum_{H \in \mathcal{H}} |H|$ of a hypergraph $\mathcal{H}$ that covers p?*

Theorems 3.2 and 6.7 provide the following generalization of Theorem 5.7.

Theorem 6.8 (Szigeti 1999). *Let $p : 2^V \to \mathbb{Z}$ be a symmetric skew–supermodular function. Then there exists a hypergraph $\mathcal{H}$ on V with $\sum_{H \in \mathcal{H}} |H| \leq \gamma$ so that $\mathcal{H}$ covers p if and only if (13) is satisfied.*

Note that only the total size of the hypergraph is guaranteed and no information is available on the size of the hyperedges. This is not a surprise in the light of the NP-completeness of the problem MINIMUM COVER OF A SYMMETRIC, SKEW-SUPERMODULAR FUNCTION BY A GRAPH. However, Bernáth and T. Király (2007) have recently observed that the hypergraph $\mathcal{H}$ in Theorem 6.8 can be chosen so that each edge except one is of size two. They also proved that the hypergraph can be chosen so that each edge is of size k or $k + 1$ for some $2 \leq k \in \mathbb{Z}$.

22.7 Open Problems

In this section we pose some open problems related to (more precisely: generalizing of) the problems of this paper.

22.7.1 Graphs

The following open problem, called *node to area local edge-connectivity augmentation* and mentioned in Ishii and Hagiwara (2006), is a natural generalization of the node to area global edge-connectivity augmentation problem: *Given a graph $G = (V, E)$, a family $\mathcal{W}$ of sets $W \subseteq V$, and a requirement function $r : \mathcal{W} \times V \to \mathbb{Z}_+$, add a minimum number of new edges to G so that the resulting graph contains $r(W, v)$ edge-disjoint paths from any area $W \in \mathcal{W}$ to any vertex $v \notin W$.*

A result on *partition constrained detachment preserving global edge-connectivity* would imply Theorems 4.14 and 4.9. To be more precise we can consider the following problem: *Given a graph $G = (V + s, E)$, a partition $\mathcal{P}$ of $\delta(s)$, a degree specification $f(s)$ and a positive integer k, decide whether G has an $f(s)$-detachment that is k-edge-connected in V and $\mathcal{P}$-allowed* (meaning that for each new vertex s_i, the edges incident to s_i belong to different members of $\mathcal{P}$).

A characterization of the existence of a *partition constrained complete splitting off satisfying a requirement function* would imply Theorem 4.13 and also Theorem 4.11 as it was observed by Frank (1999). Indeed, for a graph $G = (V + s, E)$, a requirement function $r : V \times V \to \mathbb{Z}_+$ ($r(u, v) \geq 2 \; \forall u, v \in V$) and a degree specification $f(s) = (d_1, \ldots, d_p)$ ($d_i \geq 2 \; \forall i$), let G' be obtained from G by adding p new vertices $s_1, \ldots, s_p$ and connecting each vertex s_i to s by d_i new edges, let $r'(u, v) := r(u, v)$ if $u, v \in V$ and 2 if $\{u, v\} \cap \{s_1, \ldots s_p\} \neq \emptyset$ and let $\mathcal{P} := \{V, \{s_1, \ldots, s_p\}\}$. Then G'' is a complete r'-edge-connected $\mathcal{P}$-allowed splitting off of G' if and only if G'' is a r-edge-connected $f(s)$-detachment of G.

Note that the complexity of the *local edge-connectivity augmentation problem in a graph with bipartition constraint* is not known, however, as we have seen, the problem BIPARTITION CONSTRAINED HYPERGRAPH LOCAL EDGE-CONNECTIVITY AUGMENTATION BY A GRAPH is NP-complete.

A common generalization of the above two problems is the problem of the existence of a *partition constrained detachment satisfying a requirement function*.

22.7.2 Hypergraphs

The following problem of *global edge-connectivity augmentation in hypergraphs with hyperedges of given sizes* is still open: *Given a hypergraph $\mathcal{G}$ and $k, d_1, \ldots, d_p \in \mathbb{Z}^+$ decide whether it is possible to add p hyperedges of sizes $d_1, \ldots, d_p$ to $\mathcal{G}$ to get a k-edge-connected hypergraph.* More generally we can consider the problem of covering a symmetric crossing supermodular function with hyperedges of given sizes. A solution to these problems would imply Theorems 5.3 and 6.1, respectively.

We may also consider the problem of *partition constrained complete splitting off preserving global edge-connectivity in hypergraphs*: *Given a hypergraph $\mathcal{G} = (V + s, E)$ such that $\mathcal{G}$ is k edge-connected in V and no hyperedge of size at least three is incident to s, and a partition $\mathcal{P}$ of $\delta(s)$, decide whether there exists a $\mathcal{P}$-allowed k-admissible complete splitting off at s.* A result on this problem would imply Theorem 4.13. The bipartite case was done by Cosh (2000).

The following problem of *detachment preserving global edge-connectivity in hypergraphs* is also open: *Given a hypergraph $\mathcal{G} = (V + s, E)$ such that $\mathcal{G}$ is k-edge-connected in V and no hyperedge of size at least three is incident to s, and a degree specification $f(s)$, decide whether there exists an $f(s)$-detachment such that the resulting hypergraph is k-edge-connected in V.* A result on this problem would imply Theorems 4.9 and 5.1. The case of local edge-connectivity (being a generalization of the problem HYPERGRAPH LOCAL EDGE-CONNECTIVITY AUGMENTATION BY A GRAPH) is NP-complete.

Appendix

A.1 A short proof of Theorem 4.11

The following short proof is from Szigeti (2004a). Note that, by Menger's Theorem, G is r-edge-connected if and only if $h_G^r(X) \geq 0 \ \forall X \subseteq V$, where $h_G^r(X) := d_G(X) - R(X)$. The following basic property of the function h follows from the facts that $d_G(X)$ satisfies both (2) and (3) and that $R(X)$ is skew-supermodular, and will be used frequently in this section. For any two subsets $X, Y \subseteq V$ at least one of (66) and (67) holds. If $X \cup Y = V$ then (67) always holds (with equality).

$$h_G^r(X) + h_G^r(Y) \geq h_G^r(X \cap Y) + h_G^r(X \cup Y) + 2d_G(X, Y), \tag{66}$$

$$h_G^r(X) + h_G^r(Y) \geq h_G^r(X - Y) + h_G^r(Y - X) + 2\overline{d}_G(X, Y). \tag{67}$$

Proof of the Necessity of Theorem 4.11

Let $G' := (V + \{s_1, \ldots, s_p\}, E)$ be an r-edge-connected $f(s)$-detachment of G at s. By (4), applied for $X = \{s_1, \ldots, s_p\}$, (22) is satisfied since $G'/X = G$. By (5), applied for every vertex s_i $1 \leq i \leq p$, (23) is satisfied since $G' - X = G - s$.

Proof of the Sufficiency of Theorem 4.11

Wlog. $p \geq 2$ and $\varphi \geq 2$. We will use induction on $z(G) := |V| + d_G(s)$. As we already mentioned, (22) and (23) can be reformulated as (68) and (69). Note that (70) holds.

$$h_G^r(X) \geq 0 \quad \forall X \subseteq V, \tag{68}$$

$$h_{G-s}^{r-\varphi}(X) \geq 0 \quad \forall X \subseteq V. \tag{69}$$

$$h_{G-s}^{r-\varphi}(X) = h_G^r(X) - d_G(s, X) + \varphi \quad \forall X \subseteq V. \tag{70}$$

Lemma A.1. *We may assume that*

$$\textit{every set } \emptyset \neq X \subset V \textit{ with } h_G^r(X) = 0 \textit{ is a singleton}. \tag{71}$$

Proof. Suppose there exists a set Q with $h_G^r(Q) = 0$ and $|Q| > 1$. Then let $\hat{G} := (\hat{V}, \hat{E})$ be obtained from G by contracting Q into a vertex q and let $\hat{r}(u, v) := r(u, v)$ if $u, v \in \hat{V} - q$, and $\max\{r(w, x) : w \in Q\}$ if $q \in \{u, v\}$ where $x = \{u, v\} - q$. It can be verified easily that $\hat{R}(\hat{X}) = R(X) \ \forall \hat{X} \subseteq \hat{V}$, so (68) and (69) are satisfied for $\hat{G}$ and $\hat{r}$. Since $|Q| > 1$, $z(\hat{G}) < z(G)$ and hence, by induction, $\hat{G}$ has an $\hat{r}$-edge-connected $f(s)$-detachment $\hat{G}'$. We show that the graph G' obtained from $\hat{G}'$ by "blowing up" Q is r-edge-connected and we are done. Let $X' \subseteq V'$. Using that $h_{G'}^r(Q) = h_G^r(Q) = 0$, the skew-submodularity of $h_{G'}^r$ and the fact that if X' and Q are not intersecting then $h_{G'}^r(X') \geq 0$ (because if $X' \subset Q$ then $h_{G'}^r(X') = h_G^r(X) \geq 0$ by (68) and if $Q \subseteq X'$ or $Q \cap X' = \emptyset$ then $h_{G'}^r(X') = h_{\hat{G}'}^{\hat{r}}(\hat{X}') \geq 0$ since $\hat{G}'$ is $\hat{r}$-edge-connected) we get that $h_{G'}^r(X') \geq 0$ as we wanted. □

For $T \subset \delta_G(s)$, the *T-split* of G is the $(|T|, d_G(s) - |T|)$-detachment G' of G at s where $\delta_{G'}(s_1) = T$. For $X \subseteq V$, let $e(T, X) = |T \cap \delta_G(X)|$.

Lemma A.2. *There exists* $T \subset \delta_G(s)$ *with* $|T| = 3$ *if* $f(s) = (3, 3, \ldots, 3)$ *and* $|T| = 2$ *otherwise such that the T-split* G' *of* G *satisfies* (72) *and* (73) *where* $r'(u, v) := r(u, v)$ *if* $u, v \in V$ *and* 2 *otherwise and* $V' = V \cup s_1$.

$$G' \text{ is } r'\text{-edge-connected in } V', \tag{72}$$
$$G' - s \text{ is } (r' - (\varphi - 1))\text{-edge-connected in } V', \tag{73}$$

Proof. Let $\mathcal{C}$ be defined as the minimal sets X with $h_{G-s}^{r-\varphi}(X) = 0$.

Claim. (72) and (73) are equivalent to

$$h_G^r(X) \geq 2e(T, X) - |T| \quad \forall X \subset V, \tag{74}$$
$$e(T, C) \geq 1 \qquad\qquad \forall C \in \mathcal{C}. \tag{75}$$

Proof. (72) is satisfied if and only if $0 \leq h_{G'}^{r'}(X') \ \forall X' \subset V'$. Since, for $X' \subset V' - s_1, h_{G'}^{r'}(X') = h_G^r(X) \geq 0$, (72) is equivalent to $0 \leq h_{G'}^{r'}(X') \ \forall X' \subset V'$ containing s_1 which is, by $h_{G'}^{r'}(X') = h_G^r(X) - e(T, X) + (|T| - e(T, X))$ with $X = X' - s_1$, equivalent to (74). (73) is satisfied if and only if $0 \leq h_{G'-s}^{r'-\varphi'}(X)$ $\forall X \subset V'$ not containing s_1 which is, by $h_{G'-s}^{r'-\varphi'}(X) = h_{G-s}^{r-\varphi}(X) + e(T, X) - 1$, equivalent to (75). $\square$

Claim. The following are true for $\mathcal{C}$ and for all $C \in \mathcal{C}$:

$$\text{the sets in } \mathcal{C} \text{ are pairwise disjoint} \quad \text{and} \quad d_G(s, C) \geq \varphi, \tag{76}$$
$$|\mathcal{C}| \in \{0, 2, 3\}, \quad \text{if } |\mathcal{C}| = 3 \text{ then } f(s) = (3, 3, \ldots, 3) \text{ and } h_G^r(C) = 0. \tag{77}$$

Proof. By the skew-submodularity of $h_{G-s}^{r-\varphi}(X)$, the minimality of the sets in $\mathcal{C}$, (69), (70) and (68), (76) follows. If $X \in \mathcal{C}$, then $V - X$ contains a set $Y \in \mathcal{C}$ so $|\mathcal{C}| \neq 1$. By (76) and $d_i \geq 2$, $|\mathcal{C}|\varphi \leq \sum_{C \in \mathcal{C}} d_G(s, C) \leq d_G(s) = \sum_{i=1}^{p} d_i \leq 3 \sum_{i=1}^{p} \lfloor \frac{d_i}{2} \rfloor = 3\varphi$ thus $|\mathcal{C}| \leq 3$ and if $|\mathcal{C}| = 3$ then each $d_i = 3$, that is, $f(s) = (3, 3, \ldots, 3)$ and $\forall C \in \mathcal{C}, d_G(s, C) = \varphi$, so by (70), $h_G^r(C) = 0$. $\square$

By (77), either $|\mathcal{C}| = 3$ or $|\mathcal{C}| \in \{0, 2\}$. If $|\mathcal{C}| = 3$, then, by (77), $f(s) = (3, 3, \ldots, 3)$. By (76), there exists $T \subset \delta_G(s)$ with $|T| = 3$ that satisfies (75). T also satisfies (74). Indeed, by (77), (71) and (76), $d_G(s, X) \geq \varphi \ e(T, X)$. So, by (70), (69) and $\varphi \geq 2$, $h_G^r(X) \geq d_G(s, X) - \varphi \geq \varphi(e(T, X) - 1) \geq 2(e(T, X) - 1) \geq 2e(T, X) - |T|$.

From now on $|\mathcal{C}| \in \{0, 2\}$.

Claim. There exists $T = \{su, sv\}$ that satisfies (74) and (75).

Proof. If $|\mathcal{C}| = 0$, then (75) is redundant, and, by Theorem 4.3(a), there is a pair $T = \{su, sv\}$ that is λ-admissible which is equivalent to (74). If $\mathcal{C} = \{C_1, C_2\}$, then, by (76), there is a $T = \{su, sv\}$ satisfying (75). T also satisfies (74). Otherwise, there exists $u, v \in X \subset V$ with $h_G^r(X) \leq 1$. If $C_1 \cup C_2 \subseteq X$, then, by (70), (69), (76), $h_G^r(X) \geq d_G(s, X) - \varphi \geq d_G(s, C_1 \cup C_2) - \varphi \geq 2\varphi - \varphi \geq 2$, contradiction, so wlog. $Y := C_1 - X \neq \emptyset$. Since $C_1 \in \mathcal{C}$, $h_{G-s}^{r-\varphi}(Y) \geq 1$. Let $d := d_G(s, C_1 \cap X)$. Then, by (70), $h_G^r(Y) = h_{G-s}^{r-\varphi}(Y) + d_G(s, Y) - \varphi \geq 1 + (d_G(s, C_1) - d) - \varphi = h_G^r(C_1) + 1 - d$. If (67) applies for C_1 and X, then, by (68), $h_G^r(X) + h_G^r(C_1) \geq h_G^r(X - C_1) + h_G^r(Y) + 2\overline{d}_G(X, C_1) \geq h_G^r(Y) + 2d \geq h_G^r(C_1) + 1 + d \geq h_G^r(C_1) + 2$, contradiction. So (66) applies for C_1 and X and $Z := C_1 \cup X \neq V$. Since $\mathcal{C} = \{C_1, C_2\}$, $h_{G-s}^{r-\varphi}(Z) = h_{G-s}^{r-\varphi}(V - Z) \geq 1$. Then, by (70), $h_G^r(Z) = h_{G-s}^{r-\varphi}(Z) + d_G(s, Z) - \varphi \geq 1 + (d_G(s, C_1) + 1) - \varphi = h_G^r(C_1) + 2$, so, by (68), $h_G^r(X) + h_G^r(C_1) \geq h_G^r(X \cap C_1) + h_G^r(Z) \geq h_G^r(C_1) + 2$, a contradiction that finishes the proof of the claim. $\square$

If $f(s) \neq (3, 3, \ldots, 3)$, then, by the above claim, we are done. From now on $f(s) = (3, 3, \ldots, 3)$. Then $d_G(s) = 3\varphi$.

Claim. T can be extended to $T' \subset \delta_G(s)$ with $|T'| = 3$ such that T' satisfies (74).

Proof. First suppose $\Gamma(s) = \{u, v\}$. Since $d_G(s) = 3\varphi$ and $\varphi \geq 2$, wlog. $d_G(s, u) \geq \varphi + 1$, thus there exists another copy e' of su. Then, by (70) and (69), $T' := T \cup e'$ satisfies (74). Hence $\Gamma(s) \neq \{u, v\}$. Suppose indirect that there exists a minimal set $\mathcal{M}$ of subsets of V such that for every $z_i \in \Gamma(s) - \{u, v\} \neq \emptyset$ there exists a set $M_i \in \mathcal{M}$ violating (74) for $T' := T \cup sz_i$. Then, since T satisfies (74) and by (71), $e(T', M_i) = 3$ so $\{u, v, z_i\} \subseteq M_i$ and $h_G^r(M_i) \leq 2$. Clearly, $|\mathcal{M}| \geq 1$. By (70), (69), $h_G^r(M_i) \leq 2$ and $\varphi \geq 2$, we have $|\mathcal{M}| \geq 2$. For $M_i, M_j \in \mathcal{M}$,

$$h_G^r(M_i - M_j) = 0 \text{ (so, by (71), } M_i - M_j = z_i), \tag{78}$$

$$\overline{d}_G(M_i, M_j) = 2, \tag{79}$$

$$d_G(z_i, M_i - z_i) \geq 1. \tag{80}$$

Indeed, $2 \geq h_G^r(M_i), 2 \geq h_G^r(M_j), h_G^r(M_i \cap M_j) \geq 2e(T, M_i \cap M_j) - |T| \geq 2 \times 2 - 2 = 2$ (by (74) and $\{u, v\} \subset M_i \cap M_j$), $h_G^r(M_i \cup M_j) \geq 3$ (by the minimality of $\mathcal{M}$), so (66) cannot be satisfied for M_i and M_j. Then M_i and M_j satisfy (67) implying (78) and (79). Moreover, since $\max\{R(z_i), R(M_i - z_i)\} \geq R(M_i)$ and $\min\{R(z_i), R(M_i - z_i)\} \geq 2$, we have $R(z_i) + R(M_i - z_i) \geq R(M_i) + 2$, thus $2 \leq h_G^r(z_i) + h_G^r(M_i \cap M_j) = h_G^r(z_i) + h_G^r(M_i - z_i) \leq h_G^r(M_i) - 2 + 2d_G(z_i, M_i - z_i) \leq 2d_G(z_i, M_i - z_i)$ implying (80).

Case 1 If $\mathcal{M} = \{M_1, M_2\}$. Then, by (70), (78), (69) and (79), $3\varphi = d_G(s) = d_G(s, z_1) + d_G(s, z_2) + d_G(s, M_1 \cap M_2) = h_G^r(z_1) - h_{G-s}^{r-\varphi}(z_1) + \varphi + h_G^r(z_2) - h_{G-s}^{r-\varphi}(z_2) + \varphi + d_G(s, M_1 \cap M_2) \leq 2\varphi + 2 \leq 3\varphi$. Thus $h_{G-s}^{r-\varphi}(z_1) = 0$, so $z_1 \in \mathcal{C}$, that is, (75) is violated for T, contradiction.

Case 2 If $M_1, M_2, M_3 \in \mathcal{M}$. Then, by (80), (78), (79), $1 \leq d_G(M_3 - z_3, z_3) = d_G(M_1 \cap M_2, z_3) \leq \overline{d}_G(M_1, M_2) - d_G(M_1 \cap M_2, s) \leq 2 - 2 = 0$, contradiction. The proof of the claim is finished. $\square$

Since T satisfies (75), so does T' and the proof of Lemma A.2 is complete. □

Let G' be the T-split of G from Lemma A.2. Let us denote the new vertex of G' of degree $|T|$ by t. Wlog. $d_1 \geq d_2 \geq \dots \geq d_p$. If $d_p = |T|$ then let $f'(s) := (d_1, \dots, d_{p-1})$ otherwise ($|T| = 2, d_1 \geq 4$) let $f'(s) := (d_1 - 2, d_2, \dots, d_p)$. Then $(G', f'(s))$ satisfies (72) and (73) and $z(G') < z(G)$, so by induction, G' has an r-edge-connected $f'(s)$-detachment G''. Then, in the former case G'', in the latter case the graph obtained from G'' by identifying s_1 and t, is an r-edge-connected $f(s)$-detachment of G.

A.2 A Short Proof of Theorem 4.14

I hope that the reader will find interesting the following shortened proof from Szigeti (2004b). I ask the reader not to be frightened by the technical aspects of the proof and to make figures to help to understand the proofs. In this section we call a set $X \subset V$ *tight* (resp. *dangerous*) if $d(X) = k$ (resp. $d(X) \leq k+1$). We will abbreviate k-admissible by admissible. In order to have a more convenient notation, $e \in P_j$ will also be denoted by $c(e) = j$.

Preliminaries

The following easy observations are from Frank (1992b).

Proposition A.3.

(a) $\{su, sv\}$ *is admissible if and only if there is no dangerous set containing u and v.*
(b) *For any edge su, there exist at most two dangerous sets M_1 and M_2 so that $u \in M_1 \cap M_2$ and $\{v : \{su, sv\}$ is not admissible$\} \subseteq M_1 \cup M_2$.*
(c) *For a tight set T, $\{su, sv\}$ is admissible in G if and only if it is admissible in G/T.*

The following proposition contains some technical remarks.

Proposition A.4.

(a) $d(X) - k \geq 2d(s, X) - d(s)$ $\forall X \subset V$ *where equality holds if and only if $d(V - X) = k$.*
(b) *If $k \geq 3$ and $d(X) \leq k+2$ then $G[X]$ is connected.*
(c) *If k is odd, X_1, X_2, X_3 are disjoint tight sets, $d(\bigcup_{i=1}^{3} X_i) = k+2$ and $d(X_1, X_3) = 0$, then $d(X_1, X_2) = d(X_2, X_3) = \frac{k-1}{2}$.*

Proof. (a) By (34), $d(X) - k = d(V - X) - k + d(s, X) - (d(s) - d(s, X)) \geq 2d(s, X) - d(s)$.

(b) For a set $\emptyset \neq Y \subset X$, by (2) and (34), $(k+2) + 2d(Y, X - Y) \geq d(X) + 2d(Y, X - Y) = d(Y) + d(X - Y) \geq k + k \geq k + 3$, and (b) follows.

(c) By (2) and (34), $\forall i \in \{1, 3\}$, $2k = d(X_2) + d(X_i) = d(X_2 \cup X_i) + 2d(X_2, X_i) \geq k + 2d(X_2, X_i)$, thus, by parity, $2d(X_2, X_i) \leq k - 1$. Then $3k = \sum_{i=1}^{3} d(X_i) = d(\bigcup_{i=1}^{3} X_i) + \sum_{i \neq j} 2d(X_i, X_j) \leq (k+2) + 2(k-1) + 0 = 3k$, and (c) follows. □

We present now some important properties of C_4- and C_6-obstacles.

Proposition A.5.

(a) *If $\mathcal{A}$ is a C_4-obstacle, then $d(s, A_i) \geq 1 \; \forall A_i \in \mathcal{A}$.*

(b) *If $\{A_1, A_2, A_3, A_4\}$ is a C_4-obstacle, then for each set $\emptyset \neq X \subseteq A_i$, $d(V - X) \geq k + 2$ with equality only if $d(X) = k$.*

(c) *If $\{A_1, a_2, a_3, a_4\}$ is a C_4-obstacle, then for each set X with $X \cap A_1 \neq \emptyset$ and $a_3 \in X$, $d(X) \geq k + 2$.*

(d) *If $\{A_1, \ldots, A_6\}$ is a C_6-obstacle, then for every allowed pair $\{sx, sy\}$, $G_{x,y}$ contains a C_4-obstacle.*

(e) *If $\{A_1, a_2, \ldots, a_6\}$ is a C_6-obstacle and, for a set $X \neq V$, $X \cap A_1 \neq \emptyset$, $y \in X \cap (a_3 \cup a_5)$ and $d(X) \leq k + 2$, then (e1) $d(X) = k + 2$ and (e2) $X \cup A_1$ is the union of three consecutive sets in $\mathcal{A}$.*

Proof. (a) Suppose wlog. $d(s, A_1) = 0$. Then, by (27) and (26), $d(A_1, A_2) + d(A_1, A_4) = d(A_1) = k$, so, since k is odd, wlog. $d(A_1, A_2) \geq \frac{k+1}{2}$. Then, by (34), (2) and (26), $k \leq d(A_1 \cup A_2) = d(A_1) + d(A_2) - 2d(A_1, A_2) \leq k + k - (k+1) = k - 1$, contradiction.

(b) Let $j := i + 2$ if $i \leq 2$ and $j := i - 2$ if $i \geq 3$. Then, by (34), Proposition A.5(a), (28) and (29), $k + 2 \leq d(X) + 2d(s, A_j) = d(X) - (\frac{d(s)}{2} - d(s, A_j)) + \frac{d(s)}{2} + d(s, A_j) = d(V - X)$ and (b) follows.

(c) If $a_2, a_4 \notin X$, then, by $X \cap A_1 \neq \emptyset$, (34), (26) and (27), $d(X) = d(X \cap A_1) + d(a_3) = k + k \geq k + 2$. If $a_2, a_4 \in X$, then, by Proposition A.5(b), $d(X) \geq k + 2$. Otherwise, wlog. $a_2 \notin X$ and $a_4 \in X$. By Proposition A.5(b), $d(X \cup A_1) \geq k + 2$. Then, by (2) and (34), $d(X) \geq d(X \cap A_1) + d(X \cup A_1) - d(A_1) \geq k + (k+2) - k = k + 2$.

(d) Wlog. $x \in A_1$. By (2), (30), (31), $d(A_i \cup A_{i+1}) = d(A_i) + d(A_{i+1}) - 2d(A_i, A_{i+1}) = k + k - (k - 1) = k + 1$. Then, since $\{sx, sy\}$ is admissible, $y \notin A_2 \cup A_6$ by Proposition A.3(a). $\{sx, sy\}$ is allowed so, by (33), $y \notin A_4$. Thus wlog. $y \in A_3$. Then $\{A_1 \cup A_2 \cup A_3, A_4, A_5, A_6\}$ is a C_4-obstacle in $G_{x,y}$.

(e) Let $X^* := X \cup A_1$. By (31), $d_{G-s}(X^*) \geq k - 1$ where equality holds if and only if X^* is the union of $2 < l < 6$ consecutive sets in $\mathcal{A}$. By Proposition A.4(a), $d(s, X) \leq 4$, by (32), $d(s, A_1) = 1$ and $d(s, V) = 6$ so $X^* \neq V$. By (30), $d(A_1) = k$, by $X \cap A_1 \neq \emptyset$ and (34), $d(X \cap A_1) \geq k$, so by (2), $(k+2) + k \geq d(X) + d(A_1) \geq d(X \cap A_1) + d(X^*) \geq k + d(X^*)$, so $k + 2 \geq d(X^*)$ and if equality holds then $d(X) = k + 2$. Then, by Proposition A.4(b), $G[X^*]$ is connected. Since $d_{G-s}(y, A_1) = 0$, $X' := X - (y \cup A_1) \neq \emptyset$. Then $k + 2 \geq d(X^*) = d(s, X^*) + d_{G-s}(X^*) \geq d(s, y) + d(s, X') + d(s, A_1) + d_{G-s}(X^*) \geq 1 + 1 + 1 + (k - 1)$, so $d(X^*) = k + 2$ and hence $d(X) = k + 2$ (implying (e1)), $d(s, X') = 1$ and $d_{G-s}(X^*) = k - 1$, thus (e2) is satisfied.   □

The following lemma will allow us to easily find an allowed pair. The main difficulty of the proof of Theorem 4.14 will be to show that there exists an allowed pair whose splitting off creates no C_4- and no C_6-obstacle.

Lemma A.6. *If G contains no C_4-obstacle and (37) is satisfied then each edge su belongs to an allowed pair.*

Proof. Let $S := \{sv \in E : \{su, sv\}$ is admissible$\}$. Suppose su belongs to no allowed pair. Then every $sv \in S$ and su belong to the same P_j. Then, by (37), $\frac{d(s)}{2} \geq |P_j| \geq |S| + 1$, so $|S| \leq \frac{d(s)}{2} - 1$ and if equality holds then $\frac{d(s)}{2} = |P_j|$. It also follows, by Proposition A.3(b), that there are at most two dangerous sets M_1 and M_2 so that $u \in M_1 \cap M_2$ and $\{v_i : sv_i \in \delta(s) - S\} \subseteq M_1 \cup M_2$. In fact there are exactly two, because, by Proposition A.4, $d(M_1 \cup M_2) - k \geq 2d(s, M_1 \cup M_2) - d(s) = 2(d(s) - |S|) - d(s) \geq d(s) - 2(\frac{d(s)}{2} - 1) = 2$, and if equality holds then $d(V - M_1 \cup M_2) = k$ and $|S| = \frac{d(s)}{2} - 1$. The following claim provides a contradiction.

Claim. $\{A_1 = M_1 \cap M_2, A_2 = M_1 - M_2, A_3 = V - M_1 \cup M_2, A_4 = M_2 - M_1\}$ forms a C_4-obstacle.

Proof. Note that $A_i \neq \emptyset$ $1 \leq i \leq 4$ and $\bigcup_{i=1}^4 A_i = V$. By (2), (34) and $d(M_1 \cup M_2) \geq k + 2$, $2(k+1) \geq d(M_1) + d(M_2) = d(A_1) + d(M_1 \cup M_2) + 2d(M_1, M_2) \geq k + (k + 2)$, so $d(A_1) = k, d(M_1 \cup M_2) = k + 2$ and hence $d(A_3) = k$ and $\frac{d(s)}{2} = |P_j|$ so (28) is satisfied, and $d(A_2, A_4) = d(M_1, M_2) = 0$. By (3) and (34), $2(k + 1) \geq d(M_1) + d(M_2) = d(A_2) + d(A_4) + 2d(A_1, A_3 + s) \geq k + k + 2d(A_1, A_3) + 2d(A_1, s) \geq 2k + 0 + 2$, so $d(A_2) = d(A_4) = k, d(A_1, A_3) = 0$ and $d(s, A_1) = 1$. It also follows that $\delta(A_1 \cup A_3) \cap \delta(s) = P_j$, so (26), (27) and (29) are satisfied. This completes the proof of the claim and also of Lemma A.6. □

Proof of the Necessity of Theorem 4.14

Suppose there exists a graph that has a complete allowed splitting off $\{\{e_i, f_i\} : 1 \leq i \leq \frac{d_G(s)}{2}\}$ and violates (37) or (36). Choose such a graph G with $d_G(s)$ minimum. For every $1 \leq i \leq \frac{d_G(s)}{2}, 1 \leq j \leq r, |P_j \cap \{e_i, f_i\}| \leq 1$ so (37) is satisfied, whence G contains a C_4- or a C_6-obstacle. By Proposition A.5(a) and (32), $d_G(s) \neq 0$. Then, either by (28) and (29) or by Proposition A.5(d), G_{e_1, f_1} contains a C_4-obstacle, G_{e_1, f_1} has of course a complete allowed splitting off, and $d_{G_{e_1, f_1}}(s) < d_G(s)$, contradiction.

Proof of the Sufficiency of Theorem 4.14

Induction on $|V|$. By Proposition A.3(c), we may assume that

$$\text{every tight set is a singleton.} \tag{81}$$

Wlog. $|P_1|$ is maximum. By Lemma A.6, there is an allowed pair $\{e = sx, f = sy\}$ with $sx \in P_1$. If, after splitting off this pair, we get stuck by a C_6-obstacle, then we can get rid of it by the following lemma.

Lemma A.7. *Suppose that $G' := G_{e,f}$ contains a C_6-obstacle $\mathcal{A} = \{A_1, \ldots, A_6\}$. Then there exists an edge $f' = sy'$ such that $\{e, f'\}$ is allowed in G and $G'' := G_{e,f'}$ satisfies (34), (37) and (36).*

Proof. The suitable y' will be chosen as follows: Wlog. $x \in A_1$. Then, by (30) and (81), $A_j = a_j \; \forall 2 \leq j \leq 6$. By (33), $c(sa_3) \neq c(sa_5)$ so either $c(sa_3) \neq 1$ in which case let $y' := a_3$ or $c(sa_5) \neq 1$ and then let $y' := a_5$. We show that y' will do. For each set X with $x, y' \in X \neq V$, by Proposition A.5(e), $k + 2 \leq d_{G'}(X) \leq d_G(X)$, so by Proposition A.3(a), $\{e, f'\}$ is admissible in G, thus G'' satisfies (34). Since $c(sx) = 1 \neq c(sy')$, G'' satisfies (37) and $\{e, f'\}$ is allowed in G. It remains to show that G'' satisfies (36). Since $xy \in E(G')$, either Case a: $y \in A_1$ or Case b: wlog. $y \in A_2$. In both cases we suppose indirect that G'' contains a C_4- (Case 1) or a C_6-obstacle (Case 2) $\mathcal{A}'$.

Case a: Wlog. $y' = a_5$ and $x \in A_1'$. Suppose $y' \notin A_1'$. By (30), $k+2 = d_{G'}(A_1)+2 = d_G(A_1)$. By (26) or (30), $d_G(A_1') = k$ and then, by (81), $|A_i'| = 1 \; \forall A_i' \in \mathcal{A}'$. Hence $|V| = 4$ or 6. But $|V| \geq 6$ because G' contains the C_6-obstacle $\mathcal{A}$, so $|V| = 6$. Thus $A_1' = A_1$ and hence $k = d_{G''}(A_1') = d_{G''}(A_1) = d_G(A_1)$, contradiction. Thus $y' \in A_1'$ and $d_G(A_1') = k + 2$. Since $d_{G'}(A_1') \leq d_G(A_1')$, $A_1' \cup A_1$ is the union of three consecutive sets in $\mathcal{A}$ by Proposition A.5(e).

Case a1: Then $3 = |V - (A_1 \cup A_1')| \leq |V - A_1'| = 3$ by (81) so $A_1 \subset A_1'$ thus, by (31), wlog. $A_j' = a_j \; 2 \leq j \leq 4$. By (33) for $\mathcal{A}$, there is a $w \in A_1$ with $c(sw) = c(sa_4)$ but $w \in A_1'$ and $a_4 \in A_4'$, contradiction by (29) for $\mathcal{A}'$.

Case a2: Then $a_6 \in A_1'$. Wlog. $A_2' = a_4$ and $A_3' = a_3$. Then, by (33) for $\mathcal{A}$ and $\mathcal{A}'$, $c(sa_3) = c(sa_6) \neq c(sa_3)$, contradiction, and we are done in Case a.

Case b: Then, by (30) and (81), $A_1 = a_1$ so $|V| = 6$.

Case b1: Since $d_{G''}(s, y') = 0$, wlog. $x, y' \in A_1'$ and $d_G(A_1') = k + 2$. Since $d_{G'}(A_1') \leq d_G(A_1')$, $d_{G'}(A_1') = k + 2$ and $A_1' \cup A_1$ is the union of three consecutive sets in $\mathcal{A}$ by Proposition A.5(e). Then $d_{G'}(A_1') = d_G(A_1')$ so $y' = a_5$. Thus $A_1' = \{a_5, a_6, a_1\}$. Wlog. $A_j' = a_j \; 2 \leq j \leq 4$ by (31) for $\mathcal{A}$ and (27) for $\mathcal{A}'$. By (33) for $\mathcal{A}$, $c(sa_1) = c(sa_4)$, contradiction by (29) for $\mathcal{A}'$.

Case b2: Note that, by (32), $d_G(s, a_1) = d_G(s, a_2) = 2$ and $d_G(s, a_h) = 1$ ($3 \leq h \leq 6$). Then $d_{G''}(s, a_2) = 2$ that is, by (32), a contradiction, and we are done in Case b, and the proof of Lemma A.7 is complete.   □

The following lemma allows us to get rid of a C_4-obstacle if during the splitting off we get stuck with one.

Lemma A.8. *Suppose that $G' := G_{e,f}$ contains a C_4-obstacle $\mathcal{A} := \{A_1, A_2, A_3, A_4\}$. Then there exists an allowed pair $e' = sx'$, $f' = sy'$ in G such that $G'' := G_{e',f'}$ satisfies (34), (37) and (36).*

Proof. Wlog. $x \in A_1$. Since $xy \in E(G')$, either Case a: wlog. $y \in A_2$ or Case b: $y \in A_1$. By (26) and (81), $A_j = a_j \; \forall 2 \leq j \leq 4$, in Case a: $A_1 = a_1$ so $|V| = 4$ and in Case b: $d_G(A_1) = k + 2$.

Case a: We chose e' and f' as follows. If there exists an edge $g = sa_3$ with $c(g) \neq c(e)$ then let $e' := e$, $f' := g$. Otherwise, since $\mathcal{A}$ is not a C_4-obstacle in G, there is an edge $h = sa_1$ with $c(h) \neq c(e)$ and then let $e' := sa_3$, $f' := h$. We show that this pair will do. Note that $\{x', y'\} = \{a_1, a_3\}$. For each set X with $x', y' \in X \neq V$, either $X = \{x', y'\}$ and then, by (26) and (27), $d_G(X) = d_G(a_1) + d_G(a_3) = k + k \geq k + 2$ or, by $X \neq V$, $\exists i \in \{2, 4\}$ $X = \{a_1, a_3, a_i\}$ and then, by Proposition A.5(b), $k + 2 \leq d_{G'}(X) \leq d_G(X)$. In both cases, by Proposition A.3(a), $\{e', f'\}$ is admissible in G, thus G'' satisfies (34). Since $c(e') = c(e)$, G'' satisfies (37) and $\{e', f'\}$ is allowed in G. It remains to show that G'' satisfies (36). Suppose not. Then, since $|V| = 4$, G'' contains a C_4-obstacle $\mathcal{A}' := \{A_1', A_2', A_3', A_4'\}$. Since $a_1 a_3 \in E(G'')$, wlog. $A_1' \cup A_2' = a_1 \cup a_3$ and $A_3' \cup A_4' = a_2 \cup a_4$. By Proposition A.5(a), $d_{G''}(s, A_i') \geq 1$ so $d_G(s, A_i') \geq 2$ $i \in \{1, 2\}$. By (28) and (29) for $\mathcal{A}$ in G', there exist $1 \leq l \leq r$ and $j \in \{1, 2\}$ such that for every edge $sd \in E(G')$ with $d \in A_j \cup A_{j+2}$, $c(sd) = l$. Then there exist $sd_1, sd_2 \in E(G'')$ with $d_1 \in A_j, d_2 \in A_{j+2}$ and $c(sd_1) = c(sd_2) = l$. This contradicts (29) for $\mathcal{A}'$.

Case b: Either we reduce this case to a case already seen or we show that the following edge pair e', f' will do. Let $g := sa_3$. If $c(g) \neq c(e)$, then let $e' := e$ and $f' := g$, otherwise, let $e' := g$ and $f' := f$. Since $c(e') = c(e)$, G'' satisfies (37). By Propositions A.3(a) and A.5(c), $\{e', f'\}$ is admissible in G, so allowed in G and (34) is satisfied. If G'' satisfies (36), then we are done. Otherwise, either G'' contains a C_6-obstacle, and then, by Lemma A.7, we are done, or G'' contains a C_4-obstacle $\mathcal{A}' := \{A_1', A_2', A_3', A_4'\}$. Wlog. $x', y' \in A_1'$, otherwise restarting the proof by e' and f' we are in Case a. Then, by (26) and (81), $|A_j'| = 1$ $\forall 2 \leq j \leq 4$ and $d_G(A_1') = k + 2$. By Proposition A.4(b), $G[A_1']$ is connected, so, by (27), wlog. $V - a_4 \subseteq A_1' \cup A_1$. First suppose that $A_1' \cup A_1 = V$. Then, by Propositions A.5(b) applied for G', $d_{G'}(A_1') = k + 2$ and $k = d_{G'}(V - A_1')$. The common edge of $\{e, f\}$ and $\{e', f'\}$ enters $A_1 \cap A_1'$, so $d_{G'}(V - A_1') = d_G(V - A_1')$ thus, by (81), $1 = |V - A_1'| (= 3)$, contradiction, so $A_1' \cup A_1 = V - a_4$. Then, by Proposition A.5(b) again applied for G', $d_{G'}(A_1' \cup A_1) \geq k + 2$. Then, by (2) and (34), $d_G(A_1' \cap A_1) = k$ thus, by (81), $|A_1' \cap A_1| = 1$, say $A_1' \cap A_1 = a_1$. Then it follows that $|V| = 6$, say $V = \{a_1, a_2, \ldots, a_6\}$. Note that $d_G(a_i) = k$ $1 \leq i \leq 6$. By Proposition A.5(a) for $\mathcal{A}$ and for $\mathcal{A}'$, $1 \leq d_G(sa_i)$ $1 \leq i \leq 6$ so $6 \leq d_G(s)$. The following claim provides a contradiction.

Claim. $\{a_1, a_2, \ldots, a_6\}$ forms a C_6-obstacle in G.

Proof. We start with some structural observations.

Proposition A.9. *(a)* $A_1 = \{a_1, a_5, a_6\}$, $A_i = a_i$ $2 \leq i \leq 4$, $A_1' = \{a_1, a_2, a_3\}$, *wlog.* $A_2' = a_6$, $A_3' = a_5$, $A_4' = a_4$, *(b)* $d_G(a_1, a_2) = d_G(a_2, a_3) = d_G(a_1, a_6) = d_G(a_5, a_6) = \frac{k-1}{2}$, *(c)* $\{x, y\} = \{a_1, a_5\}$.

Proof. We know that $A_1' = \{a_1, a_2, a_3\}$ and $A_1 = \{a_1, a_5, a_6\}$. Then, by (27) for $\mathcal{A}$, $d_G(a_1, a_3) = 0$ so, by Proposition A.4(c), $d_G(a_1, a_2) = d_G(a_2, a_3) = \frac{k-1}{2}$. Wlog. $A_2' = a_6$. Suppose that $A_4' = a_5$. Then, by (27) for $\mathcal{A}'$, $d_G(a_5, a_6) = 0$

so, by Proposition A.4(c), $d_G(a_1, a_5) = d_G(a_1, a_6) = \frac{k-1}{2}$. Then $k = d_G(a_1) \geq d_G(a_1, a_2) + d_G(a_1, a_5) + d_G(a_1, a_6) + d_G(a_1, s) \geq 3\frac{k-1}{2} + 1$, that is $k \leq 1$, contradiction. Thus $A_4' = a_4$ and $A_3' = a_5$, that is, (a) is satisfied. Then, by (27) for $\mathcal{A}'$, $d_G(a_1, a_5) = 0$ so, by Proposition A.4(c), $d_G(a_1, a_6) = d_G(a_5, a_6) = \frac{k-1}{2}$ and (b) is satisfied. By definition, $\{x, y\} \cap \{x', y'\} = a_1$, so $a_1 \in \{x, y\}$. Suppose $a_5 \notin \{x, y\}$. Then $\{x, y\} = \{a_1, a_6\}$. By (2), (30) and (b), $d_G(\{a_1, a_6\}) = d_G(a_1) + d_G(a_6) - 2d_G(a_1, a_6) = k + k - (k - 1) = k + 1$, hence, by Proposition A.3(a), $\{sx, sy\}$ is not admissible in G, contradiction, thus (c) is satisfied. □

By (29) for $\mathcal{A}'$, $c(sa_2) \neq c(sa_4)$ so $\delta_{G'}(A_1 \cup A_3) \cap \delta_{G'}(s) = P_l'$ in (29) for $\mathcal{A}$ for some l with $|P_l| \geq |P_l'| = \frac{d_{G'}(s)}{2} = \frac{d_G(s)}{2} - 1$. By (29) for $\mathcal{A}$, $c(sa_6) \neq c(sa_4)$ so $\delta_{G''}(A_1' \cup A_3') \cap \delta_{G''}(s) = P_{l'}$ in (29) for $\mathcal{A}'$ for some l' with $|P_{l'}| \geq |P_{l'}'| = \frac{d_{G''}(s)}{2} = \frac{d_G(s)}{2} - 1$. In particular, $c(sa_2) = c(sa_5) = l'$. By (29) for $\mathcal{A}$, $l = c(sa_3) \neq c(sa_2) = l'$ thus, by Proposition A.9(c), $e = e' = sa_1$, $f = sa_5$, $f' = sa_3$. Since $\{e, f\}$ and $\{e', f'\}$ are allowed, $l \neq 1 \neq l'$. Then, by the maximality of P_1, $|P_1| \geq |P_l| \geq \frac{d_G(s)}{2} - 1$. $d_G(s) \geq |P_1| + |P_l| + |P_{l'}| \geq 3(\frac{d_G(s)}{2} - 1)$, that is, $d_G(s) \leq 6$. Then $d_G(s) = 6$ and $|P_1| = |P_l| = |P_{l'}| = 2$, namely $P_1 = \{sa_1, sa_4\}$, $P_l = \{sa_3, sa_6\}$, $P_{l'} = \{sa_2, sa_5\}$, so (32) and (33) are satisfied. We have already seen that (30) is satisfied. By (26) and (27) for $\mathcal{A}'$ and for $\mathcal{A}$, Proposition A.9(b) and (32), $d_G(a_5, a_4) = \frac{k-1}{2} = d_G(a_3, a_4)$. Then, by Proposition A.9(b), (31) is satisfied. This completes the proof of the claim and also of Lemma A.8. □

By Lemma A.6, there exists an allowed pair, and then, by Lemmas A.7 and A.8, there exists one that does not create a C_4- or C_6-obstacle, so by induction, there exists a complete allowed splitting off and Theorem 4.14 is proved.

Acknowledgement

I thank T. Király for the fruitful discussions on this topic.

References

Bang-Jensen, J., Jackson, B.: Augmenting hypergraphs by edges of size two. Math. Program. **84**(3), 467–481 (1999)

Bang-Jensen, J., Jordán, T.: Edge-connectivity augmentation preserving simplicity. SIAM J. Discrete Math. **11**(4), 603–623 (1998)

Bang-Jensen, J., Jordán, T.: Splitting off edges within a specified subset preserving the edge-connectivity of the graph. J. Algorithms **37**, 326–343 (2000)

Bang-Jensen, J., Frank, A., Jackson, B.: Preserving and increasing local edge-connectivity in mixed graphs. SIAM J. Discrete Math. **8**(2), 155–178 (1995)

Bang-Jensen, J., Gabow, H., Jordán, T., Szigeti, Z.: Edge-connectivity augmentation with partition constraints. SIAM J. Discrete Math. **12**(2), 160–207 (1999)

Benczúr, A., Frank, A.: Covering symmetric supermodular functions by graphs. Math. Program. **84**(3), 483–503 (1999)

Bernáth, A., Király, T.: Personal communication (2007)

Cai, G.R., Sun, Y.G.: The minimum augmentation of any graph to k-edge-connected graphs. Networks **19**, 151–172 (1989)

Cheng, E., Jordán, T.: Successive edge-connectivity augmentation problems. Math. Program. **84**(3), 577–593 (1999)

Cosh, B.: Vertex splitting and connectivity augmentation in hypergraphs. Ph.D. thesis, University of London (2000)

Cosh, B., Jackson, B., Király, Z.: Local connectivity augmentation in hypergraphs is NP-complete. Submitted (2008)

Eswaran, K.P., Tarjan, E.: Augmentation problems. SIAM J. Comput. **5**, 653–665 (1976)

Fleiner, B.: Detachments of vertices of graphs preserving edge-connectivity. SIAM J. Discrete Math. **18**(3), 581–591 (2005)

Fleiner, T., Jordán, T.: Covering and structure of crossing families. Math. Program. **84**(3), 505–518 (1999)

Frank, A.: Augmenting graphs to meet edge-connectivity requirements. SIAM J. Discrete Math. **5**(1), 22–53 (1992a)

Frank, A.: On a theorem of Mader. Discrete Math. **101**, 49–57 (1992b)

Frank, A.: Personal communication (1999)

Grappe, R., Szigeti, Z.: Covering semi-monotone symmetric functions. Discrete Appl. Math. **156**, 138–144 (2008)

Ishii, T.: Minimum augmentation of edge-connectivity with monotone requirements in undirected graphs. In: Proc. of the 13th Australasian Symp. on Theory of Comp., pp. 91–100 (2007)

Ishii, T., Hagiwara, M.: Minimum augmentation of local edge-connectivity between vertices and vertex subsets in undirected graphs. Discrete Appl. Math. **154**, 2307–2329 (2006)

Jackson, B.: Some remarks on connectivity, vertex splitting and orientation in digraphs. J. Graph Theory **12**, 429–436 (1988)

Jordán, T.: Two NP-complete augmentation problems. IMADA, Odense Universitet, Preprint No. 8 (1997)

Jordán, T.: Constrained edge-splitting problems. SIAM J. Discrete Math. **17**(1), 88–102 (2003)

Jordán, T., Szigeti, Z.: Detachments preserving local edge-connectivity of graphs. SIAM J. Discrete Math. **17**(1), 72–87 (2003)

Király, T.: Merging hyperedges to meet edge-connectivity requirements. Egres Technical Reports, TR-2005-08, http://www.cs.elte.hu/egres/ (2004a)

Király, T.: Covering symmetric supermodular functions by uniform hyperedges. J. Comb. Theory, Ser. B **91**, 185–200 (2004b)

Király, Z.: Personal communication (2001)

Lovász, L.: Combinatorial Problems and Exercises. North-Holland, Amsterdam (1979)

Mader, W.: A reduction method for edge-connectivity in graphs. Ann. Discrete Math. **3**, 145–164 (1978)

Nutov, Z.: Approximating connectivity augmentation problems. In: SODA 2005, pp. 176–185 (2005)

Szigeti, Z.: Hypergraph connectivity augmentation. Math. Program. **84**(3), 519–527 (1999)

Szigeti, Z.: A short proof on the local detachment theorem. Egres Technical Reports, TR-2004-09, http://www.cs.elte.hu/egres/ (2004a)

Szigeti, Z.: On partition constrained splitting off. Egres Technical Reports, TR-2004-08, http://www.cs.elte.hu/egres/ (2004b)

Szigeti, Z.: Edge-splittings preserving edge-connectivity of graphs. Discrete Appl. Math. **156**, 1011–1018 (2008a)

Szigeti, Z.: Edge-connectivity augmentation of graphs over symmetric parity families. Discrete Math., to appear (2008b)

Watanabe, T., Nakamura, A.: Edge-connectivity augmentation problems. J. Comput. Syst. Sci. **35**, 96–144 (1987)

Some Problems on Approximate Counting in Graphs and Matroids

Dominic Welsh

Summary. I shall discuss some of the more well known counting problems associated with graphs and matroids. Except in special cases all these problems have no exact counting algorithm which runs in polynomial time unless there is a remarkable collapse of some existing classes. Hence the focus is on obtaining fast algorithms which give good approximations. The problems studied include counting forests, trees and colourings of graphs and bases of matroids.

23.1 Introduction

Most problems involving counting structures of a particular type in a given graph seem to be $\#P$-hard. One notable exception is the number of spanning trees of a connected graph when the solution is given by Kirchhoff's determinant formula. As a result much attention over the last twenty years has been given to methods of producing good approximations which can be completed in polynomial time. In this article I will try to survey some of the progress that has been made. I will also consider some similar problems about matroids where the situation is more complicated.

Most of the concepts used will be defined as they are encountered. However although the ideas of computational complexity are now fairly familiar concepts, it might be useful to review briefly and informally the basic notions, particularly with respect to randomised computation.

A function f is *polynomial time computable* or *P-time computable* if there exists an algorithm which computes f in time (= number of steps) bounded by a polynomial in the size of the input to the problem. We use *FP* to denote the class of all polynomial time computable functions, and as usual P is the class of languages (= sets of strings) *recognisable* in polynomial time. Thus, P can be identified with the class of functions in *FP* with range $\{0, 1\}$. Informally, *nondeterministic polynomial time*, *NP*, is the class of strings (objects, or theorems) which can be recognised or checked in polynomial time.

The simplest model of randomised computation is the *probabilistic Turing machine* in which the only source of randomness is a fair coin.

In a more powerful model, the *oracle coin machine*, the unbiased coin is replaced by a device which chooses transitions with probabilities p and $1 - p$, where p is the ratio of a pair of rationals presented in a special bias tape and which can be varied by the machine according to the state of the computation. Although a probabilistic machine will have difficulty simulating an oracle machine exactly, it can approximate its behaviour very accurately in polynomial time, so that the notion of polynomial time approximation algorithm is robust with respect to choice of model.

Random polynomial time, RP, is the subset of NP consisting of those strings which can be recognised in polynomial time by a probabilistic Turing machine. Obviously

$$P \subseteq RP \subseteq NP,$$

but neither of the inclusions is known to be proper. The class $\#P$ is the set of functions which count structures which are recognisable in polynomial time. In other words there is a polynomial time algorithm which will verify whether a given object has the structure needed to be included in the count. For example, it is easy to check in polynomial time whether a given assignment of colours to the vertices of a graph is in fact a proper 3-colouring in that no two adjacent vertices are the same colour. Like NP, $\#P$ has hardest problems known as $\#P$-*complete (hard) functions*. A polynomial time algorithm for solving any of these would produce an amazing collapse in the complexity hierarchy, much more surprising than the collapse of NP to P.

It is therefore extremely strong evidence of intractability when a function is shown to be $\#P$-hard. By the same token it is somewhat surprising that exact counting of some of the very easily recognisable objects to be discussed below turns out to be $\#P$-hard. For more on this see Garey and Johnson (1979) or Welsh (1993).

23.2 Approximation Schemes

Computing the number of 3-colourings of a graph G is $\#P$-hard. How well can we approximate it?

For positive numbers a and $r \geq 1$, we say that a third quantity $\hat{a}$ *approximates a within ratio r* or is an r-*approximation* to a, if

$$r^{-1}a \leq \hat{a} \leq ra.$$

In other words the ratio $\hat{a}/a$ lies in $[r^{-1}, r]$.

Suppose now we were able to find a polynomial time algorithm which gave an approximation within r to the number of 3-colourings of a graph. We would clearly have a polynomial time algorithm which would decide whether or not a graph is 3-colourable. But this is NP-hard. Thus no such algorithm can exist unless $NP = P$.

The problem of 3-colouring is just a typical example and the same argument can be applied to any function which counts objects whose existence is NP-hard to decide. In other words:

Proposition 2.1. *If $f : \Sigma^* \to \mathbb{N}$ is such that it is NP-hard to decide whether $f(x)$ is non-zero, then for any constant r there cannot exist a polynomial time r-approximation to f unless $NP = P$.*

We now turn to consider a randomised approach to counting problems and make the following definition.

Let f be a function from input strings to the natural numbers. A *randomised approximation scheme* for f is a probabilistic algorithm that takes as an input a string x and a rational number ϵ, $0 < \epsilon < 1$, and produces as output a random variable Y, such that Y approximates $f(x)$ within ratio $1 + \epsilon$ with probability $\geq 3/4$.

In other words,

$$Pr\left\{ \frac{1}{1+\epsilon} \leq \frac{Y}{f(x)} \leq 1 + \epsilon \right\} \geq \frac{3}{4}.$$

A *fully polynomial randomised approximation scheme* (fpras) for a function $f : \Sigma^* \to \mathbb{N}$ is a randomised approximation scheme which runs in time which is a polynomial function of n and ϵ^{-1}.

It is important to note that there is no significance in the constant $3/4$ appearing in this definition. Any success probability *strictly greater* than $1/2$ will suffice, for the following reason.

Suppose we have such an approximation scheme and suppose further that it works in polynomial time. Then we can boost the success probability up to $1 - \delta$ for any desired $\delta > 0$, by using the following trick of Jerrum, Valiant and Vazirani (1986). This consists of running the algorithm $O(\log \delta^{-1})$ times and taking the median of the results.

It is worth emphasising here that the existence of an fpras for a counting problem is a very strong result, it is the analogue of an *RP* algorithm for a decision problem and corresponds to the notion of tractability. However we should also note, that by an analogous argument to that used in proving Proposition 2.1 we have:

Proposition 2.2. *If $f : \Sigma^* \to \mathbb{N}$ is such that deciding if f is nonzero is NP-hard then there cannot exist an fpras for f unless NP is equal to random polynomial time RP.*

Since this is believed to be most unlikely, it makes sense only to seek out an fpras when counting objects for which the decision problem is not *NP*-hard.

Most of the randomised algorithms which give good approximation schemes are based on developing algorithms which will generate objects uniformly at random and then using what Lovász (1996) calls the *product estimator*.

This works as follows. Suppose F is the set of objects we wish to count and that we can find a chain of subsets $F_0 \subset F_1 \subset \cdots \subset F_r = F$ such that for each i:

(a) F_0 is known so $|F_0|$ is known,
(b) $|F_{i+1}|/|F_i|$ is bounded by a polynomial in the size of the input,
(c) r is polynomially bounded in the size of input,
(d) for each i, $1 \leq i \leq r$, we can generate a member of F_i uniformly (or almost uniformly) at random in time polynomial in n.

Then we use (d) a polynomial number of times to estimate the ratio

$$|F_{i-1}|/|F_i|.$$

Since products preserve good approximations, we then use

$$\prod_{i=1}^{r} |F_i|/|F_{i-1}|$$

and our knowledge of F_0 to estimate $|F|$.

As an example, suppose we try to estimate the number of forests in an n vertex graph G by this method. We would need to be able to generate a random forest of size k, for $1 \leq k \leq n - 1$. As yet no one knows how to do this in polynomial time.

Note that it is not necessary for the generation procedure to be exactly uniform. All that is really necessary is that it be close to uniform in the following sense. Throughout Σ will denote a finite alphabet which is usually taken to be $\{0, 1\}$. Σ^* denotes the set of all finite strings or words of symbols from Σ. If x is a word of Σ^*, then $|x|$ denotes the length (= number of symbols) of x.

A relation $R \subseteq \Sigma^* \times \Sigma^*$ is a *p-relation* if

(i) there exists a polynomial p such that

$$\langle x, y \rangle \in R \quad \Rightarrow \quad |y| \leq p(|x|),$$

(ii) the predicate $\langle x, y \rangle \in R$ can be tested in deterministic polynomial time.

Typically, x may be a graph G and y might be the set of edges of a forest of G, both encoded in the alphabet Σ.

A *uniform generator* for a relation R is a probabilistic Turing machine M such that the following conditions are satisfied.

(a) There exists a function ϕ such that for all inputs x

$$Pr\{M \text{ outputs } y\} = \begin{cases} 0 & \text{if } \langle x, y \rangle \notin R \\ \phi(x) & \text{if } \langle x, y \rangle \in R. \end{cases}$$

(b) If $\{y : xRy\}$ is nonempty then

$$Pr\{M \text{ accepts } x\} \geq \frac{1}{2}.$$

Thus the condition (b) demands that the machine behaves uniformly.

The machine is *polynomially time bounded* if there exists a polynomial t such that for all inputs x of length n and all positive integers n, every accepting computation halts within $t(n)$ steps.

A probabilistic machine M is an *almost uniform generator* for the relation $R \subseteq \Sigma^* \times \Sigma^*$ iff

(c) there exists a mapping $\phi : \Sigma^* \to (0, 1]$ such that for all inputs $(x, \epsilon) \in \Sigma^* \times \mathbb{R}^+$ and for all words $y \in \Sigma^*$,

$$\langle x, y \rangle \notin R \quad \Rightarrow \quad Pr\{M \text{ outputs } y\} = 0$$

$$\langle x, y \rangle \in R \quad \Rightarrow \quad (1 + \epsilon)^{-1}\phi(x) \leq Pr\{M \text{ outputs } y\} \leq (1 + \epsilon)\phi(x);$$

(d) for all inputs (x, ϵ) such that $\{y \in \Sigma^* : xRy\}$ is nonempty, the probability that M accepts (x, ϵ) is at least $\frac{1}{2}$.

An almost uniform generator is *fully polynomial* (f.p.) if its running time on input (x, ϵ) is bounded by a polynomial in x and $\log(\epsilon^{-1})$. The importance of almost uniform generation is that provided the relation R is what is known as *self reducible* (and in most natural problems this seems to be the case) the existence of a f.p. almost uniform generator implies the existence of a fpras.

The concept of self reducibility was introduced by Schnorr (1976), its formal definition is rather technical, but intuitively it represents the notion that the solution set associated with a given instance of a problem can be expressed in terms of the solution sets of a number of smaller instances of the same problem.

Example 2.3. Let R denote the relation $\langle G, A \rangle$ where A is the set of edges of a connected subgraph of a graph G. Then the set of solutions is the union of those where G is replaced by G'_e and G''_e where G'_e denotes the deletion of the edge e, and G''_e the effect of identifying its endpoints.

When R is self reducible it is easy to see how an exact counting oracle helps with uniform generation. Consider Example 2.3; if there are 35 connected subgraphs of G'_e and 65 of G''_e then at the first step of the generation algorithm branch to G'_e with probability 0.35, and to G''_e with probability 0.65. This reduces the size of the problem by one. Now proceed recursively.

A similar method based on approximate counting gives an almost uniform generator, and this forms the basis of the proof of the main theorem of Jerrum, Valiant and Vazirani (1986), which we now state.

Theorem 2.4. *If R is a self reducible relation then there exists a fully polynomial uniform generator for R iff there exists a fully polynomial randomised approximation scheme for the associated counting problem.*

Most natural problems seem to be self-reducible, though as far as I know there is no general theorem of this type. An example of a problem which is not known (or indeed likely) to be self-reducible is a planar graph colouring. This is because Khuller and Vazirani (1991) have proved:

Theorem 2.5. *Unless P is equal to NP, planar graph colouring is not self-reducible.*

Clearly this makes it very unlikely that planar graph colouring is self-reducible.

23.2.1 Quantities Which Are Impossible to Approximate

We have already seen that unless $NP = P$ there cannot exist fast approximation algorithms for quantities such as the number of 3-colourings of a graph. There is another class of problems where the objects being counted have the property that they always exist, so the decision problem is trivial, but for which approximate counting is not possible unless something very unlikely is true.

A collection of such problems is given in the following proposition.

Proposition 2.6. *Unless $NP = RP$ there cannot exist an fpras for the following:*

(a) *the number of cycles in a (directed) graph,*
(b) *the number of paths linking two specific vertices in a graph,*
(c) *the number of cliques in a graph,*
(d) *the number of maximal cliques.*

This list is not exhaustive, and in each case the proof idea is the same.

Proof (sketch). I illustrate with (a).

Blow the input up by introducing a "gadget" which transforms the input G to H so that the existence or not of a Hamilton cycle in G has such a huge effect on the number of cycles in H that it will show up in the answer obtained by applying the fpras to H. Thus we would have an RP algorithm to decide whether G was Hamiltonian.   □

For further details see Jerrum et al. (1986) and Sinclair (1988).

23.3 Rapidly Mixing Markov Chains

We turn now to sampling procedures based on Markov chain simulation. Although the mathematical analysis can be very difficult the basic idea is straightforward.

Identify the set of objects being sampled with the state space Ω of a finite Markov chain constructed in such a way that the stationary distribution of the Markov chain agrees with the probability distribution on the set of objects. Now start the Markov chain at an arbitrary state and let it converge to the stationary distribution. Provided the convergence is fast, we have a good approximation to the stationary distribution fairly quickly. However the key to the success of this approach is that the time or number of steps needed to ensure that it is sufficiently close to its stationary distribution is not too large.

The first major practical application of the theory seems to have been in the work of Broder (1988) on approximating the permanent. The paper of Jerrum and Sinclair (1989) which built on Broder's work was a major advance, particularly in their idea of relating rapid mixing with conductance.

To be more specific, suppose that P is the transition matrix of an irreducible aperiodic finite Markov chain $(X_t; 0 \leq t < \infty)$ with state space Ω. Suppose also that the chain is *reversible* in that it satisfies the detailed balance condition

$$\pi(i)P(i, j) = \pi(j)P(j, i) \quad \text{for all } i, j \in \Omega.$$

This condition implies that π is a stationary or equilibrium distribution for P and that for all X_0

$$Pr\left\{\lim_{t \to \infty} X_t = i\right\} = \pi_i$$

for all states $i \in \Omega$.

Moreover, P has eigenvalues $1 = \lambda_1 \geq \lambda_2 \geq \cdots \geq \lambda_N > -1$ where N is the number of states, and all these eigenvalues are real. The rate of convergence to the distribution π is determined by the quantity

$$\lambda_{max} = \max\{\lambda_2, |\lambda_N|\}.$$

More precisely we have the following proposition. Let $P^t(i, j)$ denote $Pr\{X_{t+h} = j \mid X_h = i\}$ and let

$$\Delta_i(t) = \frac{1}{2} \sum_j |P^t(i, j) - \pi(j)|,$$

denote what is called the *variation distance* at time t. It is clearly a measure of the separation from the stationary distribution at time t.

Now define, for $\epsilon > 0$, the *mixing time* function τ_i by

$$\tau_i(\epsilon) = \min\{t : \Delta_i(s) \leq \epsilon \; \forall s \geq t\}.$$

The following result shows the relationship between mixing times and the maximum eigenvalues. It has now become a well-known classical result, see (Lawler and Sokal 1988; Sinclair 1992; Sinclair and Jerrum 1989).

Proposition 3.1. *For $\epsilon > 0$, $\tau_i(\epsilon)$ satisfies*

$$\text{(i)} \quad \tau_i(\epsilon) \leq \frac{1}{1 - \lambda_{max}}\left(\ln \frac{1}{\pi(i)} + \ln \frac{1}{\epsilon}\right);$$

$$\text{(ii)} \quad \max_i \tau_i(\epsilon) \geq \frac{\lambda_{max}}{2(1 - \lambda_{max})} \ln\left(\frac{1}{2\epsilon}\right).$$

In order to achieve rapid convergence we need $\tau_i(\epsilon)$ be small, for all i and hence define the mixing time $\tau(\epsilon) = \max_i \tau_i(\epsilon)$.

Note. A well-known bootstrap lemma shows that it is sufficient to bound the mixing time (or mixing function) $\tau(c)$ for any c strictly less than $1/2$.

More precisely we have:

Lemma 3.2. *Suppose a Markov chain has mixing function $\tau(c) \leq T$ for some c strictly less than $1/2$. Then for any positive ϵ*

$$\tau(\epsilon) \leq \ln(\epsilon)T/\ln(2c).$$

It is useful to note the following trick which concentrates the interest on λ_2. Replace P by $P' = \frac{1}{2}(I + P)$ where I is the identity matrix. This only affects rates of convergence by a polynomial factor. All eigenvalues of P' are non-negative, and the quantity λ'_{max} of P' is $\frac{1}{2}\lambda_{max}$, so that henceforth we need only consider the second eigenvalue λ_2. Ideally we want $1 - \lambda_2$ to be large so λ_2 must be small.

The key idea of Sinclair and Jerrum (1989) was to relate this to the *conductance* Φ of the chain. This is defined by

$$\Phi = \max_{S \subseteq \Omega} \left\{ \sum_{i \in S, j \in \Omega \setminus S} P(i, j)\pi(i) \bigg/ \sum_{i \in S} \pi(i) \right\}.$$

The bracketed term is just the conditional probability of leaving a set S in the equilibrium state. In other words Φ is a measure of the ability of the chain to escape from any subset of the state space Ω.

Theorem 3.3. *The second eigenvalue λ_2 of a reversible ergodic Markov chain satisfies*

$$1 - 2\Phi \leq \lambda_2 \leq 1 - \Phi^2/2.$$

Hence for fast approximation we need large conductance, namely

$$\Phi \geq \frac{1}{\text{poly}(n)}$$

where $\text{poly}(n)$ denotes some polynomial function of the input size.

23.3.1 Canonical Paths

In order to get good lower bounds on the conductance Φ the key step is to construct a unique *canonical path* γ_{xy} between each ordered pair of states x, y in the graph Γ having vertex set Ω (which represents the state space of the Markov chain). Provided these canonical paths can be chosen in such a way that no edge (transition) is overloaded by paths, then the chain cannot contain a bottleneck (or constriction). This implies that Φ cannot be too small, for if there were a bottleneck between S and $\Omega \setminus S$ then any choice of paths would overload the edges in the bottleneck. Thus we have shifted the problem from bounding eigenvalues, through conductances, to finding a "good" set of canonical paths having the property that the maximum loading of an edge $e = (u, v)$ of Γ, measured by

$$\rho = \max_e \frac{1}{Q(e)} \sum_{\gamma_{xy}} \pi(x)\pi(y)$$

is not too large.

Here

$$Q(e) = Q(u, v) = \pi(u)P(u, v)$$

and the sum is over all canonical paths from x to y which use e.

The relationship between ρ and conductance is that it can be shown that

$$\Phi \geq 1/2\rho.$$

The difficult part is to find a good set of canonical paths. This was first done by Jerrum and Sinclair (1989) to count matchings (see Sect. 23.6).

23.3.2 The Coupling Method

Recall that we need a good upper bound on the *mixing time*

$$\tau(\varepsilon) = \max_i \tau_i(\varepsilon).$$

A *coupling* for the chain X_t is a stochastic process (Y_t, Z_t) on $\Omega \times \Omega$ with the following properties.

(i) Each of the processes Y_t and Z_t is a faithful copy of X_t for specified initial states $Y_t = y$, $Z_t = z$.
(ii) If $Y_t = Z_t$ then $Y_{t+1} = Z_{t+1}$.

Thus, viewed in isolation, each of Y_t and Z_t behaves like X_t, however Y_t and Z_t need not be independent. The basic idea is to construct a joint distribution which brings them close together. This is because of the crucial result that the expected time for the processes to couple is an upper bound on the mixing time. More precisely, if the *coupling time* is defined by

$$T^{y,z} = \min\{t : Y_t = Z_t \mid Y_0 = y, \; Z_t = z\}$$

then we have

Theorem 3.4. *The mixing and coupling times are related by*

$$\tau(\varepsilon) \leq 8 \log_2(\varepsilon^{-1}) \max_{y,z \in \Omega} E(T^{y,z}).$$

An alternative way of looking at coupling is to note that the variation distance of (X_t) from the stationary distribution is bounded above by the inequality

$$\Delta(t) \leq \max_{y,z \in \Omega} Pr\{Y_t \neq Z_t \mid Y_0 = y, \; Z_0 = z\}.$$

For a detailed account of the probability background we refer to Aldous (1982).

We illustrate the use of this coupling technique to a colouring problem in the next section.

23.4 Colouring Problems

I start with a problem about colourings first posed in Welsh (1993).

We have immediately from the *NP*-hardness of k-colouring that:

- Unless $NP = RP$ there cannot exist an fpras for counting k-colourings for any integer $k > 2$.

This does not rule out the possibility of there being an fpras for special classes of graphs. For example; There is no inherent obstacle to there being an fpras for estimating the number of 4-colourings of a planar graph.

I do not believe such a scheme exists but cannot see how to prove it. It certainly is not ruled out by any known results and I therefore pose the specific question:

Problem 4.1. Is there a fully polynomial randomised approximation scheme for counting the number of k-colourings of a planar graph for any fixed $k \geq 4$?

I conjecture that the answer is negative.

Similarly, since by Seymour's theorem (Seymour 1981) every bridgeless graph has a nowhere zero 6-flow, there is no obvious obstacle to the existence of an fpras for estimating the number of k-flows for $k \geq 6$. Thus a natural question which is in the same spirit is the following.

Problem 4.2. Show that there does not exist an fpras for estimating the number of nowhere-zero k-flows for $k \geq 6$.

This question has recently been answered by Goldberg and Jerrum (2008) who have proved a stronger result, namely:

Theorem 4.3. *Unless $NP = RP$ there is no fpras for counting the number of nowhere-zero k-flows in a graph for k strictly greater than 2.*

Another question about colourings where decision is trivial but counting is hard is the following, considered by Bartels and Welsh (1995).

#$|V|$-COLOURINGS

Input: Graph $G = (V, E)$
Task: Find the number of colourings of G using $|V|$ or fewer colours.

There is considerable evidence in Bartels and Welsh (1995) to suggest that a natural Markov chain which starts off in a proper $|V|$-colouring and moves through the set of all proper colourings using $|V|$ or fewer colours is in fact rapidly mixing. Further more substantial evidence in support of this is given by the following result of Jerrum (1995).

Consider the Markov chain (X_t) whose state space $\Omega = \Omega_k\{(G)\}$ is the set of all proper k-colourings of G and whose transition probabilities from state (colouring) (X_t) are determined by the following procedure:

(i) choose a vertex v of G and a colour c uniformly at random
(ii) recolour vertex v with colour c
(iii) if the resulting colouring X' is proper then let $X_{t+1} = X'$ otherwise let $X_{t+1} = X_t$.

It is known that provided $k \geq \Delta + 2$ this Markov chain is ergodic. Since the matrix is symmetric the (unique) stationary distribution is uniform over all proper k-colourings of G.

Theorem 4.4. *For graphs of maximum degree Δ there exists an fpras for counting k-colourings for all $k > 2\Delta + 1$.*

Because of Brooks' theorem there is no obstacle to there existing an fpras for k strictly greater than Δ.

Apart from the appeal of the result in Theorem 4.4 its proof is one of the nicest applications of the coupling technique described in Section 23.3.2.

Proof (idea). Let C be a set of k colours. Since $k > 2\Delta(G)$ there is no problem about starting the chain (Y_t, Z_t) in a pair of k-colourings. The transitions $(Y_t, Z_t) \rightarrow (Y_{t+1}, Z_{t+1})$ are defined as follows:

(a) choose a vertex v of G, uniformly at random (u.a.r.),
(b) choose a colour $c \in C$ u.a.r.,
(c) compute a permutation σ of C according to procedure CHOOSE below,
(d) in colouring Y_t recolour vertex v with colour c to obtain Y'_t,
(e) in colouring Z_t recolour vertex v with colour $\sigma(c)$ to obtain Z'_t,
(f) if Y'_t (respectively Z'_t) is a proper colouring then $Y_{t+1} = Y'_t$ (respectively $Z_{t+1} = Z'_t$) otherwise.

Let $Y_{t+1} = Y_t$ (respectively $Z_{t+1} = Z_t$).

First note that whatever method is used to select σ, since c is u.a.r. from C, so will $\sigma(c)$ be. Thus: (Y_t) and (Z_t) are both faithful copies of (X_t).

The key step is to design the procedure CHOOSE in such a way that Y_{t+1} and Z_{t+1} are stochastically closer together than Y_t and Z_t.

Procedure CHOOSE

If v has a different colour in Y_t and Z_t then take σ as the identity permutation so that v stays different.

If v has the same colour proceed as follows.

Consider the set of neighbours $N(v)$ in G. Let C_Y be the set of colours c such that $N(v)$ uses c in Y_t but not in Z_t. Define C_Z analogously. Thus $C_Y \cap C_Z = \emptyset$. Suppose without loss of generality that $|C_Y| \leq |C_Z|$. Choose any $C' \subseteq C_Z$ such that $|C'| = |C_Y|$. Let $C_Y = \{c_1, \ldots, c_r\}$ and $C' = \{c'_1, \ldots, c'_r\}$ be arbitrary orderings and take g to be the permutation $(c_1, c'_1), \ldots, (c_r, c'_r)$ which interchanges the colour sets C_Y and C' and leaves the remainder of C fixed.

Now let D_t be the set of vertices on which the colourings Y_t and Z_t *disagree*. Because of the way the processes Y_t, Z_t have been chosen, $|D_t|$ is a nonnegative random variable and a not too difficult argument gives

$$Pr\{|D_t| > 0\} \leq ne^{-t(k-2\Delta)}.$$

Thus $Pr\{D_t \neq \emptyset\} \leq \varepsilon$ provided $t \geq a^{-1}\ln(n\varepsilon^{-1})$ where $a = (k - 2\Delta)/kn$ and this gives mixing time $\tau(\varepsilon) \leq \frac{k}{k-2\Delta}n\ln(\frac{n}{\varepsilon})$. $\square$

Since Jerrum proved this theorem in 1995 there has been a lot of effort put into improving the bound on k. By Brooks' theorem there is no obvious obstacle to there being an fpras for all k strictly greater than Δ. However the best that has been achieved is a very ingenious proof by Vigoda (2000) of the following strengthening

Theorem 4.5. *For graphs of maximum degree Δ there exists an fpras for all k greater than $11\Delta/6$.*

In the seven years since Vigoda obtained his result there has been a series of partially successful attempts to improve it. Most of these results have imposed some further conditions on the graphs, such as lower bounds on the girth. For example Hayes and Vigoda (2003) have reduced the threshold on k/Δ to 1 on the assumptions that $\Delta = \Omega(\log n)$ and the girth is greater than 10. A recent survey of these and related results can be found in Frieze and Vigoda (2007).

23.5 Path Coupling

In 1997 Bubley and Dyer introduced an important method for designing couplings. Using it we will only need to define a coupling for certain pairs of initial states and not for all pairs as is the case with the standard coupling method.

This method, called *path-coupling* we will illustrate when Ω is the set of proper k-colourings of the graph $G = (V, E)$.

For any two k-colourings x, y in Ω join x, y by an edge if they differ at exactly one vertex. Let S denote this set of edges, and let $k \geq \Delta + 2$. Then, as remarked in Sect. 23.4, the Markov chain $M(G, k)$ of proper k-colourings is known to be ergodic, hence irreducible and thus the graph (Ω, S) is a connected graph.

We can now define the (path)-metric d on Ω where, for $x, y \in \Omega$, the distance $d(x, y)$ is the length of the shortest path between x and y in the graph (Ω, S).

Thus $d(x, y) = 1$ if and only if the colourings differ at exactly one vertex and so on.

We are now in a position to present the key result of Bubley and Dyer.

Theorem 5.1. *Suppose there exists a constant β strictly less than 1 and a coupling (X_t, Y_t) of the Markov chain so that for all pairs of states i, j which are adjacent in (Ω, S) the expected value $E(d(X_1, Y_1) \mid X_0 = i, Y_0 = j)$ is strictly less than $\beta d(i, j)$. Then the mixing time $\tau(1/4)$ is bounded by $2\log(D)/(1 - \beta)$ where D is the maximum distance between any pair of states i, j in the graph (Ω, S).*

To illustrate the power of the path coupling method we sketch the proof given by Frieze and Vigoda (2007) of the following weaker version of Jerrum's Theorem 4.4.

Theorem 5.2. *For graphs of maximum degree Δ there exists an fpras for counting k-colourings for all k greater than 3Δ.*

Proof. Let Ω denote the space of all k-colourings of G, not just the proper k-colourings. Form the graph with vertex set Ω and join two colourings by an edge iff they differ at exactly one vertex.

Now to apply path coupling. Consider a pair of k-colourings (X_t, Y_t) which differ only at the vertex v. The coupling is the identity coupling which means we pick a vertex z and colour c at random and try to recolour z with colour c. Clearly the only updates which can succeed or fail in exactly one chain are those in which z is adjacent to v and c is the colour of v in either X_t or Y_t. Hence there are at most 2Δ updates which increase the distance and it is easy to see that there are at least $k - \Delta$ updates which decrease the distance.

A simple argument now shows that provided k is greater than 3Δ we satisfy the conditions of Theorem 5.1 and prove Theorem 5.2.   □

Using only a slightly more complicated path coupling Frieze and Vigoda are able to obtain the full proof of Jerrum's theorem.

Proof (Jerrum's Theorem 4.4). When X_t tries to colour a neighbour z of v to the colour of v in X_t then Y_t tries to recolour z to the colour of v in Y_t. Otherwise X_t and Y_t do exactly the same. Since z cannot be successfully coloured in either case this has no effect.

This means that there are at most Δ cases in which the distance between the colourings increases.

Hence, the expected increase in the distance becomes bounded by $\frac{1}{kn} (\Delta - (k - \Delta)) \leq \frac{-1}{kn}$ provided $k \geq 2\Delta + 1$.   □

23.6 Matchings

Counting matchings and perfect matchings was one of the original problems studied in the seminal paper of Jerrum and Sinclair (1989). It is closely related to the classical problems of (a) evaluating the permanent and (b) counting monomer-dimer coverings in statistical physics. It is therefore very well covered in all existing surveys on the Markov chain method, such as (Jerrum 2003; Motwani and Raghavan 1995; Vazirani 1991). Here we just highlight the basics.

The main results of Jerrum and Sinclair (1989) can be stated as follows

Theorem 6.1.

 (i) *There exists an fpras for the number of matchings in a bipartite graph.*

(ii) *There exists an fpras for the number of perfect matchings in all $2n$-vertex bipartite graphs G which satisfy the condition*

$$\alpha(G) = \frac{m_{n-1}(G)}{m_n(G)} \leq poly(n),$$

and where $m_k(G)$ is the number of matchings of size k.

A consequence of (ii) is:

Corollary 6.2. *There exists an fpras for the permanent of $n \times n$ $(0, 1)$-matrices provided each row and column sum is at least $n/2$.*

This follows from the following two facts:

- If A is an $n \times n$ $(0, 1)$-matrix its permanent is the number of perfect matchings in the bipartite $2n$ vertex graph determined by A.
- The "density condition" on the row and column sums is sufficient to guarantee that condition (ii) of Theorem 6.1 is satisfied.

After the announcement of the results of Jerrum and Sinclair (1989) there was considerable effort to remove the denseness conditions in Corollary 6.2. However it was not until 2000 that Jerrum, Sinclair and Vigoda (2004) produced a full proof of the definitive result, namely:

Theorem 6.3. *There exists a fully polynomial randomised approximation scheme for estimating the permanent of an arbitrary square matrix with non-negative entries.*

Theorem 6.3 is proved by analysing a Markov chain on the perfect matchings and 'near-perfect' matchings of a bipartite graph, with $2n$ vertices and where by near-perfect we mean a matching of size $n - 1$.

The transitions of the Markov chain are of three types depending on the given state, namely:

(a) remove an edge at random from a perfect matching;
(b) add an edge to a near-perfect matching;
(c) exchange an edge of a near-perfect matching with another edge adjacent to one of its holes.

The chain will be ergodic and its stationary distribution will be uniform over the set of perfect matchings. this will therefore give an almost uniform sampler over the perfect matchings using the following procedure:

Simulate the chain for $\tau(\epsilon)$ steps where τ is large enough to ensure that the variation distance from the stationary distribution is sufficiently small.

Output the final state if it is a perfect matching.

Since it is proved that there is probability at least $1/(4n^2 + 1)$ of this, the above procedure, when iterated, will give a polynomial time, almost uniform sampler of the perfect matchings.

Although the broad structure of the proof is similar to that used by Jerrum and Sinclair (1989) to prove the existence of an fpras for the 'dense' permanent, the technical difficulties needed to be overcome to achieve this full proof are much more severe.

An interesting consequence of Theorem 6.3 is that several other counting problems are reducible by approximation preserving reductions to the $(0, 1)$-permanent. These were not accessible by the earlier approximation methods for restricted cases of the permanent. Two such examples are the following:

Corollary 6.4. *For an arbitrary bipartite graph G there exists an fpras for the number of labelled subgraphs of G having a specific degree sequence.*

The second example is:

Corollary 6.5. *There exists an fpras for counting the number of $(0, 1)$-flows in an arbitrary directed graph.*

Finally, the reader may wonder why the approximation schemes are restricted to matrices having non-negative entries. The authors point out that if such an extension was possible then it would be possible to construct a polynomial time algorithm which with high probability would compute the permanent exactly. Since this problem is $\#P$-hard it would imply a remarkable collapse of complexity classes.

23.7 Counting Trees, Forests and Matroid Bases

Counting spanning trees is one of the few enumerative problems in graph theory which has a P-time algorithm, namely take the determinant of a cofactor of the Kirchhoff matrix.

Counting all trees or all forests on the other hand are both $\#P$-complete problems, see Jerrum (1994) and Jaeger et al. (1990) respectively, and their approximation versions have proved frustratingly difficult. As far as I am aware, absolutely no progress has been made on approximating the number of subtrees in a graph.

For #FORESTS, Annan (1994) has shown that the following is true.

Theorem 7.1. *If $\mathcal{G}_\alpha$, $0 < \alpha \leq 1$, denotes the classes of graphs defined by $G \in \mathcal{G}_\alpha$ iff each vertex in G has at least $\alpha|V(G)|$ neighbours then there exists a polynomial time randomised algorithm which for $G \in \mathcal{G}_\alpha$ generates a forest of G uniformly at random from the set of all forests in G. Moreover the running time of this algorithm is $O(n^{1/\alpha})$.*

As a consequence, if we call G dense if $G \in \mathcal{G}_\alpha$ for some $\alpha > 0$, then we have

Corollary 7.2. *There exists an fpras for #FORESTS in dense graphs.*

What seems to be hard is deciding:

Question 7.3. Can the condition of denseness be removed in the last corollary?

Counting forests of a given size is a particular instance of the problem of estimating the number of bases in a matroid of some class.

A *matroid M* is just a pair (E, r) where E is a finite set and r is a *rank function* mapping $2^E \to \mathbb{Z}$ and satisfying the conditions

$$0 \le r(A) \le |A| \qquad\qquad (A \subseteq E),$$
$$A \subseteq B \Rightarrow r(A) \le r(B) \qquad (A, B \subseteq E),$$
$$r(A \cup B) + r(A \cap B) \le r(A) + r(B) \quad (A, B \subseteq E).$$

The edge set of any graph G with its associated rank function as defined by

$$r(A) = |V(G)| - k(A),$$

where $k(A)$ is the number of connected components of the graph $G : A$ having vertex set $V = V(G)$ and edge set A, is a matroid. This defines a very small subclass of matroids known as *graphic matroids*.

A set X is *independent* if $r(X) = |X|$, it is a *base* if it is a maximal independent subset of E.

M is *representable* over a field F if it is isomorphic with the matroid induced on the columns of some matrix over F by linear independence. It is *binary* if F can be taken to be $GF(2)$ and is *unimodular* or *regular* iff it is representable over $GF(2)$ and the reals. This is equivalent to having a representation over $\mathbb{R}$ by a totally unimodular matrix. For more details see Oxley (1992).

In its most general form, and avoiding niceties of how input is to be described, the crucial underlying question may be described as:

Problem 7.4. How difficult is it to count the number of bases of a matroid?

In terms of standard complexity this is easily answered for the following rather trivial reason. On a set of n elements there are $O(2^{2^n})$ distinct matroids and hence the input size to describe a general matroid is $O(2^n)$. Exhaustive search to count bases can be carried out in time polynomially bounded in the size of this huge input. Hence in this form the problem is in P.

However most interesting classes of matroids have what we might call "succinct representations", either as graphs, matrices, or the like. Hence the specific forms of Problems 7.4 that we wish to consider are those when the class of matroids has such a representation.

To see what is known, consider some of our earlier examples.

- Graphic matroids have as bases their spanning trees (or spanning forests in the disconnected case) and these can be counted *exactly* by Kirchhoff's determinantal formula.
- Binary matroids: counting bases is known to be #P-hard by Vertigan (1992). No nontrivial approximation algorithm is known and this is one of the key open problems in approximate counting.

Proposition 7.5. *The number of bases of regular (unimodular) matroids can be counted exactly in polynomial time.*

Proof. Given a totally unimodular matrix A coordinatising the regular matroid M, then $\det(AA^T)$ is the number of bases of M. This follows from the Binet–Cauchy expansion and the fact that all maximal minors have determinant ± 1 or zero. □

For general matroids having some reasonable presentation, Azar, Broder and Frieze (1993) prove the following negative result. Suppose that a matroid is presented by an *independence oracle* (see Robinson and Welsh 1980) which when presented by a set A determines if it is independent and in such a case provides a base of the matroid containing A. Such a call to the oracle is called a *probe*. The main result of Azar et al. (1993) is that any deterministic algorithm which relies on probes must make an exponential number in order to approximate the number of bases to within some constant ratio. However, on the positive side, it is shown in Chávez Lomelí and Welsh (1996) that for what is believed to be a majority of matroids, a randomised algorithm based on probes gives a $(1 + \varepsilon)$-approximation in polynomial time.

An interesting approach to these problems is the following:

Suppose now that we wish to find f_k, the number of forests of size k, where k may be a function of n.

The *matroid polyhedron* $P(M)$ is defined to be the convex polytope in $\mathbb{R}^E$ consisting of the intersection of half spaces

$$\sum_{e \in A} x_e \leq r(A), \quad A \subseteq E, \ x_e \geq 0.$$

The fundamental property of this polytope proved in Edmonds (1970) is

Theorem 7.6. *$P(M)$ has all its vertices integral, moreover they are the $(0, 1)$-incidence vectors of the independent sets of M.*

Geometrically it means that to count forests of a given size we want to count the integer points in the slice of $\mathbb{R}^E$ given by

$$S_G^{(k)} \equiv P(M(G)) \cap \left\{ \sum x_i = k \right\}.$$

Thus one approach to the problem of generating a random forest in G, would be to carry out random walk on the 1-skeleton of this polytope $P(M(G))$ and hope that the convergence to stationarity is rapid.

The polytope is well understood. We know the vertices and adjacency in the 1-skeleton is completely characterised in the following theorem from Hausmann and Korte (1978) who prove

Proposition 7.7. *Two distinct independent sets F_1, F_2 are adjacent iff $|F_i \Delta F_2| = 1$ or $|F_1 \Delta F_2| = 2$ and $F_1 \cup F_2$ is dependent.*

However showing random walk on this polytope is rapidly mixing is a special case of the much wider conjecture about general $(0, 1)$-polytopes made by Mihail and Vazirani (1989), see Vazirani (1991). This can be stated as follows.

Conjecture 7.8 (Polytope conjecture). For any bipartition of the vertices of a $(0, 1)$-polytope, the number of edges in the cut set is at least as large as the number of vertices in the smaller block of the partition.

The importance of this algorithmically is as follows.

Consider a class C of polytopes satisfying the conditions:

(i) There is a polynomial p such that the maximum degree of a vertex in an n-dimensional member of C is bounded by $p(n)$.
(ii) There is a P-time algorithm for enumerating the neighbours of a vertex in the polytope.
(iii) There is a P-time algorithm which outputs a vertex of the polytope.

Then there is an fpras for counting the number of vertices of polytopes in the class C.

Two important classes of polytopes which satisfy the conditions (i)–(iii) are: (a) matroid polyhedra, and (b) basis polytopes of matroids, provided they are suitably presented.

In case (b) the 1-skeleton is the *basis exchange graph*, in which two bases B_1, B_2 are adjacent vertices iff $|B_1 \oplus B_2| = 2$.

We close this section by stating a special case of the Polytope Conjecture which is purely combinatorial.

Conjecture 7.9 (Cut set expansion conjecture). The basis exchange graph $G(M)$ of any matroid has the property that for any bipartition (V_1, V_2) of the vertices of $G(M)$, the number of edges in the cut set (V_1, V_2) is at least as large as $\min(|V_1|, |V_2|)$.

Notice that this is demanding more than is needed for random walk on basis graphs to be rapidly mixing.

A new light on these problems was outlined in the work of Feder and Mihail (1992) who show that a sufficient condition for the natural random walk on the set of bases of M to be rapidly mixing is that M and its minors are *negatively correlated*. That is, they satisfy the constraint, that if B_R denotes a random base of M and e, f are distinct elements of $E(M)$, then

$$Pr\{e \in B_R \mid f \in B_R\} \leq Pr\{e \in B_R\}. \tag{1}$$

This concept goes back to 1975 when Seymour and Welsh (1975) related it with the log-concavity of various matroid quantities. It is somewhat surprising that this concept has reappeared now in an entirely new context.

Call matroids all of whose minors satisfy (1) *balanced*. Balanced binary matroids include regular matroids but "not all that much more" since any binary matroid containing a specific 8-element S_8 (see Oxley 1992 or Seymour and Welsh 1975) as a minor is not balanced. Since we can count exactly bases in regular matroids the practical improvement produced by balance is not that significant in binary matroids.

23.8 Conclusion

Several of the counting problems discussed above are particular evaluations of the 2-variable Tutte polynomial of the underlying graph or matroid. For example counting spanning trees, bases, forests, k-colourings and nowhere-zero k-flows come into this category. While the complexity of the exact computation of this polynomial is well understood (see Jaeger et al. 1990) determining for which points/regions of the (x, y)-plane there exists an fpras was largely unknown. However a significant advance has been made recently by Goldberg and Jerrum (2008). A very recent review of this area has been given by Jerrum (2007) which includes details of these improvements.

Of the open problems mentioned above, approximating the number of bases of a matroid has probably attracted the most attention and is regarded as the most important. A closely related problem is the following:

Question 8.1. Is there an fpras for the number of independent sets of a matroid?

In its general form this has the same difficulties as regards presentation of the input as does the basis problem. However apart from the partial results on approximating the number of forests of a graph described above I know of no other results.

What would be very surprising is that there existed an fpras for independent sets but not for bases. It suggests that the following statement may be true.

Conjecture 8.2. There exists an fpras for counting bases in matroids if and only if there exists an fpras for independent sets.

We close with what may be an easier problem. It certainly has not yet received much attention.

Consider the following computational question:

PLANAR SUBGRAPHS
Input: Labelled graph G
Output: Number of (labelled) planar subgraphs of G

As far as I am aware this has not been shown to be #P-complete, though it surely is.

Denise et al. (1996) give a very simple Markov chain algorithm for generating a planar subgraph approximately uniformly at random when the underlying graph is K_n. It generalises easily to the general case. The algorithm of Denise et al. (1996) works in the following way.

At time $t = 0$ let X_0 be the empty subgraph.

At each subsequent step choose an edge e of G uniformly at random. If e belongs to X_t then move to the subgraph obtained by deleting e from X_t.

If e does not belong to X_t then it is added to X_t if and only if such an addition preserves planarity.

This process defines a Markov chain on the set of all planar subgraphs of G which converges to the uniform distribution. However it is not known to converge rapidly. There is experimental evidence in Denise et al. (1996) to suggest this is the case when $G = K_n$ but as far as I am aware nothing is known about the general case.

References

Aldous, D.: Random walks on finite groups and rapidly mixing Markov chains. In: Séminaire de Probabilités XVII 1981/1982. Lecture Notes in Mathematics, vol. 986, pp. 243–297. Springer, Berlin (1982)

Annan, J.D.: A randomised approximation algorithm for counting the number of forests in dense graphs. Comb. Probab. Comput. **3**, 273–283 (1994)

Azar, Y., Broder, A.Z., Frieze, A.M.: On the problem of approximating the number of bases of a matroid. Inf. Process. Lett. **50**, 9–11 (1993)

Bartels, E., Welsh, D.J.A.: The Markov chain of colourings. In: Proc. 4th Conference on Integer Programming and Combinatorial Optimisation. Lecture Notes in Computer Science, vol. 920, pp. 373–387. Springer, Berlin (1995)

Broder, A.Z.: How hard is it to marry at random? In: Proceedings of the Eighteenth Annual ACM Symposium on Theory of Computing, pp. 50–58 (1986). Erratum in Proceedings of the 20th Annual ACM Symposium on Theory of Computing, p. 51 (1988)

Bubley, R., Dyer, M.: Path coupling, a technique for proving rapid mixing in Markov chains. In: Proceedings of the 38th Annual IEEE Symposium on Foundations of Computer Science FOCS, pp. 223–231 (1997)

Chávez Lomelí, L., Welsh, D.J.A.: Randomised approximation of the number of bases. In: Matroid Theory, Proceedings of the 1995 AMS–IMS–SIAM Joint Summer Research Conference. Contemporary Mathematics, vol. 197, pp. 371–376. Am. Math. Soc., Providence (1996)

Denise, A., Vasconcellos, M., Welsh, D.J.A.: The random planar graph. Congr. Numer. **13**, 61–79 (1996)

Dyer, M.E., Frieze, A.M.: Random walks, totally unimodular matrices and a randomised dual simplex algorithm. Math. Program. **64**, 1–16 (1994)

Edmonds, J.: Submodular functions, matroids, and certain polyhedra. In: Guy, R., Hanani, H., Sauer, N., Schönheim, J. (eds.) Combinatorial Structures and their Applications, pp. 69–87. Gordon and Breach, New York (1970)

Feder, T., Mihail, M.: Balanced matroids. In: Proceedings of 24th Annual ACM Symposium on the Theory of Computing, pp. 26–38 (1992)

Frieze, A., Vigoda, E.: A survey on the use of Markov chains to randomly sample colourings. In: Grimmett, G., McDiarmid, C. (eds.) Combinatorics, Complexity and Chance, pp. 53–71. Oxford University Press, Oxford (2007)

Garey, M.R., Johnson, D.S.: Computers and Intractability—A Guide to the Theory of NP-Completeness. Freeman, San Francisco (1979)

Goldberg, L.A., Jerrum, M.: Inapproximability of the Tutte polynomial. Inf. Comput., to appear (2008)

Grötschel, M., Lovász, L., Schrijver, A.: Geometric Algorithms and Combinatorial Optimization. Springer, Berlin (1988)

Hausmann, D., Korte, B.: Colouring criteria for adjacency on 0-1 polyhedra. In: Balinski, M.L., Hoffman, A.J. (eds.) Mathematical Programming Study, vol. 8, pp. 106–127. North-Holland, Amsterdam (1978)

Hayes, T.P., Vigoda, E.: A non-Markovian coupling for randomly sampling colourings. In: Proceedings of the 44th Annual IEEE Symposium on Foundations of Computer Science FOCS, pp. 618–627 (2003)

Jaeger, F., Vertigan, D.L., Welsh, D.J.A.: On the computational complexity of the Jones and Tutte polynomials. Math. Proc. Camb. Phil. Soc. **108**, 35–53 (1990)

Jerrum, M.R.: Counting trees in a graph is #P-complete. Inf. Process. Lett. **51**, 111–116 (1994)

Jerrum, M.R.: A very simple algorithm for estimating the number of k-colourings of a low-degree graph. Random Struct. Algorithms **7**, 157–165 (1995)

Jerrum, M.R.: Counting, Sampling and Integrating: Algorithms and Complexity. Birkhäuser, Basel (2003)

Jerrum, M.R.: Approximating the Tutte Polynomial. In: Grimmett, G., McDiarmid, C. (eds.) Combinatorics, Complexity and Chance, pp. 141–161. Oxford University Press, Oxford (2007)

Jerrum, M.R., Sinclair, A.: Approximating the permanent. SIAM J. Comput. **18**, 1149–1178 (1989)

Jerrum, M.R., Sinclair, A.J.: Fast uniform generation of regular graphs. Theor. Comput. Sci. **73**, 91–100 (1990)

Jerrum, M.R., Sinclair, A.: Polynomial-time approximation algorithms for the Ising model. SIAM J. Comput. **22**, 1087–1116 (1993)

Jerrum, M.R., Valiant, L.G., Vazirani, V.V.: Random generation of combinatorial structures from a uniform distribution. Theor. Comput. Sci. **43**, 169–188 (1986)

Jerrum, M.R., Sinclair, A., Vigoda, E.: A polynomial-time approximation algorithm for the permanent of a matrix with non-negative entries. J. Assoc. Comput. Mach. **51**, 671–697 (2004)

Kannan, R.: Markov chains and polynomial time algorithms. In: 35th Annual IEEE Symposium on Foundations of Computer Science, Santa Fe, New Mexico, pp. 656–671. IEEE Press, New York (1994)

Khuller, S., Vazirani, V.V.: Planar graph colourability is not self reducible, assuming P $\neq$ NP. Theor. Comput. Sci. **88**, 183–190 (1991)

Lawler, G.F., Sokal, A.D.: Bounds on the L^2 spectrum for Markov chains and Markov processes: a generalisation of Cheeger's inequality. Trans. Am. Math. Soc. **309**, 557–580 (1988)

Lovász, L.: Random walks on graphs: a survey. In: Miklos, D., Sos, V.T., Szonyi, T. (eds.) Combinatorics, Paul Erdös is Eighty. Bolyai Society Mathematical Studies, vol. 2, pp. 353–398. Janos Bolyai Mathematical Society, Budapest (1996)

Mihail, M., Vazirani, U.: On the magnification of 0-1 polytopes. Harvard Tech. Report TR 05-89 (1989)

Motwani, R., Raghavan, P.: Randomised Algorithms. Cambridge University Press, Cambridge (1995)

Oxley, J.G.: Matroid Theory. Oxford University Press, Oxford (1992)

Robinson, G.C., Welsh, D.J.A.: The computational complexity of matroid properties. Math. Proc. Camb. Phil. Soc. **87**, 29–45 (1980)

Schnorr, C.P.: Optimal algorithms for self-reducible problems. In: Proc. of 3rd International Colloq. on Automata, Languages and Programming, pp. 322–337 (1976)

Seymour, P.D.: Nowhere-zero 6-flows. J. Comb. Theory, Ser. B **30**, 130–135 (1981)

Seymour, P.D., Welsh, D.J.A.: Combinatorial applications of an inequality from statistical mechanics. Math. Proc. Camb. Phil. Soc. **77**, 485–497 (1975)

Sinclair, A.: Randomised algorithms for counting and generating combinatorial structures. Ph.D. thesis, Edinburgh (1988)

Sinclair, A.J.: Improved bounds for mixing rates of Markov chains and multicommodity flow. Comb. Probab. Comput. **1**, 351–370 (1992)

Sinclair, A.J., Jerrum, M.R.: Approximate counting, uniform generating and rapidly mixing Markov chains. Inf. Comput. **82**, 93–133 (1989)

Vazirani, U.: Rapidly mixing Markov chains. In: Bollobas, B. (ed.) Proceedings of Symposium in Applied Mathematics, Probabilistic Combinatorics and its Applications, vol. 44, pp. 99–121. Am. Math. Soc., Providence (1991)

Vertigan, D.L.: Private communication (1992)

Vigoda, E.: Improved bounds for sampling colourings. J. Math. Phys. **41**, 1555–1569 (2000)

Welsh, D.J.A.: Complexity: Knots, Colourings and Counting. London Mathematical Society Lecture Note Series, vol. 186. Cambridge University Press, Cambridge (1993)

Subject Index

Author Index